AF595092

Coumarin and Its Derivatives

Coumarin and Its Derivatives

Editor

Maria João Matos

MDPI • Basel • Beijing • Wuhan • Barcelona • Belgrade • Manchester • Tokyo • Cluj • Tianjin

Editor
Maria João Matos
Organic Chemistry
University of Santiago de
Compostela
Santiago de Compostela
Spain

Editorial Office
MDPI
St. Alban-Anlage 66
4052 Basel, Switzerland

This is a reprint of articles from the Special Issue published online in the open access journal *Molecules* (ISSN 1420-3049) (available at: www.mdpi.com/journal/molecules/special_issues/Coumarin_Derivatives).

For citation purposes, cite each article independently as indicated on the article page online and as indicated below:

LastName, A.A.; LastName, B.B.; LastName, C.C. Article Title. *Journal Name* **Year**, *Volume Number*, Page Range.

ISBN 978-3-0365-2775-8 (Hbk)
ISBN 978-3-0365-2774-1 (PDF)

Contents

About the Editor

Maria João Matos

Maria João Matos obtained a degree in Pharmaceutical Sciences from the University of Porto (Portugal), a MSc in Organic Chemistry and a PhD in Medicinal Chemistry from the University of Santiago de Compostela. Between 2015 and 2018, she has worked in Chemical Biology at the University of Cambridge (UK) as Postdoctoral Fellow. During those three years, Maria was also Fellow of the Newnham College, University of Cambridge.

Currently, Maria is Research Associate and Teaching Assistant at the University of Porto (Portugal) and the University of Santiago de Compostela (Spain). Her research interests include new tools for drug design, discovery and delivery, in particular in the field of neurodegenerative diseases. Maria is currently author of 116 publications in both Medicinal Chemistry and Chemical Biology. She established her scientific career in fruitful collaborations with both academic and industrial partners. Since 2018, Maria closely collaborates with the Kaertor Foundation (Spain).

Preface to "Coumarin and Its Derivatives"

Coumarins are widely distributed in nature and can be found in a large number of naturally occurring and synthetic bioactive molecules. Their unique and versatile oxygen-containing heterocyclic structure makes them a privileged scaffold in medicinal chemistry. The large conjugated system, with electron-rich and charge-transport properties, is important for the interaction of this scaffold with other molecules and ions. Therefore, a great number of coumarin derivatives have been designed, synthetized, and evaluated on different pharmacological targets. In addition, coumarin-based ion receptors, fluorescent probes, and biological stains are growing quickly and have extensive applications to monitor timely enzyme activity, complex biological events, as well as accurate pharmacological and pharmacokinetic properties in living cells. The extraction, synthesis, and biological evaluation of coumarins have become extremely attractive and rapidly developing topics. Research articles, reviews, communications, and concept papers focused on the multidisciplinary profile of coumarins, highlighting natural sources and the most recent synthetic pathways, along with the main biological applications and theoretical studies, constitute this book.

Maria João Matos
Editor

ditorial

;oumarin and Its Derivatives—Editorial

aria João Matos [1,2]

[1] CIQUP/Department of Chemistry and Biochemistry, Faculty of Sciences, University of Porto, 4169-007 Porto, Portugal; maria.matos@fc.up.pt or mariajoao.correiapinto@usc.es
[2] Departamento de Química Orgánica, Facultade de Farmacia, Universidade de Santiago de Compostela, 15782 Santiago de Compostela, Spain

Coumarins are widely distributed in nature and can be found in a large number of naturally occurring and synthetic bioactive molecules [1]. The unique and versatile oxygen-containing heterocyclic structure makes them a privileged scaffold in Medicinal Chemistry [1]. The large-conjugated system, with electron-rich and charge-transport properties, is important for the interaction of this scaffold with other molecules and ions [1]. Therefore, many coumarin derivatives have been extracted from natural sources, designed, synthetized, and evaluated on different pharmacological targets [2]. In addition, coumarin-based ion receptors, fluorescent probes, and biological stains are growing quickly and have extensive applications to monitor timely enzyme activity, complex biological events, as well as accurate pharmacological and pharmacokinetic properties in living cells [3]. The extraction, synthesis, and biological evaluation of coumarins have become extremely attractive and rapidly developing topics. A large number of research and review papers compile information on this important family of compounds in 2020 [3]. Research articles, reviews, communications, and concept papers focused on the multidisciplinary profile of coumarins, highlighting natural sources, most recent synthetic pathways, along with the main biological applications and theoretical studies, were the main focus of this Special Issue.

The anticoagulatory activity of coumarins is one of the most classic applications of this family of compounds, acenocoumarol and warfarin being the most important approved drugs. The use of one or another depends on different factors. However, the real evidence on their different results is not completely clear. Therefore, the clinical results for both molecules were studied on 2111 MPHV patients included in the nationwide PLECTRUM registry [4]. In addition, the antiplatelet aggregation profile of coumarin, esculetin and esculin, were determined by studying cyclooxygenase I (COX-I) inhibition [5].

Inflammation is another area of constant interest. Hydroxycoumarins are on the top of the list, 4-hydroxy-7-methoxycoumarin being described as an inhibitor of inflammation in LPS-activated RAW264.7 macrophages by suppressing the nuclear factor kappa B (NF-κB) and MAPK activation [6]. This simple coumarin reduced the production of nitric oxide (NO), prostaglandin E2 (PGE2), proinflammatory cytokines such as tumor necrosis factor (TNF)-α, interleukin (IL)-1β and IL-6, and the expression of inducible nitric oxide synthase (iNOS) and cyclooxygenase 2 (COX-2), being non-cytotoxic for different cell lines. Moreover, this molecule decreased phosphorylation of extracellular signal-regulated kinase (ERK1/2) and c-Jun N-terminal kinase/stress-activated protein kinase (JNK), but not that of p38 MAPK [6]. In addition, coumarins have been described as anti-inflammatory and antioxidant compounds with a potential action in inflammatory bowel disease [7]. These molecules display a protective action in intestinal inflammation by modulating different mechanisms and signaling pathways, mainly modulating immune and inflammatory responses, and protecting against oxidative stress.

Neurodegenerative diseases are another classical application of coumarins in drug discovery. The design of new hybrids, especially looking for a multitarget function, is a trend strategy. Coumarin-chalcone hybrids have been described as potent and selective

Citation: Matos, M.J. Coumarin and Its Derivatives—Editorial. *Molecules* **2021**, *26*, 6320. https://doi.org/10.3390/molecules26206320

Academic Editor: Thomas J. Schmidt

Received: 13 September 2021
Accepted: 15 October 2021
Published: 19 October 2021

monoamine oxidase B (MAO-B) inhibitors [8]. A series of fourteen new derivatives were described, an IC_{50} in the nanomolar range presenting the best compound. Theoretical approaches corroborated the interaction and selectivity of these compounds for the B isoform. Coumarin-chalcone hybrids also attracted the attention by being adenosine receptor modulators [9]. This family of G-protein-coupled receptors (GPCRs) is especially important in neurological and psychiatric disorders such as Parkinson's and Alzheimer's diseases epilepsy, and schizophrenia. The studied series proved to be interesting for the design of potent and selective human A_1 or A_3 ligands. In general, molecules bearing hydroxy groups showed more A_1 affinity, while the methoxy counterparts showed A_3 selectivity On the other hand, extracts from plants and their isolated compounds are also being used as inhibitors of enzymes involved on neurodegenerative diseases. Coumarin glycyrol and liquiritigenin, isolated from *Glycyrrhiza uralensis*, were the most promising molecules [10] The first one proved to inhibit butyrylcholinesterase (BuChE), acetylcholinesterase (AChE and MAO-B in the micromolar range, being reversible and noncompetitive inhibitors o BuChE. The second one proved to be reversible and competitive with MAO-B inhibito in the nanomolar range. Finally, curcumin–coumarin hybrids have been also describe as multitarget agents against neurodegenerative disorders [11]. From the studied serie most of the 3-(7-phenyl-3,5-dioxohepta-1,6-dien-1-yl)coumarins proved to be moderat inhibitors of *h*MAO, AChE, and BuChE, also displaying antioxidant activity (scavengin DPPH free radical). Two compounds out of this series also showed neuroprotective activit against hydrogen peroxide (H_2O_2) in the SH-SY5Y cell line. The formulation of thes derivatives in nanoparticles improved this last property.

Anticancer activities for coumarins have been also reported. Coumarin-3-carboxamid derivatives have been reported, and 4-fluoro and 2,5-difluoro benzamides presented a tivities against HepG2 and HeLa cancer cell lines comparable to doxorubicin, exhibitin low cytotoxicity against LLC-MK2 normal cell line [12]. From the combination of sin ple coumarins (osthole, umbelliferone, esculin or 4-hydroxycoumarin) with sorafenib, a antiglioma compound was also reported by studying human glioblastoma multiform (T98G) and anaplastic astrocytoma (MOGGCCM) cells lines [13].

Psoralen derivatives with electrophilic warhead variations at position 3 have bee described for their immunoproteasome inhibitory activity [14]. The studied compound proved to be slightly less active inhibiting the β5i subunit of immunoproteasome than th previously reported 7*H*-furo[3,2-g]chromen-7-one (psoralen)-based compounds with a oxathiazolone warhead. These results allowed to establish important structure–activity relationships that will guide the design of potent and selective immunoproteasome inhibitors.

As said before, several coumarins are naturally occurring molecules. Therefore there is intensive research on plants and extracts analysis. Sixty coumarin derivatives from *Artemisia capillaris* were studied for their constitutive androstane receptor (CAR) activation [15]. Amongst all the molecules studied in the in vitro CAR activation screening, 6,7-diprenoxycoumarin proved to be the most interesting for further studies. A review paper on the natural occurrence, biosynthesis, and biological properties of two 3-prenylated coumarins has been described [16]. A dihydrofuranocoumarin (chalepin) and furanocoumarin (chalepensin) are in the focus of this overview. They were isolated from the first time from the medicinal plant *Ruta chalepensis* L. (Fam: Rutaceae) but are also present in species of the genera Boenminghausenia, Clausena, and Ruta. These two natural products have been described for their anticancer, antidiabetic, antifertility, antimicrobial, antiplatelet aggregation, antiprotozoal, antiviral, and calcium antagonistic properties. The same group focused a second review on the natural origin, biosynthesis, and pharmacological activities of tetracyclic 4-substituted dipyranocoumarins, the calanolides [17]. Ultra-high-performance liquid chromatography coupled with a mass spectrometry (UHPLC-MS) methodology has been used for identifying and quantifying coumarins from a group of twenty-eight plants (roots and leaves) from Arabidopsis natural populations [18]. Simple coumarins such as scopoletin, umbelliferone and esculetin, along with their glycosides scopolin, skimmin and esculin, respectively, have been identified. Finally, the ability of different coumarins to inhibit quorum sensing when combined with small plant-derived molecules identified in various plants extracts has been described [19].

The development of new chemical tools and strategies to obtain different coumarins, and the update of the traditional ones, are a continuous field of research. Chiral tertiary amine catalyzed asymmetric [4 + 2] cyclization of 3-aroylcoumarins with 2,3-butadienoate has been described [20]. Two reviews on the synthetic strategies to obtain coumarin(benzopyrone)-fused five-membered aromatic heterocycles built on the α-pyrone moiety, one centered on five-membered aromatic rings with a single heteroatom and the other one with multiple heteroatoms, have also been published [21,22]. New 3-ethynylaryl coumarin-based dyes for DSSC applications were included in this monographic issue [23]. The synthetic pathways, spectroscopic properties and theoretical calculations were included. The structural characterization (UV-Visible spectroscopy, thermal analysis by differential scanning calorimetry and TGA, ^{1}H NMR and X-ray diffraction) of mono and dihydroxylated umbelliferone derivatives has been also described [24]. 3-Carboxylic acid and formyl-derived coumarins have been proposed as photoinitiators in the photo-oxidation or photo-reduction processes for photopolymerization upon visible light [25]. These characteristics are related to the potential of these molecules in the photocomposite synthesis and 3D printing applications [25]. Finally, *in silico* tools (i.e. MetFrag, SIRIUS version 4.8.2, CSI:FingerID and CANOPUS) have been used for the structural elucidation of ferulenol, synthetized by engineered *Escherichia coli* [26]. This study highlights the importance of 4-hydroxycoumarins as lead molecules for the chemical synthesis of several bioactive compounds and drugs.

The huge and growing range of applications of coumarins described in this Special Issue is a demonstration of the potential of this family of compounds in Organic Chemistry, Medicinal Chemistry, and different sciences related to the study of natural products. This Special Issue includes 24 articles: 18 original papers and 6 review papers. The versatility of this scaffold is also being demonstrated by the number of manuscripts revealing and highlighting its potential. Based on the current results, it may be expected that the utility of coumarins as scaffolds for drug design, as structures for chemical synthesis and as fluorescent probes, may grow in the next years. Finally, it seems that simple coumarins are still more explored than complex derivatives.

Author Contributions: Writing, review, and editing, M.J.M. All authors have read and agreed to the published version of the manuscript.

Funding: M.J.M. would like to thank Fundação para a Ciência e Tecnologia (FCT, CEECIND/02423/2018 and UIDB/00081/2020).

Institutional Review Board Statement: Not applicable.

Informed Consent Statement: Not applicable.

Conflicts of Interest: The author declares no conflict of interest.

References

1. Borges, F.; Roleira, F.; Milhazes, N.; Santana, L.; Uriarte, E. Simple Coumarins and Analogues in Medicinal Chemistry: Occurrence, Synthesis and Biological Activity. *Curr. Med. Chem.* **2005**, *12*, 887–916. [CrossRef]
2. Stefanachi, A.; Leonetti, F.; Pisani, L.; Catto, M.; Carotti, A. Coumarin: A Natural, Privileged and Versatile Scaffold for Bioactive Compounds. *Molecules* **2018**, *23*, 250. [CrossRef] [PubMed]
3. Carneiro, A.; Matos, M.J.; Uriarte, E.; Santana, L. Trending Topics on Coumarin and Its Derivatives in 2020. *Molecules* **2021**, *26*, 501. [CrossRef]
4. Menichelli, D.; Poli, D.; Antonucci, E.; Cammisotto, V.; Testa, S.; Pignatelli, P.; Palareti, G.; Pastori, D.; the Italian Federation of Anticoagulation Clinics (FCSA). Comparison of Anticoagulation Quality between Acenocoumarol and Warfarin in Patients with Mechanical Prosthetic Heart Valves: Insights from the Nationwide PLECTRUM Study. *Molecules* **2021**, *26*, 1425. [CrossRef]
5. Zaragozá, C.; Zaragozá, F.; Gayo-Abeleira, I.; Villaescusa, L. Antiplatelet Activity of Coumarins: In Vitro Assays on COX-1. *Molecules* **2021**, *26*, 3036. [CrossRef]
6. Kang, J.K.; Hyun, C.-G. 4-Hydroxy-7-Methoxycoumarin Inhibits Inflammation in LPS-activated RAW264.7 Macrophages by Suppressing NF-κB and MAPK Activation. *Molecules* **2020**, *25*, 4424. [CrossRef] [PubMed]

7. Di Stasi, L.C. Coumarin Derivatives in Inflammatory Bowel Disease. *Molecules* **2021**, *26*, 422. [CrossRef] [PubMed]
8. Moya-Alvarado, G.; Yañez, O.; Morales, N.; González-González, A.; Areche, C.; Núñez, M.T.; Fierro, A.; García-Beltrán, O. Coumarin-Chalcone Hybrids as Inhibitors of MAO-B: Biological Activity and In Silico Studies. *Molecules* **2021**, *26*, 2430. [CrossRef] [PubMed]
9. Vazquez-Rodriguez, S.; Vilar, S.; Kachler, S.; Klotz, K.-N.; Uriarte, E.; Borges, F.; Matos, M.J. Adenosine Receptor Ligands: Coumarin–Chalcone Hybrids as Modulating Agents on the Activity of hARs. *Molecules* **2020**, *25*, 4306. [CrossRef] [PubMed]
10. Jeong, G.S.; Kang, M.-G.; Lee, J.Y.; Lee, S.R.; Park, D.; Cho, M.L.; Kim, H. Inhibition of Butyrylcholinesterase and Human Monoamine Oxidase-B by the Coumarin Glycyrol and Liquiritigenin Isolated from *Glycyrrhiza uralensis*. *Molecules* **2020**, *25*, 3896. [CrossRef]
11. Quezada, E.; Rodríguez-Enríquez, F.; Laguna, R.; Cutrín, E.; Otero, F.; Uriarte, E.; Viña, D. Curcumin–Coumarin Hybrid Analogues as Multitarget Agents in Neurodegenerative Disorders. *Molecules* **2021**, *26*, 4550. [CrossRef]
12. Phutdhawong, W.; Chuenchid, A.; Taechowisan, T.; Sirirak, J.; Phutdhawong, W.S. Synthesis and Biological Activity Evaluation of Coumarin-3-Carboxamide Derivatives. *Molecules* **2021**, *26*, 1653. [CrossRef] [PubMed]
13. Sumorek-Wiadro, J.; Zając, A.; Langner, E.; Skalicka-Woźniak, K.; Maciejczyk, A.; Rzeski, W.; Jakubowicz-Gil, J. Antiglioma Potential of Coumarins Combined with Sorafenib. *Molecules* **2020**, *25*, 5192. [CrossRef] [PubMed]
14. Schiffrer, E.S.; Proj, M.; Gobec, M.; Rejc, L.; Šterman, A.; Mravljak, J.; Gobec, S.; Sosič, I. Synthesis and Biochemical Evaluation of Warhead-Decorated Psoralens as (Immuno)Proteasome Inhibitors. *Molecules* **2021**, *26*, 356. [CrossRef] [PubMed]
15. Juang, S.-H.; Hsieh, M.-T.; Hsu, P.-L.; Chen, J.-L.; Liu, H.-K.; Liang, F.-P.; Kuo, S.-C.; Chiu, C.-Y.; Liu, S.-H.; Chou, C.-H.; et al. Studies of Coumarin Derivatives for Constitutive Androstane Receptor (CAR) Activation. *Molecules* **2021**, *26*, 164. [CrossRef]
16. Nahar, L.; Al-Majmaie, S.; Al-Groshi, A.; Rasul, A.; Sarker, S.D. Chalepin and Chalepensin: Occurrence, Biosynthesis and Therapeutic Potential. *Molecules* **2021**, *26*, 1609. [CrossRef]
17. Nahar, L.; Talukdar, A.D.; Nath, D.; Nath, S.; Mehan, A.; Ismail, F.M.D.; Sarker, S.D. Naturally Occurring Calanolides: Occurrence, Biosynthesis, and Pharmacological Properties Including Therapeutic Potential. *Molecules* **2020**, *25*, 4983. [CrossRef]
18. Perkowska, I.; Siwinska, J.; Olry, A.; Grosjean, J.; Hehn, A.; Bourgaud, F.; Lojkowska, E.; Ihnatowicz, A. Identification and Quantification of Coumarins by UHPLC-MS in *Arabidopsis thaliana* Natural Populations. *Molecules* **2021**, *26*, 1804. [CrossRef]
19. Deryabin, D.; Inchagova, K.; Rusakova, E.; Duskaev, G. Coumarin's Anti-Quorum Sensing Activity Can Be Enhanced When Combined with Other Plant-Derived Small Molecules. *Molecules* **2021**, *26*, 208. [CrossRef]
20. Li, J.-L.; Wang, X.-H.; Sun, J.-C.; Peng, Y.-Y.; Ji, C.-B.; Zeng, X.-P. Chiral Tertiary Amine Catalyzed Asymmetric [4 + 2] Cyclization of 3-Aroylcoumarines with 2,3-Butadienoate. *Molecules* **2021**, *26*, 489. [CrossRef]
21. El-Sawy, E.R.; Abdelwahab, A.B.; Kirsch, G. Synthetic Routes to Coumarin(Benzopyrone)-Fused Five-Membered Aromatic Heterocycles Built on the α-Pyrone Moiety. Part 1: Five-Membered Aromatic Rings with One Heteroatom. *Molecules* **2021**, *26*, 483. [CrossRef] [PubMed]
22. El-Sawy, E.R.; Abdelwahab, A.B.; Kirsch, G. Synthetic Routes to Coumarin(Benzopyrone)-Fused Five-Membered Aromatic Heterocycles Built on the α-Pyrone Moiety. Part II: Five-Membered Aromatic Rings with Multi Heteroatoms. *Molecules* **2021**, *26*, 3409. [CrossRef]
23. Sarrato, J.; Pinto, A.L.; Malta, G.; Röck, E.G.; Pina, J.; Lima, J.C.; Parola, A.J.; Branco, P.S. New 3-Ethynylaryl Coumarin-Based Dyes for DSSC Applications: Synthesis, Spectroscopic Properties, and Theoretical Calculations. *Molecules* **2021**, *26*, 2934. [CrossRef]
24. Seoane-Rivero, R.; Ruiz-Bilbao, E.; Navarro, R.; Laza, J.M.; Cuevas, J.M.; Artetxe, B.; Gutiérrez-Zorrilla, J.M.; Vilas-Vilela, J.L.; Marcos-Fernandez, A. Structural Characterization of Mono and Dihydroxylated Umbelliferone Derivatives. *Molecules* **2020**, *25*, 3497. [CrossRef] [PubMed]
25. Rahal, M.; Graff, B.; Toufaily, J.; Hamieh, T.; Noirbent, G.; Gigmes, D.; Dumur, F.; Lalevée, J. 3-Carboxylic Acid and Formyl-Derived Coumarins as Photoinitiators in Photo-Oxidation or Photo-Reduction Processes for Photopolymerization upon Visible Light: Photocomposite Synthesis and 3D Printing Applications. *Molecules* **2021**, *26*, 1753. [CrossRef] [PubMed]
26. Klamrak, A.; Nabnueangsap, J.; Puthongking, P.; Nualkaew, N. Synthesis of Ferulenol by Engineered *Escherichia coli*: Structural Elucidation by Using the *In Silico* Tools. *Molecules* **2021**, *26*, 6264. [CrossRef]

MDPI

Review

Trending Topics on Coumarin and Its Derivatives in 2020

Aitor Carneiro [1], Maria João Matos [1,2], Eugenio Uriarte [1,3] and Lourdes Santana [1,*]

1 Departamento de Química Orgánica, Facultade de Farmacia, Universidade de Santiago de Compostela, 15782 Santiago de Compostela, Spain; aitor.carneiro@gmail.com (A.C.); mariajoao.correiapinto@usc.es (M.J.M.); eugenio.uriarte@usc.es (E.U.)
2 CIQUP/Department of Chemistry and Biochemistry, Faculty of Sciences, University of Porto, Rua Campo Alegre 687, 4169-007 Porto, Portugal
3 Instituto de Ciencias Químicas Aplicadas, Universidad Autónoma de Chile, 7500912 Santiago, Chile
* Correspondence: lourdes.santana@usc.es; Tel.: +34-881-814-936

Abstract: Coumarins are naturally occurring molecules with a versatile range of activities. Their structural and physicochemical characteristics make them a privileged scaffold in medicinal chemistry and chemical biology. Many research articles and reviews compile information on this important family of compounds. In this overview, the most recent research papers and reviews from 2020 are organized and analyzed, and a discussion on these data is included. Multiple electronic databases were scanned, including SciFinder, Mendeley, and PubMed, the latter being the main source of information. Particular attention was paid to the potential of coumarins as an important scaffold in drug design, as well as fluorescent probes for decaging of prodrugs, metal detection, and diagnostic purposes. Herein we do an analysis of the trending topics related to coumarin and its derivatives in the broad field of drug discovery.

Keywords: coumarins; biological applications; drug discovery; fluorescent probes

Citation: Carneiro, A.; Matos, M.J.; Uriarte, E.; Santana, L. Trending Topics on Coumarin and Its Derivatives in 2020. *Molecules* **2021**, *26*, 501. https://doi.org/10.3390/molecules26020501

Academic Editor: Emerson F. Queiroz

Received: 10 December 2020
Accepted: 15 January 2021
Published: 19 January 2021

1. Introduction

Coumarins are molecules that belong to a very special family. Their conjugated double ring system makes them interesting molecules for different fields of research. Coumarins can be found in industry as cosmetics and perfume ingredients, as food additives, and especially in the pharmaceutical industry in the synthesis of a large number of synthetic pharmaceutical products [1]. This last application is the main focus of our overview.

Coumarin (Figure 1) is found in nature in a wide variety of plants, particularly in high concentration in the tonka bean (*Dipteryx odorata*). It can also be found in sweet woodruff (*Galium odoratum*), vanilla grass (*Anthoxanthum odoratum*), and sweet grass (*Hierochloe odorata*), among others. This explains the great interest in the extraction and characterization techniques of natural coumarins, and in the synthesis of their derivatives. In addition, the simplicity of its chemical backbone is very attractive, as well as the reactivity of the benzene and pyrone rings. Conjugated double bonds are responsible for an electronic environment that plays a very important role in this family of compounds.

R = OCH_3, CH_3, OH, X, …

Simple coumarins **Pyranocoumarins** **Furocoumarins**

Figure 1. Basic classification of coumarins: chemical structures of the three main classes.

This review is based on the most relevant literature that comprises new data from recent research articles and overviews on the development of new therapeutic solutions and fluorescent probes based on the coumarin scaffold. The research articles and reviews organized to prepare this manuscript have been compiled from various electronic databases, including SciFinder, Mendeley, and PubMed. The latter was the main source of information, due to its specificity in the biomedical field.

2. Discussion

Searching for the word "coumarin" in Mendeley, PubMed, and SciFinder in early November 2020, and filtering by year "2020", more than a thousand references appeared. Another search was conducted in late December to include as much information as possible in this manuscript. A diversity of journals from different fields publishes research articles and reviews related to the biological interest of both plant extracts containing coumarins and/or synthetic molecules based on this scaffold. Hybrid molecules containing different pharmacophores [2,3] like piperazines or pyrazolines [4] are at the top of the list. For simplicity, in the current review the information is organized taking into account the potential pharmacological/biological applications of coumarin derivatives.

2.1. Anticancer Activity

The activity of coumarins as anticancer agents is at the top of reviews published in 2020 [5–8], as well as research papers. Potent inhibitors (Figure 2, general structure **I**) of aldo–keto reductase (AKR) presenting an iminocoumarin scaffold, with activities between 25 and 56 nM, have been described for the treatment of prostatic cancer [9]. The design of sulfamide 3-benzylcoumarin hybrids bearing an oxadiazole ring at position 7 (Figure 2, general structure **II**) has allowed the preparation of new multitarget mitogen-activated protein kinase (MEK) inhibitors and nitric oxide (NO) donors, both with antiproliferative properties [10]. In other cases, the anticancer profile has been directed to other targets. Such is the case of new inhibitors of cyclin-dependent kinases, specifically CDK9, designing hybrids that incorporate an aminopyrimidine fragment to coumarin, both pharmacophores of known activity on these therapeutic targets [11]. We highlight here compound **III** (Figure 2), with high activity and selectivity for these receptors in comparison with other kinases.

Other important targets for cancer treatment, especially lymphomas, are histone deacetylases (HDACs). A series of coumarins (Figure 2, general structure **IV**) exhibiting a hydroxamate structure similar to HDACi vorinostat (SAHA) has been published [12]. The compounds show inhibitory activity in the nanomolar range, being higher in the case of propyl or methoxypropyl derivatives.

In addition, it is worth highlighting the design of hybrids in which one part of the molecule provides fluorescent properties, and another provides therapeutic action (theranostic). Such is the case of the fusion of a 7-aminocoumarin fluorescent ring with a chalcone fragment (Figure 2, compound **V**). This molecule is an inhibitor of thioredoxin reductases (TrxRs), presenting high antitumor activity (IC_{50} = 3.6 μM), and is also used as a diagnostic agent [13]. Coumarin scaffold fluorescence is being explored extensively in biomedicine, as described at the end of this review.

The preparation of photo-triggered drug delivery systems (PTDDSs, Figure 2, general structure **VI**) has also been described, in which the chlorambucil pharmacophore is incorporated into more complex carbazole–coumarins (electron donor and electron acceptor fragments, respectively), carriers of a mitochondrial triphenylphosphonium ligand. This system allows, by irradiation, the controlled release of the chemotherapeutic agent [14].

Finally, coumarins are widely used as ligands in the formation of metal complexes, as described in a very recent review [15] focusing on their application as anticancer agents. Such is the case of complexes with platinum, palladium, gold, copper, or ruthenium, many of which are also used as described below, in the design of antimicrobial agents.

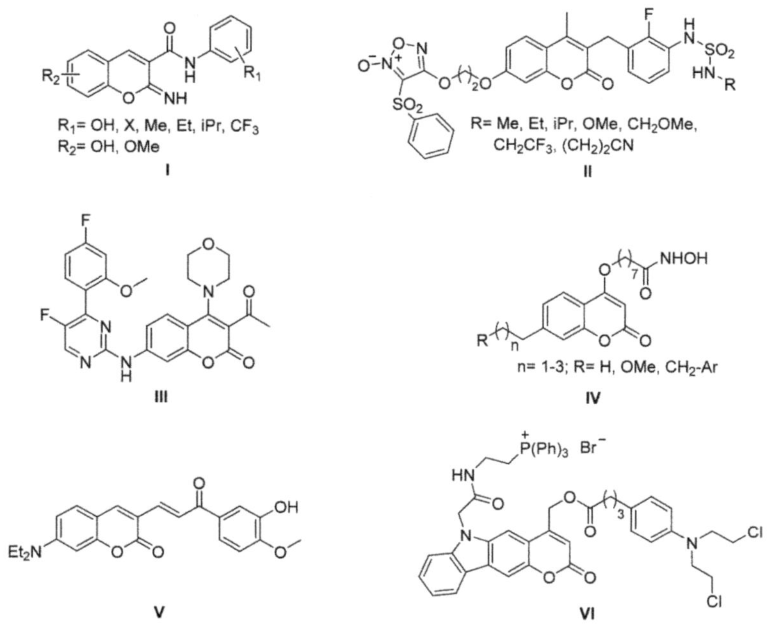

Figure 2. Structures of coumarin derivatives as anticancer agents.

Within the group of compounds with anticancer activity, coumarins exhibiting an antiglioma profile may be highlighted. Simple coumarins such as osthole, umbelliferone, esculin, and 4-hydroxycoumarin, combined with sorafenib (a kinase inhibitor drug approved for the treatment of primary kidney cancer, advanced primary liver cancer, FLT3-ITD positive acute myeloid leukemia (AML), and radioactive iodine-resistant advanced thyroid carcinoma) were studied [16]. The same group also studied a combination of the same simple coumarins with temozolomide (used in the treatment of brain tumors such as glioblastoma multiforme or anaplastic astrocytoma) [17].

2.2. Antimicrobial Activity

There is also an abundant bibliography related to the interest of coumarins as antimicrobials. Most of the projects are still inspired by the classic antibiotic novobiocin. There are several works in which antibacterial activity is found due to the presence of an azole ring introduced in different positions of the coumarin system. Articles have been published recently on the antibacterial activity of azole–coumarins, as well as 3/4/7 substituted arylcoumarins (Figure 3, general structures **VII** and **VIII**), especially active on Gram-positive and negative bacteria according to substitution patterns [15,18–20]. In other cases, thiazolidinedione–coumarin hybrids have been described (Figure 3, general structure **IX**) that show activity on methicillin-resistant *Staphylococcus aureus* (MRSA) [21].

Interestingly, coumarin metal complexes also show antibacterial activity. Such is the case of 3-arylcoumarins that present general structures **X** (Figure 3), coordinated with Re(**I**), active against MRSA in nanomolar concentrations [22]; or the complexes of general structure **XI**, a coordination of coumarin–quinoline hybrids with Cu(**I**), with activity against *Flavobacterium psychrophilum*, a Gram-negative bacterium that causes significant septicemia in fish, causing devastating economic problems in aquaculture [23].

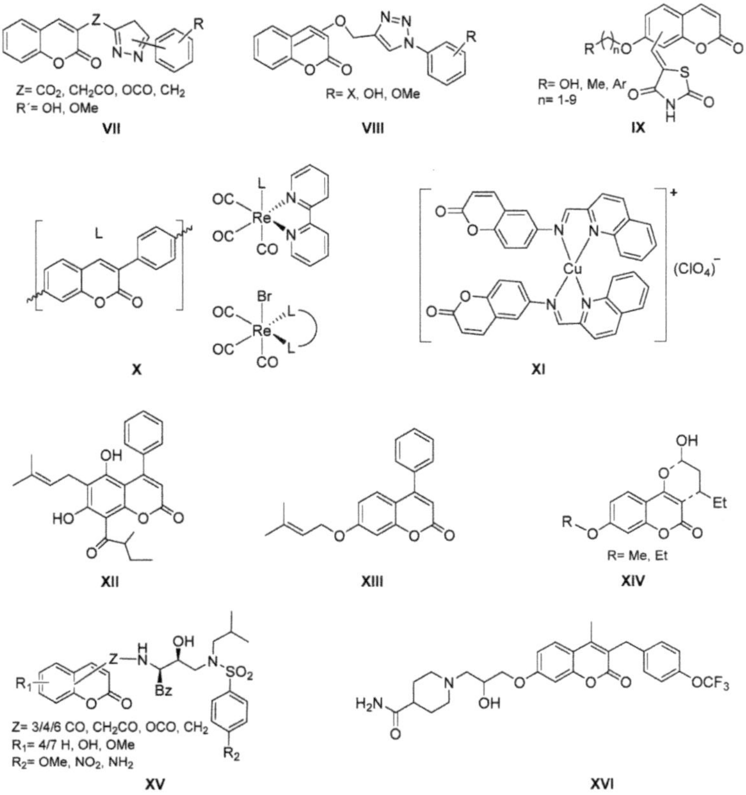

Figure 3. Structures of coumarin derivatives as antimicrobial agents.

In addition to the antibacterial activity, in a recent and comprehensive review on coumarins, activity against protozoa of the genus *Leishmania* was described [24]. The most promising compounds are prenylated, glycosylated, furan/pyranocoumarins, or simple hydroxy- or methoxy-substituted coumarins, along with the natural coumarin mammea A/BB (Figure 3, structure **XII**). Derivatives of this natural product have been prepared and substitutions at positions 6 and 8, as well as the phenyl ring at position 4, turned out to be mandatory for the studied activity. This structure–activity relationship (SAR) study led to the synthesis of the simplest and most lipophilic analogue **XIII** (Figure 3), the most promising member of the group as an antileishmanial agent. Similar structures, some also derived of the *Mammea* genus, have been evaluated against *Mycobacterium tuberculosis*, an activity that also shows simpler synthetic analogues derived from 4-hydroxycoumarin (Figure 3, general structure **XIV**) [25].

Finally, it is worth mentioning two articles reported this year on the design and preparation of coumarin derivatives with potential antiviral activity. This is the case of the dual hybrid inhibitors inspired by the antiviral activity of calanolide, known as reverse transcriptase (RT) inhibitor. With this in mind, dual inhibitors of HIV-1 RT and protease (PR) have been designed, in which the coumarin fragment responsible for RT activity is linked to the fragment of the antiretroviral darunavir, active against PR of the HIV, through different amide, carbamate, or amine linkers (Figure 3, general structure **XV**) [26]. The second case described the introduction of a piperidine ring through a linker in position 7 of the coumarin scaffold, originating compounds with outstanding activity against certain filoviruses such as Marburg virus (MARV) or Ebolavirus (EBOV). From the SAR studied, it is interesting to highlight the role of substitution in *para* position with a trifluoromethoxy group that originated compound **XVI** (Figure 3) with IC_{50} = 0.5 μM and 1.2 μM against EBOV and MARV, respectively [27].

2.3. Antioxidant and Anti-Inflammatory Activities

Although we have found very few publications related to these activities, in some cases the antioxidant activity of coumarin derivatives of both natural [28] and synthetic [29] origin has been reported. This is the case of NOs inhibitors, an activity described for coumarins that bind through different linkers to phenolic fragments capable of acting as radical scavengers (Figure 4, general structure **XVII**), hybrids that can therefore be used in the treatment of immunomodulatory diseases.

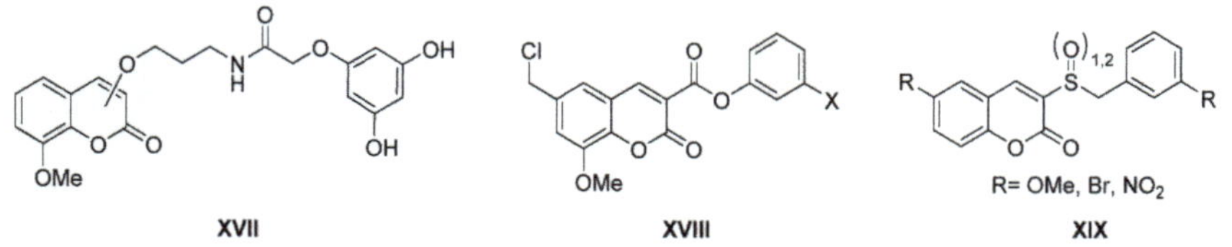

R= OMe, Br, NO_2

XVII XVIII XIX

Figure 4. Structures of coumarin derivatives as antioxidant and anti-inflammatory agents.

Regarding the anti-inflammatory activity, it is worth mentioning a review on the coumarins of natural origin (simple coumarins, prenylcoumarins, furocoumarins, coumestans, and benzocoumarins) with a detailed anti-inflammatory activity due to the activation of the nuclear factor erythroid 2-related factor 2 (Nrf2 factor) that protects cells against stress oxidative [30]. In other cases, the anti-inflammatory activity found for coumarin esters (Figure 4, general structure **XVIII**) as inhibitors of the Kallikrein-related peptidase 9 (KLK9) involved in inflammatory processes of the skin is reported [31]. Finally, the replacement of the carboxylic group by a sulfone or sulfoxide group (Figure 4, general structure **XIX**) gives rise to new inhibitors of cyclooxygenase-2 (COX-2) with activities comparable, in many cases, to indomethacin [32].

2.4. Adenosine Ligands

The affinity of the coumarin system for adenosine receptors has also been published recently. The 3-arylcoumarins (Figure 5, compound **XX**) have been described as antagonists of hA_3 receptors, showing a high affinity (in the low nanomolar range) and selectivity for this subtype [33], while the 3-aroylcoumarins (Figure 5, general structure **XXI**) have been described as dual hA_1/hA_3 antagonists in the low micromolar range [34]. These works are aligned with the already known potential of these derivatives as modulators of the different adenosine receptors, published in the last decade.

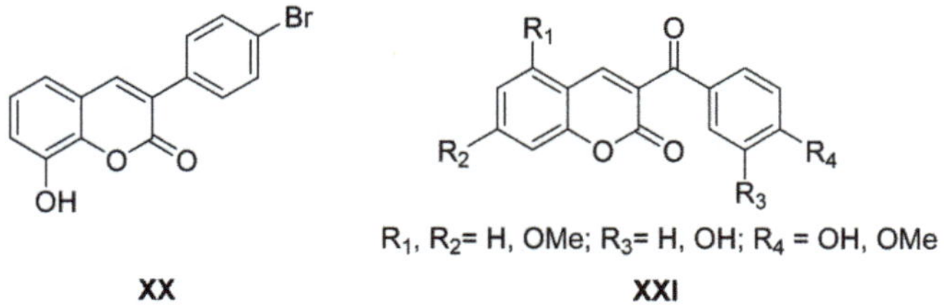

R_1, R_2= H, OMe; R_3= H, OH; R_4 = OH, OMe

XX XXI

Figure 5. Structures of coumarin derivatives as adenosine ligands.

2.5. Enzymatic Inhibitory Activity: α-Glucosidase, Carbonic Anhydrase, Tyrosinase, Sulfatase, and Xanthine Oxidase

The activity of coumarin derivatives on α-glucosidase was also reviewed in 2020 [35], in a study in which the influence of the substitution pattern was evaluated, and an important SAR was established. In addition to α-glucosidase, aldehyde dehydrogenase 1A1 (ALDH1A1) is another target for the treatment of diabetes and obesity, and 3-amidocoumarins have been described as inhibitors of this enzyme, with compound **XXII** (Figure 6) being a very promising derivative (IC_{50} = 3.87 mM) [36]. This activity has also recently been found for hybrids of coumarin and cinnamic acid, with compound **XXIII** (Figure 6) being described as a very promising derivative (IC_{50} = 12.98 mM) [37].

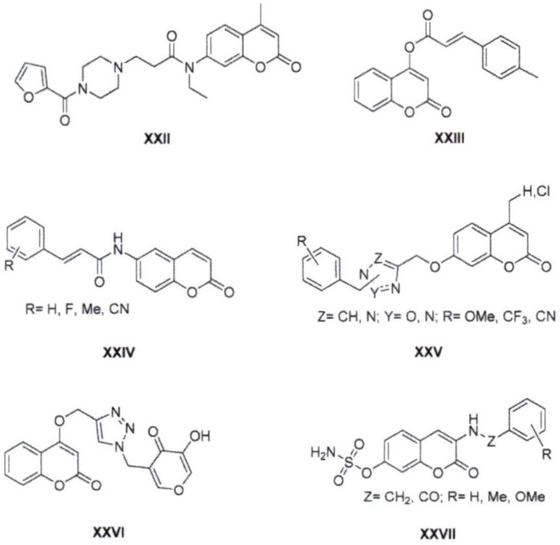

Figure 6. Structures of coumarin derivatives as enzymatic inhibitors.

Closely related to these structures, works have been published on coumarin derivatives with inhibitory activity on carbonic anhydrase **IX** and **XII**. These are coumarins that incorporate arylacrylamide substituents at position 3 (Figure 6, general structure **XXIV**) that showed inhibitory activity in the nanomolar range [38]. In other cases, anhydrase inhibitory activity was reported for hybrids connected by a methyleneoxy linker at position 7, oxadiazole heterocycles [39] that the same authors extend to the triazole ring (Figure 6, general structure **XXV**) [40].

These last structures are closely related to others that present tyrosinase inhibitory activity in the sub-micromolar range. This is the case of coumarins (umbelliferone and other phenolic analogues) that incorporate a kojic acid fragment through a triazole linker at position 4 (Figure 6, compound **XXVI**), both fragments with demonstrated tyrosinase inhibitory activity [41].

Likewise, the introduction of a sulfamate group in the coumarin scaffold originates a hybrid prototype (Figure 6, general structure **XXVII**) that presents a high inhibitory activity of the steroid sulfatase (best compound of the series with IC_{50} = 0.13 μM), which is of interest in the treatment of hormone-dependent breast cancers [42].

During 2020, 3-phenylcoumarins were also studied as xanthine oxidase inhibitors [43]. Methoxy and nitro substituents were introduced into the framework. The best compound in the series proved to be 3-(4-methoxyphenyl)-6-nitrocoumarin, with an IC_{50} = 8.4 μM, being also non-cytotoxic in B16F10 cells.

2.6. *Anti-Neurodegenerative Diseases Activity: MAO and AChE/BChE Inhibitors*

The role played by coumarin derivatives as agents that exhibit biological activities associated with neurodegenerative diseases, such as Alzheimer's disease, is very important. Throughout this year, a large number of manuscripts related to this field have been found. Due to the multidirectional nature of these diseases, there are also many works on hybrid coumarins directed at different pharmacological targets, such as monoamine oxidase B

(MAO-B) or acetylcholinesterase (AChE), amyloid aggregation, or oxidative stress, among others. Hybrids of general structure **XXVIII** (Figure 7) have been described, in which the rasagiline fragment with MAO-B inhibitory activity and neuroprotection properties is incorporated into the coumarin scaffold also with demonstrated MAO-B inhibitory activity, antioxidant, and neuroprotective properties [44]. The incorporation at position 3 of a pyridazine ring (Figure 7, general structure **XXIX**) is another case of hybrid structures as selective MAO-B inhibitors [45]. The incorporation of isoxazole-type heterocycles in carboxamide–coumarins (Figure 7, general structure **XXX**) allowed obtaining derivatives with significant inhibitory activities of AChE/BuChE and beta-secretase 1 (BACE1) [46]. In other cases, taking into account the importance of metals in the pathogenesis of Alzheimer's disease, a pyridinone fragment (Figure 7, general structure **XXXI**) with iron-chelating properties was incorporated [47].

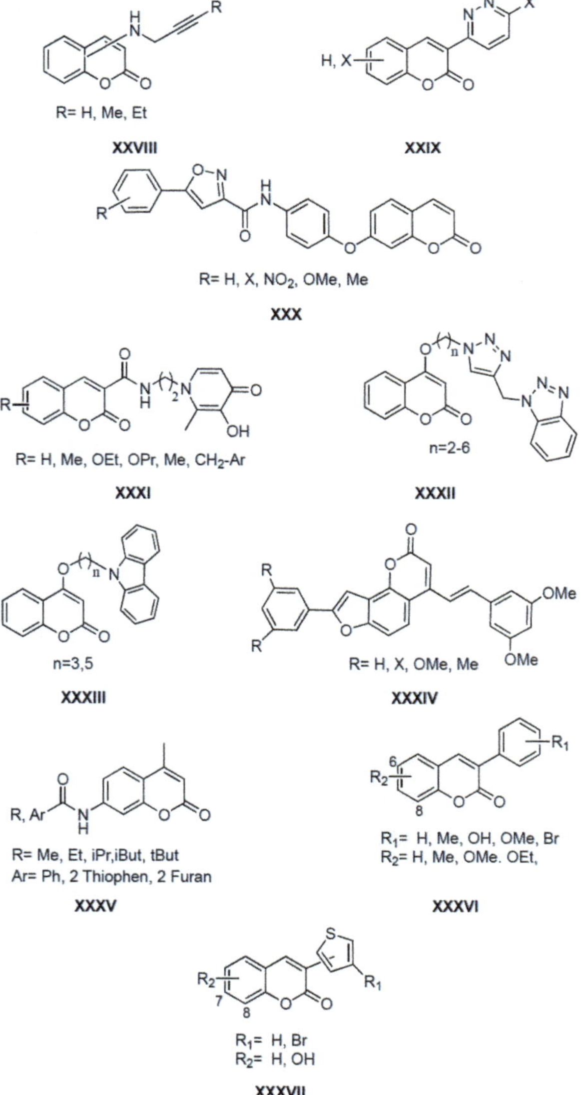

Figure 7. Structures of coumarin derivatives as anti-neurodegenerative diseases agents.

Other multitarget structures are the benzotriazole–coumarin (Figure 7, general structure **XXXII**) and carbazole–coumarin (Figure 7, general structure **XXXIII**) hybrids [48,49]. In both cases, the molecules show antioxidant activity, as well as AChE and β-amyloid aggregation inhibitory properties. Finally, it is worth mentioning another type of hybrid, this time a furocoumarin that incorporates two fragments of resveratrol (Figure 7, general structure **XXXIV**) [50]. Compounds containing this scaffold present AChE and BACE1 inhibitory properties related to the furocoumarin fragment, and antioxidant (radical scavenging) and COX-2 inhibition, related to the resveratrol [50].

The 7-amidocoumarins (Figure 7, general structure **XXXV**) have recently been published for their potential against monoamine oxidase A (*h*MAO-A), *h*MAO-B, *h*BACE1, *h*AChE, and butyrylcholinesterase (*h*BuChE) [51]. The research project is based on a screening of compounds with potential activity against Alzheimer's and Parkinson's diseases, since these multifactorial pathologies share some of their pharmacological targets. Five derivatives of the studied series were described as potent and selective *h*MAO-B inhibitors in the nanomolar range; six turned out to be *h*MAO-A inhibitors in the low micromolar range; one showed inhibitory activity of *h*BACE1, and another one *h*AChE inhibitory activity, both in the micromolar range. In addition to the enzymatic inhibition, all of the studied molecules proved to be non-cytotoxic to neurons in the motor cortex. As a main conclusion, results suggest that by modulating the substitution pattern at position 7 of the scaffold, selective or multitarget molecules can be achieved.

The 3-arylcoumarins are a family of compounds with proven activity on different targets related to neurodegenerative diseases, especially Alzheimer's and Parkinson's diseases, the two most prevalent. In the last decade, several manuscripts described very promising activities of this scaffold, both as selective and multitarget compounds. SAR studies were performed, and important conclusions were drawn based on the substitution patterns in the main scaffold. Due to the large number of molecules based on this scaffold currently synthetized and studied as *h*MAO inhibitors, in 2020 a theoretical work was published comparing different QSAR models and docking calculations, in order to predict the *h*MAO-B activity of the 3-arylcoumarins [52]. Based on the predictions, a small series of compounds was synthetized and evaluated against both *h*MAO-A and *h*MAO-B, and the most promising models were validated. Selective activities were found in the low nanomolar range against this isoenzyme for 6 and 8 methyl-substituted 3-arylcoumarins, also presenting methoxy groups or bromine atoms in different positions of the 3-phenyl ring (Figure 7, general structure **XXXVI**). These advancements may represent robust tools in the design of potent and selective derivatives.

Analogues of 3-phenylcoumarins were also published during 2020. The discovery and optimization of 3-thiophenylcoumarins (Figure 7, general structure **XXXVII**) as novel and promising agents against Parkinson's disease have been described [53]. This study explores, for the first time, the potential of these structures as in vitro and in vivo agents against this disease. The inhibitory activities of *h*MAO-A and *h*MAO-B, antioxidant profile, neurotoxicity in neurons of the motor cortex, and neuroprotection against hydrogen peroxide production were studied. The in vivo effect on locomotor activity was also evaluated by an open field test (OFT) for the most potent, selective and reversible *h*MAO-B inhibitor of the series: 3-(4′-bromothiophen-2′-yl)-7-hydroxycoumarin (IC_{50} = 140 nM). In reserpinized mice pre-treated with levodopa and benserazide, this molecule exhibited a slightly better in vivo profile than selegiline, currently a therapeutic option for Parkinson's disease. The results suggested that the 7-position substitution of the coumarin scaffold is interesting for enzyme inhibition. Furthermore, the presence of a catechol at positions 7 and 8 exponentially increases the antioxidant potential and the neuroprotective properties.

The neuroprotective effects of xanthotoxin and umbelliferone on streptozotocin (STZ)-induced cognitive dysfunction in rats were evaluated [54]. Alzheimer's disease was induced in these animals and both compounds were administrated, proving to prevent cognitive deficits in the Morris water maze and object recognition tests. In addition, both compounds reduced the activity of hippocampal AChE and the level of malondialdehyde, increasing the glutathione content. These coumarins also modulated neuronal cell death by reducing the level of proinflammatory cytokines, inhibiting the overexpression of inflammatory markers (nuclear factor κB and cyclooxygenase **II**), and upregulating the expression of NF-κB inhibitor (IκB-α). An attenuation of cognitive dysfunction by these compounds was observed. This effect can at least be attributed to the inhibition of AChE and the reduction of oxidative stress, neuroinflammation, and neuronal loss, opening a new door for these classic coumarins.

2.7. Anticoagulant Activity

The classic anticoagulant effect of specific coumarin derivatives, based on acenocoumarol and warfarin, also remains one of the classic applications for this family. During 2020, a review on this topic was published [55].

2.8. Fluorescent Probes

In addition to the interest of coumarins as a versatile scaffold in drug design, the important role that this scaffold plays as fluorescent probes to detect metals, enzymes, and biomaterials, among others, should be highlighted [56–59]. These fluorescent probes have a great imaging potential for the diagnosis of several pathologies.

Coumarins are being used in the selective detection of metals such as copper (Figure 8, general structure **XXXVIII**) [60] or its determination in drinking water (Figure 8, compound **XXXIX**) [61,62]. A recent review focuses on the detection of iron in water and its applications [63]. Other works study the fluorescence determination of the presence of silver in aqueous medium (Figure 8, compound **XL**) [64]. In the case of mercury, there are also published works in which the selective determination in water is studied (Figure 8, general structure **XLI**) [65]. In some cases, this determination is selective, but in this case, the innovative methodology can be applied over a wide pH range (Figure 8, compound **XLII**) [66].

In some cases, these metal complexes serve as probes for the detection of biothiols, as in the case of copper complexes with benzothiazoles (Figure 8, compound **XLIII**) used in the determination of cysteines [67], or coumarin–quinoline complexes used in the detection of glutathione (Figure 8, compound **XLIV**) [68]. In other cases, aromatization to form the coumarin ring is used as a fluorescence test to detect the superoxide anion (Figure 8, compound **XLV**) [69].

In addition to the determination of metals, in many cases coumarin derivatives are used as fluorescence probes for the detection of hypochlorite, as in the case of coumarin–thiophene complexes (Figure 8, compound **XLVI**) [70] or of 2-thiocoumarins in which the presence of ClO– allows the formation of a fluorescent coumarin (Figure 8, compound **XLVII**) [71].

Coumarins can also be used as photocleavable linkers in the controlled release of drugs or biomaterials. This is the case of the in vivo photolysis of the microtubule inhibitor 4-pyridinomethylcoumarin (Figure 8, compound **XLVIII**) [72]. 7-Hydroxymethyl substituted aminocoumarins are used as iron complexes (Figure 8, general structures **XLIX** and **L**), and can be used to photochemotherapeutically target the mitochondria in the treatment of cancer [60,73].

Other reviews report on the use of 7-hydroxycoumarin and its derivatives in determining the activity of cytochromes P450 (CYP) enzymes [74] as well as 7-aminocoumarin derivatives in determining amino acids from serine or cysteine proteases [58]. In other cases, they are used to detect the metabolism of mitochondrial cysteines, the oxidation of which is a measure of cellular oxidative stress (Figure 8, compound **LI**) [75]. The use of the chiral coumarin-BINOL hybrid allows the enantioselective detection of amino acids (Figure 8, compound **LII**) [76]. Finally, coumarins can be used for easy detection of bacterial carbapenamases in which the coumarin fluorophore binds to the carbapenemic structure via a reactive linker (Figure 8, general structure **LIII**) [77,78].

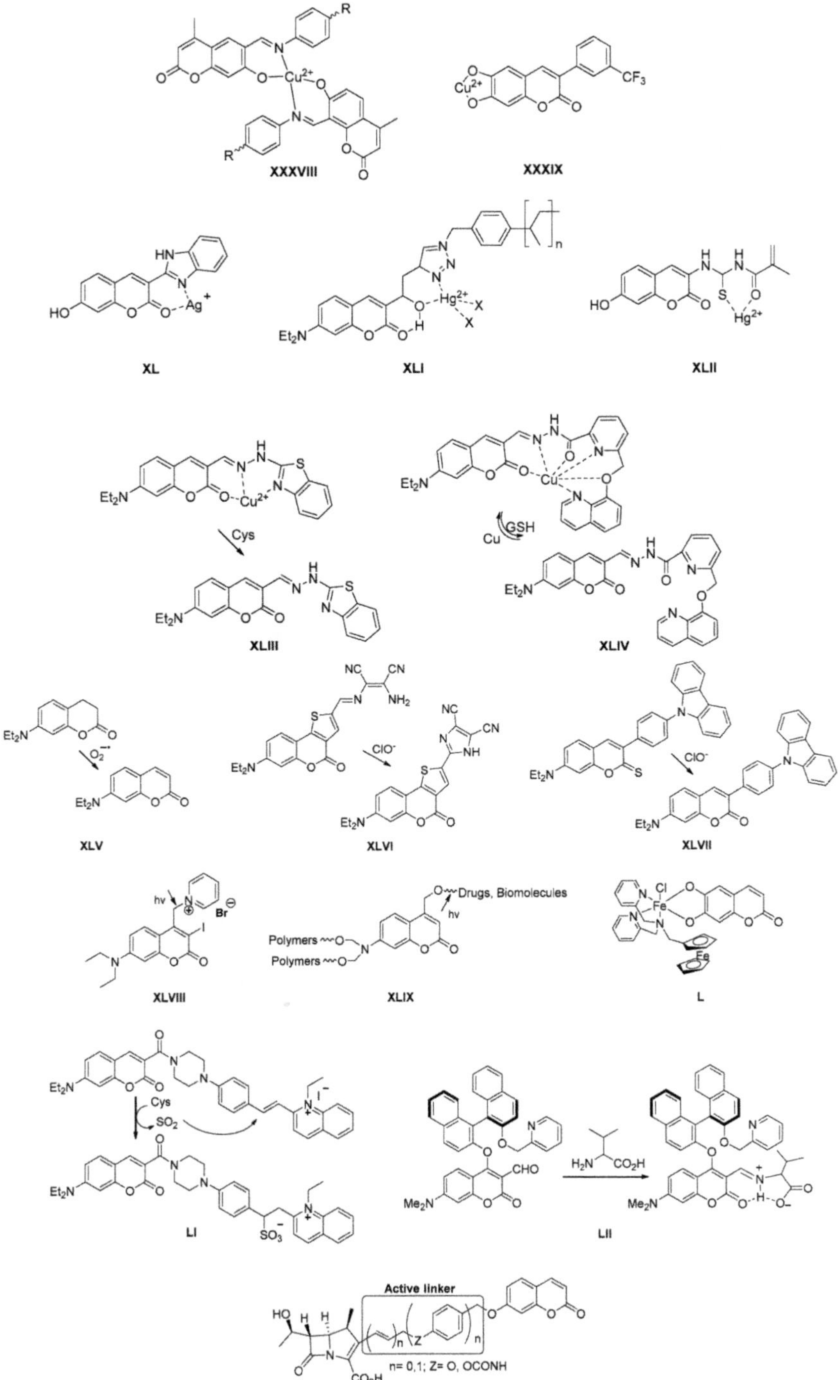

Figure 8. Structures of coumarin derivatives as fluorescent probes.

3. Perspectives

Coumarins are privileged structures for biological applications. Their conjugated double ring system allows different spots for chemical modifications, and a large number of derivatives can be obtained. Therefore, structure/activity studies appear to be the hottest emerging topic. During the year 2020, more than a thousand research articles and reviews related to coumarins could be found. This highlights the great potential that these molecules can have in different fields of research. For simplicity, our overview focused on the potential of coumarins in medicinal chemistry. The most relevant studies were included. The range of applications described in this document, and some others outside the scope of this general description (i.e., optoelectronic applications [79], polymers [80], etc.), reflect the versatility of this scaffold.

In our opinion, the potential of coumarins as fluorescent probes appears to be the most promising field of research for the next few years, since several coumarin derivatives have shown great potential in prodrug degradation (drug release) and diagnostics.

Due to the length of this general overview, synthetic strategies for obtaining new coumarins have not been discussed in detail. To find information on the most recent synthetic pathways, we recommend the manuscripts by Molnar and co-authors [81] and Kovač and co-authors [82], both from 2020. To find an overview of the wide range of biological activities of coumarins, a 2020 review by Pinto and co-authors [1] is strongly recommended. Finally, to find information on the most recent analytical methods (fundamentals, instrumentation, purification and quantification applications, optimization of experimental conditions, emerging ecological methods, etc.) we recommend the review by Xue-song and co-authors [83].

4. Conclusions

Coumarins belong to a privileged family for biomedical proposes. Their simplicity, chemical properties, and the efficiency of the synthetic routes to obtain a wide range of substitution patterns make these compounds highly attractive and versatile for medicinal and biological chemists. To date, and during 2020, more than a thousand research articles and reviews containing information on coumarins appear in the PubMed, SciFinder, and Mendeley databases. The number of research groups working on this scaffold, and the impact of the results, highlight the potential of these molecules. Special attention has been paid to the potential of coumarins in drug design, as well as to fluorescent probes. This last application seems to be the most promising field of research for the next few years, since several coumarin derivatives have shown great potential in the decaging of prodrugs (drug release) and for diagnostic purposes.

Author Contributions: Conceptualization, M.J.M. and L.S.; methodology, A.C., M.J.M., E.U., and L.S.; formal analysis, A.C., M.J.M., E.U., and L.S.; investigation, A.C., M.J.M., E.U., and L.S.; resources, M.J.M., E.U., and L.S.; writing—original draft preparation and editing, M.J.M. and L.S.; writing—review and editing, M.J.M., L.S., and E.U.; visualization, M.J.M., E.U., and L.S.; supervision, L.S.; project administration, M.J.M., E.U., and L.S.; funding acquisition, M.J.M., E.U., and L.S. All authors have read and agreed to the published version of the manuscript.

Funding: This research was funded by Xunta de Galicia (Galician Plan of Research, Innovation and Growth 2011–2015, Plan I2C, ED481B 2014/027-0, ED481B 2014/086–0 and ED481B 2018/007) and Fundação para a Ciência e Tecnologia (FCT, CEECIND/02423/2018 and UIDB/00081/2020).

Institutional Review Board Statement: Not applicable.

Informed Consent Statement: Not applicable.

Data Availability Statement: Data presented is original and not inappropriately selected, manipulated, enhanced, or fabricated.

Conflicts of Interest: The authors declare no conflict of interest.

References

1. Annunziata, F.; Pinna, C.; Dallavalle, S.; Tamborini, L.; Pinto, A. An overview of coumarin as a versatile and readily accessible scaffold with broad-ranging biological activities. *Int. J. Mol. Sci.* **2020**, *21*, 4618. [CrossRef] [PubMed]
2. Fotopoulos, I.; Hadjipavlou-Litina, D. Hybrids of coumarin derivatives as potent and multifunctional bioactive agents: A review. *Med. Chem.* **2020**, *16*, 272–306. [CrossRef] [PubMed]
3. Feng, D.; Zhang, A.; Yang, Y.; Yang, P. Coumarin-containing hybrids and their antibacterial activities. *Arch. Pharm.* **2020**, *353*, e1900380. [CrossRef] [PubMed]
4. Matiadis, D.; Sagnou, M. Pyrazoline hybrids as promising anticancer agents: An up-to-date overview. *Int. J. Mol. Sci.* **2020**, *21*, 5507. [CrossRef] [PubMed]
5. Song, X.F.; Fan, J.; Liu, L.; Liu, X.F.; Gao, F. Coumarin derivatives with anticancer activities: An update. *Arch. Pharm.* **2020**, *353*, e2000025. [CrossRef] [PubMed]
6. Akkol, E.K.; Genç, Y.; Karpuz, B.; Sobarzo-Sánchez, E.; Capasso, R. Coumarins and coumarin-related compounds in pharmacotherapy of cancer. *Cancers* **2020**, *12*, 1959. [CrossRef] [PubMed]
7. Al-Warhi, T.; Sabt, A.; Elkaeed, E.B.; Eldehna, W.M. Recent advancements of coumarin-based anticancer agents: An up-to-date review. *Bioorg. Chem.* **2020**, *103*, 104163. [CrossRef]
8. Goud, N.S.; Kumar, P.; Bharath, R.W. Recent developments of target based coumarin derivatives as potential anticancer agents. *Mini-Rev. Med. Chem.* **2020**, *20*, 1754–17668. [CrossRef]
9. Endo, S.; Oguri, H.; Segawa, J.; Kawai, M.; Hu, D.; Xia, S.; Okada, T.; Irie, K.; Fujii, S.; Gouda, H.; et al. Development of novel AKR1C3 inhibitors as new potential treatment for castration-resistant prostate cancer. *Med. Chem.* **2020**, *63*, 10396–10411. [CrossRef]
10. Wang, C.; Xi, D.; Wang, H.; Niu, Y.; Liang, L.; Xu, F.; Peng, Y.; Xu, P. Hybrids of MEK inhibitor and NO donor as multitarget antitumor drugs. *Eur. J. Med. Chem.* **2020**, *196*, 112271. [CrossRef]
11. Xu, J.; Li, H.; Wang, X.; Huang, J.; Li, S.; Liu, C.; Dong, R.; Zhu, G.; Duan, C.; Jiang, F.; et al. Discovery of coumarin derivatives as potent and selective cyclin-dependent kinase 9 (CDK9) inhibitors with high antitumour activity. *Eur. J. Med. Chem.* **2020**, *200*, 112424. [CrossRef] [PubMed]
12. Zhao, N.; Yang, F.; Han, L.; Qu, Y.; Ge, D.; Zhang, H. Development of coumarin-based hydroxamates as histone deacetylase inhibitors with antitumor activities. *Molecules* **2020**, *25*, 717. [CrossRef] [PubMed]
13. Wang, U.; Zhang, W.; Dong, J.; Gao, J. Design, synthesis, and bioactivity evaluation of coumarin-chalcone hybrids as potential anticancer agents. *Bioorg. Chem.* **2020**, *95*, 103530. [CrossRef]
14. Wang, B.Y.; Lin, Y.C.; Lai, Y.T.; Ou, J.Y.; Chang, W.W.; Chu, C. Targeted photoresponsive carbazole–coumarin and drug conjugates for efficient combination therapy in leukemia cancer cells. *Bioorg. Chem.* **2020**, *100*, 103904. [CrossRef] [PubMed]
15. Qin, H.L.; Zhang, Z.W.; Ravindar, L.; Rakesh, K.P. Antibacterial activities with the structure-activity relationship of coumarin derivatives. *Eur. J. Med. Chem.* **2020**, *207*, 112832. [CrossRef]
16. Sumorek-Wiadro, J.; Zając, A.; Langner, E.; Skalicka-Woźniak, K.; Maciejczyk, A.; Rzeski, W.; Jakubowicz-Gil, J. Antiglioma potential of coumarins combined with Sorafenib. *Molecules* **2020**, *25*, 5192. [CrossRef]
17. Sumorek-Wiadro, J.; Zając, A.; Bądziul, D.; Langner, E.; Skalicka-Woźniak, K.; Maciejczyk, A.; Wertel, I.; Rzeski, W.; Jakubowicz-Gil, J. Coumarins modulate the anti-glioma properties of temozolomide. *Eur. J. Pharmacol.* **2020**, *881*, 173207. [CrossRef]
18. Liu, H.; Xia, D.G.; Chu, Z.W.; Hu, R.; Cheng, X.; Lv, X.H. Novel coumarin-thiazolyl ester derivatives as potential DNA gyrase Inhibitors: Design, synthesis, and antibacterial activity. *Bioorg. Chem.* **2020**, *100*, 103907. [CrossRef]
19. Sutar, S.M.; Savanur, H.M.; Malunavar, S.S.; Pawashe, G.M.; Aridoss, G.; Kim, K.M.; Lee, J.Y.; Kalkhambkar, R.G. Synthesis and molecular modelling studies of coumarin and 1-aza-coumarin linked miconazole analogues and their antimicrobial properties. *ChemistrySelect* **2020**, *5*, 1322–1330. [CrossRef]
20. Lnufaie, R.; Hansa, R.K.C.; Alsup, N.; Whitt, J.; Chambers, S.A.; Gilmore, D.; Alam, M.A. Synthesis and antimicrobial studies of coumarin-substituted pyrazole derivatives as potent anti-*Staphylococcus aureus* agents. *Molecules* **2020**, *25*, 2758. [CrossRef]
21. Hu, C.F.; Zhang, P.L.; Sui, Y.F.; Lv, J.S.; Ansari, M.F.; Battini, N.; Li, S.; Zhou, C.H.; Geng, R.X. Ethylenic conjugated coumarin thiazolidinediones as new efficient antimicrobial modulators against clinical methicillin-resistant. *Bioorg. Chem.* **2020**, *94*, 103434. [CrossRef]
22. Nasiri Sovari, S.; Vojnovic, S.; Skaro Bogojevic, S.; Crochet, A.; Pavic, A.; Nikodinovic-Runic, J.; Zobi, F. Design, synthesis and in vivo evaluation of 3-arylcoumarin derivatives of rhenium (I) tricarbonyl complexes as potent antibacterial agents against methicillin-resistant *Staphylococcus aureus* (MRSA). *Eur. J. Med. Chem.* **2020**, *204*, 112533. [CrossRef] [PubMed]
23. Aldabaldetrecu, M.; Parra, M.; Soto, S.; Arce, P.; Tello, M.; Guerrero, J.; Modak, B. New Copper (I) complex with a coumarin as ligand with antibacterial activity against *Flavobacterium psychrophilum*. *Molecules* **2020**, *25*, 3183. [CrossRef] [PubMed]
24. Gonçalves, G.A.; Spillere, A.R.; das Neves, G.M.; Kagami, L.P.; von Poser, G.L.; Canto, R.F.S.; Eifler-Lima, V.L. Natural and synthetic coumarins as antileishmanial agents: A review. *Eur. J. Med. Chem.* **2020**, *203*, 112514. [CrossRef] [PubMed]
25. Pires, C.T.A.; Scodro, R.B.L.; Cortez, D.A.G.; Brenzan, M.A.; Siquiera, V.L.D.; Caleffi-Ferracioli, K.R.; Vieira, L.C.C.; Monteiro, J.L.; Corrêa, A.G.; Cardoso, R.F. Structure–activity relationship of natural and synthetic coumarin derivatives against *Mycobacterium tuberculosis*. *Future Med. Chem.* **2020**, *11*, 1533–1546. [CrossRef]

26. Zhu, M.; Ma, L.; Wen, J.; Dong, B.; Wang, Y.; Wang, Z.; Zhou, J.; Zhang, G.; Wang, J.; Guo, Y.; et al. Rational design and structure-activity relationship of coumarin derivatives effective on HIV-1 protease and partially on HIV-1 reverse transcriptase. *Eur. J. Med. Chem.* **2020**, *186*, 111900. [CrossRef]
27. Gao, Y.; Cheng, H.; Khan, S.; Xiao, G.; Rong, L.; Bai, C. Development of coumarin derivatives as potent anti-filovirus entry inhibitors targeting viral glycoprotein. *Eur. J. Med. Chem.* **2020**, *204*, 112595. [CrossRef]
28. Minhas, R.; Bansal, G.; Bansal, Y. Novel coupled molecules from active structural motifs of synthetic and natural origin as immunosuppressants. *Med. Chem.* **2020**, *16*, 544–554. [CrossRef]
29. Salar, U.; Khan, K.M.; Jabeen, A.; Faheem, A.; Naqvi, F.; Ahmed, S.; Iqbal, E.; Ali, F.; Kanwal; Perveen, S. ROS inhibitory activity and cytotoxicity evaluation of benzoyl, acetyl, alkyl ester, and sulfonate ester substituted coumarin derivative. *Med. Chem.* **2020**, *16*, 1099–1111. [CrossRef]
30. Hassanein, E.H.M.; Sayed, A.M.; Hussein, O.E.; Mahmoud, A.M. Coumarins as modulators of the Keap1/Nrf2/ARE signaling pathway. *Oxidative Med. Cell. Longev.* **2020**, *2020*, 1675957. [CrossRef]
31. Hanke, S.; Tindall, C.A.; Pippel, J.; Ulbricht, D.; Pirotte, B.; Reboud-Ravaux, M.; Heiker, J.T.; Sträter, N. Structural studies on the inhibitory binding mode of aromatic coumarinic esters to human Kallikrein-related peptidase 7. *J. Med. Chem.* **2020**, *63*, 5723–57336. [CrossRef]
32. Wang, T.; Peng, T.; Wen, X.; Wang, G.; Liu, S.; Sun, Y.; Zhang, S.; Wang, L. Design, Synthesis and evaluation of 3-substituted coumarin derivatives as anti-inflammatory agents. *Chem. Pharm. Bull.* **2020**, *68*, 443–446. [CrossRef]
33. Matos, M.J.; Vilar, S.; Vazquez-Rodriguez, S.; Kachler, S.; Klotz, K.N.; Buccioni, M.; Delogu, G.; Santana, L.; Uriarte, E.; Borges, F. Structure-based optimization of coumarin hA_3 adenosine receptor antagonists. *J. Med. Chem.* **2020**, *63*, 2577–2587. [CrossRef]
34. Vazquez-Rodriguez, S.; Vilar, S.; Kachler, S.; Klotz, K.N.; Uriarte, E.; Borges, F.; Matos, M.J. Adenosine receptor ligands: Coumarin-chalcone hybrids as modulating agents on the activity of *h*ARs. *Molecules* **2020**, *25*, 4306. [CrossRef]
35. Tafesse, T.B.; Bule, M.H.; Khoobi, M.; Faramarzi, M.A.; Abdollahi, M.; Amini, M. Coumarin-based scaffold as α-glucosidase inhibitory activity: Implication for the development of potent antidiabetic agents. *Mini-Rev. Med. Chem.* **2020**, *20*, 134–151. [CrossRef]
36. Liang, D.; Fa, Y.; Yang, Z.; Zhang, Z.; Liu, M.; Liu, L.; Jiang, C. Discovery of coumarin-based selective aldehyde dehydrogenase 1A1 inhibitors with glucose metabolism improving activity. *Eur. J. Med. Chem.* **2020**, *187*, 111923. [CrossRef]
37. Xu, X.T.; Deng, X.Y.; Chen, J.; Liang, Q.M.; Zhang, K.; Li, D.L.; Wu, P.P.; Zheng, X.; Zhou, R.P.; Jiang, Z.Y.; et al. Synthesis and biological evaluation of coumarin derivatives as α-glucosidase inhibitors. *Eur. J. Med. Chem.* **2020**, *189*, 112013. [CrossRef]
38. Swain, B.; Angeli, A.; Singh, P.; Supuran, C.T.; Arifuddin, M. New coumarin/sulfocoumarin linked phenylacrylamides as selective transmembrane carbonic anhydrase inhibitors: Synthesis and in-vitro biological evaluation. *Bioorg. Med. Chem. Lett.* **2020**, *28*, 115586. [CrossRef]
39. Thacker, P.S.; Angeli, A.; Argulwar, O.S.; Tiwari, P.L.; Arifuddin, M.; Supuran, C.T. Design, synthesis and biological evaluation of coumarin linked 1,2,4-oxadiazoles as selective carbonic anhydrase IX and XII inhibitors. *Bioorg. Chem.* **2020**, *98*, 103739. [CrossRef]
40. Thacker, P.S.; Goud, N.S.; Argulwa, O.S.; Soman, J.; Angeli, A.; Alvala, M.; Arifuddin, M.; Supuran, C.T. Synthesis and biological evaluation of some coumarin hybrids as selective carbonic anhydrase IX and XII inhibitors. *Bioorg. Chem.* **2020**, *104*, 104272. [CrossRef]
41. Ashooriha, M.; Khoshneviszadeh, M.; Khoshneviszadeh, M.; Rafiei, A.; Kardan, M.; Yazdian-Robati, R.; Emami, S. Kojic acid–natural product conjugates as mushroom tyrosinase inhibitors. *Eur. J. Med. Chem.* **2020**, *201*, 112480. [CrossRef] [PubMed]
42. Hng, Y.; Lin, M.H.; Lin, T.S.; Liu, I.C.; Lin, I.C.; Lu, Y.L.; Chang, C.N.; Chiu, P.F.; Tsai, K.C.; Chen, M.J.; et al. Design and synthesis of 3-benzylaminocoumarin-7-*O*-sulfamate derivatives as steroid sulfatase inhibitors. *Bioorg. Chem.* **2020**, *96*, 103618. [CrossRef] [PubMed]
43. Era, B.; Delogu, G.L.; Pintus, F.; Fais, A.; Gatto, G.; Uriarte, E.; Borges, F.; Kumar, A.; Matos, M.J. Looking for new xanthine oxidase inhibitors: 3-phenylcoumarins versus 2-phenylbenzofurans. *Inter. J. Biolog. Macromol.* **2020**, *162*, 774–780. [CrossRef] [PubMed]
44. Matos, M.J.; Herrera Ibatá, D.M.; Uriarte, E.; Viña, D. Coumarin-rasagiline hybrids as potent and selective *h*MAO-B inhibitors, antioxidants, and neuroprotective agents. *ChemMedChem* **2020**, *15*, 532–538. [CrossRef]
45. Rodríguez-Enríquez, F.; Costas-Lago, M.C.; Besada, P.; Alonso-Pena, M.; Torres-Terán, I.; Viña, D.; Fontenla, J.A.; Sturlese, M.; Moro, S.; Quezada, E.; et al. Novel coumarin-pyridazine hybrids as selective MAO-B inhibitors for the Parkinson's disease therapy. *Bioorg. Chem.* **2020**, *104*, 104203. [CrossRef]
46. Saeedi, M.; Rastegari, A.; Hariri, R.; Mirfazli, S.S.; Mahdavi, M.; Edraki, N.; Firuzi, O.; Akbarzadeh, T. Design and synthesis of novel arylisoxazole-chromenone carboxamides: Investigation of biological activities associated with Alzheimer's disease. *Chem. Biodivers* **2020**, *17*, e1900746. [CrossRef]
47. Jiang, X.; Guo, J.; Lv, Y.; Yao, C.; Zhang, C.; Mi, Z.; Shi, Y.; Gu, J.; Zhou, T.; Bai, R.; et al. Rational design, synthesis and biological evaluation of novel multitargeting anti-AD iron chelators with potent MAO-B inhibitory and antioxidant activity. *Bioorg. Med. Chem. Lett.* **2020**, *28*, 115550. [CrossRef]
48. Singh, A.; Sharma, S.; Arora, S.; Attri, S.; Kaur, P.; Gulati, H.K.; Bhagat, K.; Kumar, N.; Singh, H.; Singh, J.V.; et al. New coumarin-benzotriazole based hybrid molecules as inhibitors of acetylcholinesterase and amyloid aggregation. *Bioorg. Med. Chem. Lett.* **2020**, *30*, 127477. [CrossRef]

49. Shi, D.H.; Min, W.; Song, M.Q.; Si, X.X.; Li, M.C.; Zhang, Z.Y.; Liu, Y.W.; Liu, W.W. Synthesis, characterization, crystal structure and evaluation of four carbazole-coumarin hybrids as multifunctional agents for the treatment of Alzheimer's disease. *J. Mol. Struct.* **2020**, *1209*, 127897. [CrossRef]
50. Agbo, E.N.; Gildenhuys, S.; Choong, Y.S.; Mphahlele, M.J.; More, G.K. Synthesis of furocoumarin–stilbene hybrids as potential multifunctional drugs against multiple biochemical targets associated with Alzheimer's disease. *Bioorg. Chem.* **2020**, *101*, 103997. [CrossRef]
51. Rodríguez-Enríquez, F.; Viña, D.; Uriarte, E.; Laguna, R.; Matos, M.J. 7-Amidocoumarins as multitarget agents against neurodegenerative diseases: Substitution pattern modulation. *ChemMedChem* **2020**, *15*. [CrossRef] [PubMed]
52. Mellado, M.; Mella, J.; González, C.; Viña, D.; Uriarte, E.; Matos, M.J. 3-Arylcoumarins as highly potent and selective monoamine oxidase B inhibitors: Which chemical features matter? *Bioorg. Chem.* **2020**, *101*, 103964. [CrossRef] [PubMed]
53. Rodríguez-Enríquez, F.; Viña, D.; Uriarte, E.; Fontenla, J.A.; Matos, M.J. Discovery and optimization of 3-thiophenylcoumarins as novel agents against Parkinson's disease: Synthesis, in vitro and in vivo studies. *Bioorg. Chem.* **2020**, *101*, 103986. [CrossRef] [PubMed]
54. Hindam, M.O.; Sayed, R.H.; Skalicka-Woźniak, K.; Budzyńska, B.; EL Sayed, N.S. Xanthotoxin and umbelliferone attenuate cognitive dysfunction in a streptozotocin-induced rat model of sporadic Alzheimer's disease: The role of JAK2/STAT3 and Nrf2/HO-1 signalling pathway modulation. *Phytoth. Res.* **2020**, *34*, 2351–2365. [CrossRef]
55. Kasperkiewicz, K.; Ponczek, M.B.; Owczarek, J.; Guga, P.; Budzisz, E. Antagonists of vitamin K -popular coumarin drugs and new synthetic and natural coumarin derivatives. *Molecules* **2020**, *25*, 1465. [CrossRef]
56. Sunacd, X.Y.; Liubcd, T.; Sun, J.; Wan, X.J. Synthesis and application of coumarin fluorescence probes. *RSC Adv.* **2020**, *10*, 10826–10847. [CrossRef]
57. Breidenbach, J.; Bartz, U.; Gütschow, M. Coumarin as a structural component of substrates and probes for serine and cysteine proteases. *BBA-Proteins Proteom.* **2020**, *1868*, 140445. [CrossRef]
58. Raunio, H.; Pentikaeinen, O.; Juvonen, R.O. Coumarin-based profluorescent and fluorescent substrates for determining xenobiotic-metabolizing enzyme activities in vitro. *Int. J. Mol. Sci.* **2020**, *21*, 4708. [CrossRef]
59. Shen, W.; Zheng, J.; Zhou, Z.; Zhang, D. Approaches for the synthesis of *o*-nitrobenzyl and coumarin linkers for use in photocleavable biomaterials and bioconjugates and their biomedical applications. *Acta Biomater.* **2020**, *115*, 75–91. [CrossRef]
60. Ying, W.; Xiaohui, H.; Lixun, L.; Luyao, G.; Xumin, R.; Yonggang, W.; Hongchi, Z. A coumarin-containing Schiff base fluorescent probe with AIE effect for the copper (II) ion. *RSC Adv.* **2020**, *10*, 6109–6113. [CrossRef]
61. Arslan, F.N.; Geyik, G.A.; Koran, K.; Ozen, F.; Aydin, D.; Elmas, S.N.K.; Gorgulu, A.O.; Yilmaz, I. Fluorescence "turn on-off" sensing of copper (II) ions utilizing coumarin-based chemosensor: Experimental study, theoretical calculation, mineral and drinking water analysis. *J. Fluoresc.* **2020**, *30*, 317–327. [CrossRef] [PubMed]
62. Hien, N.K.; Bay, M.V.; Bao, N.C.; Vo, Q.V.; Cuong, N.D.; Thien, T.V.; Nhung, N.T.A.; Van, D.U.; Nam, P.C.; Quang, D.T. Coumarin-based dual chemosensor for colorimetric and fluorescent detection of Cu^{2+} in water media. *ACS Omega* **2020**, *5*, 21241–21249. [CrossRef]
63. Liu, J.; Guo, Y.; Dong, B.; Sun, J.; Lyu, J.; Sun, L.; Hu, S.; Xu, L.; Bai, X.; Xu, W.; et al. Water-soluble coumarin oligomer based ultra-sensitive iron ion probe and applications. *Sens. Actuator B-Chem.* **2020**, *320*, 128361. [CrossRef]
64. Jiang, X.; Yang, Y.; Li, H.; Qi, X.; Zhou, X.; Deng, M.; Lü, M.; Wu, J.; Liang, S. A Water-soluble fluorescent probe for the selective sensing of Ag^{+} and its application in imaging of living cells and nematodes. *J. Fluoresc.* **2020**, *30*, 121–129. [CrossRef] [PubMed]
65. Ngororabanga, J.M.V.; Tshentu, Z.R.; Mama, N. New highly selective colorimetric and fluorometric coumarin-based chemosensor for Hg^{2+}. *J. Fluoresc.* **2020**, *30*, 985–997. [CrossRef] [PubMed]
66. Pan, Z.; Xu, Z.; Chen, J.; Hu, L.; Li, H.; Zhang, X.; Gao, X.; Wang, M.; Zhang, J. Coumarin thiourea-based fluorescent turn-on Hg^{2+} probe that can be utilized in a broad pH range 1–11. *J. Fluoresc.* **2020**, *30*, 505–514. [CrossRef] [PubMed]
67. Khoa Hien, N.; Van Bay, M.; Diem Tran, P.; Tan Khanh, N.; Dinh Luyen, N.; Vo, Q.V.; Ung Van, D.; Cam Nam, P.; Tuan Quang, D. A coumarin derivative-Cu^{2+} complex-based fluorescent chemosensor for detection of biothiols. *RSC Adv.* **2020**, *10*, 36265–36274. [CrossRef]
68. Li, S.; Cao, D.; Meng, X.; Hu, Z.; Li, Z.; Yuan, C.; Zhou, T.; Han, X.; Ma, W. A novel fluorescent sensor for specific recognition of GSH based on the copper complex and its bioimaging in living cells. *Bioorg. Chem.* **2020**, *100*, 103923. [CrossRef]
69. Wang, Y.; Han, J.; Xu, Y.; Gao, Y.; Wen, H.; Cui, H. Taking advantage of the aromatization of 7-diethylamino-4-methyl-3,4-dihydrocoumarin in the fluorescence sensing of superoxide anion. *Chem. Commun.* **2020**, *56*, 9827–9829. [CrossRef]
70. Shi, L.; Yu, H.; Zeng, X.; Yang, S.; Gong, S.; Xiang, H.; Zhangd, K.; Shao, G. A novel ratiometric fluorescent probe based on thienocoumarin and its application for the selective detection of hypochlorite in real water samples and in vivo. *New J. Chem.* **2020**, *44*, 6232–6237. [CrossRef]
71. Nguyen, V.N.; Heo, S.; Kim, S.; Swamy, K.M.K.; Ha, J.; Park, S.; Yoon, J. A thiocoumarin-based turn-on fluorescent probe for hypochlorite detection and its application to live-cell imaging. *Sens. Actuator B-Chem.* **2020**, *317*, 128213. [CrossRef]
72. Tang, X.J.; Wu, Y.; Zhao, R.; Kou, X.; Dong, Z.; Zhou, W.; Zhang, Z.; Tan, W.; Fang, X. Photorelease of pyridines using a metal-free photoremovable protecting group. *Angew. Chem. Int. Edit.* **2020**, *59*, 18386–18389. [CrossRef]
73. Sarkar, T.; Bhattacharyya, A.; Banerjee, S.; Hussain, A. LMCT transition-based red-light photochemotherapy using a tumour-selective ferrocenyl iron (III) coumarin conjugate. *Chem. Commun.* **2020**, *56*, 7981–7984. [CrossRef]

74. Xia, Z.; Chen, D.; Song, S.; van der Vlag, R.; van der Wouden, P.E.; van der Merkerk, R.; Cool, R.H.; Hirsch, A.K.H.; Melgert, B.N.; Quax, W.J.; et al. 7-Hydroxycoumarins are affinity-based fluorescent probes for competitive binding studies of macrophage migration inhibitory factor. *J. Med. Chem.* **2020**, *63*, 11920–11933. [CrossRef]
75. Han, X.; Zhai, Z.; Yang, X.; Zhang, D.; Tang, J.; Zhu, J.; Zhu, X.; Ye, Y. A FRET-based ratiometric fluorescent probe to detect cysteine metabolism in mitochondria. *Org. Biomol. Chem.* **2020**, *18*, 1487–1492. [CrossRef]
76. Wu, X.; Wang, Q.; Dickie, D.; Pu, L. Mechanistic study on a BINOL–coumarin-based probe for enantioselective fluorescent recognition of amino acids. *J. Org. Chem.* **2020**, *85*, 6352–6358. [CrossRef]
77. Kim, J.; Kim, Y.; Abdelazem, A.Z.; Kim, H.J.; Choo, H.; Kim, H.S.; Kim, J.O.; Park, Y.J.; Min, S.J. Development of carbapenem-based fluorogenic probes for the clinical screening of carbapenemase-producing bacteria. *Bioorg. Chem.* **2020**, *94*, 103405. [CrossRef]
78. Wang, J.; Xu, W.; Xue, S.; Yua, T.; Xie, H. A minor structure modification serendipitously leads to a highly carbapenemase-specific fluorogenic probe. *Org. Biomol. Chem.* **2020**, *18*, 4029–4033. [CrossRef]
79. Kumar, A.; Baccoli, R.; Fais, A.; Cincotti, A.; Pilia, L.; Gatto, G. Substitution effects on the optoelectronic properties of coumarin derivatives. *Appl. Sci.* **2020**, *10*, 144. [CrossRef]
80. Cuevas, J.M.; Seoane-Rivero, R.; Navarro, R.; Marcos-Fernández, A. Coumarins into polyurethanes for smart and functional materials. *Polymers* **2020**, *12*, 630. [CrossRef]
81. Lončarić, M.; Gašo-Sokač, D.; Jokić, S.; Molnar, M. Recent advances in the synthesis of coumarin derivatives from different starting materials. *Biomolecules* **2020**, *10*, 151. [CrossRef] [PubMed]
82. Molnar, M.; Lončarić, M.; Kovač, M. Green chemistry approaches to the synthesis of coumarin derivatives. *Curr. Org. Chem.* **2020**, *24*, 4–43. [CrossRef]
83. Dong-wei, C.; Yuan, Z.; Xiao-Yi, D.; Yu, Z.; Guo-hui, L.; Xue-song, F. Progress in pretreatment and analytical methods of coumarins: An update since 2012—A review. *Crit. Rev. Anal. Chem.* **2020**. [CrossRef] [PubMed]

 MDPI

Review

Synthetic Routes to Coumarin(Benzopyrone)-Fused Five-Membered Aromatic Heterocycles Built on the α-Pyrone Moiety. Part 1: Five-Membered Aromatic Rings with One Heteroatom

Eslam Reda El-Sawy [1], Ahmed Bakr Abdelwahab [2] and Gilbert Kirsch [3,*]

1 National Research Centre, Chemistry of Natural Compounds Department, Dokki-Cairo 12622, Egypt; eslamelsawy@gmail.com
2 Plant Advanced Technologies (PAT), 54500 Vandoeuvre-les-Nancy, France; ahm@plantadvanced.com
3 Laboratoire Lorrain de Chimie Moléculaire (L.2.C.M.), Université de Lorraine, 57050 Metz, France
* Correspondence: gilbert.kirsch@univ-lorraine.fr; Tel.: +33-03-72-74-92-00; Fax: +33-03-72-74-91-87

Abstract: This review gives an up-to-date overview of the different ways (routes) to the synthesis of coumarin (benzopyrone)-fused, five-membered aromatic heterocycles with one heteroatom, built on the pyrone moiety. Covering 1966 to 2020.

Keywords: coumarins; benzopyrones; five-membered aromatic heterocycles; furan; pyrrole; thiophene; selenophen

Citation: El-Sawy, E.R.; Abdelwahab, A.B.; Kirsch, G. Synthetic Routes to Coumarin(Benzopyrone)-Fused Five-Membered Aromatic Heterocycles Built on the α-Pyrone Moiety. Part 1: Five-Membered Aromatic Rings with One Heteroatom. *Molecules* **2021**, *26*, 483. https://doi.org/10.3390/molecules26020483

Academic Editor: Maria João Matos
Received: 26 December 2020
Accepted: 13 January 2021
Published: 18 January 2021

1. Introduction

Coumarins are a family of benzopyrones (1,2-benzopyrones or 2*H*-[1]benzopyran-2-ones), which represent an important family of oxygen-containing heterocycles, widely distributed in nature [1–4]. Coumarins display a broad range of biological and pharmacological activities, [5,6] such as antiviral [7–10], anticancer [11–13], antimicrobial [14,15], and antioxidant [16–18] activities. On the other hand, coumarin represents an ingredient in perfumes [19], cosmetics [20], and as industrial additives [21,22]. Furthermore, coumarins play a pivotal role in science and technology as fluorescent sensors, mainly due to their interesting light-emissive characteristics, which are often responsive to the environment [23–26]. The coumarin (benzopyrane)-fused, membered aromatic heterocycles built on the α-pyrone moiety are an important scaffold. The only fused heterocycle with an α-pyrone moiety of coumarin that can be found in nature is the furan ring. One example is the naturally occurring furan 4*H*-furo[3,2-*c*]benzopyran-4-one, which provides the main core of many natural compounds of so-called coumestans. These comestans include coumestan, wedelolactone, and coumestrol. The coumestans are found in a variety of plant species that are commonly used in traditional medicine [27].

coumestan wedelolactone coumestrol

Naturally occurring furan, so-called coumestans

In order to enrich the limited versatility of the structures found in nature, synthesis of coumarin (benzopyrane)-fused, membered aromatic heterocycles has received considerable attention, including numerous reported routes.

This review gives an up-to-date overview of the different ways (routes) to the synthesis of benzopyrone-fused, five-membered aromatic heterocycles with one heteroatom, built on the pyrone moiety, from 1966 to 2020. Our main interest in this current work is to describe the components that have one heteroatom in an alicyclic-fused ring with the pyrone part of coumarin. The synthetic pathway of the investigated scaffold has provided systems containing oxygen, nitrogen, sulfur, and selenium in their core structure. The last heteroatom is less described in the output of the synthetic efforts. The fused heterocycles that contain more than one heteroatom will be detailed in the next part, which we intend to publish in the future.

Many strategies have been developed for the synthesis of the fused, five-membered aromatic heterocycle-benzopyran-4-ones. There are two main approaches to constructing these skeletons: five-membered, aromatic heterocycle construction, and pyrone-ring construction.

2. Synthesis of Benzopyrone-Fused, Five-Membered Aromatic Heterocycles

2.1. Five-Membered Aromatic Rings with One Heteroatom

2.1.1. Furans

Furobenzopyrone (or furocoumarins) comprises an important class of coumarins found in a wide variety of plants, particularly in the carrot (*Apiaceae/Umbelliferae*), legume (*Fabaceae*), and citrus families (*Rutaceae*) [27]. The chemical structure of furobenzopyrone (furocoumarins) consists of a furan ring fused with coumarin. The fusion of the furan ring to the α-pyrone moiety of coumarin forms the core structure of the three most common isomers, viz. 4*H*-furo[2,3-*c*]chromen(benzopyran)-4-one, 4*H*-furo[3,4-*c*]chromen(benzopyran)-4-one, and 4*H*-furo[3,2-*c*]chromen (benzopyran)-4-one (Figure 1).

4*H*-furo[2,3-*c*]chromen-4-one
4*H*-furo[2,3-*c*]benzopyran-4-one

4*H*-furo[3,4-*c*]chromen-4-one
4*H*-furo[3,4-*c*]benzopyran-4-one

4*H*-furo[3,2-*c*]chromen-4-one
4*H*-furo[3,2-*c*]benzopyran-4-one

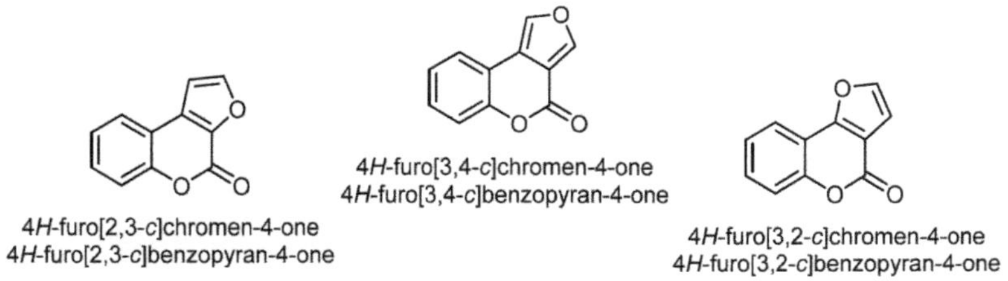

Figure 1. The three most common isomers of a furan ring fused to the α-pyrone moiety of coumarin.

4*H*-Furo[2,3-*c*]benzopyran-4-one

Furan Construction

The basic building block for the formation of 4*H*-furo[2,3-*c*]benzopyran-4-one is the 3-hydroxycoumarin (**1**) [28,29]. Pandya and coworkers [30] developed a method to synthesize some 4*H*-furo[2,3-*c*]benzopyran-4-ones starting with 3-hydroxycoumarin using the Nef reaction. Thus, the reaction of 3-hydroxycoumarin (**1**) with various 2-aryl-1-nitro ethenes **2a**,**b**, in the presence of piperidine and methanol as a solvent, followed the Nef reaction condition and afforded a series of 1-aryl-furo[2,3-*c*]benzopyran-4-ones **3a**,**b** and 1-phenyl-2-methyl-furo[2,3-*c*]benzopyran-4-one (**4**), respectively (Scheme 1). The formation of these products was explained by the reaction mechanism (Scheme 1).

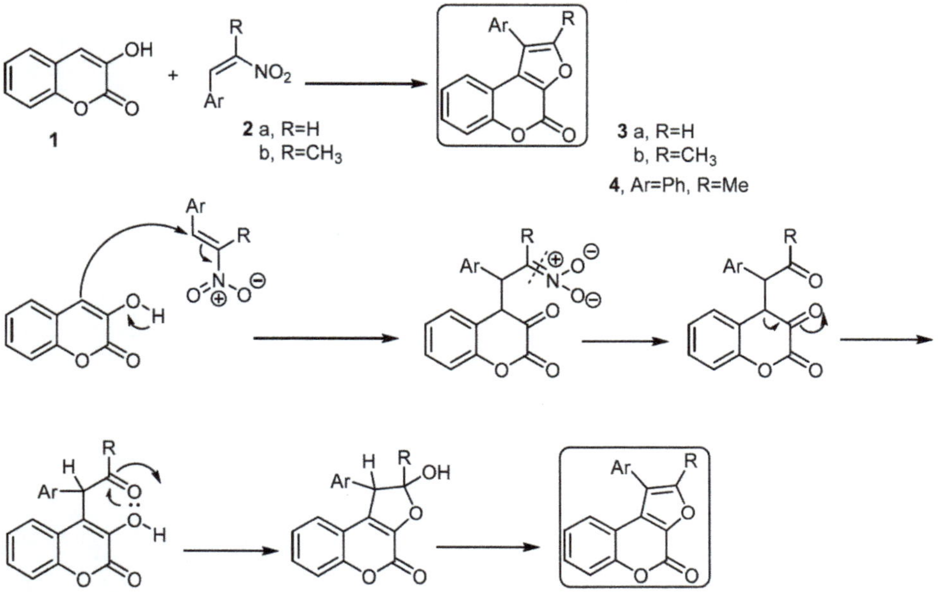

Scheme 1. The Nef reaction to synthesize furo[2,3-c]benzopyran-4-ones **3a,b** and **4**. *Reagents and conditions*: MeOH, piperidine, reflux, five outputs in 55%–61% yield.

Pyrone Construction

Dong et al., 2020 developed a novel and facile rhodium(III)-catalyzed process of sulfoxonium ylide (**5**) with hydroquinone (**6**). The carbonyl in the sulfoxonium ylide assisted the ortho-C–H functionalization of the sulfoxonium ylide, followed by intramolecular annulation with hydroquinone to afford 8-hydroxy-4*H*-furo[2,3-*c*]benzopyran-4-one (**7**) (Scheme 2) [31].

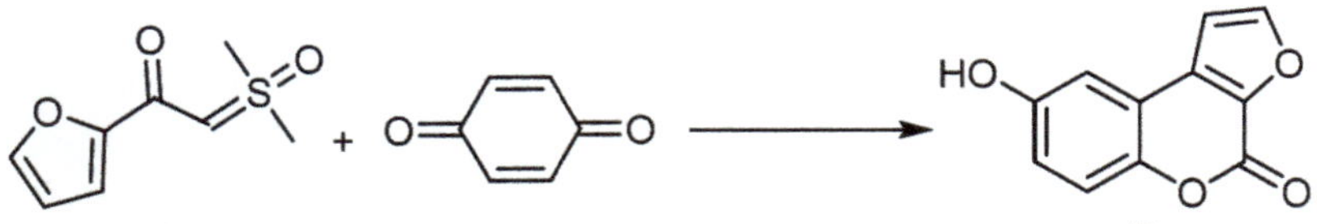

Scheme 2. Rhodium(III)-catalyzed sequential ortho-C–H oxidative arylation/cyclization of sulfoxonium ylide to afford 4*H*-furo[2,3-*c*]benzopyran-4-one (**7**). *Reagents and conditions*: $[Cp^*RhCl_2]_2$ (5 mol %), $AgBF_4$ (20 mol %), $Zn(OAc)_2$ (0.225 mmol), AcOH (0.3 mmol), and acetone (2 mL), 12 h, in a sealed Schlenk tube under N_2 at 100 °C, 25% yield.

4*H*-Furo[3,4-*c*]benzopyran-4-one

Furan and Pyrone Construction

In the literature, a large number of reports described the synthesis of 4*H*-furo[2,3-*c*] and 4*H*-furo[3,2-*c*]benzopyran-4-ones, while synthesis of the 4*H*-furo[3,4-*c*]benzopyran-4-one was reported by only one study, that of Brahmbhatt and his coworkers [32]. The first 4*H*-furo[3,4-*c*]benzopyran-4-ones (**10**) was synthesized by the demethylation–cyclization reaction of intermediates, 3-substituted-4-ethoxycarbonyl furans **9** (Scheme 3). For the demethylation and in situ lactonization steps, several reagents were tried, of which pyridine hydrochloride and HBr in acetic acid were found to be the most promising.

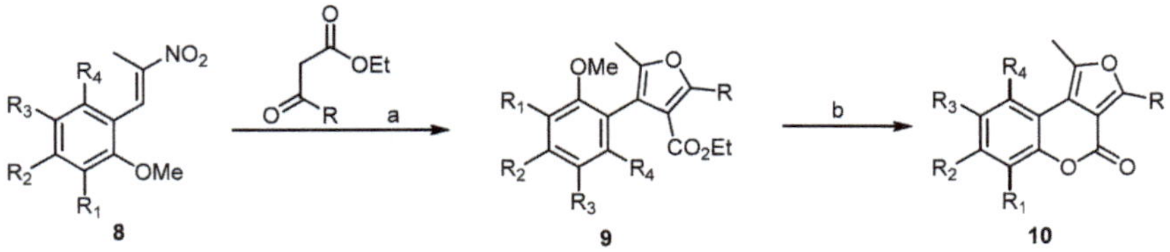

Scheme 3. Demethylation and in situ lactonization steps to prepare the first 4*H*-furo[3,4-*c*]benzopyran-4-one **10**. *Reagents and conditions:* (a) ethyl acetoacetate or ethyl benzoylacetate, piperidine, and MeOH (the Nef reaction condition); (b) HBr, AcOH concentration, 130 °C, 4 h, 16 outputs with 50%–65% yield.

4*H*-Furo[3,2-*c*]benzopyran-4-one

Furan Construction

A wide range of research has demonstrated that 4-hydroxycoumarin is the key compound for the synthesis of 4*H*-furo[3,2-*c*]benzopyran-4-ones, which can readily react with the C=C bond of the alkene, or the C≡C bond of the alkyne [33–37].

Reisch reported the condensation of 4-hydroxycoumarin (**11**) with 1-phenyl-2-propyn-1-ol (**12**) under acidic conditions (a mixture of glacial acetic and concentrated sulfuric acid) to deliver the corresponding 2-methyl-3-phenylfuro[3,2-*c*]benzopyran-4-one (**13**) (Scheme 4) [38].

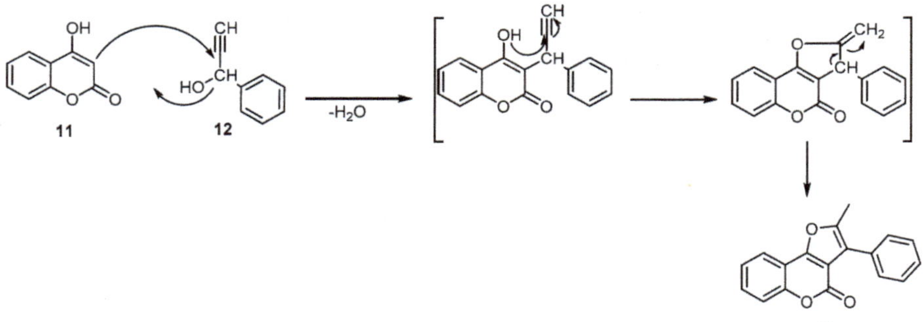

Scheme 4. Synthesis of 2-methyl-3-phenylfuro[3,2-*c*]benzopyran-4-one (**13**). *Reagents and conditions:* AcOH, conc. H_2SO_4, 110 °C, 1 h, 70% yield.

A few studies employed the aliphatic aldehydes as building blocks with 4-hydroxycoumarin (**11**) to synthesize 4*H*-furo[3,2-*c*]benzopyran-4-ones [25,39]. This method was ineffective as it gave a poor yield as well as a mixture of 2,3-dihydrofuran, 4*H*-furo[3,2-*c*]benzopyran-4-ones, and 4*H*-furo[3,2-*c*]benzopyran-4-ones [39]. Conversely, in the case of using the aromatic aldehyde as a building block, the 4*H*-furo[3,2-*c*]benzopyran-4-one was obtained [40]. Kadam et al. developed atom-efficient multicomponent reactions (MCRs) and step-efficient, one-pot synthesis of 3-(4-bromophenyl)-2-(cyclohexylamino)-4*H*-furo[3,2-*c*]benzopyran-4-one (**16**) using 4-hydroxycoumarin (**11**) with 4-bromobenzaldehyde (**14**) and cyclohexyl isocyanide (**15**) as an alkylene source (Scheme 5) [40].

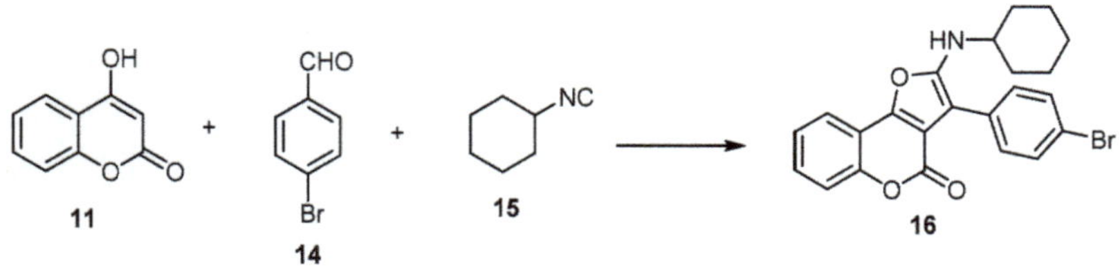

Scheme 5. Atom-efficient multicomponent reactions (MCRs) and step-efficient, one-pot synthesis of 4*H*-furo[3,2-*c*]benzopyran-4-one (**16**). *Reagents and conditions*: DMF or toluene, μw, 80 °C, 20 min, 97% yield.

4-Hydroxycoumarin derivatives have received significant attention from researchers, as these derivatives possess 1,3-dicarbonyl systems. It allows for the easy generation of α,α'-dicarbonyl radicals, which can be readily added to the C=C bond of the alkene [41]. The first example of this reaction was described in 1998, by Lee and his coworkers. They reported an efficient way to prepare 4*H*-furo[3,2-*c*]benzopyran-4-ones **19** by Ag_2CO_3/celite (Fetizon's reagent)-mediated oxidative cycloaddition of 4-hydroxycoumarin **17** to olefins, such as vinyl sulfide and phenyl propenyl sulfide. The resulting dihydrofuro[3,2-*c*]benzopyran-4-ones **18** was treated by sodium periodate in aqueous methanol to form the corresponding sulfoxides, which, upon refluxing with pyridine in carbon tetrachloride, directly delivered the 4*H*-furo[3,2-*c*]benzopyran-4-one **19** in good yields (Scheme 6) [41].

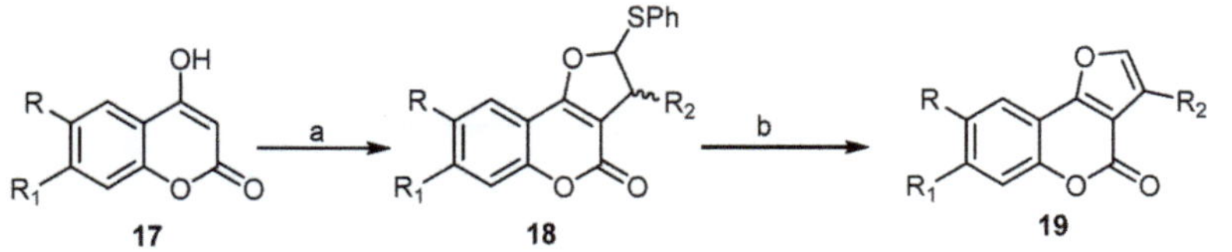

Scheme 6. A facile synthesis of 4*H*-furo[3,2-*c*]benzopyran-4-ones **19** by silver(I)/celite promoted an oxidative cycloaddition reaction. *Reagents and conditions*: (a) CH_2=CHSPh and/or CH_3CH=CHSPh, Ag_2CO_3/celite, acetonitrile, reflux, 3 h; (b) $NaIO_4$, MeOH, CCl_4, pyridine, Al_2O_3, four outputs with 71%–82% yield.

Recently, different catalytic methodologies have been developed for the synthesis of 2*H*-chromenes, and they are based on three main approaches: catalysis with (transition) metals, metal-free Brønsted catalysis, and Lewis acid/base catalysis, which includes examples of nonenantioselective organocatalysis and enantioselective organocatalysis [42–44]. Alkynes have been widely employed as building blocks for this reaction in most cases.

To date, different transition metal (Au, Pt, and Cu) catalyzed/mediated methodologies for benzopyrane synthesis have been reported [27,42,45,46]. Cheng and Hu described a one-pot cascade of an addition/cyclization/oxidation sequence using $CuCl_2$ as the oxidant and CH_3SO_3H as the acid for regioselective synthesis of 2-substituted-4*H*-furo[3,2-*c*]benzopyran-4-ones **22** from the substituted 3-alkynyl-4*H*-benzopyran-4-one **20** (Scheme 7) [47]. This strategy included the CH_3SO_3H-acid-catalyzed construction of the furan ring, followed by oxidation of **21** with $CuCl_2$ (Scheme 7) [47]. When the reaction was carried out in the presence of a catalytic amount of CuCl as a Lewis acid and atmospheric oxygen as an oxidative reagent, compound **22** was provided directly. On the other hand, the presence of 10% CuBr and an excess of $CuCl_2$ as the oxidant afforded the corresponding 3-chloro-2-substituted- 4*H*-furo[3,2-*c*]benzopyran-4-ones **23** (Scheme 7) [48].

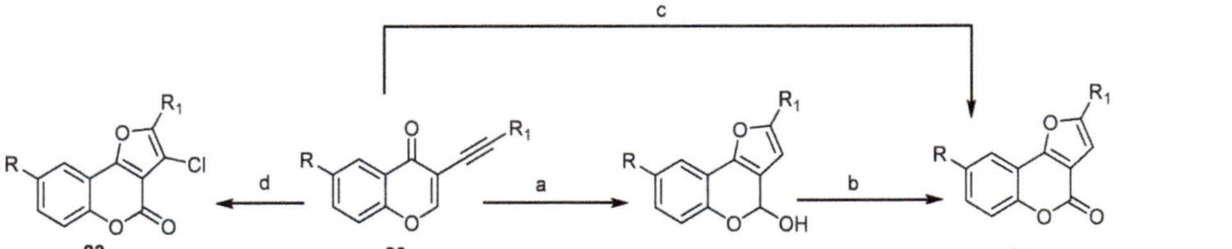

Scheme 7. Transition metal Cu catalyzed/mediated methodologies for synthesis of the 4*H*-furo[3,2-*c*]benzopyran-4-ones **22** and **23**. *Reagents and conditions*: (a) CH_3SO_3H, H_2O, DMF, 90 °C, 1–3 h; (a) $CuCl_2$, 90 °C, 20 h; (c) CuCl, O_2, DMF, H_2O, 90 °C, 10–20 h, 10 outputs with 37%–88% yield; (d) CuBr, $CuCl_2$, DMF, H_2O, 75 °C, 10 h, 13 outputs with 45%–81% yield.

Brønsted-acid-catalyzed propargylations of several organic substrates, including 1,3-dicarbonyl compounds, with alkynols have been reported [49]. In most cases, the acid catalyst is required to promote the propargylation process efficiently. Zhou and coworkers developed a one-pot $Yb(OTf)_3$ propargylation–cycloisomerization sequence of 4-hydroxy coumarin (**11**) with the propargylic alcohol (**24**) for the synthesis of a 2-benzyl-3- phenyl-4*H*-furo[3,2-*c*]chromen-4-one (**25**) skeleton using $Yb(OTf)_3$ as a Lewis acid (Scheme 8) [50].

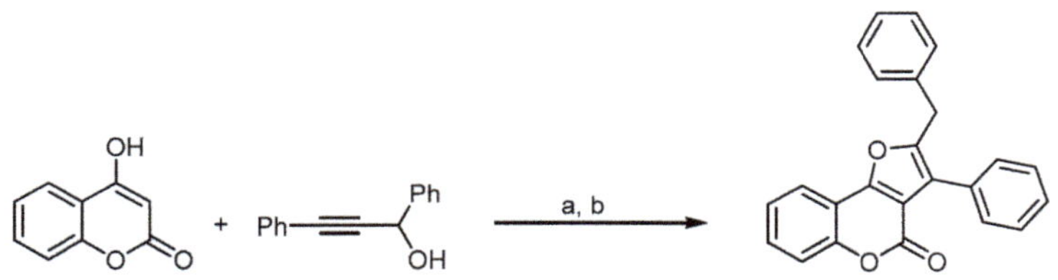

Scheme 8. One-pot synthesis of 4*H*-furo[3,2-*c*]chromen-4-one (**25**) using a $Yb(OTf)_3$-catalyzed propargylation and allenylation reaction. *Reagents and conditions*: (a) 5 mol % $Yb(OTf)_3$, CH_3NO_2, dioxane, 50 °C; (b) K_2CO_3, 70 °C, 37% yield.

Similarly, 4*H*-furo[3,2-*c*] benzopyran-4-one formation reactions proceeded in higher yields and in a one-pot manner, employing a catalytic system composed of the 16-electron allyl–ruthenium(II) complex [Ru(η3-2-C_3H_4Me)(CO)(dppf)][SbF_6] (dppf=1,1′-bis(diphenyl phosphino)ferrocene) and trifluoroacetic acid (TFA) in the reaction of 4-hydroxycoumarin (**11**), with 1-(4-methoxyphenyl)-2-propyn-1-ol (**26**) as an example. The 4*H*-furo[3,2-*c*]benzo pyran-4-one (**27**) was synthesized with a 72% yield (Scheme 9) [50–52].

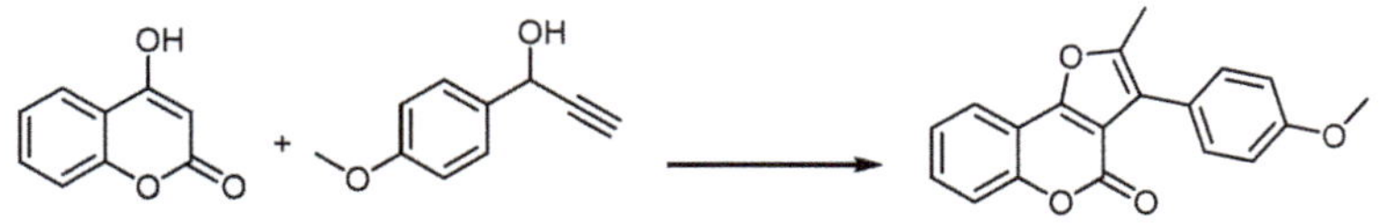

Scheme 9. The 16-electron allyl–ruthenium(II) complex in preparation of 4*H*-furo[3,2-*c*]benzopyran-4-one (**27**). *Reagents and conditions*: 16-electron allyl–ruthenium(II) complex [Ru(η3-2-C_3H_4Me)(CO)(dppf)][SbF_6] (5 mol %), trifluoroacetic acid (TFA) (50 mol %), THF, 75 °C, 5 h, 72% yield.

Extensive work has been done to investigate the utility of an aryl alkynyl ether as a furan substrate, instead of arylalkynol, in the synthesis of 4*H*-furo[3,2-*c*]benzopyran-4-one [29,35]. The treatment of 3-iodo-4-methoxycoumarin (**28**) with phenylacetylene by means of sequential Sonogashira C–C coupling conditions resulted in a high-yield formation of the 4*H*-furo[3,2-*c*]benzopyran-4-one (**30**) (Scheme 10) [53]. In this reaction, the triethylamine was used as a base to induce the S_N2-type demethylation of the Sonogashira coupling product, followed by an intramolecular attack of the enolate onto the cuprohalide π-complex of the triple bond (Scheme 10).

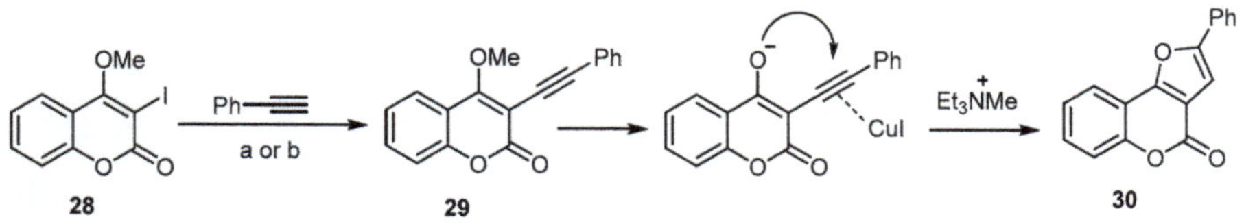

Scheme 10. Et_3N-induced demethylation–annulation of an aryl alkynyl ether in the synthesis of 4*H*-furo[3,2-*c*]benzopyran-4-one (**30**). *Reagents and conditions*: (a) alkyne (3 equiv.), 8 mol % $PdCl_2(PPh_3)_2$, 8 mol % CuI, Et_3N/DMF, 80 °C, 48 h, 82% yield; (b) alkyne (3 equiv.), 8 mol % $PdCl_2(PPh_3)_2$, 8 mol % CuI, Et_3N/MeCN, 60 °C, 15 h, 70% yield.

As a follow-up to this type of reaction, a novel and rapid assembly of an interesting class of 4*H*-furo[3,2-*c*]benzopyran-4-ones, **33**, was successfully achieved using a one-pot sequential coupling/cyclization strategy with 3-bromo-4-acetoxycoumarins **31** and dialkynlzincs **32** prepared in situ as reactive acetylides in transition-metal-catalyzed crosscoupling. The cascade transformation relies on palladium/copper-catalyzed alkynylation and intramolecular hydroalkoxylation (Scheme 11) [54].

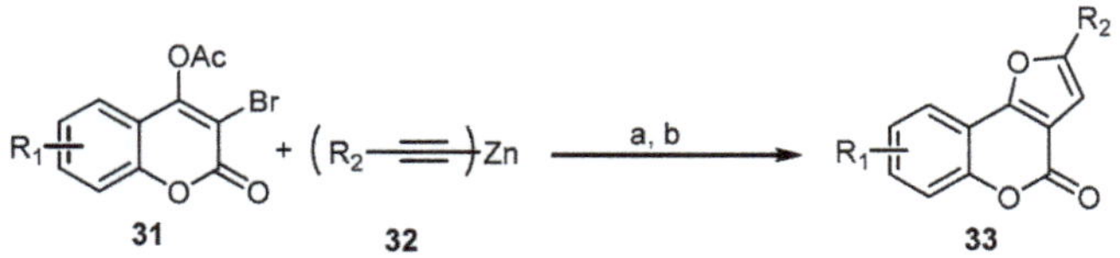

Scheme 11. A one-pot sequential coupling/cyclization strategy in the synthesis of 4*H*-furo[3,2-*c*]-benzopyran-4-ones **33**. *Reagents and conditions*: (a) $Pd(PPh_3)_2$, CuI, THF, 60 °C; (b) K_2CO_3, H_2O, 13 outputs with 51%–96% yield.

A transition-metal-free approach was developed to achieve 4-*H*-furo[3,2-*c*]benzopyran-4-ones via an iodine-promoted one-pot cyclization between 4-hydroxycoumarins **34** and acetophenones **35**. The transformation spontaneously proceeded to produce (**36**) in the presence of NH_4OAc. The possible reaction mechanism suggested for the iodine-promoted one-pot cyclization is depicted (Scheme 12) [55].

Additionally, Traven et al. [56] provided a new short way for the synthesis of 4*H*-furo-[3,2-*c*]benzopyran-4-one, employing the Fries rearrangement of 4-chloroacetoxycoumarin (**37**) to yield two products, namely 3-chloroacetyl4-hydroxycoumarin (**38**) and dihydrofuro [2,3-*c*]coumarin-3-one (**39**), in the ratio of 2:1. Compound (**38**), which underwent cyclization, led to the formation of (**39**). The latter, under reduction and dehydration conditions, afforded 4*H*-furo[3,2-*c*]chromen-4-one (**41**) (Scheme 13). A closely related reaction that allowed for the preparation of (**41**) was developed by Majumdar and Bhattacharyya [57], following a similar procedure but using chloroacetaldehyde instead of chloroactylchloride in the presence of aqueous potassium carbonate to give 3-hydroxy-2,3- dihydrofuro[3,2-*c*]benzopyran-4-one (**40**), which upon treatment with aqueous hydrochloric acid provided 4*H*-furo[3,2-*c*]benzopyran-4-one (**41**) with 72% yield (Scheme 13).

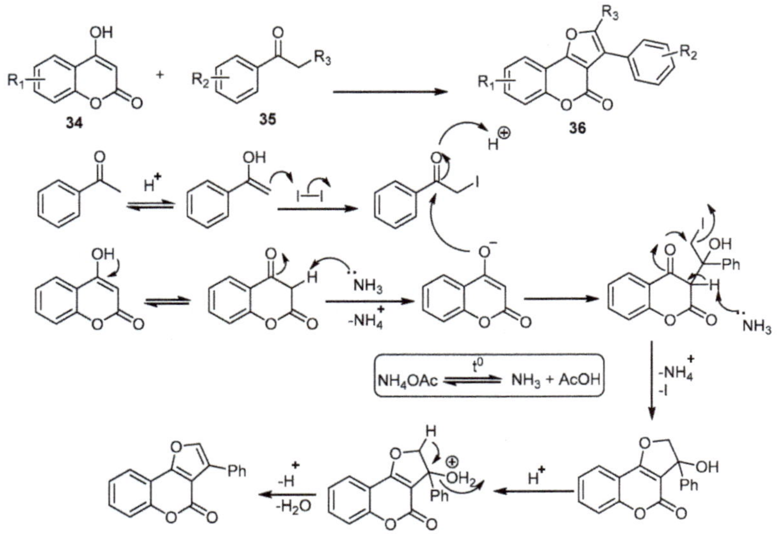

Scheme 12. Metal-free synthesis of 4-*H*-furo[3,2-*c*]benzopyran-4-ones **36**. *Reagents and conditions:* I_2, NH_4OAc, PhCl, 120 °C, 18 outputs with 28%–90% yield.

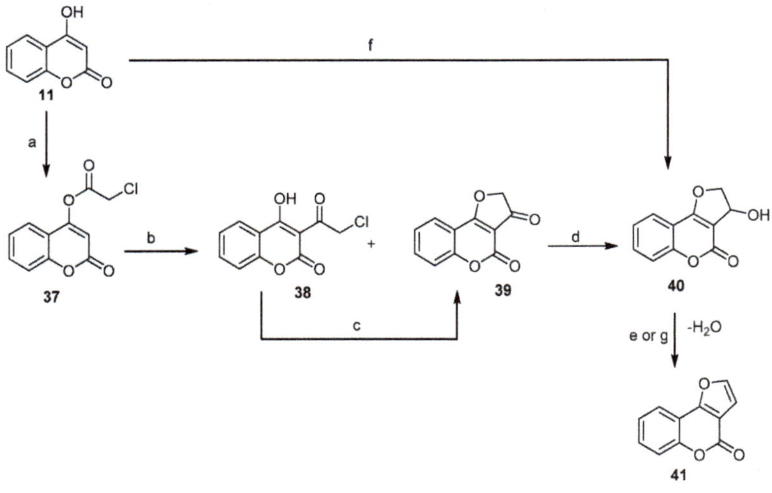

Scheme 13. Regioselective synthesis of 4*H*-furo[3,2-*c*]chromen-4-one (**41**). *Reagents and conditions:* (a) $ClCH_2COCl$, dry pyridine, 40 min, reflux, 85% yield; (b) $AlCl_3$, 140–150 °C, 60% yield; (c) $AlCl_3$, 140–150 °C, 30–40 min or K_2CO_3, acetone, 10 min, stirring, r.t., 50% yield; (d) $NaBH_4$, 85% yield; (e) H_2SO_4 (30%), EtOH, heat, 30 min, 80% yield; (f) $COCH_2Cl$, K_2CO_3, 73% yield; (g) HCl, 72% yield.

Pyrone Construction

Recently, much effort has been devoted to the development of oxidative intramolecular C–O bond-forming cyclization reactions for the synthesis of bioactive benzopyranones. These methods are limited to being used with arenes building blocks [58–60]. Fu et al. reported a ligand-enabled, site-selective carboxylation of 2-(furan-3-yl)phenols **42** under the atmospheric pressure of CO_2. It was performed through an Rh(ii)-catalyzed C–H bond activation, assisted by the ligand chelation of the phenolic hydroxyl group to afford 4*H*-furo[3,2-*c*]benzopyran-4-ones **43** (Scheme 14) [61]. This reaction indicates the role of phosphine ligands in combination with $Rh_2(OAc)_4$ in promoting the reactivity and the selectivity during C–H carboxylation. The right choice of a suitable basic catalyst is an additional critical point.

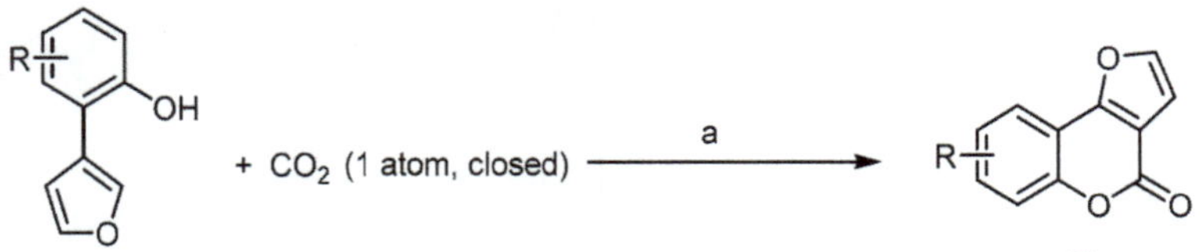

Scheme 14. Rhodium(II)-catalyzed aryl C-H carboxylation with CO_2 in the synthesis of 4*H*-furo[3,2-*c*] benzopyran-4-ones **43**. *Reagents and conditions*: (a) $Rh_2(OAc)_4$ (1 mol %), tricyclohexylphosphine PCy_3 (2 mol %), t-BuOK (4.5 equiv.), diglyme, 100 °C, 48 h, six outputs with 70%–86% yield.

2.1.2. Pyrroles

Fusion of the pyrrole ring with the pyrone ring of coumarin (benzopyrane) leads to three structural isomers, viz. chromeno[3,4-*b*]pyrrol-4(3*H*)-one, chromeno[3,4-*c*]pyrrol-4(2*H*)-one, and chromeno[4,3-*b*]pyrrol-4(1*H*)-one (Figure 2).

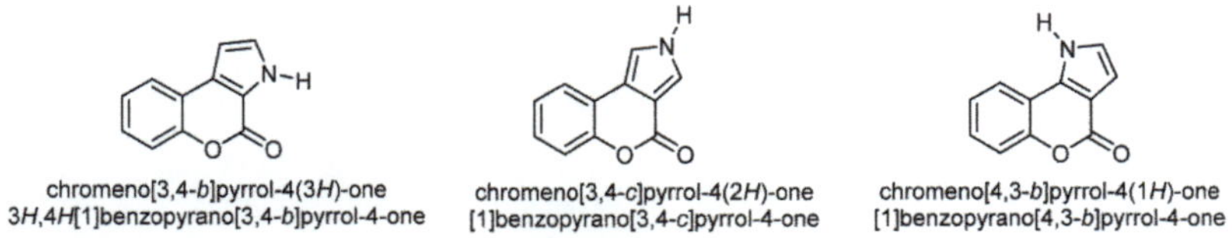

Figure 2. The three common isomers of the pyrrole ring fused to the α-pyrone moiety of coumarin.

3*H*,4*H*[1]Benzopyrano[3,4-*b*]pyrrol-4-one

Pyrrole Construction

The 3-Aminocoumarin (**44**) is considered the starting compound for the preparation of fused 3*H*,4*H*[1]benzopyrano[3,4-*b*]pyrrol-4-ones. The amino group represents the key moiety of this cyclization process in the reaction with different reagents [30,62–64]. Compound **45** was prepared by the reaction of 3-aminocoumarin (**44**) with different carbonyls via Fischer indole synthesis after being diazotized and reduced to coumarin-3-yl-hydrazine [62]. The compound 1-Aryloxy-4-chlorobut-2-ynes reacted with 3-aminocoumarin (**44**) to afford **47** through amino-Claisen rearrangement [64]. Condensation of (**44**) with α-halo ketones, followed by cyclization catalyzed by TFA, led to the formation of **49** [63], while **50** was prepared under Nef conditions using 2-aryl-1- nitro ethenes [30] (Scheme 15).

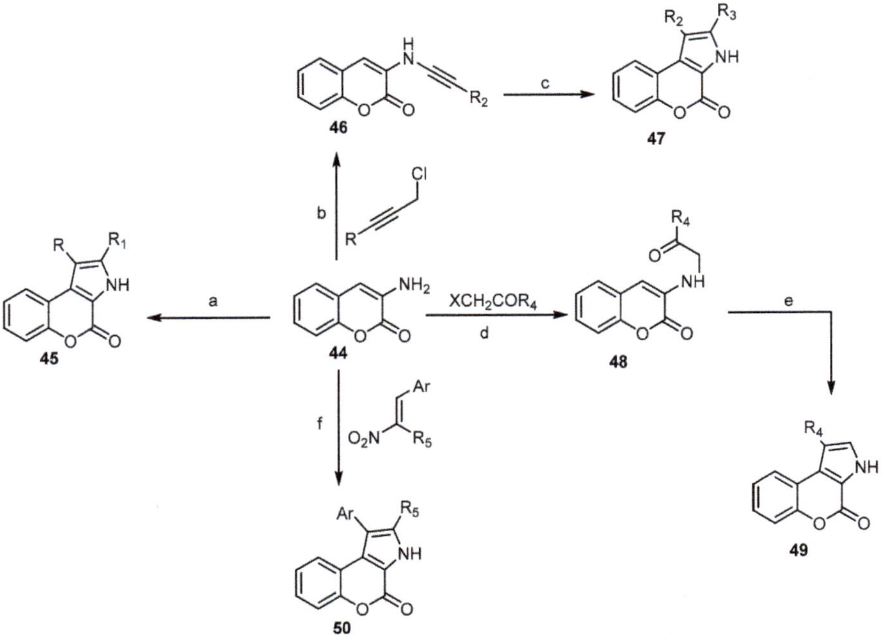

Scheme 15. Different pathways to synthesize *3H,4H*[1]benzo- pyrano[3,4-*b*]pyrrol-4-ones **45**, **47**, **49**, and **50** via 3-aminocoumarin. *Reagents and conditions*: (a) i: $NaNO_2$, HCl, −30 °C, ii: $SnCl_2$, −10 °C, HCl, 2 h, iii: carbonyl compounds, polyphosphoric acid, 1 h, 130 °C, seven outputs with 33%–51% yield; (b) anhydride ethyl methyl ketone, K_2CO_3, NaI, reflux, 24–30 h, five outputs with 51%–63% yield; (c) *N,N*-dimethylaniline, reflux, 6–9 h, five outputs with 90%–94% yield; (d) KI (cat.), reflux, 5 h; (e) TFA (cat.), AcOH, 12 h, reflux, four outputs with 28%–86% yield; (f) MeOH, piperidine, reflux, five outputs with 55%–61% yield.

3*H*,4*H*[1]Benzopyrano[3,4-*e*]pyrrol-4-one

Pyrrole Construction

A one-pot, three-component reaction of phenylsulphinyl-2*H*-benzopyran-2-one (**51**), phenylglycine (**52**), and benzaldehyde (**53**) led to the formation of 1,3-diphenyl[l]benzopyrano[3,4-*e*]pyrrol-4-one (**55**) (Scheme 16) [65].

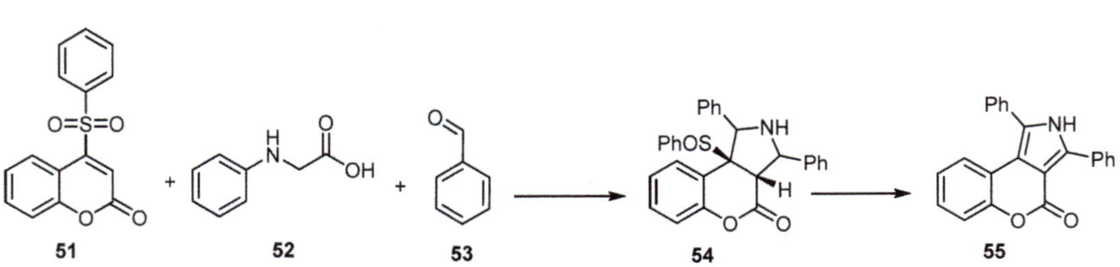

Scheme 16. Synthesis of 1,3-diphenyl[l]benzopyrano[3,4-*e*]pyrrol-4-one (**55**). *Reagents and conditions*: DMF, stirring, 120 °C, 24 h, 51% yield.

Xue et al. developed an efficient and straightforward synthetic protocol for the preparation of [1]benzopyrano[3,4-*e*]pyrrol-4-ones **59** through $FeCl_3$-promoted, three-component reactions between substituted 2-(2-nitrovinyl)phenols **58**, acetylene dicarboxylate (**57**), and amines **58** (Scheme 17) [66]. This reaction involved the sequential $FeCl_3$-mediated nucleophilic addition of acetylenedicarboxylates, amines, and 2-(2-nitrovinyl)phenols, following intramolecular transesterifcation to form a coumarin core. This strategy offers a complementary approach to substituted pyrrolo[3,4-*c*]coumarin compounds, with advantages that include a variety of cheap and readily available reactants and a wide range of substrates with dense or flexible substitution patterns [66].

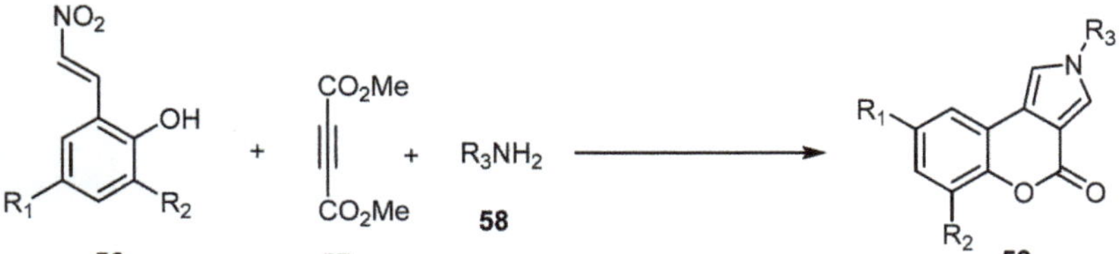

Scheme 17. Synthesis of [1]benzopyrano[3,4-*e*]pyrrol-4-ones **59** by a $FeCl_3$-promoted, three-component reaction. *Reagents and conditions*: $FeCl_3$, toluene, 110 °C, 6 h, 17 outputs with 62%–92% yield.

Alizadeh et al. reported a sequential three-component reaction of salicylaldehydes **60**, β-keto esters 61, and *p*-toluenesulfonylmethyl isocyanide (TosMIC) (**62**) via [1,3] acyl shift to give 2-acyl[1] benzopyrano[3,4-*e*]pyrrol-4-ones **63** (Scheme 18) [67]. A simple workup procedure, mild reaction conditions, lack of side products, and good yields of 62%–95% are the main aspects of this method.

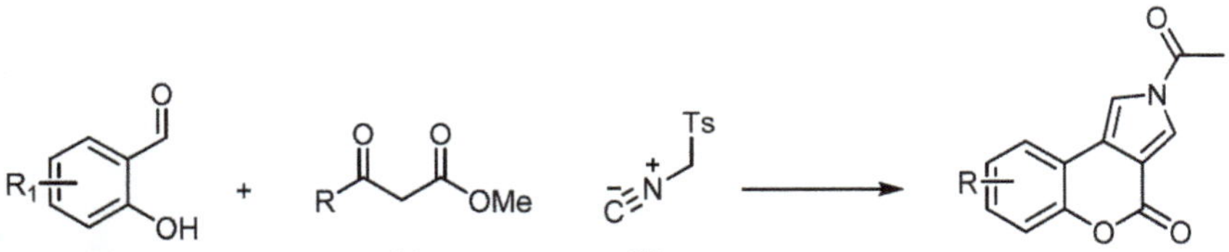

Scheme 18. A sequential three-component reaction to synthesize [1]benzopyrano[3,4-*e*]pyrrol-4-ones **63**. *Reagents and conditions*: TEA, piperidine, DMF, r.t., 8 h, 10 outputs with 62%–95% yield.

Recently, Khavasi and his coworkers investigated the reactivity, chemo-, region-, and diastreo-selectivity of *p*-toluenesulfonylmethyl isocyanide (TosMIC) (**62**) in Van Leusen-type [3 + 2] cycloaddition reactions with the 3-acetylcoumarins **64** to give [1]benzopyrano [3,4-*e*]pyrrol-4-ones **65** (Scheme 19) [68]. This method offers several advantages, such as being inexpensive, providing good to excellent yields, producing short reaction times, high atom economy, and ease of product isolation under catalyst-free conditions without any activation at ambient temperature.

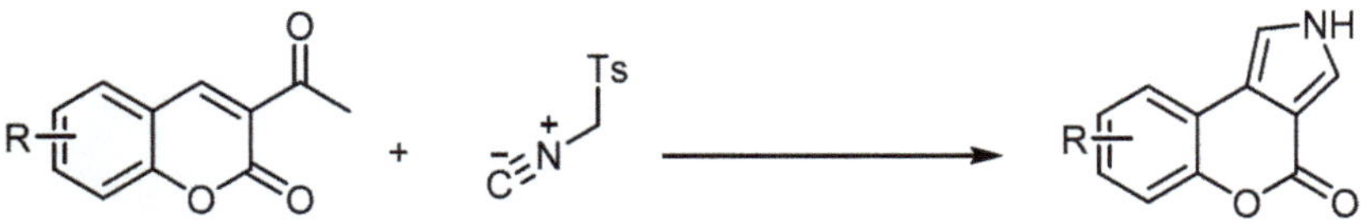

Scheme 19. The Van Leusen protocol for the synthesis of [1]benzopyrano[3,4-*e*]pyrrol-4-ones **65**. *Reagents and conditions*: K_2CO_3, EtOH, 15 min, r.t, three outputs with 72%–86% yield.

[1]Benzopyrano[4,3-*b*]pyrrol-4-one

Pyrrole Construction

Many synthetic protocols have been reported for the synthesis of [l]benzopyrano [4,3-*b*]pyrrole-4(1*H*)-ones, including the reaction of β-nitroalkenes **67** and 4-phenylamino coumarins **66** under solvent-free conditions to afford **68** (Scheme 20) [69]. Moreover, the reaction of the 4-aminocoumarin (**69**), amines **70**, and glyoxal monohydrates **71** in the presence of nanocrystalline $CuFe_2O_4$ [70], or $KHSO_4$, led to the formation of [l]benzopyrano [4,3-*b*]pyrrole-4(1*H*)-ones **72** (Scheme 21) [71]. The synthesis of 71 using nanocrystalline $CuFe_2O_4$ discloses a rapid, high-yielding, green synthetic protocol for a variety of chromeno [4,3-*b*]pyrrol-4(1*H*)-one derivatives by assembling the basic building blocks in an aqueous medium using nano $CuFe_2O_4$ as the efficient, magnetically recoverable catalyst [70].

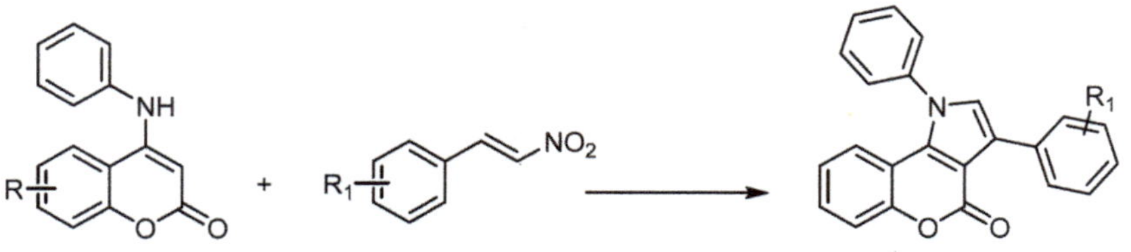

Scheme 20. Synthetic protocol to synthesize [l]benzopyrano[4,3-*b*]pyrrole-4(1*H*)-ones **68**. *Reagents and conditions*: TsOH.H_2O, solvent-free, 24 outputs with 6%–77% yield.

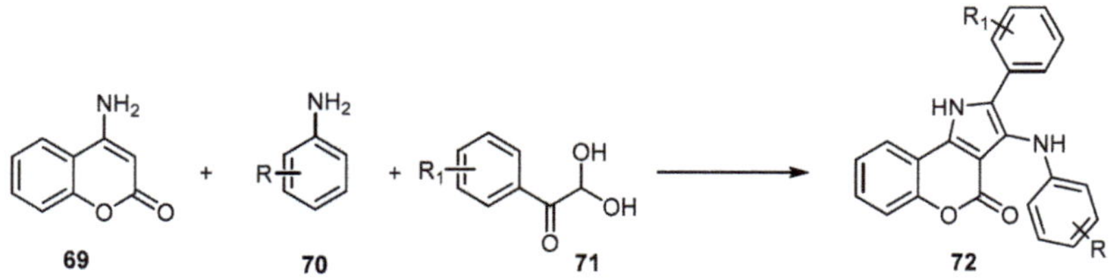

Scheme 21. Synthetic protocol to synthesize [l]benzopyrano[4,3-*b*]pyrrole-4(1*H*)-ones **72**. *Reagents and conditions*: $CuFe_2O_4$, H_2O, 70 °C, 25 outputs with 68%–94% yield; or $KHSO_4$, toluene, reflux, 23 outputs with 37%–93% yield.

Many articles have found that the 4-chlorocoumarin is the key compound for the preparation of various [l]benzopyrano[4,3-*b*]pyrrol-4-ones by Knorr- or Fischer–Fink-type reactions [72,73]. Albrola et al. indicated the preparation of N(α)-(2-oxo-2*H*-l-benzopyran-4-yl)Weinreb-α-aminoamides **75** from 4-chlorocoumarin (**73**) and different α-aminoacids **74**. The reaction of **75** with various organometallic compounds, followed by cyclization, led to the formation of [l]benzopyrano[4,3-*b*]pyrrol-4-ones **76** (Scheme 22) [74,75].

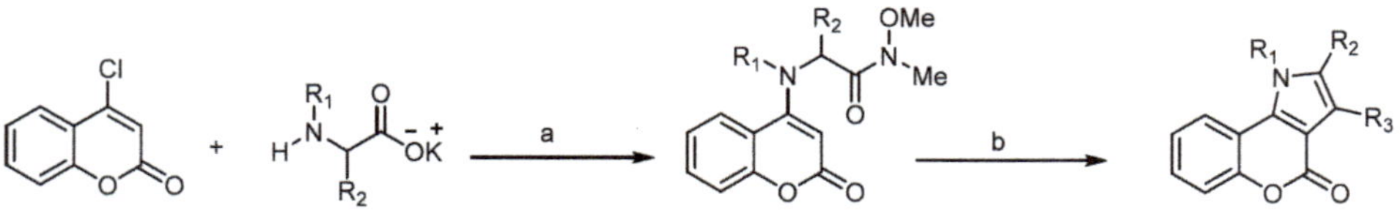

Scheme 22. Synthesis of [l]benzopyrano[4,3-*b*]pyrrol-4-ones **76** from 4-chlorocoumarin. *Reagents and conditions*: (a) i: EtOH, TEA, HCl.N(Me)OMe, ii: CH_2Cl_2, TEA, DCC; (b) i: organometallic compounds (R_3M), THF, N_2, -H_2O, ii: NaOEt, EtOH, r.t., 2 h, reflux, 1 h, nine outputs with 27%–92% yield.

On the other hand, the 4-amino-2*H*-benzopyran-2-one is employed as a key compound for the preparation of [l]benzopyrano[4,3-*b*]pyrrol-4-one [76,77]. Peng et al. synthesized a series of [l]benzopyrano[4,3-*b*]pyrrol-4-ones **79** via a palladium-catalyzed oxidative annulation reaction of 4-amino-2*H*-benzopyran-2-ones **77**, with electron-withdrawing or electron-donating groups with different alkynes **78** (Scheme 23) [76]. The method utilizes simple and readily available enamines and alkynes, and employs direct Pd(II)-catalyzed oxidative annulation to synthesize [l]benzopyrano[4,3-*b*]pyrrol-4-ones in high yields of 72%–99%.

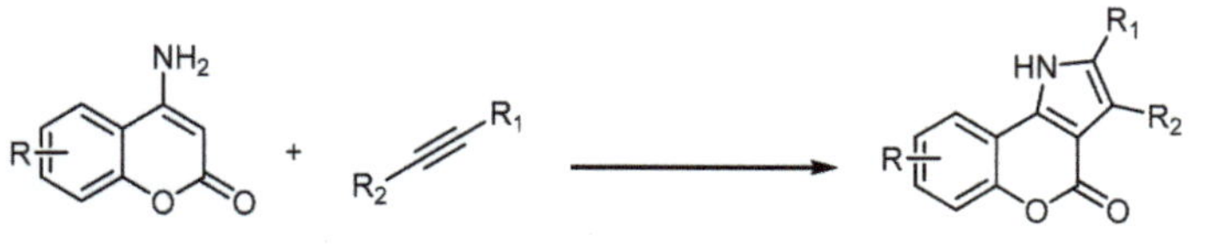

Scheme 23. The scope of 4-aminocoumarins in the preparation of [l]benzopyrano[4,3-*b*]pyrrol-4-ones **79**. *Reagents and conditions*: $Pd(OAc)_2$, oxidant, DMSO, 100 °C, 12 outputs with 72%–99% yield.

Recently, Yang et al. reported a one-pot, two-step reaction of 4-amino-2*H*-benzopyran-2-one (**69**) with arylglyoxal monohydrates 80 and *p*-toluenesulfonates **81** to afford a series of 3-alkoxy-substituted [l]benzopyrano[4,3-*b*]pyrrol-4-ones **82** and **83**, respectively (Scheme 24) [78]. On the other hand, Yahyavi et al. described the Knoevenagle treatment of the arylglyoxals **84** with active methylene compounds and consequently an iodine-activated Michael-type reaction with 4-aminocoumarin (**69**) in a one-pot manner to afford disubstituted [l]benzopyrano[4,3-*b*]pyrrol-4-ones **85** (Scheme 24) [79].

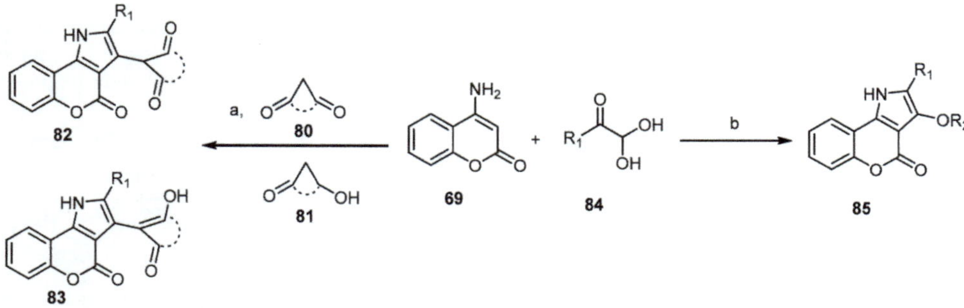

Scheme 24. Various arylglyoxals in the synthesis of [l]benzopyrano[4,3-*b*]pyrrol-4-ones **82**, **83**, and **85**. *Reagents and conditions*: (a) i: AcOH, reflux, 40 min; ii: an appropriate alkyl *p*-toluenesulfonate ($TsOR_2$), 1,8-diazabicyclo[5.4.0]undec-7-ene (DBU), toluene, reflux, 1.5 h, 14 outputs with 73%–89% yield; (b) R_1=Ph, 4-$CH_3OC_6H_4$, 4-$CH_3C_6H_4$, 4-FC_6H_4, 4-ClC_6H_4, 4-BrC_6H_4, 2-thienyl; dimedone, 2-hydroxy-1,4-naphtoquinone barbituric acid, 1,3 dimethyl barbituric acid, I_2, DMSO, stirring, 100 °C, 7 h, 15 outputs with 15%–80% yield.

2.1.3. Thiophenes

Fusion of the thiophene ring with the pyrone ring of coumarin(benzopyrane) leads to three structural isomers, viz. 4*H*-thieno[2,3-*c*]chromen(benzopyran)-4-one, 4*H*-thieno[3,4-*c*]chromen(benzopyran)-4-one, and 4*H*-thieno[3,2-*c*] chromen(benzopyran)-4-one (Figure 3)

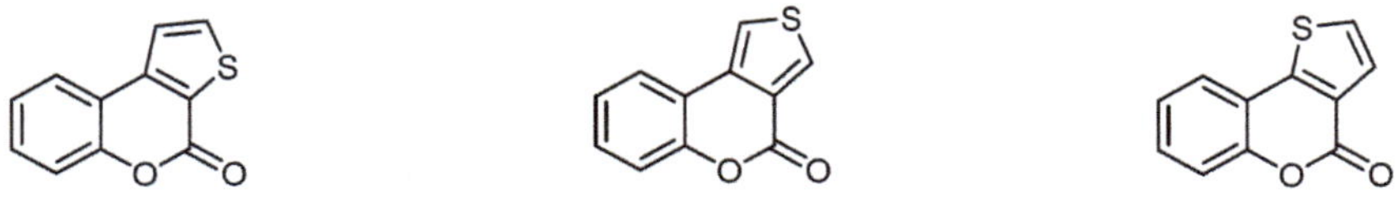

4*H*-thieno[2,3-*c*]chromen-4-one
4*H*-thieno[2,3-*c*]benzopyran-4-one

4*H*-thieno[3,4-*c*]chromen-4-one
4*H*-thieno[3,4-*c*]benzopyran-4-one

4*H*-thieno[3,2-*c*]chromen-4-one
4*H*-thieno[3,2-*c*]benzopyran-4-one

Figure 3. The three common isomers of the thiophene ring fused to the α-pyrone moiety of coumarin.

4*H*-Thieno[2,3-*c*]benzopyran-4-one

Thiophene Construction

The one-pot cascade addition/condensation/intramolecular cyclization sequence of chromone (**86**) with ethyl 2-mercaptoacetate (**87**) using a complexing ligand 1,8-diazabicyclo [5.4.0]undec-7-ene (DBU) in 1,4-dioxane led to the formation of 4*H*-thieno[2,3-*c*]benzopyran-4-one (**88**) (Scheme 25) [80].

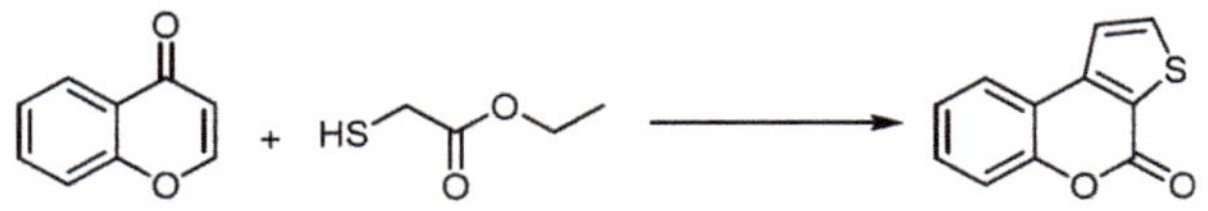

86 **87** **88**

Scheme 25. Synthesis of 4*H*-thieno[2,3-*c*]benzopyran-4-one (**88**). *Reagents and conditions:* 1,8-diazabicyclo[5.4.0]undec-7-ene (DBU), 1,4-dioxane, 60 °C, 12 h, N_2, 95% yield.

Pyrone Construction

The Suzuki–Miyaura cross coupling of bromoarylcarboxylates and *o*-hydroxy(methoxy) arylboronic acids is one of the methods that plays an important role in the preparation of 4*H*-thieno[2,3-*c*] **91** and 4*H*-thieno[3,4-*c*]benzopyran-4-ones **92** (Scheme 26) [81–83].

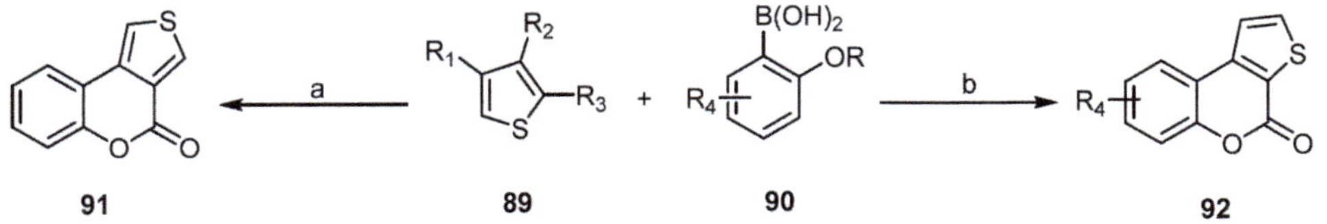

91 **89** **90** **92**

Scheme 26. 4*H*-thieno[3,4-*c*]benzopyran-4-ones **91** and **92** via a Suzuki–Miyaura cross coupling reaction. *Reagents and conditions:* (a) $Pd(PPh_3)_4$ (10 mol %), Cs_2CO_3 (4 equiv.), DME, H_2O, MW 125 °C, 15 min, 86% yield; (b) i: $Pd(PPh_3)_4$ (5 mol %), K_3PO_4 (1.5 equiv.), 1,4-dioxane, 90 °C, 4 h; ii: (a) BBr_3, CH_2Cl_2, (b) KO*t*Bu, H_2O, five outputs with 75%–81% yield [83].

4*H*-Thieno[3,4-*c*]benzopyran-4-one

Thiophene Construction

4*H*-thieno[3,4-*c*]benzopyran-4-ones (**94**) were meanly prepared through the Gewald reaction (Scheme 27) [84–90]. Low yields of 37% and 48% were observed using these methods.

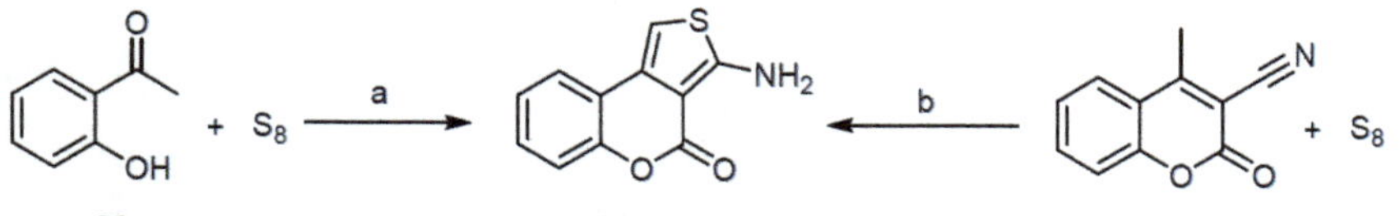

Scheme 27. The Gewald reaction to synthesize 4*H*-thieno[3,4-*c*]benzopyran-4-one (**94**). *Reagents and conditions*: (a) $CNCH_2COOEt$, base; (b) NH_3, EtOH, 48% yield [85], 37% yield [90].

Yu et al. created a new technique for the preparation of 4*H*-thieno[3,4-*c*] [1]benzopyran-4(4*H*)-ones **97** by applying a chemoselective reaction of thioamides **95** with α-bromoacetophenones **96** (Scheme 28) [91].

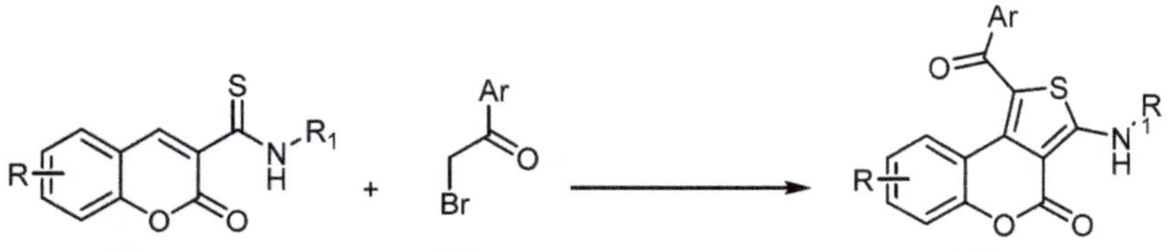

Scheme 28. Synthesis of 4*H*-thieno[3,4-*c*] [1]benzopyran-4(4*H*)-ones **97** via [4 + 1] annulations. *Reagents and conditions*: NaOH, diethyl azodicarboxylate (DEAD), MeCN, r.t., 20 min, 17 outputs with 68%–95% yield.

A rhodium-catalyzed intramolecular transannulation reaction of alkynyl thiadiazoles **98** provided 4*H*-thieno[3,4-*c*][1]benzopyran-4(4*H*)-ones **99**. A plausible reaction mechanism proposed for the Rh-catalyzed intramolecular transannulation of alkynyl thiadiazoles is outlined in (Scheme 29) [92].

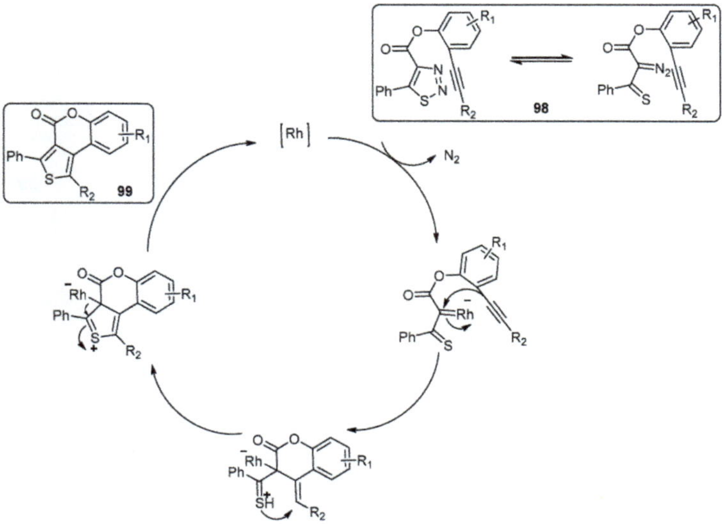

Scheme 29. A plausible reaction mechanism proposed for the preparation of 4*H*-thieno[3,4-*c*][1] benzopyran-4(4*H*)-ones **99**. *Reagents and conditions*: $[Rh(COD)Cl]_2$ (5 mol %), 1,1′-Ferrocenediyl-bis(diphenylphosphine) (DPPF) (12 mol %), PhCl, 130 °C, 30 min, three outputs with 75%–90% yield.

4*H*-Thieno[3,2-*c*][1]benzopyran-4(4H)-one

Thiophene Construction

Numerous articles state the use of 4-chloro-2-*H*-benzopyran-3-carboxaldehydes **101** as a key compound in the preparation of a series of 4*H*-thieno[3,2-*c*][1]benzopyran-4(4*H*)-ones. This compound was prepared by a Vilsmeier–Haack reaction, and the cyclization process was performed through a rection with thioglycolate or dithiane to produce **102** and **103**, respectively (Scheme 30) [25,26,93,94].

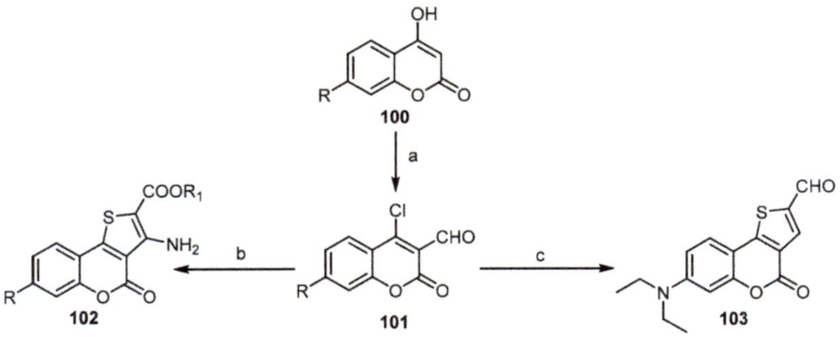

Scheme 30. The synthesis of 4*H*-thieno[3,2-*c*][1]benzopyran-4(4H)-ones **102** and **103** via the cyclization of Vilsmeier–Haack products. *Reagents and conditions*: (a) $POCl_3$, DMF, 60 °C, overnight; (b) $SHCH_2COOR_1$, EtOH, base, 90% yield [25], 88% yield [94]; (c) 1,4-dithiane-2,5-diole, K_2CO_3, acetone, stirring, 1 h, r.t., 45 °C, 3 h, 85% yield.

Pyrone Construction

The palladium-catalyzed oxidative carbonylation of 2-(thiophen-3-yl)phenol (**104**) under acid-base-free and mild conditions yielded the corresponding 4*H*-thieno[3,2-*c*][1] benzopyran-4(4*H*)-one (**105**) (Scheme 31) [58].

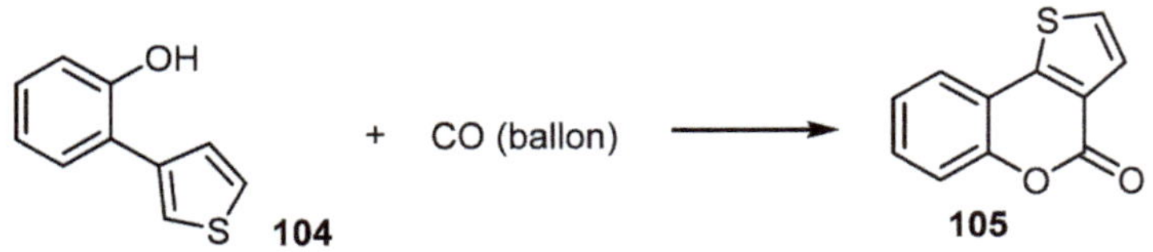

Scheme 31. One-pot synthesis of 4*H*-thieno[3,2-*c*][1]benzopyran-4(4*H*)-one (105) by CO insertion into phenol. *Reagents and conditions*: 10 mol % $Pd(OAc)_2$, AgOAc, CH_3CN, 80 °C, 48 h, 62% yield.

2.1.4. Selenophene

The fusion of the selenophene ring with the pyrone ring of coumarin leads to two structural isomers, viz. 4*H*-selenophen[2,3-*c*]chromen(benzopyran)-4-one and 4*H*-selenophen[3,2-*c*]chromen(benzopyran)-4-one (Figure 4).

4*H*-selenophen[2,3-*c*]chromen-4-one
4*H*-selenophen[2,3-*c*]benzopyran-4-one

4*H*-selenophen[3,2-*c*]chromen-4-one
4*H*-selenophen[3,2-*c*]benzopyran-4-one

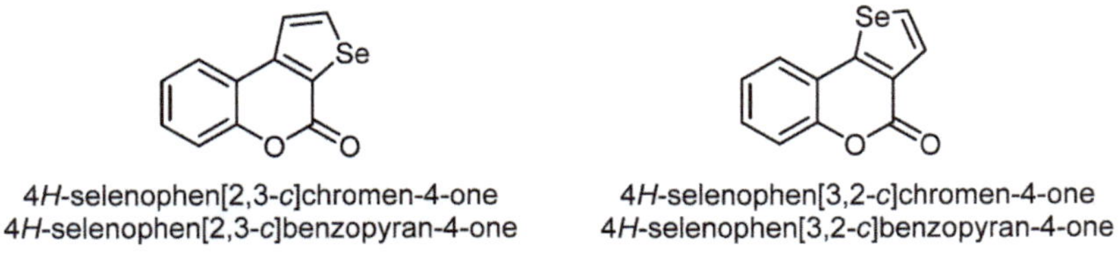

Figure 4. The two common isomers of the selenophene ring fused to the α-pyrone moiety of coumarin.

4*H*-Selenophen[2,3-*c*] and [3,2-*c*]benzopyran-4-ones

A few articles have discussed the possibility of synthesizing the fused selenophen-chromen-(benzopyran)-4-one moiety. A simple method for the synthesis of substituted 4*H*-selenopheno[2,3-*c*] benzopyran-4-ones **108** is by the treatment of 3-ethynylcoumarins **107** with selenium (IV) oxide and concentrated hydrobromic acid at room temperature [16,95]. Similarly, 4*H*-selenopheno-[3,2-*c*]benzopyran-4-ones **111** was prepared under the same conditions using 4-ethynylcoumarins **110**. The alkenyl derivatives were obtained from bromocoumarin (**106**) and 4-(trifluoromethane-sulfonyl)coumarin (**109**) by Sonogashira coupling. All the reaction steps were carried out in situ from the starting materials and until the end product (Scheme 32) [16,95].

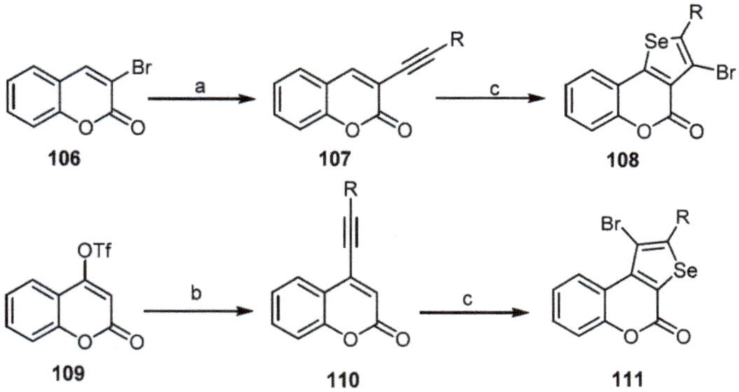

Scheme 32. *Reagents and reaction conditions*: a: $PdCl_2$ (10 mol %), Ph_3P (20 mol %), CuI (10 mol %), terminal acetylene (1.5 equiv.), NMP, Et_3N, 55 °C, 20 h; b: $(Ph_3P)4Pd$ (5 mol %), CuI (20 mol %), terminal acetylene (1.5 equiv.), DMF, Et_3N, r.t., 20 h; **c**: SeO_2 (2 equiv.), conc. HBr, dioxane, r.t., 24–48 h, compounds **108**, four outputs with 68%–75% yield; compounds 111, four outputs with 64%–70% yield.

Additionally, Kirsch and his coworkers reported the synthesis of selenopheno[2,3-*c*]benzopyran-4-ones (**115**) via a multistep reaction. This reaction cascade started by Vilsmeier formylation of 4-hydroxycoumarine (**111**). The formylated product reacted with hydroxyl amine to afford (**113**) according to the reaction conditions, which subsequently transformed into 3-cyano-4-coumarinselenol (**114**), by refluxing with selenium and sodium-borohydride in ethanol (Scheme 33). It was the precursor of selenopheno[2,3-*c*]benzopyran-4-ones (**115**) as it was reactive towards a series of haloacids, such as chloroacetonitrile, ethyl chloroacetoacetate, and chloroacetamide [96].

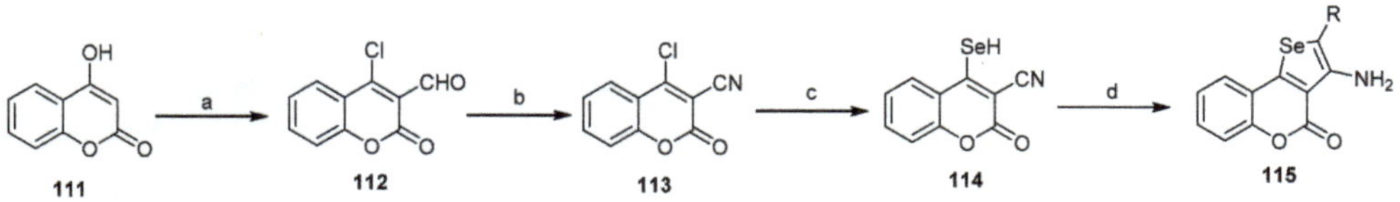

Scheme 33. Synthesis of selenopheno[2,3-*c*]benzopyran-4-ones (**115**) from 4-hydroxycoumarin. *Reagents and reaction conditions*: a: $POCl_3$, DMF, 84% yield; b: $NH_2OH.HCl$, 80% yield; c: Se, $NaBH_4$, EtOH, 85% yield; d: $ClCH_2CN(R)$, DMF, stirring, r.t., 2 h, three outputs with 80%–88% yield.

In conclusion, since coumarins have versatile applications, the synthesis of different structures of the coumarin-based scaffold was attempted. Among all the heterocycles built on the α-pyrone moiety of coumarin, the furan ring is the only available structure in nature. Thus, it has inspired a lot of researchers to replace the oxygen atom with other heteroatoms. Wide varieties of heterocycles were constructed by a synthetic pathway to introduce furans, pyrroles, thiophenes, and selenophenes as a fused ring that is characterized by a single heteroatom to the α-pyrone moiety of coumarin. The fused heterocycles that contain more than one heteroatom will be described in the next part, which we intend to publish in the future.

Author Contributions: E.R.E.-S. and G.K. collected publications and sorted them. E.R.E.-S. analyzed the literature and wrote the first draft. A.B.A. and G.K. developed the concept of the review. E.R.E.-S., A.B.A., and G.K. wrote the final draft and corrected the manuscript. All authors have read and agreed to the published version of the manuscript.

Funding: The APC was funded by MDPI.

Conflicts of Interest: The authors declare no conflict of interest.

References

1. Molnar, M.; Lončarić, M.; Kovač, M. Green Chemistry Approaches to the Synthesis of Coumarin Derivatives. *Curr. Org. Chem.* **2020**, *24*, 4–43. [CrossRef]
2. Matos, M.J.; Santana, L.; Uriarte, E.; Abreu, O.A.; Molina, E.; Yordi, E.G. Coumarins—An Important Class of Phytochemicals. *Phytochem. Isol. Characterisation Role Hum. Health* **2015**. [CrossRef]
3. Lončarić, M.; Gašo-Sokač, D.; Jokić, S.; Molnar, M. Recent Advances in the Synthesis of Coumarin Derivatives from Different Starting Materials. *Biomolecules* **2020**, *10*, 151. [CrossRef] [PubMed]
4. Salem, M.A.; Helal, M.H.; Gouda, M.A.; Ammar, Y.A.; El-Gaby, M.S.A.; Abbas, S.Y. An Overview on Synthetic Strategies to Coumarins. *Synth. Commun.* **2018**, *48*, 1534–1550. [CrossRef]
5. Irfan, A.; Rubab, L.; Rehman, M.U.; Anjum, R.; Ullah, S.; Marjana, M.; Qadeer, S.; Sana, S. Coumarin Sulfonamide Derivatives: An Emerging Class of Therapeutic Agents. *Heterocycl. Commun.* **2020**, *26*, 46–59. [CrossRef]
6. Venugopala, K.N.; Rashmi, V.; Odhav, B. Review on Natural Coumarin Lead Compounds for Their Pharmacological Activity. *BioMed. Res. Int.* **2013**, *2013*, 1–14. [CrossRef] [PubMed]
7. Mishra, S.; Pandey, A.; Manvati, S. Coumarin: An Emerging Antiviral Agent. *Heliyon* **2020**, *6*, e03217. [CrossRef]
8. Chattopadhyay, D.S.K. Pharmacological Potentiality of Coumarins as Anti-Viral Agent. *Int. J. Res. Anal. Rev.* **2018**, *5*, 11.
9. Bizzarri, B.M.; Botta, L.; Capecchi, E.; Celestino, I.; Checconi, P.; Palamara, A.T.; Nencioni, L.; Saladino, R. Regioselective IBX-Mediated Synthesis of Coumarin Derivatives with Antioxidant and Anti-Influenza Activities. *J. Nat. Prod.* **2017**, *80*, 3247–3254. [CrossRef]
10. Hassan, M.Z.; Osman, H.; Ali, M.A.; Ahsan, M.J. Therapeutic Potential of Coumarins as Antiviral Agents. *Eur. J. Med. Chem.* **2016**, *123*, 236–255. [CrossRef]
11. Menezes, J.C.; Diederich, M. Translational Role of Natural Coumarins and Their Derivatives as Anticancer Agents. *Future Med. Chem.* **2019**, *11*, 1057–1082. [CrossRef] [PubMed]
12. Kaur, M.; Kohli, S.; Sandhu, S.; Bansal, Y.; Bansal, G. Coumarin: A Promising Scaffold for Anticancer Agents. *Anticancer Agents Med. Chem.* **2015**, *15*, 1032–1048. [CrossRef] [PubMed]
13. Majnooni, M.B.; Fakhri, S.; Smeriglio, A.; Trombetta, D.; Croley, C.R.; Bhattacharyya, P.; Sobarzo-Sánchez, E.; Farzaei, M.H.; Bishayee, A. Antiangiogenic Effects of Coumarins against Cancer: From Chemistry to Medicine. *Molecules* **2019**, *24*, 4278. [CrossRef] [PubMed]
14. Bhagat, K.; Bhagat, J.; Gupta, M.K.; Singh, J.V.; Gulati, H.K.; Singh, A.; Kaur, K.; Kaur, G.; Sharma, S.; Rana, A.; et al. Design, Synthesis, Antimicrobial Evaluation, and Molecular Modeling Studies of Novel Indolinedione–Coumarin Molecular Hybrids. *ACS Omega* **2019**, *4*, 8720–8730. [CrossRef]
15. Al-Majed, Y.K.; Kadhum, A.H.; Al-Amier, A.A.; Mohama, A. Coumarins: The Antimicrobial Agents. *Syst. Rev. Pharm.* **2017**, *8*, 62–70. [CrossRef]
16. Arsenyan, P.; Vasiljeva, J.; Shestakova, I.; Domracheva, I.; Jaschenko, E.; Romanchikova, N.; Leonchiks, A.; Rudevica, Z.; Belyakov, S. Selenopheno[3,2-c]- and [2,3-c]Coumarins: Synthesis, Cytotoxicity, Angiogenesis Inhibition, and Antioxidant Properties. *C. R. Chim.* **2015**, *18*, 399–409. [CrossRef]
17. Annunziata, F.; Pinna, C.; Dallavalle, S.; Tamborini, L.; Pinto, A. An Overview of Coumarin as a Versatile and Readily Accessible Scaffold with Broad-Ranging Biological Activities. *Int. J. Mol. Sci.* **2020**, *21*, 4618. [CrossRef]
18. Boisde, P.M.; Meuly, W.C. Coumarin. In *Kirk-Othmer Encyclopedia of Chemical Technology*; American Cancer Society: Atlanta, GA, USA, 2000; ISBN 978-0-471-23896-6.
19. Api, A.M.; Belmonte, F.; Belsito, D.; Biserta, S.; Botelho, D.; Bruze, M.; Burton, G.A.; Buschmann, J.; Cancellieri, M.A.; Dagli, M.L.; et al. RIFM Fragrance Ingredient Safety Assessment, Coumarin, CAS Registry Number 91-64-5. *Food Chem. Toxicol.* **2019**, *130*, 110522. [CrossRef]
20. Coumarin | Cosmetics Info. Available online: https://cosmeticsinfo.org/ingredient/coumarin (accessed on 18 September 2020).
21. Lončar, M.; Jakovljević, M.; Šubarić, D.; Pavlić, M.; Buzjak Služek, V.; Cindrić, I.; Molnar, M. Coumarins in Food and Methods of Their Determination. *Foods* **2020**, *9*, 645. [CrossRef]
22. Krüger, S.; Winheim, L.; Morlock, G.E. Planar Chromatographic Screening and Quantification of Coumarin in Food, Confirmed by Mass Spectrometry. *Food Chem.* **2018**, *239*, 1182–1191. [CrossRef]

23. Sun, X.; Liu, T.; Sun, J.; Wang, X. Synthesis and Application of Coumarin Fluorescence Probes. *RSC Adv.* **2020**, *10*, 10826–10847. [CrossRef]
24. Sarih, N.M.; Ciupa, A.; Moss, S.; Myers, P.; Slater, A.G.; Abdullah, Z.; Tajuddin, H.A.; Maher, S. Furo[3,2-c]Coumarin-Derived Fe^{3+} Selective Fluorescence Sensor: Synthesis, Fluorescence Study and Application to Water Analysis. *Sci. Rep.* **2020**, *10*, 7421. [CrossRef] [PubMed]
25. Akchurin, I.O.; Yakhutina, A.I.; Bochkov, A.Y.; Solovjova, N.P.; Traven, V.F. Synthesis of Novel Push-Pull Fluorescent Dyes—7-(Diethylamino)Furo[3,2-c]Coumarin and 7-(Diethylamino)Thieno[3,2-c]Coumarin Derivatives. *Heterocycl. Commun.* **2018**, *24*, 85–91. [CrossRef]
26. Shi, L.; Yu, H.; Zeng, X.; Yang, S.; Gong, S.; Xiang, H.; Zhang, K.; Shao, G. A Novel Ratiometric Fluorescent Probe Based on Thienocoumarin and Its Application for the Selective Detection of Hypochlorite in Real Water Samples and in Vivo. *New J. Chem.* **2020**, *44*, 6232–6237. [CrossRef]
27. Cadierno, V. Chapter 4—Metal-Catalyzed Routes for the Synthesis of Furocoumarins and Coumestans. In *Green Synthetic Approaches for Biologically Relevant Heterocycles*; Brahmachari, G., Ed.; Elsevier: Amsterdam, The Netherlands, 2015; pp. 77–100. ISBN 978-0-12-800070-0.
28. Ram, R.; Krupadanam, S.; Srimannarayana, G. Synthesis of 2-Methyl-9-Oxo-9H-Furo[2,3-c]Benzopyrans and 2-Methyl-3H,4H[1]Benzopyrano[3,4-b]Pyrrol-4-Ones. *Synth. Commun.* **1998**, *28*, 2421–2428. [CrossRef]
29. Majumdar, K.C.; Khan, A.T.; Das, D.P. Facile Regioselective Synthesis of 2-Methyl Furo [3,2-c][1] Benzopyran-4-One 2-Methyl Furo[2,3-c] [1]Benzopyran-4-One. *Synth. Commun.* **1989**, *19*, 917–930. [CrossRef]
30. Pandya, M.K.; Chhasatia, M.R.; Vala, N.D.; Parekh, T.H. Synthesis of Furano[2,3-c] /Pyrrolo[2,3-c]Coumarins and Synthesis of 1(H)-[1]Benzopyrano[3,4-b][1]Benzopyrano[3′,4′-d] Furan-7(H)-Ones /1(H)-[1]Benzopyrano[3,4-b][1]Benzopyrano [3′,4′-d]Pyrrole-7(H)-Ones. *J. Drug Deliv. Ther.* **2019**, *9*, 32–42. [CrossRef]
31. Dong, Y.; Yu, J.-T.; Sun, S.; Cheng, J. Rh(III)-Catalyzed Sequential Ortho -C–H Oxidative Arylation/Cyclization of Sulfoxonium Ylides with Quinones toward 2-Hydroxy-Dibenzo[b,d]Pyran-6-Ones. *Chem. Commun.* **2020**, *56*, 6688–6691. [CrossRef]
32. Brahmbhatt, D.I.; Gajera, J.M.; Patel, C.N.; Pandya, V.P.; Pandya, U.R. First Synthesis of Furo[3,4-c]Coumarins. *J. Heterocycl. Chem.* **2006**, *43*, 1699–1702. [CrossRef]
33. Wang, X.; Nakagawa-Goto, K.; Kozuka, M.; Tokuda, H.; Nishino, H.; Lee, K.-H. Cancer Preventive Agents. Part 6: Chemopreventive Potential of Furanocoumarins and Related Compounds. *Pharm. Biol.* **2006**, *44*, 116–120. [CrossRef]
34. Rajitha, B.; Geetanjali, Y.; Somayajulu, V. Synthesis of Coumestrol Analogs as Possible Antifertility Agents. *Indian J. Chem* **1986**, *25B*, 872–873.
35. Al-Sehemi, A.G.; El-Gogary, S.R. Synthesis and Photooxygenation of Furo[3,2-c]Coumarin Derivatives as Antibacterial and DNA Intercalating Agent. *Chin. J. Chem.* **2012**, *30*, 316–320. [CrossRef]
36. Gorbunov, Y.O.; Mityanov, V.S.; Melekhina, V.G.; Krayushkin, M.M. Synthesis of Novel 4H-Furo[3,2-c]Pyran-4-Ones and 4H-Furo[3,2-c]Chromen-4-Ones. *Russ. Chem. Bull.* **2018**, *67*, 304–307. [CrossRef]
37. Panja, S.K.; Ghosh, J.; Maiti, S.; Bandyopadhyay, C. Synthesis of Furocoumarins and Biscoumarins by an Isocyanide-Induced Multicomponent Reaction: Isocyanide as a Masked Amine. *J. Chem. Res.* **2012**, *36*, 222–225. [CrossRef]
38. Reisch, J. Die Synthese von Hydroxy- und Furo-cumarinderivaten. *Archiv. Pharm.* **1966**, *299*, 798–805. [CrossRef]
39. Rahman, M.-U.; Khan, K.-Z.; Siddiqi, Z.S.; Zaman, A. A Convenient Route for 3-Formyl-4-Hydroxycoumarin and Reactions of 3-Bromo-4-Hydroxycoumarin. *J. Chem. Sect. Org. Chem. Incl. Med. Chem. Indian* **1990**, *29*, 941–943.
40. Kadam, A.; Bellinger, S.; Zhang, W. Atom- and Step-Economic Synthesis of Biaryl-Substituted Furocoumarins, Furoquinolones and Furopyrimidines by Multicomponent Reactions and One-Pot Synthesis. *Green Process. Synth.* **2013**, *2*, 491–497. [CrossRef]
41. Lee, Y.R.; Kim, B.S.; Wang, H.C. Silver(I)/Celite Promoted Oxidative Cycloaddition of 4-Hydroxycoumarin to Olefins. A Facile Synthesis of Dihydrofurocoumarins and Furocoumarins. *Tetrahedron* **1998**, *54*, 12215–12222. [CrossRef]
42. Majumdar, N.; Paul, N.D.; Mandal, S.; de Bruin, B.; Wulff, W.D. Catalytic Synthesis of 2H-Chromenes. *ACS Catal.* **2015**, *5*, 2329–2366. [CrossRef]
43. Nebra, N.; Díaz-Álvarez, A.E.; Díez, J.; Cadierno, V. Expeditious Entry to Novel 2-Methylene-2,3-Dihydrofuro[3,2-c] Chromen-2-Ones from 6-Chloro-4-Hydroxychromen-2-One and Propargylic Alcohols. *Molecules* **2011**, *16*, 6470–6480. [CrossRef]
44. Raffa, G.; Rusch, M.; Balme, G.; Monteiro, N. A Pd-Catalyzed Heteroannulation Approach to 2,3-Disubstituted Furo[3,2-c]Coumarins. *Org. Lett.* **2009**, *11*, 5254–5257. [CrossRef] [PubMed]
45. Medina, F.G.; Marrero, J.G.; Macías-Alonso, M.; González, M.C.; Córdova-Guerrero, I.; Teissier García, A.G.; Osegueda-Robles, S. Coumarin Heterocyclic Derivatives: Chemical Synthesis and Biological Activity. *Nat. Prod. Rep.* **2015**, *32*, 1472–1507. [CrossRef] [PubMed]
46. Nealmongkol, P.; Tangdenpaisal, K.; Sitthimonchai, S.; Ruchirawat, S.; Thasana, N. Cu(I)-Mediated Lactone Formation in Subcritical Water: A Benign Synthesis of Benzopyranones and Urolithins A–C. *Tetrahedron* **2013**, *69*, 9277–9283. [CrossRef]
47. Cheng, G.; Hu, Y. One-Pot Synthesis of Furocoumarins through Cascade Addition–Cyclization–Oxidation. *Chem. Commun.* **2007**, 3285–3287. [CrossRef]
48. Cheng, G.; Hu, Y. Two Efficient Cascade Reactions to Synthesize Substituted Furocoumarins. *J. Org. Chem.* **2008**, *73*, 4732–4735. [CrossRef] [PubMed]
49. Sanz, R.; Miguel, D.; Martínez, A.; Álvarez-Gutiérrez, J.M.; Rodríguez, F. Brønsted Acid Catalyzed Propargylation of 1,3-Dicarbonyl Derivatives. Synthesis of Tetrasubstituted Furans. *Org. Lett.* **2007**, *9*, 727–730. [CrossRef] [PubMed]

50. Huang, W.; Wang, J.; Shen, Q.; Zhou, X. Yb(OTf)3-Catalyzed Propargylation and Allenylation of 1,3-Dicarbonyl Derivatives with Propargylic Alcohols: One-Pot Synthesis of Multi-Substituted Furocoumarin. *Tetrahedron* **2007**, *63*, 11636–11643. [CrossRef]
51. Cadierno, V. A Novel Propargylation/Cycloisomerization Tandem Process Catalyzed by a Ruthenium(II)/Trifluoroacetic Acid System: One-Pot Entry to Fully Substituted Furans from Readily Available Secondary Propargylic Alcohols and 1,3-Dicarbonyl Compounds. *Adv. Synth. Catal.* **2007**, *349*, 382–394. [CrossRef]
52. Cadierno, V.; García-Garrido, S.E.; Gimeno, J.; Nebra, N. Atom-Economic Transformations of Propargylic Alcohols Catalyzed by the 16-Electron Allyl-Ruthenium(II) Complex [Ru(H3-2-C3H4Me)(CO)(Dppf)][SbF6] (Dppf=1,1′-Bis(Diphenylphosphino)Ferrocene). *Inorg. Chim. Acta* **2010**, *363*, 1912–1934. [CrossRef]
53. Conreaux, D.; Belot, S.; Desbordes, P.; Monteiro, N.; Balme, G. Et3N-Induced Demethylation−Annulation of 3-Alkynyl-4-Methoxy-2-Pyridones and Structurally Related Compounds in the Synthesis of Furan-Fused Heterocycles. *J. Org. Chem.* **2008**, *73*, 8619–8622. [CrossRef]
54. Lei, C.; Li, Y.; Mh, X. One-Pot Synthesis of Furocoumarins via Sequential Pd/Cu-Catalyzed Alkynylation and Intramolecular Hydroalkoxylation. *Org. Biomol. Chem.* **2010**, *8*, 3073–3077. [CrossRef]
55. Pham, P.H.; Nguyen, Q.T.D.; Tran, N.K.Q.; Nguyen, V.H.H.; Doan, S.H.; Ha, H.Q.; Truong, T.; Phan, N.T.S. Metal-Free Synthesis of Furocoumarins: An Approach via Iodine-Promoted One-Pot Cyclization between 4-Hydroxycoumarins and Acetophenones: Metal-Free Synthesis of Furocoumarins: An Approach via Iodine-Promoted One-Pot Cyclization between 4-Hydroxycoumarins and Acetophenones. *Eur. J. Org. Chem.* **2018**, *2018*, 4431–4435. [CrossRef]
56. Traven, V.F.; Kravtchenko, D.V.; Chibisova, T.A.; Shorshnev, S.V.; Eliason, R.; Wakefield, D.H. A New Short Way to Furocoumarins. *Heterocycl. Commun.* **1996**, *2*. [CrossRef]
57. Majumdar, K.C.; Bhattacharyya, T. Regioselective Synthesis of Furo[3,2-c][1]Benzopyran-4-One and Furo[3,2-c]Quinolin-4-One. *J. Chem. Res.* **1997**, 244–245. [CrossRef]
58. Lee, T.-H.; Jayakumar, J.; Cheng, C.-H.; Chuang, S.-C. One Pot Synthesis of Bioactive Benzopyranones through Palladium-Catalyzed C–H Activation and CO Insertion into 2-Arylphenols. *Chem. Commun.* **2013**, *49*, 11797. [CrossRef]
59. Xiao, B.; Gong, T.-J.; Liu, Z.-J.; Liu, J.-H.; Luo, D.-F.; Xu, J.; Liu, L. Synthesis of Dibenzofurans via Palladium-Catalyzed Phenol-Directed C-H Activation/C-O Cyclization. *J. Am. Chem. Soc.* **2011**, *133*, 9250–9253. [CrossRef]
60. Moon, Y.; Kim, Y.; Hong, H.; Hong, S. Synthesis of Heterocyclic-Fused Benzofurans via C–H Functionalization of Flavones and Coumarins. *Chem. Commun.* **2013**, *49*, 8323. [CrossRef]
61. Fu, L.; Li, S.; Cai, Z.; Ding, Y.; Guo, X.-Q.; Zhou, L.-P.; Yuan, D.; Sun, Q.-F.; Li, G. Ligand-Enabled Site-Selectivity in a Versatile Rhodium(Ii)-Catalysed Aryl C–H Carboxylation with CO_2. *Nat. Catal.* **2018**, *1*, 469–478. [CrossRef]
62. Khan, M.A.; De Brito Morley, M.L. Condensed Benzopyrans III. 3H,4H [1] Benzopyrano [3,4- b] Pyrrol-4-Ones. *J. Heterocycl. Chem.* **1978**, *15*, 1399–1401. [CrossRef]
63. Soman, S.S.; Thaker, T.H.; Rajput, R.A. Novel Synthesis and Cytotoxic Activity of Some Chromeno[3,4-b]Pyrrol-4(3H)-Ones. *Chem. Heterocycl. Comp.* **2011**, *46*, 1514. [CrossRef]
64. Majumdar, K.; Chattopadhyay, B. Amino-Claisen versus Oxy-Claisen Rearrangement: Regioselective Synthesis of Pyrrolocoumarin Derivatives. *Synthesis* **2008**, *2008*, 921–924. [CrossRef]
65. Gtigga, R.; Vipond, D. 4-phenylsulphinyl-and 4-Phenylsulphonylcoumarins as components in cycloaddition reactions. *Temahedron* **1989**, *45*, 7587–7592.
66. Xue, S.; Yao, J.; Liu, J.; Wang, L.; Liu, X.; Wang, C. Three-Component Reaction between Substituted 2-(2-Nitrovinyl)-Phenols, Acetylenedicarboxylate and Amines: Diversity-Oriented Synthesis of Novel Pyrrolo[3,4-c]Coumarins. *RSC Adv.* **2016**, *6*, 1700–1704. [CrossRef]
67. Alizadeh, A.; Ghanbaripour, R.; Zhu, L.-G. An Approach to the Synthesis of 2-Acylchromeno[3,4-c]Pyrrol-4(2H)-One Derivatives via a Sequential Three-Component Reaction. *Synlett* **2013**, *24*, 2124–2126. [CrossRef]
68. Shaabani, A.; Sepahvand, H.; Bazgir, A.; Khavasi, H.R. Tosylmethylisocyanide (TosMIC) [3+2] Cycloaddition Reactions: A Facile Van Leusen Protocol for the Synthesis of the New Class of Spirooxazolines, Spiropyrrolines and Chromeno[3,4-c]Pyrrols. *Tetrahedron* **2018**, *74*, 7058–7067. [CrossRef]
69. Padilha, G.; Iglesias, B.A.; Back, D.F.; Kaufman, T.S.; Silveira, C.C. Synthesis of Chromeno[4,3-b]Pyrrol-4(1 H)-Ones, from β-Nitroalkenes and 4-Phenylaminocoumarins, under Solvent-Free Conditions. *ChemistrySelect* **2017**, *2*, 1297–1304. [CrossRef]
70. Saha, M.; Pradhan, K.; Das, A.R. Facile and Eco-Friendly Synthesis of Chromeno[4,3-b]Pyrrol-4(1H)-One Derivatives Applying Magnetically Recoverable Nano Crystalline $CuFe_2O$ 4 Involving a Domino Three-Component Reaction in Aqueous Media. *RSC Adv.* **2016**, *6*, 55033–55038. [CrossRef]
71. Chen, Z.; Yang, X.; Su, W. An Efficient Protocol for Multicomponent Synthesis of Functionalized Chromeno[4,3-b]Pyrrol-4(1H)-One Derivatives. *Tetrahedron Lett.* **2015**, *56*, 2476–2479. [CrossRef]
72. Alberola, A.; Calvo, L.; González-Ortega, A.; Encabo, A.P.; Sañudo, M.C. Synthesis of [1]Benzopyrano[4,3-b]Pyrrol-4(1H)-Ones from 4-Chloro-3-Formylcoumarin. *Synthesis* **2001**, *2001*, 1941–1948. [CrossRef]
73. Navarro, R.A.; Bleye, L.C.; González-Ortega, A.; Ruíz, C.S. Synthesis of 1H-[1]Benzopyrano[4,3-b]Pyrrole and 4hthieno[3,2-c][1]Benzopyran Derivatives. Functionalisation by Aromatic Electrophilic Substitution. *Heterocycles* **2001**, *55*, 2369–2386.
74. Alberola, A.; Álvaro, R.; Ortega, A.G.; Sádaba, M.L.; Carmen Sañudo, M. Synthesis of [1]Benzopyrano[4,3-b]Pyrrol-4(1H)-Ones from N(α)-(2-Oxo-2H-1-Benzopyran-4-Yl)Weinreb α-Aminoamides. *Tetrahedron* **1999**, *55*, 13211–13224. [CrossRef]

75. Alberola, A.; Alvaro, R.; Andres, J.M.; Calvo, B.; González, A. Synthesis of [1]Benzopyrano[4,3-b]Pyrrol-4(1H)-Ones from 4-Chlorocoumarin. *Synthesis* **1994**, 279–281. [CrossRef]
76. Peng, S.; Wang, L.; Huang, J.; Sun, S.; Guo, H.; Wang, J. Palladium-Catalyzed Oxidative Annulation C-H/N-H Functionalization: Access to Substituted Pyrroles. *Adv. Synth. Catal.* **2013**, *355*, 2550–2557. [CrossRef]
77. Lin, C.-H.; Yang, D.-Y. Synthesis of Coumarin/Pyrrole-Fused Heterocycles and Their Photochemical and Redox-Switching Properties. *Org. Lett.* **2013**, *15*, 2802–2805. [CrossRef]
78. Yang, X.; Jing, L.; Chen, Z. An Efficient Method for One-Pot Synthesis of 3-Alkoxy-Substituted Chromeno[4,3-b]Pyrrol-4(1H)-One Derivatives. *Chem. Heterocycl. Comp.* **2018**, *54*, 1065–1069. [CrossRef]
79. Yahyavi, H.; Heravi, M.M.; Mahdavi, M.; Foroumadi, A. Iodine-Catalyzed Tandem Oxidative Coupling Reaction: A One-Pot Strategy for the Synthesis of New Coumarin-Fused Pyrroles. *Tetrahedron Lett.* **2018**, *59*, 94–98. [CrossRef]
80. Yang, Y.; Qi, X.; Liu, R.; He, Q.; Yang, C. One-Pot Transition-Metal-Free Cascade Synthesis of Thieno[2,3-c]Coumarins from Chromones. *RSC Adv.* **2016**, *6*, 103895–103898. [CrossRef]
81. Vishnumurthy, K.; Makriyannis, A. Novel and Efficient One-Step Parallel Synthesis of Dibenzopyranones via Suzuki−Miyaura Cross Coupling. *J. Comb. Chem.* **2010**, *12*, 664–669. [CrossRef]
82. Janosik, T.; Bergman, J. *Five-Membered Ring Systems: Thiophenes and Se/Te Analogues in Progress in Heterocyclic Chemistry*; Gordon, W.G., John, A.J., Eds.; Elsevier: Amsterdam, The Netherlands, 2007; Volume 18.
83. Iaroshenko, V.O.; Ali, S.; Mkrtchyan, S.; Gevorgyan, A.; Babar, T.M.; Semeniuchenko, V.; Hassan, Z.; Villinger, A.; Langer, P. Design and Synthesis of Condensed Thienocoumarins by Suzuki–Miyaura Reaction/Lactonization Tandem Protocol. *Tetrahedron Lett.* **2012**, *53*, 7135–7139. [CrossRef]
84. Fondjo, E.S.; Tsemeugne, J.; De Dieu Tamokou, J.; Djintchui, A.N.; Kuiate, J.R.; Sondengam, B.L. Synthesis and Antimicrobial Activities of Some Novel Thiophene Containing Azo Compounds. *Heterocycl. Commun.* **2013**, *19*. [CrossRef]
85. Al-Mousawi, S.M.; El-Apasery, M.A. Synthesis of Some Monoazo Disperse Dyes Derived from Aminothienochromene. *Molecules* **2013**, *18*, 8837–8844. [CrossRef] [PubMed]
86. Djeukoua, K.S.D.; Fondjo, E.S.; Tamokou, J.-D.; Tsemeugne, J.; Simon, P.F.W.; Tsopmo, A.; Tchieno, F.M.M.; Ekom, S.E.; Pecheu, C.N.; Tonle, I.K.; et al. Synthesis, Characterization, Antimicrobial Activities and Electrochemical Behavior of New Phenolic Azo Dyes from Two Thienocoumarin Amines. *Arkivoc* **2020**, *2019*, 416–430. [CrossRef]
87. Gewald, K. Zur Reaktion von α-Oxo-mercaptanen mit Nitrilen. *Angew. Chem.* **1961**, *73*, 114. [CrossRef]
88. Hoberg, H. Halogenmethyl-aluminiumverbindungen. *Angew. Chem.* **1961**, *73*, 114–115. [CrossRef]
89. Gewald, K. Heterocyclen aus CH-aciden Nitrilen, VII. 2-Amino-thiophene aus α-Oxo-mercaptanen und methylenaktiven Nitrilen. *Chem. Ber.* **1965**, *98*, 3571–3577. [CrossRef]
90. Sopbué Fondjo, E.; Döpp, D.; Henkel, G. Reactions of Some Anellated 2-Aminothiophenes with Electron Poor Acetylenes. *Tetrahedron* **2006**, *62*, 7121–7131. [CrossRef]
91. Yu, L.-S.-H.; Meng, C.-Y.; Wang, J.; Gao, Z.; Xie, J.-W. Substrate-Controlled Diastereoselectivity Switch in the Formation of Dihydrothieno[3,4-c]Coumarins via [4+1] Annulations. *Adv. Synth. Catal.* **2019**, *361*, 526–534. [CrossRef]
92. Kim, J.E.; Lee, J.; Yun, H.; Baek, Y.; Lee, P.H. Rhodium-Catalyzed Intramolecular Transannulation Reaction of Alkynyl Thiadiazole Enabled 5,n-Fused Thiophenes. *J. Org. Chem.* **2017**, *82*, 1437–1447. [CrossRef]
93. Weibenfels, M.; Hantschmann, A.; Steinfuhrer, T.; Birkner, E. Synthesis of 7-Substituted Ones Thieno[3,2-c]Coumarin-3-Carboxlyic Acid. *Z. Chem.* **1989**, *29*, 166. [CrossRef]
94. El-Dean, A.M.K.; Zaki, R.M.; Geies, A.A.; Radwan, S.M.; Tolba, M.S. Synthesis and Antimicrobial Activity of New Heterocyclic Compounds Containing Thieno[3,2-c]Coumarin and Pyrazolo[4,3-c]Coumarin Frameworks. *Russ. J. Bioorg. Chem.* **2013**, *39*, 553–564. [CrossRef]
95. Domracheva, I.; Kanepe-Lapsa, I.; Jackevica, L.; Vasiljeva, J.; Arsenyan, P. Selenopheno Quinolinones and Coumarins Promote Cancer Cell Apoptosis by ROS Depletion and Caspase-7 Activation. *Life Sci.* **2017**, *186*, 92–101. [CrossRef] [PubMed]
96. Abdel-Hafez, S.H.; Elkhateeb, A.; Gobouri, A.A.; Azab, I.H.E.; Kirsch, G. An Efficient Route for Synthesis and Reactions of Seleno-[2, 3-c]Coumarin. *Heterocycles* **2016**, *92*, 1054–1062. [CrossRef]

Review

Synthetic Routes to Coumarin(Benzopyrone)-Fused Five-Membered Aromatic Heterocycles Built on the α-Pyrone Moiety. Part II: Five-Membered Aromatic Rings with Multi Heteroatoms

Eslam Reda El-Sawy [1], Ahmed Bakr Abdelwahab [2] and Gilbert Kirsch [3,*]

1 National Research Centre, Chemistry of Natural Compounds Department, Dokki, Cairo 12622, Egypt; eslamelsawy@gmail.com
2 Plant Advanced Technologies (PAT), 54500 Vandoeuvre-les-Nancy, France; ahm@plantadvanced.com
3 Laboratoire Lorrain de Chimie Moléculaire (L.2.C.M.), Université de Lorraine, 57078 Metz, France
* Correspondence: gilbert.kirsch@univ-lorraine.fr; Tel.: +33-0372-749-200; Fax: +33-0372-749-187

Abstract: Coumarins are natural heterocycles that widely contribute to the design of various biologically active compounds. Fusing different aromatic heterocycles with coumarin at its 3,4-position is one of the interesting approaches to generating novel molecules with various biological activities. During our continuing interest in assembling information about fused five-membered aromatic heterocycles, and after having presented mono-hetero-atomic five-membered aromatic heterocycles in Part I. The current review Part II is intended to present an overview of the different synthetic routes to coumarin (benzopyrone)-fused five-membered aromatic heterocycles with multi-heteroatoms built on the pyrone ring, covering the literature from 1945 to 2021.

Keywords: coumarins; benzopyrones; pyrazole; imidazole; thiazole; oxazole; triazole; thiadiazole

Citation: El-Sawy, E.R.; Abdelwahab, A.B.; Kirsch, G. Synthetic Routes to Coumarin(Benzopyrone)-Fused Five-Membered Aromatic Heterocycles Built on the α-Pyrone Moiety. Part II: Five-Membered Aromatic Rings with Multi Heteroatoms. *Molecules* **2021**, *26*, 3409. https://doi.org/10.3390/molecules26113409

Academic Editor: Maria João Matos

Received: 11 May 2021
Accepted: 31 May 2021
Published: 4 June 2021

Publisher's Note: MDPI stays neutral with regard to jurisdictional claims in published maps and institutional affiliations.

1. Introduction

The fusion of the pyrone ring with the benzene nucleus gives rise to a class of heterocyclic compounds known as benzopyrone [1]. Coumarin is one of the benzopyrones (1,2-benzopyrones or 2H-[1] benzopyran-2-ones) and represents an important family of oxygen-containing heterocycles widely distributed in nature [1]. The incorporation of another heterocyclic moiety into coumarin enriches the properties of the parent structure and the resulting compounds may exhibit promising properties [2–5]. Certain derivatives of 3,4-heterocycle-fused coumarins play an important role in medicinal chemistry and have been extensively used as versatile building blocks in organic synthesis [2,3,5–7]. Many examples of biologically active coumarins containing 3,4-heterocycles-fused were cited in the literature [2,3,8] including antimicrobial [9–12], antiviral [13,14], anticancer [15–17], antioxidant [18,19], and anti-inflammatory [20,21] activities.

The development of synthetic pathways towards active coumarins containing heterocyclics has attracted great interest from researchers [5]. Significant efforts have been focused on developing new methodologies to enrich structural libraries and reduce the number of synthetic steps of novel coumarin derivatives [5,22].

In proceeding to our interest in coumarin(benzopyrone)-fused five-membered aromatic heterocycles built on the α-pyrone ring, which was recently issued in Part I [22]. The present review, Part II, describes the components which have multi heteroatom in an aromatic fused ring with the pyrone part of coumarin. The synthetic pathways of the investigated scaffolds provided systems containing oxygen, nitrogen, and sulfur in their core structure.

2. Synthesis of Benzopyrone-Fused Five-Membered Aromatic Heterocycles

2.1. Five-Membered Aromatic Rings with Two Heteroatoms

2.1.1. Two Identical Heteroatoms (N-N)

Pyrazole

Fusion of pyrazole ring with the pyrone ring of coumarin results in formation of two structural isomers, namely chromeno[4,3-*c*]pyrazol-4(2*H*)-one, chromeno[4,3-*c*]pyrazol-4(1*H*)-one, (1*H*-benzopyrano[4,3-*c*]pyrazole), and chromeno[3,4-*c*]pyrazol-4(2*H*)-one, chromeno[3,4-*c*]pyrazol-4(3*H*)-one, (1*H*-benzopyrano[3,4-*c*]pyrazole) (Figure 1).

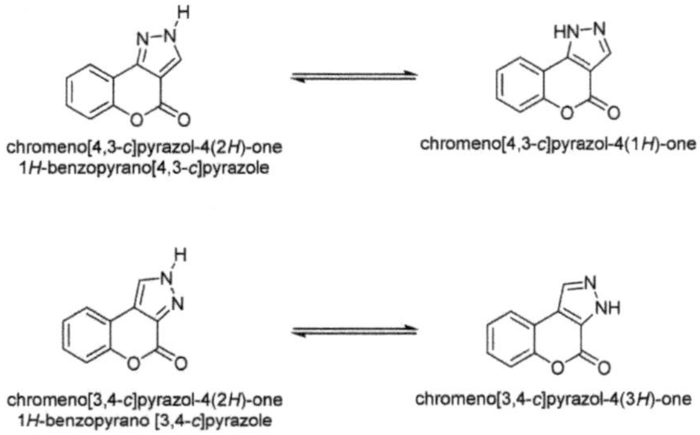

chromeno[4,3-c]pyrazol-4(2*H*)-one
1*H*-benzopyrano[4,3-*c*]pyrazole

chromeno[4,3-c]pyrazol-4(1*H*)-one

chromeno[3,4-c]pyrazol-4(2*H*)-one
1*H*-benzopyrano [3,4-c]pyrazole

chromeno[3,4-c]pyrazol-4(3*H*)-one

Figure 1. The isomers of fused chromeno-pyrazole.

Chromeno[4,3-c]pyrazol-4(2*H*)-one; (1*H*-Benzopyrano[4,3-c]pyrazole)

Many synthetic protocols reported the synthesis of chromeno[4,3-*c*]pyrazol-4(2*H*)-one including the pyrazole and/or the pyrone-ring construction. The synthesis of the pyrazole ring in the literature started from coumarin, 4-hydroxy, 3-aldehyde, or 3-acetyl coumarin in addition to the chromone derivatives.

Pyrazole Construction

Shawali and his co-workers described the 1,3-dipolar additions of diphenylnitrilimine (DPNI) to coumarin (**1**) to afford 1,3-diphenyl-3a,9b-dihydro-4-oxo-1*H*-chromeno[4,3-*c*]pyrazol-4(1*H*)one (**2**) [23]. Upon dehydrogenation of 2 with lead tetra-acetate the corresponding 1,3-diphenyl-chromeno[4,3-c]pyrazol-4(1H)one (3) was obtained (Scheme 1).

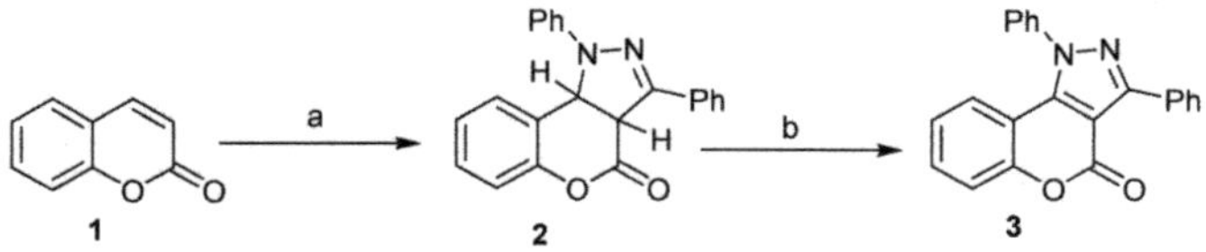

Scheme 1. Cycloaddition of diphenylnitrilimine to coumarin. Reagents and conditions. (**a**) diphenylnitrilimine (DPNI), benzene, TEA, compound **2**, 65% yield, compound **5**, 65% yield; (**b**) lead tetraacetate, dichloromethane, r.t., 12 h, 83% yield.

3-Aminochromeno[4,3-*c*]pyrazol-4(1*H*)one (**8**) was prepared through multi-step reactions starting from 4-hydroxycoumarin (**4**) [24]. Vilsmeier–Haack formylation of **4** developed 4-chlorocoumarin-3-carboxaldehyde (**5**). The reaction of **5** with hydroxylamine hydrochloride followed with phosphorus oxychloride gave the corresponding 4-chlorocoumarin -3-carbonitrile (**7**). Compound **7** under the reaction with hydrazine hydrate provided the target compound, 3-aminochromeno[4,3-*c*]pyrazol-4(1*H*)one (**8**) (Scheme 2).

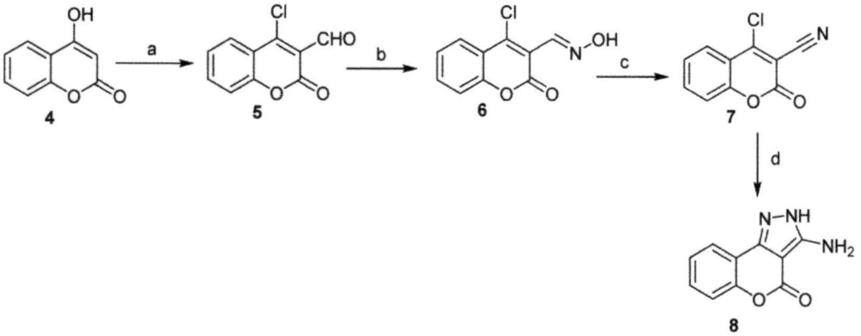

Scheme 2. Synthesis of 3-aminochromeno[4,3-*c*]pyrazol-4(1*H*)one (**8**). Reagents and conditions. (**a**) $POCl_3$, DMF, $CHCl_3$; (**b**) NH_2OH. HCl, AcONa, EtOH; (**c**) $POCl_3$; (**d**) N_2H_4. H_2O, EtOH, 80% yield.

The preparation of chromen[4,3-*c*]pyrazol-4-ones **9** from the reaction of 3-formyl-4-chlorocoumarin (**5**) with the appropriate aryl or alkyl hydrazine hydrochloride in the presence of base was intensively investigated (Scheme 3) [10,18,19,25–30]. Compound **9a** was employed as starting material to enrich the derivatives of chromen[4,3-*c*]pyrazol-4-ones **10–12** through the reaction with benzyl bromides [26], alkyl sulfonyl chlorides [28], or *N*-piperazine sulfonyl chlorides [30] (Scheme 4).

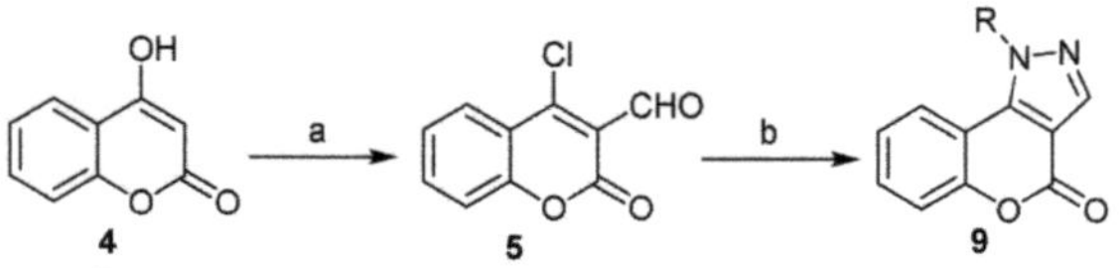

Scheme 3. Synthesis of chromeno[4,3-*c*]pyrazol-4(1*H*)ones **9**. Reagents and conditions. (**a**) $POCl_3$, DMF, $CHCl_3$; (**b**) NH_2NHR, EtOH, TEA or NaOAc, 25 °C, 2 h, five outputs with 67–89% yield [18], 14 outputs with 38–69% yield [25].

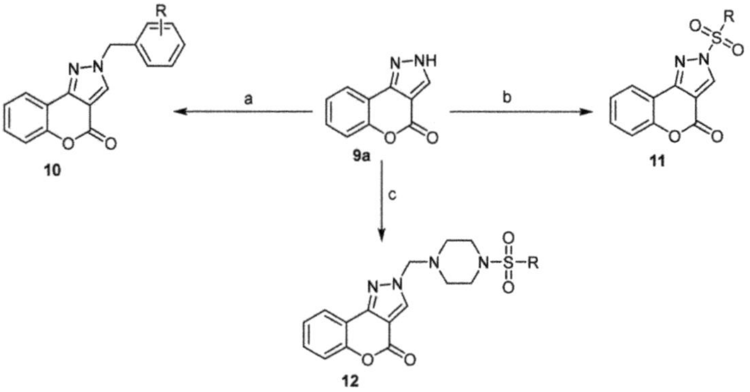

Scheme 4. Synthesis of 2-substituted-1*H*-chromeno[4,3-*c*]pyrazol-4(1*H*)ones **10–12**. Reagents and conditions. (**a**) Benzyl bromides, DMF, Cs_2CO_3, 100 °C, 10–12 h, 13 outputs; (**b**) benzene(alkyl)sulfonyl chlorides, DCM, TEA, 0 °C, 6–8 h, 28 outputs with 33–52% yield; (**c**) *N*-alkyl sulfonyl piperazines, HCHO, EtOH, rt, 4–6 h, 26 outputs with 37–65% yield.

Steinfiihrer et al., synthesized chromeno[4,3-*c*]pyrazol-4(2*H*)-ones **16** in a three-step reaction starting from 4-hydroxycoumarin derivatives **13**. The intermediate, 4-azido-3-coumarincarboxaldehydes **15** was produced in situ from 4-chlorocoumarin-3-carboxaldehydes **14** which subsequently reacted with some hydrazine derivatives to deliver the final products (Scheme 5) [31].

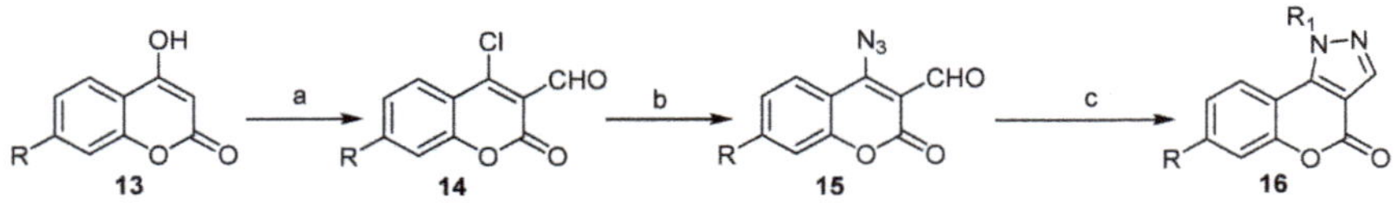

Scheme 5. Synthesis of chromeno[4,3-*c*]pyrazol-4(2*H*)-ones **16**. Reagents and conditions. (**a**) $POCl_3$, DMF, $CHCl_3$, 4 outputs with 65–80% yield; (**b**) NaN_3, DMF, 4 outputs with 50–80% yield; (**c**) R_1NHNH_2, DMF, 40–50 °C, six outputs with 70–90% yield.

Additionally, 3-coumarinyl alkyl ketones, such as 4-hydroxy-3-acetylcoumarin (**17a**) [32], and 3-coumarinyl phenyl ketone (**17b**) [12] were the starting materials in the preparation of chromeno[4,3-*c*]pyrazol-4(2*H*)-ones **18** and **19** by the base-catalyzed reaction with hydrazine hydrate (Scheme 6).

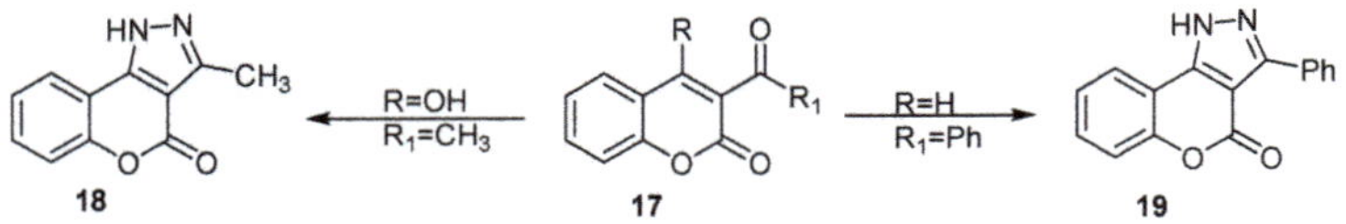

Scheme 6. 3-Coumarinyl alkyl ketone in the synthesis of chromeno[4,3-*c*]pyrazol-4(2*H*)-ones **18** and **19**. Reagents and conditions. $NH_2NH_2.H_2O$, EtOH, TEA, **18**: 75% yield, **19**: 58% yield.

The transformation of three-substituted chromone to chromeno[4,3-*c*]pyrazol-4(2*H*)-ones was within the scope of interest of Ibrahim's research group [33–36]. In 2008, they studied the ring transformation of chromone-3-carboxylic acid (**20**) under the reaction with 2-cyanoacetohydrazide or the hydrazine hydrate as the nucleophile to give the corresponding chromeno[4,3-*c*]pyrazol-4(2*H*)-one (**9a**) (Scheme 7) [33]. Furthermore, they examined the chemical reactivity of chromone-3-carboxamide (**21**) towards hydrazine hydrate or phenylhydrazine and they could construct the pyrazole within the scaffold (Scheme 7) [34]. Furthermore, 3-cyano-2,6-dimethyl chromone (**23**) was allowed to react with the nucleophile *S*-benzyldithiocarbazate to afford the corresponding 6,8-dimethyl chromeno[4,3-*c*]pyrazol-4(2*H*)-one (**24**) (Scheme 7) [35,36].

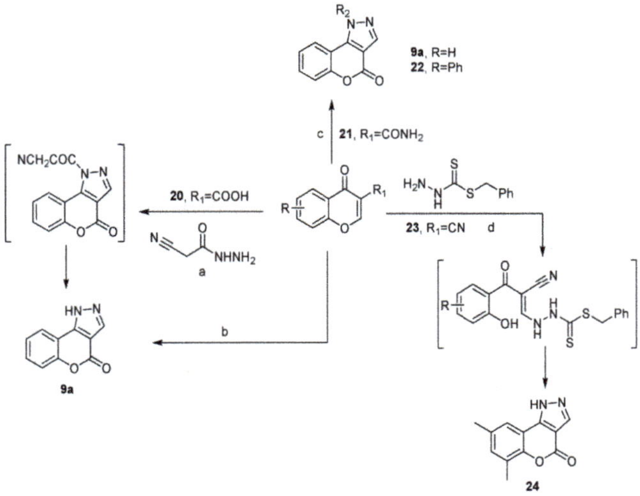

Scheme 7. Synthesis of chromeno[4,3-*c*]pyrazol-4(2*H*)-ones **9a**, **22**, and **24**. Reagents and conditions. (**a**) AcOH, 2 h, reflux, 48% yield; (**b**) NH_2NH_2. H_2O, AcOH, 2 h, reflux, 48% yield; (**c**) NH_2NH_2. H_2O, or NH_2NHPh, EtOH, 2 h, reflux, R=H: 56%, R=Ph: 46% yield; (**d**) DMF, 30 min, reflux, 63% yield.

Another synthetic pathway was performed by the dissociation of a large molecule in an acidic medium. [1]Benzopyrano[4,3-c][1,5]benzodiazepin-7(8*H*)-one (**25**) was prone to lose *o*-phenylenediamine in a reaction with some *N*-binucleophiles, such as hydrazine and methyl hydrazine that led exclusively to chromeno[4,3-*c*]pyrazol-4(2*H*)-ones **9a** and **26** (Scheme 8) [37].

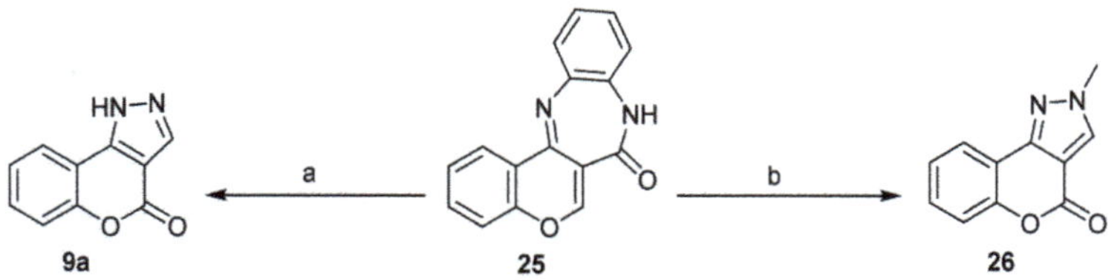

Scheme 8. [1]Benzopyrano[4,3-c][1,5]benzodiazepin-7(8*H*)-one in the synthesis of chromeno[4,3-*c*]pyrazol-4(2*H*)-ones **9a** and **26**. Reagents and conditions. (**a**) $NH_2NH_2.H_2O$, AcOH, reflux, 10 min, 48% yield; (**b**) NH_2NHCH_3, AcOH, reflux, 10 min, 48% yield.

Pyrone Construction

Lokhande et al. introduced a simple and convenient method for the synthesis of fused chromeno[4,3-*c*]pyrazol-4(2*H*)-one. It was accomplished using iodine catalyzed oxidative cyclization of 3-(2-hydroxyaryl)-1-phenyl-1*H*-pyrazole-4-carbaldehydes **27** or 3-(2-(allyloxyaryl)-1-phenyl-1*H*-pyrazole -4-carbaldehydes **28** in dimethyl-sulfoxide that supplied the corresponding 2-phenyl-pyrazolo[4,3-*c*]coumarins **29** (Scheme 9) [38]. The reaction was initially performed with (5%) of iodin in dimethyl-sulfoxide in the presence of H_2SO_4 at 60 °C using known methods, but the reaction did not take place. To overcome this problem, the reaction was carried out by increasing the iodine ratio to 10% and raising the temperature to 120 °C, which gave good results. This pathway has an advantage over the previous pyrazole construction methods which were relatively unstable and gave a mixture of isomeric 1-aryl and 2-arylpyrazolo[4,3-*c*]coumarins [25,39].

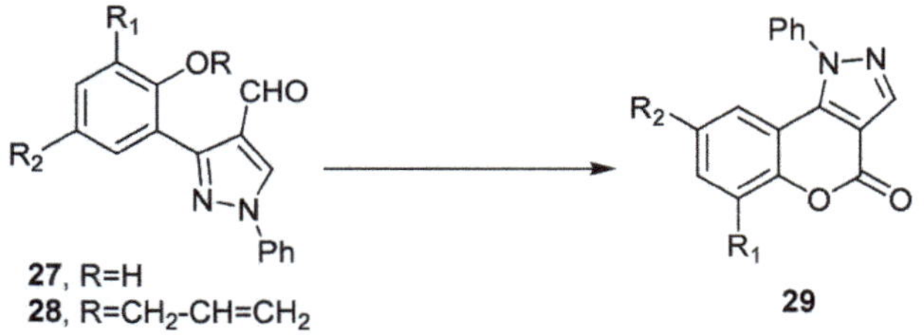

Scheme 9. Synthesis of 2-phenyl-chromeno[4,3-*c*]pyrazol-4(1*H*)ones **29**. Reagents and conditions. R = H, or CH_2-CH=CH_2, 10 mol%, DMSO, conc. H_2SO_4, 120 °C, 5 h, seven outputs with 88–93% yield.

Raju and his co-workers described the synthetic strategy for chromeno[4,3-*c*]pyrazol-4(1*H*)ones **34** and 2-tosyl-chromeno[4,3-*c*]pyrazol-4(1*H*)ones **35.** The pathway started from a reaction between salicylaldehydes **30** and *p*-toluenesulfonyl hydrazide **31.** It proceeded towards salicylaldehyde tosylhydrazone **32** which was in situ reacted with 3-oxobutanoates **33** in presence of lanthanum tris(trifluoromethanesulfonate) to deliver the desired coumarin (Scheme 10) [40].

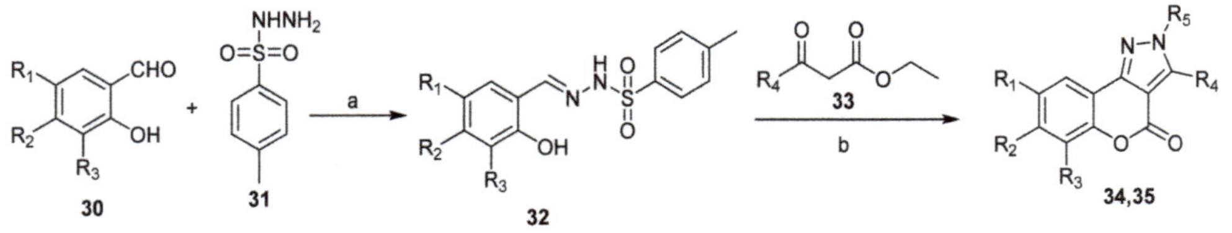

Scheme 10. Synthesis of chromeno[4,3-*c*]pyrazol-4(1*H*)ones **34** and **35**. Reagents and conditions. (**a**) CH_3CN, 2 h, r.t.; (**b**) $La(OTf)_3$, 130 °C, 8 h; 25 outputs with 22–78% yield.

Chromeno[3,4-c]pyrazol-4(2*H*)-one; (1*H*-Benzopyrano[3,4-c]pyrazole)
Pyrazole Construction

The literature reported numerous synthetic routes for chromeno[3,4-*c*]pyrazol-4(2*H*)-one. A simple pathway beginning from the 1,3-dipolar cycloaddition of diphenylnitrilimine (DPNI) to 3-ethoxy carbonyl coumarin (**36**) to build the pyrazole moiety **37**. Treatment of the ester (**37**) with an aqueous solution of potassium hydroxide (10%) gave the corresponding acid (**38**). Decarboxylation of **37** was accompanied by dehydrogenation which facilitated the attainment of 1,3-diphenyl-chromeno[3,4-*c*]pyrazol-4(1*H*)one (**39**) in a considerable yield (Scheme 11) [41].

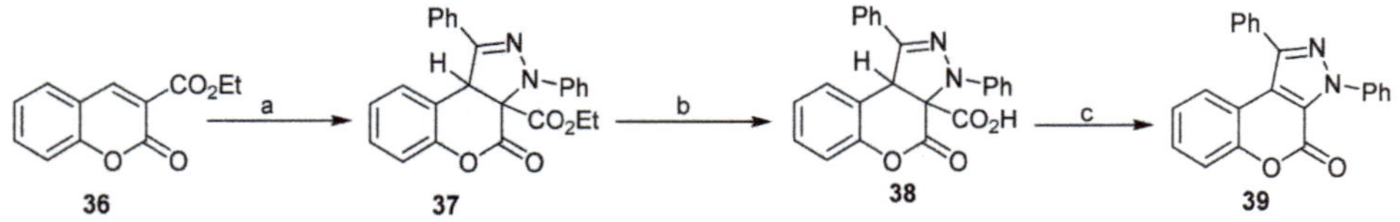

Scheme 11. Synthesis of 1,3-diphenyl-chromeno[3,4-*c*]pyrazol-4(1*H*)one (**39**). Reagents and conditions. (**a**) Diphenylnitrilimine (DPNI), benzene, TEA, 65% yield; (**b**) KOH (10%), 1 h, reflux, 80% yield; (**c**) heat, 98% yield.

On the other hand, thermal conversion of 4-diazo-methyl coumarins **40** into the corresponding chromeno[3,4-*c*]pyrazol-4(2*H*)-ones **41** in toluene followed the pathway depicted in Scheme 12 [42,43]. The presence of the alkyl substituent at the peri position to the attached diazo-methyl group in coumarins markedly destabilized the diazo-methyl group and facilitated the pyrazole isomerization.

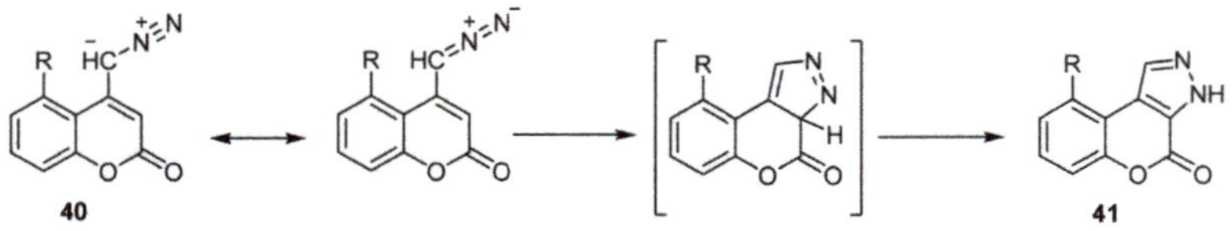

Scheme 12. Thermal conversion of 4-diazo-methyl coumarin. Reagents and conditions. Toluene, heat, stirring, R = H: 85% yield, R = Me: 98% yield.

Hydrazonyl halides played an important role in the preparation of chromeno[3,4-*c*]pyrazol-4(2*H*)-ones [11]. This coumarin was accessible from the reaction of hydrazonyl halides **42** with 3-acetylcoumarin (**43**) or 3-benzoylcoumarin (**17b**) in the presence of triethylamine. Subsequently, the dihydro-products **44** were refluxed in aqueous potassium hydroxide (10%) solution, and toluene successively to give chromeno[3,4-*c*]pyrazol-4(2*H*)-ones **45** (Scheme 13). Alternatively, compounds **45** were synthesized via cycloaddition of the hydrazonyl halides **42** to the 3-phenylsuphonylcoumarin (**46**), or the 3-bromocoumarin (**47**). The thermal elimination of benzenesulfinic acid or hydrogen bromide from the corresponding cyclo-adducts **48** resulted in the target **45** (Scheme 13).

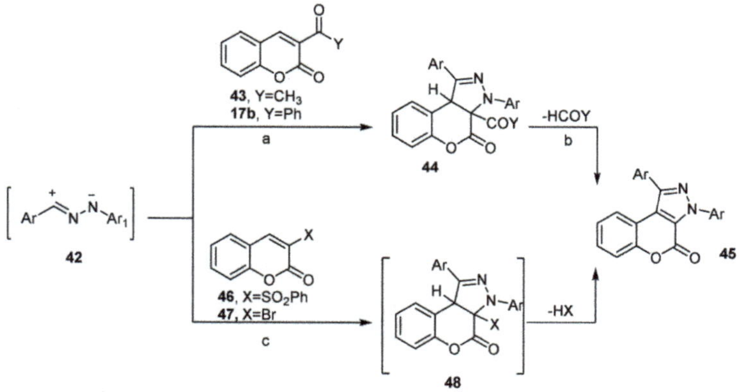

Scheme 13. The role of hydrazonyl halides in the preparation of chromeno[3,4-*c*]pyrazol-4(2*H*)-ones. Reagents and conditions. (**a**) Benzene, reflux, TEA, two outputs Y = CH_3 (51%), Y = Ph (48%); (**b**) i: KOH (10%), 12 h, reflux; ii: toluene, 2 h, reflux; (**c**) benzene, reflux, TEA, two outputs Ar = C_6H_4F-*p*, Ar_1 = $C_6H_4NO_2$-*p* (48%), Ar = $C_6H_3Cl_2$-2,4, Ar_1 = $C_6H_4NO_2$-*p* (51%).

Imidazole

Fusion of the imidazole ring with the pyrone ring of coumarin leads to one structural isomer, namely chromeno[3,4-*d*]imidazol-4-one, chromeno[3,4-*d*]imidazole-4(3*H*)-one, (1*H*-benzopyrano[3,4-*d*]imidazole) (Figure 2).

Figure 2. The common isomer of fused chromeno-imidazole.

Chromeno[3,4-d]imidazol-4-one; (1*H*-Benzopyrano[3,4-d]imidazole)
Imidazole Construction

Few synthetic routes to prepare benzo-pyrano-imidazoles were reported. The main synthetic approach involved the condensation of 3,4-diaminocoumarin (**49**) with different reagents [7,44]. Kitan and his coworkers elaborated a simple synthetic route of novel 1*H*-benzopyrano[3,4-*d*]imidazole-4-one starting from the in situ prepared 3,4-diaminocoumarin (**49**) by catalytic hydrogenation of 4-amino-3-nitrocoumarin. The condensation of 3,4-diaminocoumarin (**49**) with different acids including formic acid, acetic acid, or formaldehyde in presence of concentrated hydrochloric acid established 1*H*-benzopyrano[3,4-*d*]imidazoles **50** (Scheme 14) [7]. On the other hand, different series of 3-*N*-(4-aminocoumarin-3-yl)aroylamides **51** and 3-*N*-arylidenamino-4-aminocoumarins **52** were prepared from 3,4-diaminocoumarin **49**. Cyclization of **51** and **52** under heating, or in the presence of lead tetraacetate resulted in the fused 2-substituted 4*H*-chromeno[3,4-*d*]oxazol-4-ones **53** (Scheme 14) [44].

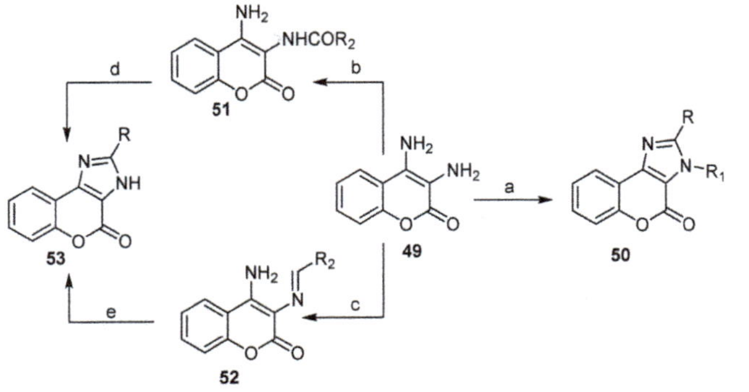

Scheme 14. Simple synthetic routes of novel 1*H*-benzopyrano[3,4-*d*]imidazoles **50** and **53**. Reagents and conditions. (**a**) i: Formic acid, conc. HCl, heating, 12 h, R = R_1 = H, 87% yield; ii: glacial AcOH, conc. HCl, heating, 12 h, R = CH_3, R_1 = H, 92% yield; iii: HCHO, EtOH, conc. HCl, heating, 2 h, R = H, R_1 = CH_3, 82% yield; (**b**) $(AcO_2)_2O$ or RCOCl, benzene, heat, 40 °C, 1 h, 4 outputs with 50–84% yield; (**c**) appropriate aldehyde, EtOH, reflux, 1 h, 4 outputs with 59–81% yield; (**d**) oil bath, 310–320 °C, 30 min; (**e**) lead tetraacetate, benzene, stirring, r.t., 2 h, 4 outputs with 59–91% yields.

Trimarco's research group developed a new method to synthesize substituted [1]benzopyrano[3,4-*d*]imidazol-4(3*H*)-ones bearing hydrogen or alkyl groups on N-3, and alkyl or amino substituents on C-2 in good yields [45]. 2-Alkyl-[1]benzopyrano[3,4-*d*]imidazol-4(3*H*)-ones **57** were obtained from acetamidines **56** carrying a 3-nitrocoumarin group at N-1 by reduction with sodium borohydride/palladium 10% (Scheme 15) [45]. Benzopyranoimidazoles **58** of an amino substituent on C-2 were obtained by refluxing **56** in excess of triethyl phosphite (Scheme 15) [45].

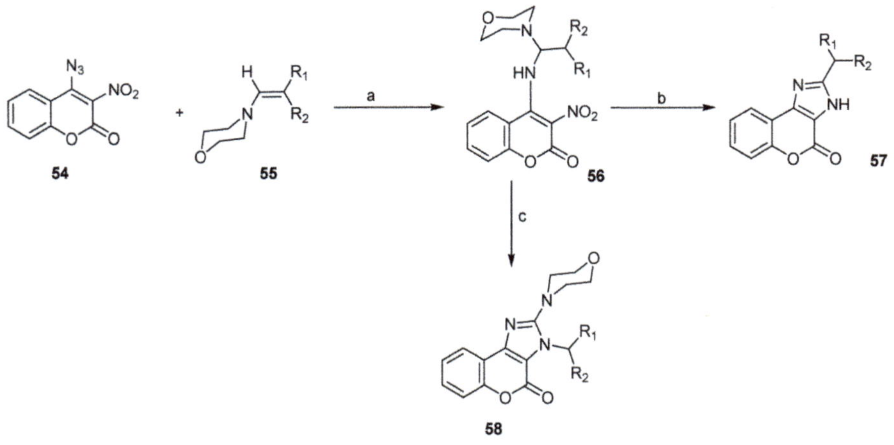

Scheme 15. [1]Benzopyrano[3,4-d]imidazol-4(3H)-ones **57** and **58** bearing hydrogen or alkyl groups on N-3, and alkyl or amino substituents on C-2. Reagents and conditions. i, R_1 = Ph, R_2 = H; ii, R_1 = 4-BrC_6H_4, R_2 = H; iii, R_1 = CH_2Ph, R_2 = H; iv, R_1 = CH_2CH_3, R_2 = H; v, R_1 = R_2 = CH_3; (**a**) CH_2Cl_2, −40°C, five outputs with 67–89% yields; (**b**) $NaBH_4$, Pd/C, MeOH, H_2O, five outputs with 55–65% yield; (**c**) N_2, $P(OC_2H_3)_3$, reflux, five outputs with 57–65% yields.

2.1.2. Two Different Heteroatoms

Thiazole and Isothiazole

4*H*-chromeno[3,4-*d*]thiazol-4-one (1*H*-benzopyrano[3,4-*d*]thiazole), 4*H*-chromeno[3,4-*c*]isothiazol-4-one (1*H*-benzopyrano[3,4-*c*]isothiazole), and 4*H*-chromeno[3,4-*d*]isothiazol-

4-one (1*H*-benzopyrano[3,4-*d*]isothiazole) are isomers of the fused five-member ring (containing N and S atoms) with the pyrone ring of coumarin (Figure 3).

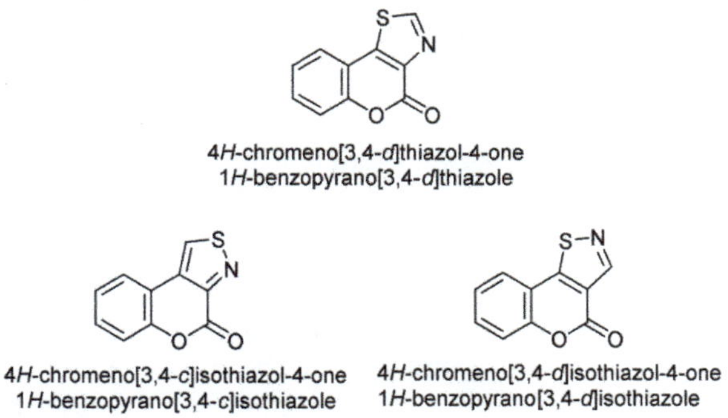

4*H*-chromeno[3,4-*d*]thiazol-4-one
1*H*-benzopyrano[3,4-*d*]thiazole

4*H*-chromeno[3,4-*c*]isothiazol-4-one
1*H*-benzopyrano[3,4-*c*]isothiazole

4*H*-chromeno[3,4-*d*]isothiazol-4-one
1*H*-benzopyrano[3,4-*d*]isothiazole

Figure 3. The common three isomers of fused the thiazole and isothiazole ring to the α-pyrone moiety of the coumarin.

4*H*-Chromeno[3,4-d]thiazol-4-one
Thiazole Construction

The reaction of 4-chloro-3-nitrocoumarin (**59**) with carbon disulfide in ethanol in the presence of sodium hydrogen sulfate produced 4*H*-chromeno[3,4-*d*]thiazol-4-one (**60**) (Scheme 16) [7].

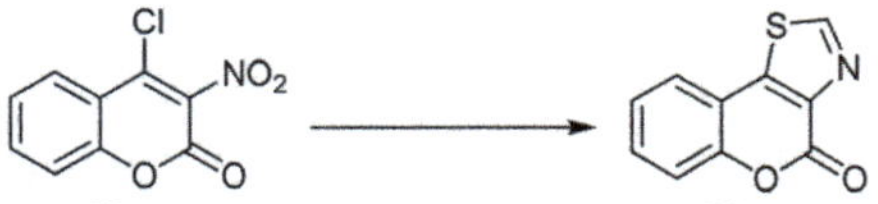

Scheme 16. Synthesis of 4*H*-chromeno[3,4-*d*]thiazol-4-one (**60**). Reagents and conditions. EtOH, $NaHSO_3$, pH 3, reflux, 3 h, 78% yield.

In situ prepared 4-aroylthio-3-nitrocoumarins **62** which was obtained from the reaction of 4-mercapto-3-nitrocoumarin (**61**) with different aroyl chlorides were allowed to cyclize in the presence of iron and glacial acetic acid that gave 2-aryl-4*H*-benzopyrano[3,4-*d*]thiazol-4-ones **63** (Scheme 17) [44].

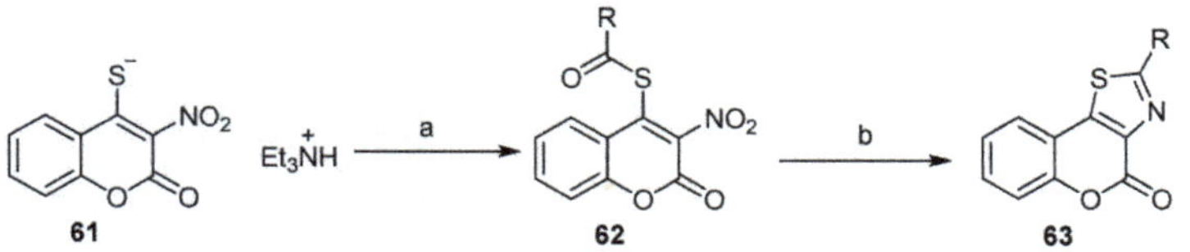

Scheme 17. Synthetic pathway of 2-aryl-4*H*-chromeno[3,4-*d*]thiazol-4-ones **63**. Reagents and conditions. (**a**) THF, stirring, 30 min, five outputs with 40–73% yield; (**b**) Fe, conc. AcOH, reflux, 60 °C for 1 h, 90 °C for 3–4 h, five outputs with 45–95% yield.

Anwar et al. employed a facile, green, and efficient, one-pot multicomponent reaction (MCR) catalyzed by iron(III) chloride to synthesize the coumarin annulated 2-aminothiazole (**65**) (Scheme 18) [46].

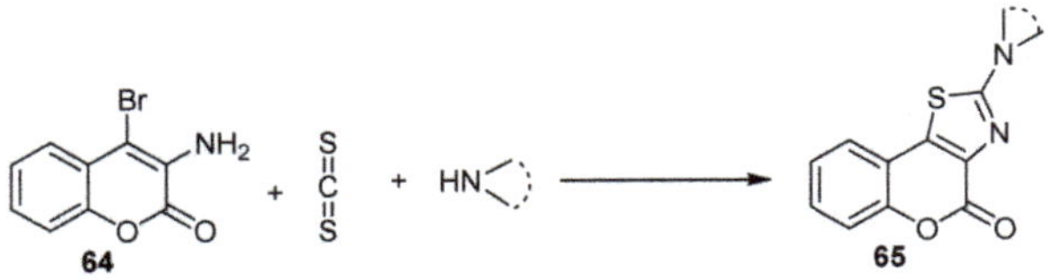

Scheme 18. Synthetic pathway of 2-amino-4*H*-chromeno[3,4-*d*]thiazol-4-ones **65**. Reagents and conditions. CS_2 (1.2 mmol) and K_2CO_3 (3.0 mmol), $FeCl_3$ (1.5 mmol), 110 °C, 6–8 h, N_2, pyrrolidine 64% yield, *N*-ethylaniline 74% yield.

Transition metal-free oxidative coupling for the formation of C–S bonds was employed to synthesize different 4*H*-chromeno[3,4-*d*]thiazol-4-ones. The key point in C–H bond activation of 3-(benzylamino)-4-bromo-substituted chromenone derivatives **66** was the presence of sodium sulfide as a source of sulfur, and iodine as a catalyst. The terminal oxidation was performed by H_2O_2 to form various 2-phenyl-4*H*-chromeno(3,4-*d*)thiazol-4-one derivatives **67** (Scheme 19) [47].

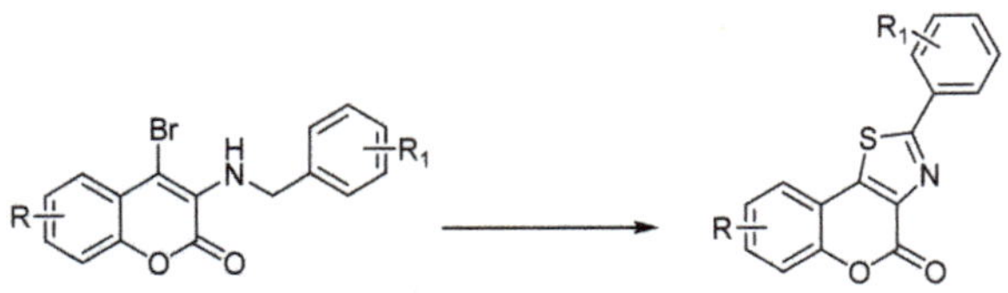

Scheme 19. Oxidative cross-coupling reaction to synthesis of 4*H*-chromeno[3,4-*d*]thiazol-4-ones **67**. Reagents and conditions. Na_2S (3 equiv.), I_2 (20 mmol%), 30% aq. H_2O_2 (5 equiv), DMF, 120 °C, 22 outputs with 65–89% yields.

4*H*-Chromeno[3,4-c]isothiazol-4-one
Pyrone Construction

The 1,3-dipolar cycloaddition reactions of nitrile sulfides (RC-N≡S-) played a particular role in the synthesis of 4*H*-chromeno[4,3-*c*]isothiazole [48–50]. For example, heating of acetylenic oxathiazolone (**68**) in xylene afforded 4-oxo-3-phenyl-4*H*-chromeno[4,3-*c*]isothiazole (**69**) [48]. The initial step of the reaction was decarboxylation of the oxathiazolone followed by intramolecular 1,3-dipolar cycloaddition of the resulting nitrile sulfide (**70**) to the adjacent alkyne (Scheme 20).

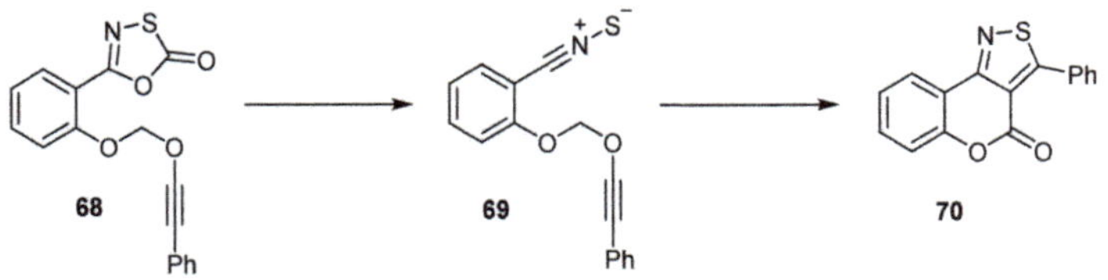

Scheme 20. Synthesis of 4-oxo-3-phenyl-4*H*-chromeno[4,3-*c*]isothiazole (**70**). Reagents and conditions. Xylene, heat, 16 h, 70% yield.

In 2010, Fordyce et al. improved a synthetic approach of 4*H*-chromeno[4,3-*c*]isothiazole as a result of the 1, 3-dipolar cycloaddition reactions of *o*-hydroxybenzonitrile sulfide (**72**), generated by microwave-assisted decarboxylation of oxathiazolone (**71**) [51]. The reaction of the *o*-hydroxyphenyloxathiazolone (**73**) with dimethyl acetylenedicarboxylate DMAD (1:2) in ethyl acetate supplied methyl 4-oxo-4*H*-chromeno[4,3-*c*]isothiazole-3-carboxylate (**74**) (Scheme 21).

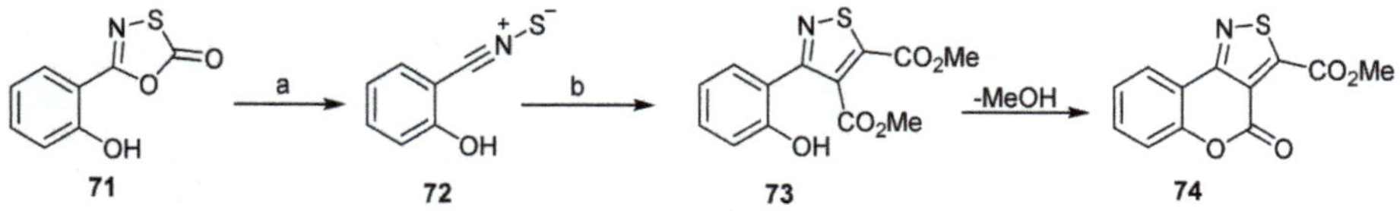

Scheme 21. Synthesis of methyl 4-oxo-4*H*-chromeno[4,3-*c*]isothiazole-3-carboxylate (**74**). Reagents and conditions. (**a**) Heat, -CO_2, 84% yield; (**b**) EtOAc, DMAD,10 min, 160 °C, microwave, 94% yield.

In 2017, Lee and his coworkers reported an efficient intramolecular Rh-catalyzed transannulation of thiadiazoles linked to cyanoalkoxycarbonyl. The ring closure of compound 75 was catalyzed by 1,1′-bis(diphenylphosphino) ferrocene DPPF to form the corresponding 8-substituted-3-phenyl-4*H*-chromeno[4,3-*c*]isothiazol-4-ones **76** in good yields of 90–99% (Scheme 22) [52].

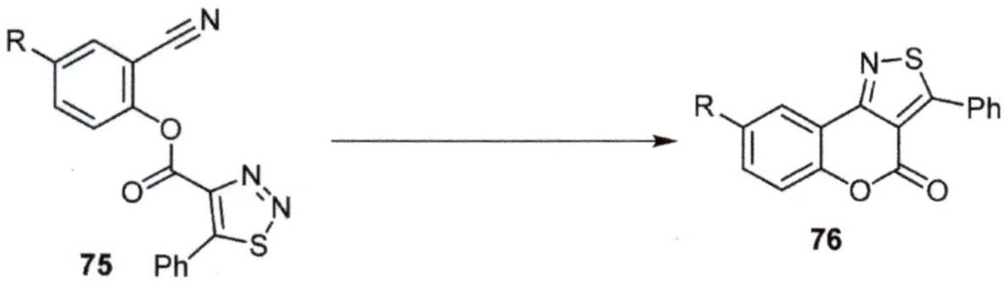

Scheme 22. Synthesis of 8-substituted-3-phenyl-4*H*-chromeno[4,3-*c*]isothiazol-4-ones **76**. Reagents and conditions. An amount of 5.0 mol% [Rh(COD)Cl]$_2$, 12.0 mol% DPPF, PhCl (1.0 mL), N_2, 80 °C; R = H 90% yield; R = Me 99% yield; R = Br 90% yield.

4*H*-Chromeno[3,4-d]isothiazol-4-one

Up to our knowledge, only one article discussed the synthesis of 4*H*-chromeno[3,4-*d*]isothiazol-4-one [53]. The reported method included cyclization of 4-mercapto-2-oxo-2*H*-chromene-3-carboxamide (**77**) on heating with bromine in ethyl acetate to form 2-hydroxy-4*H*-chromeno[3,4-*d*]isothiazol-4-one (**78**) (Scheme 23).

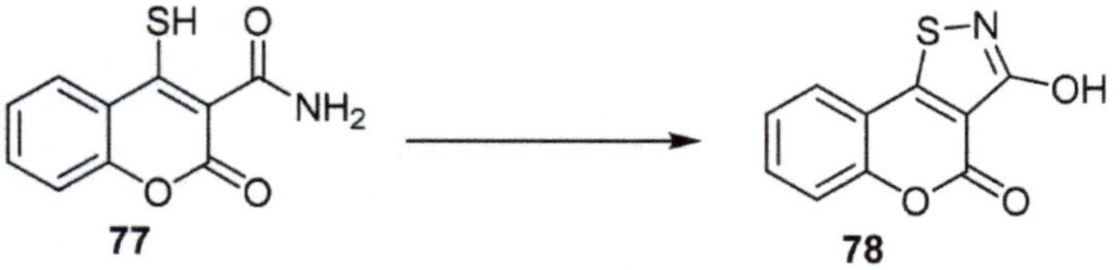

Scheme 23. Synthesis of 4*H*-chromeno[3,4-*d*]isothiazol-4-one (**78**). Reagents and conditions. Br_2, EtOAc, heat, 4 h, 96% yield.

Oxazole and Isoxazole

Fusion of a five-member ring containing (N and O atoms) with the pyrone ring of coumarin leads to four structural isomers, viz. 4*H*-chromeno[3,4-*d*]oxazol-4-one, 4*H*-chromeno[4,3-*d*]oxazol-4-one, 4*H*-chromeno[3,4-*d*]isoxazol-4-one and 4*H*-chromeno[4,3-*c*]isoxazol-4-one (Figure 4).

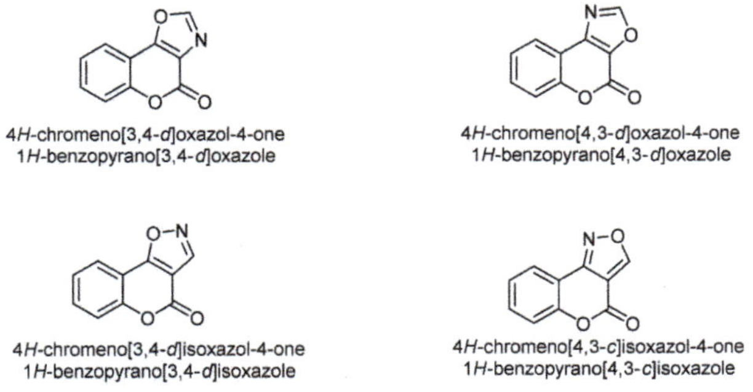

4H-chromeno[3,4-d]oxazol-4-one
1H-benzopyrano[3,4-d]oxazole

4H-chromeno[4,3-d]oxazol-4-one
1H-benzopyrano[4,3-d]oxazole

4H-chromeno[3,4-d]isoxazol-4-one
1H-benzopyrano[3,4-d]isoxazole

4H-chromeno[4,3-c]isoxazol-4-one
1H-benzopyrano[4,3-c]isoxazole

Figure 4. The common four isomers of fused oxazole and isoxazole ring to the α-pyrone moiety of the coumarin.

4*H*-Chromeno[3,4-d]oxazol-4-one
Oxazole Construction

Rhodium(II)-catalyzed reactions of 3-diazo-2,4-chromenedione (**79**) with several nitriles, such as acetonitrile, chloroacetonitrile and phenylacetonitrile established 2-substituted-4*H*-chromeno[3,4-*d*]oxazol-4-ones **80** (Scheme 24) [54]. The 3-diazo-2,4-chromenedione (**79**) was prepared by the diazo-transfer reaction of the corresponding 4-hydroxycoumarin (**4**) with mesyl azide according to Taber's method [55].

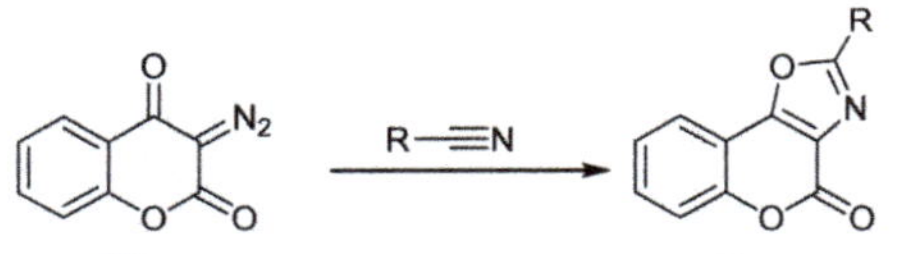

R—≡N

79 80

Scheme 24. 3-Diazo-2,4-chromenedione in the preparation of 4*H*-chromeno[3,4-*d*]oxazol-4-ones **72**. Reagents and conditions. $Rh_2(OAc)_4$, 60 °C, 5 h, three outputs, R = CH_3 50% yield, R = CH_2Cl 95% yield, R = CH_2Ph 73% yield.

3-Nitro-4-hydroxycoumarin was crucial for the preparation of 2-substituted-4*H* -chromeno[3,4-*d*]oxazol-4-one. In 1961, Dallacker et al. reported the preparation of 3-nitrocoumarin (**73**) from 4-hydroxycoumarin (**4**) upon nitration with nitric acid in glacial acetic acid. Reduction of **81** by Raney nickel in presence of propionic anhydride afforded the corresponding *N*-(4-hydroxy-2-oxo-2*H*-chromen-3-yl)- propionamide (**82**). Intramolecular cyclization of amide **82** by heating in acetic anhydride afforded 2-ethyl-4*H*-chromeno[3,4-*d*]oxazol-4-one (**83**) (Scheme 25) [56].

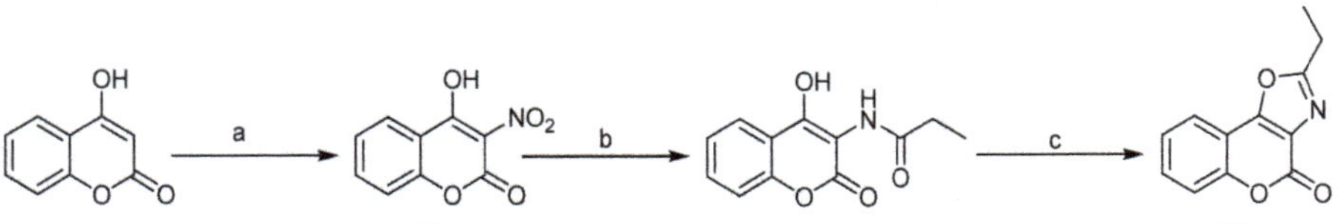

4 81 82 83

Scheme 25. 3-Nitrocoumarin in the preparation of 4*H*-chromeno[3,4-*d*]oxazol-4-one (**83**). Reagents and conditions. (**a**) HNO_2, glacial AcOH; (**b**) Raney Ni, $CH_3CH_2COOCOCH_2CH_3$; (**c**) acetic acid anhydride, heat, 96% yield.

In 2018, Litinas and his coworkers summarized the previous scheme in a green chemistry methodology [57]. They described the reaction of 4-hydroxy-3-nitrocoumarin (**81**) with acids in a one-pot reaction in the presence of PPh_3 and P_2O_5 under microwave irradiation. Another one-pot two-step reaction was accomplished in the presence of

Pd/C and hydrogen, followed by treatment with P_2O_5 under microwave irradiation (Scheme 26) [57].

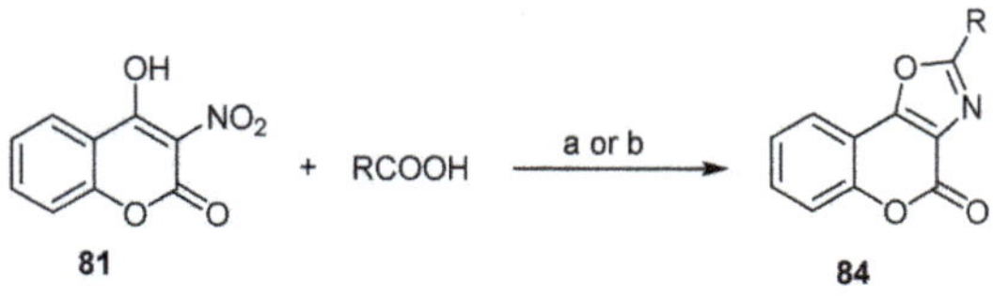

Scheme 26. One-pot synthesis of 2-substituted 4*H*-chromeno[3,4-*d*]oxazol-4-ones **84**. Reagents and conditions: (**a**) PPh_3 (2.5 equiv), P_2O_5 (4 equiv), MW irradiation, 130 °C or 140 °C, 1.5 h; (**b**) 5 mole % Pd/C (10%), H_2 1 atm, r.t., 1–3 h, then P_2O_5 (4 equiv), MW irradiation, 130 °C, 1 h, ten outputs with up to 91% yield.

Pyrone and Oxazole Construction

Wilson and his coworkers reported a high yield six-step synthesis of 7-hydroxy -2,6-dimethylchromeno[3,4-*d*]oxazol-4-one (**91**) from commercially available 2,4-dihydroxy-3-methyl-acetophenone [58]. The chemoselective benzylation of 2,4-dihydroxy3 -methylacetophenone (**85**) gave the corresponding 4-benzyloxy derivative (**86**). Compound **86** was converted into the 4-hydroxycoumarin derivative (**87**) using diethyl carbonate and sodium hydride. Nitration of **87** with fuming nitric acid in chloroform at room temperature afforded the nitro derivative (**88**). Reduction of **88** using zinc in refluxing acetic acid afforded 3-acetamido-4,7-dihydroxycoumarin (**89**). Cyclization of **89** was achieved using pyridine-buffered $POCl_3$ in THF under the Robinson–Gabriel mechanism. Finally, 7-hydroxy-2,6-dimethylchromeno[3,4-*d*]oxazol-4-one (**91**) was attainable through the debenzylation of oxazole **90** (Scheme 27) [58].

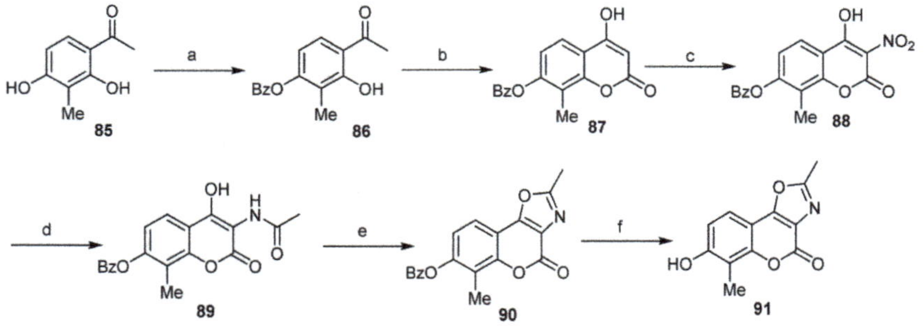

Scheme 27. Synthesis of 7-hydroxy-2,6-dimethylchromeno[3,4-*d*]oxazol-4-one (**91**). Reagents and conditions: (**a**) BnCl, K_2CO_3, KI, acetone, 56 °C, 88% yield; (**b**) NaH, $CO(OEt)_2$, toluene, 110 °C, 76%yield; (**c**) HNO_3, H_2SO_4, $CHCl_3$, room temperature, 93% yield; (**d**) Zn, AcOH, 110 °C, 86% yield; (**e**) $POCl_3$, pyridine, THF, 66 °C, 87% yield; (**f**) 10% Pd/C, H_2, THF/CH_2Cl_2, room temperature, 74% yield.

4*H*-Chromeno[4,3-d]oxazol-4-one

Oxazole Construction

Regarding 4*H*-chromeno[4,3-*d*]oxazole-4-one, only one article mentioned the preparation of such a fused system [59]. In 2004, Ray and Paul reported the synthesized 4*H*-chromeno[4,3-*d*]oxazole-4-one (**92**) via the reaction of 4-hydroxycoumarin (**4**) with formamide under reflux (Scheme 28).

Scheme 28. Synthesis of 4*H*-chromeno[4, 3-*d*]oxazol-4-one (**84**). Reagents and conditions. $HCONH_2$, 160 °C, 84% yield.

4*H*-Chromeno[3,4-d]isoxazol-4-ones
Isoxazole Construction

4-Chromanone (**93**) was found to be one of the key compounds for the preparation of fused coumarin-isoxazole. 4-chromanone (**93**) was treated by a lower alkyl oxalate such as ethyl oxalate, in the presence of a suitable base (e.g., sodium amide, sodium methoxylate, or sodium hydride) in an anhydrous reaction medium. The obtained 4-oxo-chroman-3-glyoxylate (**94**) was refluxed with hydroxylamine hydrochloride in ethanol to create the fused isoxazole ring and ethyl 4*H*-chromeno[3,4-*d*]isoxazole-3-carboxylate (**95**) was formed (Scheme 29) [60].

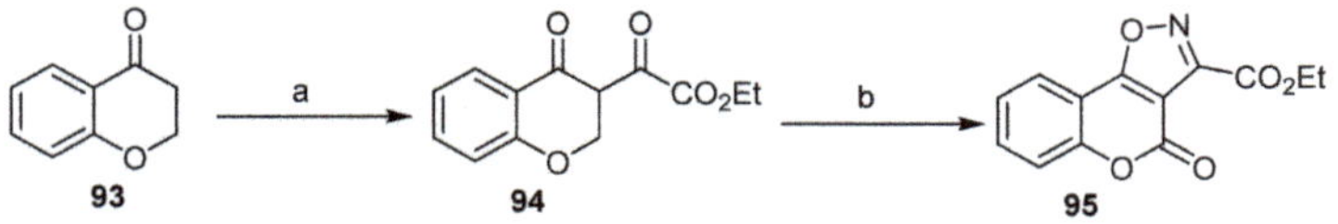

Scheme 29. 4-Chromanone in the synthesis of ethyl 4*H*-chromeno[3,4-*d*]isoxazol-3-carboxylate (**95**). Reagents and conditions. (**a**) Toluene, NaH, N_2, stirring, r.t., 94% yield; (**b**) $NH_2OH.HCl$, EtOH, reflux, 10 h, 91% yield.

Additionally, Sosnovskikh et al. studied the chemical transformation of the 3-cyanochromone (**96**) when reacted with hydroxylamine hydrochloride in basic condition. The cyclization at the CN group linked to the opened pyrone ring gave the corresponding 2-amino-3-carbamoylchromone (**98**). The re-cyclization of **98** exploiting another molecule of hydroxylamine led to the formation of 2-amino-4*H*-chromeno[3,4-*d*]isoxazol-4-one (**99**) (Scheme 30) [61].

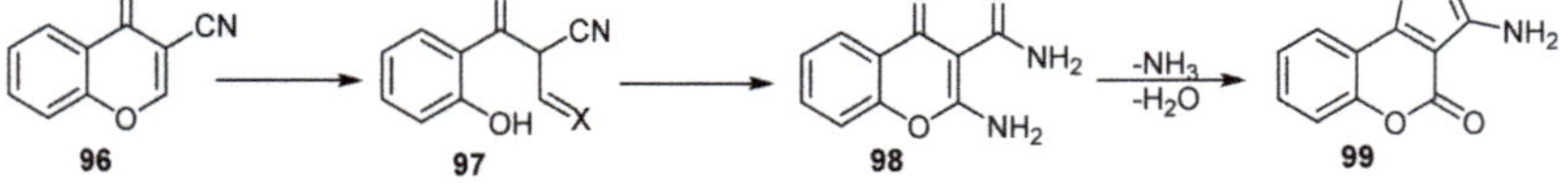

Scheme 30. Synthesis of 2-amino-4*H*-chromeno[3,4-*d*]isoxazol-4-one (**99**). Reagents and conditions. NH_2OH. HCl, NaOH, Ethanol, reflux, 45% yield.

On the other hand, the reactivity of 3-ethoxycarbonyl-2-methyl substituted 5,6,7,8-tetrafluorochromone (**100**) toward hydroxylamine in an alkaline medium was explored [62]. Wherever the pyrone ring was firstly opened, hydroxylamine was involved in the isoxazole ring formation of **101**. Moreover, the cyclization to the corresponding 2-methyl-4H-6,7,8,9-tetrafluorochromeno[3,4-d]isoxazol-4-one (**102**) was performed by sulfuric acid treatment (Scheme 31) [62].

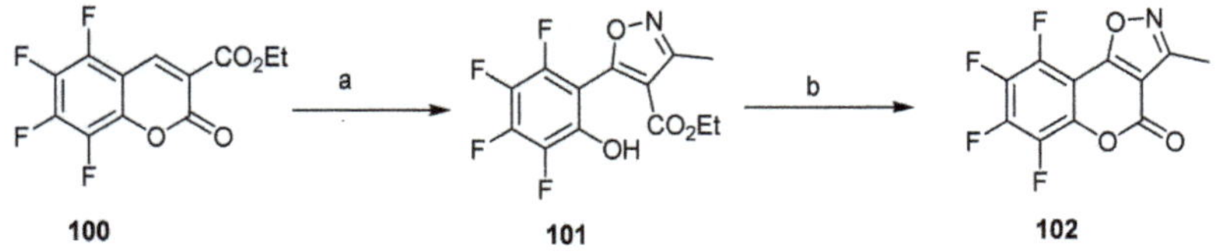

Scheme 31. Synthesis of polyfluorochromeno[3,4-*d*]isoxazol-4-one (**102**). Reagents and conditions. (**a**) NH_2OH. HCl, TEA, MeOH, 2 h, rt, 85% yield; (**b**) H_2SO_4, H_2O, reflux, 1.5 h, 82% yield.

4*H*-Chromeno[4,3-c]isoxazol-4-one
Isoxazole Construction

The synthesis of 4*H*-chromeno[4,3-*c*]isoxazol-4-one starting from 4-azido-3-hydroxycoumarin was scarcely reported. Vilsmeier–Haack reagent supplied the system with chloride and formyl moieties (**14**). The chloride was replaced by azido when compound **14** was treated by NaN_3 to produce 4-azido-3-coumarin carboxaldehyde **15**. This last decomposed thermally to be depleted from nitrogen and spontaneously cyclized to 4*H*-chromeno[4,3-*c*]isoxazol-4-ones **103** (Scheme 32) [31].

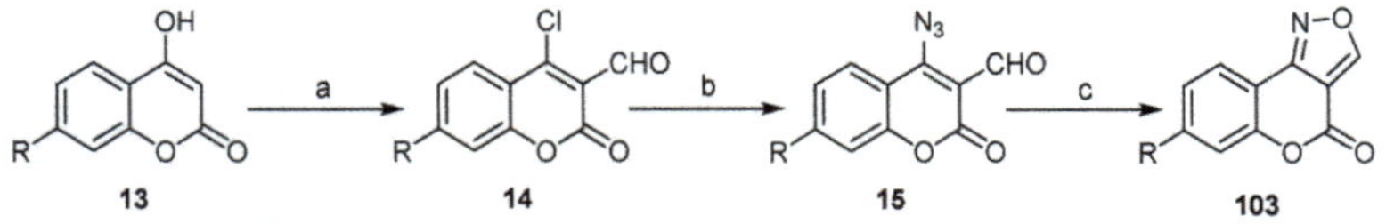

Scheme 32. Synthetic pathway of 4*H*-chromeno[4,3-*c*]isoxazol-4-ones **103**. Reagents and conditions. (**a**) $POCl_3$, DMF, H_2O, four outputs with 65–80% yields; (**b**) NaN_3, DMF, four outputs with 50–80% yield; (**c**) DMF, heat, 60–90 °C, -N_2, four outputs with 30–65% yields.

2.2. Five-Membered Aromatic Rings with Three Heteroatoms

2.2.1. Three Identical Heteroatoms

Triazole

Fusion of the triazole ring with the pyrone ring of coumarin leads to one structural isomer (chromeno[3,4-d][1,2,3]triazol-4(9b*H*)-one) (Figure 5).

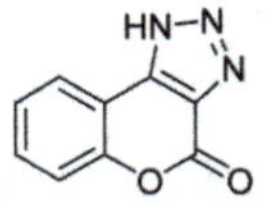

chromeno[3,4-*d*][1,2,3]triazol-4(9b*H*)-one
benzopyrone[3,4-*d*][1,2,3]triazole

Figure 5. The common isomer of the fused chromeno-triazole system.

Chromeno[3,4-d][1,2,3]triazol-4(9b*H*)-one
Triazole Construction

Dean and Park indicated the elimination of the sulphenic acid moiety during the addition of the corresponding 3-(4-methylphenylsulphinyl)coumarin (**104**) to sodium azide forming 4*H*-chromeno[3,4-d][1,2,3]triazol-(3H)4-one (**105**) (Scheme 33) [63].

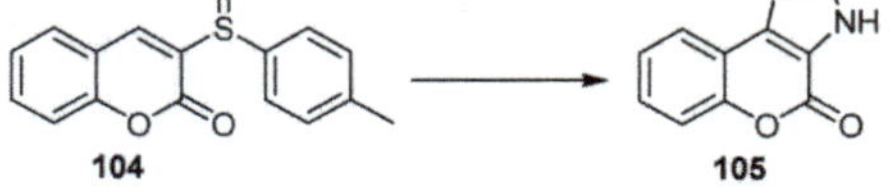

Scheme 33. Synthesis of 4*H*-chromeno[3,4-d][1,2,3]triazol-(3H)4-one (**105**). Reagents and conditions. NaN_3, DMF, 95 °C, 5 h, N_2, 84% yield.

A low to moderate yield of 4*H*-chromeno[3,4-d][1,2,3]triazol-(3*H*)4-ones **107** was obtained when 1,5-dipolar electro-cyclization took place within 4-azidocoumarins **106** that was induced by *t*-butoxide (Scheme 34) [64].

Scheme 34. 4-Azidocoumarins in the preparation of 4*H*-chromeno[3,4-d][1,2,3]triazol-(3H)4-ones **107**. Reagents and conditions. DMF or DMSO, *t*-BuOK, 50–60 °C, 5 h, stirring, six outputs with 12–41% yields.

A wide range of research demonstrated that 3-nitrocoumarin was the key compound for the synthesis of fused coumarin-triazole [65–67]. Vaccaro and his coworkers subjected the 3-nitrocoumarins **108** to a [3 + 2] cycloaddition with trimethylsilyl azide ($TMSN_3$) under a solvent-free condition (SFC). Tetrabutylammonium fluoride (TBAF) acted as a catalyst during the reaction to supply a series of 4*H*-chromeno[3,4-d][1,2,3]triazol-(3H)4-ones **109** (Scheme 35) [65]. This method confirmed that ammonium halogen TBAF salt can be efficaciously employed as a non-metallic catalyst for activating the silicon–nitrogen bond under SFC.

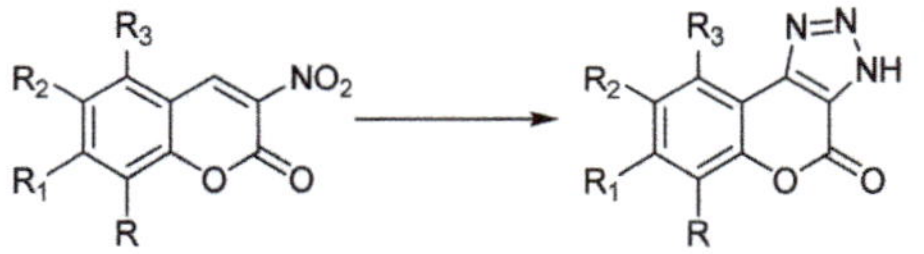

Scheme 35. TBAF-catalyzed cycloadditions of 3-nitrocoumarins with $TMSN_3$ under SFC. Regents and conditions. An amount of (10 mol%) TBAF, $TMSN_3$ (2 equiv), 50 °C, eleven outputs with 76–94% yields.

In 2012, the formation of 4*H*-chromeno[3,4-d][1,2,3]triazol-(3*H*)4-ones **111** was described through a catalyst-free 1,3-dipolar cycloaddition of 3-nitrocoumarins **110** to sodium azide (Scheme 36). It was found that good yields were obtained in the presence of electron-withdrawing substituent on the aryl ring of 3-nitrocoumarins **110**. The reaction gave the best yield in DMSO at 80 °C after three lower temperatures attempts [66]. By applying this reaction, a novel group of 4H-chromeno[3,4-d][1,2,3]triazol-(3H)4-ones was achieved using microwave-assisted green chemistry procedures (Scheme 36) [67].

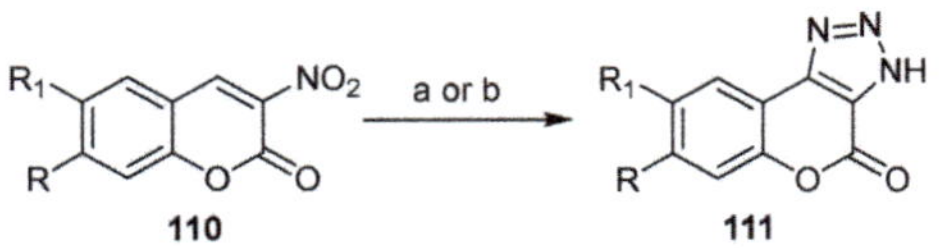

Scheme 36. 1,3-Dipolar cycloaddition of 3-nitrocoumarins with sodium azide. Reagents and conditions. a) NaN_3, DMSO, 80 °C, 1 h, five outputs with 66–89% yields; b) NaN_3 (1.2 equiv), DMF, Pyrex microwave vial equipped with a magnetic stir bar, stirred for 10 s, 160 C, 1 min, R = H, R_1 = CH_3 94% yield; R = OCH_3, R_1 = H 69% yield.

2.2.2. Three Different Heteroatoms

Thiadiazole

Fusion of thiadiazole ring with the pyrone ring of coumarin furnishes one structural isomer, namely 4*H*-chromeno[3,4-c][1,2,5]thiadiazol-4-one (Figure 6).

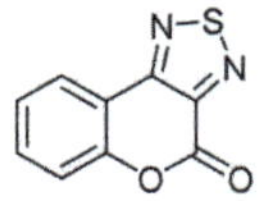

Figure 6. The common isomer of the fused chromeno-thiadiazole system.

4*H*-Chromeno[3,4-c][1,2,5]thiadiazol-4-one

Thiadiazole Construction

Synthesis of the 4*H*-chromeno[3,4-c][1,2,5]thiadiazol-4-one was reported by only one research article that belongs to Smirnov and his co-workers [68]. The reaction of 3,4-diaminocoumarin (**49**) with thionyl chloride in pyridine gave 4*H*-chromeno[3,4-*c*][l, 2,5]thiadiazol-4-one (**112**) (Scheme 37).

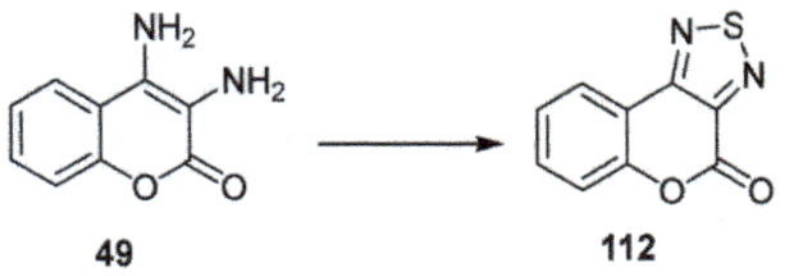

Scheme 37. Synthetic pathway to 4*H*-chromeno[3,4-*c*][l,2,5]thiadiazol-4-one (**112**). Reagents and conditions. $SOCl_2$, pyridine, stirring for 3 h, r.t., 87% yield.

In summary, coumarins are one of the heterocyclic structures of great interest in the development of valuable biologically active structures. Since coumarins have versatile applications, synthesis trials of different structures of the coumarin-based scaffold were attempted. The different synthetic routes to synthesize coumarin (benzopyrone)-fused five-membered aromatic heterocycles with multi-heteroatoms built on the pyrone ring were discussed in this review to shed light on the evolution in synthetic methods. We found that the starting scaffolds for this preparation were mainly 4-hydroxy, 4-amino, 3,4-diamino, and 3-nitro derivatives of coumarin. Moreover, other various methods of building the pyrone ring from simple functionalized compounds were discussed. To date, 4*H*-chromeno[4,3-*d*]thiazol-4-one, 4*H*-chromeno[3,4-c][1,2,5]selenadiazol-4-one, and 4H-chromeno[3,4-c][1,2,5]oxadiazol-4-one (Figure 7) have not been feasible by any synthetic procedures.

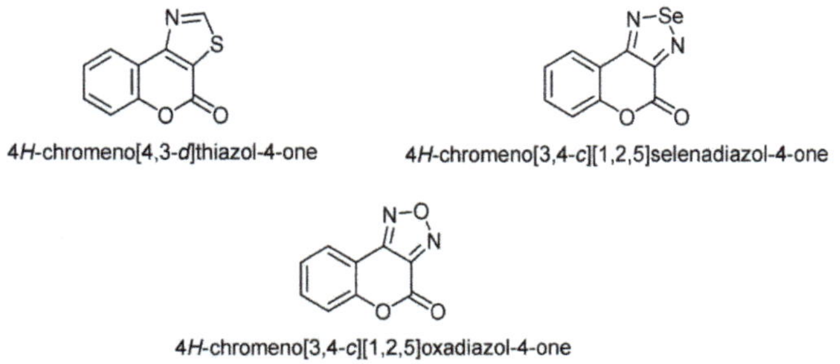

Figure 7. Coumarin-fused five-membered aromatic heterocycles have not been feasible by any synthetic procedures.

Author Contributions: E.R.E.-S. and G.K. collected publications and sorted them. E.R.E.-S. analyzed the literature and wrote the first draft. A.B.A. and G.K. developed the concept of the review. E.R.E.-S., A.B.A. and G.K. wrote the final draft and corrected the manuscript. All authors have read and agreed to the published version of the manuscript.

Funding: This research received no external funding.

Conflicts of Interest: The authors declare no conflict of interest.

References

1. Sethna, S.M.; Shah, N.M. The Chemistry of Coumarins. *Chem. Rev.* **1945**, *36*, 1–62. [CrossRef]
2. Medina, F.G.; Marrero, J.G.; Macías-Alonso, M.; González, M.C.; Córdova-Guerrero, I.; García, A.G.T.; Osegueda-Robles, S. Coumarin heterocyclic derivatives: Chemical synthesis and biological activity. *Nat. Prod. Rep.* **2015**, *32*, 1472–1507. [CrossRef]
3. Molnar, M.; Lončarić, M.; Kovač, M. Green Chemistry Approaches to the Synthesis of Coumarin Derivatives. *Curr. Org. Chem.* **2020**, *24*, 4–43. [CrossRef]
4. Salehian, F.; Nadri, H.; Jalili-Baleh, L.; Youseftabar-Miri, L.; Bukhari, S.N.A.; Foroumadi, A.; Küçükkilinç, T.T.; Sharifzadeh, M.; Khoobi, M. A review: Biologically active 3,4-heterocycle-fused coumarins. *Eur. J. Med. Chem.* **2021**, *212*, 113034. [CrossRef]
5. Geetika, P.; Subhash, B. Review on Synthesis of Bio-Active Coumarin-Fused Heterocyclic Molecules. *Curr. Org. Chem.* **2020**, *24*, 2566–2587.
6. Lončar, M.; Jakovljevi, Ć.M.; Šubarić, D.; Pavlić, M.; Služek, V.B.; Cindrić, I.; Molnar, M. Coumarins in Food and Methods of Their Determination. *Foods* **2020**, *9*, 645. [CrossRef]
7. Trkovnik, M.; Kalaj, V.; Kitan, D. Synthesis of new heterocyclocoumarins from 3,4-diamino- and 4-chloro-3-nitrocoumarins. *Org. Prep. Proced. Int.* **1987**, *19*, 450–455. [CrossRef]
8. Önder, F.C.; Durdaği, S.; Sahin, K.; Özpolat, B.; Ay, M. Design, Synthesis, and Molecular Modeling Studies of Novel Coumarin Carboxamide Derivatives as eEF-2K Inhibitors. *J. Chem. Inf. Model.* **2020**, *60*, 1766–1778. [CrossRef] [PubMed]
9. Sahoo, C.R.; Sahoo, J.; Mahapatra, M.; Lenka, D.; Sahu, P.K.; Dehury, B.; Padhy, R.N.; Paidesetty, S.K. Coumarin derivatives as promising antibacterial agent(s). *Arab. J. Chem.* **2021**, *14*, 102922. [CrossRef]
10. Mulwad, V.; Shirodkar, J. Synthesis of Some of the Antibacterial Compounds from 4-Hydroxycoumarins: Part II. *Ind. J. Chem.* **2002**, *41B*, 1263–1267.
11. Abunada, N.M.; Hassaneen, H.M.; Abu Samaha, A.S.M.; Miqdad, O.A. Synthesis and antimicrobial evaluation of some new pyrazole, pyrazoline and chromeno[3,4-c]pyrazole derivatives. *J. Braz. Chem. Soc.* **2009**, *20*, 975–987. [CrossRef]
12. El-Saghier, A.M.M.; Naili, M.B.; Rammash, B.K.; Saleh, N.A.; Kreddan, K.M. Synthesis and antibacterial activity of some new fused chromenes. *Arkivoc* **2007**, *2007*, 83–91. [CrossRef]
13. Mishra, S.; Pandey, A.; Manvati, S. Coumarin: An emerging antiviral agent. *Heliyon* **2020**, *6*, e03217. [CrossRef]
14. Reddy, M.V.R.; Rao, M.R.; Rhodes, D.; Hansen, M.S.T.; Rubins, K.; Bushman, F.D.; Venkateswarlu, Y.; Faulkner, D.J. Lamellarin α 20-Sulfate, an Inhibitor of HIV-1 Integrase Active against HIV-1 Virus in Cell Culture. *J. Med. Chem.* **1999**, *42*, 1901–1907. [CrossRef] [PubMed]
15. Boger, D.L.; Soenen, D.R.; Boyce, C.W.; Hedrick, M.P.; Jin, Q. Total Synthesis of Ningalin B Utilizing a Heterocyclic Azadiene Diels−Alder Reaction and Discovery of a New Class of Potent Multidrug Resistant (MDR) Reversal Agents. *J. Org. Chem.* **2000**, *65*, 2479–2483. [CrossRef] [PubMed]
16. Neagoie, C.; Vedrenne, E.; Buron, F.; Mérour, J.-Y.; Roşca, S.; Bourg, S.; Lozach, O.; Meijer, L.; Baldeyrou, B.; Lansiaux, A.; et al. Synthesis of chromeno[3,4-b]indoles as Lamellarin D analogues: A novel DYRK1A inhibitor class. *Eur. J. Med. Chem.* **2012**, *49*, 379–396. [CrossRef]
17. Rajabi, M.; Hossaini, Z.; Khalilzadeh, M.A.; Datta, S.; Halder, M.; Mousa, S.A. Synthesis of a new class of furo[3,2-c]coumarins and its anticancer activity. *J. Photochem. Photobiol. B Biol.* **2015**, *148*, 66–72. [CrossRef] [PubMed]
18. Al-Ayed, A.S. Synthesis of New Substituted Chromen[4,3-c]pyrazol-4-ones and Their Antioxidant Activities. *Molecules* **2011**, *16*, 10292–10302. [CrossRef]
19. Hamdi, N.; Fischmeister, C.; Puerta, C.; Valerga, P. A rapid access to new coumarinyl chalcone and substituted chromeno[4,3-c]pyrazol-4(1H)-ones and their antibacterial and DPPH radical scavenging activities. *Med. Chem. Res.* **2010**, *20*, 522–530. [CrossRef]
20. Hosni, H.M.; Abdulla, M. Anti-inflammatory and analgesic activities of some newly synthesized pyridinedicarbonitrile and benzopyranopyridine derivatives. *Acta Pharm.* **2008**, *58*, 175–186. [CrossRef]
21. Khan, I.A.; Kulkarni, M.V.; Gopal, M.; Shahabuddin, M.; Sun, C.-M. Synthesis and biological evaluation of novel angularly fused polycyclic coumarins. *Bioorg. Med. Chem. Lett.* **2005**, *15*, 3584–3587. [CrossRef] [PubMed]
22. El-Sawy, E.; Abdelwahab, A.; Kirsch, G. Synthetic Routes to Coumarin(Benzopyrone)-Fused Five-Membered Aromatic Heterocycles Built on the α-Pyrone Moiety. Part 1: Five-Membered Aromatic Rings with One Heteroatom. *Molecules* **2021**, *26*, 483. [CrossRef]
23. Shawali, A.S.; Elanadouli, B.E.; Albar, H.A. Cycloaddition of diphenylnitrilimine to coumarins. The synthesis of 3a,9b-dihydro-4-oxo-1H-benzopyrano [4,3-c]pyrazole derivatives. *Tetrahedron* **1985**, *41*, 1877–1884. [CrossRef]
24. El-Dean, A.M.K.; Zaki, R.M.; Geies, A.A.; Radwan, S.M.; Tolba, M.S. Synthesis and antimicrobial activity of new heterocyclic compounds containing thieno[3,2-c]coumarin and pyrazolo[4,3-c]coumarin frameworks. *Russ. J. Bioorg. Chem.* **2013**, *39*, 553–564. [CrossRef]
25. Strakova, I.; Petrova, M.; Belyakov, S.; Strakovs, A. Reactions of 4-Chloro-3-formylcoumarin with Arylhydrazines. *Chem. Heterocycl. Compd.* **2003**, *39*, 1608–1616. [CrossRef]

26. Yin, Y.; Wu, X.; Han, H.-W.; Sha, S.; Wang, S.-F.; Qiao, F.; Lü, A.-M.; Lv, P.-C.; Zhu, H.-L. Discovery and synthesis of a novel series of potent, selective inhibitors of the PI3K?: 2-alkyl-chromeno[4,3-c]pyrazol-4(2H)-one derivatives. *Org. Biomol. Chem.* **2014**, *12*, 9157–9165. [CrossRef] [PubMed]
27. Chekir, S.; Debbabi, M.; Regazzetti, A.; Dargère, D.; Laprévote, O.; Ben Jannet, H.; Gharbi, R. Design, synthesis and biological evaluation of novel 1,2,3-triazole linked coumarinopyrazole conjugates as potent anticholinesterase, anti-5-lipoxygenase, anti-tyrosinase and anti-cancer agents. *Bioorg. Chem.* **2018**, *80*, 189–194. [CrossRef]
28. Yin, Y.; Hu, J.-Q.; Wu, X.; Sha, S.; Wang, S.-F.; Qiao, F.; Song, Z.-C.; Zhu, H.-L. Design, synthesis and biological evaluation of novel chromeno[4,3-c]pyrazol-4(2H)-one derivates containing sulfonamido as potential PI3Kα inhibitors. *Bioorg. Med. Chem.* **2019**, *27*, 2261–2267. [CrossRef] [PubMed]
29. Lu, L.; Sha, S.; Wang, K.; Zhang, Y.-H.; Liu, Y.-D.; Ju, G.-D.; Wang, B.; Zhu, H.-L. Discovery of Chromeno[4,3-c]pyrazol-4(2H)-one Containing Carbonyl or Oxime Derivatives as Potential, Selective Inhibitors PI3Kα. *Chem. Pharm. Bull.* **2016**, *64*, 1576–1581. [CrossRef]
30. Yin, Y.; Sha, S.; Wu, X.; Wang, S.-F.; Qiao, F.; Song, Z.-C.; Zhu, H.-L. Development of novel chromeno[4,3-c]pyrazol-4(2H)-one derivates bearing sulfonylpiperazine as antitumor inhibitors targeting PI3Kα. *Eur. J. Med. Chem.* **2019**, *182*, 111630. [CrossRef] [PubMed]
31. Steinfiihrer, T.; Hantschmann, A.; Pietsch, M.; WeiDenfels, M. Heterocyclisch [c]-Anellierte Cumarine Aus 4-Azido-3-Cumarincarbaldehyden. *Liebigs Ann. Chem.* **1992**, *1992*, 23–28. [CrossRef]
32. Mustafa, A.; Hishmat, O.H.; Nawar, A.A.; Khalil, K.H.M.A. Pyrazolo-cumarine und Pyrazolyl-cumarone. *Eur. J. Org. Chem.* **1965**, *684*, 194–200. [CrossRef]
33. Ibrahim, M.A. Ring transformation of chromone-3-carboxylic acid under nucleophilic conditions. *Arkivoc* **2008**, *2008*, 192. [CrossRef]
34. Ibrahim, M.A. Ring Transformation of Chromone-3-Carboxamide under Nucleophilic Conditions. *J. Braz. Chem. Soc.* **2013**, *24*, 1754–1763. [CrossRef]
35. Ibrahim, M.; Badran, A.; El-Gohary, N.; Hashiem, S. Studies on the Chemical Reactions of Some 3-Substituted-6,8-dimethylchromones with Nucleophilic Reagents. *J. Heterocycl. Chem.* **2018**, *55*, 2315–2324. [CrossRef]
36. Ibrahim, M.A.; El-Gohary, N.M.; Said, S. Reactivity of 6-Methylchromone-3-carbonitrile Towards Some Nitrogen Nucleophilic Reagents. *Heterocycles* **2018**, *96*, 690. [CrossRef]
37. Trimeche, B.; Gharbi, R.; Martin, M.-T.; Nuzillard, J.M.; Mighri, Z.; El Houla, S. Reactivity of [1]benzopyrano[4,3-c][1,5]benzodiazepin-7(8H)-ones towards some N-binucleophiles. *J. Chem. Res.* **2004**, *2004*, 170–173. [CrossRef]
38. Lokhande, P.; Hasanzadeh, K.; Konda, S.G. A novel and efficient approach for the synthesis of new halo substituted 2-arylpyrazolo[4,3-c] coumarin derivatives. *Eur. J. Chem.* **2011**, *2*, 223–228. [CrossRef]
39. Stadlbauers, W.; Hojas, G. Ring closure reactions of 3-arylhydrazonoalkyl-quinolin-2-ones to 1-aryl-pyrazolo[4,3-c]quinolin-2-ones. *J. Heterocycl. Chem.* **2004**, *41*, 681–690. [CrossRef]
40. Hariprasad, K.S.; Anand, A.; Rathod, B.B.; Zehra, A.; Tiwari, A.K.; Prakasham, R.S.; Raju, B.C. Neoteric Synthesis and Biological Activities of Chromenopyrazolones, Tosylchromenopyrazolones, Benzoylcoumarins. *ChemistrySelect* **2017**, *2*, 10628–10634. [CrossRef]
41. Fathi, T.; An, N.D.; Schmitt, G.; Cerutti, E.; Laude, B. Regiochemistry of the cycloadditions of diphenylnitrilimine to coumarin, 3-ethoxycarbonyl and 3-acetyl coumarins. *Tetrahedron* **1988**, *44*, 4527–4536. [CrossRef]
42. Ito, K.; Maruyama, J. 4-Diazomethylcoumarins and Related Stable Heteroaryldiazomethanes. Thermal Conversion into Condensed Pyrazoles. *J. Heterocycl. Chem.* **1988**, *25*, 1681–1687. [CrossRef]
43. Ito, K.; Maruyama, J. A Facile Intramolecular Cyclization of 4-Diazomethylcoumarins. A Convenient Route to Benzopyrano[3,4-c]pyrazol-4(3H)-ones. *Heterocycles* **1984**, *22*, 1057. [CrossRef]
44. Colotta, V.; Catarzi, D.; Varano, F.; Cecchi, L.; Filacchioni, G.; Martini, C.; Giusti, L.; Lucacchini, A. Tricyclic heteroaromatic systems. Synthesis and benzodiazepine receptor affinity of 2-substituted-1-benzopyrano[3,4-d]oxazol-4-ones, -thiazol-4-ones, and -imidazol-4-ones. *IL Farm.* **1998**, *53*, 375–381. [CrossRef]
45. Beccalli, E.M.; Contini, A.; Trimarco, P. 3-Nitrocoumarin Amidines: A New Synthetic Strategy for Substituted [1]Benzopyrano[3,4-d]imidazol-4(3H)-ones. *Eur. J. Org. Chem.* **2003**, *2003*, 3976–3984. [CrossRef]
46. Anwar, S.; Paul, S.B.; Majumdar, K.C.; Choudhury, S. Green, One-Pot, Multicomponent Synthesis of Fused-Ring 2-Aminothiazoles. *Synth. Commun.* **2014**, *44*, 3304–3313. [CrossRef]
47. Belal, M.; Khan, A.T. Oxidative cross coupling reaction mediated by I_2/H_2O_2: A novel approach for the construction of fused thiazole containing coumarin derivatives. *RSC Adv.* **2015**, *5*, 104155–104163. [CrossRef]
48. Brownsort, P.A.; Paton, R.; Sutherland, A.G. Intramolecular cycloaddition reactions involving nitrile sulphides. *Tetrahedron Lett.* **1985**, *26*, 3727–3730. [CrossRef]
49. Brownsort, P.A.; Michael Paton, A. Nitrile Sulphides. Part 7. Synthesis of [I]Benzopyrano[4,3-c]Isothiazoles and Isothiazolo[4,3-c]Quinolines. *J. Chem. Soc. Perkin Trans.* **1987**, 2339–2344. [CrossRef]
50. Brownsort, P.A.; Paton, R.M.; Sutherland, A.G. Nitrile Sulphides. Part 1O. Intramolecular 1.3-Dipolar Cycloadditions. *J. Chem. Soc. Perkin Trans.* **1989**, 1679–1686. [CrossRef]
51. Fordyce, E.A.; Morrison, A.J.; Sharp, R.D.; Paton, R.M. Microwave-induced generation and reactions of nitrile sulfides: An improved method for the synthesis of isothiazoles and 1,2,4-thiadiazoles. *Tetrahedron* **2010**, *66*, 7192–7197. [CrossRef]

52. Seo, B.; Kim, H.; Kim, Y.G.; Baek, Y.; Um, K.; Lee, P.H. Synthesis of Bicyclic Isothiazoles through an Intramolecular Rhodium-Catalyzed Transannulation of Cyanothiadiazoles. *J. Org. Chem.* **2017**, *82*, 10574–10582. [CrossRef] [PubMed]
53. Checchi, S.; Pecori Vettori, L.; Vincieri, F. 4-Hydroxycoumarins. VIII. Substitution Reactions of 4-Chloro-3-Cyanocoumarins. *Gazz. Chim. Ital.* **1968**, *98*, 1488–1502.
54. Lee, Y.R.; Suk, J.Y.; Kim, B.S. Rhodium(II)-catalyzed reactions of 3-diazo-2,4-chromenediones. First one-step synthesis of pterophyllin 2. *Tetrahedron Lett.* **1999**, *40*, 6603–6607. [CrossRef]
55. Taber, D.F.; Ruckle, R.E.; Hennessy, M.J. Mesyl azide: A superior reagent for diazo transfer. *J. Org. Chem.* **1986**, *51*, 4077–4078. [CrossRef]
56. Dallacker, F.; Kratzer, P.; Lipp, M. Derivate des 2.4-Pyronons und 4-Hydroxy-cumarins. *Eur. J. Org. Chem.* **1961**, *643*, 97–109. [CrossRef]
57. Balalas, T.D.; Stratidis, G.; Papatheodorou, D.; Vlachou, E.-E.; Gabriel, C.; Hadjipavlou-Litina, D.J.; Litinas, K.E. One-pot Synthesis of 2-Substituted 4H-Chromeno[3,4-d]oxazol-4-ones from 4-Hydroxy-3-nitrocoumarin and Acids in the Presence of Triphenylphosphine and Phosphorus Pentoxide under Microwave Irradiation. *SynOpen* **2018**, *2*, 0105–0113. [CrossRef]
58. Gammon, D.W.; Hunter, R.; Wilson, S.A. An efficient synthesis of 7-hydroxy-2,6-dimethylchromeno[3,4-d]oxazol-4-one—a protected fragment of novenamine. *Tetrahedron* **2005**, *61*, 10683–10688. [CrossRef]
59. Ray, S.; Paul, S.K. Studies on Reaction of Hydroxycoumarin Compounds with Formamide. *Cheminform* **2005**, *36*. [CrossRef]
60. Freedman, J.; Milwaukee, W. 3-Substituted-4H(1)Benzopyrano(3,4-d) Isoxazoles 1971. U.S. Patent 3,553,228, 5 January 1971.
61. Sosnovskikh, V.Y.; Moshkin, V.S.; Kodess, M.I. A reinvestigation of the reactions of 3-substituted chromones with hydroxylamine. Unexpected synthesis of 3-amino-4H-chromeno[3,4-d]isoxazol-4-one and 3-(diaminomethylene)chroman-2,4-dione. *Tetrahedron Lett.* **2008**, *49*, 6856–6859. [CrossRef]
62. Saloutin, V.; Skryabina, Z.; Bazyl', I.; Kisil', S. Interaction of 3-ethoxycarbonyl(carboxy)-substituted 5,6,7,8-tetrafluorochromones with N-nucleophiles: Synthesis of fluorocoumarins. *J. Fluor. Chem.* **1999**, *94*, 83–90. [CrossRef]
63. Dean, F.M.; Park, B.K. Activating groups for the ring expansion of coumarin by diazoethane: Benzoyl, pivaloyl, arylsulphonyl, arylsulphinyl, and nitro. *J. Chem. Soc. Perkin Trans.* **1976**, *1*, 1260–1268. [CrossRef]
64. Ito, K.; Hariya, J. Electrocyclization of 4-Azidocoumarins Leading to Benzopyrano[3,4-d]-1,2,3-triazol-4-ones. *Heterocycles* **1987**, *26*, 35. [CrossRef]
65. D'Ambrosio, G.; Fringuelli, F.; Pizzo, F.; Vaccaro, L. TBAF-catalyzed [3 + 2]cycloaddition of TMSN3 to 3-nitrocoumarins under SFC: An effective green route to chromeno[3,4-d][1,2,3]triazol-4(3H)-ones. *Green Chem.* **2005**, *7*, 874–877. [CrossRef]
66. Wang, T.; Hu, X.-C.; Huang, X.-J.; Li, X.-S.; Xie, J.-W. Efficient synthesis of functionalized 1,2,3-triazoles by catalyst-free 1,3-dipolar cycloaddition of nitroalkenes with sodium azide. *J. Braz. Chem. Soc.* **2012**, *23*, 1119–1123. [CrossRef]
67. Schwendt, G.; Glasnov, T. Intensified synthesis of [3,4-d]triazole-fused chromenes, coumarins, and quinolones. *Mon. Für Chem. Chem. Mon.* **2017**, *148*, 69–75. [CrossRef]
68. Savel'Ev, V.L.; Samsonova, O.L.; Troitskaya, V.S.; Vinokurov, V.G.; Lezina, V.P.; Smirnov, L.D. Synthesis of 4H-[1]benzopyrano[3,4-c][1,2,5]thyadiazol-4-one and its reactions with some nucleophilic and electrophilic agents. *Chem. Heterocycl. Compd.* **1988**, *24*, 805–810. [CrossRef]

Review

Naturally Occurring Calanolides: Occurrence, Biosynthesis, and Pharmacological Properties Including Therapeutic Potential

Lutfun Nahar [1,*], Anupam Das Talukdar [2], Deepa Nath [3], Sushmita Nath [4], Aman Mehan [5], Fyaz M. D. Ismail [4] and Satyajit D. Sarker [4,*]

1. Laboratory of Growth Regulators, Institute of Experimental Botany ASCR & Palacký University, Šlechtitelů 27, 78371 Olomouc, Czech Republic
2. Department of Life Science and Bioinformatics, Assam University, Silchar, Assam 788011, India; adtddt@gmail.com
3. Department of Botany, Gurucharan College, Silchar, Assam 788004, India; deepa.nath@gmail.com
4. Centre for Natural Products Discovery, School of Pharmacy and Biomolecular Sciences, Liverpool John Moores University, James Parsons Building, Byrom Street, Liverpool L3 3AF, UK; sushmitanath84@gmail.com (S.N.); fyaz.ismail@gmail.com (F.M.D.)
5. School of Clinical Medicine, University of Cambridge, Cambridge CB2 OSP, UK; ahm41@cam.ac.uk

* Correspondence: drnahar@live.co.uk (L.N.); S.Sarker@ljmu.ac.uk (S.D.S.)

Academic Editors: Maria João Matos and Pascal Richomme
Received: 30 September 2020; Accepted: 26 October 2020; Published: 28 October 2020

Abstract: Calanolides are tetracyclic 4-substituted dipyranocoumarins. Calanolide A, isolated from the leaves and twigs of *Calophyllum lanigerum* var. *austrocoriaceum* (Whitmore) P. F. Stevens, is the first member of this group of compounds with anti-HIV-1 activity mediated by reverse transcriptase inhibition. Calanolides are classified pharmacologically as non-nucleoside reverse transcriptase inhibitors (NNRTI). There are at least 15 naturally occurring calanolides distributed mainly within the genus *Calophyllum*, but some of them are also present in the genus *Clausena*. Besides significant anti-HIV properties, which have been exploited towards potential development of new NNRTIs for anti-HIV therapy, calanolides have also been found to possess anticancer, antimicrobial and antiparasitic potential. This review article provides a comprehensive update on all aspects of naturally occurring calanolides, including their chemistry, natural occurrence, biosynthesis, pharmacological and toxicological aspects including mechanism of action and structure activity relationships, pharmacokinetics, therapeutic potentials and available patents.

Keywords: calanolides; pseudocalanolides; calanolide A; *Calophyllum*; Calophyllaceae; anti-HIV; reverse transcriptase; non-nucleoside reverse transcriptase inhibitors (NNRTIs)

1. Introduction

Calanolides are tetracyclic 4-substituted dipyranocoumarins, and their C-ring contains a *gem*-dimethyl group (Figure 1), e.g., (+)-calanolide A (**1**), (−)-calanolide B (costatolide) (**14**) (Figure 2). The discovery of calanolides from the leaves and twigs of the tree *Calophyllum lanigerum var. austrocoriaceum* (Whitmore) P. F. Stevens, collected from Sarawak, Malaysia in 1987 happened during one of the largest anti-HIV screening programs conducted by the National Cancer Institute (NCI) during 1987–1996. In that program, over 30,000 plant extracts were screened utilizing an in vitro cell-based anti-HIV screen that could determine the degree of HIV-1 replication in treated infected lymphoblastic cells versus that in treated uninfected control cells [1,2]. Calanolide A (**1**) (Figure 1), which can be described as a 11,12-dihydro-2*H*,6*H*,10*H*-dipyrano[2,3-f:2′,3′-h]chromen-2-one substituted by a hydroxyl (–OH) group at C-12, methyl groups at positions 6, 6, 10 and 11 and a propyl group at C-4 (the 10*R*,11*S*,12*S* stereoisomer), was isolated as the first member of anti-HIV compounds, calanolides, as a potential novel therapeutic option for the treatment of HIV infections. However, a subsequent attempt to recollect this plant sample failed and the collection of other specimens of the same species (not necessarily the same variety), afforded only a negligible amount of calanolide A (**1**). In fact, calanolides are among the first plant-based compounds to demonstrate potential anti-HIV-1 activity. Later, an extract of the latex of *C. teysmanii* showed significant anti-HIV activity in the screening, but the major active compound was (−)-calanolide B (**14**, also known as costatolide), regrettably not calanolide A (**1**) (Figure 2). The anti-HIV activity of (−)-calanolide B (**14**) was less potent than that of calanolide A (**1**), possibly because of difference in stereochemistry at the chiral centers. To date calanolides A-F and some of their methyl, acetyl and dihydro derivatives have been reported mainly from various *Calophyllum* species (Figure 2). Among these, the structures of calanolides C (**6**) and D (**7**), as reported initially by Kashman et al. [1] from *C. lanigerum*, were revised and renamed as pseudocalanolides C (**8**) and D (**9**) [3] (Figure 2). However, the true calanolides C (**6**) and D (**7**) were later reported from *C. brasiliense* Cambess. [4–6].

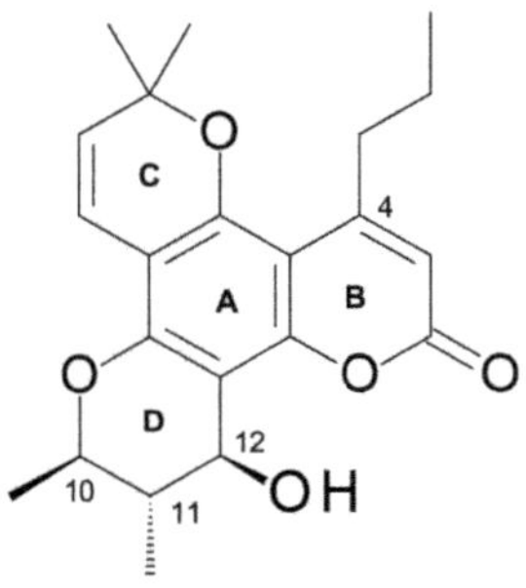

Figure 1. Rings A, B, C and D, and carbon numbering in (+)-calanolide A (**1**).

The first isolation of calanolides from *C. lanigerum var. austrocoriaceum*, involved multiple steps, starting with the extraction of dried fruits and twigs of this plant with a 1:1 mixture of dichloromethane and methanol, followed by a sequential solvent partitioning process involving various solvents. The *n*-hexane and CCl_4 fractions emerged as the active fractions [1]. Repeated vacuum liquid chromatography (VLC) on silica gel, eluting with a mixture of *n*-hexane and ethyl acetate afforded crude calanolides, which were further purified by HPLC, employing normal phase for calanolide A (**1**), calanolide B (**4**) and pseudocalanolide D (**9**) [reported incorrectly as calanolide D (**7**)], while reversed-phase for 12-acetoxycalanolide A (**2**), 12-methoxycalanolide A (**3**), 12-methoxycalanolide B (**5**), pseudocalanolide C (**8**) [reported as calanolide C (**6**)] and calanolide E (**10**). The structures of these compounds were determined by a combination of UV, IR, NMR and MS spectroscopic methods, and all spectroscopic data were published [1]. The absolute stereochemistry of calanolides A (**1**) and B (**4**) was confirmed by a modified Mosher's method.

There is a review [7] and a book chapter on calanolides [8], published about six years ago, that mainly cover anti-HIV activity, and the literature published until early 2014. This present review is not on the genus *Calophyllum*, the family Calophyllaceae or pyranocoumarins a such, but it exclusively focuses on various aspects of naturally occurring calanolides. This review is significantly different from any other previous articles on calanolides in its approach and coverage, and is a comprehensive update on naturally occurring calanolides, encompassing their chemistry, natural occurrence, biosynthesis, pharmacological and toxicological aspects including mechanism of action and structure activity relationships, pharmacokinetics, therapeutic potentials and available patents.

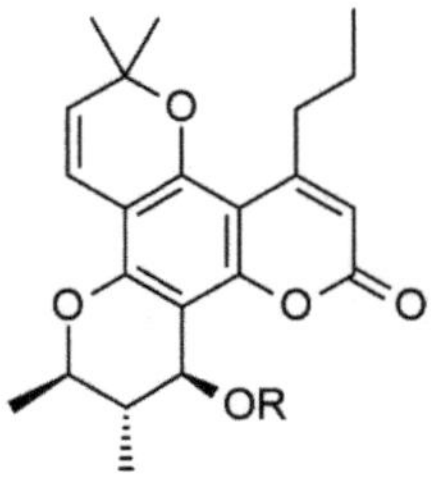

Calanolide A (**1**) R = H
12-*O*-Acetylcalanolide A (**2**) R = Ac
12-*O*-Methylcalanolide A (**3**) R = Me

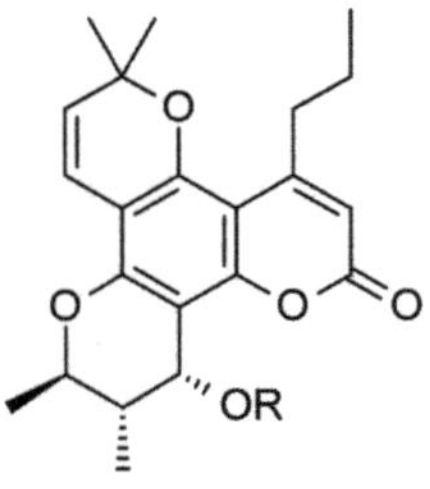

Calanolide B (**4**) R = H
12-*O*-Methylcalanolide B (**5**) R = Me

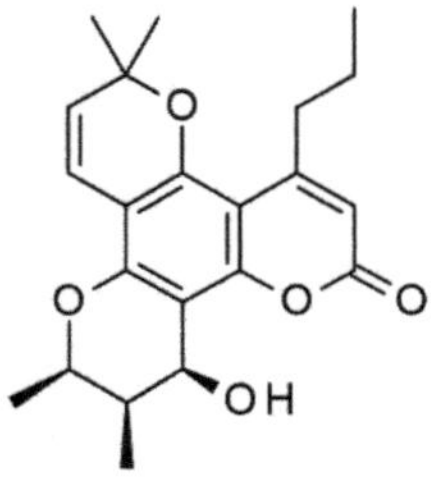

Calanolide C (**6**)

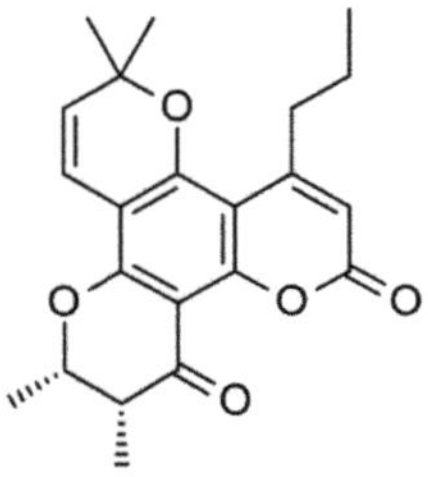

Calanolide D (**7**)

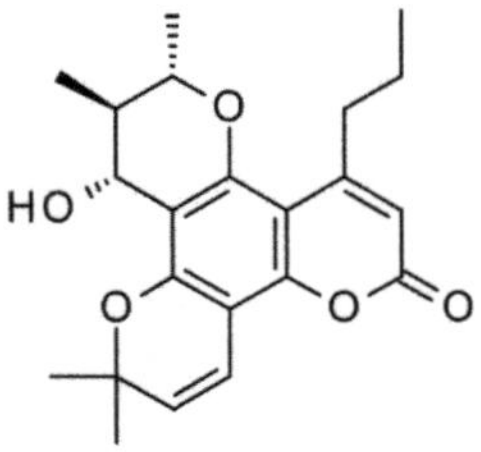

Revised to Pseudocalanolide C (**8**)

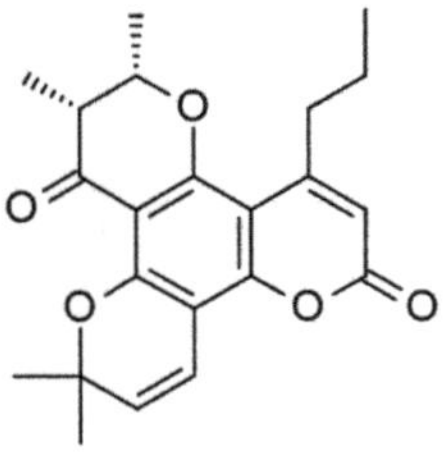

Revised to Pseudocalanolide D (**9**)

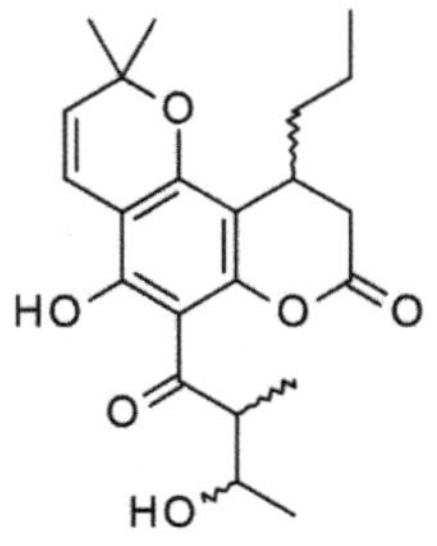

Calanolide E (**10**)

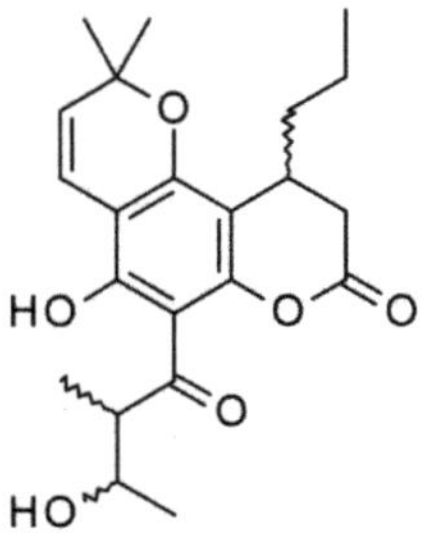

Calanolide E1 (**11**)
Calanolide E2 (**12**)

Figure 2. *Cont.*

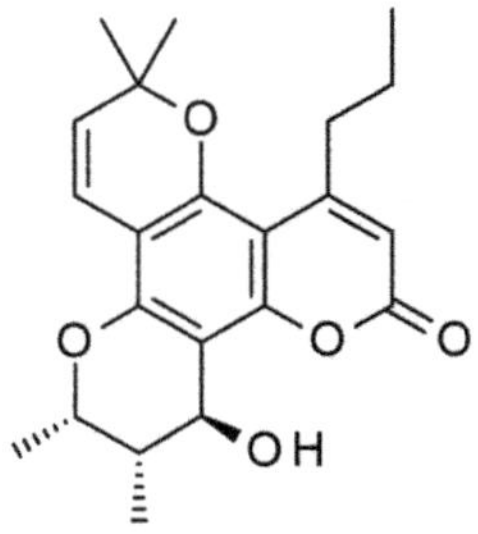

Calanolide F (**13**)
(10-*epi*-calanolide A)

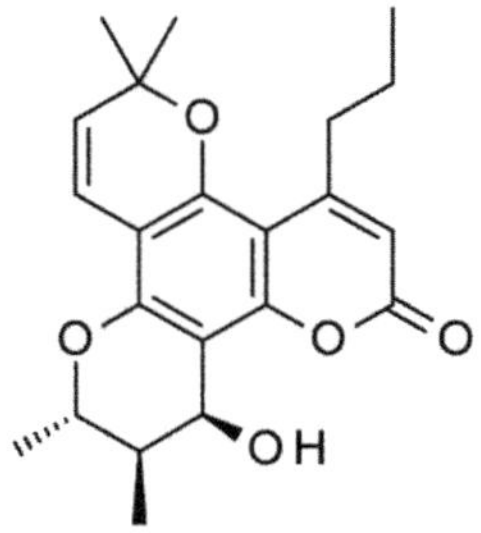

Costatolide (**14**)
[10,11-Di-*epi*-calanolide A or (-)-calanolide B]

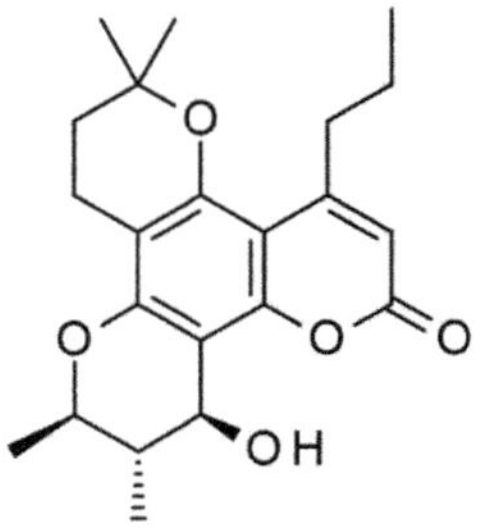

7,8-Dihydrocalanolide A (**15**)

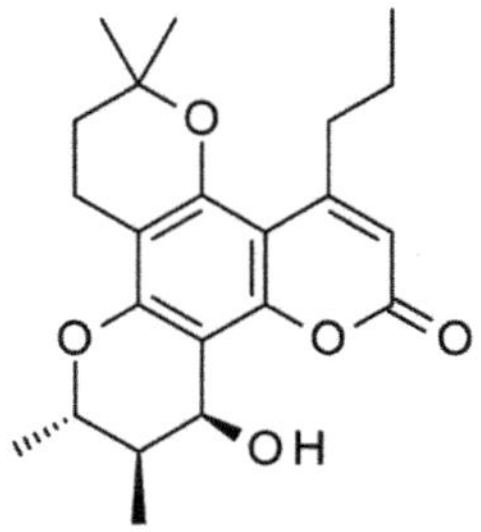

Dihydrocostatolide (**16**)

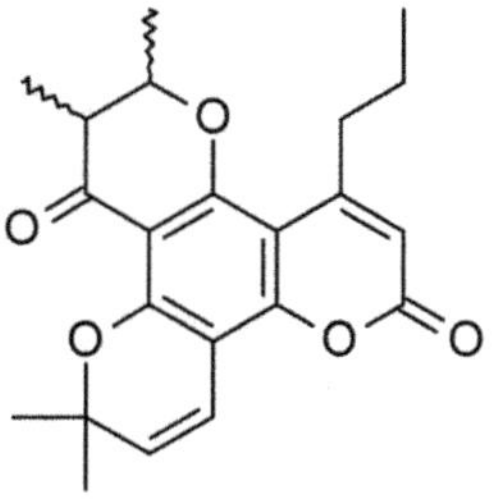

Tomentolide B (**17**)

Figure 2. Naturally occurring calanolides

2. Occurrence

Calanolides, calanolide A (**1**) being the first member of these 4-substituted pyranocoumarins isolated from *C. lanigerum var. austrocoriaceum*, are almost exclusively distributed within the genus *Calophyllum* L., which comprises a large group of ca. 200 species of tropical trees distributed in the Indo-Pacific region, but was also reported from one species (*Clausena excavate* Brum. f.) of the closely related genus *Clausena* [9–12] (Table 1). Calanolide A (**1**) and other calanolides were subsequently isolated from other *Calophyllum* species, e.g., *C. brasiliense* Cambess. [4,13,14], *C. inophyllum* L. [6], *C. teysmanii* Miq. [2] and *C. wallichianum* Planch. & Triana [15]. In a chemotaxonomic study on the *Calophyllum* species, the presence of calanolides was detected in the extracts of *C. inophyllum*, *C. lanigerum* var. *austrocoriaceum*, *C. mole* King, *C. nodosum*, aff. *Pervillei* Vesque., *C. soulattri* Burm. f., *C. tacamahaca* Willd. and *C. teysmanii* [9] (Table 1).

Table 1. Naturally occurring calanolides, their sources and properties.

Calanolides	Sources	Physical State	Mol. Formula	Mol. Weight	Optical Rotation [a]$_D$	UV λ_{max} (MeOH) nm	References
Calanolide A (**1**)	*Calophyllum lanigerum var. austrocoriaceum* *Calophyllum brasiliense* *Calophyllum inophyllum* *Calophyllum teysmannii* *Clausena excavata*	Oil	$C_{22}H_{26}O_5$	370.44	[a]$_D$ +60° (c, 0.7 in $CHCl_3$)	228, 284 and 325	[1,9,11,16] [4,17,18] [6,11] [19] [10]
12-*O*-Acetylcalanolide A (**2**)	*Calophyllum lanigerum var. austrocoriaceum*	Oil	$C_{24}H_{28}O_6$	412.48	[a]$_D$ +20° (c, 0.5 in $CHCl_3$)	228, 284 and 325	[1]
12-*O*-Methylcalanolide A (**3**)	*Calophyllum lanigerum var. austrocoriaceum*	Oil	$C_{23}H_{28}O_5$	384.47	[a]$_D$ +32° (c, 0.8 in $CHCl_3$)	228, 284 and 325	[1]
Calanolide B (**4**)	*Calophyllum lanigerum var. austrocoriaceum* *Calophyllum brasiliense* *Calophyllum teysmannii var. inophylloide*	Oil	$C_{22}H_{26}O_5$	370.44	[a]$_D$ +8° (c, 1.0 in acetone)	228, 284 and 325	[1] [5,17] [9]
12-*O*-Methyl-calanolide B (**5**)	*Calophyllum lanigerum var. austrocoriaceum*	Oil	$C_{23}H_{28}O_5$	384.47	[a]$_D$ +34° (c, 0.5 in $CHCl_3$)	228, 284 and 325	[1]
Calanolide C (**6**)	*Calophyllum brasiliense*	Oil	$C_{22}H_{26}O_5$	370.44	-	-	[4,5]
Calanolide D (**7**)	*Calophyllum brasiliense*	Amorphous solid	$C_{22}H_{24}O_5$	368.42	-	-	[6]
Calanolide E (**10**)	*Calophyllum lanigerum var. austrocoriaceum* *Calophyllum membranaceum* *Calophyllum molle* *Calophyllum polyanthum* *Calophyllum teysmannii var. inophylloide* *Calophyllum wallichianum*	Amorphous powder	$C_{22}H_{28}O_6$	388.50	[a]$_D$ +30° (c, 0.7 in acetone)	-	[1,9,20] [21] [9] [22] [20] [15]
Calanolide E1 (**11**)	*Calophyllum lanigerum var. austrocoriaceum* *Calophyllum brasiliense* *Calophyllum molle*	Amorphous powder	$C_{22}H_{28}O_6$	388.50	-	-	[9,20] [9,14] [9]
Calanolide E2 (**12**)	*Calophyllum lanigerum var. austrocoriaceum* *Calophyllum brasiliense* Cambess. *Calophyllum membranaceum* *Calophyllum molle* *Calophyllum polyanthum* *Calophyllum teysmannii var. inophylloide*	Amorphous powder	$C_{22}H_{28}O_6$	388.50	-	-	[9,20] [14] [21] [9] [22,23] [9,20]

Table 1. *Cont.*

Calanolides	Sources	Physical State	Mol. Formula	Mol. Weight	Optical Rotation $[a]_D$	UV λ_{max} (MeOH) nm	References
Calanolide F (**13**)	*Calophyllum lanigerum* var. *austrocoriaceum*	Amorphous powder	$C_{22}H_{26}O_5$	370.44	$[a]_D$ −51.5 (c, 0.3 in $CHCl_3$)	227, 283, 322	[9,20]
	Calophyllum teysmannii var. *inophylloide*						[9,20]
Costatolide (**14**) [(−)-Calanolide B]	*Calophyllum brasiliense*	Crystals (M.p. 181–183°)	$C_{22}H_{26}O_5$	370.44	$[a]_D$ −19.9 (c, 0.42 in $CHCl_3$)	228, 284 and 325	[4]
	Calophyllum costatum						[24]
	Calophyllum inophyllum L.						[24]
	Calophyllum teysmannii var. *inophylloide*						[9,19,25]
7,8-Dihydrocalanolide A (**15**)	*Calophyllum lanigerum* var. *austrocoriaceum*	Amorphous solid	$C_{22}H_{28}O_5$	372.46	Negative optical rotation	-	[25]
Dihydrocostatolide (**16**)	*Calophyllum costatum*	Amorphous solid	$C_{22}H_{28}O_5$	372.46	-	-	[26]
Pseudocalanolide C (**8**) [incorrectly named as calanolide C (**6**)]	*Calophyllum lanigerum* var. *austrocoriaceum*	Amorphous solid	$C_{22}H_{26}O_5$	370.44	$[a]_D$ +68° (c, 0.7 in $CHCl_3$)	-	[1,3,9]
Pseudocalanolide D (**9**) [incorrectly named as calanolide D (**7**)]	*Calophyllum lanigerum* var. *austrocoriaceum*	Amorphous solid	$C_{22}H_{24}O_5$	368.43	$[a]_D$ +60° (c, 0.5 in $CHCl_3$)	-	[1,3]
Tomentolide B (**17**)	*Calophyllum tomentosa*	Amorphous solid (M.p. 158–160°)	$C_{22}H_{24}O_5$	368.43	Racemic mixture	-	[1,3,9]

Bernabe-Antonio et al. [5] reported the production of calanolides in a callus culture of *C. brasiliense*, where different concentrations and combinations of plant growth regulators were tested in leaf and seed explants to establish callus cultures capable of producing calanolides. Higher calanolides B (**4**) and C (**6**) production was observed in calluses from seed explants than those developed from leaves. In continuation of the search for new natural anti-HIV compounds, and at the same time to find new botanical sources of calanolides, McKee et al. [20] purified calanolide E2 (**12**), and calanolide F (**13**) from the extracts of *C. lanigerum var. austrocoriaceum* and *C. teysmanii var. inophylloide* (King.) P. F. Stevens. Later, costatolide (**14**), also known as (−)-calanolide B, was reported as an anti-HIV compound present in *C. cerasiferum* Vesque and *C. inophyllum* L. [24]. Calanolides A (**1**), and C (**6**), and costatolide (**14**) were isolated from the leaves of *C. brasiliense*, and their anti-HIV potential was evaluated [4].

3. Biosynthesis

Calanolides are biosynthesized from the parent simple coumarin 7-hydroxycoumarin, also known as umbelliferone (Schemes 1–3). The biosynthesis of umbelliferone in plants starts from the amino acid L-phenylalanine, and proceeds through the formation of *trans*-cinnamic acid, *p*-coumaric acid, 2-hydroxy-*p*-coumaric acid, 2-glucosyloxy-*p*-coumaric acid, and 2-glucosyloxy-*p*-*cis*-coumaric acid with the help of various enzymes like cinnamate 4-hydroxylase, 4-coumarate-CoA ligase, 4-coumaroyl 2′-hydroxylase and so on [27]. The biosynthesis of dipetalolactone, a pyranocoumarin, and subsequent conversion to the 3-propyl-intermediate for calanolides may proceed through two routes, one through conversion of umbelliferone to osthenol (Scheme 1), and the other via formation of 5,7-dihdroxycoumarin (Scheme 2). Reactions are generally mediated by p450 monooxygenase and other non-p450 enzymes [28]. 3-Propyl-intermediate is converted to the precursor compound for calanolides A–C (**1, 4** and **6**), utilizing the Wagner-Meerwein rearrangement reaction, and the precursor compound is believed to be converted to calanolides with the help of p450 monooxygenase enzyme (Scheme 3). Published studies on the biosynthesis of calanolides are rather limited and only two publications are available on this topic to date [28,29]. Therefore, detailed knowledge of specific enzymes involved in the biosynthesis of calanolides is still in its infancy.

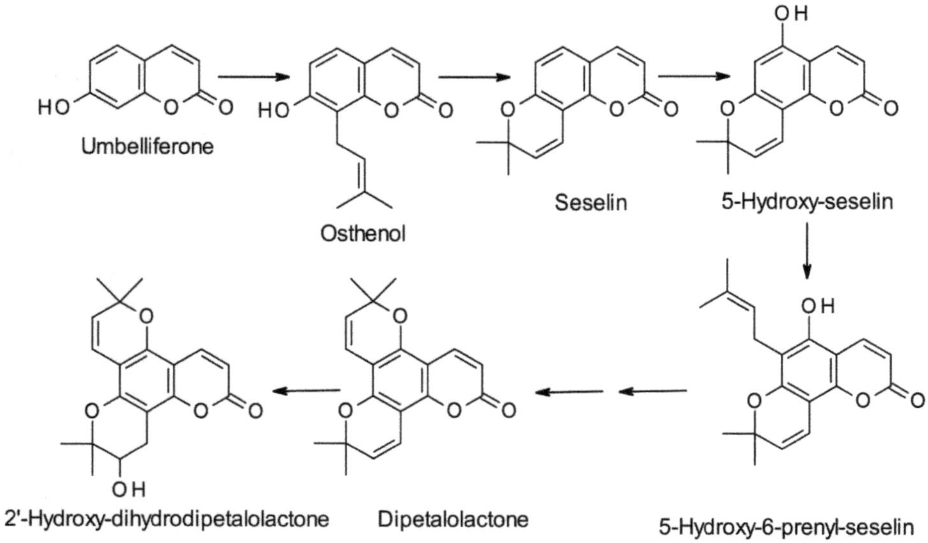

Scheme 1. Plausible biosynthetic route to 2′-hydroxydihydrodipetalolactone from umbelliferone *via* formation of 5-hydroxy-6-prenylseselin.

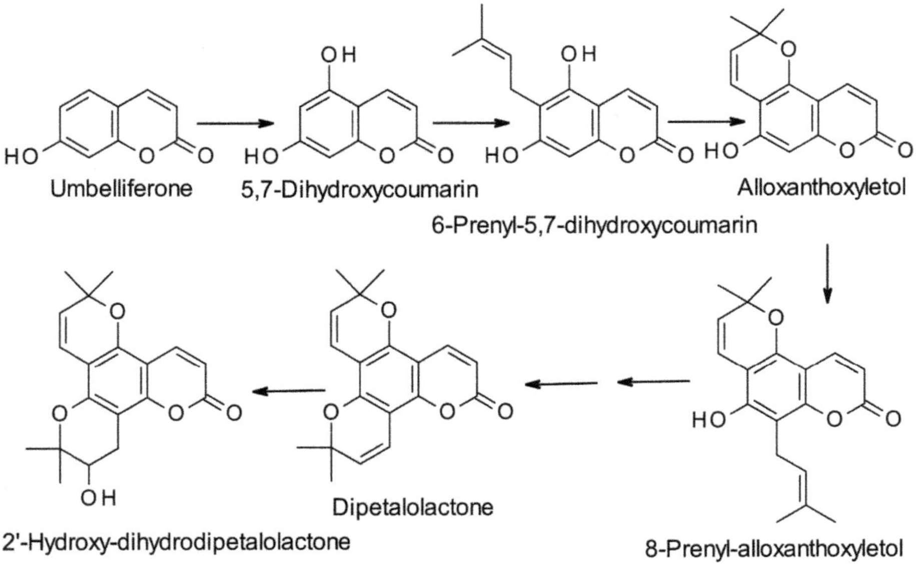

Scheme 2. Plausible biosynthetic route to 2′-hydroxydihydrodipetalolactone from umbelliferone *via* formation of 8-prenylalloxanthoxyletol.

In a recent study, the influence of soil nutrients, e.g., Ca^{2+} and K^+, on the biosynthesis of pharmacologically active calanolides in the seedlings of *C. brasiliense* was studied [29]. It was observed that the use of K^+ deficient modified Hoagland solution (MHS) could induce a 15, 4.2 and 4.3-fold decrease of calanolides B (**4**), C (**6**), and apetalic acid concentrations in the leaves of the seedlings, respectively. On the other hand, Ca^{2+} deficient MHS could lead to a decrease of 4.3 and 2.4-fold for calanolides B (**4**) and C (**6**), respectively. This study demonstrated that, like many other plant secondary metabolites, the biosynthesis of calanolides, albeit genetically controlled, may also be affected by environmental conditions, e.g., soil nutrients (minerals).

As genes dictate biosynthesis of secondary metabolites, a study was conducted to identify candidate genes that regulate to the biosynthesis of calanolides in *C. brasiliense* [28]. The unigene dataset constructed in this study could offer an insight for further molecular studies of *C. brasiliense*, particularly for characterizing candidate genes responsible for the biosynthesis of angular and linear pyranocoumarins. The candidate genes, e.g., UN36044, UN28345 and UN34582, identified in the transcriptome of the leaves, stem and roots of *C. brasiliense* might be involved in the biosynthesis of calanolides, which are essentially modified angular pyranocoumarins. Candidate unigenes in the transcriptome dataset were screened using mainly homology-based BLAST and phylogenetic analyses. It is worthy of mention that the BLAST programs are widely used for searching protein and DNA databases for optimizing sequence similarities [30]. For protein comparisons, several definitional, algorithmic and statistical refinements allow substantial decrease in the execution time of the BLAST programs and enhancement of their sensitivity to weak similarities.

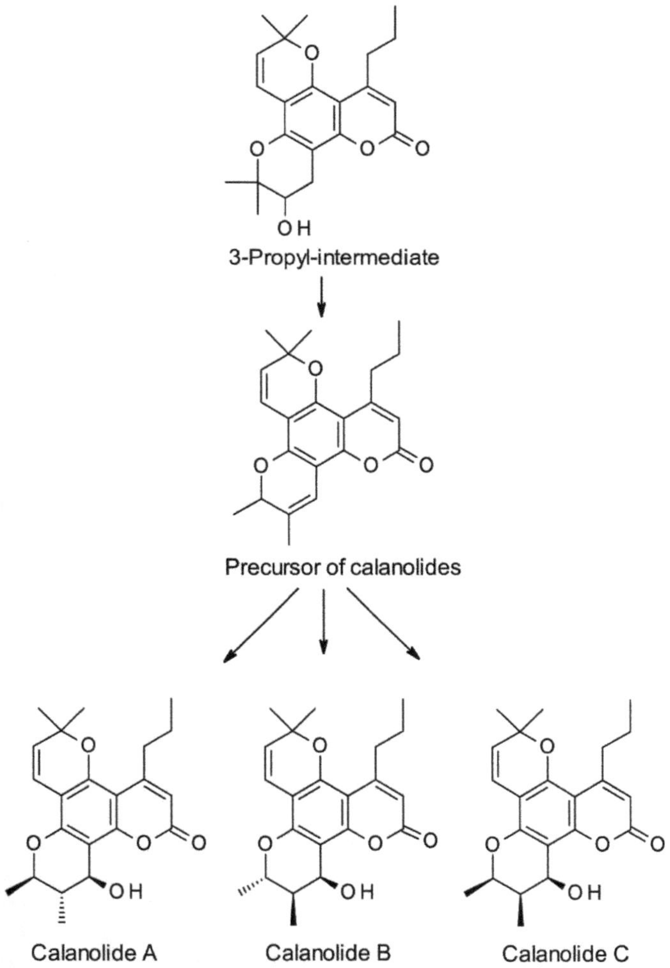

3-Propyl-intermediate

Precursor of calanolides

Calanolide A Calanolide B Calanolide C

Scheme 3. Plausible biosynthetic route to calanolides A (**1**), B (**4**) and C (**6**) from the intermediate, 2′-hydroxydihydrodipetalolactone.

4. Pharmacological Properties

Although well-known for non-nucleoside reverse transcriptase inhibitory activity offering anti-HIV potential, calanolides have also been shown to possess various other pharmacological properties (Figure 3). The following sub-sections deal with anticancer, anti-HIV, antimycobacterial and antiparasitic activity of naturally occurring calanolides. As much of the published pharmacological studies, both in vitro and in vivo including human trials, on naturally occurring calanolides are about their anti-HIV property, over the years, significant amounts of information have become available on their mechanism of action, structure-activity-relationships, synergistic and/or additive property and their potential in anti-HIV combination therapy, which have been discussed adequately under individual headings within the anti-HIV sub-section. All other pharmacological properties of these compounds as outlined in different publications still require further investigations to establish their realistic therapeutic potential. Also, in silico pharmacological activity and toxicity studies on these pyranocoumarins have just begun to emerge in recent years.

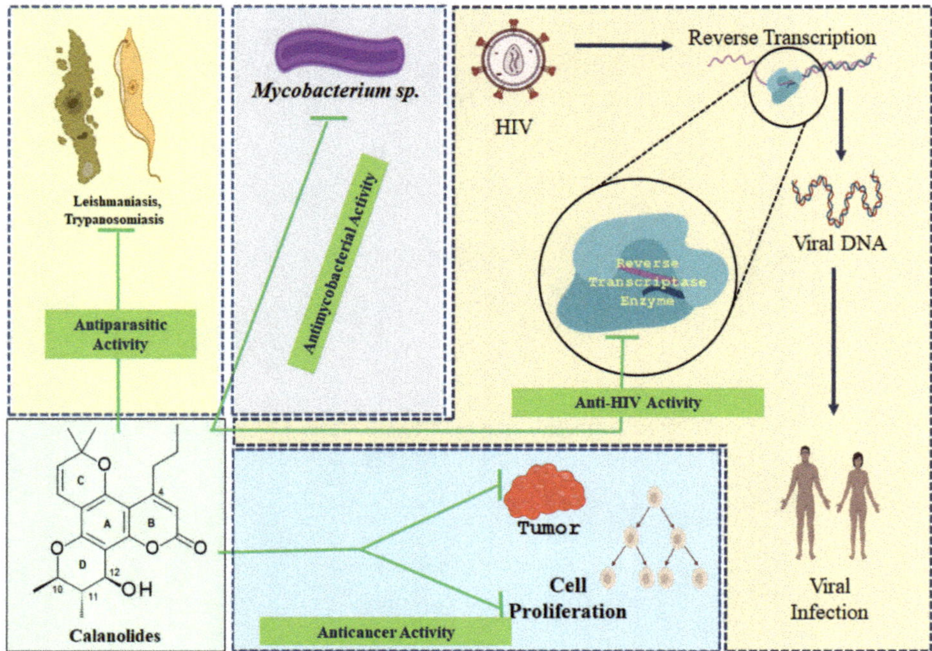

Figure 3. A pictorial summary of pharmacological properties of naturally occurring calanolides.

4.1. Anticancer Activity

In the later part of 1980s, as a part of the initiative of the United States National Cancer Institute (NCI), plant samples from the Malaysian flora were collected for routine screening for potential cytotoxicity against a collection of cancer cell lines as well as for possible anti-HIV activity. One of the samples, the leaves and twigs of the tree *C. lanigerum* var. *austrocoriaceum*, despite not being active against any of the cancer cell lines tested, showed inhibitory activity of viral replication when tested against HIV-1 virus [31,32]. However, later, calanolide A (**1**) and calanolide C (**6**) were shown to possess antiproliferative or antitumor-promoting property through inhibition of TPA-induced EBV-EA activation in Raji cell lines [13]. The phorbol ester, 12-*O*-tetradecanoylphorbol-13-acetate (TPA) is a potent stimulator of differentiation and apoptosis in myeloid leukemia cells. Calanolide A (**1**) was found to be more active (IC_{50} = 290 mol ratio/32 pmol TPA) than its 10,11-*cis*-isomer, calanolide C (**6**) (IC_{50} = 351 mol ratio/32 pmol TPA). It was inferred that 4-substituted pyranocoumarins like calanolides might possess potential as cancer chemopreventive agents or antitumor-promoters. A recent study with the crude ethanolic extract of the leaves of *C. inophyllum* revealed its potential as a cytotoxic agent (IC_{50} 120 μg/mL) against the breast cancer cell line MCF-7 [33]; it was also found to possess antiproliferative and apoptotic properties. However, no definitive proof was provided to establish which of the secondary metabolites biosynthesized by this plant, calanolides being one major class, were responsible for the putative anticancer activity. Although not calanolides, a few other 4-substituted coumarins, isolated from *C. brasiliense*, were tested against human leukemia HL-60 cells with some promising results [34], which might highlight the need for more comprehensive studies with all major 4-susbtitued coumarins, including calanolides, to find antileukemia lead compounds for new anticancer drug development. Calanolide A (**1**), isolated from a chloroform extract of *Clausena excavata*, was found to induce toxicity to the cells used in a syncytium assay for anti-HIV activity [10].

The efficacy of calanolide A (**1**) in AIDS-associated cancer was evaluated in silico utilizing an integrated approach combining network-based systems biology, molecular docking and molecular dynamics [35]. Molecular targets were screened and only the targets, e.g., HRAS, that are common to HIV and sarcoma, HIV and lymphoma, and HIV and cervical cancer, were utilized in this study. Calanolide A (**1**) was found to form a stable complex with the screened target HRAS, which is a small G protein in the RAS subfamily of the RAS superfamily of small GTPases, and is considered as a proto-oncogene; when mutated, this proto-oncogene has the potential to cause normal cells to become cancerous.

4.2. Anti-HIV Activity

Calanolide A (**1**), an anti-HIV non-nucleoside reverse transcriptase inhibitor (NNRTI), paved the way for the discovery and synthesis of a series of 4-substituted angular pyranocoumarins with potential anti-HIV property [1,36]. NNRTIs are a class of anti-HIV drugs that prevent healthy T-cells in the body from becoming infected with HIV. Kashman et al. [1] first reported this new class of anti-HIV agents from the tropical rainforest tree, *C. lanigerum*. Calanolide A (**1**), 12-acetoxycalanolide A (**2**), 12-methoxycalanolide A (**3**), calanolide B (**4**), 12-methoxycalanolide B (**5**), pseudocalanolide C (**8**), pseudocalanolide D (**9**) and calanolide E (**10**) (Figure 2) were isolated through an anti-HIV bioassay-guided isolation. Calanolides A (**1**) and B (**4**) were found to be protective against HIV-1 replication and cytopathicity with EC_{50} values of 0.1 μM and 0.4 μM, respectively. However, both compounds were inactive against HIV-2, which is known as less pathogenic than HIV-1 and mainly found in West African countries. The other compounds showed a low level of anti-HIV-1 activity. This study involving purified bacterial recombinant reverse transcriptases established that the calanolides are indeed HIV-1 specific reverse transcriptase inhibitors. A comparative report on the anti-HIV potentials of calanolide A (**1**), costatolide (**14**) and dihydrocostatolide (**16**) against a series of HIV isolates of different cellular phenotypes was published by Buckheit et al. [26], which clearly demonstrated that calanolide A (**1**) was the best anti-HIV candidate among the three calanolides tested.

Two analogs of calanolide A (**1**), i.e., costatolide (**14**) and dihydrocostatolide (**16**), were shown to possess anti-HIV property similar to that of calanolide A (**1**) [26] and could be ascribed to the class of NNRTIs. In fresh human cells, costatolide (**14**) and dihydrocostatolide (**16**) could significantly inhibit the low-passage clinical virus strains, including those representative of the various HIV-1 clade strains, syncytium-inducing and non-syncytium-inducing isolates, and T-tropic and monocyte-tropic isolates [26,37]. In continuation of the search for new natural anti-HIV compounds, McKee et al. [20] purified calanolide E2 (**12**), and calanolide F (**13**) from extracts of *C. lanigerum* var. *austrocoriaceum* and *C. teysmanii* var. *inophylloide* (King.) P. F. Stevens, and calanolide E2 (**12**) emerged as one of the most active anti-HIV compounds. Later, costatolide (**14**) was reported as an anti-HIV compounds present in *C. cerasiferum* Vesque and *C. inophyllum* L. [24], while calanolides A (**1**), and C (**6**), and costatolide (**14**), isolated from the leaves of *C. brasiliense*, were shown to possess anti-HIV potential [4]. Comparative anti-HIV activities of some naturally occurring calanolides, e.g., calanolide A (**1**), costatolide (**14**) and dihydrocostatolide (**16**), against various strains of HIV are available in the article by Buckheit, et al. [26].

4.2.1. Activity Against Drug Resistant Strains of HIV-1

Interestingly, calanolide A (**1**) was not only found to be active against standard strains of HIV-1, but it was also active against the resistant strains, eAZT-resistant G-9106 strain of HIV-1 and pyridinone-resistant A17 strain [1,38]. The activity against the pyridinone-resistant A17 strain was of interest as this strain is highly resistant to most of the HIV-1 specific NNRTIs, for example, TIBO, BI-RG-587 and L693,593. Later, it was established that pyranocoumarin **1** could interact with HIV-1 reverse transcriptase within the previously defined common binding site for nonnucleoside inhibitors [38]. An assessment of the inhibition patterns of the chimeric reverse transcriptases containing complementary segments of HIV-1 and HIV-2 reverse transcriptases established that there was a segment between residues 94 and 157 in HIV-1 reverse transcriptase that was crucial for inhibition by calanolide A (**1**) [39]. However, it was assumed that there might be a second segment, essential for specifying susceptibility to the drug, between amino acids 225 and 427 in HIV-1 reverse transcriptase. A couple of years later, it was noted that calanolide A (**1**) was active against virus isolates resistant to 1-[(2-hydroxyethoxy)methyl]-6-(phenylthio)thymine and its derivative, [1-benzyloxymethyl-5-ethyl-6-(alpha-pyridylthio)uracil] [40]. Furthermore, this pyranocoumarin (**1**) showed activity against HIV with the two most common NNRTI-related mutations, K103N and Y181C, and was found to select for a mutation that does not cause cross-resistance with any other NNRTIs under investigation. It was postulated that substitution at codon Y188H of reverse transcriptase could be associated with 30-fold resistance to calanolide A (**1**) in vitro [41]. The compound is essentially inactive against all strains of the less common HIV type 2. It is necessary to carry out appropriate in vivo experimentations, either in animal models or in human clinical trials, to understand the true potential of any putative drug candidate. In vivo anti-HIV activity of (+)-calanolide A (**1**) was assessed in a hollow fibre mouse model [42], and it was observed that this compound could suppress virus replication in two unique, but separate physiologic compartments following oral or parenteral administration.

Calanolides were found to possess an enhanced antiviral activity against one of the most prevalent NNRTI-resistant viruses that is engendered by the Y181C amino acid change in reverse transcriptase as well as with reverse transcriptases that possess the Y181C change together with AZT-resistant mutations [26,37]. Calanolides could also be active against viruses containing Y181C and K103N dual mutations, which are generally highly resistant to other known non-nucleus reverse transcriptase inhibitors. Anti-HIV activity of naturally occurring calanolides against drug-resistant strains of HIV have made these compounds promising structural templates for new anti-HIV drug development.

4.2.2. Calanolides in Anti-HIV-1 Combination Therapy

For the treatment of HIV infections, use of combination therapy comprising several anti-HIV drugs has become a common practice in recent years. The synergistic effects of calanolide A (**1**), costatolide (**14**) and dihydrocostatolide (**16**) [26] in combination with established anti-HIV drugs, e.g., azidothymidine (AZT), indinavir, nelfinavir and saquinavir, are available in the literature [26]. Synergistic effects were observed in both cultured cells and animal models when calanolides were used in combination with other anti-HIV agents [43]. Both calanolide A (**1**) and costatolide (**14**) were found to be effective in combination therapy for HIV infections [44]; in combination with NNRTIs, costatolide (**14**) could only synergistically inhibit HIV type 1 with UC38, whilst calanolide A (**1**) in combination with one of the NNRTIs helped this drug to retain activity against virus isolates with the single Y181C mutation [26,41,44,45].

A combination of (+)-calanolide A (**1**) and nevirapine (marketed under the trade name viramune among others for the treatment and prevention HIV-1 infection) was found to possess an additive to weakly synergistic effect in blocking replication of HIV-1 in an in vitro tissue culture assay [41], indicating the possibility of using (+)-calanolide A (**1**) in anti-HIV-1 combination therapy. In an in vivo study using a hollow fibre mouse model [42], the synergistic potential of (+)-calanolide A (**1**) in combination therapy with AZT, a well-known anti-retroviral medication, was further established. A more comprehensive study on the anti-HIV activity of (+)-calanolide A (**1**) and its analogs, e.g., costatolide (**14**), dihydrocostatolide (**16**) and (+)-12-oxo-calanoldie A, in combination with other inhibitors of HIV-1 replication was published about a decade ago [37,46]. Calanolides were found to display synergistic antiviral interactions with other nucleoside and non-nucleoside reverse transcriptase inhibitors and protease inhibitors. In addition, additive interactions were also observed with calanolides when used with other anti-HIV drugs. It was concluded that the utility of convergent and divergent combination therapies using reverse transcriptase inhibitors and protease inhibitors in combination with (+)-calanolide A (**1**) or one of its analogues could be clinically relevant. Budihas et al. [47] demonstrated significant synergy between β-thujaplicinol and calanolide A (**1**).

4.2.3. Structure-Activity-Relationships (SAR)

Among the naturally occurring calanolides, calanolide A (**1**) is one of the most potent anti-HIV compounds and has been the focus of various studies including the study of its possible mechanism of action, structural modifications, pharmacokinetics and toxicity [9,48–51]. The structures of naturally occurring calanolides mainly differ in their stereochemistry at various chiral centers (C-10, C-11 and C-12) on the ring D (Figures 1 and 2). McKee et al. [20] reported that calanolide-type compounds with a 12β hydroxyl group (as in compound **1**) generally possess anti-HIV activity. While calanolide A (**1**) and costatolide (**14**) were found to be active, (+)-calanolide C (**6**) was inactive in the in vitro anti-HIV assay [4,24]. The inactivity of (+)-calanolide C (**6**) despite possessing the pharmacophoric ring D, as well as a propyl group on C-4, could be due to the β-*cis* orientation of methyl groups on C-10 and C-11.

Like any other optically active drug molecules, optical activity plays an important role in the anti-HIV activity of calanolides. It has long been established that (+)-calanolide A (**1**) and (−)-calanolide B (**14**) are potent HIV-1 inhibitors, whilst (−)-calanolide A and (+)-calanolide B (**4**) are inactive against the virus [52]. It should be mentioned here that (+)-calanolide A (**1**) is the natural product, but its enantiomer (−)-calanolide A was prepared from the naturally occurring (−)-costatolide (**14**), isolated from *C. costatum*. Similarly, to establish structure-activity-relationships of calanolides, several analogs of calanolides have been synthesized to date, and tested in anti-HIV assays [53]. Although the synthesis of calanolides and the anti-HIV activity of synthetic calanolide analogs are not within the scope of this review, a few examples are given here in the context of structure-activity-relationships. One of the first attempts in this area was from Galinis et al. [53], where $\Delta^{7,8}$ olefinic bonds within (+)-calanolide A (**1**) and (−)-calanolide B (**14**) were reduced, and C-12 hydroxyl group in (−)-calanolide B (**14**) was modified to investigate variations in anti-HIV activity compared to parent calanolides. In this study, none of the 14 derivatives was found to possess superior activity to parent calanolides but revealed some preliminary structure-activity requirements for anti-HIV potencies. Later, in order to identify the structural features of naturally occurring (+)-calanolide A (**1**) necessary for its anti-HIV activity and to prepare synthetic analogues, oxo-derivatives (+)-, (−)- and (±)-12-oxocalanolides, were synthesized and tested in vitro using a biochemical reverse transcriptase inhibition assay for determining anti-HIV activity with a promising outcome [48]. In a review article covering various aspects of anti-HIV 4-substitued coumarins with an alkyl or a phenyl group as the substituent, isolated from the genus *Calophyllum*, summarized that all *trans* configurations (10*R*, 11*S*, 12 *S*), as in (+)-calanolide A (**1**) and (+)-inophyllum B (a 4-phenyl-substituted pyranocoumarin), are essential for the best anti-HIV activity [54].

Most of the SAR studies involving calanolides for their anti-HIV activities concentrated on the three chiral centers at C-10, C-11 and C-12 of (+)-calanolide A (**1**) [55,56]. As the number of naturally occurring calanolides are rather limited (calanolides A–F) (Figure 2), the SAR studies were often carried out with natural calanolides as well as their synthetic analogs. Of the diastereomers, compounds containing 10,11¬-*trans*-methylation and 12-(*S*)-OH chirality (Figure 2) displayed the most potent activity with EC_{50} values in between 0.18 and 2.0 μM [55]. It was also observed that either the enantiomers (12-*R*-OH) or epimeric alcohols, e.g., calanolide C (**6**) could not produce any noticeable anti-HIV effect. It could be concluded that the relative stereochemistry at C-10 and C-11 are essential structural features for potent anti-HIV activity of calanolides, and at the same time, the *S* configuration at C-12 as well as the presence of a heteroatom, e.g., O, at C-12 are necessary for anti-HIV effects.

In order to assess the importance of the presence of 11-methyl functionality on calanolide A for its anti-HIV activity, the activity of the semi-synthetic racemic mixture of 11-demethyl-calanolide A was compared with the anti-HIV activity of its parent compound, (±)-calanolide A [57]. The in vitro HIV-1 reverse transcriptase inhibitory activity of these compounds was determined with the isotope 3H assay, which is a thymidine incorporation assay that often utilizes a strategy wherein a radioactive nucleoside, 3H-thymidine, is incorporated into new strands of chromosomal DNA during mitotic cell division; a scintillation beta-counter is used to measure the radioactivity in DNA recovered from the cells in order to determine the extent of cell division that has occurred in response to a test agent.

The cytotoxicity and inhibition of cytopathic effect of (±)-calanolide A and (±)-11-demethyl-calanolide A were studied in HIV-1 IIIB infected MT-4 cell cultures by the MTT staining method. Both compounds inhibited HIV-1 reverse transcriptase in vitro with IC_{50} value of 3.028 μM/L and 3.965 μM/L, respectively, for (±)-11-demethyl-calanolide and (±)-calanolide A. They also inhibited cytopathic effect in HIV-1 IIIB infected MT-4 cell cultures with IC_{50} values of 1.081 and 1.297 μM/L, respectively. The outcome form this study indicated that (±)-11-demethyl-calanolide had a slightly more potent anti-HIV activity than (±)-calanolide A, suggesting the methyl functionality at C-11 in calanolide A (**1**) might not be an essential structural feature for anti-HIV activity. With the help of synthetic analogues a few other structural features that could impact on the anti-HIV activity of calanolides could be identified. Some of those are summarized below:

(i) $\Delta^{11,12}$ Olefination diminishes activity.
(ii) A C-12 hetero atom is essential for the activity.
(iii) Relative potencies of C-12 ketone, thiol, azide, amine, and acetylated derivatives suggest stringent spatial and stereochemical requirements around C-12.
(iv) The enantiomers of 12-oxocalanolide A, synthetic intermediates containing one fewer chiral center, still retain anti-HIV potency in the cytopathic assays.
(v) The oxygen substituent can either be in the plane of the aromatic system or possess *S* configuration.
(vi) Optical activity is important. For example, (+)-12-oxocalanolide A and (±)-12-oxocalanolide A have similar (but not same) anti-HIV activity, but (−)-12-oxocalanolide A is much less active.
(vii) The racemic form, for example, (±)-12-oxocalanolide A, is more active than its pure (+)-enantiomer, (+)-12-oxocalanolide A, which suggests a possible synergistic effect in the combination of the two enantiomers.
(viii) Hydrogenation at C-7 and C-8 of calanolides has little effect on the anti-HIV activity, e.g., the dihydro derivatives of calanolides A (**1**) and B (**4**) possess the same activity as the parent calanolides.
(ix) Modifications at C-4 substituent can affect the anti-HIV activity of calanolides. For example, a methyl substituent at C-4 (as in cordatolides), instead of a propyl function as in calanolides reduces the anti-HIV potency.
(x) Both the surface area of the substituted group attached on C-10, S-R3, and the distance between atoms O-13 and X-14 (O, N, S), L, of the calanolide analogues play important roles in determining the inhibitory activity of HIV-1 [56].

With the advent of various modern computational tools and mathematical models, it is now possible to study quantitative structure activity relationships (QSAR) in silico, and to predict the potential of any drug candidates for any therapeutic application [58]. A Caco-2 cell permeability QSAR model has recently been used to study various HIV-1 reverse transcriptase inhibitors, including (+)-calanolides A (**1**) and B (**4**), both of which showed a high degree of permeability [59]. This parallel computational screening method incorporated approaches of intestinal absorption prediction, receptor affinity estimation, inhibitor shape similarity, lipophilicity, and index-based lipophilic efficiency analyses. Calanolide A (**1**), among a few other HIV-1 reverse transcriptase inhibitors, emerged as one of the prioritized hits, as a result of guided prioritization task by the better binding affinity, crystal ligand similarity, permissible log*P* value and top lipophilic ligand efficiency scores.

4.2.4. Mechanism of Action

The evaluation of the activity of (+)-calanolide A (**1**) against reverse transcriptase and nonnucleoside reverse transcriptase inhibitor-resistant viruses and enzyme kinetic studies for reverse transcriptase inhibition suggest that this coumarin possibly interacts with the HIV-1 reverse transcriptase in a fashion mechanistically different from other known NNTRIs. The biochemical mechanism of inhibition of HIV-1 reverse transcriptases by calanolide A (**1**) was studied using two primer systems, ribosomal RNA and homopolymeric rA-dT(12-18) [60]. Calanolide A (**1**) was found to bind near the active site of the enzyme and interfered with dNTP binding; it inhibited HIV-1 reverse transcriptase in a synergistic fashion with nevirapine, further distinguishing it from the general class of NNRTIs. It was also observed that at certain concentrations, this compound could bind HIV-1 reverse transcriptase in a mutually exclusive manner with respect to both the pyrophosphate analog, phosphonoformic acid and the acyclic nucleoside analogue 1-ethoxymethyl-5-ethyl-6-phenylthio-2-thiouracil. It was concluded that calanolide A (**1**) could share some binding domains with both phosphonoformic acid and 1-ethoxymethyl-5-ethyl-6-phenylthio-2-thiouracil. It might interact with reverse transcriptase near both the pyrophosphate binding site and the active site of the enzyme. Later, the same group of researchers studied possible mechanism of action of action of calanolide A (**1**) against the HIV type 1 including a variety of laboratory strains, with EC_{50} values of 0.10–0.17 µM [60]. Calanolide (**1**) could inhibit promonocytotropic and lymphocytotropic isolates from patients in various stages of HIV disease, and drug-resistant strains, and was found to act early in the infection process like the known HIV reverse transcriptase inhibitor 2′,3′-dideoxycytidine. It could selectively inhibit recombinant HIV type 1 reverse transcriptase but not cellular DNA polymerases or HIV type 2 reverse transcriptase. Auwerx et al. [50] studied the possible role of Thr139 in the HIV-1 reverse transcriptase sensitivity to (+)-calanolide A (**1**). As T139I reverse transcriptase proved to be resistant to (+)-calanolide A (**1**), represents a catalytically efficient enzyme, and requires only a single transition point mutation (ACA→ATA) in codon 139 could provide some explanation as to why mutant T139I reverse transcriptase virus strains, but not the other strains containing other amino acid changes at this position, predominantly emerge in cell cultures under (+)-calanolide A (**1**) pressure.

Calanolides are non-nucleoside reverse transcriptase inhibitors and mediate their inhibitory effect in two different template primer systems: primed ribosomal RNA template, and homopolymeric poly rA-oligo$T_{12\text{-}18}$ primer. Calanolide A (**1**) was found to inhibit reverse transcriptase by involving two binding sites, and the action is because of the bi-bi ordered mechanism of reverse transcriptase, requiring primer binding prior to polymerization [55]. Calanolide A (**1**) can bind HIV-1 reverse transcriptase in a mutually exclusive manner with the pyrophosphate analogues phosphoformic acid or 1-ethoxymethyl-5-ethyl-6-phenylthio-2-thiouracil. This indicates that calanolide A (**1**) can interact with reverse transcriptase near the pyrophosphate binding site as well as the active site. Unlike general non-nucleoside reverse transcriptase inhibitors, calanolide A (**1**) appears to be at least partially competitive inhibitor of dNTP binding. Clinical and laboratory assessment on viral load and CD4 count indicated that antiviral effects of calanolide A (**1**) appeared to be dose-dependent and maximized on day 14 or 16. Viral life-cycle studies indicated that calanolide A (**1**) could act early in the infection process, similar to the known HIV reverse transcriptase inhibitor 2′,3′-dideoxycytidine. In enzyme inhibition assays, calanolide A (**1**) could potently and selectively inhibit recombinant HIV type 1 reverse transcriptase but not cellular DNA polymerases or HIV type 2 reverse transcriptase within the concentration range tested.

4.3. Antimycobacterial Activity

The antibacterial (against *Bacillus cereus, B. pumilius, B. subtilis, Escherichia coli, Pseudomonas aeruginosa, Salmonella typhi, Staphylocossus aureus and Vibrio cholerae*) and antifungal (against *Alternaria tenuissima, Aspergillus fumigatus, Aspergillus niger, Candida albicans and Candida tropicalis*) properties of *Calophyllum* species and their bioactive secondary metabolites, including calanolides, are already known [6,15,61–67]. Kudera et al. [66] reported in vitro growth inhibitory activity of *C. inophyllum* extract against diarrhea-causing microorganisms, e.g., *Clostridium difficile infant, Clostridium perfringens, Enterococcus faecalis, Escherichia coli, Listeria monocytogenes* and *Salmonella enterica*. The extract was particularly active against *C. perfringens* and *L. monocytogenes* (MIC = 128 µg/mL). Later, calanolide E (**10**) was isolated from *C. wallichianum* and tested for its anti-*Bacillus* activity against *Bacillus cereus, B. megaterium, B. pumilus* and *B. subtilis* [20]. However, calanolide E (**10**) was not bactericidal on the tested Bacillus species, and at the tested concentration.

Based on the initial findings on promising antimicrobial properties of calanolides and *Calophyllum* extracts, efforts have recently been directed to the study on the effect of these compounds on the acid-fast bacillus *Mycobacterium tuberculosis* that causes tuberculosis [17,68,69]. As over the years several antibiotic resistant and multidrug-resistant *M. tuberculosis* strains have emerged, and complicated the existing treatment modalities for tuberculosis, and there has been a recent increase in incidents of tuberculosis globally observed, the need for new effective, safe and affordable antimycobacterial drugs has become paramount. *Calophyllum brasiliense* extract was reported to be active against *M. tuberculosis* (IC_{50} 3.02–3.64 μg/mL), and a follow up HPLC analysis of the active extract provided evidence of presence of calanolides and the antimycobacterial activity induced by *C. brasilliense* was attributed mainly to calanolides A (**1**) and B (**4**) [17]. Earlier, Xu et al. [68] demonstrated that calanolide A (**1**), from Colombian *C. lanigerum*, was active against both drug-susceptible and drug-resistant strains of *Mycobacterium tuberculosis*, e.g., H37Ra (ATCC 25177), H37Rv (ATCC 27294), CSU 19, CSU 33, H37Rv-INH-R (ATCC 35822), CSU 36, CSU 38 and H37Rv-EMB-R (ATCC 35837). Efficacy evaluations in macrophages established that this pyranocoumarin could inhibit intracellular replication of *M. tuberculosis* at concentrations below the minimum inhibitory concentration (MIC) determined in vitro. It was postulated that calanolide A (**1**), like the antitubercular drug rifampicin, could rapidly inhibit RNA and DNA synthesis followed by an inhibition of protein synthesis, and could lead to the generation of a new class of pyranocoumarin-based antitubercular drugs. In this study, the natural calanolides A (**1**), B (**4**) and D (**7**), as well as their semisynthetic analogues were tested, and (+)-calanolide A (**1**) and the semisynthetic analogue, 7,8-dihydrocalanolide B emerged as most effective against tuberculosis with the MIC value of 3.13 μg/mL. While (−)-calanolide B (**14**) was moderately effective, calanolide D (**7**) was found inactive at the highest tested concentration of 12.5 μg/mL. In fact, calanolides, especially calanolide A (**1**), is unique in a sense that these compounds have anti-HIV property and were found to be active against *M. tuberculosis* (MIC = 3.1 μg/mL) and an array of drug-resistant strains (MIC = 8–16 μg/mL). The antimycobacterial activity of calanolide A (**1**) is comparable to that of the well-known anti-tubercular drug isoniazid, and effective against rifampicin- and streptomycin-resistant *M. tuberculosis* strains. A recent patent described potent antimycobacterial property of calanolides and their analogs and provided a method of using these compounds for the treatment and prevention of mycobacterial infections [70].

4.4. Antiparasitic Activity

Traditionally, natural products, especially in crude forms, have long been used to treat various parasitic diseases, like babesiosis, leishmaniasis, malaria, trypanosomiasis and so on. Recently, leishmaniasis and trypanosomiasis have been in research focus of natural products researchers, aiming at discovering new drug candidates to treat these neglected diseases [71,72]. Extracts of *C. brasiliense* and *C. inophyllum* and calanolides were shown effective against intracellular parasites causing American trypanosomiasis and leishmaniasis [6]. In a recent study, Silva et al. (2020) [14] demonstrated that the MeOH extract from stem bark of *C. brasiliense* was active against amastigote forms of *Trypanosoma cruzi* and *Leishmania infantum*. Bioactivity-guided purification of the extra afforded calanolides E1 (**11**) and E2 (**12**), which were found to be active against *T. cruzi* (EC_{50} values of 12.1 and 8.2 μM, respectively) and *L. infantum*, (EC_{50} values of 37.1 and 29.1 μM, respectively) in vitro. Calanolide E1 (**11**) displayed the best selectivity index (SI) with values >24.4 to *T. cruzi* and >6.9 to *L. infantum* in comparison to calanolide E2 (**12**). It was concluded that these coumarins could be utilized as scaffolds for the design and development of novel drug candidates to treat Leishmaniasis and Chagas diseases.

5. Toxicological Aspects Including Pharmacokinetics

Among the naturally occurring calanolides (Figure 2), calanolide A (**1**), a specific nonnucleoside inhibitor of human immunodeficiency virus type 1 (HIV-1) reverse transcriptase, first isolated from a tropical tree *C. lanigerum* that grows abundantly in the Malaysian rain forest, is the most-studied compound in terms of its pharmacology, toxicology and synthesis. A series of animal studies [43] involving mice, rats and dogs established that calanolide A (**1**) is generally well-tolerated at oral doses of up to 150 mg/kg in rats and 100 mg/kg in dogs, and possesses a good safety profile [73,74]. Calenolides A (**1**), B (**4**) and C (**6**) were found to be nontoxic in mice (LD_{50} = 1.99 g/kg), and no alternation on hepatocytes could be observed during the histological study of the mice treated with the highest dose applied [74]. During a study looking at the anti-HIV efficacy and toxicity of calanolides when used in combination with other anti-HIV drugs, no noticeable toxicity could be detected [46].

In the very first study on the safety and pharmacokinetics of calanolide A (**1**) in healthy HIV-negative human volunteers revealed that the toxicity of calanolide A (**1**) was minimal in the majority of subjects treated with four successive single dose, 200, 400, 600 and 800 mg. While there were no acute serious or life-threatening adverse effects were observed, among the usual minor adverse effects, dizziness, oily taste, headache, eructation, and nausea were noticed, but were of minimal clinical significance. These adverse effects were non-dose-dependent [73]. In this study, it was found that calanolide A (**1**) was rapidly absorbed following administration, with time to maximum concentration of drug in plasma (T_{max}) values, depending on the doses, occurring between 2.4 and 5.2 h. It was noted that the levels of calanolide A (**1**) in human plasma were higher than would have been predicted from animal studies, but the safety profile was benign. However, taking calanolide A (**1**) with food was found to generate significant variability in pharmacokinetics, but with no detectable interaction with food. Later, these findings were further confirmed by another similar study carried out by Eiznhamer et al. [75]. Calanolide A (**1**), the first member of the new family of NNRTIs, was found to have long elimination half-life, the benign toxicity profile, to achieve trough plasma levels approximating the EC_{90} of calanolide A (**1**) for HIV-1, to have the potential for twice daily dosing, and to offer the unique HIV-1 resistance profile could make this compound an attractive candidate for further clinical studies. It was reported that after oral administration, (+)-calanolide A (**1**) was generally well tolerated and indication of any safety concern could be observed [48]. Its plasma concentrations in humans were higher than anticipated from animal data. The AUC and C_{max} values increased with increasing dose, and it appeared that therapeutic levels could easily be achieved in humans.

A comparative study on the relative pharamacokinetic parameters and bioavailability of calanolide A (**1**) and its synthetic analogue dihydrocalanolide A (**15**) was reported [76]. This study compared the intravenous pharmacokinetics of the dihydro analog relative to the parent compound, calanolide A (**1**), and determined the relative oral bioavailability of each drug in CD2F1 mice. Both compounds displayed similar pharmacokinetic parameters, but the oral bioavailability of the dihydro analogue was considerably better (almost 3.5-fold) than calanolide A (**1**). Moreover, the relative ability of calanolide A (**1**) and its dihydro analog to change to their inactive epimer forms, (+)-calanolide B (**4**) and (+)-dihydrocalanolide B, respectively, was also determined; while conversion of active calanolides to inactive forms occurred in vitro especially under acidic conditions, no epimers of either compound were observed in plasma of mice after administration of either (+)-calanolide A (**1**) or (+)-dihydrocalanolide A (**15**). It was suggested that the selection of the dihydro derivative of calanolide A (**1**) could be a reasonable choice for further preclinical development and possible Phase I clinical evaluation as an oral drug candidate for the treatment of HIV infection. Calanolide A (**1**) was shown to be distributed readily into the brain and lymph [55]. The distribution and elimination pattern of calanolide A (**1**) and its 7,8-dihydro derivative were found to be similar, but the apparent volume of distribution (Vd) and oral clearance of these compounds were significantly different after oral administration. It was also demonstrated that calanolide A (**1**) is generally well tolerated in doses up to 600 mg. As evident from animal studies, the gastrointestinal intolerance for this compound is not severe, but the most common adverse events as observed in human trails of calanolide A (**1**) include an oily after taste and transient dizziness [55]. The calculated half-life of calanolide A (**1**) from 800 mg dosing was reported to be 20 h [55,73].

During the study directed to the evaluation of antitubercular property of calanolides and their semisynthetic analogues, the pharmacokinetic data indicated that the (+)-calanolide A (**1**) concentrations in plasma could be comparable to the observed in vitro MICs against *M. tuberculosis* [68]. Both calanolides A (**1**) and B (**4**) metabolized by cytochrome P450 CYP3A, and their blood levels could be enhanced if co-administered with ritonavir. Usach et al. [77] reported the safety, tolerability and pharmacokinetics profiles of calanolide A (**1**), as a result of a comprehensive Phase I clinical trial.

6. Therapeutic Potential

Naturally occurring calanolides and their synthetic or semi-synthetic analogs have undergone several pre-clinical and clinical trials for their anti-HIV activity, aiming at novel anti-HIV drug development [2,16,55,78]. In fact, calanolide A (**1**) was at an advanced stage of development as an anti-HIV drug about a decade ago [78]. Buckheit [79] reviewed therapeutic potential of non-nucleoside reverse transcriptase inhibitors like calanolides as anti-HIV and commented on strategies for the treatment modalities for HIV infections. In fact, NNRTIs opened a new avenue of treatment of HIV infections, as previously this therapeutic area was predominantly covered by nucleoside reverse transcriptase inhibitors and protease inhibitors. Soon after the discovery of calanolides as a potential ant-HIV agents by the NCI/NIH, Sarawak Medichem Pharmaceuticals (Sarawak, Malaysia) synthesized calanolide A (**1**) and started developing calanolide A (**1**) as a clinical drug for the treatment of HIV infections. It was a joint venture between the Sarawak State Government and Medichem Research Inc.

During 2001–2005, an interventional clinical trial was conducted on human volunteers [80], where patients were randomized to receive (+)-calanolide A (**1**) or placebo for 21 days. All patients could elect to receive an open-label, 3-month course of approved retroviral therapy (up to triple-drug therapy) to be selected by, and administered under the care of, the patients' physicians. If the patient had no insurance coverage or did not wish to utilize his/her insurance for anti-HIV medications, Sarawak MediChem Pharmaceuticals provided these medications at no charge for up to three months. The trial was primarily aimed at the assessment of the safety and effectiveness of (+)-calanolide A (**1**) in HIV-infected patients who had never taken anti-HIV drugs. In 2006, Craun Research (Kuching, Malaysia), a company established by the Sarawak Government, acquired Sarawak MediChem, and in 2016, Craun Research announced the completion of Phase I clinical trials for calanolide A (**1**) with doses of 200 to 800 mg, which initially started in 2013 [77]. In 2017, F18 (10-chloromethyl-11-demethyl-12-oxo-calanolide A), a synthetic structural analog of calanolide A (**1**) was shown to have more potent anti-HIV activity than original molecule, calanolide A (**1**) [81,82]. This compound showed better druggable profile with 32.7% oral bioavailability in rat, tolerable oral single-dose toxicity in mice, and suppressed both the wild type HIV-1 and Y181C mutant HIV-1 at an EC_{50} of 7.4 nM and 0.46 nM, respectively [83]. Furthermore, it was shown that two enantiomers F18, (*R*)-F18 and (*S*)-F18, had quite similar anti-HIV property, but (*R*)-F18 was more potent than (*S*)-F18 against wild type virus, K101E mutation and P225H mutation pseudoviruses [81]. However, calanolides, particularly calanolide A (**1**) remains as an investigational anti-HIV drug and has not yet been approved by the FDA or any other drugs regulatory bodies for their commercial pharmaceutical production.

7. Patents

In 1999, calanolides and related antiviral compounds were patented by the Board of Trustees of the University of Illinois [84]. The patent covered novel antiviral compounds, calanolides, and their derivatives that could be isolated from plants of the genus *Calophyllum* in accordance with the specified method. The patent also included the uses of these compounds and their derivatives alone or in combination with other antiviral agents in compositions, such as pharmaceutical compositions, to inhibit the growth or replication of a virus, such as a retrovirus, in particular a human immunodeficiency virus, specifically HIV-1 or HIV-2. Later, another patent, owned by Parker Hughes Institute, was reported, which described the novel uses of calanolides as Tec family/BTK (Bruton's tyrosine kinase) inhibitors, methods for their identification, and pharmaceutical compositions [85]. It can be mentioned here that the BTK inhibitors inhibit the enzyme BTK, which is a crucial part of the B-cell receptor signaling pathway, and these inhibitors have emerged as a new therapeutic target in a variety of malignancies, e.g., chronic lymphocytic leukemia and small lymphocytic lymphoma [86].

8. Conclusions

Non-nucleoside reverse transcriptase inhibitors (NNRTIs), efavirenz, nevirapine and delavirdine, have become one of the cornerstones of highly active anti-retroviral therapy for HIV infections. Calanolides, as they belong to this pharmacological class of NNRTIs, and because of their high safety margins and favorable pharmacokinetic profiles, are ideal candidates for novel anti-HIV drug development. While several analogues of the naturally occurring calanolides have been synthesized, a good number of preclinical and clinical trials have been conducted to date, and there are a few patents published, further work is still required to commercially bring any of the calanolide candidates, natural or synthetic, to anti-HIV drug market. As calanolides show an excellent synergistic and additive profile in combination with other anti-HIV drugs, it is assumed that calanolides can be considered for use in combination therapy for HIV infections.

Author Contributions: All authors contributed equally to the data collection and compilation of information. Additionally, L.N. and S.D.S. played the lead role in writing, editing and submission of this manuscript. All authors have read and agreed to the published version of the manuscript.

Funding: This research was funded by the European Regional Development Fund—Project ENOCH (No. CZ. 02.1.01/0.0/0.0/16_019/0000868).

Conflicts of Interest: The authors declare no conflict of interest.

References

1. Kashman, Y.; Gustafson, K.R.; Fuler, R.W.; Cardellina, J.H.; McMahon, J.B.; Currens, M.J.; Buckheit, R.W.; Hughes, S.H.; Cragg, G.M.; Boyd, M.R. HIV inhibitory natural products 7. The calanolides, a novel HIV-inhibitory class of coumarin derivatives from the tropical rain forest tree, *Calophyllum lanigerum*. *J. Med. Chem.* **1992**, *35*, 2735–2743. [CrossRef] [PubMed]
2. Cragg, G.M.; Newman, D.J. Plants as a source of anti-cancer and anti-HIV agents. *Ann. Appl. Biol.* **2003**, *143*, 127–133. [CrossRef]
3. McKee, T.C.; Cardellina, J.H.; Dreyer, G.B.; Boyd, M.R. The pseudocalanolides—Structure revision of Calanolide C. and Calanolide D. *J. Nat. Prod.* **1995**, *58*, 916–920. [CrossRef] [PubMed]
4. Huerta-Reyes, M.; Basualdo, M.D.C.; Abe, F.; Jimenez-Estrada, M.; Soler, C.; Reyes-Chilpa, R. HIV-1 inhibitory compounds from *Calophyllum brasiliense* leaves. *Biol. Pharm. Bull.* **2004**, *27*, 1471–1475. [CrossRef]
5. Bernabé-Antonio, A.; Estrada-Zúñiga, M.E.; Buendía-González, L.; Reyes-Chilpa, R.; Chávez-Ávila, V.M.; Cruz-Sosa, F. Production of anti-HIV-1 calanolides in a callus culture of *Calophyllum brasiliense* (Cambes). *Plant Cell Tissue Organ Cult.* **2010**, *103*, 33–40. [CrossRef]
6. Gomez-Verjan, J.C.; Gonzalez-Sanchez, I.; Estrella-Parra, E.; Reyes-Chilpa, R. Trends in the chemical and pharmacological research on the tropical trees *Calophyllum brasiliense* and *Calophyllum inophyllum*, a global context. *Scientometrics* **2015**, *105*, 1019–1030. [CrossRef] [PubMed]
7. Brahmachari, G.; Jash, S.K. Naturally occurring calanolides: An update on their anti-HIV potential and total syntheses. *Recent Patents Biotechnol.* **2014**, *8*, 3–16. [CrossRef] [PubMed]
8. Brahmachari, G. Naturally occurring calanolides: Chemistry and biology. In *Bioactive Natural Products: Chemistry and Biology*; Brahmachari, G., Ed.; Wiley-VCH: Verlag, UK, 2014; pp. 349–374.
9. McKee, T.C.; Covington, C.D.; Fuller, R.W.; Bokesch, H.R.; Young, S.; Cardellina, J.H.; Kadushin, M.R.; Soejarto, D.D.; Stevens, P.F.; Cragg, G.M.; et al. Pyranocoumarins from Tropical Species of the Genus Calophyllum: A Chemotaxonomic Study of Extracts in the National Cancer Institute Collection1. *J. Nat. Prod.* **1998**, *61*, 1252–1256. [CrossRef]
10. Sunthitikawinsakul, A.; Kongkathip, N.; Kongkathip, B.; Phonnakhu, S.; Daly, J.W.; Spande, T.F.; Nimit, Y.; Napaswat, C.; Kasisit, J.; Yoosook, C. Anti-HIV limonoid: First isolation from Clausena excavate. *Phytotherap. Res.* **2003**, *17*, 1101–1103. [CrossRef]
11. De Clercq, E. Current lead natural products for the chemotherapy of human immunodeficiency virus (HIV) infection. *Med. Res. Rev.* **2000**, *20*, 323–349. [CrossRef]

12. Ishikawa, T. Anti-HIV-1 Active Calophyllum Coumarins: Distribution, Chemistry, and Activity. *Heterocycles* **2000**, *53*, 453. [CrossRef]
13. Ito, C.; Itoigawa, M.; Mishina, Y.; Filho, V.C.; Enjo, F.; Tokuda, H.; Nishino, H.; Furukawa, H. Chemical Constituents of *Calophyllum brasiliense* 2. Structure of Three New Coumarins and Cancer Chemopreventive Activity of 4-Substituted Coumarins. *J. Nat. Prod.* **2003**, *66*, 368–371. [CrossRef] [PubMed]
14. Silva, L.G.; Gomes, K.S.; Costa-Silva, T.A.; Romanelli, M.M.; Tempone, A.G.; Sartorelli, P.; Lago, J.H.G. Calanolides E1 and E2, two related coumarins from *Calophyllum brasiliense* Cambess. (Clusiaceae), displayed in vitro activity against amastigote forms of *Trypanosoma cruzi* and *Leishmania infantum*. *Nat. Prod. Res.* **2020**, 1–5. [CrossRef]
15. Tee, K.H.; Ee, G.C.L.; Ismail, I.S.; Karunakaran, T.; Teh, S.S.; Jong, V.Y.M.; Nor, S.M.M. A new coumarin from stem bark of *Calophyllum wallichianum*. *Nat. Prod. Res.* **2018**, *32*, 2565–2570. [CrossRef]
16. Singh, I.P.; Bharate, S.B.; Bhutani, K.K. Anti-HIV natural products. *Curr. Sci.* **2005**, *89*, 269–290.
17. Gómez-Cansino, R.; Espitia-Pinzón, C.I.; Campos-Lara, M.G.; Guzmán-Gutiérrez, S.L.; Segura-Salinas, E.; Echeverría-Valencia, G.; Torras-Claveria, L.; Cuevas-Figueroa, X.M.; Reyes-Chilpa, R. Antimycobacterial and HIV-1 Reverse Transcriptase Activity of Julianaceae and Clusiaceae Plant Species from Mexico. *Evidence-Based Complement. Altern. Med.* **2015**, *2015*, 1–8. [CrossRef] [PubMed]
18. Garcia-Zebadua, J.C.; Reyes-Chilpa, R.; Huerta-Reyes, M.; Castillo-Arellano, J.I.; Santillan-Hernandez, S.; Vazaquez-Astudillo, B.; Mendoza-Espinoza, J.A. The tropical tree *Calophyllum brasiliense*: A botanical, chemical and pharmacological review. *Vita Rev. Facul. Quimica Farmaceut.* **2014**, *21*, 126–145.
19. Gustafson, K.R.; Bokesch, H.R.; Fuller, R.W.; Cardellina, J.H.; Kadushin, M.R.; Soejarto, D.D.; Boyd, M.R. Calanone, a novel coumarin from *Calophyllum teysmannii*. *Tetrahedron Lett.* **1994**, *35*, 5821–5824. [CrossRef]
20. McKee, T.C.; Fuller, R.W.; Covington, C.D.; Cardellina, J.H.; Gulakowski, R.J.; Krepps, B.L.; McMahon, J.B.; Boyd, M.R. New Pyranocoumarins Isolated from *Calophyllum lanigerum* and *Calophyllum teysmannii* 1. *J. Nat. Prod.* **1996**, *59*, 754–758. [CrossRef]
21. Zou, J.; Jin, D.; Chen, W.; Wang, J.; Liu, Q.; Zhu, X.; Zhao, W. Selective Cyclooxygenase-2 Inhibitors from *Calophyllum membranaceum*. *J. Nat. Prod.* **2005**, *68*, 1514–1518. [CrossRef]
22. Ma, C.-H.; Chen, B.; Qi, H.-Y.; Li, B.-G.; Zhang, G.-L. Two Pyranocoumarins from the Seeds of *Calophyllum polyanthum*. *J. Nat. Prod.* **2004**, *67*, 1598–1600. [CrossRef] [PubMed]
23. Chen, J.-J.; Xu, M.; Luo, S.-D.; Wang, H.-Y.; Xu, J.-C. Chemical constituents of *Calophyllum polyanthum*. *Acta Botan. Yunnanica* **2001**, *23*, 521–526.
24. Spino, C.; Dodier, M.; Sotheeswaran, S. Anti-HIV coumarins from calophyllum seed oil. *Bioorganic Med. Chem. Lett.* **1998**, *8*, 3475–3478. [CrossRef]
25. Yang, S.S.; Cragg, G.M.; Newman, A.D.J.; Bader, J.P. Natural Product-Based Anti-HIV Drug Discovery and Development Facilitated by the NCI Developmental Therapeutics Program. *J. Nat. Prod.* **2001**, *64*, 265–277. [CrossRef] [PubMed]
26. Buckheit, R.W.; White, E.L.; Fliakas-Boltz, V.; Russell, J.; Stup, T.L.; Kinjerski, T.L.; Osterling, M.C.; Weigand, A.; Bader, J.P. Unique Anti-Human Immunodeficiency Virus Activities of the Nonnucleoside Reverse Transcriptase Inhibitors Calanolide A, Costatolide, and Dihydrocostatolide. *Antimicrob. Agents Chemother.* **1999**, *43*, 1827–1834. [CrossRef] [PubMed]
27. Nahar, L.; Sarker, S.D. *Chemistry for Pharmacy Students: General, Organic and Natural Product Chemistry*, 2nd ed.; Wiley and Sons: Chichester, UK, 2019.
28. Gómez-Robledo, H.-B.; Cruz-Sosa, F.; Bernabé-Antonio, A.; Guerrero-Analco, A.; Olivares, J.L.; Alonso-Sanchez, A.; Villafán, E.; Ibarra-Laclette, E. Identification of candidate genes related to calanolide biosynthesis by transcriptome sequencing of *Calophyllum brasiliense* (Calophyllaceae). *BMC Plant Biol.* **2016**, *16*, 177. [CrossRef]
29. Castillo-Arellano, J.I.; Osuna-Fernández, H.R.; Mumbru-Massip, M.; Gómez-Cancino, R.; Reyes-Chilpa, R. The biosynthesis of pharmacologically active compounds in *Calophyllum brasiliense* seedlings is influenced by calcium and potassium under hydroponic conditions. *Bot. Sci.* **2019**, *97*, 89–99. [CrossRef]
30. Altschul, S.F.; Madden, T.L.; Schäffer, A.A.; Zhang, J.; Zhang, Z.; Miller, W.; Lipman, D.J. Gapped BLAST and PSI-BLAST: A new generation of protein database search programs. *Nucleic Acids Res.* **1997**, *25*, 3389–3402. [CrossRef]
31. Crag, G.M.; Newman, D.J. Natural products drug discovery in the next millennium. *Pharm. Biol.* **2001**, *39*, 8–17.

32. Wilson, E.O. What is nature worth? *Wilson Quarter.* **2002**, *26*, 36–37.
33. Jaikumar, K.; Sheikh, N.M.M.; Anand, D.; Saravanan, P. Anticancer activity of *Calophyllum inophyllum* L. ethanolic leaf extract in MCF human breast cell lines. *Int. J. Pharm. Sci. Res.* **2016**, *7*, 3330–3335.
34. Ito, C.; Murata, T.; Itoigawa, M.; Nakao, K.; Kaneda, N.; Furukawa, H. Apoptosis inducing activity of 4-substituted coumarins from *Calophyllum brasiliense* in human leukaemia HL-60 cells. *J. Pharm. Pharmacol.* **2006**, *58*, 975–980. [CrossRef]
35. Omer, A.; Singh, P. An integrated approach of network-based systems biology, molecular docking, and molecular dynamics approach to unravel the role of existing antiviral molecules against AIDS-associated cancer. *J. Biomol. Struct. Dyn.* **2016**, *35*, 1547–1558. [CrossRef] [PubMed]
36. Hanna, L. Calanolide A: A natural non-nucleoside reverse transcriptase inhibitor. *BETA Bull. Exp. Treat. AIDS Publ. San Francisco AIDS Found.* **1999**, *12*, 8–9.
37. Xu, Z.-Q.; Norris, K.J.; Weinberg, D.S.; Kardatzke, J.; Wertz, P.; Frank, P.; Flavin, M.T. Quantification of (+)-calanolide A, a novel and naturally occurring anti-HIV agent, by high-performance liquid chromatography in plasma from rat, dog and human. *J. Chromatogr. B Biomed. Sci. Appl.* **2000**, *742*, 267–275. [CrossRef]
38. Boyer, P.L.; Currens, M.J.; McMahon, J.B.; Boyd, M.R.; Hughes, S.H. Analysis of nonnucleoside drug-resistant variants of human immunodeficiency virus type 1 reverse transcriptase. *J. Virol.* **1993**, *67*, 2412–2420. [CrossRef]
39. Hizi, A.; Tal, R.; Shaharabany, M.; Currens, M.J.; Boyd, M.R.; Hughes, S.H.; McMahon, J.B. Specific inhibition of the reverse transcriptase of human immunodeficiency virus type 1 and the chimeric enzymes of human immunodeficiency virus type 1 and type 2 by nonnucleoside inhibitors. *Antimicrob. Agents Chemother.* **1993**, *37*, 1037–1042. [CrossRef]
40. Buckheit, R.W.; Fliakasboltz, V.; Yeagybargo, S.; Weislow, O.; Mayers, D.L.; Boyer, P.L.; Hughes, S.H.; Pan, B.C.; Chu, S.H.; Bader, J.P. Resistance to 1-[(2-hydroxyethoxy)methyl]-6-(phenylthiol)thymine derivatives is generated by mutations ad multiple sites in the HIV-1 reverse-transcriptase. *Virology* **1995**, *210*, 186–193. [CrossRef]
41. Quan, Y.; Motakis, D.; Buckheit, R.; Xu, Z.-Q.; Flavin, M.T.; Parniak, M.A.; Wainberg, M.A. Sensitivity and resistance to (+)-calanolide A of wild-type and mutated forms of HIV-1 reverse transcriptase. *Antivir. Ther.* **1999**, *4*, 203–209.
42. Xu, Z.-Q.; Hollingshead, M.G.; Borgel, S.; Elder, C.; Khilevich, A.; Flavin, M.T. In vivo anti-HIV activity of (+)-calanolide a in the hollow fiber mouse model. *Bioorganic Med. Chem. Lett.* **1999**, *9*, 133–138. [CrossRef]
43. Xu, Z.-Q.; Flavin, M.T.; Jenta, T.R. Calanolides, the naturally occurring anti-HIV agents. *Curr. Opin. Drug Discov. Dev.* **2000**, *3*, 155–166.
44. Buckheit, R.W.; Kinjerski, T.L.; Fliakasboltz, V.; Russell, J.D.; Stup, T.L.; Pallansch, L.A.; Brouwer, W.G.; Dao, D.C.; Harrison, W.A.; Schultz, R.J.; et al. Structure-activity and cross resistance evaluations of a series of human-deficiency-virus type-1 specific compounds related to oxanthin carboxanilide. *Antimicrob. Agents Chemotherap.* **1995**, *39*, 2718–2727. [CrossRef] [PubMed]
45. Buckheit, R.; Fliakas-Boltz, V.; Russell, J.; Snow, M.; Pallansch, L.; Yang, S.; Bader, J.; Khan, T.; Zanger, M. A Diarylsulphone Non-Nucleoside Reverse Transcriptase Inhibitor with a Unique Sensitivity Profile to Drug-Resistant Virus Isolates. *Antivir. Chem. Chemother.* **1996**, *7*, 243–252. [CrossRef]
46. Buckheit, J.R.W.; Russell, J.D.; Xu, Z.-Q.; Flavin, M. Anti-HIV-1 Activity of Calanolides Used in Combination with other Mechanistically Diverse Inhibitors of HIV-1 Replication. *Antivir. Chem. Chemother.* **2000**, *11*, 321–327. [CrossRef]
47. Budihas, S.R.; Gorshkova, I.; Gaidmakov, S.; Wamiru, A.; Bona, M.K.; Parniak, M.A.; Crouch, R.J.; McMahon, J.B.; Beutler, J.A.; le Grice, S.F.J.; et al. Selective inhibition of HIV-1 reverse transcriptase-associated ribonuclease H activity by hydroxylated tropolones. *Nucleic Acid Res.* **2005**, *33*, 1249–1256. [CrossRef]
48. Xu, Z.Q.; Buckheit, R.W.; Stup, T.L.; Flavin, M.T.; Khilevich, A.; Rizzo, J.D.; Lin, L.; Zembower, D.E. In vitro anti-human deficiency virus (HIV) activity of the chromanone derivative, 12-oxocalanolide A, a novel NNRTI. *Bioorg. Med. Chem. Lett.* **1998**, *8*, 2179–2184. [PubMed]
49. Sorbera, L.A.; Leeson, P.; Castaner, J. Calanolide A: Antiviral for AIDS, reverse transcriptase inhibitor. *Drugs Future* **1999**, *24*, 235–245. [CrossRef]
50. Auwerx, J.; Rodríguez-Barrios, F.; Ceccherini-Silberstein, F.; San-Félix, A.; Velázquez, S.; de Clercq, E.; Camarasa, M.-J.; Perno, C.F.; Gago, F.; Balzarini, J. The Role of Thr139 in the Human Immunodeficiency Virus Type 1 Reverse Transcriptase Sensitivity to (+)-Calanolide A. *Mol. Pharmacol.* **2005**, *68*, 652–659. [CrossRef]

51. Reyes-Chilpa, R.; Huerta-Reyes, M. Natural compounds from plants of the Clausiaceae family: Type 1 human immunodeficiency virus inhibitors. *Interciencia* **2009**, *34*, 385–392.
52. Cardellina, J.H.; Bokesch, H.R.; McKee, T.C.; Boyd, M.R. Resolution and comparative anti-HIV evaluation of the enantiomers of calanolides A and B. *Bioorganic Med. Chem. Lett.* **1995**, *5*, 1011–1014. [CrossRef]
53. Galinis, D.L.; Fuller, R.W.; McKee, T.C.; Cardellina, J.H.; Gulakowski, R.J.; McMahon, J.B.; Boyd, M.R. Structure–Activity Modifications of the HIV-1 Inhibitors (+)-Calanolide A and (−)-Calanolide B1. *J. Med. Chem.* **1996**, *39*, 4507–4510. [CrossRef] [PubMed]
54. Ishikawa, T. Chemistry of Anti HIV-1 Active Calophyllum Coumarins. *J. Synth. Org. Chem. Jpn.* **1998**, *56*, 116–124. [CrossRef]
55. Yu, D.; Suzuki, M.; Xie, L.; Morris-Natschke, S.L.; Lee, K.-H. Recent progress in the development of coumarin derivatives as potent anti-HIV agents. *Med. Res. Rev.* **2003**, *23*, 322–345. [CrossRef] [PubMed]
56. Qiu, K.X.; Xie, H.D.; Guo, Y.P.; Huang, Y.; Liu, B.; Li, W. QSAR studies on the calanolide analogues as anti-HIV-1 agents. *Chin. J. Struct. Chem.* **2010**, *29*, 1477–1482.
57. Peng, Z.-G.; Chen, H.-S.; Wang, L.; Liu, G. Anti-HIV activities of HIV-1 reverse transcriptase inhibitor racemic 11-demethyl-calanolide A. *Acta Pharm. Sin.* **2008**, *43*, 456–460.
58. Sarker, S.D.; Nahar, L. An Introduction to Computational Phytochemistry. *Comput. Phytochem.* **2018**, *1146*, 1–41.
59. Patel, R.D.; Kumar, S.P.; Patel, C.N.; Shankar, S.S.; Pandya, H.A.; Solanki, H. Parallel screening of drug-like natural compounds using Caco-2 cell permeability QSAR model with applicability domain, lipophilic ligand efficiency index and shape property: A case study of HIV-1 reverse transcriptase inhibitors. *J. Mol. Struct.* **2017**, *1146*, 80–95. [CrossRef]
60. Currens, M.J.; Gulakowski, R.J.; Mariner, J.M.; Moran, R.A.; Buckheit, R.W.; Gustafson, K.R.; McMahon, J.B.; Boyd, M.R. Antiviral activity and mechanism of action of calanolide A against the human deficiency virus type-1. *J. Pharmacol. Experim. Therapeut.* **1996**, *279*, 645–651.
61. Ha, M.H.; Nguyen, V.T.; Nguyen, K.Q.C.; Cheah, E.L.C.; Heng, P.W.S. Antimicrobial activity of Calophyllum inophyllum crude extracts obtained by pressurised liquid extraction. *Asian J. Trad. Med.* **2009**, *4*, 141–146.
62. Alkhamaiseh, S.I.; Taher, M.; Ahmad, F. The Phytochemical Contents and Antimicrobial Activities of Malaysian *Calophyllum rubiginosum*. *Am. J. Appl. Sci.* **2011**, *8*, 201–205. [CrossRef]
63. Saravanan, R.; Dhachinamoorthi, D.; Senthikumar, K.; Thamizhvanan, K. Antimicrobial activity of various extracts from various parts of *Calophyllum inophyllum* L. *J. Appl. Pharm. Sci.* **2011**, *1*, 102–106.
64. Adewuyi, A.; Fasusi, O.H.; Oderinde, R.A. Antibacterial activities of acetonides prepared from the seed oils of *Calophyllum inophyllum* and *Pterocarpus osun*. *J. Acute Med.* **2014**, *4*, 75–80. [CrossRef]
65. Léguillier, T.; Lecsö-Bornet, M.; Lemus, C.; Rousseau-Ralliard, D.; Lebouvier, N.; Hnawia, E.; Nour, M.; Aalbersberg, W.; Ghazi, K.; Raharivelomanana, P.; et al. The Wound Healing and Antibacterial Activity of Five Ethnomedical *Calophyllum inophyllum* Oils: An Alternative Therapeutic Strategy to Treat Infected Wounds. *PLoS ONE* **2015**, *10*, e0138602. [CrossRef]
66. Kudera, T.; Rondevaldova, J.; Kant, R.; Umar, M.; Skřivanová, E.; Kokoska, L. In vitro growth-inhibitory activity of *Calophyllum inophyllum* ethanol leaf extract against diarrhoea-causing bacteria. *Trop. J. Pharm. Res.* **2017**, *16*, 2207. [CrossRef]
67. Oo, W.M. Pharmacological properties of *Calophyllum inophyllum*—Updated review. *Int. J. Photochem. Photobiol.* **2018**, *2*, 28–32.
68. Xu, Z.Q.; Barrow, W.W.; Suling, W.J.; Westbrook, L.; Barrow, E.; Lin, Y.M.; Flavin, M.T. Anti-HIV natural product (+)-calanolide A is active against both drug-susceptible and drug-resistant strains of *Mycobacterium tuberculosis*. *Bioorg. Med. Chem.* **2004**, *12*, 1199–1207. [CrossRef] [PubMed]
69. Bueno, J.; Coy, E.D.; Stashenko, E. Antimycobacterial natural products—An opportunity for the Colombian biodiversity. *Rev. Espanola Quimiot.* **2011**, *24*, 175–183.
70. Xu, Z.-Q.; Lin, Y.M.; Flavin, M.T. Method for Treating and Preventing Mycobacterium infections. U.S. Patent 6,268,393, 31 July 2001.
71. Souto, E.B.; Dias-Ferreira, J.; Craveiro, S.A.; Severino, P.; Sanchez-Lopez, E.; Garcia, M.L.; Silva, A.M.; Souto, S.B.; Mahant, S. Therapeutic Interventions for Countering Leishmaniasis and Chagas's Disease: From Traditional Sources to Nanotechnological Systems. *Pathogens* **2019**, *8*, 119. [CrossRef] [PubMed]
72. Ismail, F.M.D.; Nahar, L.; Zhang, K.; Sarker, S.D. Antimalarial and antiparasitic natural products. In *Medicinal Natural Products—A Disease-Focused Approach*; Sarker, S.D., Nahar, L., Eds.; Elsevier: London, UK, 2020; pp. 115–151.

73. Creagh, T.; Ruckle, J.L.; Tolbert, D.T.; Giltner, J.; Eiznhamer, D.A.; Dutta, B.; Flavin, M.T.; Xu, Z.-Q. Safety and Pharmacokinetics of Single Doses of (+)-Calanolide A, a Novel, Naturally Occurring Nonnucleoside Reverse Transcriptase Inhibitor, in Healthy, Human Immunodeficiency Virus-Negative Human Subjects. *Antimicrob. Agents Chemother.* **2001**, *45*, 1379–1386. [CrossRef]
74. César, G.-Z.J.; Gil-Alfonso, M.-G.; Marius, M.-M.; Elizabeth, E.-M.; Ángel, C.-B.M.; Maira, H.-R.; Guadalupe, C.-L.M.; Manuel, J.-E.; Ricardo, R.-C. Inhibition of HIV-1 reverse transcriptase, toxicological and chemical profile of *Calophyllum brasiliense* extracts from Chiapas, Mexico. *Fitoterapia* **2011**, *82*, 1027–1034. [CrossRef]
75. Eiznhamer, D.A.; Creagh, T.; Ruckle, J.L.; Tolbert, D.T.; Giltner, J.; Dutta, B.; Flavin, M.T.; Jenta, T.; Xu, Z.-Q. Safety and Pharmacokinetic Profile of Multiple Escalating Doses of (+)-Calanolide A, a Naturally Occurring Nonnucleoside Reverse Transcriptase Inhibitor, in Healthy HIV-Negative Volunteers. *HIV Clin. Trials* **2002**, *3*, 435–450. [CrossRef]
76. Newman, R.A.; Chen, W.; Madden, T.L. Pharmaceutical Properties of Related Calanolide Compounds with Activity against Human Immunodeficiency Virus. *J. Pharm. Sci.* **1998**, *87*, 1077–1080. [CrossRef] [PubMed]
77. Usach, I.; Melis, V.; Peris, J.E. Non-nucleoside reverse transcriptase inhibitors: A review on pharmacokinetics, pharmacodynamics, safety and tolerability. *J. Int. AIDS Soc.* **2013**, *16*, 1–14. [CrossRef] [PubMed]
78. Singh, I.P.; Bodiwala, H.S. Recent advances in anti-HIV natural products. *Nat. Prod. Rep.* **2010**, *27*, 1781–1800. [CrossRef]
79. Buckheit, R.W. Non-nucleoside reverse transcriptase inhibitors: Perspectives on novel therapeutic compounds and strategies for the treatment of HIV infection. *Expert Opin. Investig. Drugs* **2001**, *10*, 1423–1442. [CrossRef]
80. ClinicalTrials.gov. The Safety and Effectiveness of (+)-Calanolide A in HIV-Infected Patients Who Have Never Taken Anti-HIV Drugs. NIH US National Library of Medicine. 2005. Available online: https://clinicaltrials.gov/ct2/show/NCT00005120 (accessed on 21 August 2020).
81. Zhang, L.L.; Xue, H.; Li, L.; Lu, X.F.; Chen, Z.W.; Liu, G. HPLC enantioseparation, absolute configuration determination and anti-HIV activity of (+/-)-F19 enantiomers. *Yaoxue Xuebao* **2015**, *50*, 733–737.
82. Wu, X.; Zhang, Q.; Guo, J.; Jia, Y.; Zhang, Z.; Zhao, M.; Yang, Y.; Wang, B.; Hu, J.; Sheng, L.; et al. Metabolism of F18, a Derivative of Calanolide A, in Human Liver Microsomes and Cytosol. *Front. Pharmacol.* **2017**, *8*, 479. [CrossRef]
83. Xue, H.; Lu, X.; Zheng, P.; Liu, L.; Han, C.; Hu, J.; Liu, Z.; Ma, T.; Li, Y.; Wang, L.; et al. Highly Suppressing Wild-Type HIV-1 and Y181C Mutant HIV-1 Strains by 10-Chloromethyl-11-demethyl-12-oxo-calanolide A with Druggable Profile. *J. Med. Chem.* **2010**, *53*, 1397–1401. [CrossRef] [PubMed]
84. Boyd, M.R.; Cardellina, J.H., II; Gustafson, K.R.; McMahon, J.B.; Fuller, R.W.; Cragg, G.M.; Kashman, Y.; Soejarto, D. Calanolide and related antiviral compounds, compositions, and uses thereof. U.S. Patent 5,859,049, 12 January 1999.
85. Uckun, F.M.; Sudbeck, E. Calanolides for Inhibiting BTK. Official Gazette of the United States Patents and Trademark Office Patents. U.S. Patent 6,306,897, 23 October 2001.
86. Aalipour, A.; Advani, R.H. Bruton's tyrosine kinase inhibitors and their clinical potential in the treatment of B-cell malignancies: Focus on ibrutinib. *Ther. Adv. Hematol.* **2014**, *5*, 121–133. [CrossRef]

molecules

MDPI

Review

Chalepin and Chalepensin: Occurrence, Biosynthesis and Therapeutic Potential

Lutfun Nahar [1,*], Shaymaa Al-Majmaie [2], Afaf Al-Groshi [2], Azhar Rasul [3] and Satyajit D. Sarker [2,*]

[1] Laboratory of Growth Regulators, Institute of Experimental Botany ASCR and Palacký University, Šlechtitelů 27, 78371 Olomouc, Czech Republic

[2] Centre for Natural Products Discovery, School of Pharmacy and Biomolecular Sciences, Liverpool John Moores University, James Parsons Building, Byrom Street, Liverpool L3 3AF, UK; bio.shaymaa@yahoo.com (S.A.-M.); afaf.gerushi@gmail.com (A.A.-G.)

[3] Cell and Molecular Biology Lab, Department of Zoology, Government College University, Faisalabad 38000, Pakistan; drazharrasul@gmail.com

* Correspondence: drnahar@live.co.uk (L.N.); S.Sarker@ljmu.ac.uk (S.D.S.); Tel.: +44-(0)-1512312096 (S.D.S.)

Abstract: Dihydrofuranocoumarin, chalepin (**1**) and furanocoumarin, chalepensin (**2**) are 3-prenylated bioactive coumarins, first isolated from the well-known medicinal plant *Ruta chalepensis* L. (Fam: Rutaceae) but also distributed in various species of the genera *Boenminghausenia*, *Clausena* and *Ruta*. The distribution of these compounds appears to be restricted to the plants of the family Rutaceae. To date, there have been a considerable number of bioactivity studies performed on coumarins **1** and **2**, which include their anticancer, antidiabetic, antifertility, antimicrobial, antiplatelet aggregation, antiprotozoal, antiviral and calcium antagonistic properties. This review article presents a critical appraisal of publications on bioactivity of these 3-prenylated coumarins in the light of their feasibility as novel therapeutic agents and investigate their natural distribution in the plant kingdom, as well as a plausible biosynthetic route.

Keywords: *Ruta chalepensis*; Rutaceae; chalepin; chalepensin; bioactivity; biosynthesis

Citation: Nahar, L.; Al-Majmaie, S.; Al-Groshi, A.; Rasul, A.; Sarker, S.D. Chalepin and Chalepensin: Occurrence, Biosynthesis and Therapeutic Potential. *Molecules* **2021**, *26*, 1609. https://doi.org/10.3390/molecules26061609

Academic Editor: Maria João Matos

Received: 24 February 2021
Accepted: 12 March 2021
Published: 14 March 2021

1. Introduction

Chalepin (**1**; mol formula: $C_{19}H_{22}O_4$; mol weight 314) and chalepensin (**2**; mol formula: $C_{16}H_{14}O_3$; mol weight 254) (Figure 1) are, respectively, a dihydrofuranocoumarin and a furanocoumarin, with a prenylation at C-3 of the coumarin core structure. These coumarins, as the names imply, were first isolated from *Ruta chalepensis* L. (Fam: Rutaceae), but are also found in other *Ruta* species, e.g., *R. angustifolia* and a few other plants of the genus *Clausena* (Fam: Rutaceae), e.g., *Clausena anisata* (Willd.) Hook. F. ex Benth. [1–4]. While chalepin (**1**), also known as heliettin, is optically active, chalepensin (**2**), also known as xylotenin, does not possess any optical activity. Although these coumarins are rather rare in the sense that there are not many 3-prenylated naturally occurring furanocoumarins reported to date, there are quite a good number of bioactivity studies carried out on these compounds. The present review critically appraises publications on bioactivity of these 3-prenylated furanocoumarins in the light of their feasibility as novel therapeutic agents and covers their natural distribution in the plant kingdom, as well as a plausible biosynthetic route.

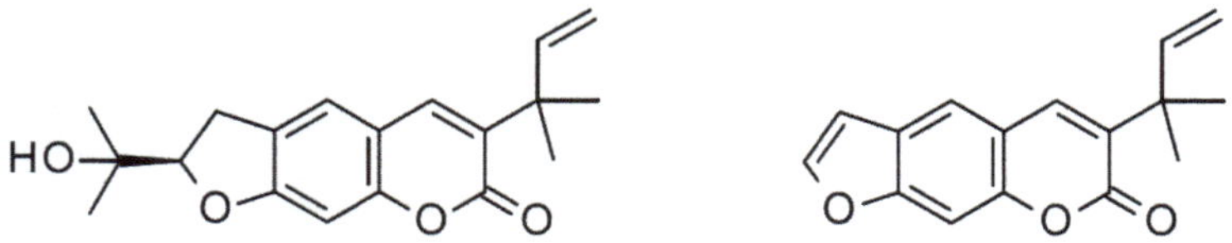

1 **2**

Figure 1. Structures of chalepin (**1**) and chalepensin (**2**).

2. Distribution

First isolated from *Ruta chalepensis* more than half a century ago, chalepin (**1**) and chalepensin (**2**) have been further reported mainly from various species of the genera *Clausena* and *Ruta* of the family Rutaceae [4,5]. It appears that these compounds exclusively occur in the family Rutaceae [1–20], and predominantly within these two genera. However, *Boenminghausenia albiflora* var. *japonica (Hook.)* Rchb. Ex Meisn and *B. sessilicarpa* H. Lev. also produce chalepensin (**2**) [6,20] and this genus is phylogenetically close to the genus *Ruta* [21]. Chalepensin (**2**) was further found in the leaves of *Esenbeckia alata* (Karst and Triana) Tr. and Pl. [9], while *E. grandiflora* Mart. was reported to produce chalepin (**1**) [10]. Interestingly, the genus *Esenbeckia* Kunth. is a part of a small group of phylogenetically distant Rutaceae including the genera *Clausena* and *Ruta*, where 3-prenylated coumarins like **1** and **2** are generally produced [9]. Thus, co-occurrence of these 3-prenylated furanocoumarins in these genera might have some chemotaxonomic implications, at least at the family level, within the family Rutaceae. The distribution of these two coumarins (**1** and **2**) is summarized in Table 1. Within the source plants these compounds are well distributed almost in all parts, leaves, stem, flowers and fruits. Although not chalepensin (**2**) itself, a series of 5-*O*-prenylated chalepensin derivatives were reported from *Dorstenia foetida* Schweinf., a medicinal plant from the family Moraceae, distributed in various countries in the Middle-East Asia [22].

Table 1. Distribution of chalepin (**1**) and chalepensin (**2**) in the plant kingdom.

Plant Names	Family	Chalepin (1)	Chalepensin (2)	References
Boenminghausenia albiflora var. *japonica* (Hook.) Rchb. Ex Meisn.	Rutaceae	−	+	[6]
Boenminghausenia sessilicarpa H. Lev.	Rutaceae	−	+	[20]
Clausena anisata (Willd.) Hook. F. ex Benth.	Rutaceae	+	−	[4,7]
Clausena emarginata C. C. Huang	Rutaceae	+	−	[4,7]
Clausena indica (Dalz.) Oliver	Rutaceae	+	+	[1]
Clausena lansium (Lour.) Skeels	Rutaceae	+	+	[8]
Esenbeckia alata (Karst & Triana) Tr. & Pl.	Rutaceae	−	+	[9]
Esenbeckia grandiflora Mart.	Rutaceae	+	−	[10]
Ruta angustifolia L. Pers	Rutaceae	+	−	[3,11,12]
Ruta chalepensis L.	Rutaceae	+	+	[5,13–15]
Ruta graveolens L.	Rutaceae	+	+	[16,17]
Ruta montana L.	Rutaceae	−	+	[18]
Stauranthus perforatus Liebm.	Rutaceae	+	+	[19]

+ = Found; − = Not found.

3. Biosynthesis

Like all other coumarins, the biosynthesis of chalepin (**1**) and chalepensin (**2**) begins from the simple coumarin umbelliferone, which is formed from the amino acid L-phenylalanine through the formation of *trans*-cinnamic acid, *p*-coumaric acid, 2-hydroxy-*p*-coumaric acid, 2-glucosyloxy-*p*-coumaric acid and 2-glucosyloxy-*p*-*cis*-coumaric acid aided by different enzymes, e.g., cinnamate 4-hydroxylase and 4-coumarate-CoA ligase, 4-coumaroyl 2′-hydroxylase (Figure 2) [23,24]. Sharma et al. [25] studied the biosynthesis of chalepin (**1**) in *Ruta graveolens*. They suggested that 3-(1,1-dimethylallyl)-umbelliferone could be the key intermediate for the biosynthesis of chalepin (**1**), and the dihydrofuran moiety in chalepin (**1**) is formed via prenylation, aided by dimethylallyldiphosphate, at C-6 of the core coumarin skeleton followed by oxidative cyclization with neighboring hydroxyl function at C-7. Generally, prenyltransferases (6-prenyltransferase was identified in *R. graveolens* as a plastidic enzyme) are considered the enzymes involved in the biosynthesis of furano-/dihydrofuranocoumarins through umbelliferone prenylation. Further oxidation of chalepin (**1**) could lead to the formation of the furanocoumarin chalepensin (**2**) in a similar fashion as observed in the conversion of marmesin to psoralen [26]. In fact, biosynthesis of chalepin (**1**) resembles that of 3-prenylated furanocoumarin, rutamarin (acetyl-chalepin) [26]. At this moment, it is not clear from the literature if the prenylation at C-3 takes precedence over that on C-6. In fact, the published information on the biosynthesis of these coumarins **1** and **2** is rather extremely limited, and much work, especially using radioisotopes is much needed to explore other possible routes to the biosynthesis of these compounds.

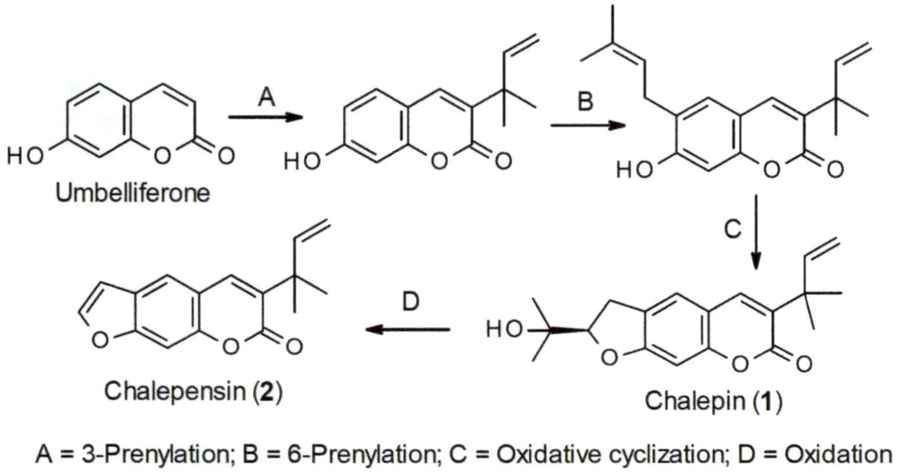

A = 3-Prenylation; B = 6-Prenylation; C = Oxidative cyclization; D = Oxidation

Figure 2. Putative biosynthetic route for the formation of chalepin (**1**) and chalepensin (**2**).

4. Bioactivity

The general, *Clausena* and *Ruta*, the main sources of chalepin (**2**) and chalepensin (**2**), are well known for their uses in traditional medicines, and different studies have established their bioactivities [27,28]. Chalepin (**1**) and chalepensin (**2**) have emerged as two major bioactive components in many of those plants through bioassay-guided isolation protocols, and their bioactivities include antimicrobial, anti-inflammatory, anticancer, antiviral and many more. In this section, using several subsections, a critical appraisal is presented on bioactivities of these two compounds (**1** and **2**) reported in the literature to date (Table 2) [29–51]. Most of the reported bioactivity studies on these compounds involved predominantly in vitro assays and only a handful of in vivo and in silico studies. However, there is no report on any systematic preclinical or clinical trial with these compounds involving human volunteers available in the literature to date.

Table 2. Reported bioactivities of chalepin (**1**) and chalepensin (**2**).

Bioactivity	Chalepin (1)	Chalepensin (2)	References
Antidiabetic	+	NR	[8]
Antifertility	+	+	[13,29]
Antimicrobial	+	+	[4,30–32]
Antiplatelet aggregation	NR	+	[17]
Antiprotozoal	+	+	[5,33–38]
Antiviral	+	NR	[39–41]
Calcium antagonist	NR	+	[6]
Cytotoxicity (potential anticancer and antitumor)	+	+	[3,7,11,12,42]
Spasmolytic	+	NR	[43]
Effect on drug metabolizing enzymes	NR	+	[44–49]
Mutagenicity and other toxicities	+	NR	[50,51]

NR = No report available; + = Active.

4.1. Antidiabetic Activity

Among the bioactive compounds isolated from the stem bark of *Clausena lansium* (Lour.) Skeels, chalepin (**1**) exhibited antidiabetic properties, exerted through dose-dependent stimulated (glucose-mediated) insulin release in vitro from INS-1 cells (rat insulinoma cell line) [8]. Chalepin (**1**) showed 138% insulin secretory response in vitro at the concentration of 0.1 mg/mL. INS-1 cells are widely used as rat islet β-cell models for screening for antidiabetic properties of plant extracts or purified compounds. They express muscarinic M1 and M3 receptors, which are activated by carbachol to promote insulin release. Chalepensin (**2**) does not appear to have gone through any antidiabetic screening yet. It is known that insulin secretion involves a sequence of events in β-cells that lead to fusion of secretory granules with the plasma membrane; it is secreted primarily in response to glucose, while other nutrients such as free fatty acids and amino acids can augment glucose-induced insulin secretion.

4.2. Antifertility Activity

During the assessment of the extracts of *R. chalepensis* var *latifolia* for antifertility activity in rodents, chalepin (**1**) and chalepensin (**2**) were discovered as the major active antifertility principles in the extracts [13]. Despite these compounds showing antifertility activity, most of the tested animals developed cystic and atretic follicles in their ovaries and glomerulocapsular adhesion and segmental fusion in the kidneys. However, no brain toxicity was observed with these compounds. Kong et al. [29] assessed the antifertility activity of the chloroform extracts of the roots, stem and leaves of *R. graveolens* L. in rats and fractionation of the extracts afforded coumarin **2** as the active component with moderate toxicity. Time-dosing experiments showed that this coumarin (**2**) could act at the early stages of pregnancy. The observed antifertility activity of **1** and **2** [13,29] could provide some scientific evidence in support of the traditional uses of *R. chalepensis* as an abortifacient.

4.3. Antimicrobial Property

Antimicrobial assay-guided analysis of a root extract of *Clausena anisata* (Willd.) Hook. f. Benth., a well-known medicinal plant used traditionally for the treatment of parasitic infections, influenza, abdominal pain and constipation, afforded chalepin (**1**) as an antibacterial agent, particularly effective against *Bacillus subtilis* with a zone of inhibition of 16 mm as opposed to 15 mm of the positive control cifrofloxacin [4]. This coumarin was also found active against two other pathogenic bacterial strains, *Staphylococcus aureus* and *Pseudomonas aeruginosa*. Chalepensin (**2**), on the other hand, was reported to possess antifungal property and was found to inhibit the growth of the fungal strains, *Candida albicans* and *Cryptococcus neoformans* [30]. However, interestingly, none of these coumarins showed any antimicrobial activity at tested concentrations (50-100 μg/mL) against a range of microorganisms, e.g., *Bacillus subtilis, Mycobacterium smegmatis, Staphylococcus aureus* and *Candida albicans*, using a modified microtitre-plate assay as reported by El Sayed et al. [52]. Chalepensin (**2**), isolated from *R. chalepensis*, was assessed for antibacterial activity against *Streptococcus mutans* using the method of colony forming units counts in solid medium culture and reduction of tetrazolium salt MTT [3-(4,5-dimethylthiazol-2-yl)-2,5-diphenyltetrazolium bromide] in liquid medium [31] and was shown to significantly inhibit the growth of this bacterial strain with an MIC (minimum inhibitory concentration) of 7.8 μg/mL.

In the most recent study [32] on anti-MRSA (methicillin-resistant *Staphylococcus aureus*) activity of several compounds, mainly coumarins and flavonoids, isolated from *R. chalepensis* grown in Iraq, both chalepin (**2**) and chalepensin (**3**) showed significant antimicrobial activities against the MRSA strains, ATCC 25923, SA-1199B, XU212, MRSA-274819 and EMRSA-15 with MIC values ranging between 32 and 128 μg/mL. In that study, two other furanocoumarins, bergapten and isopimpineline, which do not have a 3-prenylation as in **1** and **2**, were found inactive at tested concentrations. Based on this finding, it was suggested that the prenylation at C-3 of the coumarin nucleus might be a key determinant of anti-MRSA activity. Chalepensin (**2**) was found to be more active than chalepin (**1**) and was subjected to in silico studies to gain an insight into the extent at which this compound (**2**) is able to bind to MRSA proteins and also their drug−like physicochemical characters. In silico studies on compound **2** showed that this compound could have high GI absorption and no violation of the Lipinski rules. It was also shown that chalepensin (**2**) could bind with certain MRSA protein targets, predominantly through hydrogen bonding as well as van de Waals forces. It was suggested that this coumarin could be utilized as a structural template for generating structural analogs and developing potential anti-MRSA therapeutic agents.

4.4. Antiprotozoal Activity

One of the major traditional medicinal uses of *R. chalepensis* and other *Ruta* species is their efficacy as antiparasitic agents [28], particularly as an anthelmintic medication. This traditional medicinal use of *R. chalepensis* has prompted antiparasitic activity screening of its extracts and isolated major compounds, including chalepin (**1**) and chalepensin (**2**). Antiprotozoal activity of chalepin (**1**) and chalepensin (**2**), obtained from *R. chalepensis* following a bioassay-guided protocol, against *Entamoeba histolytica*, which is a causative organism of ameoebiasis, was reported as a meeting presentation, but no further full scientific report was published [33]. Both coumarins showed >90% growth inhibition against *E. histolytica*, an anaerobic parasitic amoebozoan, at a concentration of 150 μg/mL with IC_{50} values of 28.67 and 38.71 μg/mL, respectively, for compounds **1** and **2**. However, in a previous study [5], conducted by the same group, evaluated antiprotozoal activity of plants used in northwest Mexican traditional medicine, particularly *Lippia graveolens* Kunth. and *R. chalepensis*, against *E. histolytica*, and chalepensin (**2**) was found to be the main antiprotozoal component in *R. chalepensis*. Earlier, Kundu and Roy [34] carried out in silico studies involving chalepin (**1**) and glyceraldehyde-3-phosphate dehydrogenase (GAPDH) of the pathogenic protozoa *E. histolytica*. It can be noted that GAPDH is a major glycolytic enzyme (~37 kDA), which catalyzes the sixth step of glycolysis, and an attractive drug target like *E. histolytica* lacks a functional citric acid cycle and exclusively depends on glycolysis for its energy needs. Chalepin (**1**) was predicted as a GAPDH inhibitor and structural modifications offering additional polar interactions were suggested to improve potency.

Trypanosoma cruzi, a species of parasitic euglenoids, characteristically can bore tissue in another organism and feed on blood and lymph, causing diseases like Chagas disease (also known as American trypanosomiasis) in humans, that affects more than 7 million people worldwide, with Latin American countries being most affected. In recent years, a renewed interest has been observed in the search for antitrypanosomal natural products, especially from higher plants. In many in vitro as well as in silico studies, glycosomal glyceraldehyde-3-phosphate dehydrogenase (gGAPDH) from *T. cruzi* has been used as a target molecule for screening compounds for potential antitrypanosomal activity [35]. In an in silico study with various natural products, chalepin (**1**) emerged as a hit molecule for antitrypanosomal drug discovery [36], and subsequently, a series of 3-piperonylcoumarins were synthesized and tested for their inhibitory activity against gGAPDH. Chalepin (**1**) was shown in silico to possess the highest binding affinity to gGAPDH (IC_{50} = 55.5 μM) among the natural coumarins screened and the best inhibitor of gGAPDH [36,37]. Earlier, during an in vitro screening of natural coumarins for trypanocidal or antitrypanosomal activity, chalepin (**1**) was found to be the most active coumarin with an IC_{50} value of 64 μM [38]. However, to the best of our knowledge, there is no report available to date on antitrypanosomal property of chalepensin (**2**).

4.5. Antiviral Activity

Chalepin (**1**), isolated from *R. graveolens*, along with its 28 synthetic analogs were tested for their inhibitory activity on the Epstein–Barr virus (EBV, also known as human herpes virus 4) lytic replication activity [41]. It was noted that most of the synthesized analogs were more active than their parent or precursor, (-)-chalepin (**1**). EVP is a human gamma-herpes virus that infects more than 90% of the human population globally, and preferentially infects B lymphocytes and epithelial cells causing various diseases like Hodgkin's disease, Burkitt's lymphoma, nasopharyngeal carcinoma and gastric carcinoma in humans. Thus, inhibition of EBV lytic replication is considered as one of the pragmatic strategies for the treatment of some these diseases.

Chalepin (**1**), isolated from the leaves of *R. angustifolia*, displayed significant inhibitory activity (IC_{50} = 1.7 μg/mL) against hepatitis C virus replication and was found to be more potent than the positive control ribavirin (IC_{50} = 2.8 μg/mL), a well-known antiviral drug used for the treatment of hepatitis C and other viral diseases [39]. In continuation of their study, they have recently reported enhancement of antihepatitis C virus activity of chalepin (**1**) in combination with conventional antiviral drugs including cyclosporine A, daclatasvir, ribavirin, simeprevir and telaprevir [40]. It was found that chalepin (**1**) could enhance antihepatitis C activities of these conventional drugs with a synergistic combination index of <1. It could be considered as an excellent finding as the need for new and effective drugs for treating hepatitis C is of paramount importance. It can be mentioned that hepatitis C virus infects around 71 million people globally, causes severe liver disease, e.g., liver cancer and deaths; the WHO (World Health Organization) estimated that in 2016, about 400,000 people died from hepatitis C, mainly from liver cirrhosis and liver cancer [53,54].

Like many other antiviral coumarins including some 3-substitued ones, it can be assumed that chalepin (**1**) might offer antiviral activity through inhibition of various proteins that are involved in the transcription/translation processes essential for viral life cycle at different stages, and via modulation of host cell signaling, NF-kB (nuclear factor κB), and inflammatory redox-sensitive pathways and thus blocking viral replication [54]. However, clearly, further research is necessary to understand and establish definite mode of antiviral action mechanism of chalepin (**1**). However, there is no data available on any antiviral property of chalepensin (**2**) to date.

4.6. Cytotoxicity (Potential Anticancer and Antitumor Activity)

Cancer is one of the major causes of human mortality and morbidity. Currently available cancer treatment options or modalities are rather limited, and often suffer from severe side effects. Therefore, the search for new, effective, safe and affordable anticancer drugs is a part of many major modern drug discovery initiatives worldwide. Natural products have long been considered one of the major contributors in the continuing search for new anticancer molecules for safer and more effective anticancer drug development, and evidently, have already provided several successful anticancer drugs, e.g., taxol, vincristine and vinblastine [55]. The most common starting point in the search for anticancer molecules is the screening compounds for cytotoxicity against various human cancer cell lines because cytotoxicity is regarded as one of the major characteristics of anticancer agents. In order to assess anticancer potential of chalepin (**1**) and chalepensin (**2**), cytotoxicity of these compounds has been assessed against different human cancer cell lines in vitro, and some mechanistic studies on how they kill the cancer cells have also been published, showing anticancer and antitumor potential of these compounds (Table 2).

Chalepin (**1**), isolated from *Clausena emarginata* C. C. Huang, has been found to possess significant cytotoxicity against five human cancer cell lines including human leukemia (HL-60), hepatocarcinoma (SMMC-7721), lung carcinoma (A-549), breast cancer (MCF-7) and colon adenocarcinoma (SW-480) with IC_{50} values comparable to that of the positive control, doxorubicin [7]. Chalepin (**1**), isolated from *Ruta angustifolia* Pers., was demonstrated to induce apoptosis through phosphatidylserine externalizations and DNA fragmentation in breast cancer cell line, MCF-7 [12,42]; this compound was considerably cytotoxic to MCF-7 cells, moderately cytotoxic to the epithelial human breast cancer cells (MDA-MB231), but not cytotoxic to normal cells, MRC-5 (Medical Research Council cell strain 5) in the SRB (sulforhodamine B) assay [56]. MRC-5 is a diploid cell culture line comprising fibroblasts, first developed from the lung tissue of a 14-week-old aborted Caucasian male fetus. It can be mentioned here that apoptosis is a process by which cell commit suicide and is eliminated from the system; induction of apoptosis, a cell toxicity pathway, is considered as one of the early-stage mechanism for compounds to exert anticancer activity. This differential cytotoxicity against cancer cells and noncancerous cells might make this compound an ideal candidate, or at least a structural template, for anticancer drug development.

Earlier, in order to understand how chalepin (**1**) could exert its anticancer potential, a study conducted by Richardson et al. [11], revealed that this compound could dose-dependently exhibit cell cycle arrest at S phase, suppress nuclear factor kappa B (NFκB) pathway, signal transducer and activation of transcription 3 phosphorylation and extrinsic apoptotic pathway in human non-small cell lung cancer cell line A-549. Cell cycle analysis using the flow cytometry confirmed that chalepin (**1**) could inhibit cell cycle at S phase (synthesis phase), which is the phase of the cell cycle, where DNA is replicated and occurs between the G_1 and G_2 phases. Since accurate duplication of the genome is essential for successful cell division to take place, the processes involved in the S phase are tightly regulated and widely conserved. A significant accumulation of cells in the S phase was observed after chalepin (**1**) treatment (45 μg/mL) for 48 (accumulation 27.7%) and 72 h (accumulation 25.4%), whereas the accumulation was only about 4% for the untreated cells [11]. It is well known that there is a remarkable link between cell cycle and cancer, as cell cycle appears to be the machinery that controls cell proliferation, and uncontrolled cell proliferation happens in cancer. The suppression of the NF-κB pathway by chalepin (**1**) was shown to be through modulation of the p65 subunit of NF-κB, where the phosphorylation of p65 and the translocation of the p65 subunit to nucleus were inhibited [11]. It can be noted that the NF-κB pathway is generally induced by carcinogens and inflammatory agents. Thus, suppression of NF-κB pathway by chalepin (**1**) could suggest its potential as an anticancer agent.

Caspase 8 is implicated to the activation of the intrinsic apoptotic pathway, and enhancement of caspase 8 activity can be exploited to identify compounds with plausible anticancer activity. In chalepin (**1**) treated cells, a significantly increased level of caspase 8 activity was noticed, when compared to the control; after 48 and 72 h of incubations, chalepin (**1**) (45 μg/mL) enhanced caspase 8 activity, respectively, by 5-fold and 8.6-fold [11].

This group of researchers also demonstrated that chalepin (**1**) and chalepensin (**2**) could induce mitochondrial mediated apoptosis in lung carcinoma cells (A-549), with chalepin (**1**) being more cytotoxic than chalepensin (**2**) [3]; chalepin (**1**) exhibited selective cytotoxicity against A-549 cells with an IC_{50} value of 8.69 μg/mL (27.64 μM). Chalepin (**1**) was mildly toxic to the normal cell line with an IC_{50} value of 23.4 μg/mL. Chalepensin (**2**) exhibited considerable cytotoxic property against A-549 cell line with IC_{50} value of 18.5 μg/mL, while the cytotoxicity (IC_{50} = 23.4 μg/mL) of this coumarin against noncancerous MRC-5 human lung fibroblast cell line was of moderate level as was with chalepin (**1**). Chalepin (**1**) showed morphological changes, typical for apoptosis, e.g., plasma membrane blebbing, cell vacuolization, echinoid spiking, chromatin condensation, formation of apoptotic bodies, cell shrinkage and nuclear fragmentation. Both coumarins (**1** and **2**) were found to downregulate inhibitors of apoptosis such as Bcl-2, survivin, Bcl-xl and cFLIP. They also triggered release of cytochrome c and activated caspases 9 and 3 to induce apoptosis. Chalepensin (**2**) was shown to possess cytotoxicity against colon (H-T29), lung (A-549), breast (MCF-7), kidney (A-498), and pancreatic (PACA-2) cancer cell lines [3].

Wu et al. [17] screened 19 compounds isolated from *Ruta chalepensis*, including chalepensin (**2**), for their potential cytotoxicity against KB (keratin forming tumor), Hela, DLD (colorectal adenocarcinoma) and Hepa tumor cell lines, but chalepensin (**2**) was found to be inactive against any of these cell lines at tested concentrations. From the available literature data, it is obvious that chalepin (**1**) is more cytotoxic than chalepensin (**2**). However, considerably more work has been carried out with chalepin (**1**) than with chalepensin (**2**) to date, and further comparative work may be necessary to gain a better insight into their anticancer potential.

4.7. Miscellaneous Activities

Spasmolytic activities of chalepin (**1**) and a few other coumarins, isolated from *Boenninghausenia albiflora* (Hook.) Rchb. Ex Meisn., were reported by Rizvi et al. [43]. Effects of aqueous extracts of *R. graveolens* and its ingredients, chalepensin (**2**) being one of them, on major drug metabolizing enzymes, cytochrome P450, uridine diphosphate (UDP)-glucuronosyltransferase and reduced nicotinamide adenine dinucleotide (phosphate) (NAD(P)H)-quinone oxidoreductase, were evaluated in mice [15]. The repeated administration of *R. graveolens* extract, rich in rutin and chalepensin (**2**), could induce hepatic CYP1a and CYP2b activities in a dose-dependent fashion. It was observed that male mice were more responsive than female mice to the extract-medicated induction of UGT (uridine glucuronosyltransferase). Earlier, the same group of researchers [48] showed mechanism-based inhibition of CYP1a1 and CYP3A4 by chalepensin (**2**), while this compound was also found to inhibit human CYP1a2, CYP2a13, CYP2c9, CYP2d6 and CYP2e1.

In order to study the in vivo effect of chalepensin (**2**), Lo et al. [44] assessed its effect on multiple hepatic P450 enzymes in C57BL/6JNarl mice, and observed that this coumarin, after oral administration (10 mg/kg) in mice for 7 days, could decrease hepatic coumarin 7-hydroxylation by CYP2a, and increase 7-pentoxyresorufin *O*-dealkylation by CYP2b, without affecting the activities of other CYP enzymes. It was further observed that the suicidal inhibition of CYP2a5 and the constitutive androstane receptor (CAR) mediated CYP2b9/10 induction simultaneously happened in chalepensin (**2**)-treated mice. Previously, Ueng et al. [46,47] and Lo et al. [45] carried out related extensive studies on mechanism-based inhibition of CYP enzymes by chalepensin (**2**) in various in vitro and in vivo models. However, there is no report on such activities of chalepin (**1**) available in the published literature to date.

In a study conducted by Wu et al. [17], chalepensin (**2**) at 100 μg/mL concentration displayed significant antiplatelet aggregation activity, induced by arachidonic acid and collagen. This coumarin, isolated from *Boenninghausenia albiflora* var. *japonica,* was also reported to possess calcium antagonistic property [6].

5. Mutagenicity and Other Toxicities

The mutagenicity of chalepin (**1**) was assessed at the HGPRT locus (AzGr) in Chinese hamster V79 cells [50], and this compound was found to be mutagenic. Chalepin (**1**), isolated from *Clausena aniseta*, a well-known medicinal plant from West Africa, showed anticoagulant (blood-thinning) activity when administered to rats in a single dose [51], and it could depress aniline hydroxylase activity. Ethylmorphine demethylase, hepatic DNA, reduced glutathione and glucose-6-phosphatase were unaffected by chalepin (**1**) treatment at a dose of 50 mg/kg for 3 days prior to sacrifice. This coumarin also resulted in α-1-globulin increase and a decrease in β-globulin content of the serum. Intraperitoneal treatment with chalepin (100 mg/kg) for 2 days caused death of 4 rats out of 100 within 48 h of treatment. Livers of dead rats showed generalized necrosis of hepatocytes. Chalepin (**1**) induced alterations in the serum protein pattern within this period. Liver lesions were observed in chalepin treated animals and were characterized by mild necrosis of hepatocytes. However, no report on mutagenicity of chalepensin (**2**) is available to date.

6. Drugability' of Chalepin (1) and Chalepensin (2)

"Drugability" can simply be defined as the ability of a compound to be used as a pharmaceutical drug. In order for a molecule to be developed as a drug, it must have certain physicochemical characteristics, which can be measured or predicted by various experimental or mathematical models. The Lipinski rule of five, formulated in 1997 by Christopher A. Lipniski, can be used, albeit not conclusively, to predict whether a compound could be an ideal candidate as a drug molecule, i.e., whether a compound possesses "druglikeness" or not [57]. This rule states that an orally active drug does not have more than one violation of the following criteria: a molecular mass less than 500 Daltons, no more than five hydrogen donors, no more than 10 hydrogen bonds and an octanol-water partition coefficient (log *P*) that does not exceed five. Sometimes an additional criterion, "molar refractivity should be between 40–130" is also added to the above rule. If we consider these criteria in relation to chalepin (**1**) and chalepensin (**2**), both compounds tend to follow Lipinski rule of five, and there is no violation of this rule whatsoever (Table 3), which suggests that these compounds possess "druglikeness" or "drugability" and have the potential for further development as commercial drugs. However, it must be noted that this rule of five was originally presented to aid the development of orally bioavailable drugs and was not intended for guiding the medicinal chemistry in the development of all small-molecule drugs. Moreover, there is hardly any reliable experimental bioavailability data available on these coumarins (**1** and **2**) to make any connections between bioavailability and the predicted values for the criteria shown in Table 3.

Table 3. "Druglikeness" of chalepin (**1**) and chalepensin (**2**) *.

Criteria	Chalepin (1)	Chalepensin (2)
Molar mass	314	254
Hydrogen bond donor	1	0
Hydrogen bond acceptors	4	3
Log *P*	3.72	4.32
Molar refractivity	86.6 cm^3	72.5 cm^3
Lipinski rule of 5 violation	0	0

* Data obtained from ChemSpider (www.chemspider.com, (accessed on 24 February 2021)) and DrugBank (https://go.drugbank.com/drugs/DB02205, (accessed on 24 February 2021)).

7. Conclusions

The present work generated the first comprehensive and critical review of published literature on chalepin (**1**) and chalepensin (**2**), revealing various bioactivities of these compounds and their potential as new therapeutic agents. Among the activities, it appeared that antiprotozoal, antiviral and particularly anticancer activities bear promises for these compounds for further consideration for development as therapeutic agents, when considered in the light of nonviolation of the Lipinski rule of five and low level of toxicities. However, there is no report on any systematic preclinical or clinical trial with these compounds involving human volunteers available in the literature to date. Therefore, further studies, including controlled preclinical and clinical trials, are still needed before we can comment on the true therapeutic potential of these compounds.

Author Contributions: All authors contributed equally to collation of relevant information from extensive literature search. Additionally, L.N. and S.D.S. prepared, edited and submitted the manuscript as corresponding authors. All authors have read and agreed to the published version of the manuscript.

Funding: Lutfun Nahar (L.N.) gratefully acknowledges the financial support of the European Regional Development Fund—Project ENOCH (No. CZ.02.1.01/0.0/0.0/16_019/0000868).

Data Availability Statement: All relevant data have been presented as an integral part of this man.

Acknowledgments: Lutfun Nahar gratefully acknowledges the financial support of the European.

Conflicts of Interest: The authors declare no conflict of interest.

References

1. Joshi, B.S.; Gawad, D.H. Isolation of some furanocoumarins from *Clausena indica* and identity of chalepensin with xylotenin. *Phytochemistry* **1971**, *10*, 480–481. [CrossRef]
2. Mohr, N.; Budzikiewicz, H.; El-Tawil, B.A.H.; El-Beih, F.K.A. Further furoquinoline alkaloids from Ruta chalepensis. *Phytochemistry* **1982**, *21*, 1838–1839. [CrossRef]
3. Richardson, J.S.M.; Sethi, G.; Lee, G.S.; Malek, S.N.A. Chalepin: Isolated from *Ruta angustifolia* L. Pers induces mitochondrial mediated apoptosis in lung carcinoma cells. *BMC Complementary Altern. Med.* **2016**, *16*, 389. [CrossRef]

4. Tamene, D.; Endale, M. Antibacterial activity of coumarins and carbazole alkaloids from roots of *Clausena anisata*. *Adv. Pharmacol. Sci.* **2019**, *2019*, 5419854. [CrossRef]
5. Quintanilla-Licea, R.; Mata-Cardenas, B.D.; Vargas-Villarreal, J.; Bazaldua-Rodriguez, A.F.; Angeles-Hernandez, I.K.; Garza-Gonzalez, J.N.; Hernandez-Garcia, M.E. Antiprotozoal activity against *Entamoeba histolytica* of plants used in northeast Mexican traditional medicine. Bioactive compounds from *Lippia graveolens* and *Ruta chalepensis*. *Molecules* **2014**, *19*, 21044–21065. [CrossRef] [PubMed]
6. Yamahara, J.; Miki, S.; Murakami, H.; Matsuda, H.; Fujimura, H. Screening test for calcium antagonists in natural products. The active principles of *Boebminghausenia albiflora* var. *japonica*. *Yakuhaku Zasshi* **1987**, *107*, 823–826. [CrossRef] [PubMed]
7. Hu, S.; Guo, J.-M.; Zhang, W.-H.; Zhang, M.-M.; Liu, Y.-P.; Fu, Y.-H. Chemical constituents from stems and leaves of *Clausena emarginata*. *China J. Chin. Mater. Med.* **2019**, *44*, 2096–2101.
8. Adebajo, A.C.; Iwalewa, E.O.; Obuotor, E.M.; Ibikunle, G.F.; Omisore, N.O.; Adewunmi, C.O.; Obaparusi, O.O.; Klaes, M.; Adetogun, G.E.; Schmidt, T.J.; et al. Pharmacological properties of the extract and some isolated compounds of Clausena lansium sten bark: Anti-trichomonal, antidiabetic, anti-inflammatory, hepatoprotective and antioxidant effects. *J. Ethnopharmacol.* **2009**, *122*, 10–19. [CrossRef]
9. Garcia-Beltran, O.; Areche, C.; Cassels, B.K.; Suarez, L.E.C. Coumarins isolated from *Esenbeckia alata* (Rutaceae). *Biochem. Syst. Ecol.* **2014**, *52*, 39–40. [CrossRef]
10. Oliveira, F.M.; Santana, A.E.G.; Conserva, L.M.; Maia, J.G.S.; Guilhon, G.M.P. Alkaloids and coumarins from Esenbeckia species. *Phytochemistry* **1996**, *41*, 647–649. [CrossRef]
11. Richardson, J.S.M.; Aminudin, N.; Malek, A.; Nurestri, S. A Compound from *Ruta angustifolia* L. Pers exhibits cell cycle arrest at s phase, suppresses nuclear factor-kappa B (NF-kappa B) pathway, signal transducer and activation of transcription 3 (STAT3) phosphorylation and extrinsic apoptotic pathway in non-small cell lung cancer carcinoma (A549). *Pharmacogn. Mag.* **2017**, *13*, S489–S498.
12. Fakai, M.I.; Malek, S.N.A.; Karsani, S.A. Induction of apoptosis by chalepin through phosphatidylserine externalisations and DNA fragmentation in breast cancer cells (MCF7). *Life Sci.* **2019**, *220*, 186–193. [CrossRef]
13. Ulubelen, A.; Ertugrul, L.; Birman, H.; Yigit, R.; Erseven, G.; Olgac, V. Antifertility effects of some coumarins isolated from *Ruta chalepensis* and *R. chalepensis* var. *latifolia* in rodents. *Phytother. Res.* **1994**, *8*, 233–236. [CrossRef]
14. Elbeih, F.K.; Eltawil, B.A.H.; Baghlaf, A.O. Constituents of local plants.12. Coumarin and chalepensin, a further constituent of *Ruta chalepensis* L. *J. Chin. Chem. Soc.* **1981**, *28*, 237–238. [CrossRef]
15. Al-Majmaie, S.; Nahar, L.; Sharples, G.P.; Wadi, K.; Sarker, S.D. Isolation and antimicrobial activity of rutin and its derivatives from *Ruta chalepensis* (Rutaceae) growing in Iraq. *Rec. Nat. Prod.* **2019**, *13*, 64–70. [CrossRef]
16. Orlita, A.; Sidwa-Gorycka, M.; Paszkiewicz, M.; Malinski, E.; Mumirska, J.; Siedlecka, E.M.; Lojkowska, E.; Stepnowski, P. Application of chitin and chitosan as elicitors of coumarins and furoquinoline alkaloids in *Ruta graveolens* l. (Common rue). *Biotechnol. Appl. Biochem.* **2008**, *51*, 91–96. [CrossRef] [PubMed]
17. Wu, T.S.; Shi, L.S.; Wang, J.J.; Iou, S.C.; Chang, H.C.; Chen, Y.P.; Kuo, Y.H.; Chang, Y.L.; Teng, C.M. Cytotoxic and antiplatelet aggregation principles of *Ruta graveolens*. *J. Chin. Chem. Soc.* **2003**, *50*, 171–178. [CrossRef]
18. Ulubelen, A.; Doganca, S. Constituents of the aerial parts of *Ruta montana*. *Fitoterapia* **1991**, *62*, 279.
19. Anaya, A.L.; Macias-Rubalcava, M.; Cruz-Ortega, R.; Garcia-Santana, C.; Sanchez-Monterrubio, P.N.; Hernandez-Bautista, B.E.; Mata, R. Allelochemicals from *Stauranthus perforatus*, a Rutaceous tree of the Yucatan Peninsula, Mexico. *Phytochemistry* **2004**, *66*, 487–494. [CrossRef] [PubMed]
20. Yang, Q.-Y.; Tian, X.-Y.; Fang, W.-S. Bioactive coumarins from Boenninghausenia sessilicarpa. *J. Asian Nat. Prod. Res.* **2007**, *9*, 59–65. [CrossRef]
21. Wei, L.; Xiang, X.-G.; Wang, Y.-Z.; Li, Z.-Y. Phylogenetic relationships and evolution of the Androecia in Ruteae (Rutaceae). *PLoS ONE* **2015**, *10*, e0137190. [CrossRef]
22. Heinke, R.; Franke, K.; Porzel, A.; Wessjohann, L.A.; Ali, N.A.A.; Schmidt, J. Furanocoumarins from *Dorstenia foetida*. *Phytochemistry* **2011**, *72*, 929–934. [CrossRef]
23. Sarker, S.D.; Nahar, L. Progress in the chemistry of naturally occurring coumarins. *Prog. Chem. Org. Nat. Prod.* **2017**, *106*, 241–304. [PubMed]
24. Nahar, L.; Sarker, S.D. *Chemistry for Pharmacy Students: General, Organic and Natural Product Chemistry*, 2nd ed.; Wiley and Sons: Chichester, UK, 2019.
25. Sharma, R.B.; Raj, K.; Kapil, R.S. Biosynthesis of chalepin in Ruta graveolens. *Indian J. Chem. Sect. B—Org. Chem. Incl. Med. Chem.* **1998**, *37*, 247–251.
26. Grundon, M.F. Biosynthesis of aromatic hemiterpenes. *Tetrahedron* **1978**, *34*, 143–161. [CrossRef]
27. Gali, L.; Bedjou, F.; Velikov, K.P.; Ferrari, G.; Donsi, F. High-pressure homogenization-assisted extraction of bioactive compounds from Ruta chalepensis. *J. Food Meas. Charact.* **2020**, *14*, 2800–2809. [CrossRef]
28. Günaydin, K.; Savci, S. Phytochemical studies on *Ruta chalepensis* (LAM) Lamarck. *Nat. Prod. Res.* **2005**, *19*, 203–210. [CrossRef]
29. Kong, Y.C.; Lau, C.P.; Wat, K.H.; Ng, K.H.; But, P.P.; Cheng, K.F.; Waterman, P.G. Antifertility principles of Ruta graveolens. *Planta Med.* **1989**, *55*, 176–178. [CrossRef]
30. Orabi, K.Y.; El Sayed, K.A. Biocatalytic studies of the furanocoumarins angelicin and chalepensin. *Nat. Prod. Commun.* **2007**, *2*, 565–569. [CrossRef]

31. Gomez-Flores, R.; Gloria-Garza, M.A.; de la Garza-Ramos, M.A.; Quintanilla-Licea, R.; Tamez-Guerra, P. Antimicrobial effect of chalepensin against *Streptococcus mutans*. *J. Med. Plants Res.* **2016**, *10*, 631–634.
32. Al-Majmaie, S.; Nahar, L.; Rahman, M.M.; Nath, S.; Saha, P.; Talukdar, A.D.; Sharples, G.P.; Sarker, S.D. Anti-MRSA constituents from *Ruta chalepensis*, and in silico studies on two of most active compounds, chalepensin and 6-hydroxy-rutin 3′,7-dimethyl ether. *Molecules* **2021**, in press. [CrossRef] [PubMed]
33. Quintanilla-Licea, R.; Mata-Cardenas, B.D.; Vargas-Villarreal, J.; Bazaldua-Rodriguez, A.F.; Verde-Star, M.J. Antiprotozoal activity against *Entamoeba histolytica* of furocoumarins isolated from *Ruta chalepensis*. *Planta Med.* **2016**, *82*, 424. [CrossRef]
34. Kundu, S.; Roy, D. Computational study of glyceraldehyde-3-phosphate dehydrogenase of *Entamoeba histolytica:* Implications for structure-based drug design. *J. Biomol. Struct. Dyn.* **2007**, *25*, 25–33. [CrossRef] [PubMed]
35. de Marchi, A.A.; Castilho, M.S.; Nascimento, P.G.B.; Archanjo, F.C.; Del Ponte, G.; Oliva, G.; Pupo, M.T. New 3-piperonylcoumarins as inhibitors of glycosomal glyceraldehyde-3-phosphate dehydrogenase (gGAPDH) from *Trypanosoma cruzi*. *Bioorg. Med. Chem.* **2004**, *12*, 4823–4833. [CrossRef]
36. Menzes, I.R.A.; Lopes, J.C.D.; Montanari, C.A.; Oliva, G.; Pavao, F.; Castilho, M.S.; Vieira, P.A.; Pupo, M.T. 3D QSAR studies on binding affinities of coumarin natural products for glycosomal GAPDH of *Trypanosoma cruzi*. *J. Comput. Aided Mol. Des.* **2003**, *17*, 277–290. [CrossRef] [PubMed]
37. Pavao, F.; Castilho, M.S.; Pupo, M.T.; Dias, R.L.A.; Correa, A.G.; Fernandes, J.B.; da Silva, M.F.G.F.; Mafezoli, J.; Vieira, P.C.; Oliva, G. Structure of *Trypanosoma cruzi* glycosomal glyceraldehyde-3-phosphate dehydrogenase complexed with chalepin, a natural inhibitor, at 1.95 angstrom resolution. *FEBS Lett.* **2002**, *520*, 13–17. [CrossRef]
38. Vieira, P.C.; Mafezoli, J.; Pupo, M.T.; Fernandez, J.B.; da Silva, M.F.G.F.; de Albuquerque, S.; Oliva, G.; Pavao, F. Strategies for the isolation and identification of trypanocidal compounds from the Rutales. *Pure Appl. Chem.* **2000**, *73*, 617–622. [CrossRef]
39. Wahyuni, T.S.; Widyawaruyanti, A.; Lusida, M.; Fuad, A.; Soetjipto, A.; Fuchino, H.; Kawahara, N.; Hayashi, Y.; Aoki, C.; Hotta, H. Inhibition of hepatitis C virus replication by chalepin and pseudane IX isolated from *Ruta angustifolia* leaves. *Fitoterapia* **2014**, *99*, 276–283. [CrossRef]
40. Wahyuni, T.S.; Permanasari, A.A.; Widyawaruyanti, A.; Hafid, A.F.; Fuchino, H.; Kawahara, N.; Hotta, H. Enhancement of anti-hepatitis C virus activity by the combination of chalepin from *Ruta angustifolia* and current antiviral drugs. *Southeast. Asian J. Trop. Med. Public Health* **2020**, *51*, 18–25.
41. Lin, Y.S.; Wang, Q.; Gu, Q.; Zhang, H.A.; Jiang, C.; Hu, J.Y.; Wang, Y.; Yan, Y.; Xu, J. Semisynthesis of (-)-rutamarin derivatives and their inhibitory activity against Esptein-Barr virus lytic replication. *J. Nat. Prod.* **2017**, *80*, 53–60. [CrossRef]
42. Fakai, M.I.; Karsani, S.A.; Malek, S.N.A. Chalepin and rutamarin isolated from *Ruta angustifolia* inhibit cell growth in selected cancer cell lines (MCF7, MDA-MB-231, HT29, and HCT116). *J. Inf. Syst. Technol. Manag.* **2017**, *2*, 8–17.
43. Rizvi, S.H.; Shoeb, A.; Kapil, R.S.; Popli, S.P. Medicinal plants 7. Spasmolytic coumarins from *Boenninghausenia Albiflora*. *Indian J. Pharm. Sci.* **1979**, *41*, 205–206.
44. Lo, W.S.; Lim, Y.P.; Chen, C.C.; Hsu, C.C.; Soucek, P.; Yun, C.H.; Xie, W.; Ueng, Y.F. A dual function of the furanocoumarin chalepensin in inhibiting CYP2a and inducing CYP2b in mice: The protein stabilization and receptor-mediated activation. *Arch. Toxicol.* **2012**, *86*, 1927–1938. [CrossRef] [PubMed]
45. Lo, W.-S.; Chang, Y.-P.; Chen, C.-C.; Chung, Y.-T.; Liu, T.-Y.; Murayama, N.; Yamazaki, H.; Soucek, P.; Chai, G.Y.; Chi, C.W. Metabolism-enhanced inhibition of human and mouse cytochrome P4502A enzymes by chalepensin in vitro and in vivo. *Drug Metab. Rev.* **2011**, *43*, 44–45.
46. Ueng, Y.-F.; Chen, C.-C.; Chung, Y.-T.; Liu, T.-Y.; Chang, Y.-P.; Lo, W.-S.; Murayama, N.; Yamazaki, H.; Soucek, P.; Chau, G.-Y. Mechanism-based inhibition of cytochrome P450 (CYP)2A6 by chalepensin in recombinant systems, in human liver microsomes and in mice in vivo. *Br. J. Pharmacol.* **2011**, *163*, 1250–1262. [CrossRef] [PubMed]
47. Ueng, Y.-F.; Chen, C.-C.; Chung, Y.-T.; Liu, T.-Y.; Chang, Y.-P.; Lo, W.-S. Corrigendum. *Br. J. Pharmacol.* **2012**, *165*, 2808.
48. Ueng, Y.-F.; Chen, C.-C.; Yamazaki, H.; Kiyotani, K.; Chang, Y.P.; Lo, W.S.; Li, D.T.; Tsai, P.L. Mechanism-based inhibition of CYP1A1 and CYP3A4 by the furanocoumarin chalepensin. *Drug Metab. Pharmacokinet.* **2013**, *28*, 229–238. [CrossRef]
49. Ueng, Y.-F.; Chen, C.-C.; Huang, Y.-L.; Lee, I.-J.; Yun, C.-H.; Chen, Y.-H.; Huang, C.-C. Effects of aqueous extracts of *Ruta graveolens* and its ingredients in cytochrome P450, uridine diphosphate (UDP)-glucuronosyltransferase, and reduced nicotinamide adenine dinucleotide (phosphate) (NAD(P)H)-quinone oxidoreductase in mice. *J. Food Drug Anal.* **2015**, *23*, 516–528. [CrossRef]
50. Uwaifo, A.O. Analysis of structure activity relationship in the mutagenicity of aflatoxin B-1 and 4 other furocoumarins in Chinese hamster V-9 cells. *Life Sci. Adv.* **1984**, *3*, 62–70.
51. Emerole, G.; Thabrew, M.I.; Anosa, V.; Okorie, D.A. Structure-activity-relationship in the toxicity of some naturally occurring coumarins-chalepin, imperatorin and oxypeucedanine. *Toxicology* **1981**, *20*, 71–80. [CrossRef]
52. El Sayed, K.; Al-Said, M.S.; El-Feraly, F.S.; Ross, S.A. New quinoline alkaloids from *Ruta chalepensis*. *J. Nat. Prod.* **2000**, *63*, 995–997. [CrossRef]
53. World Health Organization (WHO) Hepatitis C. 2020. Available online: https://www.who.int/news-room/fact-sheets/detail/hepatitis-c (accessed on 30 January 2021).
54. Mishra, S.; Pandey, A.; Manvati, S. Coumarin: An emerging antiviral agent. *Heliyon* **2020**, *6*, e03217. [CrossRef] [PubMed]
55. Sarker, S.D.; Nahar, L.; Miron, A.; Guo, M. *Medicinal Natural Products—A Disease-focused Approach*; Sarker, S.D., Nahar, L., Eds.; Elsevier: London, UK, 2020; Volume 55, pp. 45–75.

56. Vicahi, V.; Kanyawim, K. Sulforhodamine B colorimetric assay for cytotoxicity screening. *Nat. Protoc.* **2006**, *1*, 1112–1116. [CrossRef] [PubMed]
57. Lipinski, C.A.; Lombardo, F.; Dominy, B.W.; Feeney, P.J. Experimental and computational approaches to estimate solubility and permeability in drug discovery and development settings. *Adv. Drug Deliv. Rev.* **2001**, *46*, 3–26. [CrossRef]

Review

Coumarin Derivatives in Inflammatory Bowel Disease

Luiz C. Di Stasi

Laboratory of Phytomedicines, Pharmacology, and Biotechnology (PhytoPharmaTech), Department of Biophysics and Pharmacology, Institute of Biosciences, São Paulo State University (UNESP), 18618-689 Botucatu, SP, Brazil; luiz.stasi@unesp.br

Abstract: Inflammatory bowel disease (IBD) is a non-communicable disease characterized by a chronic inflammatory process of the gut and categorized into Crohn's disease and ulcerative colitis, both currently without definitive pharmacological treatment and cure. The unclear etiology of IBD is a limiting factor for the development of new drugs and explains the high frequency of refractory patients to current drugs, which are also related to various adverse effects, mainly after long-term use. Dissatisfaction with current therapies has promoted an increased interest in new pharmacological approaches using natural products. Coumarins comprise a large class of natural phenolic compounds found in fungi, bacteria, and plants. Coumarin and its derivatives have been reported as antioxidant and anti-inflammatory compounds, potentially useful as complementary therapy of the IBD. These compounds produce protective effects in intestinal inflammation through different mechanisms and signaling pathways, mainly modulating immune and inflammatory responses, and protecting against oxidative stress, a central factor for IBD development. In this review, we described the main coumarin derivatives reported as intestinal anti-inflammatory products and its available pharmacodynamic data that support the protective effects of these products in the acute and subchronic phase of intestinal inflammation.

Keywords: inflammatory bowel disease; coumarin; isocoumarin; Crohn's disease; ulcerative colitis; glutathione; oxidative stress; complementary therapies; intestinal inflammation

Citation: Di Stasi, L.C. Coumarin Derivatives in Inflammatory Bowel Disease. *Molecules* **2021**, 26, 422. https://doi.org/10.3390/molecules26020422

Academic Editor: Maria João Matos
Received: 28 November 2020
Accepted: 23 December 2020
Published: 15 January 2021

1. Introduction

Currently, the search and discovery of new drugs with efficacy, safety, and quality control to prevent and treat non-communicable diseases is a huge challenge for the chemical and pharmaceutical sciences as well as medicine. This task is not only a challenge in the present time but also to guarantee health quality for the next generations. Non-communicable diseases, also known as chronic diseases, are persistent illnesses generally without a pharmacological cure, that tend to be of long duration, requiring long-term and systematic treatment approaches, and the result from multifactorial etiological factors. In general, patients with non-communicable diseases live throughout their lives with several symptoms, continuously using drugs to relieve them. Non-communicable diseases are a group of chronic disorders including cardiovascular diseases, diabetes, multiple sclerosis, obesity, arthritis, asthma, Parkinson's and Alzheimer's diseases, cancer, and inflammatory bowel disease (IBD), which have a high impact on the health system. According to a recent study, 56 million people died in 2017 and non-communicable diseases account for more than 73.4% of these global deaths, i.e., 41.1 million people [1]. Based on this, the search for new drug development to relieve symptoms and mainly to prevent non-communicable diseases is an important approach to improve several world health problems and patients' life quality.

The discovery of new drugs is based on synthetic chemistry, partial synthesis or modification of active molecules of synthetic or natural origin, and bioprospection of natural products, particularly from fungi and plant species. The research with natural products was overshadowed by the advent of the new technologies, synthesis of several chemical active compounds, and international regulatory systems for biodiversity access established by the United Nations Convention on Biological Diversity. However, all current difficulties did not reduce the importance of the world biological biodiversity, particularly from tropical areas, as an inexhaustible and magnificent source of new medicines, which should be carefully and legally studied, respecting international regulatory systems and the traditional knowledge from local communities.

Natural products from plant and fungi origin are the source of several drugs with wide applications and pharmacological importance. Some of these compounds have defined the way of science and modern medicine as well as represented the basis of the treatment of several serious diseases and health problems affecting the world's people. The antibiotic penicillin discovered in *Penicillium* genus fungi; morphine, an opioid compound useful as a pain reliever, isolated of the opium plant (*Papaver somniferum*), and acetylsalicylic acid, a lead compound of the non-steroidal anti-inflammatory drugs, which is related to salicin, obtained from plants belonging genera *Salix* and *Populus*, are some emblematic examples of the natural products that have changed the history of medicine. Even with advanced modern medicine and biotechnology, the most recent discoveries of lead compounds include two products of plant origin, artemisinin and taxol, an antimalarial and an antineoplastic agent, respectively. The research related to the discovery of artemisinin from *Artemisia annua* received the 2015 Nobel Prize in Physiology or Medicine. Artemisinin completely changes the control of malaria and represents a new class of antimalarial drugs, whereas taxol, isolated from species belonging to the genus *Taxus*, particularly *Taxus brevifolia*, represents a new class of anti-cancer drugs. The number of lead compounds obtained from nature is high, showing that natural products play a key role in human health surveillance and represent the support basis of drug research and discovery.

Plant-based products are rich in several chemical classes of compounds, among which the alkaloids, terpenoids, tannins, and phenol and polyphenol compounds stand out, which are potentially useful to prevent and treat several disorders, particularly non-communicable diseases. Phenol and polyphenolic compounds, one of the most important classes of secondary metabolites from plants, include a plethora of different classes of molecules with high pharmacological value, among which the flavonols, flavanones, flavones, anthocyanidins, xanthones, stilbenes, catechins, quinones, and coumarins may be highlighted. These compounds represent an important source of new molecules with several pharmacological properties and are widespread in vegetables commonly consumed daily as dietary foods and spices. Dietary intake of several plants containing these compounds contributes to the plasma bioavailability of active molecules, which are useful both to improve immune response and act as preventative products for several non-communicable diseases. Nowadays, it has been considered that a properly used nutritional approach might be a part of the treatment of non-communicable diseases, particularly patients with Crohn's disease and ulcerative colitis, two chronic inflammatory disorders of the gut [2]. The pharmacological properties of phenol and polyphenol compounds against the inflammatory processes of the gut have been exhaustively reported, focusing on flavonoids [3,4], proanthocyanidins and anthocyanins [5,6], and catechins [7]. However, there is a lack of data and analysis of the potential use and application of coumarin and their derivatives as preventative and curative compounds in non-communicable diseases, particularly in inflammatory bowel diseases (IBDs).

In this review, we aim to update and systematize the available knowledge on the pharmacological activities of coumarin derivatives in the various in vivo experimental models of intestinal inflammation and in vitro studies to provide data and insights to further preclinical, clinical, and molecular studies, demonstrating the main and potential active coumarins useful to prevent or treat inflammatory bowel diseases as well as its main mechanisms of action and signaling pathways.

2. Coumarin and the Main Coumarin Derivatives

Coumarins, also known as benzopyrones, comprise a class of cinnamic acid-derived phenolic compounds found in fungi, bacteria, and plant species, particularly in edible, medicinal, and spice plants from different botanical families [8]. Coumarins are secondary plant heterocyclic metabolites composed of fused benzene and α-pyrone rings (Figure 1), and they occur widely in different parts of plants, such as roots, seeds, nuts, flowers, and fruits either as heterosides or in free form [8,9]. The term coumarin originated from de name "cumaru" the local name for the Brazilian teak plant (*Dipteryx odorata* Wild.) from the Fabaceae botanical family, in the traditional medicine of the Brazilian Amazon forest. *Dipteryx odorata* is an endemic plant of Central America and the North of South America, widespread in the Amazon Forest region, from which the coumarin was firstly isolated by Vogel in 1820 [10]. Its seeds, named tonka beans, are the natural source of coumarin, a compound widely used by the perfumery companies to replace vanilla, particularly as a fixative and enhancing agent in perfumes as well as added to toilet soap, detergents, toothpaste, tobacco, and alcoholic products [9]. Moreover, the isocoumarin, also recognized as 1*H*-benzopyran, is an isomer of the basic structure of coumarin, in which the orientation of the lactone ring is reversed (Figure 1). From the different substituted groups on the basic structure of isocoumarin, several subclasses and isocoumarin derivatives are also found in plant species such as paepalantine, capillarin, and thumberginol A (Figure 1).

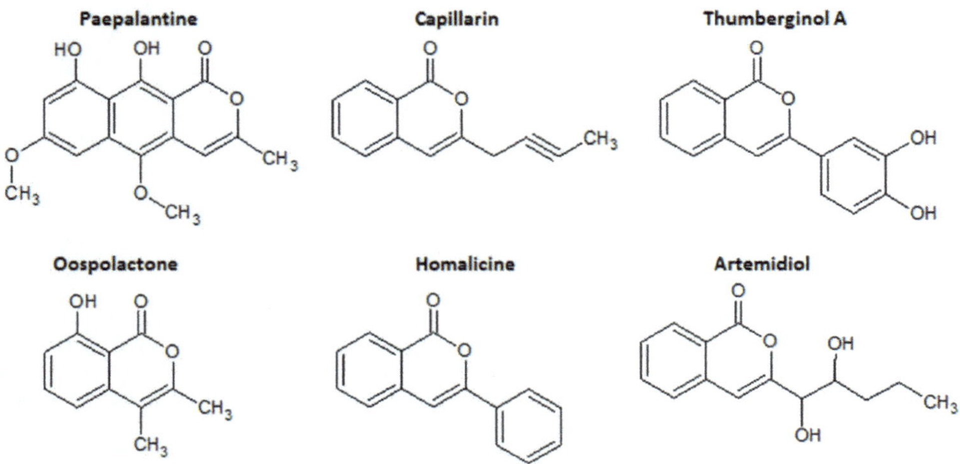

Figure 1. Basic structures of coumarins and isocoumarins and the main isocoumarin derivatives. Chemical structures were drawn using ACD/ChemSketch software.

Coumarins are categorized into four main subtype classes of compounds: simple coumarins, furanocoumarins, pyranocoumarins, and the pyrone-substituted coumarins. Simple coumarins are composed of molecules with hydroxyl, alkoxyl, and alkyl substitution patterns on the basic structure and their glucosides [11]. Simple coumarin class represents the main class of coumarin derivatives with intestinal anti-inflammatory properties, particularly esculetin, esculin, 4-hydroxycoumarin, osthole, and 4-methylesculetin (Figure 2).

Furanocoumarins are composed of coumarins derivatives in which a furan ring is fused with the basic structure of coumarin via C6-C7 or C7-C8 [10,11], generating linear furanocoumarins (fusion via C6-C7) such as psolaren, imperatorin, and xanthotoxin or angular furanocoumarins (fusion via C7-C8) such as isobergapten and angelicin (Figure 3). Similarly, in the pyranocoumarins, a six-membered pyran ring is fused with the benzene ring of the basic structure of coumarins via C6-C7 or C7-C8 [10,11]. Decursin (a linear pyranocoumarin) and seselin (an angular pyranocoumarin) are some examples of pyranocoumarins (Figure 3), which have no intestinal anti-inflammatory activity.

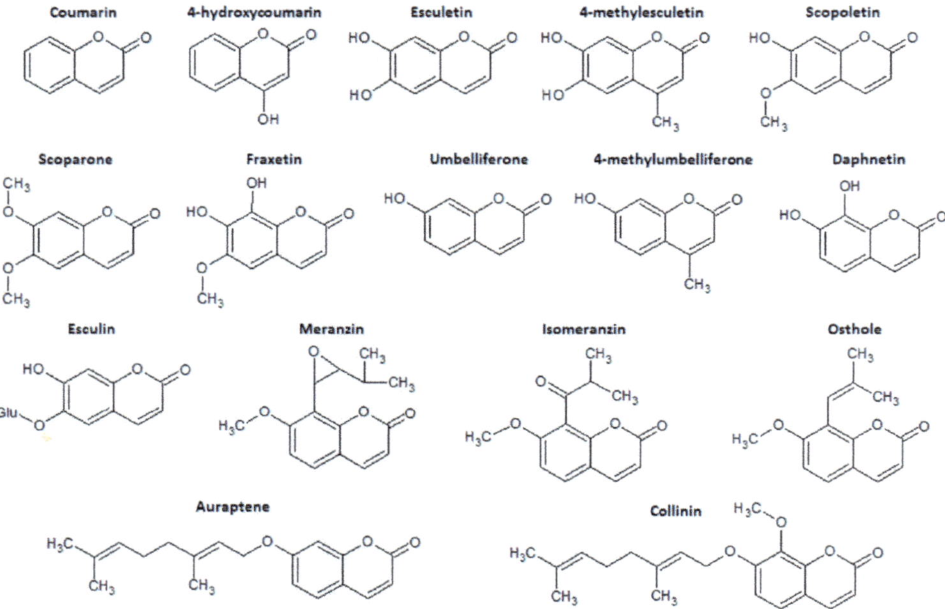

Figure 2. Chemical structures of the main simple coumarin derivatives with intestinal anti-inflammatory activity. Chemical structures were drawn using ACD/ChemSketch software.

Finally, pyrone-substituted coumarins are coumarin derivatives containing different chemical radicals fused with the pyran ring of coumarin [10,11]. Pyrone-substituted coumarins include natural and synthetic coumarins such as warfarin and dicoumarol (Figure 3). Pyrano-substituted coumarins have no intestinal anti-inflammatory activity but comprise some compounds with high pharmacological relevance. Warfarin is a derivative of dicoumarol, a pyrano-substituted coumarin isolated from hay species (*Melilotus alba* and *Melilotus officinalis*) after natural oxidation by several fungi, mainly *Penicillium nigricans* and *Penicillium jensi* found in moldy hay [12]. Dicoumarol and warfarin were first used as rodenticides due to their ability to promote internal hemorrhage in rodents [12]. The anticoagulant properties of dicoumarol and warfarin were the basis for the development of anticoagulant drugs to prevent stroke in patients with cardiovascular diseases, mainly atrial fibrillation and valvular heart disease, and to prevent and treat vein thrombosis and pulmonary embolism [13].

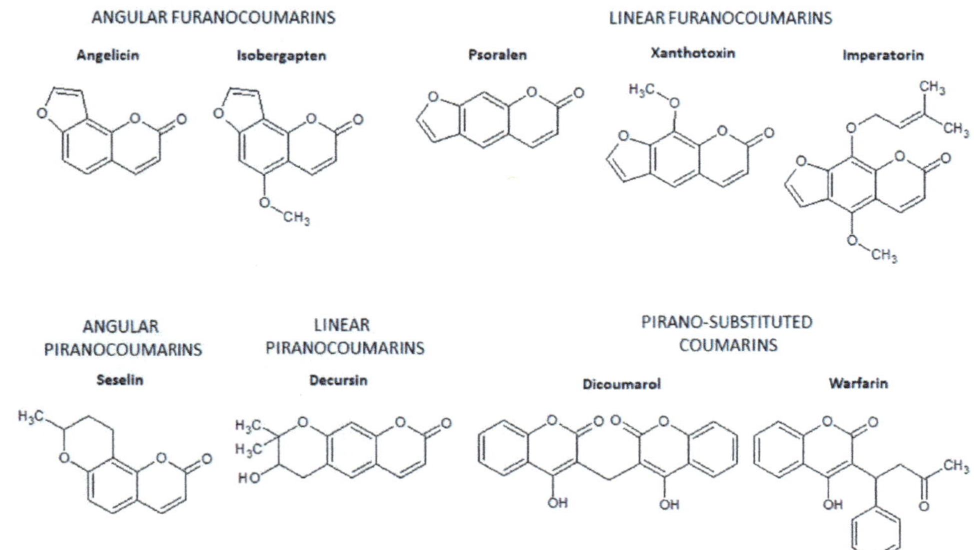

Figure 3. Chemical structures of angular and linear furanocoumarins, angular and linear pyranocoumarins, and pyrano-substituted coumarins. Chemical structures were drawn using ACD/ChemSketch software.

3. Inflammatory Bowel Diseases: General Aspects

Inflammatory bowel disease (IBD) consists of Crohn's disease (CD) and ulcerative colitis (UC), two relapsing chronic inflammatory processes of the gastrointestinal tract, which are part of a group of immune-mediated inflammatory diseases, without a definitive pharmacological treatment and cure [14]. Patients with CD or UC live with several harmful effects in their daily physical, social, and professional activities because these diseases produce limiting effects such as changes in intestinal habits with several evacuations, abdominal pain, diarrhea, bleeding, perianal fistulas and other extraintestinal manifestations.

IBD is a disease that is increasing globally, affecting some 6 to 8 million people in the world and presenting a prevalence rate of 84.3 people (79.2 to 89.9) per 100,000 population in 2017 [15]. Although it is a disorder with low mortality and with a death rate of 0.51 per 100,000 population, IBD is growing exponentially around the world, showing the highest prevalence rates in North America and the United Kingdom and other European countries such as Norway, Poland, and Slovakia, whereas lower prevalence rate has been reported in several countries of Africa, South America, and Southeast Asia [15]. There is a direct relationship between the high prevalence rate of IBD and the industrialization level of a specific country, but the prevalence and incidence rates notably are also rising in newly industrialized countries [16]. The prevalence and incidence rate increment in developing or newly industrialized countries is associated with the industrialization process and migration of population from rural to urban areas, which promote changes in the lifestyle and the people choices related to diet, daily activities, and social behaviors. These changes suggest rates of IBD prevalence and incidence should thrive in parallel to those in the industrialized countries [16].

The IBD etiology is unclear, but several triggering factors have been related to its occurrence and development, including dysregulated immune response, dysfunctional intestinal barrier function, genetic, and environmental aspects. Currently, the pathogenesis of IBD involves a dysregulated autoimmune response and increased intestinal permeability related to gut dysbiosis, which is accelerated by exposure to environmental factors in individuals who have a pre-existing high-risk genotype [17]. The complex and inexact etiology of IBD is a limiting factor in the discovery of new pharmacological and complementary therapies, and the development of preventive strategies useful to define a general protocol of treatment and definitive management of IBD patients. The multifactorial aspects of IBD also explain the high frequency of patients who are refractory to current pharmacological treatments, including conventional drugs such as aminosalicylates, glucocorticoids, immunosuppressants, and biological therapies based on monoclonal antibodies [18]. Current pharmacological treatment of IBD is based on the relief or to create a time of deep remission of symptoms. However, the long-term use of these drugs, produces serious side effects, reducing patient adherence to pharmacological treatment.

IBD also includes a lot of risk factors with an imbricated relationship among them. A series of interactions among risk factors, which do not act in isolation because none of the risk factors alone is sufficient for IBD development, have been suggested [19]. Risk factors for IBD development include intrinsic and extrinsic factors. The intrinsic factors involve genetic predisposition and familiar history as well as the intestinal microbiota, whereas extrinsic risk factors include smoking, appendectomy, hygiene, infections, use of antibiotics and other drugs such as NSAIDs (Non-steroidal anti-inflammatory drugs) and oral contraceptives, a diet with lower fiber and higher fat, vitamin D deficiency as well as lifestyle and social behavior, mainly high stress, sleep privation and lower physical activity [19]. Genetic and epigenetic studies have been extensively used as an important source of data, which are important for a better prediction of IBD course, identification of loci and candidate genes yielding valuable insights into the pathogenesis of IBD and disease pathways, which can be relevant in the clinical practice [20,21].

Intestinal microbiota, which has a key role in the pathogenesis of IBD, is an intrinsic factor that can be modulated by a series of products able to differentially affect distinct microorganisms, including functional food products, mainly probiotic, prebiotic, and symbiotic, natural products such as polyphenol compounds and standardized phytomedicines. In health conditions, intestinal microbiota via fermentation of dietary components produces a series of metabolites, mainly short-chain fatty acids (SCFAs), which are a source of energy for colonocytes and bacteria, and play several protective effects in the body after prompt absorption (Figure 4). On the other hand, the management of extrinsic factors, including changes in lifestyle, social behavior, and diet options as well as a lower exposition to other extrinsic factors can represent an important approach to reduce IBD development.

The combinatory action among genetic predisposition, external environmental factors, and intestinal microbiota is essential to the development of the dysregulated immune response and dysfunctional intestinal barrier [22], which are responsible for IBD development (Figure 4). Both CD and UC patients exhibit a dysfunctional intestinal epithelial barrier with increased permeability as well as an exaggerated immune response in the gastrointestinal tract towards the intestinal microbiota, which is not appropriately controlled, leading to intestinal inflammation [3,22]. The increase of intestinal permeability has been recognized as an early feature of the intestinal inflammatory process, which reduces the intestinal barrier function, a key factor to maintain intestinal homeostasis [23]. In this process, several factors and mediators are involved, particularly the zonulin pathway activation, a key process to control intestinal permeability, suggesting zonulin as a biomarker of gut permeability as well as a key target for the action of intestinal anti-inflammatory products [23–25]. The dysregulated immune response includes an innate immune response, the first-line defense against any damage promoted by pathogens, which is mediated by different cell types including macrophages, neutrophils, monocytes, dendritic, epithelial, and endothelial cells [3,22]. These cell types are responsible for phagocytosis, elimination of pathogens, production of several cytokines, and development of barrier and transport functions (Figure 4). The response to pathogens includes prompt participation of the antigen-presenting cells (APCs), which mediate the differentiation of T-cells into effector T helper (Th) cells, including Th1, Th2, and Th17 cell types, and regulator T-cells (Treg) (Figure 4), which are constituents of the adaptative immune response [3,22]. These different cell types are responsible for the synthesis and dysregulated release of a series of immunologic and inflammatory mediators with wide importance in the pathogenesis of

IBD, including a plethora of chemokines and cytokines. The pathophysiology of intestinal inflammation is very complex, including a wide number of signaling pathways and endogenous mediators as illustrated in Figure 4, where are particularly indicated the main targets for the action of coumarin derivatives.

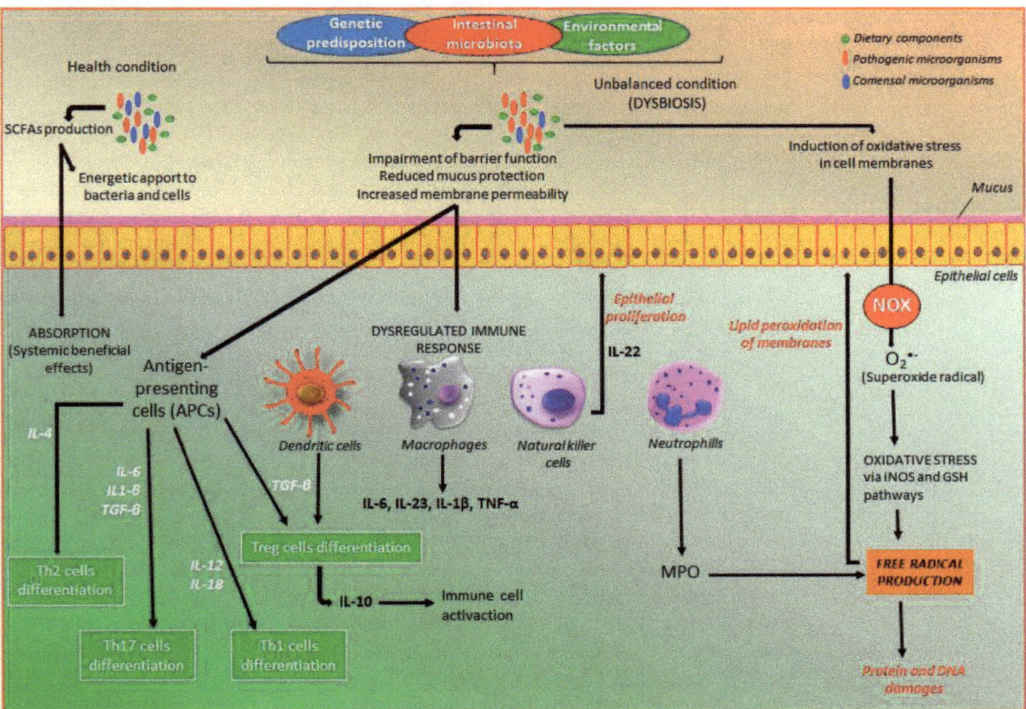

Figure 4. The main pathways of the intestinal inflammatory process or the action of coumarin derivatives. GSH, glutathione; IL, interleukin; iNOS, inducible nitric oxide synthase; MPO, myeloperoxidase; SCFAs, short-chain fatty acids; TGF, transforming growth factor; Th, T helper cells; Treg, T regulatory cells; TNF, tumor necrosis factor.

Moreover, in the gastrointestinal tract, there is homeostasis of intestinal microbiota with immune cells responsible for a balance of host defense and immune tolerance, which can be shift leading to a dysbiosis process that plays a key role in the pathogenesis of IBD [17]. Although dysregulated immune response associated with both genetic predisposition and intestinal microbiota participates in the intestinal inflammatory process, oxidative stress characterized by an excessive release of reactive oxygen and nitrogen species (ROS/RNS) plays a key role in IBD pathogenesis [26,27]. The excess of oxidative mediators reacts with cell membrane fatty acids, proteins, and DNA impairing their functions (Figure 4). The production of superoxide, the main source of free radicals, is required to kill bacteria, a process that occurs especially in neutrophils and other cell types such as epithelial cells from the intestine. From superoxide production, several free radicals are produced via nitric oxide synthase and glutathione-related enzymes (Figure 4).

4. Intestinal Anti-Inflammatory Activity of Coumarin Derivatives

The dysregulated immune and oxidative response triggered by intestinal inflammation is an imbricated and complex interaction involving a series of endogenous mediators from different signaling pathways and receptors, such as nuclear factor-kappa b (NF-κB), nuclear factor erythroid 2 (NEF2)-related factor 2 (Nrf2), peroxisome proliferator-activated receptor gamma (PPAR-γ), pregnane X receptor (PXR), hypoxia-inducible factor (HIF), several enzymes, especially cyclooxygenase 2 (COX-2), mitogen-activated protein kinases (MAPKs), and HIF-prolyl hydroxylases (PHDs) as well as mediators of intestinal epithelial barrier function such as zona occludens 1 (ZO-1), occludin, mucins (MUC1, MUC2, MUC 3, MUC4), and E-cadherin. These different pathways will be discussed in this review to explain the effects of some coumarin derivatives and their partially elucidated mechanisms of action. However, until now the majority of coumarin derivatives produce intestinal anti-inflammatory activity acting as modulators of oxidative stress and immune response, whereas a little number of coumarin derivatives were reported to act by other signaling pathways, which are also partially involved in oxidative stress modulation.

The intestinal anti-inflammatory activity of different coumarin derivatives was described using different experimental models of intestinal inflammation, mainly trinitrobenzene sulphonic acid (TNBS) and dextran sodium sulfate (DSS) in rats or mice as well as several in vitro studies with distinct cell types. Although coumarin is a class of natural and synthetic compounds with high chemical diversity, the intestinal anti-inflammatory has been limited to three subclasses of coumarin derivative, i.e., isocoumarins having one active compound namely paepalantine, a lot of simple coumarin derivatives, and the furanocoumarin imperatorin.

4.1. *Effects of Coumarin Derivatives on Oxidative Stress*

A recent review reported that different antioxidant compounds act by six general mechanisms: inhibiting free radical production by activated oxygen metabolites, changing the structural organization of free radical, producing a local decrease of oxygen concentration, interacting with organic radicals, chelating metal ions, and converting peroxides to stable and inactive products [28]. These processes reduce the availability of free radical species, improving the response against oxidative stress, which are the main properties of the coumarin derivatives with intestinal anti-inflammatory activity.

Oxidative stress is considered an imbalance in which excessive levels of oxygen free radicals such as reactive oxygen species (ROS) and reactive nitrogen species (RNS) are present in the biological system in the face of inadequate availability of the antioxidants, which are capable to destroy these harmful products from metabolic processes [29]. A free radical is a molecule of the normal metabolism of oxygen with an unpaired or odd number of electrons, which is highly reactive and able to react with lipids from the cell membranes, proteins, and nucleic acids, affecting their structure and functions. In an inflammation process, ROS and RNS production is a prompt defense response of the body to kill several invading pathogens as well as to regulate the immune response via pro-inflammatory chemotaxis induction into the site of inflammatory processes as well as modulate the interactions and activation of the immune cell types [30]. However, when occurs an imbalance between free radical production and endogenous antioxidant response, the reaction of reactive oxygen and nitrogen species with host lipids, proteins, and nucleic acids generate oxidative stress, triggering a series of molecular and cellular events such as tissue damage and fibrosis [30], which are related to the origin and development of several chronic diseases, including IBD.

The reactive oxygen species and the endogenous mediators of antioxidative response are represented in Figure 5. The main reactive species in the biological system responsible for oxidative stress is the superoxide radical anion ($O_2^{\bullet-}$), which is formed by the addition of a single electron in molecular oxygen (O_2) by the action of reduced nicotinamide adenine dinucleotide phosphate (NADPH) oxidase (NOX) enzyme [26,27]. $O_2^{\bullet-}$ is the source of hydrogen peroxide (H_2O_2), formed by the action of superoxide dismutase (SOD) enzyme, which catalyzes the dismutation of $O_2^{\bullet-}$ into H_2O_2. Although H_2O_2 is not a free radical, it is a molecule highly reactive [29,30] and represents a key substrate for free radical production. H_2O_2 via Fenton reaction generates hydroxyl radical (OH^-), which is the most reactive and harmful metabolite of oxygen metabolism. Simultaneously, H_2O_2 by the action of myeloperoxidase (MPO) produces hypochlorous acid (HOCl), a potent antioxidant with antimicrobial activity and useful as a defense against several infectious pathogens [30,31].

MPO is the most toxic enzyme found in the granules of neutrophils and monocytes, two important cell types that participate in the intestinal inflammatory process with a key role in the innate immune response to pathogens [30]. MPO generates reactive intermediates, inducing oxidative peroxidation of lipids, and proteins and DNA damage. During the inflammatory response, MPO is released from neutrophils and monocytes, catalyzing the formation of HOCl from H_2O_2 (Figure 5). HOCl produced by the action of MPO influences the conversion of glutathione (GSH) to oxidized glutathione (GSSG), which disrupts the cellular redox balance, reducing the antioxidant GSH pool (Figure 5) and increasing the susceptibility to oxidative stress [30]. Generally, MPO activity is strongly increased in experimental models of intestinal inflammation, whereas simultaneously is possible to observe a depletion of GSH [32].

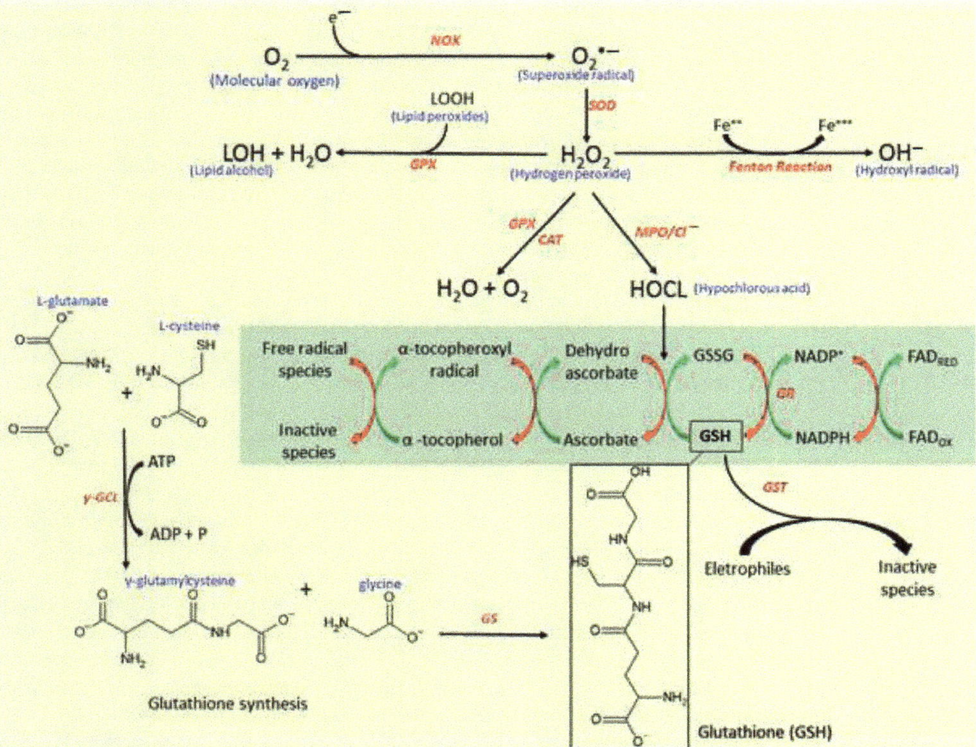

Figure 5. Free radical production and glutathione (GSH) antioxidant system. CAT, catalase; GPX, glutathione peroxidase, GR, glutathione reductase; GS, glutathione synthetase; GST, glutathione S-transferase; NOX, NADPH oxidase; MPO, myeloperoxidase; SOD, superoxide dismutase; γ-GLC, γ-glutamylcysteine ligase. In these oxidative processes, the following activities were reported: A. Counteraction of GSH depletion by paepalantine, coumarin, 4-hydroxycoumarin, esculetin, 4-methylesculetin, daphnetin, esculin, scopoletin, scoparone, and fraxetin; B. Inhibition of MPO activity by paepalantine, esculetin, 4-methylesculetin, daphnetin, and esculin; C. Scavenging activity of free radical by paepalantine, daphnetin, esculin, scopoletin, scoparone, and fraxetin; D. Inhibition of lipid peroxidation by esculetin, and daphnetin; E. Inhibition of GPX activity and expression by 4-methylesculetin; F. Increase of GST and GR activity and expression by 4methylesculetin. Chemical structures were drawn using ACD/ChemSketch software.

GSH, a tripeptide formed by glutamic acid, cysteine, and glycine, is the key endogenous antioxidant that participates in antioxidant response. GSH is produced by a reaction with two steps (Figure 5). Firstly, a residue of glutamic acid (Glu) binding with cysteine via γ-glutamylcysteine ligase (γ-GCL) action, producing γ-glutamylcysteine [33]. Secondly, γ-glutamylcysteine reacts with a residue of glycine under the action of glutathione synthetase (GS) enzyme to produce GSH [33]. GSH participates in antioxidant response acting as a free radical scavenger, reducing dehydroascorbate to ascorbate, which regenerates α-tocopherol from the α-tocopherol radical oxidation, and serving as a co-substrate for several antioxidant enzymes, mainly glutathione S-transferase (GST) and glutathione peroxidase (GPx) [34]. High pools of GSH is essential to modulate oxidative stress, and GSH also can be regenerated by oxidation of its disulfide-oxidized dimer (GSSG) by the action of glutathione reductase (GR) (Figure 5). This reaction occurs at expense of the reduction of NADPH from the pentose phosphate pathway [34].

The modulation of oxidative stress for the body's defense against free radical tissue damage and macromolecules oxidation is modulated by antioxidants, which are classified into nonenzymatic antioxidants consisting of micronutrient components, and enzymatic endogenous system [17,29]. The nonenzymatic antioxidant system includes several small molecules, mainly GSH, vitamin E, vitamin C, β-carotene, retinol, uric acid, and ubiquinol as well as several microelements like selenium, iron, zinc, copper, and manganese [29]. Vitamins act as donors and acceptors of ROS and the micronutrients act as cofactors, which regulate the activities of the antioxidant enzymes [29]. Endogenous enzymatic antioxidants involve superoxide dismutase (SOD), catalase (CAT), glutathione peroxidase (GPX), glutathione S-transferase (GST), glutathione reductase (GR), and γ-glutamyl transferase (γ-GT) enzymes [32], which act by different pathways to reduce the free radical availability and control oxidative stress.

In this process, CAT can break down two H_2O_2 molecules generating one molecular oxygen and two molecules of water [35], reducing H_2O_2 pool availability, which is used as a substrate for the production of OH^- (Figure 5). In association with SOD and CAT, GPX, a dependent enzyme of micronutrient selenium, plays an important role in the reductions of H_2O_2 and lipid peroxides (LOOH) to produce water and lipid alcohol (LOH), contributing to modulation of oxidative stress and avoiding direct tissue damage [36]. Moreover, GST can catalyze the transfer of a GSH group to organic and inorganic electrophiles, reducing these compounds into unreactive products [34].

In another reaction pathway (Figure 6), nitric oxide synthase (NOS) controls the reaction of $O_2^{\bullet -}$ with nitric oxide radical ($NO^\bullet$) to produce free radical peroxynitrite ($ONOO^-$), which is responsible for nitrosylation of proteins and oxidation of lipoproteins [31]. $NO^\bullet$ is produced via enzymatic oxidation of L-arginine to L-citrulline by the action of constitutive and inducible NOS. Besides the mediator in blood pressure, $NO^\bullet$ participates in the immune and inflammatory responses with biocidal activity against several microorganisms and induces damages on the proteins and DNA [31,37]. Toxic and oxidative effects of $NO^\bullet$ results from its oxidation, generating highly reactive species, such as nitrite (NO_2^-) and peroxynitrite ($ONOO^-$).

NO_2^- is produced by $NO^\bullet$ autooxidation forming nitrous anhydride (N_2O_3), an intermediate in this conversion recognized as a potent nitrosating agent [31], which can also be used to produce nitrogen dioxide by the action of the MPO enzyme (Figure 6). Carbon dioxide (CO_2) reacts catalytically with $ONOO^-$ to produce nitroperoxycarbonate ($ONOOCO_2$), which via homolysis of the O-O bonds, carbon trioxide ($CO_3^{\bullet -}$), and nitrite dioxide ($NO_2^\bullet$) radicals are produced [31]. Moreover, when $ONOO^-$ decomposes in the absence of CO_2, the $NO_2^\bullet$ and $OH^\bullet$ radicals production take place, whereas, in the presence of CO_2, $CO_3^{\bullet -}$ and $NO_2^\bullet$ radicals are produced, and in this process (Figure 6). MPO also participate, affecting tyrosine nitration when NO_2^- is used as a co-substrate, with consequent production of $NO_2^\bullet$, a reactive free radical [31]. Based on its diverse action as a marker of neutrophil infiltration, inflammatory process, and oxidative stress, MPO represents a potential target for the development of synthetic and natural compounds against several diseases, including atherosclerosis, acute coronary syndromes, ischemic heart disease, and IBD [30,38,39].

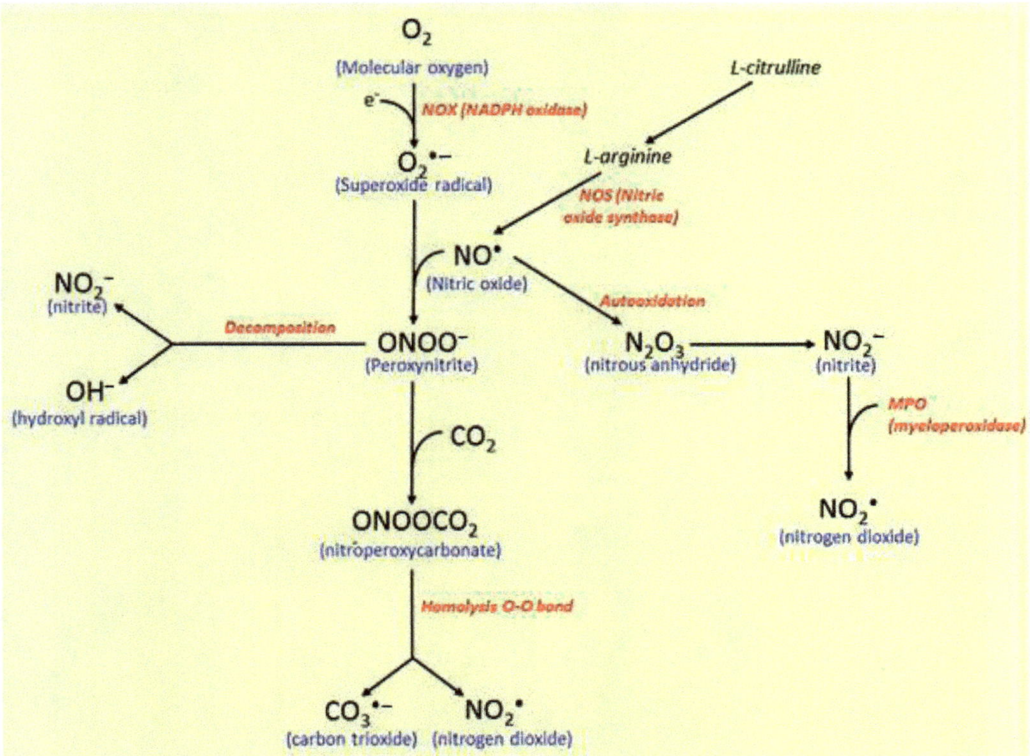

Figure 6. Nitric oxide synthase (NOS) pathway of free radical production. NOX, NADPH oxidase; NOS, nitric oxide synthase; MPO, myeloperoxidase. In these oxidative processes, the following activities were reported: A. Inhibition of iNOS activity by paepalantine, esculetin, esculin, auraptene, and collinin; B. Reduction of NO release by isomeranzin and esculin.

The major regulator of the endogenous antioxidant system is the nuclear factor erythroid 2 (NEF2)-related factor 2 (Nrf2) that protects cells from several stressors agents, such as ROS, RNS, and environmental damage [40]. In physiological conditions, Nrf2 binds with cullin 3 (cul3) and Kelch-like ECH-associated protein 1 (keap1), a key repressor of the Nrf2 signaling pathway (Figure 7), preventing the translocation of Nrf2 to the nucleus [41]. This complex, after ubiquitination, promotes Nrf2 degradation via proteolysis. Under oxidative stress conditions, Nrf2-keap1 complex is uncoupled and a free Nrf2 is translocated into the nucleus, where binds with small Maf (sMaf) proteins [42]. The heterodimer binds with antioxidant response elements (ARE) target genes (Figure 7), regulating the expression of several antioxidant-related endogenous genes, including the enzymes CAT, GPX, SOD, GST, γ-GCL, GR, NADPH quinone oxidoreductase, and heme oxygenase [41]. Based on this, Nrf2 is a key mediator of the antioxidant defense system as well as an important target for the action of new synthetic and natural compounds, including coumarin derivatives such as esculetin.

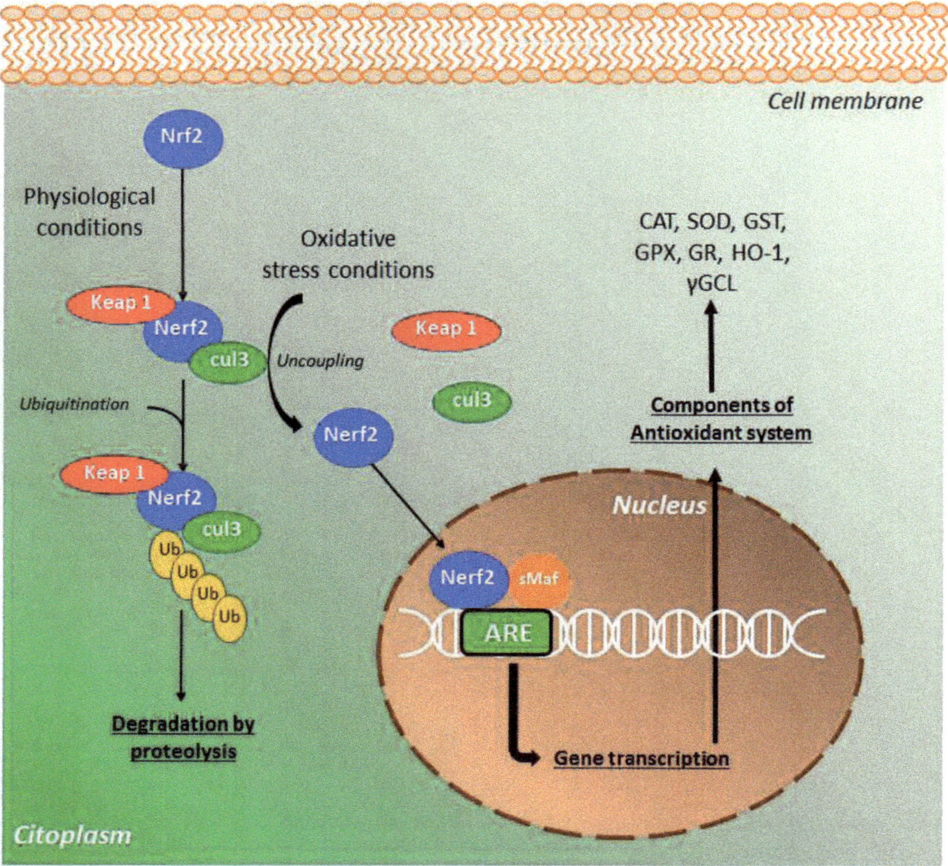

Figure 7. The nuclear factor erythroid 2 (NEF2)-related factor 2 (Nrf2) signaling pathway of oxidative stress. ARE, antioxidant element of response; CAT, catalase; cul3, cullin 3; GST, glutathione S-transferase; GPX, glutathione peroxidase; GR, glutathione reductase; HO-1, heme-oxygenase 1; Keap1, Kelch-like ECH-associated protein 1; Nrf2, nuclear factor erythroid 2-related factor 2; sMaf, small Maf proteins; Ub, ubiquitin; γGCL, γ-glutamylcysteine ligase. In these oxidative processes, the following activities were reported: A. Upregulation of Nrf2 by esculetin, 4-methylesculetin, daphnetin, and esculin.

4.1.1. The Isocoumarin Paepalantine

Paepalantine (9,10-dihydroxy-5,7-dimethoxy-1*H*-naptho(2,3c)pyran-1-one), was the first plant-derived coumarin studied in an experimental model of intestinal inflammation [43]. Paepalantine (Figure 1) is an isocoumarin previously isolated from the capitula of the Brazilian endemic *Paepalanthus bromelioides* plant from the Eriocalulaceae botanical family [44], which produced protective effects in the acute and relapse phases of the intestinal inflammation induced by TNBS in rats [43]. The protective effects observed after oral administration of the 5 and 10 mg/kg were similar to those promoted by the 25 mg/kg of sulphasalazine, a 5-aminosalicylate currently used to treat human IBD, i.e., paepalantine produced intestinal anti-inflammatory activity at doses 2.5 and 5.0-times lower than a reference drug [43]. Intestinal anti-inflammatory activity of the paepalantine was related to prevention of the GSH depletion (Figure 5) and inhibition of the colonic NOS activity (Figure 6), which was upregulated by the inflammatory process, suggesting that intestinal anti-inflammatory activity is related to its antioxidant properties [43]. Paepalantine also inhibited HOCl production in rat neutrophils, reducing oxidative stress by the inhibition of MPO activity and scavenging HOCl (Figure 5) [45]. Moreover, the antioxidant properties of paepalantine were evidenced by its potent scavenging properties in the 1,1-diphenyl-2-picrylhydrazyl (DPPH) and superoxide radicals as well as by its ability to protect mitochondria from hydroperoxide accumulation and mitochondrial membrane lipid peroxidation [46,47]. The inhibition of iNOS activity by paepalantine was recently corroborated through in vitro studies with LPS-stimulated macrophages [48]. In this study, paepalantine binds with the NOS enzyme through several structural amino acids just on the active site of the enzyme, reducing its enzymatical activity [48].

4.1.2. Coumarin and 4-hydroxycoumarin

Intestinal anti-inflammatory activity of coumarin and its derivative 4-hydroxycoumarin (Figure 2) were evaluated in the acute and subchronic phases of the TNBS-induced intestinal inflammation model in rats [49]. In this study, damage score and extension of tissue lesion induced by TNBS were significantly reduced after oral administration of coumarin (25 mg/kg) and 4-hydroxycoumarin (10 and 25 mg/kg), and these protective effects were accompanied by a counteraction of GSH depletion and inhibitory MPO activity (Figure 5) [49]. Although 4-hydroxycoumarin produced effects at lower doses when compared with coumarin, is not possible to suggest that the OH substitution in C4 was directly related to the improvement of its effects in the acute or sub-chronicle protocols of intestinal inflammation induced by TNBS in rats.

4.1.3. Esculetin (6,7-dihydroxycoumarin) and 4-methyl Esculetin

Esculetin and 4-methylesculetin (Figure 2) oral administration produced antioxidant protective effects in rats with TNBS-induced intestinal inflammation [50]. While esculetin counteracted GSH depletion at the dose of 10 mg/kg, with no effects on MPO activity, 4-methylesculetin produced significantly positive effects on the GSH levels (2.5 and 5 mg/kg) and MPO activity (5 and 10 mg/kg) in the acute phase of intestinal inflammation (Figure 5). In the sub-chronicle protocol, when coumarins were administered after induction of intestinal inflammation, the effects of 4-methylesculetin were evidenced on the GSH level and MPO activity, whereas esculetin was inactive to restore the basal value of these mediators [50]. Moreover, the inhibitory concentration of 4-methylesculetin on the lipid peroxidation membranes was approximately twice lower than esculetin [50]. A comparative analysis of data suggesting that 6,7-dihydroxylated coumarins when substituted at C4 with a methyl group had an improvement of effects on the MPO activity. Reduction of damage score and MPO activity was also demonstrated after the intrarectal administration of esculetin at 100 and 200 μM in rats with intestinal inflammation previously induced by TNBS [51]. The antioxidant property of esculetin was also corroborated by its action inhibiting iNOs activity (Figure 6) and modulating Nrf2 (Figure 7) signaling pathway [51,52].

Besides intestinal anti-inflammatory activity related to antioxidant properties counteracting GSH depletion, inhibiting MPO activity and lipid peroxidation [50], the effects of 4-methylesculetin (6,7-dihydroxy-4-methylcoumarin) in acute and subchronic phases of TNBS-induced intestinal inflammation was evaluated in comparison with effects of sulphasalazine and prednisolone in rats as well as in RAW264.7, Caco-2 and splenocytes culture cells [53]. Similar to previously reported, 4-methylesculetin improved clinical, histopathological, and biochemical parameters, such as GSH levels and MPO activity (Figure 5) in both acute and subchronic phases of the TNBS-induced intestinal inflammation model [53]. In a recent study in the DSS-induced intestinal inflammation in mice, 4-methylesculetin also improved histopathological indicators of intestinal inflammation, reduced MPO activity, and markedly counteracted GSH depletion [54].

Considering that the intestinal anti-inflammatory activity of 4-methylesculetin was closely related to several mediators of oxidative stress, an interesting study was carried out to investigate the molecular mechanisms involved in these antioxidant properties [32]. In the TNBS model of intestinal inflammation, treatments with 5 and 10 mg/kg methylesculetin significantly decreased damage score, lesion extension, and diarrhea incidence [32]. These protective effects were accompanied by an inhibition of the MPO and GPx activities and simultaneous increment of GST and GR activities, with no effects on the SOD and CAT activities [32]. Moreover, treatment with 4-methylesculetin was able to prevent downregulation of GR and Nrf2, with no effects on the GRX, GST gene expression, suggesting that GR is a target enzyme for the action of 4-methylesculetin. Molecular interaction between 4-methylesculetin and GR using UV-vis absorbance spectroscopy, fluorescence measurements, saturation transfer difference nuclear magnetic resonance, and computational modeling were performed to identify this interaction [32]. These analyses showed that 4-methylesculetin forms a complex with GR with more than one binding site close to the FAD cofactor, which was reduced by NADPH, whereas equivalents were transferred to a redox-active GSSG, stabilizing the 4-methylesculetin-GR complex with a consequent increment of the GR activity [32]. Based on this, authors demonstrate that 4-methylesculetin acts by different antioxidant mechanisms, i.e., controlling the imbalance between MPO activity and GSH production with an increment of GSH availability, upregulating the GST activity with consequent increase of electrophiles inactivation, upregulating GR activity via stabilization of its enzymatic activity, and upregulating Nrf2 expression that leads to a GR regeneration with consequent GSH maintenance levels [32].

4.1.4. Daphnetin (7,8-dihydroxycoumarin)

A comparative study with several coumarin derivatives in TNBS-model of intestinal inflammation daphnetin (Figure 2) demonstrated a protective effect of daphnetin (Figure 2) in intestinal inflammation after oral administration of the lower doses (2.5 and 5.0 mg/kg) [55]. Daphentin counteracted GSH depletion and inhibited MPO activity as well as showing a potent ROS scavenging property (Figure 5) [55]. Among coumarin derivatives, daphnetin is one of the most studied compounds, with a series of pharmacological activities that corroborate its use in the inflammatory process, mainly acting on the oxidative stress and other signaling pathways of the intestinal inflammatory process, which it will bellow discussed. Antioxidant and anti-inflammatory activities have been reported by different studies, in which daphnetin was reported as a potent antioxidant compound inhibiting lipid peroxidation, scavenging free radical generation, and upregulating the Nrf2 signaling pathway [8,56,57].

4.1.5. Esculin (7-hydroxy-6-O-glucosylcoumarin)

Esculin (Figure 2) promoted protective effects on the DSS- and TNBS-induced intestinal inflammation, counteracting GSH depletion, and inhibiting MPO activity [55,58]. Esculin relieved intestinal inflammatory clinical indicators and histopathological damage promoted by DSS, effects that were accompanied by a downregulation of iNOS expression [58]. In vitro studies with RAW264.7 cells stimulated by LPS demonstrated esculin reducing NO generation as well as the gene expression and protein level of iNOS [58]. Several studies corroborated the antioxidant properties of esculin and its use in different inflammatory processes, mainly acting as a potent scavenging agent [8,56], reducing MPO activity [56], NO production, and iNOS levels [59], as well as markedly activated the Nrf2 signaling pathway related to oxidative stress [60,61].

4.1.6. Other Simple Antioxidant Coumarin Derivatives

A comparative and preliminary study using several simple coumarins with different substitutions in the basic ring of coumarins, including scopoletin, scoparone, fraxetin, 4-methylumbelliferone, esculin, and daphnetin (Figure 2) demonstrated differential intestinal anti-inflammatory and antioxidant properties in a TNBS-induced intestinal inflammation in rats [55]. Among these coumarin derivatives, 4-methylumbelliferone produced no effects on the clinical (damage score, extension of lesion, diarrhea, and length/weight colon ratio) and biochemical parameters such as GSH level and MPO activity. Oral administration of scopoletin (5 and 25 mg/kg), scoparone (5 and 10 mg/kg), and fraxetin (5 and

10 mg/kg) were able to counteract GSH depletion induced by intestinal inflammation with no effects on the MPO activity [55]. On the other hand, oral administration of 25 mg/kg of esculin and 2.5 and 5.0 mg/kg of daphnetin counteracted GSH depletion and inhibited MPO activity, showing daphnetin with a protective effect against intestinal inflammatory process at lower doses when compared with the other coumarin derivatives [55]. Although all coumarin derivatives acted as a radical scavenger, only fraxetin and daphnetin inhibited in vitro assay of lipid peroxidation in the cell membrane with lower inhibitory concentrations [55]. The results corroborate the hypothesis that dihydroxylated coumarins with vicinal diol functionality such as fraxetin, esculetin, and daphnetin exhibit potent ROS scavenging when compared to other coumarin derivatives [8,55].

Osthole (Figure 2) in dinitrobenzene sulphonic acid model of intestinal inflammation improved the histopathological damage and some clinical indicators of intestinal inflammation and acted as an antioxidant product, reducing malondialdehyde levels and MPO activity, increasing GPX, CAT, SOD, and GST levels, and counteracting GSH depletion [62]. In the DSS-model of intestinal inflammation in mice osthole showed protective effects on intestinal inflammation improving clinical parameters and histological damages as well as reducing MPO activity and downregulating colon TNF-α and serum TNF-α levels [63].

Antioxidant simple coumarin derivatives with intestinal anti-inflammatory activity also include isomeranzin, auraptene, and collinin (Figure 2). Isomeranzin was reported as able to inhibit NO release in RAW264.7 cells [64], whereas auraptene and collinin intestinal anti-inflammatory effects were associated with reduced iNOS levels [65]. The intestinal anti-inflammatory activities of isomeranzin, auraptene, and collinin were related to other signaling pathways of the inflammatory process [64,66].

4.2. Effects of Coumarin Derivatives on Aarachidonic Acid Metabolism

Several metabolites of the arachidonic acid metabolism pathway have pro-inflammatory properties, indicating that inhibitory action on these metabolites production can be an important target for the development of intestinal anti-inflammatory drugs. Arachidonic acid is produced from membrane phospholipids by the action of several phospholipases, manly phospholipase A2 (Figure 8). The metabolism of arachidonic acid includes several enzymes, mainly cyclooxygenase 1 and 2 (COX-1 and COX-2), and lipoxygenase 5 and 12 (LOX-5 and LOX-12), which play a relevant role in intestinal inflammation [67]. COX-1 is constitutively expressed in several cell types and produces diverse eicosanoids such as thromboxane A2 (TXA_2) and prostaglandins I2 (PGI_2), which have platelet and cytoprotective effects, respectively. COX-2 is a cyclooxygenase induced under inflammatory stimuli and the main source of pro-inflammatory prostaglandins, such as PGI_2 and PGE_2, and its inhibition by different chemical agents has been considered beneficial to control the intestinal inflammatory process [3,67]. On the other hand, lipoxygenases, mainly LOX-5 is a key enzyme for the production of leukotriene B4 (LTB_4), the major pro-inflammatory metabolite of arachidonic acid that contributes to the perpetuation of intestinal inflammation [68].

Some coumarin derivatives produced different inhibitory action on arachidonic acid metabolism, improving response against intestinal inflammation. Esculetin was able to reduce the COX-2 levels in the colon of rats with intestinal inflammation induced by TNBS [51] as well as inhibited LTB_4 and TXB_2 generation via an inhibitory action on the LOX-5 activity [8,69]. Similar effects were reported to esculin, daphnetin, osthole, imperatorin, auraptene, collinin, and fraxetin [8,63,65,69–72].

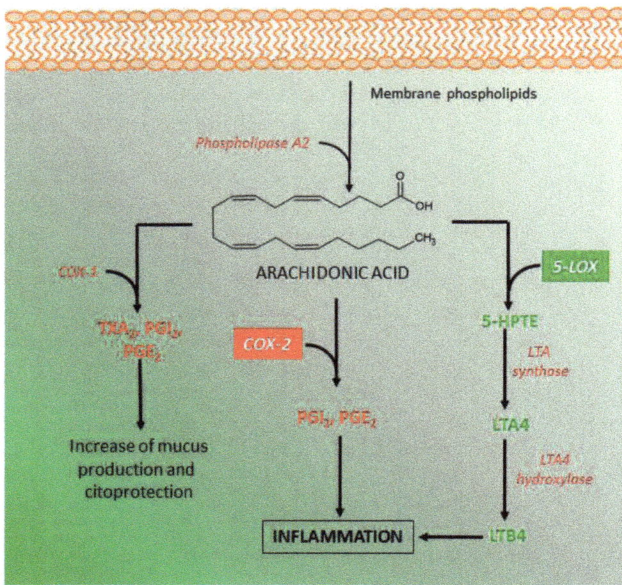

Figure 8. Arachidonic acid metabolism and its main pro-inflammatory mediators. 5-LOX, 5-lipooxygenase; COX-1, cyclooxygenase 1; COX-2 cyclooxygenase 2; LTA, leukotriene A; LTB, leukotriene B; PGI_2, prostaglandin I2, PGE_2, prostaglandin E2, TXA_2, thromboxane A2. Inhibitory action on the arachidonic acid metabolism was demonstrated by treatment with esculetin, esculin, daphnetin, osthole, imperatorin, auraptene, collinin, and fraxetin.

4.3. *Effects of Coumarin Derivatives on the Immune Response*

The modulation of the immune response has been reported as an important action of coumarin derivatives to control intestinal inflammation in experimental models and in vitro studies. Generally, the available data indicated the ability of several coumarins produce effects on the production and release of immune mediators, but the general mechanism of action to produce these responses was not fully investigated. However, several intestinal anti-inflammatory coumarin derivatives described probably modulated the immune response acting on the other signaling pathways here described, particularly those including nuclear signaling pathways.

Besides the antioxidant properties, paepalantine (Figure 1) was demonstrated to inhibit the production of pro-inflammatory cytokines TNF-α and IL-6 in human gastric carcinoma cells and murine macrophages RAW264.7 line [48]. Although in the acute phase of the TNBS model, 4-methylesculetin (Figure 2) produced no effects on the IL-1β, TNF-α, MMP-2, and MMP-9 protein levels, in vitro studies demonstrated that 4-methylesculetin inhibits the production of IL-1β in LPS-stimulated RAW264.7 cells, IL-8 in IL1-β-stimulated Caco-2 cells, and INF-γ and IL-2 in concanavalin a-stimulated splenocytes [53]. 4-methylesculetin treatment in DSS-model of intestinal inflammation reduced IL-6 colon levels, with no effects on the IL-17 and TNF-α colon levels [54]. Esculin (Figure 2) relieved intestinal inflammatory clinical indicators and histopathological damage promoted by DDS, effects that were accompanied by a downregulation of IL-1β, TNF-α, and reduction of IL-1β and TNF-α protein levels [59]. Esculetin (Figure 2) in a model of psoriasis-like skin diseasedramatically suppressed pro-inflammatory cytokine releases such as TNF-α, IL-6, IL-22, IL-23, IL-17α, and INF-γ [73]. A recent and interesting study in mice treated with daphnetin was carried out using different approaches to describe the protection of daphnetin (Figure 2) on the DSS-induced intestinal inflammation model [74]. The evaluation of immune and inflammatory response in this experimental model demonstrated daphnetin avoid intestinal inflammation progression, which was related to an improvement of histopathological damage induced by DSS and modulation of pro-inflammatory mediators, downregulating colon TNF-α, IL-6, IL-1β, IL-21, IL-23, CXCL1, and CXCL2 expression, and increasing IL-10 [74]. Osthole at 100 mg/kg by intraperitoneal route attenuated several clinical indicators of the intestinal inflammation as well as the histopathological lesions and alterations induced by TNBS [75]. The protective clinical effects were accompanied by a significant reduction of IL-1β, TNF-α, IL-6, CXCL10, and COX-2 gene expression as well as by an improvement of the intestinal barrier function, upregulating claudin-1 and ZO-1 mRNA [75]. In another set of evaluations using a model experimental of intestinal inflammation induced by dinitrobenzene sulphonic acid, osthole reduced TNF-α and increased IL-10, with no effects on the INF-γ levels [62]. Isomeranzin treatment with an oral dose of 30 mg/kg, isomeranzin attenuated several clinical and histopathological indicators of DSS- and TNBS intestinal inflammation as well as decreased serum IL-6 and TNF-α expression and colon IL1-β, IL-6, TNF-α, and iNOS mRNA expression [64].

4.4. *Effects of Coumarin Derivatives on the Nuclear Signaling Pathways*

Several drugs and natural products, including coumarin derivatives, produce intestinal anti-inflammatory activity acting on the transcription factors, nuclear receptors, and enzymes related to the inflammatory response, particularly nuclear factor-kappa b (NF-κB), peroxisome proliferator-activated receptor gamma (PPAR-γ), mitogen-activated protein kinases (MAPKs), pregnane X receptors (PXRs), rexinoid X receptors (RXRs). Other receptors such as glucocorticoid receptor (GR), farnesoid X receptor (FXR), estrogen receptor (ER), liver X receptor (LXR) regulate the inflammatory response in several diseases such as atherosclerosis, obesity, diabetes, multiple sclerosis, cancer, and IBD [76], showing that these nuclear signaling pathways are key targets for the action of new intestinal anti-inflammatory compounds.

4.4.1. NF-κB and PPAR-γ Signaling Pathways

The transcription factor kappa B (NK-κB) has a central role in the intestinal inflammatory processes, triggering a high pro-inflammatory cytokines production. NK-κB signaling pathway (Figure 9) can be activated either canonical or noncanonical pathways, however, the majority of products and studies were focused on the canonical signaling pathway [77,78]. In the canonical NK-κB signaling pathway, the NK-κB heterodimer consists of the subunits p50 and p65/Rel A, which is inactive in the cytoplasm when binding with inhibitors of protein kappa B (IκB). The IκB inhibitory enzymatic complex (IKK) is composed of a regulatory IKK gamma (IKKγ) subunit and two enzymatically active subunits, IKK alpha (IKKα) and beta (IKKβ) [79]. In the canonical NK-κB signaling pathway, IKK activation occurs by specific membrane ligands such as cytokines, bacteria, bacteria metabolites, viruses, and growth factors [80]. Under this stimulation, IKKβ is activated leading to IκB phosphorylation with consequent ubiquitination and proteasome degradation [74–77], whereas IKKα is phosphorylated to activate noncanonical NK-κB pathway (Figure 9), causing p100 processing and formation of p52/RelB dimers instead of p50 and p65/Real [77,80]. The released NK-κB is promptly translocated into the nucleus to activate specific response elements in DNA, triggering a transcriptional activity with high production of diverse inflammatory mediators, mainly TNF-α, IL-1β, COX-2, IL-6, IL-8, IL-12, and IL-23 (Figure 9) [77–81].

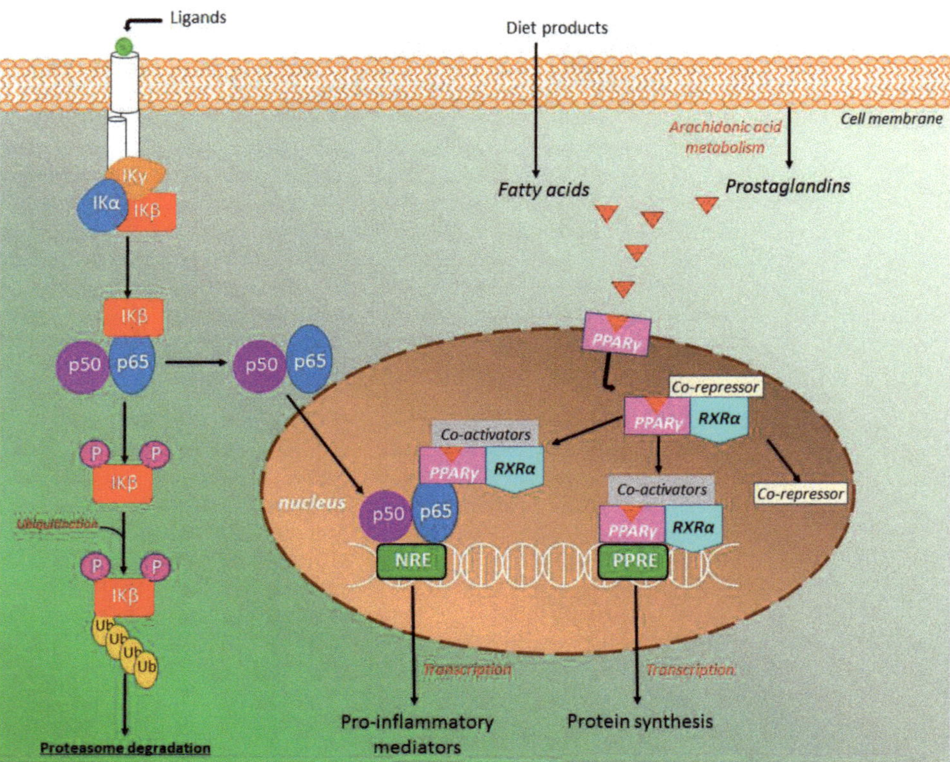

Figure 9. The nuclear factor-kappa b (NF-κB) and peroxisome proliferator-activated receptor gamma (PPAR-γ) signaling pathway in the intestinal inflammatory process. Modulation of NF-κB and PPAR-γ signaling pathways was demonstrated by treatment with esculetin, esculin, osthole, and isomeranzin.

In several studies, the peroxisome proliferator-activated receptor gamma (PPAR-γ) has been associated with the inflammatory response coordinated by the NF-κB signaling pathway [82,83]. PPAR-γ and other PPARs, such as PPARα and PPARβ are a group of nuclear receptors that modulates glucose metabolism, adipogenesis, fatty acid synthesis as well as inhibit the NF-κB inflammatory response [83]. It has been reported that PPAR-γ deletion induces an increment of the inflammatory process in the DSS model of intestinal inflammation, whereas its activation represses the nuclear localization of NF-κB [81,83], showing NF-κB-dependent response of PPAR-γ as a target for the action of intestinal anti-inflammatory compounds. PPAR-γ is a heterodimer complex with retinoid X receptor alpha (RXRα) generally binding with a co-repressor and expressed in several cells that participates in the intestinal inflammatory response such as dendritic cells, macrophages, and monocytes [83]. Under receptor activation by ligands, the co-repressor molecule displaced, whereas PPAR-γ/RXRα free complex binding with coactivator molecules [83].This activated complex binding to PPAR-γ response elements (PPRE) inducing transcription and protein synthesis (Figure 9). It has been reported that activated PPAR-γ/RXRα/coactivators complex can also bind with NF-κB repressing its transcriptional function with consequent reduction of pro-inflammatory cytokines production and consequent anti-inflammatory effects [81–85]. Several exogenous and endogenous PPAR-γ ligands have been reported, including fatty acids (linoleic, palmitoleic, and oleic acids), eicosanoids (eicosapentaenoic and docosahexaenoic acids, and prostaglandins), thiazolidinediones (rosiglitazone and pioglitazone), non-steroidal anti-inflammatory drugs (indomethacin and ibuprofen) as well as short-chain fatty acids, mainly butyrate and propionate, which are produced from the fermentative process of dietary fiber and other food products by intestinal microbiota [83,86,87].

Several coumarin derivatives produced intestinal anti-inflammatory activity acting on the NF-κB and PPAR-γ signaling pathway in both in vivo and in vitro studies. Antioxidant esculetin (Figure 2) treatment of the human pancreatic cell lines resulted in a significant reduction of NF-κB levels via its binding with Keap1 regulator of the Nrf2 signaling pathway, attenuating the NF-κB activation [88]. Moreover, esculetin reduced the NF-κB p65 levels in the cell nucleus of human NB4 leukemic cell lines [52].

Esculin (Figure 2) was also able to decrease nuclear protein levels p65 from NF-κB signaling pathway both rectal tissue from the animal with DSS-induced intestinal inflammation and RAW264.7 cells [59]. Moreover, esculin suppressed the phosphorylation of IκBα, the major step of NFκB accumulation in the cell nucleus [58]. Finally, the authors elegantly demonstrated that inhibition of NFκB activation by esculin was partially mediated by the PPAR-γ stimulation, promoting nuclear localization of PPAR-γ (Figure 9) and the regulation on NFκB activation [58]. Osthole (Figure 2) was also evaluated in the DSS-model of intestinal inflammation in mice and its protective effects on intestinal inflammation were related to a downregulation of the NFκB p65 and IκB gene expression with a simultaneous effect increasing IκBα protein levels (Figure 9), suggesting that osthole at doses of 20 mg/kg inhibited NFκB activation [63]. Isomeranzin (Figure 2) treatment reduced the phosphorylation of ERK and p65 in DSS- and TNBS-induced intestinal inflammation models. In vitro studies was demonstrated isomeranzin inhibiting NF-κB activation via prevention of TRAF6 ubiquitination, a signal transductor of NF-Kb [64].

4.4.2. MAPK Signaling Pathway

Mitogen-activated protein kinase (MAPK) signaling exerts several effects on cell function, including cell growth, proliferation, differentiation and survival, as well as is closely implicated in IBD, influencing the progression and perpetuation of intestinal inflammation [89,90]. MAPK activation is a response to several extracellular stimuli such as environmental stress, hormones, growth factors, and cytokines that via different kinase receptors, pathogen-associated molecular patterns, and danger-associated molecular patterns recruit pattern recognition receptors to induce a cell response [89,91]. The MAPK signaling pathway includes three groups of protein kinases, i.e., the extracellular signal-regulated kinases (ERKs), the c-Jun N-terminal kinases (JNKs), and the p38 MAPKs [89]. Phosphorylation, which occurs in a specific amino acid sequence of each group of MAPK, is pivotal for their activation [90]. Each group of MAPK is activated by different kinase pathways using distinct interlinked kinase components, as elegantly described [86]. After activation, MAPK is translocated to the nucleus to phosphorylate a series of transcription factors responsible for the expression of several genes and protein synthesis of mediators related to the inflammatory response [89–91].

MAPK signaling pathway has been related to the action of intestinal anti-inflammatory drugs, including aminosalicylates, glucocorticoids, and immunomodulators [92] as well as the target for the action of several coumarin derivatives. In mouse peritoneal macrophages, osthole (Figure 2) treatment significantly attenuated the production of the pro-inflammatory cytokines via suppressive effects on the p38 phosphorylation, suggesting its protective effects in TNBS-induced intestinal inflammation was related to the MAPK signaling pathway [75]. A similar evaluation of osthole was performed using the dinitrobenzene sulphonic acid model in rats, DSS-induced intestinal inflammation in mice, and murine macrophages [62,63]. Oral administration of 50 mg/kg of osthole reduced phosphorylation of the MAPK/p38 protein, promoting protective effects in the intestinal inflammatory process [62,63]. In vitro studies demonstrated osthole significantly reduced phosphorylation of p38/MAPK with no effects on the phosphorylation of the ERK and JNK [60], corroborating the data previously reported [62]. Differentially, isomeranzin (Figure 2) treatment in LPS-stimulated murine macrophages reduced phosphorylation of ERK with no effects of the JNK and p38 MAPKs [64].

4.4.3. HIF-1α Signaling Pathway

The hypoxia-inducible factor 1 alpha (HIF-1α) is an innovative target for the action of new drugs with anti-inflammatory activity (Figure 10). Several studies with HIF-1α were performed in the last years as an attempt to explain how cells sense and to adapt to oxygen availability. These studies were recognized by the Nobel Prize of Physiology or Medicine in 2019 awarded to Kaelin, Ratcliffe, and Semenza.

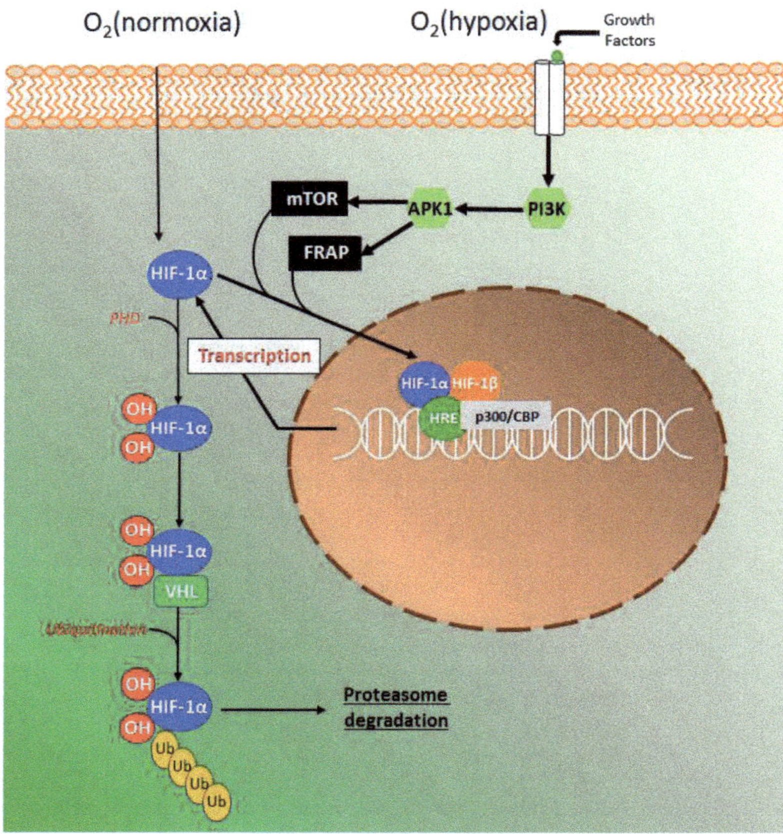

Figure 10. The hypoxia-inducible factor 1 alpha (HIF-1α) signaling pathway of intestinal inflammation as the target for the action of esculetin.

Using human colon carcinoma HCT116 cells, esculetin was demonstrated to induce the hypoxia-inducible factor 1 alpha (HIF-1α), promote the secretion of vascular endothelial grown factor (VEGF), and inhibit HIF prolyl hydroxylases (PHD) activity [51]. This elegant study also suggested that catechol moiety in esculetin is required for HPH inhibition via competition with ascorbate and 2-ketoglutarate [51], given that several compounds containing catechol moieties such as quercetin and caffeic acid tend to activate HIF-1α [93,94]. In the intestinal inflammatory process, the epithelial cells provide barrier and transport functions, which are modulated by a series of physiological and morphological events such as mucus production, microvilli, and tight junctions. On the other hand, the high vascularization of intestinal tissue contributes to the counteraction of the high oxygen gradient from luminal anaerobic conditions to oxygenated tissue [95]. It has been considered that in acute and chronic inflammation, oxygen delivery, and oxygen availability or hypoxia is a key factor to trigger an inflammatory response [26,96]. The hypoxia signaling pathway is mainly coordinated by the HIF-1α stabilization, and in normoxia conditions, proline residues are hydroxylated by PHD action producing a complex with the Von Hippel-Landau (VHL) protein [97]. The complex HIF-VHL binds with ubiquitin, leading to proteasomal degradation of HIF-1α (Figure 10). Hypoxia signaling induces growth factors, such as transforming growth factor β (TGF-β) and VEGF binds with membrane-related tyrosine kinase receptors triggering a signaling pathway of phosphatidylinositol 3-kinase (PIP3K) with consequent serine/threonine-specific protein kinase 1(Akt1) phosphorylation (Figure 10) [98]. Under hypoxia, the activity of PHD is suppressed while phosphorylated Akt promotes the phosphorylation of mammalian target of rapamycin (mTOR) and FKBP-rapamycin associate protein (FRAP), regulating HIF-1α [98]. HIF-1α subunits translocate into the nucleus to bind with HIF-1β subunit and heterodimer HIF-α:HIF-β transcription factor complex then locate to the hypoxia-response elements (HRE) target genes (Figure 10), resulting in their transcriptional upregulation with the participation of coactivator p300/CREB binding protein (p300/CBP) [97].

4.4.4. The Pregnane X Signaling Pathway

The pregnane X nuclear signaling pathway has been also reported as a target for the action of intestinal anti-inflammatory products, including coumarin derivatives such as imperatorin [99]. Nuclear pregnane X receptors (PXRs) are well-recognized for their function in the modulation of drug metabolism, acting as a flexible ligand for several products including drugs, natural and dietary products, hormones, and environmental pollutants [100]. Predominantly expressed in the intestine and liver, PXR after activation forms a heterodimer with the retinoid X receptor (RXR) [76]. This heterodimer binding to specific PXR response elements to control the gene expression of several proteins [76]. PXR agonists were demonstrated to attenuate intestinal inflammatory symptoms and to reduce intestinal permeability [101], improving epithelial barrier function via suppression of NF-κB expression that encoding pro-inflammatory cytokines [102]. PXR activation is a relevant antagonist of NF-κB transcriptional activity in the intestine during intestinal inflammation [103]. Imperatorin (Figure 2) mediated PXR activation suppressing the nuclear translocation of NF-κB and down-regulating pro-inflammatory production in DSS-induced intestinal inflammation in mice [99].

4.5. Effects of Coumarin Derivatives Intestinal Microbiota

Intestinal microbiota modulation by dietary products, mainly probiotic, prebiotic, and other natural products to improve SCFAs and other bacteria metabolites production from the fermentative process is an important approach to prevent IBD as well as to relieve symptoms of the intestinal inflammatory process. However, among all coumarin derivatives evaluated in several studies related to intestinal inflammation, only daphnetin (Figure 2) was demonstrated to act on the intestinal microbiota [74]. Daphnetin reversed DSS-induced gut dysbiosis, reducing *Bacteroides,* and increasing *Firmicutes,* which are the major SCFAs-producing bacteria [74]. Moreover, it was demonstrated that daphnetin was able to recovery zona occludens, occludin, mucin, and E-cadherin function compromised by DSS-induced intestinal inflammation, improving the intestinal epithelial integrity [74]. Using an elegant approach of the microbiota-transfer by cohousing untreated with daphnetin-treated mice, the authors reported an improvement of the clinical parameters, bacteria biodiversity, and immune response in the colon of cohousing DSS-untreated animals, when compared with DSS-inflamed mice singly housed [74]. Finally, to corroborate these data and intestinal microbiota importance in the maintenance of intestinal function, fecal microbiota from daphnetin-treated mice was transfer to mice depleted of intestinal microbiota, and the results demonstrated a remarkable improvement of disease manifestations, immune and inflammatory response when compared with the animal has received the vehicle, clearly showing that protective effects of daphnetin in intestinal inflammation, besides of its effects on the oxidative stress and immune response, were directly related of the regulation of intestinal integrity and tissue homeostasis modulated by intestinal microbiota [74]. Recently, daphnetin was also demonstrated to improve the altered intestinal microbiota composition of the glucocorticoid-induced osteoporosis rats, attenuating the intestinal barrier dysfunction [104]. Although daphnetin is the only coumarin whose intestinal anti-inflammatory activity has been directly associated with intestinal microbiota modulation, other natural and synthetic coumarin derivatives and plant extracts containing coumarins [105–111] were able to differentially modulate some pathogenic intestinal bacteria, but with no direct evidence and correlation with intestinal anti-inflammatory activity.

5. Conclusions and Perspectives

This review provided a general overview of the various coumarin derivatives with potential therapeutic applications on the intestinal inflammatory processes highlighting the ones for which the mechanism of action is at least partially defined and can serve for the design of further preclinical and clinical studies to support the use and application of coumarin derivatives as complementary therapies against IBD. In general, the mechanisms of action of coumarin derivatives observed in experimental models of intestinal inflammation and in vitro studies are similar to those described for other natural products such as flavonoids, anthocyanidins, and catechins. Although several coumarin derivatives such as paepalantine, 4-methylesculetin, daphnetin, esculetin, and osthole produce intestinal anti-inflammatory effects in lower doses when compared with other phenol compounds, it is no possible to attribute advantages in the use of these coumarins due to the lack of clinical trials and more detailed studies on efficacy and safety with these compounds. Protective effects of coumarin derivatives are related to antioxidant properties, similar to those produced by several phenolic compounds. However, some coumarins also interact with several endogenous macromolecules, different cell types, and signaling pathways as well as in innovative molecular targets. On the other hand, further studies are needed into the effects of some coumarin derivatives on the course of the disease, mechanisms of action, ability to modulated intestinal microbiota and intestinal permeability, and safety for use. Clinical trials in patients with IBD are very important to generate data for a potential application of coumarins derivatives as a complementary therapy for this chronic disease. There is scientific evidence here reported to support the suggestions of some coumarin derivatives as candidates for further pre-clinical studies and clinical trials, particularly those better studied, mechanism of action partially defined and with protective effects in lower doses, such as esculetin, 4-methylesculetin, osthole, and daphnetin.

Funding: The research in the Laboratory of Phytomedicines, Pharmacology, and Biotechnology (PhytoPharmaTech) has been supported by the São Paulo Research Foundation (FAPESP) and National Council for Scientific and Technological Development (CNPq).

Conflicts of Interest: The author declares no conflict of interest.

References

1. GDB 2017 Burden Disease Collaborators. Global, regional, and national age-sex specific mortality for 282 causes of death in 195 countries and territories, 1980–2017: A systematic analysis for the Global Burden of Disease Study 2017. *Lancet* **2018**, *392*, 1736–1788. [CrossRef]
2. Jarmakiewicz-Czaja, S.; Pia, D.; Filip, R. The influence of nutrients on Inflammatory Bowel Diseases. *J. Nutr. Metab.* **2020**, *2020*, 2894169. [CrossRef] [PubMed]
3. Vezza, T.; Rodríguez-Nogales, A.; Algieri, F.; Utrilla, M.P.; Rodriguez-Cabezaz, M.H.; Gávez, J. Flavonoids in Inflammatory Bowel Disease: A review. *Nutrients* **2016**, *8*, 211. [CrossRef] [PubMed]
4. Hoensch, H.P.; Weigmann, B. Regulation of the intestinal immune system by flavonoids and its utility in chronic inflammatory bowel disease. *World J. Gastroenterol.* **2018**, *24*, 877–881. [CrossRef] [PubMed]
5. González-Quilen, C.; Rodríguez-Gallego, E.; Beltrán-Debón, R.; Pinent, M.; Ardévol, A.; Blay, M.T.; Terra, X. Health-promoting properties of proanthocyanidins for intestinal dysfunction. *Nutrients* **2020**, *12*, 130. [CrossRef] [PubMed]
6. Li, S.; Wu, B.; Fu, W.; Reddivari, L. The anti-inflammatory effects of dietary anthocyanins against ulcerative colitis. *Int. J. Mol. Sci.* **2019**, *20*, 2588. [CrossRef]

7. Fan, F.; Sang, L.; Jiang, M. Catechins and their therapeutic benefits to Inflammatory Bowel Disease. *Molecules* **2017**, *22*, 484. [CrossRef]
8. Hoult, J.R.S.; Paya, M. Pharmacological and biochemical actions of simple coumarins: Natural products with therapeutic potential. *Gen. Pharmacol. Vasc. Syst.* **1996**, *27*, 713–722. [CrossRef]
9. Lake, B.G. Coumarin metabolism, toxicity and carcinogenicity: Relevance for human risk assessment. *Food Chem. Toxicol.* **1999**, *37*, 423–453. [CrossRef]
10. Jain, P.K.; Joshi, H. Coumarin: Chemical and pharmacological profile. *J. Appl. Pharm. Sci.* **2012**, *2*, 236–240.
11. Akkol, E.K.; Genç, Y.; Karpuz, B.; Sobarzo-Sánchez, E.; Capasso, R. Coumarins and coumarin-related compounds in pharmacotherapy of Cancer. *Cancers* **2020**, *12*, 1959. [CrossRef] [PubMed]
12. Wardrop, D.; Keeling, D. The story of the discovery of heparin and warfarin. *Br. J. Haematol.* **2008**, *141*, 757–763. [CrossRef] [PubMed]
13. Lim, G.B. Warfarin: From rat poison to clinical use. *Nat. Rev. Cardiol.* **2017**. [CrossRef] [PubMed]
14. Ghosh, S.; Pariente, B.; Mould, D.R.; Schreiber, S.; Petersson, J.; Hommes, D. New tools and approaches for improved management of inflammatory bowel diseases. *J. Crohn's Colitis* **2014**, *8*, 1246–1253. [CrossRef]
15. GBD 2017 Inflammatory Bowel Disease Collaborators. The global, regional, and national burden of inflammatory bowel disease in 195 countries and territories, 1990–2017: A systematic analysis for the Global Burden of Disease Study 2017. *Lancet Gastroenterol. Hepatol.* **2020**, *5*, 17–30. [CrossRef]
16. Kaplan, G.G. The global burden of IBD: From 2015 to 2025. *Nat. Rev. Gastroenterol. Hepatol.* **2015**, *12*, 720–727. [CrossRef]
17. Ghouri, Y.A.; Tahan, V.; Shen, B. Secondary causes of inflammatory bowel diseases. *World J. Gastroenterol.* **2020**, *26*, 3998–4017. [CrossRef]
18. Di Stasi, L.C.; Costa, C.A.R.A.; Witaicenis, A. Products for the treatment of inflammatory bowel disease: A patent review (2013–2014). *Expert Opin. Ther. Pat.* **2015**, *25*, 629–642. [CrossRef]
19. Ananthakrishnan, A.N. Epidemiology and risk factors for IBD. *Nat. Rev. Gastroenterol. Hepatol.* **2015**, *12*, 205–217. [CrossRef]
20. Annese, V. Genetics and epigenetics of IBD. *Pharmacol. Res.* **2020**, *159*, 104892. [CrossRef]
21. Ellinghaus, D.; Bethune, J.; Petersen, B.; Frankle, A. The genetics of Crohn's disease and ulcerative colitis—*Status quo* and beyond. *Scand. J. Gastroenterol.* **2015**, *50*, 13–23. [CrossRef] [PubMed]
22. de Souza, S.P.; Fiocchi, C. Immunopathogenesis of IBD: Current state of the art. *Nat. Rev. Gastroenterol. Hepatol.* **2016**, *13*, 13–27. [CrossRef] [PubMed]
23. Arrieta, M.C.; Madsen, K.; Doyle, J.; Meddings, J. Reducing small intestinal permeability attenuates colitis in the IL10 gene-deficient mouse. *Gut* **2009**, *58*, 41–48. [CrossRef] [PubMed]
24. Fasano, A. All disease begins in the (leaky) gut: Role of zonulin-mediated gut permeability in the pathogenesis of some chronic inflammatory diseases. *F1000Research* **2020**, *9*, 69. [CrossRef] [PubMed]
25. Caviglia, G.P.; Dughera, F.; Ribaldone, D.G.; Rosso, C.; Abate, M.L.; Pellicano, R.; Bresso, F.; Smedile, A.; Saracco, G.M.; Astegiano, M. Serum zonulin in patients with infammatory bowel disease: A pilot study. *Min. Med.* **2019**, *110*, 95–100.
26. Rezaie, A.; Parker, R.D.; Abdollahi, M. Oxidative stress and pathogenesis of inflammatory bowel disease: An epiphenomenon or the cause? *Dig. Dis. Sci.* **2007**, *52*, 2015–2021. [CrossRef]
27. Moura, F.A.; Andrade, K.Q.; Santos, J.C.F.; Araújo, O.R.P.; Goulart, M.O.F. Antioxidant therapy for treatment of inflammatory bowel disease: Does it work? *Redox Biol.* **2015**, *6*, 617–639. [CrossRef]
28. Ivanova, A.; Gerasimova, E.; Gazizullina, E. Study of antioxidant properties of agents from perspective of their action mechanism. *Molecules* **2020**, *25*, 4251. [CrossRef]
29. Opara, E.C. Oxidative stress. *Dis. Mon.* **2006**, *52*, 183–198. [CrossRef]
30. Chami, B.; Martin, N.J.J.; Dennis, J.M.; Witting, P.K. Myeloperoxidase in the inflamed colon: A novel target for treating inflammatory bowel disease. *Arch. Biochem. Biophys.* **2018**, *645*, 61–71. [CrossRef]
31. Dedon, P.C.; Tannenbaum, S.R. Reactive nitrogen species in the chemical biology of inflammation. *Arch. Biochem. Biophys.* **2004**, *423*, 12–22. [CrossRef] [PubMed]
32. Tanimoto, A.; Witaicenis, A.; Caruso, I.P.; Piva, H.M.R.; Araujo, G.C.; Moraes, F.R.; Fossey, M.C.; Cornélio, M.L.; Souza, F.P.; Di Stasi, L.C. 4-methylesculetin, a natural coumarin with intestinal anti-inflammatory activity, elicits a glutathione antioxidant response by different mechanisms. *Chem. Biol. Int.* **2020**, *315*, 108876. [CrossRef] [PubMed]
33. Couto, N.; Wood, J.; Barber, J. The role of glutathione reductase and related enzymes on cellular redox homeostasis network. *Free Radic. Biol. Med.* **2016**, *95*, 27–42. [CrossRef] [PubMed]
34. Leong, P.K.; Ko, K.M. Induction of the glutathione antioxidant response/glutathione redox cycling by nutraceuticals: Mechanism of protection against oxidant-induced cell death. *J. Nutrac. Food Sci.* **2016**, *1*, 1–8.
35. Nandi, A.; Yan, L.; Jana, C.K.; Das, N. Role of catalase in oxidative stress- and age-associated degenerative diseases. *Oxidative Med. Cell. Longev.* **2019**, *2019*, 9613090. [CrossRef]
36. Arthur, J.R. The glutathione peroxidases. *Cell. Mol. Life Sci.* **2000**, *57*, 1825–1835. [CrossRef]
37. Chen, B.; Deen, W.M. Analysis of the effects of cell spacing and liquid depth on nitric oxide and its oxidation products in cell cultures. *Chem. Res. Toxicol.* **2001**, *14*, 135–147. [CrossRef]
38. Loria, V.; Dato, I.; Graziani, F.; Biasucci, L.M. Myeloperoxidase: A new biomarker in ischemic heart disease and acute coronary syndromes. *Mediat. Inflamm.* **2008**, *2008*, 135625. [CrossRef]

39. Karakas, M.; Koening, W. Myeloperoxidase production by macrophage and risk of atherosclerosis. *Curr. Atheroscler. Rep.* **2012**, *14*, 277–283. [CrossRef]
40. Huang, Y.; Li, W.; Su, Z.; Kong, A.T. The complexity of the Nrf2 pathway: Beyond the antioxidant response. *J. Nutr. Biochem.* **2015**, *26*, 1401–1413. [CrossRef]
41. Iranshahy, M.; Iranshahi, M.; Abtahi, S.R.; Karimi, G. The role of nuclear factor erythroid 2-related factor 2 in hepatoprotective activity of natural products: A review. *Food Chem. Toxicol.* **2018**, *120*, 261–276. [CrossRef] [PubMed]
42. Yamamoto, T.; Suzuki, T.; Kobayashi, A.; Wakabayashi, J.; Mahler, J.; Motohashi, H.; Yamamoto, M. Physiological significance of reactive cysteine residues of keap1 in determing Nrf2 activity. *Mol. Cell. Biol.* **2008**, *28*, 2758–2770. [CrossRef] [PubMed]
43. Di Stasi, L.C.; Camuesco, D.; Nieto, A.; Vilegas, W.; Zarzuelo, A.; Gálvez, J. Intestinal anti-inflammatory activity of paepalantine, an isocumarin isolated from the capitula of *Paepalanthus bromelioides*, in the trinitrobenzene sulphonic acid model of rat colitis. *Planta Med.* **2004**, *70*, 315–320.
44. Vilegas, W.; Roque, N.F.; Salatino, A.; Giesbrecht, A.M.; Davino, S. Isocoumarin from *Paepalanthus bromelioides*. *Phytochemistry* **1990**, *29*, 2299–2301. [CrossRef]
45. Kitagawa, R.R.; Raddi, M.S.G.; Khalil, N.M.; Vilegas, W.; Fonseca, L.M. Effects of the isocoumarin paepalantine on the luminol and lucigenin amplified chemiluminescence of rat neutrophils. *Biol. Pharm. Bull.* **2003**, *26*, 905–908. [CrossRef] [PubMed]
46. Cálgaro-Helena, A.F.; Devienne, K.F.; Rodrigues, T.; Dorta, D.J.; Raddi, M.S.G.; Vilegas, W.; Uyemura, S.A.; Santos, A.C.; Curti, C. Effects of isocoumarins isolated from *Paepalanthus bromelioides* on mitochondria: Uncoupling, and induction/inhibition of mitochondrial permeability transition. *Chem. Biol. Interact.* **2006**, *161*, 155–164. [CrossRef]
47. Devienne, K.F.; Cálgaro-Helena, A.F.; Dorta, D.J.; Prado, I.M.R.; Raddi, M.S.G.; Vilegas, W.; Uyemura, S.A.; Santos, A.C.; Curti, C. Antioxidant activity of isocoumarins isolated from *Paepalanthus bromelioides* on mitochondria. *Phytochemistry* **2007**, *68*, 1075–1080. [CrossRef]
48. Ardisson, J.S.; Gonçalves, R.C.R.; Rodrigues, R.P.; Kitagawa, R.R. Antitumour, immunomodulatory activity and in silico studies of naphthopyranones targeting iNOS, a relevant target for the treatment of *helicobacter pylori* infection. *Biomed. Pharmacother.* **2018**, *107*, 1160–1165. [CrossRef]
49. Luchini, A.C.; Orsi, P.R.; Cestari, S.H.; Seito, L.N.; Witaicenis, A.; Pellizzon, C.H.; Di Stasi, L.C. Intestinal anti-inflammatory activity of coumarin and 4-hydroxycoumarin in the trinitrobenzene sulphonic acid model of rat colitis. *Biol. Pharm. Bull.* **2008**, *31*, 1343–1350. [CrossRef]
50. Witaicenis, A.; Seito, L.N.; Di Stasi, L.C. Intestinal anti-inflammatory activity of esculetin and 4-methylesculetin in the trinitrobenzene sulphonic acid model of rat colitis. *Chem. Biol. Interact.* **2010**, *186*, 211–218. [CrossRef]
51. Yum, S.; Jeong, S.; Lee, S.; Kim, W.; Nam, J.; Jung, Y. HIF-prolyl hydroxylase is a potential molecular target for esculetin-mediated anti-colitic effects. *Fitoterapia* **2015**, *103*, 55–62. [CrossRef] [PubMed]
52. Rubio, V.; García-Pérez, A.I.; Herráez, A.; Diez, J.C. Different roles of Nrf2 and NF-κB in the antioxidant imbalance produced by esculetin or quercetin on NB4 leukemia cells. *Chem. Biol. Interact.* **2018**, *294*, 158–166. [CrossRef] [PubMed]
53. Witaicenis, A.; Luchini, A.C.; Hiruma-Lima, C.A.; Felisbino, S.L.; Garrido-Mesa, N.; Utrilla, P.; Gávez, J.; Di Stasi, L.C. Supression of TNBS-induced colitis in rats by 4-methylesculetin, a natural coumarin: Cmparison with prednisolone and sulphasalazine. *Chem. Biol. Interact.* **2012**, *195*, 76–85. [CrossRef] [PubMed]
54. Witaicenis, A.; Oliveira, E.C.S.; Tanimoto, A.; Zorzella-Pezavento, S.F.G.; Oliveira, S.L.; Sartori, A.; Di Stasi, L.C. 4-metylesculetin, a coumarin derivative, ameliorates dextran sulfate sodium-induced intestinal inflammation. *Chem. Biol. Interact.* **2018**, *280*, 59–63. [CrossRef]
55. Witaicenis, A.; Seito, L.N.; Chagas, A.S.; Almeida-Junior, L.D.; Luchini, A.C.; Rodrigues-Orsi, P.; Cestari, S.H.; Di Stasi, L.C. Antioxidant and intestinal anti-inflammatory effects of plant-derived coumarin derivatives. *Phytomedicine* **2014**, *21*, 240–246. [CrossRef]
56. Hassanein, E.H.M.; Sayed, A.M.; Hussein, O.E.; Mahmoud, A.M. Coumarins as modulators of the keap1/Nrf2/ARE signaling pathway. *Oxidative Med. Cell. Longev.* **2020**, *2000*, 1675957. [CrossRef]
57. Lv, H.; Zhu, C.; Wei, W.; Lv, X.; Yu, Q.; Deng, X.; Ci, X. Enhanced keap1-Nrf2/Trx-1 axis by daphnetin protects against oxidative stress-driven hepatotoxicity via inhibiting ASK1/JNK and Txnip/NLRP3 inflammasome activation. *Phytomedicine* **2020**, *71*, 153241. [CrossRef]
58. Tian, X.; Peng, Z.; Luo, S.; Zhang, S.; Li, B.; Zhou, C.; Fan, H. Aesculin protects against DSS-induced colitis through activating PPARγ and inhbiting NF-κB pathway. *Eur. J. Pharmacol.* **2019**, *857*, 172453. [CrossRef]
59. Li, W.; Wang, Y.; Wang, X.; He, Z.; Liu, F.; Zhi, W.; Zhanf, H.; Niu, X. Esculin attenuates endotoxin shock induced by lipopolysaccharide in mouse and NO production in vitro through inhibition of NF-κB activation. *Eur. J. Pharmacol.* **2016**, *791*, 726–734. [CrossRef]
60. Liu, A.; Shen, Y.; Du, Y.; Chen, J.; Pei, F.; Fu, W.; Qiao, J. Esculin prevents lipolysaccharide/D-galactosamine-induced acute liver injury in mice. *Microb. Pathog.* **2018**, *125*, 418–422. [CrossRef]
61. Kim, K.H.; Park, H.; Park, H.J.; Choi, K.; Sadikot, R.T.; Cha, J.; Joo, M. Glycosilation enables aesculin to activate Nrf2. *Sci. Rep.* **2016**, *6*, 29956. [CrossRef] [PubMed]
62. Khairy, H.; Saleh, H.; Badr, A.M.; Marie, M.S. Therapeutic efficacy of osthole against dinitrobenzene sulphonic acid induced-colitis in rats. *Biomed. Pharmacother.* **2018**, *100*, 42–51. [CrossRef] [PubMed]

63. Fan, H.; Gao, Z.; Ji, K.; Li, X.; Wu, J.; Liu, Y.; Wang, X.; Liang, H.; Liu, Y.; Li, X.; et al. The in vitro and in vivo anti-inflammatory effect of osthole, the major natural coumarin from *Cnidium monnieri* (L.) Cuss, via the blocking of the activation of the NF-κB and MAPK/p38 pathways. *Phytomedicine* **2019**, *58*, 152864. [CrossRef] [PubMed]
64. Xu, G.; Feng, L.; Song, P.; Xu, F.; Li, A.; Wang, Y.; Shen, Y.; Wu, X.; Luo, Q.; Wu, X.; et al. Isomeranzin suppress inflammation by inhibiting M1 macrophage polarization through the NF-κB and ERK pathway. *Int. Immunopharmacol.* **2016**, *38*, 175–185. [CrossRef]
65. Kohno, H.; Suzuki, R.; Curini, M.; Epifano, F.; Maltese, F.; Gonzales, S.P.; Tanaka, T. Dietary administrations with prenyloxy-coumarins, auraptene and collinin, inhibits colitis-related colon carcinogenesis in mice. *Int. J. Cancer* **2006**, *118*, 2936–2942. [CrossRef]
66. Kawabata, K.; Murakami, A.; Ohigashi, H. Auraptene decreases the activity of matrix metalloproteinases in dextran sulfate sodium-induced ulcerative colitis in ICR mice. *Biosci. Biotech. Biochem.* **2006**, *70*. [CrossRef] [PubMed]
67. Stenson, W.F. The universe of arachidonic acid metabolites in inflammatory bowel disease: Can we tell the good from the bad? *Curr. Opin. Gastroenterol.* **2014**, *30*, 347–351. [CrossRef] [PubMed]
68. Nielsen, H.; Ronne, A.; Elmgreen, J. Abnormal metabolism of arachidonic acid in chronic inflammatory bowel disease: Enhanced release of leukotriene B4 from activated neutrophils. *Gut* **1987**, *28*, 181–185. [CrossRef] [PubMed]
69. Fylaktakidou, K.C.; Hadjipavlou-Litina, D.J.; Litinas, K.E.; Nicolaides, D.N. Natural and synthetic coumarin derivatives with anti-inflammatory/antioxidant activities. *Curr. Pharm. Des.* **2004**, *10*, 3813–3833. [CrossRef]
70. Sekiya, K.; Okuda, H.; Arichi, S. Selective inhibtion of platelet lipoxygenase by esculetin. *Biochim. Biophys. Acta (BBA) Lipids Lip. Metab.* **1982**, *713*, 68–72.
71. Zhu, H.; Huang, J. Anti-inflammatory effects of scopoletin. *Zhongcaoyao* **1989**, *15*, 462–465.
72. Huang, G.J.; Deng, J.S.; Liao, J.C.; Hou, W.C.; Wang, S.Y.; Sung, P.J.; KUo, Y.H. Inducible nitric oxide synthase and cyclooxygenase-2 participate in anti-inflammatory activity of imperatorin from *Glehnia littoralis*. *J. Agric. Food Chem.* **2012**, *60*, 1673–1681. [CrossRef] [PubMed]
73. Chen, Y.; Zhang, Q.; Liu, H.; Lu, C.; Liang, C.L.; Qiu, F.; Han, L.; Dai, Z. Esculetin ameliorates psoriasis-like skin disease in mice by inducing $CD4^+Foxp3^+$ regulatory T cells. *Front. Immunol.* **2018**, *9*, 2092. [CrossRef] [PubMed]
74. Ji, J.; Ge, X.; Chen, Y.; Zhu, B.; Wu, Q.; Zhang, J.; Shan, J.; Cheng, H.; Shi, L. Daphnetin ameliorates experimental colitis by modulating microbiota compositions and T_{reg}/T_h17 balance. *FASEB J.* **2019**, *33*, 9308–9322. [CrossRef] [PubMed]
75. Sun, W.; Cai, Y.; Zhang, X.; Chen, H.; Lin, Y.; Li, H. Osthole pretreatment alleviates TNBS-induced colitis in mice via both cAMP/PKA-dependent and independent pathways. *Acta Pharmacol. Sin.* **2017**, *38*, 1120–1128. [CrossRef] [PubMed]
76. Cheng, J.; Shah, Y.M.; Gonzalez, F.J. Pregnane X receptor as a target for treatment of inflammatory bowel disorders. *Trends Pharmacol. Sci.* **2012**, *33*, 323–330. [CrossRef] [PubMed]
77. McDaniel, D.K.; Eden, K.; Ringel, V.M.; Allen, I.C. Emerging roles for non-canonical NF-κB signaling in the modulation of Inflammatory Bowel Disease pathobiology. *Inflamm. Bowel Dis.* **2016**, *22*, 2265–2279. [CrossRef]
78. Rogler, G.; Brand, K.; Vogl, D.; Page, S.; Hofmeister, R.; Andus, T.; Knuechel, R.; Baeuerle, P.A.; Scholmerich, J.; Gross, V. NuclearFactor κB is activated in macrophages and epithelial cells of inflamed intestinal mucosa. *Gastroenterology* **1988**, *115*, 357–369. [CrossRef]
79. Spehlman, M.E.; Eckmann, L. Nuclear factor-kappa B in intestinal protection and destruction. *Curr. Opin. Gastroenterol.* **2009**, *25*, 92–99. [CrossRef] [PubMed]
80. Karrash, T.; Jobin, C. NF-κB and the intestine: Friend or foe? *Inflamm. Bowel Dis.* **2008**, *14*, 114–124. [CrossRef] [PubMed]
81. Fu, Y.; Ma, J.; Shi, X.; Song, X.; Yang, Y.; Xiao, S.; Li, J.; Gu, W.; Huang, Z.; Zhang, J.; et al. A novel pyrazole-containing indolizine derivative suppresses NF-κB activation and protects against TNBS-induced colitis via a PPAR-γ-dependent pathway. *Biochem. Pharmacol.* **2017**, *135*, 126–138. [CrossRef] [PubMed]
82. Shah, Y.M.; Morimura, K.; Gonzalez, F.J. Expression of peroxisome proliferator-activated receptor-γ in macrophage. *Am. J. Physiol. Gastrointest. Liver Physiol.* **2007**, *292*, G657–G666. [CrossRef] [PubMed]
83. Croasdell, A.; Duffney, P.F.; Kim, N.; Lacy, S.H.; Sime, P.J.; Phipps, R.P. PPARγ and the innate immune system mediate the resolution of inflammation. *PPAR Res.* **2015**, *2015*, 549691. [CrossRef] [PubMed]
84. Sánchez-Hidalgo, M.; Villegas, M.I.; Alarcón de la Lastra, C. Rosiglitazone, an agonist of perosisome proliferator-activated receptor gamma, reduces chronic colonic inflammation is rats. *Biochem. Pharmacol.* **2005**, *69*, 1733–1744. [CrossRef]
85. Hou, Y.; Moreau, F.; Chadee, K. PPARγ is an E3 ligase that induces the degradation of NF-κB/p65. *Nat. Commun.* **2012**, *3*, 1300. [CrossRef]
86. Korbecki, J.; Bobinski, R.; Dutka, M. Self-regulation of the inflammatory response by peroxisome proliferator-activated receptors. *Inflamm. Res.* **2019**, *68*, 443–458. [CrossRef]
87. Kumar, J.; Rani, K.; Datt, C. Molecular link between dietary fibre, gut microbiota and health. *Mol. Biol. Rep.* **2020**, *47*, 6229–6237. [CrossRef]
88. Arora, R.; Sawney, S.; Saini, V.; Steffi, C.; Tiwari, M.; Saluja, D. Esculetin induces antiproliferative and apoptotic response in pancreatic cancer cells by directly binding to keap1. *Mol. Cancer* **2016**, *15*, 64. [CrossRef]
89. Coskun, M.; Olsen, J.; Seidelin, J.B.; Nielsen, O.H. MAP kinases in inflammatory bowel disease. *Clin. Chim. Acta* **2011**, *412*, 513–520. [CrossRef]
90. Broom, O.J.; Widjaya, B.; Troelsen, J.; Olsen, J.; Nielsen, O.H. Mitogen activated protein kinases: A role in inflammatory bowel disease? *Clin. Exp. Immunol.* **2009**, *158*, 272–280. [CrossRef] [PubMed]

91. Kyriakis, J.M.; Avruch, J. Mammalian MAPK signal transduction pathways activated stress and inflammation: A 10-year update. *Physiol. Rev.* **2012**, *92*, 689–737. [CrossRef] [PubMed]
92. Quaglio, A.E.V.; Castilho, A.C.S.; Di Stasi, L.C. Experimental evidence of MAP kinase gene expression on the response of intestinal anti-inflammatory drugs. *Life Sci.* **2015**, *136*, 60–66. [CrossRef] [PubMed]
93. Jeon, H.; Kim, H.; Choi, D.; Kim, D.; Park, S.; Kim, Y.; Kim, Y.M.; Jung, Y. Quercetin activates an angiogenic pathway, hypoxia inducible factor (HIF)-1-vascular endothelial growth factor, by inhibiting HIF-prolyl hydroxylase: A structural analysis of quercetin for inhibiting HIF-prolyl hydroxylases. *Mol. Pharmacol.* **2007**, *71*, 1676–1684. [CrossRef]
94. Choi, D.; Han, J.; Lee, Y.; Choi, J.; Han, S.; Hong, S.; Jeon, H.; Kim, Y.M.; Jung, Y. Caffeic acid phenethyl ester is a potent inhibitor of HIF prolyl hydroxylase: Structural analysis and pharmacological implication. *J. Nutr. Biochem.* **2010**, *21*, 809–817. [CrossRef] [PubMed]
95. Robinson, A.; Keely, S.; Karhausen, J.; Gerich, M.E.; Furuta, G.T.; Colgan, S.P. Mucosal protection by hypoxia-inducible factor prolyl hydroxylase inhibition. *Gastroenterology* **2008**, *134*, 145–155. [CrossRef] [PubMed]
96. Tambuwala, M.M.; Cummins, E.P.; Lenihan, C.R.; Kiss, J.; Stauch, M.; Scholz, C.C.; Fraisl, P.; Lasitschka, F.; Mollenhauer, M.; Saunders, S.P.; et al. Loss of prolyl hydroxylase-1 protects against colitis through reduced epithelial cell apoptosis and increased barrier function. *Gastroenterology* **2010**, *139*, 2093–2101. [CrossRef]
97. Lee, J.W.; KO, J.; Ju, C.; Eltzschig, H.K. Hypoxia signaling in human diseases and therapeutic targets. *Exp. Mol. Med.* **2019**, *51*, 68. [CrossRef]
98. Zhang, Z.; Yao, L.; Yang, J.; Wang, Z.; Du, G. PI3K/Akt and HIF-1 signaling pathway in hypoxia-ischemia. *Mol. Med. Rep.* **2018**, *18*, 3547–3554. [CrossRef]
99. Liu, M.; Zhang, G.; Zheng, C.; Song, M.; Liu, F.; Huang, X.; Bai, S.; Huang, X.; Lin, C.; Zhu, C.; et al. Activating the pregnane X receptor by imperatorin attenuates dextran sulphate sodium-induced colitis in mice. *Br. J. Pharmacol.* **2018**, *175*, 3563–3580. [CrossRef]
100. Gao, J.; Xie, W. Pregnane X receptor and constitutive androstane receptor at the crossroads of drug metabolism and energy metabolism. *Drug Metab. Dispos.* **2010**, *38*, 2091–2095. [CrossRef]
101. Shah, Y.M.; Ma, X.; Morimura, K.; Kim, I.; Gonzalez, F.J. Pregnane X receptor activation ameliorates DSS-induced inflammatory bowel disease via inhibition of NF-κB target gene expression. *Am. J. Physiol. Gastrointes. Liver Physiol.* **2007**, *292*, G1114–G1122. [CrossRef] [PubMed]
102. DeMeo, M.T.; Mutlu, E.A.; Keshavarzian, A.; Tobin, M.C. Intestinal permeation and gastrointestinal disease. *J. Clin. Gastroenterol.* **2002**, *34*, 385–396. [CrossRef] [PubMed]
103. Deuring, J.J.; Li, M.; Cao, W.; Chen, S.; Wang, W.; de Haar, C.; van der Woude, C.J.; Peppelenbosch, M. Pregnane X receptor activation constrains mucosal NF-κB activity in active inflammatory bowel disease. *PLoS ONE* **2019**, *14*, e0221924.
104. Wag, Y.; Chen, J.; Chen, J.; Dong, C.; Yan, X.; Zhu, Z.; Lu, P.; Song, Z.; Liu, H.; Chen, S. Daphnetin ameliorates glucocorticoid-induced osteoporosis via activation of Wnt/GSK-3β/β-catenin signaling. *Toxicol. Appl. Pharmacol.* **2020**, *409*, 115333.
105. Lee, J.-H.; Kim, Y.-G.; Cho, S.S.; Ryu, S.Y.; Cho, M.H.; Lee, J. Coumarins reduce biofilm formation and virulence of *Escherichia coli* O157:H7. *Phytomedicine* **2014**, *21*, 1037–1042. [CrossRef]
106. Martínez Aguilar, Y.; Rodriguez, F.S.; Saavedra, M.A.; Hermosilla Espinosa, R.; Yero, O.M. Secondary metabolites and in vitro antibacterial activity of extracts from *Anacardium occidentale* L. (cashew tree) leaves. *Rev. Cuba. Plantas Med.* **2012**, *17*, 320–329.
107. Widelski, J.; Luca, S.V.; Skiba, A.; Chinou, I.; Marcout, L.; Wolfender, J.-L.; Skalicka-Wozniak, K. Isolation and antimicrobial activity of coumarin derivatives from fruits of *Peucedanum luxurians* Tamamsch. *Molecules* **2018**, *23*, 1222. [CrossRef]
108. Saho, J.; Kumar Mekap, S.; Sudhir, K.P. Synthesis, spectral characterization of some new 3-heteroaryl-azo-4-hydroxy coumarin derivatives and their antimicrobial evaluation. *J. Taibah Univ. Sci.* **2015**, *9*, 187–195. [CrossRef]
109. López-Rojas, P.; Janeczko, M.; Kubinski, K.; Amestt, Á.; Maslyk, M.; Estévez-Braun, A. Synthesis and antimicrobial activity of 4-substituted 1, 2, 3-triazole-coumarin derivatives. *Molecules* **2018**, *23*, 199. [CrossRef]
110. Zhang, Z.; Xu, H.; Zhao, H.; Geng, Y.; Ren, Y.; Guo, L.; Shi, J.; Xu, Z. *Edgeworthia gardneri* (Wall.) Meisn. Water extract improves diabetes and modulates gut microbiota. *J. Ethnopharmacol.* **2019**, *239*, 111854. [CrossRef]
111. Zhao, X.-Q.; Guo, S.; Lu, Y.-Y.; Hua, Y.; Zhang, F.; Yan, H.; Shang, E.-X.; Wang, H.-Q.; Zhang, W.-H.; Duan, J.-A. *Lyceum barbarum* L. leaves ameliorate type 2 diabetes in rats by modulating metabolic profiles and gut microbiota composition. *Biomed. Pharmacother.* **2020**, *121*, 109559. [CrossRef] [PubMed]

Article

Antiplatelet Activity of Coumarins: In Vitro Assays on COX-1

Cristina Zaragozá *, Francisco Zaragozá, Irene Gayo-Abeleira and Lucinda Villaescusa

Pharmacology Unit, Biomedical Sciences Department, University of Alcalá, Alcalá de Henares, 28871 Madrid, Spain; francisco.zaragoza@uah.es (F.Z.); irene.gayo@uah.es (I.G.-A.); lucinda.villaescusa@uah.es (L.V.)
* Correspondence: cristina.zaragoza@uah.es

Abstract: Atherosclerotic cardiovascular disease is the leading cause of death in developed countries. Therefore, there is an increasing interest in developing new potent and safe antiplatelet agents. Coumarins are a family of polyphenolic compounds with several pharmacological activities, including platelet aggregation inhibition. However, their antiplatelet mechanism of action needs to be further elucidated. The aim of this study is to provide insight into the biochemical mechanisms involved in this activity, as well as to establish a structure–activity relationship for these compounds. With this purpose, the antiplatelet aggregation activities of coumarin, esculetin and esculin were determined in vitro in human whole blood and platelet-rich plasma, to set the potential interference with the arachidonic acid cascade. Here, the platelet COX activity was evaluated from 0.75 mM to 6.5 mM concentration by measuring the levels of metabolites derived from its activity (MDA and TXB_2), together with colorimetric assays performed with the pure recombinant enzyme. Our results evidenced that the coumarin aglycones present the greatest antiplatelet activity at 5 mM and 6.5 mM on aggregometry experiments and inhibiting MDA levels.

Keywords: coumarin; esculin; esculetin; antiplatelet activity; impedance aggregometry; COX; polyphenols

Citation: Zaragozá, C.; Zaragozá, F.; Gayo-Abeleira, I.; Villaescusa, L. Antiplatelet Activity of Coumarins: In Vitro Assays on COX-1. *Molecules* **2021**, *26*, 3036. https://doi.org/10.3390/molecules26103036

Academic Editor: Maria João Matos

Received: 15 April 2021
Accepted: 14 May 2021
Published: 19 May 2021

1. Introduction

Platelets present a wide variety of functions in the blood circulation, with a key role in the development of the atherosclerotic process and the subsequent physiopathology of the cardiovascular disease [1]. Once attached to the vascular endothelium and activated, the platelets release a broad range of molecules such as chemokines, proinflammatory agents and different substances able to modulate a biological response that will promote the interaction among platelets, endothelial cells and leukocytes [2]. These cell interactions trigger a local inflammatory response, which is mainly responsible for the atherosclerotic process [3]. Platelet adhesion to the luminal vascular surface occurs after exposure of the endothelium caused by a lesion or detachment of an atherosclerotic plaque. Platelet aggregation represents the initial stage in the formation of a blood clot that can lead, to a greater or lesser extent, to the vascular occlusion and eventually result in thromboembolic disease such as stroke or myocardial infarction [4].

An irreversible stage of platelet aggregation is mainly induced by the secretion of substances from the platelet granules content. This event has also been observed in vitro as a response to the addition of high concentrations of agonists. The process includes the formation of metabolites mostly derived from arachidonic acid (cyclic endoperoxides and TXA_2) and the secretion of the content from lysosomes and dense and α-granules in platelets [5]. Coumarins (2*H*-1-benzopyran-2-ones), the lactones of the 2-hydroxy-*Z*-cinnamic acids, are phenolic compounds with complex structures that differ substantially across the family [6] and are extensively distributed in the plant kingdom, especially in the families *Apiaceae, Asteraceae* and *Rutaceae* [7]. Naturally occurring coumarins, even though all of them contain the coumarin moiety, are structurally different and can be classified according to their chemical structure in the following groups: simple coumarins, furanocoumarins, dihydro-furanocoumarins, phenylcoumarins, pyranocoumarins and dicoumarins [8].

Simple coumarins are usually substituted at position 7 (C-7) with a hydroxyl but can be also hydroxylated at positions 6 and 8. These hydroxyl groups can be sometimes methylated or substituted with sugar molecules, in which case they are referred to as glycosylated or heterosidic coumarins [9]. The presence of the different substituents in the main structure largely influences the biological activity of the resulting compound [10].

Coumarin (2*H*-1-benzopyran-2-one) (Figure 1) has been under research due to its interesting and wide-ranging bioactivities, inclusive of anti-inflammatory [11,12], antioxidant [7], antimicrobial [13–15], antiproliferative [16,17] and anticoagulant properties [18]. The vitamin K antagonists in clinical use are structurally derived from 4-hydroxycoumarin and share a common mechanism of action in that they noncompetitively inhibit the vitamin K epoxide reductase complex, which is essential in the recycling of vitamin K in the liver. As vitamin K serves as a cofactor in the activation of clotting factors II, VII, IX and X, the inhibition of its recycling results in strong anticoagulation activity [19].

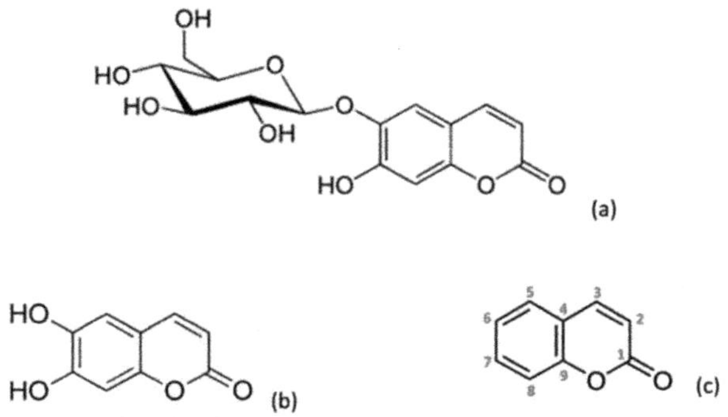

Figure 1. Chemical structure of the different coumarins assayed: esculin (**a**), esculetin (**b**) and coumarin (**c**).

Esculin (7-hydroxy-6-[(2 S, 3 R, 4 S, 5 S, 6 R)-3,4,5-trihydroxy-6-(hydroxymethyl) oxane-2-yl] oxychromen-2-one) (Figure 1) is a coumarin derivative found in *Aesculus hippocastanum* L. (horse chestnut) [20] that has demonstrated promising anti-inflammatory, antioxidant and free radical scavenging properties. This compound was effective in diminishing the elevated blood creatinine levels in diabetic mice, which ameliorated diabetes-induced renal dysfunction through a reduction on the activation levels of caspase-3 in the mice kidney [21]. Likewise, esculin showed a protective effect against lipid metabolism disorders in diabetic rats in a dose-dependent manner. The authors of this study proposed that the possible mechanism might be associated with the inhibition of AGE (advanced glycation end products) formation [22].

Conversely, esculetin (6,7-dihydroxychromen-2-one) (Figure 1) is the aglycone of the heteroside esculin. This compound has been thoroughly investigated because of its anti-inflammatory activity, which is conducted through several mechanisms that include the inhibition of ICAM-1 release, the decrease of NO and PGE_2 levels in synovial fluid, myocardial protection or the inhibition of proinflammatory cytokines during the interaction between adipocytes and macrophages [23]. Some evidence for the potential of this aglycone to decrease oxidative stress has also been demonstrated [24], together with the presence of antidiabetic [25], antibacterial [26] and antitumor activities [27].

Despite the significant number of studies based on these types of chemical compounds and the diverse biological activities described for coumarin, esculetin and esculin, the mechanisms of action remain partially unknown. This research work focuses on demonstrating the antiplatelet activity of these coumarins and shedding some light on their mechanism of action. Due to the important role of the cyclooxygenase (COX) enzyme in platelet aggregation, it has been hypothesized that the potential interaction with COX is a possible mechanism through which coumarins could exert their antiplatelet function.

2. Results

2.1. *Antiaggregant Effect of Coumarins by Impedance Platelet Aggregometry*

The percentage of platelet aggregation in whole blood (WB) and platelet-rich plasma (PRP) after activation by adenosine diphosphate (ADP) or arachidonic acid (AA) was calculated. Maximal aggregation (100%) was considered when ADP or AA were used in absence of any other compound. All the assayed phenolic compounds were tested at different concentrations: 0.75, 1.5, 3.0, 5.0, 6.5 mM. These concentrations are similar to the daily dose clinically used for the flavonoid diosmin (Daflon® 500 mg) [28]. The percentages of aggregation for coumarin, esculin and esculetin are shown in Figure 2.

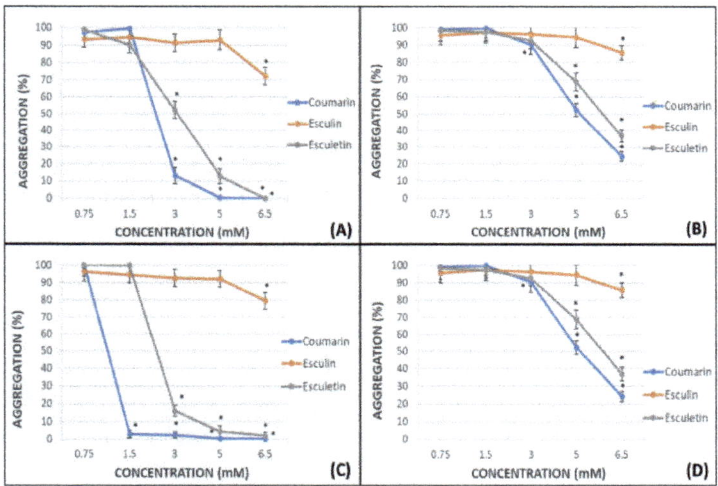

Figure 2. Graphical representation of the percentage of platelet aggregation for coumarin, esculin and esculetin. Panels (**A**) (WB samples) and (**C**) (PRP samples) shows results in AA-induced platelet aggregation. Panels (**B**) (WB samples) and (**D**) (PRP samples) shows results in ADP-induced platelet aggregation. Results are expressed as the mean and standard deviation for 10 donors. Error bars represent the standard deviation. * $p < 0.05$: statistically significant differences in platelet aggregation between samples with and without the tested phenolic compound.

In general, the antiplatelet effect of the screened phenolic compounds was observed at concentrations equal to or higher than 3 mM and was more potent in samples subjected to AA-induced platelet aggregation (Figure 2A,C) than in those subjected to ADP-induced platelet aggregation (Figure 2B,D), reaching on WB AA-induced experiments a IC50 for coumarin and esculetin of 2.45 mM and 3.07 mM respectively, and 5.12 mM and 5.82 mM on ADP experiments. The IC50 on PRP AA-induced samples were 1.12 mM for coumarin and 2.48 mM and 5.08 mM and 5.97 mM for ADP-induced assays.

The effects of coumarin and esculetin as antiaggregant agents were especially relevant in all the experiments performed. The complete inhibition of the platelet aggregation was achieved in AA-induced activated PRP samples after addition of 1.5 mM of coumarin (Figure 2C).

2.2. *Platelet MDA Levels and COX Activity*

Considering that the extent of inhibition was greater in the impedance aggregometry assays when AA was used as activator (Figure 2A,C) [29], the MDA levels were quantified in AA-induced activated PRP samples. Maximal activity of COX (100%) was set as that obtained for AA-treated samples in the absence of phenolic compound. Indomethacin was used as a positive control at the doses shown below (Figure 3).

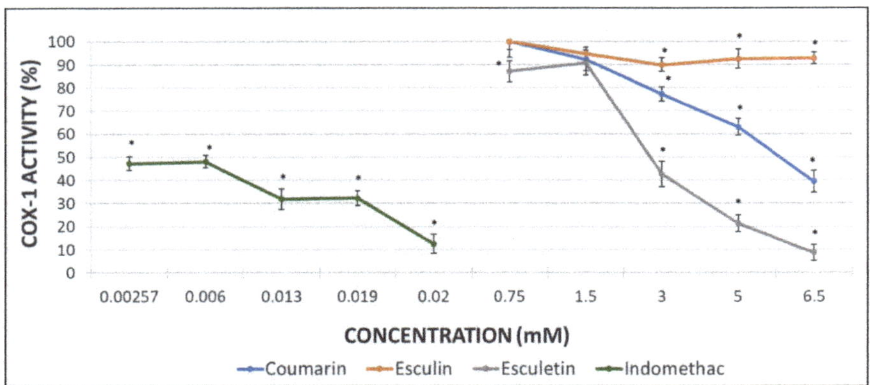

Figure 3. Graphical representation of the percentage of COX activity in AA-induced activated PRP samples after addition of increasing concentrations of indomethacin, as positive control, and assayed coumarins. Results are expressed as the mean and standard deviation for 10 donors. Error bars represent the standard deviation. * $p < 0.05$: significant differences regarding COX activity with and without the examined substances.

The inhibition of COX activity was confirmed in presence of indomethacin (Figure 3). Coumarin and esculetin showed their ability to inhibit COX-1 activity, while esculin did not affect it significantly (Figure 3). The IC50 for coumarin and esculetin were 5.93 mM and 2.76 mM for coumarin and esculetin, respectively.

2.3. *COX-1 Inhibitory Assay*

With the aim to determine the direct effect of the compounds here investigated on the COX-1 activity, analyses with the pure human recombinant enzyme (h-COX-1) were performed. Results were expressed as the percentage of COX-1 activity in the presence of indomethacin (as a positive control) or after incubation with the different tested molecules.

As it can be observed in Figure 3, indomethacin diminished almost completely the activity of the recombinant COX-1 enzyme. Coumarin produced a decrease of 49% in h-COX-1, while esculetin barely reached a drop of 42%. By contrast, esculin demonstrated the greatest COX-1 inhibitory effect, with a 74% of enzyme inhibition rate (Figure 4) and IC50 of 4.49 mM.

2.4. *TXB_2 Levels as COX-1 Activity Indicator*

TXB_2 quantification was conducted in WB and used as an indicator of the COX-1 activity since it is assumed that the administration of a COX inhibitor will decrease the levels of TXB_2. The effect of indomethacin as a positive control was analysed at different concentrations. Meanwhile, the calcium ionophore (CI) (25 mM) was employed as a platelet aggregation inducer to evaluate the potential effect of the assayed phenolic compounds.

The results demonstrated the inhibition of the TXB_2 production by indomethacin at all the concentrations tested. On the contrary, any of the phenolic compounds investigated exerted any effect on the levels of this metabolite (Figure 5).

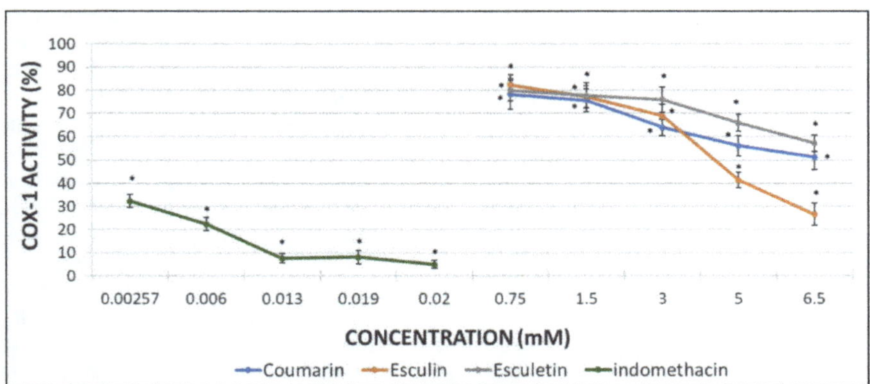

Figure 4. Graphical representation of the percentage of h-COX-1 activity after indomethacin and tested coumarins addition in AA-induced activated samples. Results are expressed as the mean and standard deviation for 10 donors. Error bars represent the standard deviation. * $p < 0.05$: significant differences on h-COX-1 activity with and without the examined substances.

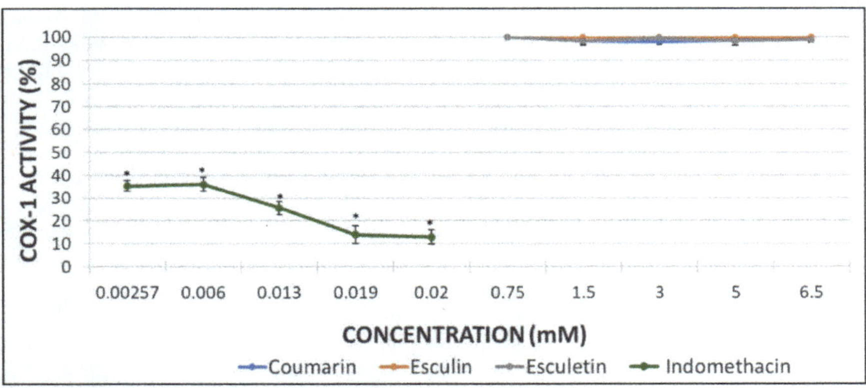

Figure 5. Graphical representation of the percentage of TXB_2 production at different concentrations of indomethacin and coumarins after CI-induced aggregation. Results are expressed as the mean and standard deviation for 10 donors. Error bars represent the standard deviation. * $p < 0.05$: significant differences between the basal TXB_2 production with and without the examined substances.

3. Discussion

In this research study, the potency of the aglycones coumarin and esculetin as antiplatelet aggregation agents was evidenced by the impedance aggregometry assays. The effect observed was similar in WB and PRP, albeit the antiaggregant activity was superior in the latter. In both cases, the effect was more remarkable in AA-induced than in ADP-induced activated samples. Coumarin and esculetin showed an inhibition of 87% and 52%, respectively, in AA-induced samples at a concentration of 3 mM, whereas the inhibition was complete with highest concentrations.

With respect to coumarin, this compound was able to completely inhibit platelet agglutination in AA-induced PRP samples when a 1.5 mM solution was used. Regarding the effect of esculetin, the inhibition rate achieved a 16% at 3 mM, and was complete at 6.5 mM. The esculetin, in its heterosidic form as esculin, demonstrated a minimal decrease in platelet aggregation at 6.5 mM that was null at lower concentrations in WB and PRP, in either AA- or ADP-induced activated samples. Thus, our results revealed that the presence of the catechol group in esculetin favours the antiplatelet activity, while this activity is lost when the C-6 hydroxyl group is replaced by a sugar, such as in esculin.

The higher antiaggregant potency of aglycones vs. heterosides was confirmed in the COX-1 activity assay from the measurement of MDA when AA was employed as aggregant agent in PRP samples. Esculetin showed a 90% of inhibition of COX-1 activity at 6.5 mM, which was similar to the inhibition reached with the positive control indomethacin. Unlike esculetin, coumarin just showed a 60% of inhibition at the maximal concentration assayed and no significant effect was found for esculin.

However, the results obtained in the COX-1 inhibitory assay performed with the human pure recombinant enzyme were somewhat different. In this case, the heteroside esculin procured the greatest inhibitory effect on h-COX-1 since the enzyme activity decreased up to the 26%. Nevertheless, coumarin produced a 50% enzyme inhibition in a similar way to the inhibition showed by the MDA measurement assay. Esculetin, that had previously shown a 90% of COX-1 inhibition, returned a 57% of enzyme inhibition when tested with the pure enzyme. Regarding the indomethacin (positive control), the results with the pure enzyme seem to be more relevant than those related to the MDA production, since the enzyme inhibition reached a 95%.

Surprisingly, any of the coumarins investigated had an impact in TXB_2 levels. Indomethacin, for its part, prevented TXB_2 production in a 95%. Hence, it can be assumed that this molecule significantly inhibits COX-1 activity.

Coumarin and esculetin presented a dose-dependent effect on the platelet aggregation, the COX-1 activity and the pure enzyme. Higher concentrations of coumarins than those selected in this research work might exert a larger effect, but the low solubility of these compounds is a major limitation. Even though DMSO is an optimal solvent for these molecules, the use of higher amounts could affect the platelets integrity and is discouraged [30,31]. Previous experiments in our laboratory showed how a higher volume than 2 μL of DMSO could damage platelets contained in 1 mL of WB [32].

Polyphenols are naturally present in plants as O- and C-glycosides, while aglycones are not found in fresh plants but can occur after processing [33]. In general, the oral bioavailability of polyphenols is considerably limited [34,35]. As a consequence of enzyme hydrolysis, the heterosides lose the glycosidic moiety before reaching the bloodstream and can then pass through the cell membranes [36].

As previously indicated, our in vitro results show that in the specific case of coumarins, the aglycones present a greater antiplatelet effect than their heterosidic parent compounds.

Esculin did not show any activity in our in vitro experiments apart from those performed with the pure enzyme, which suggests the inability of this compound to access the platelet interior. This fact could be explained by the presence of a sugar ring in its chemical structure. Considering that glycoxidation favours the biological activity of coumarins [37], it could be hypothesized that esculin would present a similar activity to its aglycone in vivo.

The smaller size of the coumarin and esculetin structures could ease their transport across the platelet membrane, and hence, produce a higher effect on the COX activity. However, our results support the feasibility of and need for future studies on the interaction of the coumarins with blood platelet membrane. Notwithstanding, these two molecules were not able to inhibit TXB_2 production. Thereby, the results here presented point to a mechanism of action at a different level that would possibly involve TXA_2 receptors.

4. Materials and Methods

4.1. Selected Compounds

The selected phenolic compounds coumarin, esculin and esculetin were purchased from Sigma-Aldrich (Sigma-Aldrich Chemical, Madrid, Spain) and dissolved in DMSO (dimethylsulphoxide) (Dismadel, Madrid, Spain) to a final concentration of 0.5 mM. This concentration is similar to the daily dose clinically used for the flavonoid diosmin (Daflon® 500 mg) [28] and was established considering the structural similarity (presence of the benzo-α-pyrone core), the almost identical physicochemical properties and their comparable molecular weight between this drug and the compounds here investigated [38]. Further dilutions were performed to reach 0.75, 1.5, 3.0, 5.0, 6.5 mM for the different assays. To avoid altering the platelet configuration, the lowest volume of DMSO (Dismadel S.L., Madrid, Spain) that could ensure the dissolution of the compounds (2 μL) was added to the blood samples [32].

4.2. *Study Cohort, Inclusion and Exclusion Criteria*

Ten healthy volunteers (seven women and three men; aged 22.2 ± 1.2 [mean ± SD] years), none of which had undergone platelet function or complement activation treatment during the previous year, were recruited to furnish blood for every assay of this work. Participants were not included if they were smokers or showed any sign of kidney, lung, heart, or autoimmune disease, any chronic or acute infection, diabetes mellitus, a history of tumours, immunodeficiency or thrombocytopathy, hypercholesterolemia or were undergoing immunosuppressant, steroids, or nonsteroidal anti-inflammatory drug (NSAID) treatment. They were excluded if they had undergone any other treatment that could affect the platelet activity during the six months prior to the assay, anovulants included.

Written informed consent was signed by every participant. The study protocol was carried out in strict accordance with the guidelines of the 1975 Declaration of Helsinki, under approval of the Biomedical Ethics Committee of the University of Alcalá.

4.3. *Peripheral Blood Extraction*

Peripheral blood was collected by an antecubital puncture in sodium citrate-containing (3.8% wt/vol) Vacutainer® tubes (Dismadel S.L., Madrid, Spain), discarding the first 2 mL. All extractions were performed at the Haematology Service of the Principe de Asturias Hospital, Alcalá de Henares (Madrid, Spain). Sodium citrate was selected as the anticoagulant instead of heparin, ethylenediaminetetraacetic acid (EDTA), or D-phenylalanyl-L-prolyl-L-arginine chloromethyl ketone (PPACK) given its lesser impact on the complement activation pathways [39].

4.4. *Blood Samples Preparation*

Platelet aggregation assays were performed in both, WB and platelet-rich plasma PRP [40]. The removal of blood cell components in PRP allows to better evaluate the effect of a specific compound on the platelets. Two different platelet activity inducers were employed: AA (Sigma-Aldrich Chemical, Madrid, Spain) 0.5 mM [41] and ADP (Sigma-Aldrich Chemical, Madrid, Spain) 5 μM [42]. The use of two different substances (AA and ADP) that promote platelet aggregation through different pathways, together with their administration on the different types of samples (WB and PRP), provides with a better understanding on the level of action of the different compounds under research. It was considered that the screened compounds could present different activities depending on the medium or the aggregation inducer.

4.4.1. WB Samples Preparation

Samples were kept at room temperature until use. They were homogenized in a plastic beaker and aliquots of 500 μL were distributed in aggregation Chronolog polyethylene cuvettes (Labmedics, Oxfordshire, UK) as soon as possible. After that, 500 μL of physiological saline solution (PSS) (Dismadel S.L., Madrid, Spain) were added to each cuvette. The diluted samples were incubated at 37 °C for 1 h with the selected compounds, or DMSO in the case of the control sample, in a thermostatic bath Unitronic 320 Selecta (Tecnylab, Madrid, Spain) to increase their solubility. Sample preparation and subsequent assays were performed in the first 3 h after blood extraction.

4.4.2. PRP Samples Preparation

Blood samples were subjected to centrifugation for 10 min at 1200 rpm twice in a centrifuge Jouan B-3.11 (Tecnylab, Madrid, Spain) and PRP was obtained by collecting the supernatant. Platelet counts were normalized to 200,000 platelets/μL PRP. Briefly, PRP was dissolved in the hematologic solvent Diluid 601 (Biolab Diagnostics, Barcelona, Spain) and counting was performed in a Neubauer cell chamber using a binocular microscope NIKON (Izasa S.L., Madrid, Spain). This method avoids some potential errors linked to automated cell counters, such as detection of bubbles as particles, the counting of cell components other than platelets or counting groups of platelets present in the sample. The refringence and morphology of platelets under the light microscope facilitated their unambiguous identification. Once the number of platelets in the original PRP samples was known, the calculated volume was transferred to the cuvettes and PSS was added until a final volume of 1 mL. Next, samples were incubated with the assayed compound or DMSO as a control for 5 min at 37 °C.

4.5. Impedance Platelet Aggregometry Assay

The procedure was carried out in a Chrono-Log 500 Lumi-Aggregometer (Labmedics, Oxfordshire, UK) connected to an Omnioscribe II data-logger, according to the manufacturer's instructions. Only plastic material was used to be in contact with the samples. The experimental method is based on the measurement of the change in the electrical impedance (the resistance to the electric current) between two electrodes when platelet aggregation is induced by an agonist [40]. Thereby, the electrodes immersed in the WB or PRP samples continuously stirred at 1200 rpm become covered by a platelet monolayer. The impedance remains constant in the absence of an aggregation agent. On the contrary, the addition of an aggregant promotes the adhesion and agglutination of the platelets in the electrodes and produces an increase in the impedance that can be used as a measurement of the platelet aggregation.

4.6. MDA Quantification and COX Activity Assessment

Plasma MDA levels reflect COX activity and can be used as a qualitative test of platelet function or to quantify the effect of COX inhibitors. MDA is a product of the arachidonic acid metabolism in platelets that can be measured by spectrophotometric techniques. Absorbance is recorded at 532 nm to ensure that it is entirely due to the released MDA. The molar extinction coefficient of MDA at 532 nm is 1.56×10^5 [29].

MDA analysis was performed in AA-induced activated PRP samples [31], since the MDA absorbance levels in the ADP-induced activated samples were much lower than those corresponding to total COX activity.

Calibration curves were created by preparing a set of solutions with known concentrations of MDA and measured in a UV/VIS Philips PU 8700 spectrophotometer at 532 nm to later extrapolate the MDA levels in PRP. The curves were prepared in PSS and PRP without aggregant to establish the possible MDA release by nonactivated platelets.

Curves ranged from 100–1000 nM to 1–10 μM and had regression coefficients near 1 (0.997 y 0.994, respectively) (Figure 6).

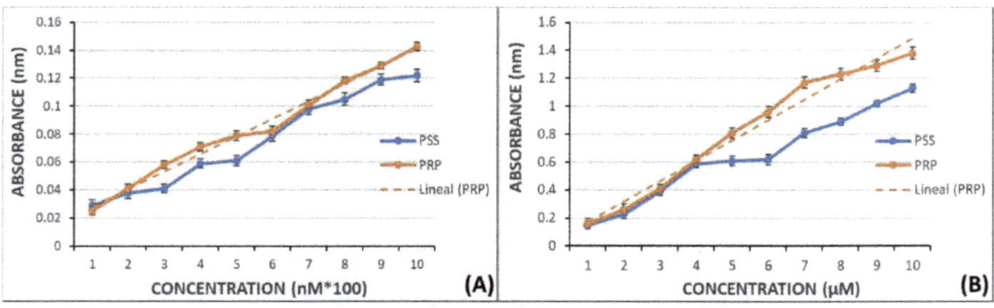

Figure 6. Graphical representation of MDA absorbance in PSS y PRP. Panel (**A**): in the range from 100 nM to 1000 nM and linear fitting in PRP samples (r = 0.997). Panel (**B**): in the range from 1 μM to 10 μM and linear fitting in PRP samples (r = 0.994).

Test samples were processed following a similar procedure to the samples used to obtain the calibration curves. A volume of 375 μL of 40% trichloroacetic acid (ATA) (Sigma-Aldrich Chemical, Madrid, Spain) was added to propylene tubes containing the AA-induced activated PRP samples and gently mixed to mediate the protein precipitation. The tubes were covered to prevent oxidation. PSS was added up to a final volume of 2 mL, like in the samples used in the calibration curves. After centrifugation at 3500 rpm for 10 min, the supernatant was filtered through glass wool. One more centrifugation step was performed in similar conditions and then 0.12 M tiobarbituric acid (TBA) (Sigma-Aldrich Chemical, Spain) was added in a relation of 0.2 volumes per volume of acid supernatant.

Next, the capped tubes were heated in a water bath at 100 °C for 15 min. Once cooled down, the spectrophotometric measures were obtained at 532 nm.

MDA concentrations in control samples were considered as 100% of COX-1 activity and the concentrations in the test samples were expressed as the percentage of COX-1 activity. Indomethacin (Sigma-Aldrich Chemical, Madrid, Spain) was employed as positive control at the following concentrations: 0.00257, 0.006, 0.013, 0.019, 0.02 mM. These concentrations were set considering that the therapeutic concentration is 0.004 mg/mL [43]. The assayed compounds were dissolved in DMSO (2 μL) and added to the reaction medium at the concentrations previously mentioned: 0.75, 1.5, 3.0, 5.0, 6.5 mM.

4.7. Procedure for the COX-1 Inhibitory Assay

Human purified COX-1 enzyme with a purity of 95% was purchased from Vitro S.A. (Madrid, Spain). The enzyme was supplied in 10KU vials prepared in Tris-HCl 80 Mm and 1% Tween 20. The enzyme unit (EU) is defined as the amount of enzyme required to produce a change of 0.001 mn^{-1} in the optical density at a wavelength of 610 nm. According to the manufacturer, the vials were stored frozen and kept in the dark on ice during the assays. One hundred EU (4 μL) were added to the samples. AA was used as enzyme substrate to replicate the aggregation inducer employed in the impedance aggregometry and MDA quantification experiments.

The enzyme activity was determined by a chromogenic method based on the oxidation of N,N,N′,N′-tetramethyl-p-phenylenediamine (TMPD) [44]. Among the substances produced during AA-induced platelet activation, the prostaglandin G_2 (PGG_2) is quickly reduced to PGH_2 by the platelet enzyme COX-1. Because of this reduction, the TMPD is oxidized in a directly proportional amount to the enzyme activity.

The experimental studies were carried out in solutions containing the pure enzyme incubated with the test compounds and AA as the enzyme substrate. Absorbance produced by TMPD was measured at 610 nm in a Biotek ELx800 Absorbance Microplate Reader (Izasa Scientific, Madrid, Spain).

The screened compounds were prepared in DMSO at the concentrations abovementioned for the previous assays. Indomethacin, a selective COX-1 inhibitor, was used as control drug at the therapeutic concentrations of 0.001, 0.0025, 0.005, 0.0075, 0.01mg/mL [43].

4.8. Enzyme Immunoassay for the Quantitative Determination of TXB_2

TXA_2 is produced from AA oxidation and physiologically active. However, it is rapidly hydrolysed (average life of 30 s) to form TXB_2, a stable and biologically inactive metabolite [45]. TXB_2 concentration, as measured by immunoassay, is maximal at 20–30 min and declines thereafter [46], being considered as a measurement of the TXA_2 levels. For this reason, the samples were incubated in WB and calcium ionophore A23187 (CI) (Sigma Aldrich, Madrid, Spain) was added at a concentration of 25 mM [32] to trigger platelet activation that produces TXB_2. In this way, it can be considered as an indirect measure of the COX-1 activity inhibition [46].

TXB_2 levels were determined by a specific enzyme immunoassay kit (TXB_2 Biotrak Enzymeimmunoassay System, Amersham Biosciences, Little Chalfont, Buckinghamshire, UK) according to the manufacturer´s protocol. This kit possesses a high sensitivity (0.2 pg) and the standard curve ranges from 0.5 to 64 pg. The absorbance values were obtained in a

Biotek ELx800 Absorbance Microplate Reader (Izasa Scientific, Madrid, Spain) coupled to an automatic microplate washer Biotek ELx50 (Izasa Scientific, Madrid, Spain).

Similarly to prior assays, indomethacin was used as control drug because of its ability to inhibit COX [43].

To perform the procedure, the total volume of WB from donors was distributed in equal aliquots of 1 mL and incubated in a thermostatic bath (Unitronic 320 Selecta) (Izasa Scientific, Madrid, Spain) at 37 °C for 1 h with 2 μL of the test solutions or DMSO as control. After that time, 2 μL of CI 25 mM were added and incubation maintained for a further 30 min cycle. The reaction was terminated by introducing the samples on dry ice. Then, the samples were centrifuged (centrifuge Jouan 3.11) (Tecnylab, Madrid, Spain) at 4000 rpm for 10 min and the supernatant collected and subjected to the enzyme-linked immunosorbent technique.

4.9. Statistical Analysis

All results are expressed as the mean ± standard deviation (SD) of values obtained in each experiment. Since most variables did not fulfil the normality hypothesis, the Wilcoxon test was used to analyse the variance of paired groups. The level of significance was set at $p < 0.05$. Statistical analysis was performed using SPSS-27.0 software (SPSS-IBM, Armonk, NY, USA).

Author Contributions: Conceptualization, C.Z. and F.Z.; methodology, C.Z. and L.V.; software, L.V.; formal analysis, F.Z., L.V. and C.Z.; investigation, L.V. and C.Z.; resources, L.V. and C.Z.; data curation, L.V.; writing—original draft preparation, F.Z.; writing—review and editing, C.Z., I.G.-A. and L.V.; supervision, F.Z.; funding acquisition, C.Z. and F.Z. All authors have read and agreed to the published version of the manuscript.

Funding: This research was funded by University of Alcalá de Henares-Reig Jofré Art. 83 LOU, grant number (154/2018) titled: "Asesoramiento en materia de medicamentos y programas de investigación, desarrollo e innovación".

Institutional Review Board Statement: This assay was leaded following the guidelines of the 1975 Declaration of Helsinki, with approval of the Biomedical Ethics Committee of the University of Alcalá. General Ethic Code of the University of Alcalá approved by the Governing Council on 22 June 2017.

Informed Consent Statement: Informed consent was obtained from all subjects involved in the study. Written informed consent has been obtained from the patients to publish this paper.

Acknowledgments: We thank the Haematology Service of the Principe de Asturias Hospital for their help with blood extractions. We appreciate the technical support.

Conflicts of Interest: The authors declare no conflict of interest. The funders had no role in the design of the study; in the collection, analyses, or interpretation of data; in the writing of the manuscript, or in the decision to publish the results.

Sample Availability: Samples of the compounds are not available from the authors.

References

1. Stratmann, J.; Karmal, L.; Zwinge, B.; Miesbach, W. Platelet Aggregation Testing on a Routine Coagulation Analyzer: A Method Comparison Study. *Clin. Appl. Thromb. Hemost.* **2019**, *25*. [CrossRef]
2. Zaragozá, C.; Villaescusa, L.; Monserrat, J.; Zaragozá, F.; Álvarez-Mon, M. Potential Therapeutic Anti-Inflammatory and Immunomodulatory Effects of Dihydroflavones, Flavones, and Flavonols. *Molecules* **2020**, *25*, 1017. [CrossRef]
3. Papapanagiotou, A.; Daskalakis, G.; Siasos, G.; Gargalionis, A.; Papavassiliou, G.A. The Role of Platelets in Cardiovascular Disease: Molecular Mechanisms. *CPD* **2016**, *22*, 4493–4505. [CrossRef]
4. Paniccia, R.; Priora, R.; Alessandrello Liotta, A.; Abbate, R. Platelet Function Tests: A Comparative Review. *VHRM* **2015**, 133. [CrossRef]
5. Golebiewska, E.M.; Poole, A.W. Secrets of Platelet Exocytosis–What Do We Really Know about Platelet Secretion Mechanisms? *Br. J. Haematol.* **2014**, *165*, 204–216. [CrossRef]
6. Rajabi, M.; Hossaini, Z.; Khalilzadeh, M.A.; Datta, S.; Halder, M.; Mousa, S.A. Synthesis of a New Class of Furo[3,2-c]Coumarins and Its Anticancer Activity. *J. Photochem. Photobiol. B Biol.* **2015**, *148*, 66–72. [CrossRef]

7. Hassanein, E.H.M.; Sayed, A.M.; Hussein, O.E.; Mahmoud, A.M. Coumarins as Modulators of the Keap1/Nrf2/ARE Signaling Pathway. *Oxidative Med. Cell. Longev.* **2020**, *2020*, 1–25. [CrossRef]
8. Sarker, S.D.; Nahar, L. Progress in the Chemistry of Naturally Occurring Coumarins. *Prog. Chem. Org. Nat. Prod.* **2017**, *106*, 241–304. [CrossRef]
9. Durazzo, A.; Lucarini, M.; Souto, E.B.; Cicala, C.; Caiazzo, E.; Izzo, A.A.; Novellino, E.; Santini, A. Polyphenols: A Concise Overview on the Chemistry, Occurrence, and Human Health. *Phytother. Res.* **2019**, *33*, 2221–2243. [CrossRef]
10. Robledo-O'Ryan, N.; Matos, M.J.; Vazquez-Rodriguez, S.; Santana, L.; Uriarte, E.; Moncada-Basualto, M.; Mura, F.; Lapier, M.; Maya, J.D.; Olea-Azar, C. Synthesis, Antioxidant and Antichagasic Properties of a Selected Series of Hydroxy-3-Arylcoumarins. *Bioorganic Med. Chem.* **2017**, *25*, 621–632. [CrossRef]
11. Kirsch, G.; Abdelwahab, A.; Chaimbault, P. Natural and Synthetic Coumarins with Effects on Inflammation. *Molecules* **2016**, *21*, 1322. [CrossRef]
12. Ed Nignpense, B.; Chinkwo, K.A.; Blanchard, C.L.; Santhakumar, A.B. Polyphenols: Modulators of Platelet Function and Platelet Microparticle Generation? *Int. J. Mol. Sci.* **2019**, *21*, 146. [CrossRef]
13. Widelski, J.; Luca, S.; Skiba, A.; Chinou, I.; Marcourt, L.; Wolfender, J.-L.; Skalicka-Wozniak, K. Isolation and Antimicrobial Activity of Coumarin Derivatives from Fruits of Peucedanum Luxurians Tamamsch. *Molecules* **2018**, *23*, 1222. [CrossRef]
14. Alshibl, H.M.; Al-Abdullah, E.S.; Haiba, M.E.; Alkahtani, H.M.; Awad, G.E.A.; Mahmoud, A.H.; Ibrahim, B.M.M.; Bari, A.; Villinger, A. Synthesis and Evaluation of New Coumarin Derivatives as Antioxidant, Antimicrobial, and Anti-Inflammatory Agents. *Molecules* **2020**, *25*, 3251. [CrossRef]
15. Venugopala, K.N.; Rashmi, V.; Odhav, B. Review on Natural Coumarin Lead Compounds for Their Pharmacological Activity. *BioMed Res. Int.* **2013**, *2013*, 1–14. [CrossRef]
16. Shokoohinia, Y.; Jafari, F.; Mohammadi, Z.; Bazvandi, L.; Hosseinzadeh, L.; Chow, N.; Bhattacharyya, P.; Farzaei, M.; Farooqi, A.; Nabavi, S.; et al. Potential Anticancer Properties of Osthol: A Comprehensive Mechanistic Review. *Nutrients* **2018**, *10*, 36. [CrossRef]
17. Venkata Sairam, K.; Gurupadayya, B.M.; Chandan, R.S.; Nagesha, D.K.; Vishwanathan, B. A Review on Chemical Profile of Coumarins and Their Therapeutic Role in the Treatment of Cancer. *Curr. Drug Deliv.* **2016**, *13*, 186–201. [CrossRef]
18. Spyropoulos, A.C.; Al-Badri, A.; Sherwood, M.W.; Douketis, J.D. Periprocedural Management of Patients Receiving a Vitamin K Antagonist or a Direct Oral Anticoagulant Requiring an Elective Procedure or Surgery. *J. Thromb. Haemost.* **2016**, *14*, 875–885. [CrossRef]
19. Stehle, S.; Kirchheiner, J.; Lazar, A.; Fuhr, U. Pharmacogenetics of Oral Anticoagulants: A Basis for Dose Individualization. *Clin. Pharmacokinet.* **2008**, *47*, 565–594. [CrossRef]
20. Liu, A.; Shen, Y.; Du, Y.; Chen, J.; Pei, F.; Fu, W.; Qiao, J. Esculin Prevents Lipopolysaccharide/D-Galactosamine-Induced Acute Liver Injury in Mice. *Microb. Pathog.* **2018**, *125*, 418–422. [CrossRef]
21. Kang, K.S.; Lee, W.; Jung, Y.; Lee, J.H.; Lee, S.; Eom, D.-W.; Jeon, Y.; Yoo, H.H.; Jin, M.J.; Song, K.I.; et al. Protective Effect of Esculin on Streptozotocin-Induced Diabetic Renal Damage in Mice. *J. Agric. Food Chem.* **2014**, *62*, 2069–2076. [CrossRef]
22. Wang, Y.-H.; Liu, Y.-H.; He, G.-R.; Lv, Y.; Du, G.-H. Esculin Improves Dyslipidemia, Inflammation and Renal Damage in Streptozotocin-Induced Diabetic Rats. *BMC Complement Altern. Med.* **2015**, *15*, 402. [CrossRef]
23. Liang, C.; Ju, W.; Pei, S.; Tang, Y.; Xiao, Y. Pharmacological Activities and Synthesis of Esculetin and Its Derivatives: A Mini-Review. *Molecules* **2017**, *22*, 387. [CrossRef]
24. Lee, M.-J.; Chou, F.-P.; Tseng, T.-H.; Hsieh, M.-H.; Lin, M.-C.; Wang, C.-J. *Hibiscus* Protocatechuic Acid or Esculetin Can Inhibit Oxidative LDL Induced by Either Copper Ion or Nitric Oxide Donor. *J. Agric. Food Chem.* **2002**, *50*, 2130–2136. [CrossRef]
25. Prabakaran, D.; Ashokkumar, N. Protective Effect of Esculetin on Hyperglycemia-Mediated Oxidative Damage in the Hepatic and Renal Tissues of Experimental Diabetic Rats. *Biochimie* **2013**, *95*, 366–373. [CrossRef]
26. Lee, J.-H.; Kim, Y.-G.; Cho, H.S.; Ryu, S.Y.; Cho, M.H.; Lee, J. Coumarins Reduce Biofilm Formation and the Virulence of Escherichia Coli O157:H7. *Phytomedicine* **2014**, *21*, 1037–1042. [CrossRef] [PubMed]
27. Yang, J.; Xiao, Y.-L.; He, X.-R.; Qiu, G.-F.; Hu, X.-M. Aesculetin-Induced Apoptosis through a ROS-Mediated Mitochondrial Dysfunction Pathway in Human Cervical Cancer Cells. *J. Asian Nat. Prod. Res.* **2010**, *12*, 185–193. [CrossRef] [PubMed]
28. Ulloa, J.H. Micronized Purified Flavonoid Fraction (MPFF) for Patients Suffering from Chronic Venous Disease: A Review of New Evidence. *Adv. Ther.* **2019**, *36*, 20–25. [CrossRef]
29. Garelnabi, M.; Gupta, V.; Mallika, V.; Bhattacharjee, J. Platelets Oxidative Stress in Indian Patients with Ischemic Heart Disease. *J. Clin. Lab. Anal.* **2010**, *24*, 49–54. [CrossRef]
30. Liu, M.; Chen, A.; Wang, Y.; Wang, C.; Wang, B.; Sun, D. Improved Solubility and Stability of 7-Hydroxy-4-Methylcoumarin at Different Temperatures and PH Values through Complexation with Sulfobutyl Ether-β-Cyclodextrin. *Food Chem.* **2015**, *168*, 270–275. [CrossRef]
31. Zhao, J.; Yang, J.; Xie, Y. Improvement Strategies for the Oral Bioavailability of Poorly Water-Soluble Flavonoids: An Overview. *Int. J. Pharm.* **2019**, *570*, 118642. [CrossRef]
32. Zaragozá, C.; Monserrat, J.; Mantecón, C.; Villaescusa, L.; Zaragozá, F.; Álvarez-Mon, M. Antiplatelet Activity of Flavonoid and Coumarin Drugs. *Vasc. Pharmacol.* **2016**, *87*, 139–149. [CrossRef]
33. Austermann, K.; Baecker, N.; Stehle, P.; Heer, M. Putative Effects of Nutritive Polyphenols on Bone Metabolism In Vivo-Evidence from Human Studies. *Nutrients* **2019**, *11*, 871. [CrossRef]

34. Hu, M.; Wu, B.; Liu, Z. Bioavailability of Polyphenols and Flavonoids in the Era of Precision Medicine. *Mol. Pharm.* **2017**, *14*, 2861–2863. [CrossRef]
35. Luca, S.V.; Macovei, I.; Bujor, A.; Miron, A.; Skalicka-Woźniak, K.; Aprotosoaie, A.C.; Trifan, A. Bioactivity of Dietary Polyphenols: The Role of Metabolites. *Crit. Rev. Food Sci. Nutr.* **2020**, *60*, 626–659. [CrossRef] [PubMed]
36. Serreli, G.; Deiana, M. *In Vivo* Formed Metabolites of Polyphenols and Their Biological Efficacy. *Food Funct.* **2019**, *10*, 6999–7021. [CrossRef]
37. Park, S.; Moon, K.; Park, C.-S.; Jung, D.-H.; Cha, J. Synthesis of Aesculetin and Aesculin Glycosides Using Engineered Escherichia Coli Expressing Neisseria Polysaccharea Amylosucrase. *J. Microbiol. Biotechnol.* **2018**, *28*, 566–570. [CrossRef]
38. Stringlis, I.A.; de Jonge, R.; Pieterse, C.M.J. The Age of Coumarins in Plant–Microbe Interactions. *Plant Cell Physiol.* **2019**, *60*, 1405–1419. [CrossRef]
39. Martel, C.; Cointe, S.; Maurice, P.; Matar, S.; Ghitescu, M.; Théroux, P.; Bonnefoy, A. Requirements for Membrane Attack Complex Formation and Anaphylatoxins Binding to Collagen-Activated Platelets. *PLoS ONE* **2011**, *6*, e18812. [CrossRef]
40. Dell'Agli, M.; Maschi, O.; Galli, G.V.; Fagnani, R.; Dal Cero, E.; Caruso, D.; Bosisio, E. Inhibition of Platelet Aggregation by Olive Oil Phenols via CAMP-Phosphodiesterase. *Br. J. Nutr.* **2008**, *99*, 945–951. [CrossRef]
41. Zhao, L.; Liu, D.; Liu, B.; Hu, H.; Cui, W. Effects of Atorvastatin on ADP-, Arachidonic Acid-, Collagen-, and Epinephrine-Induced Platelet Aggregation. *J. Int. Med. Res.* **2017**, *45*, 82–88. [CrossRef] [PubMed]
42. Frère, C.; Kobayashi, K.; Dunois, C.; Amiral, J.; Morange, P.-E.; Alessi, M.-C. Assessment of Platelet Function on the Routine Coagulation Analyzer Sysmex CS-2000i. *Platelets* **2018**, *29*, 95–97. [CrossRef] [PubMed]
43. Lucas, S. The Pharmacology of Indomethacin: Headache. *Headache J. Head Face Pain* **2016**, *56*, 436–446. [CrossRef]
44. Loke, W.M.; Sing, K.L.M.; Lee, C.-Y.J.; Chong, W.L.; Chew, S.E.; Huang, H.; Looi, W.F.; Quek, A.M.L.; Lim, E.C.H.; Seet, R.C.S. Cyclooxygenase-1 Mediated Platelet Reactivity in Young Male Smokers. *Clin. Appl. Thromb. Hemost.* **2014**, *20*, 371–377. [CrossRef]
45. Gryglewski, R.J. Prostacyclin among Prostanoids. *Pharmacol. Rep.* **2008**, *60*, 3–11.
46. Young, J.M.; Panah, S.; Satchawatcharaphong, C.; Cheung, P.S. Human Whole Blood Assays for Inhibition of Prostaglandin G/H Synthases-1 and-2 Using A23187 and Lipopolysaccharide Stimulation of Thromboxane B2 Production. *Inflamm. Res.* **1996**, *45*, 246–253. [CrossRef]

Article

Antiglioma Potential of Coumarins Combined with Sorafenib

Joanna Sumorek-Wiadro [1], Adrian Zając [1], Ewa Langner [2], Krystyna Skalicka-Woźniak [3], Aleksandra Maciejczyk [1], Wojciech Rzeski [1,2] and Joanna Jakubowicz-Gil [1,*]

1 Department of Functional Anatomy and Cytobiology, Maria Curie-Sklodowska University, Akademicka 19, 20-033 Lublin, Poland; joanna.sumorek@gmail.com (J.S.-W.); a.zajac.umcs@gmail.com (A.Z.); olamaciejczyk77@gmail.com (A.M.); rzeskiw@hektor.umcs.lublin.pl (W.R.)
2 Department of Medical Biology, Institute of Rural Health, Institute of Agricultural Medicine, Jaczewskiego 2, 20-950 Lublin, Poland; ewa.langner@gmail.com
3 Independent Laboratory of Natural Products, Medical University of Lublin, Chodzki 1, 20-093 Lublin, Poland; kskalicka@pharmacognosy.org
* Correspondence: jjgil@poczta.umcs.lublin.pl

Academic Editor: Maria João Matos
Received: 28 September 2020; Accepted: 6 November 2020; Published: 8 November 2020

Abstract: Coumarins, which occur naturally in the plant kingdom, are diverse class of secondary metabolites. With their antiproliferative, chemopreventive and antiangiogenetic properties, they can be used in the treatment of cancer. Their therapeutic potential depends on the type and location of the attachment of substituents to the ring. Therefore, the aim of our study was to investigate the effect of simple coumarins (osthole, umbelliferone, esculin, and 4-hydroxycoumarin) combined with sorafenib (specific inhibitor of Raf (Rapidly Accelerated Fibrosarcoma) kinase) in programmed death induction in human glioblastoma multiforme (T98G) and anaplastic astrocytoma (MOGGCCM) cells lines. Osthole and umbelliferone were isolated from fruits: *Mutellina purpurea* L. and *Heracleum leskowii* L., respectively, while esculin and 4-hydroxycoumarin were purchased from Sigma Aldrich (St. Louis, MO, USA). Apoptosis, autophagy and necrosis were identified microscopically after straining with specific fluorochromes. The level of caspase 3, Beclin 1, PI3K (Phosphoinositide 3-kinase), and Raf kinases were estimated by immunoblotting. Transfection with specific siRNA (small interfering RNA) was used to block Bcl-2 (B-cell lymphoma 2), Raf, and PI3K expression. Cell migration was tested with the wound healing assay. The present study has shown that all the coumarins eliminated the MOGGCCM and T98G tumor cells mainly via apoptosis and, to a lesser extent, via autophagy. Osthole, which has an isoprenyl moiety, was shown to be the most effective compound. Sorafenib did not change the proapoptotic activity of this coumarin; however, it reduced the level of autophagy. At the molecular level, the induction of apoptosis was associated with a decrease in the expression of PI3K and Raf kinases, whereas an increase in the level of Beclin 1 was observed in the case of autophagy. Inhibition of the expression of this protein by specific siRNA eliminated autophagy. Moreover, the blocking of the expression of Bcl-2 and PI3K significantly increased the level of apoptosis. Osthole and sorafenib successfully inhibited the migration of the MOGGCCM and T98G cells.

Keywords: osthole; umbelliferone; esculin; 4-hydroxycoumarin; sorafenib; apoptosis; autophagy

1. Introduction

Coumarins, classified as secondary metabolites, constitute a large group of ubiquitous compounds in the plant world. Depending on the chemical structure, simple coumarins, pyranocoumarins, and furanocoumarins can be distinguished [1,2].

Coumarin derivatives exhibit a wide spectrum of biological activity. Research conducted to date indicates a beneficial effect of coumarins on the central nervous system (analgesic, anticonvulsant, antidepressant, and sedative) and the circulatory system (anticoagulant and antihypertensive effect). They show antioxidant, antibacterial, antifungal, anti-inflammatory, antiallergic, and antiviral activity. Their antitumor activity is particularly important as well. These compounds have been shown to act at various stages of carcinogenesis. They exhibit chemopreventive properties as well as cytotoxic and antiproliferative activity against cancer cells. In addition, they limit angiogenesis and prevent the formation of metastases to other tissues [1–5]. Clinical studies have shown that coumarins have promising activity against several types of cancer, such as breast cancer, lung cancer, malignant melanoma, prostate cancer and renal cell carcinoma [6,7]. Simple coumarin derivatives improved the health condition of patients and did not show any toxic properties. Renal cell carcinoma patients tolerated a wide spectrum of coumarin doses, and the most common side effect was nausea associated with the intense aroma of the compound [7,8]. Interestingly, previous studies have also shown that coumarins may be used not only in the treatment of cancer but also in the treatment of the side effects of radiation therapy, such as radiogenic sialadenitis and mucositis [9].

The cytotoxicity of coumarins towards cancer cells depends on their chemical structure; therefore, knowledge of the effect of various substituents on the antitumor properties of these compounds will ensure in more effective plans of therapeutic strategies. Special attention has been paid to simple coumarins, e.g., 4-hydroxycoumarin, umbelliferone, esculin, and osthole, differing in their location or the type of attached substituents (Figure 1). 4-hydroxycoumarin and umbelliferone (7-hydroxycoumarin) are isomers with a hydroxyl moiety located at the C4 and C7 positions of the coumarin ring, respectively. Esculin (6,7-dihydroxycoumarin 6-glucoside) is an analogue of umbelliferone with an additional glycosidic moiety at the C6 position. In turn, osthole (7-metoxy-8-isopentenyl-coumarin) has a methoxy moiety at the C7 position and an isoprenyl substituent at the C8 position [10].

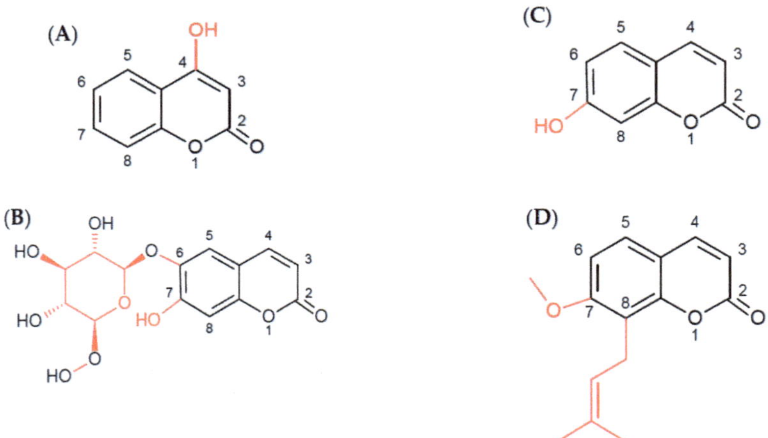

Figure 1. Structure of 4-hydroxycoumarin (**A**), esculin (**B**), umbelliferone (**C**) and osthole (**D**).

It has been shown that all these hydroxycoumarins have the ability to reduce the proliferation, adhesion, and migration of cancer cells. Esculin interferes with the adhesion of U87 glioblastoma cells by modulating the function of integrins [11]. 4-hydroxycoumarin disorganizes the actin cytoskeleton in B16-F10 melanoma cells and reduces the potential of this tumor to metastasize to the lungs, as shown in mice [12–14]. Umbelliferone, in turn, reduces the migration of laryngeal cancer (RK33) and rat breast adenocarcinoma (RBA) cells [15,16]. 7-hydroxycoumarin has cytotoxic properties against many human cell lines such as leukemia (HL-60) and lung (A549 and H727), kidney (ACHN), and breast cancers (MCF-7) [17]. This compound inhibits the G1 phase cell cycle in human renal cell carcinoma cells (786-O, OS-RC-2, and ACHN) by reducing the expression of proteins that positively regulate the cell cycle (CDK2, CyclinE1, CDK4, and CyclinD1). Moreover, it modulates the expression of proteins involved in apoptosis (Bax and Bcl-2) and in proliferation (Ki67) [18].

Similar to hydroxycoumarins, osthole inhibits the migration (MCF-7) and invasiveness of breast cancer cells (MDA-MB-231BO) [19]. Additionally, it inhibits proliferation (in lung cancer: A549, leukemia: P-388 D1, breast carcinoma: MCF-7 and MDA-MB 231, medulloblastoma: TE671 and larynx carcinoma: RK33) by inducing apoptosis and inhibiting the cell cycle in the G2/M phase. At the molecular level, it is associated with a decrease in the expression of proteins involved in the cell cycle (CDK2, CyclinB1) and activation of apoptotic proteins (caspase 3, caspase 9, caspase 8, p53 protein) [20–24].

Anaplastic astrocytoma (AA, grade III) and glioblastoma multiforme (GBM, grade IV) are malignant tumors of the central nervous system. At the molecular level, they are characterized by the presence of mutations within genes, the products of which are involved in enhancement of intracellular signal transmission from the cell membrane to the nucleus. This applies in particular to the prosurvival pathways responsible for the regulation of cell proliferation and differentiation: Ras/MEK/ERK (Ras-Ras protein, MEK—mitogen-activated protein kinase, ERK—extracellular signal-regulated kinase) and PI3K/Akt/mTOR (PI3K-phosphoinositide 3-kinase, Akt/PKB-protein kinase B, mTOR- mammalian target of rapamycin kinase). It has been described that blocked signal transmission may be beneficial in enhancement of glioma cell sensitivity. It is also known that combination therapy, especially with the use of natural compounds, can increase the anticancer potential of clinically used pharmacotherapy [25,26].

Therefore, in our research, the antiglioma effect of simple coumarins (4-hydroxycoumarin, umbelliferone, esculin, and osthole) in combination therapy with sorafenib (Raf kinase inhibitor) was evaluated in terms of programmed cell death induction and migratory potential. At the molecular level, these processes were confirmed by the level of caspase 3, Beclin 1, Raf, and PI3K expression. Direct involvement of these proteins in apoptosis, autophagy, and mobility was studied by blocking their expression by specific siRNA.

2. Results

2.1. *Effect of Simple Coumarins (Osthole, Esculin, Umbelliferone, or 4-Hydroxycoumarin) in Combination with Sorafenib on Apoptosis, Necrosis, and Autophagy Induction*

Our research shows that all the coumarins effectively eliminated tumor cells by apoptosis. Osthole was the most effective compound in both cell lines, as it induced this type of death in 40% and 30% in AA and GBM, respectively (Figure 2A,B). Moreover, the coumarin-induced autophagy (10%) in the MOGGCCM cell line. The application of umbelliferone, esculin, and 4-hydroxycoumarin led to apoptosis in approximately 15% of cells in the T98G cell line (Figure 2B). Slightly different results were obtained in the case of the MOGGCCM cell line (Figure 2A). It turned out to be less sensitive to the action of umbelliferone and 4-hydroxycoumarin, which initiated apoptosis at a level lower than 7%.

Sorafenib had no significant effect on the induction of apoptosis but initiated autophagy in approx. 15% of the T98G cells (Figure 2B). The simultaneous application with the coumarins did not potentiate such an anticancer activity effectively. In the MOGGCCM line, sorafenib diminished the antiglioma potential of osthole, inducing apoptosis in ca. 25% and autophagy in 1% of cells in comparison to the single application of the coumarin. Similar effects were obtained upon the application of esculin, which in combination with sorafenib also showed lower proapoptotic activity. Interestingly, the simultaneous treatment with the hydroxycoumarins and sorafenib was more effective, causing apoptosis in up to 17% of cells.

The experiments carried out on primary human skin fibroblasts (HSF) showed that osthole and esculin (alone and in combination with sorafenib) did not exert cytotoxic effects against normal cells. Different results were obtained for the hydroxycoumarins, as they showed a strong necrotic effect (nearly 20%), which was additionally enhanced by the addition of sorafenib. For this reason, osthole was chosen for further experiments.

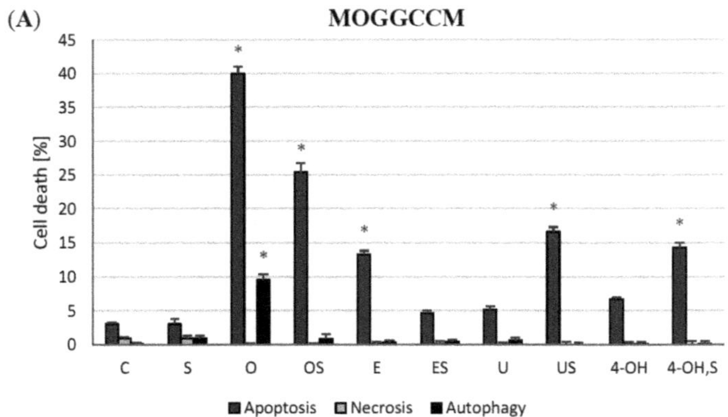

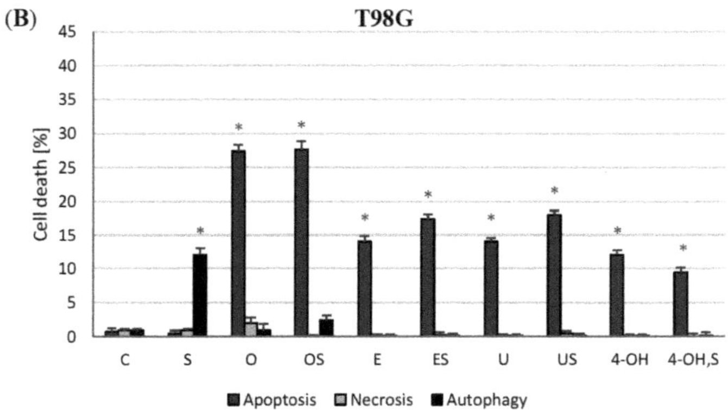

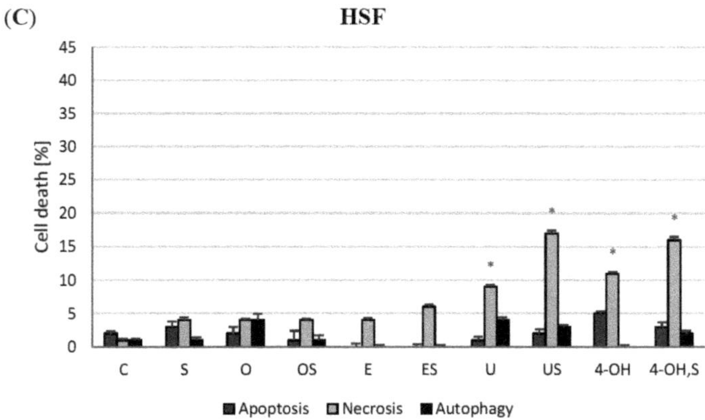

Figure 2. Effect of sorafenib and osthole, esculin, umbelliferone or 4-hydroxycoumarin administered separately or simultaneously on apoptosis, necrosis, and autophagy induction in the MOGGCCM (**A**), T98G (**B**) and HSF (Human Skin Fibroblasts) (**C**) cell lines. C—control, S—sorafenib, O—osthole, E—esculin, U—umbelliferone, 4-hydroxycoumarin; * $p < 0.05$.

2.2. Effect of Osthole and Sorafenib on the Migration Potential of Neoplastic Cells

Inhibition of tumor cell migration plays an important role in anticancer therapy. The wound healing test showed that the treatment with osthole, sorafenib, and the combination of these compounds significantly decreased the migration potential of the AA and GBM cells (Figure 3). The combination therapy was the most effective, as it lowered this activity by approx. 70% compared to the control in both cell lines.

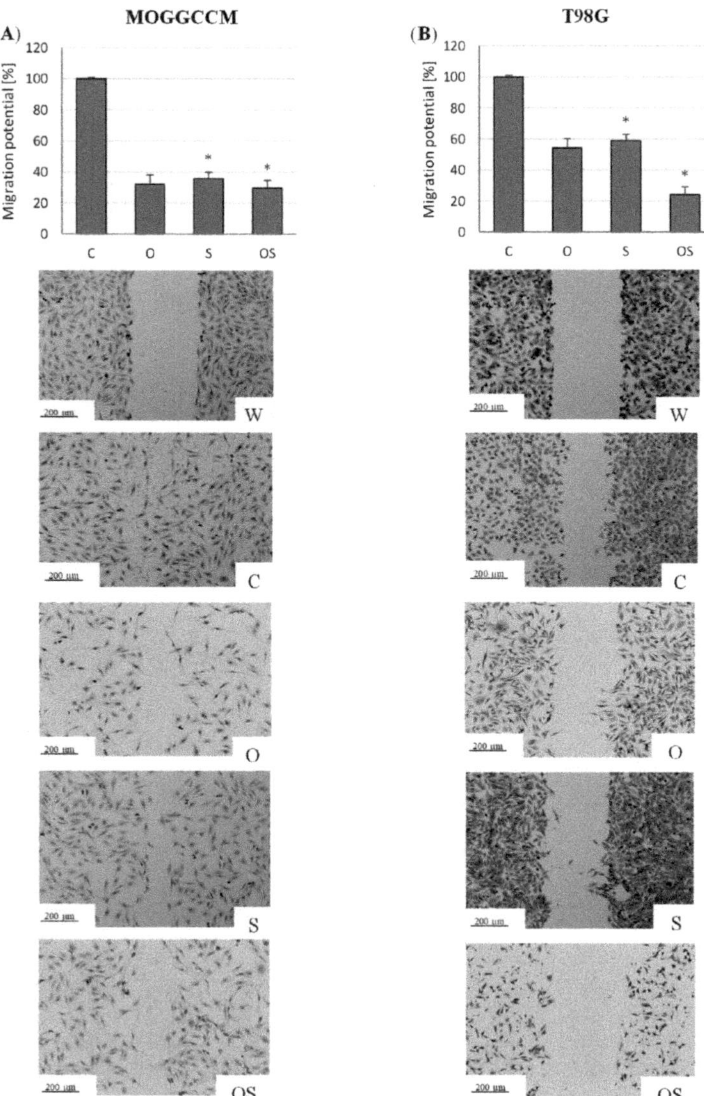

Figure 3. Migration potential of the MOGGCCM (**A**) and T98G (**B**) cells upon osthole and sorafenib treatment presented as the percent of cells within the wound. W—wound, C—control, O—osthole, S—sorafenib; * $p < 0.05$.

2.3. *Effect of Osthole and Sorafenib on the Expression of Cell Death Marker Proteins*

2.3.1. Expression of Caspase-3, PI3K, and Raf Kinases

Caspase 3 is a member of the cysteine-aspartic acid protease family playing a key role in the execution phase of apoptosis. Our studies showed that the use of osthole alone and in combination with sorafenib led to an increase in caspase 3 expression in both cell lines (Figure 4A,B). The best effects (a 14% increase) were obtained upon the administration of osthole in combination with sorafenib in the T98G line. Moreover, the treatment with sorafenib alone reduced the level of this protein by 25% in the MOGGCCM line and by 55% in the T98G line.

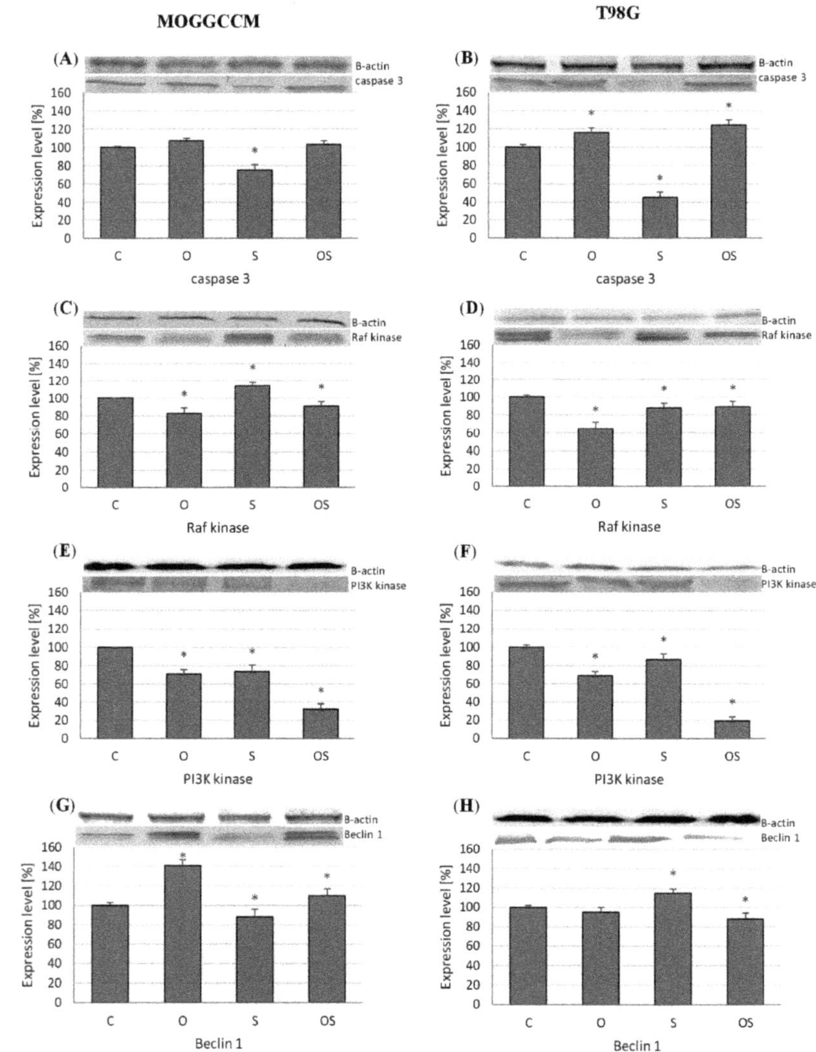

Figure 4. Effect of osthole, sorafenib and combined treatment with both drugs on the expression of caspase 3 (**A**,**B**), PI3K (Phosphoinositide 3-kinase) (**C**,**D**), Raf (Rapidly Accelerated Fibrosarcoma) (**E**,**F**), and Beclin 1 (**G**,**H**) in the MOGGCCM (**A**,**C**,**E**,**F**) and T98G (**B**,**D**,**G**,**H**) cell lines. C—control, O—osthole, S—sorafenib; * $p < 0.05$.

PI3K and Raf kinases are also involved in the course of apoptosis and promote the survival of tumor cells. The Western blot analysis showed that osthole alone and in combination with sorafenib decreased the level of Raf kinase (Figure 4C,D). In both cell lines, the treatment with osthole exerted the greatest effect, as it reduced the expression of this protein by 20% in MOGGCCM and 35% in T98G. Moreover, the application of sorafenib increased the level of this protein by 15% in the GBM cells. Better effects were evident in the case of PI3K (Figure 4E,F). The treatment with osthole, sorafenib, and the combination of both compounds decreased the level of this protein. The simultaneous application of the drugs was the most effective, as it caused an over 70% decrease in PI3K expression in both cell lines. The worst effect was observed upon the application of sorafenib alone, which reduced the PI3K levels by 25% in MOGGCCM and 15% in T98G.

2.3.2. Level of Beclin 1

Beclin 1 induces the formation of an autophagosome, thereby initiating the process of autophagy. In the MOGGCCM line (Figure 4G), overexpression of this protein was visible after the treatment with osthole alone (a 40% increase) and in combination with sorafenib (over a 10% increase). In turn, the use of sorafenib alone was associated with a slight decrease in the expression. Completely different results were obtained in the T98G cell line (Figure 4H). Sorafenib increased the level of Beclin 1 by ca. 15%. The coumarin (alone and in combination with sorafenib) inhibited the expression of this protein.

2.4. Apoptosis, Autophagy, and Necrosis Induction Upon Inhibition of PI3, Beclin 1, and Bcl-2 Expression

2.4.1. Blocking PI3K Expression

The neoplastic transformation of gliomas is associated with excessive activation of the Ras/Raf/MEK/ERK and PI3K/Akt/mTOR pathways. Therefore, the studied cell lines were incubated with anti-PI3K siRNA and cells with blocked PI3K expression were additionally incubated with sorafenib. A significant increase in the sensitivity of the MOGGCCM and T98G cells to the induction of apoptosis was then observed (Figure 5A,B). The treatment with osthole and sorafenib, alone and in combination, induced apoptosis in at least 80% of cells. Sorafenib was the most effective agent leading to the death of almost all cancer cells (97%) in both cell lines. No autophagy or necrosis was observed in both cell lines.

2.4.2. Inhibition of Beclin 1 and Bcl-2 Expression

The antiapoptotic protein Bcl-2 is responsible for the regulation of both apoptosis and autophagy. In a complex with Beclin 1, it inhibits autophagy and, after dissociation, disrupts apoptosis. Blocking the expression of Bcl-2 and Beclin 1 proteins in the AA cells inhibited autophagy and significantly increased apoptosis (Figure 5C,E). Osthole in combination with sorafenib exerted the most potent effect and induced apoptosis in over 70% of the cells. An increase in the apoptotic potential (up to 50%) was also observed in the GBM cells with blocked Bcl-2 expression upon the osthole treatment. In turn, the transfection had no effect on apoptosis induction in the treatment with sorafenib alone and in combination with the coumarin. However, it inhibited autophagy. The GBM cells with the blocked expression of Beclin 1 were less sensitive to induction of programmed death (Figure 5D,F). Apoptosis (30%), but not autophagy, was observed only after the sorafenib treatment.

2.5. Chou-Talalay Method—Effect of Combination Therapy

Drug interactions were determined using the isobologram, dose reduction, and combination index method derived from the median-effect principle proposed by Chou and Talalay [27]. As it turned out, the combination of osthole with sorafenib in the T98G line had a synergistic effect (Figure 6D–F). It was stronger when the higher dose was used, which is extremely important in anticancer therapy (for IC97, CI = 0.4). Moreover, the combination treatment significantly reduced the doses of both drugs (DRI > 1), which would have to be higher in a single application to yield the same effect. We observed different results in the MOGGCCM line, where the doses used had an additive effect (CI ≈ 1), while the higher drug concentrations were already antagonistic (Figure 6A–C). The effectiveness of the combination therapy was estimated on the basis of the ability to induce apoptosis. We also observed autophagy in this cell line, which was inhibited by the simultaneous treatment with osthole. For this reason, the use of combinations of these compounds is also appropriate in AA cells.

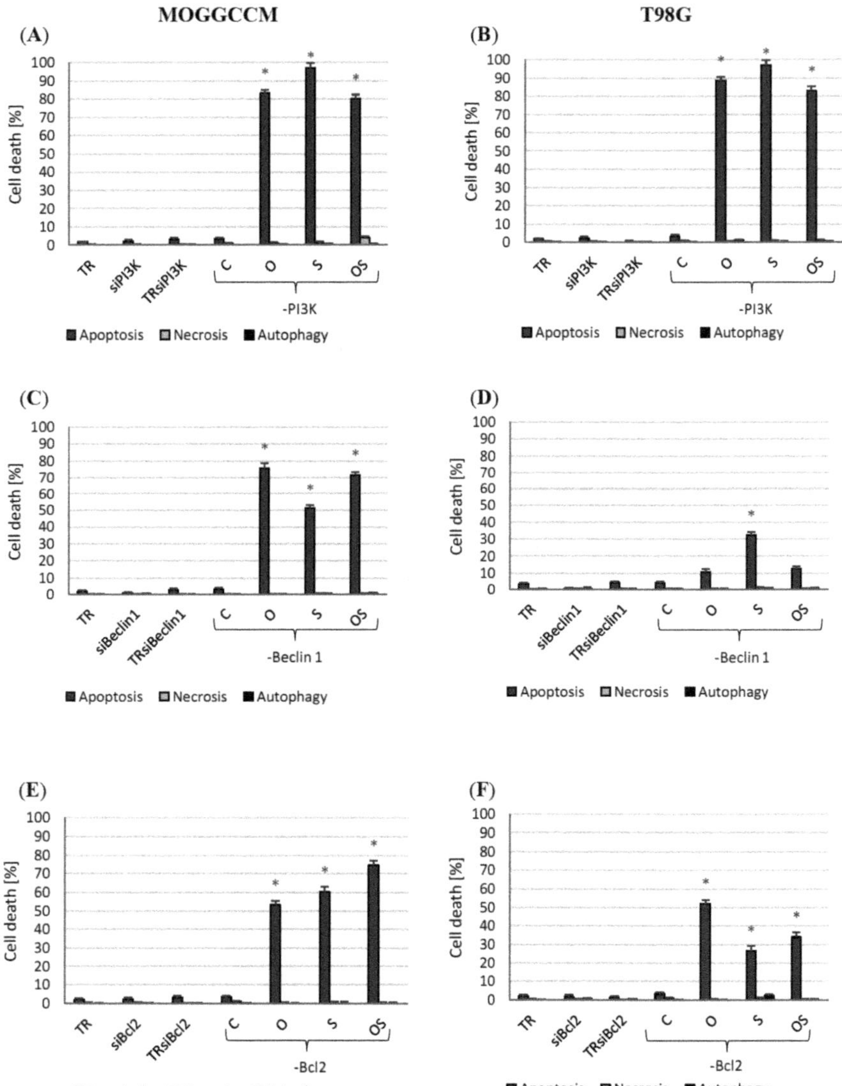

Figure 5. Level of apoptosis, autophagy, and necrosis in the MOGGCCM (**A**,**C**,**E**) and T98G (**B**,**D**,**F**) cells with inhibited PI3K (**A**,**B**), Beclin 1 (**C**,**D**), and Bcl-2 (B-cell lymphoma 2) (**E**,**F**) expression by specific siRNA (small interfering RNA) upon the osthole and sorafenib treatment. C—control, O—osthole, S—sorafenib; * $p < 0.05$.

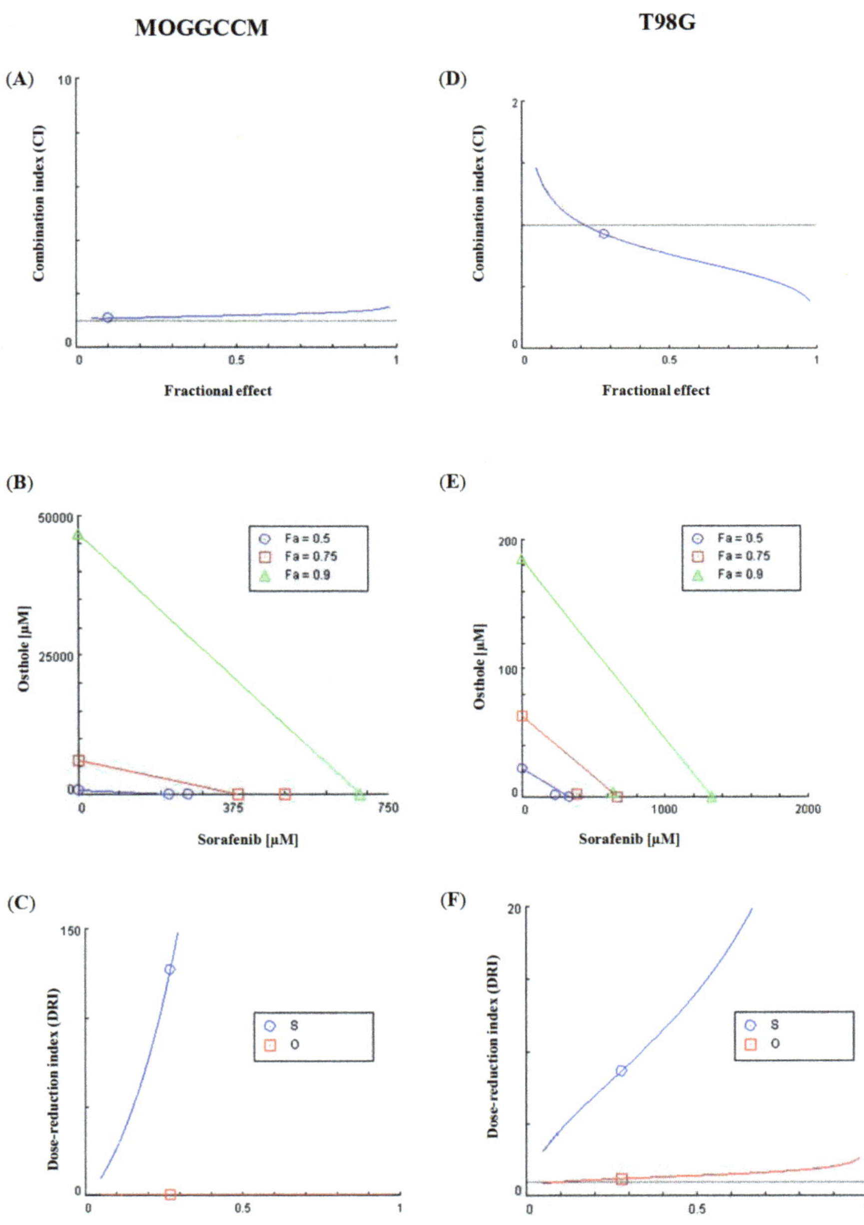

Figure 6. Osthole (O) and sorafenib (S) combination treatment in the MOGGCCM (**A–C**) and T98G (**D–F**) cell line. (**A**,**D**) Combination index (CI) plot: the combination index is plotted as a function of Fa (fractional effect, line of blue color). (**B**,**E**) Isobologram for the combination: classic isobologram at IC50, IC75, and IC90. (**C**,**F**) Fa-DRI (dose-reduction index) plot (Chou-Martin plot).

3. Discussion

Gliomas, i.e., tumors of the central nervous system, account for approx. 70% of all brain tumors. Due to their infiltrative nature, they are practically impossible to remove surgically [1,25].

Many resistance mechanisms are activated at the molecular level. An example is the Ras/Raf/MEK/ERK pathway. It has been shown that inhibition of this pathway reduces the survival, proliferation, migration, and metabolism of cancer cells [28–30]. It also reduces cancer resistance to the chemotherapy. Our previous studies have shown that sorafenib, i.e., a Raf kinase inhibitor, increases the apoptotic activity of Temozolomide and quercetin [31,32]. It has also been proved that simple coumarins: 4-hydroxycoumarin, umbelliferone, esculin, and osthole can play an important role in cancer therapy [33]. Therefore, in our studies, we used a combination of sorafenib as the Raf kinase inhibitor and the coumarins.

Our experiments showed that the anticancer properties of coumarin derivatives are closely related to their chemical structure. Osthole had the strongest apoptotic activity, alone and in combination with sorafenib. This compound has a methoxy and isopentenyl moiety attached to the C7 and C8 positions, respectively (Figure 1D). The other coumarins (4-hydroxycoumarin, umbelliferone, and esculin) are monohydroxy derivatives (Figure 1A–C). Additionally, esculin has a glycosidic substituent. It has been shown that the cytotoxic effect of coumarins is stronger with the increase in the hydrophobicity of the substituent, which is ensured by the isoprenyl group in osthole [34]. It has also been reported that the length of the substituted aliphatic chain has a great influence on the antitumor activity of the compound [35,36]. The enhanced lipophilicity of the alkyl group contributes to improvement of the ability of compounds to penetrate the cell membrane [37]. In the present experiment, there were no significant differences in the antitumor activity of umbelliferone and 4-hydroxycoumarin in both cell lines. Research conducted by Budzisz et al. showed similar effects. It was found that both hydroxycoumarins inhibited cell proliferation in a gastric carcinoma cell line with similar effectiveness [34]. Thus, the site of attachment of the hydroxy moiety (at the C4 or C7 position) does not significantly affect the proapoptotic properties of the coumarins. Moreover, the presence of an additional glycosidic moiety (esculin) did not change the properties of the compounds in the T98G line and increased the proapoptotic activity in the AA cells. Our experiments confirm observations described by other authors who reported that the anticancer effect of coumarins depends on both the chemical structure and the cell line used. The human carcinoma KB cell line was more sensitive to esculin treatment, while umbelliferone was more effective in HL60 cells [37]. In addition, the combined treatment with sorafenib decreased the sensitivity of the AA cells to the hydroxycoumarins (4-hydroxycoumarin and umbelliferone) treatment and increased the exposure to esculin.

Osthole, alone and in combination with sorafenib, induced apoptosis in approx. 30% of the glioblastoma cells. Our previous research showed that Temozolomide (TMZ)—a drug currently used to treat gliomas, induced apoptosis in 12% of the GBM cells and 5% of the AA cells [38,39]. Interestingly, another chemotherapeutic agent, also used in the treatment of gliomas, Bevacizumab (BEV), similarly to TMZ, reduces the GBM viability by approx. 15% [40]. Thus, our results suggest that the efficacy of osthole with sorafenib may be much higher than that of TMZ and BEV.

In our experiments, at the molecular level, the induction of apoptosis by osthole and sorafenib was accompanied by an increase in caspase 3 expression. On the other hand, the coumarin induced programmed death type I and II in the MOGGCCM cells. Moreover, the combination of both compounds reduced the number of apoptotic cells and completely inhibited autophagy. In this case, the elimination of the process of autophagy is desirable, as it can inhibit cell death in conditions of nutrient deficiency. Cells in the center of tumors are metabolically stressed in this manner and therefore, they use autophagy as a survival mechanism [41]. Blocking this process significantly increases the effectiveness of anticancer therapies used [42]. We also observed that the treatment with osthole alone was associated with overexpression of Beclin 1, which enhanced apoptosis in addition to autophagy. This was accompanied by an increased level of caspase 3. These results suggesting that Beclin 1 has both proautophagous and proapoptotic functions are consistent with studies conducted by Fururya et al. [43] and Huang et al. [44]. They showed that an increase in the expression of Beclin 1 in human gastric (MKN28) and glioma (U87) cells led to apoptosis by increasing the activity of caspase 3, 7, and 8. Interestingly, the treatment with osthole (alone and in combination with sorafenib) in the T98G cells with blocked Beclin 1 expression inhibited not only autophagy but also apoptosis. It has been shown that Beclin 1 can affect cell survival by interacting with Bcl-2 or Bcl-xL proteins. In turn, Bcl-2 may act as an antiautophagy protein by forming a complex with Beclin 1 [45]. Thus, after silencing the expression of the autophagy marker, the Bcl-2 protein inhibited apoptosis. Different effects were obtained in the AA cells, where instead of apoptosis reduction, we noted a significant increase in the percentage of apoptotic cells. Similar results were noticed after silencing the expression of the Bcl-2 protein. The percentage of apoptotic cells increased significantly after the combined treatment with sorafenib of the MOGGCCM cells. We did not observe autophagy. As with the Beclin 1 blocking, the T98G line was less sensitive to the treatment following the inhibition of Bcl-2 expression. The induction of apoptosis was also observed at that time, but the best effects were achieved only by the application of osthole.

Glioblastoma cells often have mutations in the PTEN and PI3K genes, resulting in continued Akt/PKB kinase activity. This enzyme performs antiapoptotic functions, reducing the susceptibility of cells to inducers of this process and thus enabling tumor growth [25]. We noticed that the use of the combination therapy decreased PI3K protein expression in both cell lines, which correlated with the induction of programmed death. Moreover, blocking the expression of this protein significantly increased the effectiveness of the drugs used. Then, the treatment with sorafenib eliminated almost 100% of cancer cells. The coumarin (alone and in combination with sorafenib) exerted slightly worse effects, inducing over 80% cell death. Interestingly, blocking only PI3K kinase did not decrease the survival rate of the cancer cells. As it turned out, the inhibitors of this protein were also not cytotoxic in the single application; however, when combined with mTOR inhibitors, they significantly increased their effectiveness in eliminating gliomas [46]. A possible explanation is that PI3K inhibition induces other pathways that promote cancer cell survival [47]. Therefore, blocking PI3K expression in combination with osthole or sorafenib gave such good results. Our results suggest that the Ras/Raf/MEK/ERK and PI3K/Akt/mTOR pathways play a key role in the pathogenesis of grade III and IV gliomas.

We observed that the coumarin treatment led to a decrease in Raf kinase expression in both cell lines. At that time, the morphological analysis showed apoptosis, which is consistent with other literature reports. Raf kinase has antiapoptotic functions, hence a decrease in its activity promotes induction of apoptosis [48]. In turn, sorafenib led to the overexpression of this protein in the MOGGCCM line. The autophagy observed was a protective and adaptive response to the inhibition of the Raf/Raf/MEK/ERK pathway. This mechanism was described in many types of cancer, including brain cancer (B76, AM38, and BT40) [49], pancreas (MiaPaCa2 and BxPC3) [50], and melanoma (e.g., A375P, SKMEL5, 1205Lu, MEL624) [51]. It has also been shown that supporting targeted therapy through the use of autophagy inhibitors significantly increased the effectiveness of treatment [49,50]. Therefore, it can be assumed that, in addition to its proapoptotic properties, osthole has antiautophagy activity.

Gliomas have high migratory and invasive potential. Therefore, when planning new therapeutic strategies, drugs that inhibit the translocation of cancer cells should be considered. Our results indicated that the combination of osthole with sorafenib can be used for this purpose.

4. Materials and Methods

4.1. Cells and Culture Conditions

Human glioblastoma multiforme cells (T98G, European Collection of Cell Cultures) and human anaplastic astrocytoma cells (MOGGCCM, European Collection of Cell Cultures) were grown in a 1:3 mixture of DMEM (Dulbecco's modified Eagle medium) and nutrient mixture F-12 Ham (Ham's F-12, Sigma). Both cell lines were cultured in medium supplemented with 10% FBS (fetal bovine serum) (Sigma), 100 U/mL penicillin (Sigma), and 100 mg/mL streptomycin (Sigma) at 37 °C in a humidified atmosphere consisting of 5% CO_2 and 95% air.

A primary culture of human skin fibroblasts (HSF), which was carried out in the same conditions as the tumor lines, was used in the study as well.

4.2. Coumarin Isolation

Osthole was obtained for the experiments after isolation from a petroleum extract of *Mutellina purpurea* L. fruits, with a method described previously [22] in the Independent Laboratory of Natural Products Chemistry, Medical University, Lublin, Poland. Two-phase solvent systems made of n-heptane, ethyl acetate, methanol, and water (HEMWat) with a volume ratio 3:2:3:2 were chosen as the most proper system for purification of target compounds (K = 1.8). After injection of 600 mg of a crude oily extract, 2 mg of the target compound were obtained. The identification of the isolated compound was carried out by comparison of the retention time and UV-DAD spectra with those obtained by standards in the same conditions. The purity of osthole was 99% (established with the HPLC-DAD method).

Umbelliferone was purified from fruits of *Heracleum leskowii* L. (Apiaceae) collected in the Medicinal Plant Garden, Department of Pharmacognosy with Medicinal Plant Unit, Medical University, Lublin, Poland in summer 2009, in accordance with a method published previously [15]. Briefly, the fruits were air-dried at room temperature and powdered, and a batch (100 g) was extracted with 100 mL of methanol under reflux for 30 min. After filtration, the procedure was repeated twice. The filtrates were combined and concentrated with a rotary evaporator to remove the solvent. The dried crude extract (13 g) was stored in a refrigerator until further separation. The Spectrum High-Performance Countercurrent Chromatograph (HPCCC) apparatus delivered by Dynamic Extractions (Slough, UK) was employed in the present study. The integrated analytical column (22 mL) was first entirely filled with the upper stationary phase. Then the apparatus was rotated at 200× *g* and the lower mobile phase was pumped into the column at a flow rate of 1.0 mL/min. After hydrodynamic equilibrium was reached, each time 30 mg of the extract dissolved in 1 mL of the two-phase solvent system was loaded onto the column through a 1 mL injection valve. When optimal conditions were determined, the procedure was transferred to a semipreparative integrated column (137 mL volume). The mobile phase was pumped at a flow rate of 6.0 mL/min and 180 mg of the extract was dissolved in 6 mL of the two-phase solvent system and loaded onto the column through a 6 mL injection valve. The solid-phase retention was 70%. The effluent from the column was continuously monitored with a UV detector at 320 nm (Ecom, Prague, Czech Republic). A mixture of n-heptane, ethyl acetate, methanol and water at a ratio of 1:2:1:2 was chosen for further experiments. Umbelliferone was isolated after 20 min. After injection of 180 mg of the crude extract, 1.8 mg of umbelliferone with 99% purity (according to HPLC analysis) was purified. All solvents for HPCCC and HPLC analysis were delivered by Avantor Performance Materials Poland S.A. (Gliwice, Poland—formerly POCh).

Esculin and 4-hydroxycoumarin were delivered by Sigma Aldrich.

4.3. Drug Treatment

Sorafenib (Nexavar, BAY 43-9006) (1 μM), osthole (150 μM), and hydroxycoumarins: 4-hydroxycoumarin (Sigma), umbelliferone, and esculin (Sigma) at the final concentrations of 200 μM were used in the experiments. The drugs were dissolved in DMSO (Sigma) to the final concentration not exceeding 0.01%. The doses were chosen based on previous studies [32,33]. The cancer cells were treated with the coumarins or with sorafenib separately or in combination for 24 h. As controls, T98G and MOGGCCM cells were incubated only with 0.01% of DMSO.

4.4. Fluorescence Microscopy (Apoptosis, Necrosis, Autophagy Identification)

For identification of apoptosis and necrosis, a solution of propidium iodide (Sigma, St. Louis, MO, USA) and Hoechst 33342 (Sigma, St. Louis, MO, USA) in distilled water in a ratio of 3:2:5 were used. The cells were stained upon incubation with the appropriate combinations of the drug. After addition of 2.5 μL of the mixture to 1 mL of the medium, the tumor cells were incubated at 37 °C for 5 min. The morphological analysis was performed under a fluorescence microscope (Nikon E—800, Tokyo, Japan). It showed bright blue fluorescence characteristic of apoptotic cells. Necrotic cells emitted red-pink fluorescence. Five percent acridine orange was used to identify autophagous cells incubated in the dark for 15 min. The dye induced the red glow of the autophagous bodies. At least 1000 cells in randomly selected microscopic fields were counted under the microscope. Each experiment was conducted in triplicate.

4.5. Cell Migration Test

Tumor cell migration was assessed by means of the wound assay model [31]. The cell lines were grown at 2.5 × 105 in standard conditions (37 °C, 95% humidity, 5% CO_2) in 4 cm-diameter culture dishes (NuncTM, ThermoFisher, Rochester, NY, USA). The next day, the cell monolayer was scratched with the pipette tip (P300), the medium and dislodged cells were aspirated, and the plates were rinsed twice with PBS. Next, fresh culture medium was applied and the number of cells that had migrated into the wound area after 24 h was estimated in the control and drug-treated cultures. The plates were stained with the May–Grünwald–Giemsa method. The observation was performed with the use of a BX51 microscope (Olympus, UK), and the distances between the scratch fronts were estimated using the CellSans program. The results were presented as the migratory potential expressed as the percent of cells within the wound.

4.6. Western Blotting Analysis

The expression of cellular proteins was evaluated by Western blotting. After treatment for 24 h, cells grown in Falcon flasks (5 mL) were lysed in buffer (125 mM Tris-HCl pH 6.8; 4% SDS; 10% glycerol; 100 mM dithiothreitol). The cells prepared in this way were boiled for 10 min and then centrifuged at 12,000× *g* centrifugal force for 10 min; next, the supernatants were collected. The protein concentration in the cell-free extracts was determined with the Bradford method [52]. Equal amounts of protein (80 μg) from each sample were separated on SDS-PAGE (SDS poliacrylamide gel electrophoresis) [53] and transferred onto an Immmobilon P membrane (Sigma). After blocking with 3% low fat milk for 1 h, the membranes were incubated overnight with primary antibodies: rabbit anti-caspase 3 (Sigma, 1:1000), anti-Beclin 1 (Santa Cruz Biotechnology, 1:500), anti-Raf (Santa Cruz Biotechnology, 1:500), and anti-PI3K (Santa Cruz Biotechnology,1:500). After three washes with PBS enriched with 0.05% Triton X-100 (Sigma), the membranes were incubated with secondary antibodies conjugated with alkaline phosphatase (AP) for 2 h. Alkaline phosphatase substrates: 5-bromo-4-chloro-3-indolylphosphate (BCIP) and nitro-blue tetrazolium (NBT) (Sigma) in *N*,*N*-dimethylformamide (DMF, Sigma) were used for visualization of proteins (Bcl-2, Beclin 1, and caspase 3). The results were analyzed qualitatively on the basis of the band thickness, width, and color depth. The quantitative analysis of protein bands was performed using the Bio-Profil Bio-1D Windows Application V.99.03 program. The data were normalized relative to β-actin (Sigma, working dilution 1:2000). Three independent experiments were performed.

4.7. Cell Transfection

The cells at a density of 2 × 10^5 were incubated for 24 h at 37 °C in a CO_2 incubator to reach 60–80% of confluence. The cells were washed with a DMEM:Ham's F-12 (3;1) mixture without serum and antibiotics and aspirated. Next, a blocking mixture containing 2 μL of specific anti-PI3K, anti-Bcl-2, or anti-BCN1 siRNA (Santa Cruz Biotech Dallas, TX, USA), 2 μL of Transfection Reagent (Santa Cruz Biotech, Dallas, TX, USA), and 250 μL of Transfection Medium (Santa Cruz Biotech) was added. The cells were incubated for 5 h at 37 °C, 5% CO_2, and 95% humidity. Next, the medium was supplemented with medium containing 20% of fetal bovine serum and 200 μg/mL of antibiotics. After 18 h, the medium was replaced with a fresh one (containing 10% FBS, 100 U/mL penicillin, and 100 mg/mL streptomycin) and the transfected cells were used for further studies (incubation with osthole and sorafenib alone and in combination as well as determination of cell death).

4.8. Statistical Analysis

A one-way anova test followed by Dunnett's multiple comparison analysis was used for statistical evaluation. $p < 0.05$ of data presented as mean ± standard deviation (SD) was taken as the criterion of significance.

4.9. Chou-Talalay Method

The combination index (CI) and the dose reduction index (DRI) were calculated with the method developed by Chou and Talalay [23] using the Compusyn software and the original data of programmed cell death induction in the MOGGCCM and T98G cells upon the sorafenib or osthole treatment [32,33]. CI < 1, CI = 1, and CI > 1 indicate a synergistic, additive, and antagonistic effect, respectively. The DRI represents the fold reduction of compounds as a result of the synergistic combination compared to the concentration of the drug alone required to reach the same effect.

5. Conclusions

Due to their widespread availability and the broad spectrum of biological activity, coumarins have enormous pharmacological potential. Their anticancer properties, which depend on the chemical structure of the compounds, deserve special attention. Our research has shown that the presence of an isoprenyl moiety (osthole) significantly increases this activity, compared to the other coumarins. It also sensitizes glioma cells with a decreased level of PI3K to apoptosis induction in combination therapy with sorafenib. Therefore, the present results may therefore constitute a basis for further research on the development of new anticancer therapies.

Author Contributions: Conceptualization, J.J.-G. and J.S.-W.; formal analysis, J.S.-W. and W.R.; investigation, J.S.-W., A.Z., E.L., K.S.-W., A.M. and W.R.; resources, A.Z., E.L.; writing—original draft preparation, J.S.-W.; writing—review and editing, J.J.-G.; visualization, A.M.; supervision, W.R. and J.J.-G.; All authors have read and agreed to the published version of the manuscript.

Funding: This research was funded by Maria Curie-Sklodowska University.

Conflicts of Interest: The authors declare no conflict of interest.

References

1. Asif, M. Pharmacologically potentials of different substituted coumarin derivatives. *Chem. Int.* **2015**, *1*, 1–11. [CrossRef]
2. Sumorek-Wiadro, J.; Zając, A.; Maciejczyk, A.; Jakubowicz-Gil, J. Furanocoumarins in anticancer therapy—For and against. *Fitoterapia* **2020**, *142*, 104492. [CrossRef] [PubMed]
3. Kirsch, G.; Abdelwahab, A.B.; Chaimbault, P. Natural and synthetic coumarins with effects on inflammation. *Molecules* **2016**, *21*, 1322. [CrossRef] [PubMed]
4. Venugopala, K.N.; Rashmi, V.; Odhav, B. Review on natural coumarin lead compounds for their pharmacological activity. *BioMed Res. Int.* **2013**, 963248. [CrossRef]
5. Xu, L.Z.; Wu, Y.L.; Zhao, X.Y.; Zhang, W. The Study on Biological and Pharmacological Activity of Coumarins. In Proceedings of the 2015 Asia-Pacific Energy Equipment Engineering Research Conference, Zhuhai, China, 13–14 June 2015.
6. Venkata-Sairam, K.; Gurupadayya, B.M.; Chandan, R.S.; Nagesha, D.K.; Vishwanathan, B. A Review on Chemical Profile of Coumarins and their Therapeutic Role in the Treatment of Cancer. *Curr. Drug Deliv.* **2016**, *13*, 186–201. [CrossRef]
7. Lacy, A.; O'Kennedy, R. Studies on coumarins and coumarin-related compounds to determine their therapeutic role in the treatment of cancer. *Curr. Pharm. Des.* **2004**, *10*, 3797–3811. [CrossRef]
8. Marshall, M.E.; Butler, K.; Fried, A. Phase I evaluation of coumarin (1,2-benzopyrone) and cimetidine in patients with advanced malignancies. *Mol. Biother.* **1991**, *3*, 170–178. [PubMed]

9. Grötz, K.A.; Wüstenberg, P.; Kohnen, R.; Al-Nawas, B.; Henneicke-von Zepelin, H.H.; Bockisch, A.; Kutzner, J.; Naser-Hijazi, B.; Belz, G.G.; Wagner, W. Prophylaxis of radiogenic sialadenitis and mucositis by coumarin/troxerutine in patients with head and neck cancer–a prospective,randomized, placebo-controlled, double-blind study. *Br. J. Oral. Maxillofac. Surg.* **2001**, *39*, 34–39. [CrossRef] [PubMed]
10. Hassanein, E.H.M.; Sayed, A.M.; Hussein, O.E.; Mahmoud, A.M. Coumarins as Modulators of the Keap1/Nrf2/ARE Signaling Pathway. *Oxid. Med. Cell. Longev.* **2020**, 1675957. [CrossRef]
11. Mokdad-Bzeouich, I.; Kovacic, H.; Chebil, L.; Ghoul, M.; Chekir-Ghedira, L.; Luis, J. Esculin and its oligomer fractions inhibit adhesion and migration of U87 glioblastoma cells and in vitro angiogenesis. *Tumor Biol.* **2015**, *37*, 3657–3664. [CrossRef]
12. Velasco-Velázquez, M.A.; Agramonte-Hevia, J.; Barrera, D.; Jiménez-Orozco, A.; García-Mondragón, M.J.; Mendoza-Patiño, N.; Landa, A.; Mandoki, J. 4-Hydroxycoumarin disorganizes the actin cytoskeleton in B16-F10 melanoma cells but not in B82 fibroblasts, decreasing their adhesion to extracellular matrix proteins and motility. *Cancer Lett.* **2003**, *198*, 179–186. [CrossRef]
13. Salinas-Jazmín, N.; de la Fuente, M.; Jaimez, R.; Pérez-Tapia, M.; Pérez-Torres, A.; Velasco-Velázquez, M.A. Antimetastatic, antineoplastic, and toxic effects of 4-hydroxycoumarin in a preclinical mouse melanoma model. *Cancer Chemother. Pharmacol.* **2010**, *65*, 931–940. [CrossRef]
14. Velasco-Velázquez, M.V.; Salinas-Jazmín, N.; Mendoza-Patiño, N.; Mandoki, J.J. Reduced paxillin expression contributes to the antimetastatic effect of 4-hydroxycoumarin on B16-F10 melanoma cells. *Cancer Cell Int.* **2008**, *8*, 8. [CrossRef] [PubMed]
15. Kielbus, M.; Skalicka-Wozniak, K.; Grabarska, A.; Jeleniewicz, W.; Dmoszynska-Graniczka, M.; Marston, A.; Polberg, K.; Gawda, P.; Klatka, J.; Stepulak, A. 7-substituted coumarins inhibit proliferation and migration of laryngeal cancer cells in vitro. *Anticancer Res.* **2013**, *33*, 4347–4356. [PubMed]
16. Lü, H.Q.; Niggemann, B.; Zänker, K.S. Suppression of the proliferation and migration of oncogenicras-dependent cell lines, cultured in a three-dimensional collagen matrix, by flavonoid-structured molecules. *J. Cancer Res. Clin. Oncol.* **1996**, *122*, 335–342. [CrossRef]
17. Musa, M.A.; Cooperwood, J.S.; Khan, M.O. A review of coumarin derivatives in pharmacotherapy of breast cancer. *Curr. Med. Chem.* **2008**, *15*, 2664–2679. [CrossRef]
18. Wang, X.; Huang, S.; Xin, X.; Ren, Y.; Weng, G.; Wang, P. The antitumor activity of umbelliferone in human renal cell carcinoma via regulation of the p110γ catalytic subunit of PI3Kγ. *Acta Pharm.* **2019**, *69*, 111–119. [CrossRef]
19. Hung, C.M.; Kuo, D.H.; Chou, C.H.; Su, Y.C.; Ho, C.T.; Way, T.D. Osthole suppresses hepatocyte growth factor (HGF)-induced epithelial-mesenchymal transition via repression of the c-Met/Akt/mTOR pathway in human breast cancer cells. *J. Agric. Food Chem.* **2011**, *59*, 9683–9690. [CrossRef]
20. Xu, X.; Zhang, Y.; Qu, D.; Jiang, T.; Li, S. Osthole induces G2/M arrest and apoptosis in lung cancer A549 cells by modulating PI3K/ Akt pathway. *J. Exp. Clin. Cancer Res.* **2011**, *30*, 33. [CrossRef]
21. Yang, D.; Gu, T.; Wang, T.; Tang, Q.; Ma, C. Effects of osthole on migration and invasion in breast cancer cells. *Biosci. Biotechnol. Biochem.* **2010**, *74*, 1430–1434. [CrossRef]
22. Jarząb, A.; Grabarska, A.; Kielbus, M.; Jeleniewicz, W.; Dmoszyńska-Graniczka, M.; Skalicka-Woźniak, K.; Sieniawska, E.; Polberg, K.; Stepulak, A. Osthole induces apoptosis, suppresses cell-cycle progression and proliferation of cancer cells. *Anticancer Res.* **2014**, *34*, 6473–6648. [CrossRef]
23. Chou, S.Y.; Hsu, C.S.; Wang, K.T.; Wang, M.C.; Wang, C.C. Antitumor effects of osthol from Cnidium monnieri: An in vitro and in vivo study. *Phytother. Res.* **2007**, *21*, 226–230. [CrossRef]
24. Shokoohinia, Y.; Jafari, F.; Mohammadi, Z.; Bazvandi, L.; Hosseinzadeh, L.; Chow, N.; Bhattacharyya, P.; Farzaei, M.H.; Farooqi, A.A.; Nabavi, S.M.; et al. Potential anticancer properties of osthol: A comprehensive mechanistic review. *Nutrients* **2018**, *10*, 36. [CrossRef]
25. Jakubowicz-Gil, J. Inhibitors of PI3K-Akt-PKB-mTOR pathway in glioma therapy. *Post. Biol. Kom.* **2009**, *36*, 189–201.
26. Schwartzbaum, J.A.; Fisher, J.L.; Aldape, K.D.; Wrensch, M. Epidemiology and molecular pathology of glioma. *Nat. Clin. Pract. Neurol.* **2006**, *2*, 494–516. [CrossRef]
27. Chou, T.C. Drug combination studies and their synergy quantification using the Chou-Talalay method. *Cancer Res.* **2010**, *70*, 440–446. [CrossRef]
28. DU, W.; Pang, C.; Xue, Y.; Zhang, Q.; Wei, X. Dihydroartemisinin inhibits the Raf/ERK/MEK and PI3K/AKT pathways in glioma cells. *Oncol. Lett.* **2015**, *10*, 3266–3270. [CrossRef]

29. Chappell, W.H.; Steelman, L.S.; Long, J.M.; Kempf, R.C.; Abrams, S.L.; Franklin, R.A.; Bäsecke, J.; Stivala, F.; Donia, M.; Fagone, P.; et al. Ras/Raf/MEK/ERK and PI3K/PTEN/Akt/mTOR inhibitors: Rationale and importance to inhibiting these pathways in human health. *Oncotarget* **2011**, *2*, 135–164. [CrossRef]
30. McCubrey, J.A.; Steelman, L.S.; Chappell, W.H.; Abrams, S.L.; Franklin, R.A.; Montalto, G.; Cervello, M.; Libra, M.; Candido, S.; Malaponte, G.; et al. Ras/Raf/MEK/ERK and PI3K/PTEN/Akt/mTOR cascade inhibitors: How mutations can result in therapy resistance and how to overcome resistance. *Oncotarget* **2012**, *3*, 1068–1111. [CrossRef]
31. Jakubowicz-Gil, J.; Bądziul, D.; Langner, E.; Wertel, I.; Zając, A.; Rzeski, W. Temozolomide and sorafenib as programmed cell death inducers of human glioma cells. *Pharmacol. Rep.* **2017**, *69*, 779–787. [CrossRef]
32. Jakubowicz-Gil, J.; Langner, E.; Bądziul, D.; Wertel, I.; Rzeski, W. Quercetin and sorafenib as a novel and effective couple in programmed cell death induction in human gliomas. *Neurotox. Res.* **2014**, *26*, 64–77. [CrossRef]
33. Sumorek-Wiadro, J.; Zając, A.; Bądziul, D.; Langner, E.; Skalicka-Woźniak, K.; Maciejczyk, A.; Wertel, I.; Rzeski, W.; Jakubowicz-Gil, J. Coumarins modulate the anti-glioma properties of temozolomide. *Eur. J. Pharmacol.* **2020**, *881*, 173207. [CrossRef]
34. Budzisz, E.; Brzezinska, E.; Krajewska, U.; Rozalski, M. Cytotoxic effects, alkylating properties and molecular modelling of coumarin derivatives and their phosphonic analogues. *Eur. J. Med. Chem.* **2003**, *38*, 597–603. [CrossRef]
35. Devji, T.; Reddy, C.; Woo, C.; Awale, S.; Kadota, S.; Carrico-Moniz, D. Pancreatic anticancer activity of a novel geranylgeranylated coumarin derivative. *Bioorg. Med. Chem. Lett.* **2011**, *21*, 5770–5773. [CrossRef]
36. Jun, M.; Bacay, A.F.; Moyer, J.; Webb, A.; Carrico-Moniz, D. Synthesis and biological evaluation of isoprenylated coumarins as potential anti-pancreatic cancer agents. *Bioorg. Med. Chem. Lett.* **2014**, *24*, 4654–4658. [CrossRef] [PubMed]
37. Yang, X.; Xu, B.; Ran, F.; Wang, R.; Wu, J.; Cui, J. Inhibitory effects of 40 coumarins compounds against growth of human nasopharyngeal carcinoma cell line KB and human leukemia cell line HL-60 in vitro. *Mod. Chin. Med.* **2006**, *8*, 8–13.
38. Jakubowicz-Gil, J.; Langner, E.; Wertel, I.; Piersiak, T.; Rzeski, W. Temozolomide, quercetin and cell death in the MOGGCCM astrocytoma cell line. *Chem. Biol. Interact.* **2010**, *188*, 190–203. [CrossRef]
39. Jakubowicz-Gil, J.; Langner, E.; Bądziul, D.; Wertel, I.; Rzeski, W. Apoptosis induction in human glioblastoma multiforme T98G cells upon temozolomide and quercetin treatment. *Tumour. Biol.* **2013**, *34*, 2367–2378. [CrossRef] [PubMed]
40. Guarnaccia, L.; Navone, S.E.; Trombetta, E.; Cordiglieri, C.; Cherubini, A.; Crisà, F.M.; Rampini, P.; Miozzo, M.; Fontana, L.; Caroli, M.; et al. Angiogenesis in human brain tumors: Screening of drug response through a patient-specific cell platform for personalized therapy. *Sci. Rep.* **2018**, *8*, 8748. [CrossRef]
41. Jin, S.; White, E. Tumor suppression by autophagy through the management of metabolic stress. *Autophagy* **2008**, *4*, 563–566. [CrossRef]
42. Amaravadi, R.K.; Kimmelman, A.C.; Debnath, J. Targeting Autophagy in Cancer: Recent Advances and Future Directions. *Cancer Discov.* **2019**, *9*, 1167–1181. [CrossRef] [PubMed]
43. Furuya, D.; Tsuji, N.; Yagihashi, A.; Watanabe, N. Beclin 1 augmented cis-diamminedichloroplatinum induced apoptosis via enhancing caspase-9 activity. *Exp. Cell Res.* **2005**, *307*, 26–40. [CrossRef]
44. Huang, X.; Qi, Q.; Hua, X.; Li, X.; Zhang, W.; Sun, H.; Li, S.; Wang, X.; Li, B. Beclin 1, an autophagy-related gene, augments apoptosis in U87 glioblastoma cells. *Oncol. Rep.* **2014**, *31*, 1761–1767. [CrossRef] [PubMed]
45. Erlich, S.; Mizrachy, L.; Segev, O.; Lindenboim, L.; Zmira, O.; Adi-Harel, S.; Hirsch, J.; Stein, R.; Pinkas-Kramarski, R. Differential interactions between Beclin 1 and Bcl-2 family members. *Autophagy* **2007**, *3*, 561–568. [CrossRef]
46. Fan, Q.W.; Knight, Z.A.; Goldenberg, D.D.; Yu, W.; Mostov, K.E.; Stokoe, D.; Shokat, K.M.; Weiss, W.A. A dual PI3 kinase/mTOR inhibitor reveals emergent efficacy in glioma. *Cancer Cell.* **2006**, *9*, 341–349. [CrossRef]
47. Cheng, C.K.; Fan, Q.W.; Weiss, W.A. PI3K signaling in glioma–animal models and therapeutic challenges. *Brain Pathol.* **2009**, *19*, 112–120. [CrossRef]
48. Chang, F.; Steelman, L.S.; Shelton, J.G.; Lee, J.T.; Navolanic, P.M.; Blalock, W.L.; Franklin, R.; McCubrey, J.A. Regulation of cell cycle progression and apoptosis by the Ras/Raf/MEK/ERK pathway (Review). *Int. J. Oncol.* **2003**, *22*, 469–480. [CrossRef]

49. Mulcahy-Levy, J.M.; Zahedi, S.; Griesinger, A.M.; Morin, A.; Davies, K.D.; Aisner, D.L.; Kleinschmidt-DeMasters, B.K.; Fitzwalter, B.E.; Goodall, M.L.; Thorburn, J.; et al. Autophagy inhibition overcomes multiple mechanisms of resistance to BRAF inhibition in brain tumors. *eLife* **2017**, *6*, e19671. [CrossRef] [PubMed]
50. Kinsey, C.G.; Camolotto, S.A.; Boespflug, A.M.; Guillen, K.P.; Foth, M.; Truong, A.; Schuman, S.S.; Shea, J.E.; Seipp, M.T. Protective autophagy elicited by RAF→MEK→ERK inhibition suggests a treatment strategy for RAS-driven cancers. *Nat. Med.* **2019**, *25*, 620–627. [CrossRef] [PubMed]
51. Ma, X.H.; Piao, S.F.; Dey, S.; McAfee, Q.; Karakousis, G.; Villanueva, J.; Hart, L.S.; Levi, S.; Hu, J.; Zhang, G. Targeting ER stress-induced autophagy overcomes BRAF inhibitor resistance in melanoma. *J. Clin. Investig.* **2014**, *124*, 1406–1417. [CrossRef]
52. Bradford, M.M. A rapid and sensitive method for quantitation of microgram quantities of protein utilizing the principle of protein-dye binding. *Anal. Biochem.* **1976**, *72*, 248–254. [CrossRef]
53. Laemmli, U.K. Cleavage of structural proteins during the assembly of the head of bacteriophage T4. *Nature* **1970**, *227*, 680–685. [CrossRef]

Sample Availability: Samples of the coumarins tested are available from the authors.

Article

Curcumin–Coumarin Hybrid Analogues as Multitarget Agents in Neurodegenerative Disorders

Elías Quezada [1], Fernanda Rodríguez-Enríquez [2], Reyes Laguna [2], Elena Cutrín [3], Francisco Otero [3], Eugenio Uriarte [1,4] and Dolores Viña [2,*]

1 Department of Organic Chemistry, Faculty of Pharmacy, Universidade de Santiago de Compostela, 15782 Santiago de Compostela, Spain; elias.quezada@usc.es (E.Q.); eugenio.uriarte@usc.es (E.U.)
2 Center for Research in Molecular Medicine and Chronic Disease (CIMUS), Department of Pharmacology, Pharmacy and Pharmaceutical Technology, Universidade de Santiago de Compostela, 15782 Santiago de Compostela, Spain; nana.enriquez@gmail.com (F.R.-E.); mdelosreyes.laguna@usc.es (R.L.)
3 Department of Pharmacy and Pharmaceutical Technology, Faculty of Pharmacy, Universidade de Santiago de Compostela, 15782 Santiago de Compostela, Spain; elena.cutrin@rai.usc.es (E.C.); francisco.otero@usc.es (F.O.)
4 Instituto de Ciencias Químicas Aplicadas, Universidad Autónoma de Chile, Santiago 7500912, Chile
* Correspondence: mdolores.vina@usc.es; Tel.: +34-881-815-424

Abstract: Neurodegenerative diseases have a complex nature which highlights the need for multitarget ligands to address the complementary pathways involved in these diseases. Over the last decade, many innovative curcumin-based compounds have been designed and synthesized, searching for new derivatives having anti-amyloidogenic, inhibitory of tau formation, as well as anti-neuroinflammation, antioxidative, and AChE inhibitory activities. Regarding our experience studying 3-substituted coumarins with interesting properties for neurodegenerative diseases, our aim was to synthesize a new series of curcumin–coumarin hybrid analogues and evaluate their activity. Most of the 3-(7-phenyl-3,5-dioxohepta-1,6-dien-1-yl)coumarin derivatives **11–18** resulted in moderated inhibitors of hMAO isoforms and AChE and BuChE activity. Some of them are also capable of scavenger the free radical DPPH. Furthermore, compounds **14** and **16** showed neuroprotective activity against H_2O_2 in SH-SY5Y cell line. Nanoparticles formulation of these derivatives improved this property increasing the neuroprotective activity to the nanomolar range. Results suggest that by modulating the substitution pattern on both coumarin moiety and phenyl ring, ChE and MAO-targeted derivatives or derivatives with activity in cell-based phenotypic assays can be obtained.

Keywords: curcumin; curcumin–coumarin hybrids; neuroprotection; monoamine oxidase inhibition; cholinesterase inhibition; scavenging activity

Citation: Quezada, E.; Rodríguez-Enríquez, F.; Laguna, R.; Cutrín, E.; Otero, F.; Uriarte, E.; Viña, D. Curcumin–Coumarin Hybrid Analogues as Multitarget Agents in Neurodegenerative Disorders. *Molecules* **2021**, *26*, 4550. https://doi.org/10.3390/molecules26154550

Academic Editors: Luciana Mosca and Jose Luis Lavandera

Received: 4 June 2021
Accepted: 26 July 2021
Published: 28 July 2021

1. Introduction

In neurodegenerative diseases, a loss of nerve cells is observed in the brain and spinal cord, leading to sensory dysfunction (dementia) or loss of function (ataxia). Mitochondrial dysfunction, oxidative stress, protein misfolding, neuroinflammation, and finally apoptosis have been recognized by different studies as pathological causes of neurodegenerative diseases such as Parkinson's disease (PD), Alzheimer's disease (AD), sclerosis multiple (MS), and amyotrophic lateral sclerosis (ALS). Currently, commercially available and approved drugs for these disorders only temporarily relieve symptoms but do not significantly alter disease progression. The development of new treatment strategies remains in the preclinical and clinical stages. Due to the complex nature of neurodegenerative diseases, it seems necessary to design multitarget ligands to address the complementary pathways involved in these diseases [1,2].

Curcumin is a dietary polyphenol presented in the curry spice turmeric. Numerous studies describe its therapeutic potential for neurodegenerative diseases, including AD and

PD, due to its powerful antioxidant, anti-protein aggregation, and anti-inflammatory properties [3,4]. However, curcumin exhibits instability, poor bioavailability, and low cellular uptake, which limits the interest of its use in these disorders [5]. To address this problem, new nanoformulations such as liposomes, solid-lipid nanoparticles, micelles, polymer nanoparticles, and polymer conjugates have been developed [6,7]. With the same objective and also to improve its activity, in recent years, many compounds derived from curcumin have been designed and synthesized. Some of them have shown anti-amyloidogenic activity, inhibitory of the formation of tau, as well as anti-neuroinflammatory, antioxidant, and inhibitory of acetylcholinesterase (AChE) [8,9].

Coumarins are natural or synthetic compounds with diverse biological activities. Many synthetic coumarin derivatives have been designed to obtain new drugs with potential activity in neurodegenerative diseases. Coumarin moiety has the potential to achieve monoamine oxidase (MAO) inhibitory activity (e.g., MAO-A and MAO-B inhibitors), AChE, β- and γ-secretase inhibition. Some of these compounds display potent antioxidant activity and, therefore, could protect cells from neurodegeneration [10]. In the last few years, our group has described different series of 3-substituted coumarins displaying these properties [11–17].

MAO inhibition by coumarins may also prevent oxidative stress, through inhibition of neurotransmitters degradation, leading a neuroprotective effect. It has been described that systemic injection of a MAO inhibitor decreases 6-hydroxydopamine-induced oxidative stress [18].

Considering the above described and with the aim to improve the properties of curcumin and coumarin for the treatment of neurodegenerative diseases such as AD or PD, we have synthesized a series of curcumin–coumarin hybrid analogues (Figure 1) to study their activity as MAO and AChE inhibitors, free radical scavengers as well as their neuroprotective activity against hydrogen peroxide (H_2O_2). In addition, to facilitate their passage through cell membranes and therefore improve their neuroprotective activity, some of the derivatives have been formulated into nanoparticles.

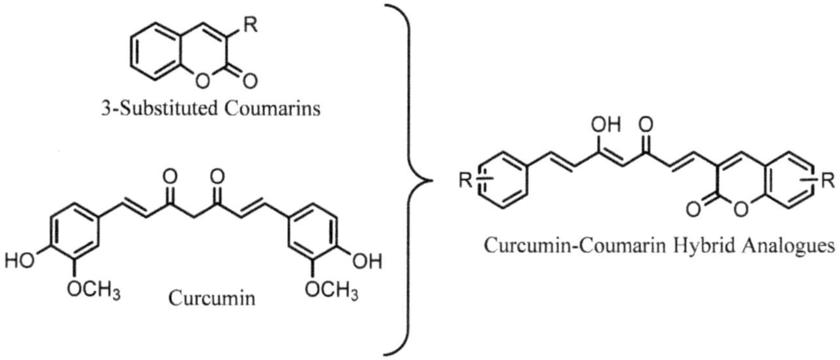

Figure 1. Overview of design of new curcumin–coumarin hybrid analogues.

2. Results

2.1. Synthesis of Coumarins **5** *and* **9**

For the synthesis of coumarin **5**, firstly, pyrogallol was treated with K_2CO_3 and dichlorodiphenylmethane in CH_3CN, obtaining a protected catechol **2**. In a second step, the protected catechol was reacted with magnesium chloride, triethylamine, and *para*-formaldehyde to afford the protected *ortho*-hydroxybenzaldehyde **3**. Then, *ortho*-hydroxybenzaldehyde **3** was reacted with sodium hydride and (trimethylsilyl)propioloyl chloride to obtain the silylated ester **4**. Silylated ester **4** was reacted with 1,4-dizabicyclo [2.2.2]octane (DABCO) in THF under reflux, resulting in the desired coumarin **5** (Scheme 1).

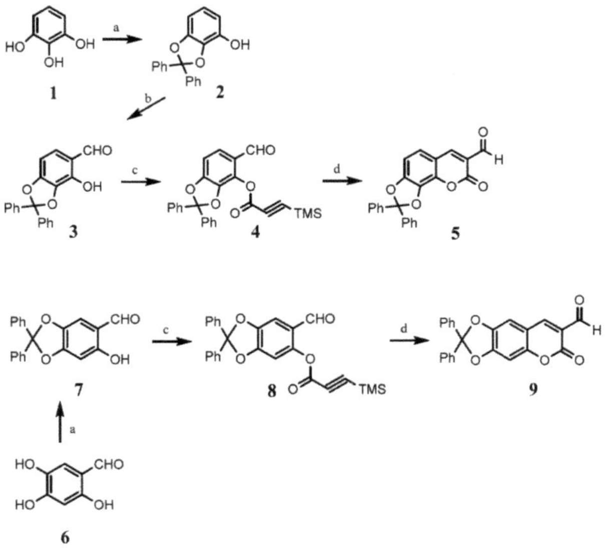

Scheme 1. (a) dichlorodiphenylmethane, $(Ph)_2O$, 180 °C, 30 min; (b) $MgCl_2$, Et_3N, $(CH_2O)_n$, THF reflux, 4 h; (c) (trimethylsilyl)propioloyl chloride, NaH, THF, reflux, 10 h; (d) DABCO, THF, reflux, 12 h.

Coumarin **9** was synthesized from 2,4,5-trihydroxybenzaldehyde (**6**) following a similar procedure to that described above (Scheme 1).

2.2. *Synthesis of a Series of Curcumin–Coumarin Hybrid Analogues* **11–18**

These compounds were obtained through direct coupling of an acethylacetone–B_2O_3 complex with the corresponding formylcoumarin previously obtained (**5** or **9**) and the adequate substituted benzaldehyde (**10a–d**) in the presence of tributyl borate and *n*-butylamine (Scheme 2). The deprotection of the phenol groups was carried out in two steps, firstly acidium medium was used to hydrolyze tris(methoxymethoxy) groups (OMOM) followed by hydrolysis of the diphenylbenzodioxole group to obtain compounds **11–18**. Compounds **11–18** were stored at −20 °C and in the dark.

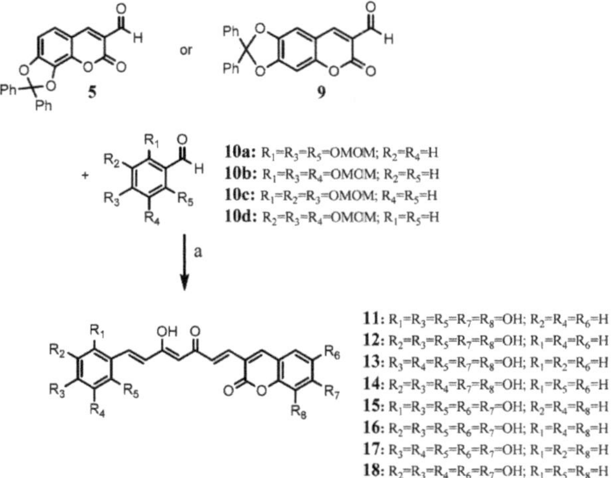

Scheme 2. (a) i: 2,4-Pentanedione, B_2O_3, EtOAc, 40 °C, 2 h; $(BuO)_3B$; $nBuNH_2$, EtOAc, 25–40 °C 22 h; ii: HCl, 60 °C, 1 h; iii: H_2, Pd/C, EtOH, rt, 48 h.

2.3. *Nanoparticles Formulations*

Curcumin and curcumin–coumarin hybrid analogue loaded PLGA nanoparticles were prepared by an interfacial deposition method. All nanoparticles showed a narrow size distribution with mean diameters between 141–168 nm and PDI of 0.121–0.153, and a Zeta potential of −20 to −26 mV (dispersed in purified water). The encapsulation efficiency was similar for all the drugs assayed, obtaining percentages of encapsulation of 56, 53, and 55% for curcumin and curcumin–coumarin hybrid analogues **14** and **16,** respectively.

2.4. *In Vitro Activity*

2.4.1. Cholinesterase Inhibition

As seen in Table 1, compounds **15** and **17** at 100 μM concentration inhibit the activity of both AChE and butyrylcholinesterase (BuChE) by approximately 50%. Therefore, their activity is lower than that presented by curcumin on AChE. Compound **12** resulted in the most selective derivative with activity only on BuChE.

Table 1. Percentage inhibition of human cholinesterases (hAChE and hBuChE) and human monoamine oxidases (hMAO-A and hMAO-B).

Compound	hAChE % Inh 100 μM	hBuChE % Inh 100 μM	hMAO-A % Inh 100 μM	hMAO-B % Inh 100 μM
11	2.96% ± 0.10%	2.90% ± 0.10%	nd	nd
12	5.70% ± 0.40%	43.26% ± 2.90%	60.91% ± 4.09%	45.76% ± 3.07%
13	38.47% ± 2.58%	36.08% ± 2.42%	nd	nd
14	5.62% ± 0.40%	8.69% ± 0.60%	54.33% ± 3.64%	45.81% ± 3.07%
15	47.69% ± 3.20%	50.94% ± 3.41%	nd	nd
16	17.33% ± 1.16%	21.37% ± 1.43%	58.18% ± 3.90%	78.93% ± 5.59%
17	42.88% ± 2.87%	46.63% ± 3.12%	nd	nd
18	5.14% ± 0.34%	10.38% ± 0.70%	51.50% ± 3.45%	55.72% ± 3.74%
Curcumin	64.83% ± 4.34%	35.46% ± 2.38%	84.20% ± 5.61%	92.60% ± 6.17%

Results are expressed as the mean ± e.e.m (n = 3). nd: not determined. At concentration > 100 μM, compounds precipitate.

2.4.2. Monoamine Oxidase Inhibition

Most of the curcumin–coumarin hybrid analogues herein evaluated were not selective, inhibiting both MAO isoforms in a similar percentage (Table 1). The most potent inhibitor was compound **16** with IC_{50} (hMAO-B) = 26.18 ± 1.76 μM, which also exhibited greater selectivity over MAO-B. In any case, its inhibitory activity turned out to be lower than that shown by curcumin: IC_{50} (hMAO-A) = 10.18 ± 0.68 μM and IC_{50} (hMAO-B) = 1.78 ± 0.12 μM. For compounds **11, 13, 15, 17**, their activity on the MAO isoforms could not be determined because they react with the Amplex Red reagent.

2.4.3. Scavenging Activity

As seen in Figure 2, most of the curcumin–coumarin hybrid analogues **11–18** showed moderate activity as free radical scavengers. All of them resulted less active than curcumine or vitamin C. In general, among the curcumin–coumarin hybrid analogues, the compounds with the highest activity are those with three hydroxyl groups in contiguous positions of the phenyl substituent. Compounds **11, 15,** and **16** that did not present this characteristic resulted in being the least active.

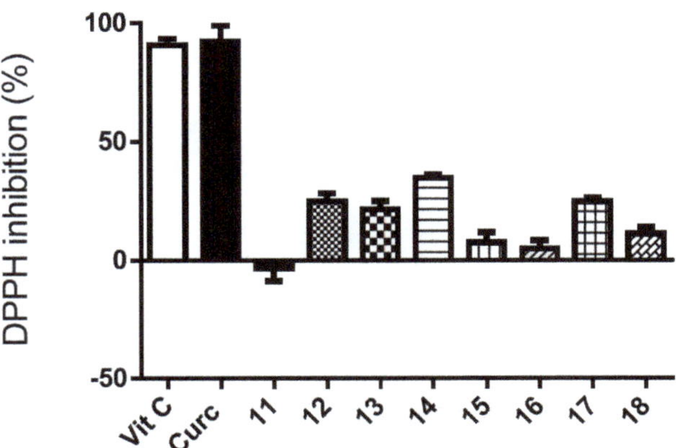

Figure 2. Percentage of neutralization of radical DPPH· by curcumin, curcumin–coumarin hybrid analogues **11–18** (100 μM) and vitamin C used as reference (100 μM). Each value is the mean ± s.e.m. of 3 experiments (*n* = 3).

2.4.4. Neuroprotective Activity against H_2O_2

The neuroprotective activity of these curcumin–coumarin hybrid analogues **11–18** was evaluated in two different cell models, primary culture of rat motor cortex neurons and SH-SY5Y cell line. Neither curcumin nor any of the curcumin–coumarin hybrid analogues s **11–18** (10 μM) showed a protective effect against hydrogen peroxide (H_2O_2) in the primary culture of rat motor cortex (data not shown). However, **14** and **16** showed a significant increase of viability on the SH-SY5Y cell line treated with H_2O_2 (Figure 3). Because of this neuroprotective effect, compounds **14** and **16** were formulated in biodegradable nanoparticles, and their neuroprotective activity against H_2O_2 was also evaluated in the SH-SY5Y cell line. In this formulation, derivatives **14** and **16** at low concentration (10 nM) presented a statistically significant neuroprotective activity. As can be seen in Figure 4, the activity of compound **16** turned from a neurotoxic effect when the cultures were treated at 1 μM concentration to a neuroprotective effect at 10 nM concentration.

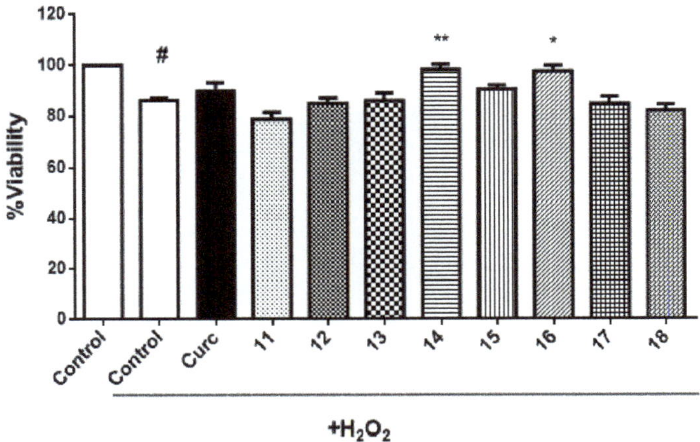

Figure 3. Neuroprotective effects of curcumin and curcumin–coumarin hybrid analogues **11–18** (10 μM) on SH-SY5Y cells. The results are expressed as % viability versus the control group (treated with DMSO 1%, or DMSO 1%, and H_2O_2 100 μM). Each value is the mean ± s.e.m of at least five experiments. # $p < 0.0001$ versus the control group without H_2O_2 treatment. * $p < 0.05$, ** $p < 0.005$ versus DMSO + H_2O_2 treated group.

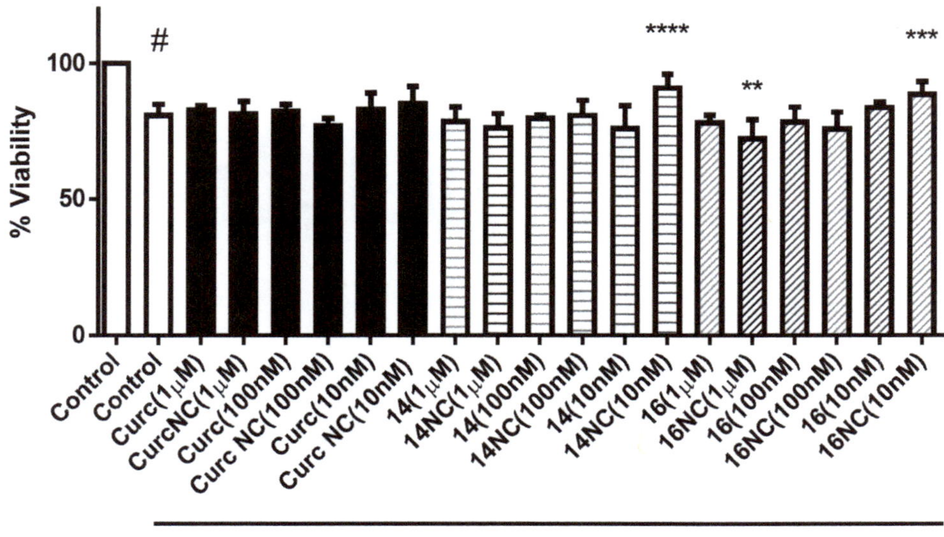

Figure 4. Neuroprotective effects on SH-SY5Y cells of different concentrations of curcumin and curcumin–coumarin hybrid analogues **14** and **16** and their nanoparticle formulations (NC). Each value is the mean ± s.e.m of at least 5 experiments. # $p < 0.0001$ versus the control group (without H_2O_2 treatment), ** $p < 0.005$, *** $p < 0.001$, **** $p < 0.0001$ versus cells treated with H_2O_2.

3. Discussion

Coumarins **5** and **9** were obtained by a similar route based on reactions described in the literature and both in good yield (68.3% and 81.3%, respectively). However, 3-(7-phenyl-3,5-dioxohepta-1,6-dien-1-yl)coumarin derivatives, namely as curcumin–coumarin hybrid analogues **11–18**, were obtained in low yields (5% approximately). This can be probably explained due to the low reactivity of the carbonyl of the formyl group at position 3 of coumarins. This fact was corroborated when obtaining in the same reaction the corresponding 1,7-biscoumarin-3,5-dioxohepta-1,6-dienyl derivatives in very low yield (data not shown). Furthermore, in the same reaction, we appreciated the formation of the corresponding 1,7-bisphenyl-3,5-dioxohepta-1,6-dienyl derivatives, previously described [19–22], and obtained in higher yields than the curcumin–coumarin hybrid analogues **11–18**. Despite the poor yield for curcumin–coumarin hybrid **11–18**, they could be easily detected and isolated because of their red color.

Regarding curcumin, the curcumin–coumarin hybrid analogues **11–18** conserve two aromatic systems in their structure, replacing a phenyl ring with a coumarin and maintain the length and flexibility of the central link region. These characteristics have been identified as a key for the derivatives to maintain the interest of curcumin in neurodegenerative diseases [23]. Additionally, different substitution partners on both aromatic systems, coumarin moiety, and phenyl ring have been studied.

MAO plays an important role in the homeostasis of neurotransmitters in the brain. MAO-B inhibitors are being used in combination with L-dopa to manage PD. However, the beneficial effects of MAO-B inhibitors in PD are not only associated with maintaining dopamine levels but also with their neuroprotective properties [24]. The occurrence of activated MAO-B in the brains of patients with AD has also been evidenced. Furthermore, MAO-A has a different appearance in different parts of the brains of patients with AD. MAO-A is increased in the hypothalamus and frontal pole, revealing that activated MAO-A in neurons is involved in the pathology of this disease as a predisposing factor. In addition, increased MAO-A activity appears more significant in the glia of patients with AD [25].

The above described demonstrates the interest in MAO inhibitors for the treatment of these diseases. Compounds **12, 14, 16,** and **18** showed moderate inhibitory activity on both MAO isoforms (Table 1). Hydroxyl substituents at positions 6 and 7 of the coumarin nucleus (compounds **16** and **18**) afforded more potent derivatives on the MAO-B isoform than substitution at positions 7 and 8 (compounds **12** and **14**). However, the position of the hydroxyl groups on the phenyl ring does not appear to significatively modify the activity of these compounds on MAO-B. The opposite behavior is observed in the activity of MAO-A. Compounds **12** and **16** resulted in the most potent derivatives, both bearing hydroxyl groups at positions 4, 6, and 7 of the phenyl ring. Only compound **16** exhibited moderate MAO-B selectivity [selectivity index (SI) = IC_{50} hMAO-A/IC_{50} hMAO-B; SI = 3.82].

Acetylcholine levels are regulated mainly by AChE but also by BuChE. Role of BuChE is less important than AChE in healthy brains. However, the AChE activity remains unchanged or even decreases in AD, while BuChE progressively increases, suggesting that inhibition of both enzymes may be considered a valid approach for AD therapy, increasing levels of AChE [26]. Curcumin–coumarin hybrid analogues **13**, **15,** and **17** showed similar activity on both AChE and BuChE, while compound **12** resulted in selectively inhibiting BuChE activity (Table 1).

Among the studied curcumin–coumarin hybrid analogues, only compound **12** showed potential to inhibit both degradation of acetylcholine (via BuChE inhibition) and monoamines (via non-selective MAO inhibition) (Table 1).

Curcumin can protect neurons against inflammation, oxidative stress, apoptosis, or mitochondrial dysfunction [27,28]. It has been described that curcumin concentrations up to 20 μM increase viability in different cell models treated with H_2O_2 [29]. However, the effect of curcumin on SH-SY5Y cells is both dose and time-dependent. Approximately 40 μM concentration and 24 h exposure are the critical parameters at which the cell viability significantly decreases [30]. Other authors describe even lower concentrations (10 μM) to decreases SH-SY5Y proliferation and 20 μM to cause apoptosis [31]. Considering the controversies found in the literature, we studied the neuroprotective effects of low concentrations ($\leq$10 μM) of curcumin and curcumin–coumarin hybrid analogues on two different neuronal models, primary culture of rat motor cortex neurons and SH-SY5Y cell line. While neither of the curcumin–coumarin hybrid analogues **11–18** nor curcumin protected rat motor cortex neurons against H_2O_2 (data not shown), compounds **14** and **16** showed neuroprotective effects at 10 μM concentration on SH-SY5Y cells (Figure 3). Compound **14** also showed scavenger activity (Figure 2) which could justify, at least partially, its neuroprotective activity. However, compound **16** lacks this activity, but it is the most potent MAO-B inhibitor, indicating that different mechanisms may be implicated in this neuroprotective activity.

Based on the statistically significant increase in viability on SH-SY5Y cells treated with H_2O_2 produced by compounds **14** and **16** at 10 μM concentration, both compounds were formulated in biodegradable nanoparticles. This formulation improves the neuroprotective effect at 10 nM concentration on SH-SY5Y cells treated with H_2O_2 (Figure 4). Furthermore, compound **16** formulated in nanoparticles goes from having a neurotoxic effect at 1 μM concentration to a statistically significant neuroprotective effect at 10 nM concentration. To explain this neuroprotective effect, it is necessary to resort to the hormesis, which is shared by several phytochemical compounds, including curcumin [32]. Hormesis is defined as a stimulation of cellular protection at low doses while it is inhibited at high doses of the compound, resulting in an inverted J or U-shaped dose-response curve as it can be obtained for compounds **14** and **16**. However, at concentrations used in this work, neuroprotective activity was not found for coumarin.

4. Materials and Methods

4.1. Materials and Instrumentation

All reactions utilizing air- or moisture-sensitive reagents were carried out in flame-dried glassware under an argon atmosphere, unless otherwise stated. Hexane, CH_2Cl_2,

THF, Et_2O, Et_3N, and n-$BuNH_2$ were distilled prior to use according to the standard protocols. Other reagents were purchased and used as received without further purification unless otherwise stated. Reactions were magnetically stirred and monitored by thin-layer chromatography (TLC). Analytical TLC was performed on plates precoated with silica gel (Merck 60 F254, 0.25 mm). Compounds were visualized with UV light and/or by staining with ethanolic phosphomolybdic acid (PMA) followed by heating on a hot plate. Flash chromatography (FC) was performed with silica gel (35–60 mesh) under pressure. Melting points were determined in a Reichert Kofler thermopan or in capillary tubes in a Buchi 510 apparatus and are uncorrected. NMR spectra were recorded on Bruker AMX 250 (^{1}H, 250 MHz; ^{13}C, 62.9 MHz) spectrometer in $CDCl_3$ or DMSO-d_6 with TMS as the internal standard. Chemical shifts (δ) are given in ppm and coupling constants (J) in Hz. Multiplicity is indicated as follows: s, singlet; d, doublet; m, multiplet; bs, broad singlet. Elemental analyses were performed on Thermo-Finnigan Flash 1112 CHNS/O analyzer (Supplementary Materials).

4.2. Chemical Synthesis

4.2.1. Synthesis of 2,2-Diphenylbenzo[1,3]dioxol-4-ol (**2**)

Dichlorodiphenylmethane (9.65 mL, 50.31 mmol) was added to a stirred mixture of pyrogallol (1, 4.23 g, 33.54 mmol) in diphenyl ether (25 mL), and the reaction mixture was heated at 180 °C for 30 min. The mixture was cooled to room temperature, and petroleum ether (50 mL) was added to give a solid compound [33,34]. Then the solid was filtered and purified by column chromatography using CH_2Cl_2 to yield **2** as a white solid (9.65 g, 99.2%). m.p.: 165 °C. ^{1}H-NMR (250 MHz, $CDCl_3$, δ ppm): 7.54 (4H, m), 7.37 (6H, m), 6.71 (1H, t, J = 6.7 Hz), 6.53 (1H, d, J = 6.4 Hz), 6.46 (1H, d, J = 6.4 Hz), 4.98 (1H, br). ^{13}C-NMR (62.9 MHz, $CDCl_3$, δ ppm): 148.2, 139.9, 139.3, 133.8, 129.1 (2C), 128.2 (4C), 126.3 (4C), 122.1, 116.1, 110.8, 101.9. Anal. Calcd. (%) for [$C_{19}H_{14}O_3$]: C, 78.61; H, 4.86; found (%): C, 78.58; H, 4.83.

4.2.2. Synthesis of 4-Hydroxy-2,2-diphenylbenzo[1,3]dioxol-5-carbaldehyde (**3**)

To a dry THF solution (300 mL) of the 2,2-diphenylbenzo[1,3]dioxol-4-ol (**2**) (6.2 g, 21.35 mmol), anhydrous magnesium chloride (4.065 g, 42.70 mmol), triethylamine (5.95 mL, 4.32 g, 42.70 mmol) and paraformaldehyde (1.923 g, 64.05 mmol) were added. The reaction mixture was heated to reflux under Ar atmosphere for 4 h, and monitored by TLC (hexane:ethyl acetate = 8:2). After complete consumption of the phenol, the reaction mixture was cooled and diluted with diethyl ether (100 mL). The organic layer was washed successively with HCl (1 M, 2 × 100 mL) and H_2O (2 × 100 mL), and then dried (Na_2SO_4) [35]. The product was purified by column chromatography using hexane:ethyl acetate (98:2) to yield **3** as a white solid (5.6 g, 82.5%). m.p.: 159 °C. ^{1}H-NMR (250 MHz, $CDCl_3$, δ ppm): 11.07 (1H, bs), 9.68 (1H, s), 7.59 (4H, m), 7.38 (6H, m), 7.12 (1H, d, J = 8.2 Hz), 6.62 (1H, d, J = 8.2 Hz). ^{13}C-NMR (62.9 MHz, $CDCl_3$, δ ppm): 194.9, 154.2, 145.3, 139.1 (2C), 133.6, 130.2, 129.3 (2C), 128.2 (4C), 126.0 (4C), 119.2, 118.09, 101.9. Anal. Calcd. (%) for [$C_{20}H_{14}O_4$]: C, 75.46; H, 4.43; found (%): C, 75.44; H, 4.40.

4.2.3. Synthesis of 5-Formyl-2,2-diphenylbenzo[1,3]dioxol-4-yle (trimethylsilyl)propiolate (**4**)

Sodium hydride (0.942 g, 23.55 mmol, 60%, washed with hexane) was suspended in anhydrous THF (25 mL), then cooled to 0 °C. A solution of **3** (2.50 g, 7.85 mmol) in anhydrous THF (25 mL) was dropwise added and the suspension was stirring for 1 h. Then, trimethylsilylpropioloyl chloride (3.78 g, 23.55 mmol) [36] in THF (10 mL) was dropwise added. The mixture was refluxed for 10 h. The mixture was quenched with ice-cold water and extracted three times with EtOAc. The combined organic layers were washed with brine, dried over Na_2SO_4, and evaporated [37]. The residue was purified by column chromatography on SiO_2 hexane:ethyl acetate (2:98) to obtain silylated ester **4** as a colorless syrup (2.8 g, 80.7%). ^{1}H-NMR (250 MHz, $CDCl_3$, δ ppm): 9.93 (1H, s), 7.53 (4H, m), 7.39 (7H, m), 6.87 (1H, d, J = 8.2 Hz), 0.27 (9H, s). ^{13}C-NMR (62.9 MHz, CDC13, δ ppm): 186.8,

153.6, 149.1, 145.6, 139.5, 138.4 (2C), 129.5 (2C), 128.2 (4C), 127.1, 126.2 (4C), 123.5, 120.3, 106.8, 98.1, 92.8, −1.1. Anal. Calcd. (%) for [$C_{26}H_{22}O_5Si$]: C, 70.57; H, 5.01; found (%): C, 70.56; H, 5.00.

4.2.4. Synthesis of 2′,2′-Diphenyl-1,3-dioxol[h]coumarin-3-carbaldehyde (**5**)

A mixture of silylated ester **4** (2.8 g, 6.33 mmol), and DABCO (1.42 g, 12.66 mmol) in THF (150 mL) was refluxed under Ar atmosphere. After 12 h, the mixture was diluted with CH_2Cl_2, washed with HCl (10%) and brine, and dried over Na_2SO_4 [38]. The solvent was evaporated to leave a residue, which was purified by silica-gel chromatography (CH_2Cl_2) to afford the 3-formylcoumarin **5** as a yellow solid (1.6 g, 68.3%). m.p.: 181 °C. ^{1}H-NMR (250 MHz, $CDCl_3$, δ ppm): 10.18 (1H, s), 8.31 (1H, s), 7.59 (4H, m), 7.40 (6H, m), 7.23 (1H, d, J = 8.3 Hz), 6.94 (1H, d, J = 8.3 Hz). ^{13}C-NMR (62.9 MHz, $CDCl_3$, δ ppm): 187.4, 159.3, 153.7, 146.2, 139.0, 138.6 (2C), 133.5, 129.6 (2C), 128.4 (4C), 126.4, 126.0 (4C), 120.6, 118.5, 114.3, 107.0. Anal. Calcd. (%) for [$C_{23}H_{14}O_5$]: C, 74.59; H, 3.81; found (%): C, 74.56; H, 3.80.

4.2.5. Synthesis of 6-Hydroxy-2,2-diphenylbenzo[1,3]dioxol-5-carbaldehyde (**7**)

Following the procedure previously described to obtain compound **2**, dichlorodiphenylmethane (7.4 mL, 38.92 mmol) was reacted with **6** (4.0 g, 25.95 mmol) in diphenyl ether (25 mL), to yield **7** as a white solid (8.0 g, 96.8% yield) [33,39]. m.p.: 128 °C. ^{1}H-NMR (250 MHz, $CDCl_3$, δ ppm): 11.79 (1H, br), 9.59 (1H, s), 7.58 (4H, m), 7.39 (6H, m), 6.90 (1H, s), 6.55 (1H, s). ^{13}C-NMR (62.9 MHz, $CDCl_3$, δ ppm): 193.6, 161.4 (2C), 154.6, 141.0, 139.1 (2C), 129.4 (2C), 128.3 (4C), 126.1 (2C), 126.0 (2C), 109.4, 98.4, 90.1. Anal. Calcd. (%) for [$C_{20}H_{14}O_4$]: C, 75.46; H, 4.43; found (%): C, 75.42; H, 4.42.

4.2.6. Synthesis of 6-Formyl-2,2-diphenylbenzo[1,3]dioxol-5-yle(trimethylsilyl)propiolate (**8**)

Following the procedure previously described to obtain compound **4**, compound **7** (2.5 g, 7.85 mmol) was reacted with sodium hydride (0.942 g, 23.55 mmol, 60%, washed with hexane), and trimethylsilylpropioloyl chloride (3.78 g, 23.55 mmol) [36] to obtain **8** as a colorless syrup (2.5 g, 72.0%) [37]. ^{1}H-NMR (250 MHz, $CDCl_3$, δ ppm): 9.96 (1H, s), 7.50 (5H, m), 7.34 (6H, m), 6.91 (1H, s), 0.20 (9H, s). ^{13}C-NMR (62.9 MHz, $CDCl_3$, δ ppm): 187.0, 154.3, 151.0, 145.9, 140.9, 138.2 (2C), 129.2 (2C), 128.4 (4C), 126.9 (4C), 125.8, 122.5, 118.2, 107.1, 97.9, 85.8. Anal. Calcd. (%) for [$C_{26}H_{22}O_5Si$]: C, 70.57; H, 5.01; found (%): C, 70.55; H, 4.99.

4.2.7. Synthesis of 2′,2′-Diphenyl-1,3-dioxol[g]coumarin-3-carbaldehyde (**9**)

Following the procedure previously described to obtain compound **5**, a mixture of the propionic ester **8** (2.5 g, 5.65 mmol), and DABCO (1.26 g, 11.30 mmol) afforded the 3-formylcoumarin **9** as a yellow solid (1.7 g, 81.3%) [38]. m.p.: 173 °C. ^{1}H-NMR (250 MHz, $CDCl_3$, δ ppm): 9.97 (1H, s), 8.12 (1H, s), 7.44 (4H, m), 7.38 (6H, m), 6.92 (1H, s), 6.69 (1H, s). ^{13}C-NMR (62.9 MHz, $CDCl_3$, δ ppm): 191.8, 160.6, 152.1, 151.2, 145.3, 143.5, 138.9 (2C), 129.5 (2C), 128.4 (4C), 126.0 (4C), 122.4, 119.2, 108.3, 103.8, 98.4. Anal. Calcd. (%) for [$C_{23}H_{14}O_5$]: C, 74.59; H, 3.81; found (%): C, 74.55; H, 3.79.

4.2.8. Synthesis of 7,8-Dihydroxy-3-(7-(2′,4′,6′-trihydroxyphenyl)-3,5-dioxohepta-1,6-dien-1-yl)coumarin (**11**)

2,4-Pentanedione (0.25 g, 2.5 mmol) and boric anhydride (0.121 g, 1.75 mmol) were dissolved in EtOAc (5 mL) and stirred for 2 h at 40 °C. Coumarin **5** (0.925 g, 2.5 mmol), tris(methoxymethoxy)benzaldehyde **10a** [40,41] (0.715 g, 2.5 mmol) and tributyl borate (2.3 g, 10 mmol) were added and the reaction mixture was stirred for 0.5 h. Then a solution of n-butylamine (0.182 g, 2.5 mmol) in EtOAc (2.5 mL) was dropwise added over a period of 30 min, the mixture was stirred for a further 18 h at room temperature and 4 h at 40 °C. The mixture was hydrolyzed by the addition of 0.4 N HCl (10 mL) and heating to 60 °C for 1 h. The organic layer was separated, and the aqueous layer was extracted three times with EtOAc. The combined organic layers were washed with water and dried over Na_2SO_4.

Evaporation of the solvent left a red-brown powder [42]. The solid was dissolved in 80 mL of ethanol and treated with 10% Pd/C (0.6 g, 33 wt. % of starting material) [43]. The system was purged several times with hydrogen and stirred under hydrogen for 48 h. The reaction mixture was then purged with Ar and filtered through Celite washing with CH_3OH. The filtrate was evaporated and the dark red solid was purified by column chromatography using CH_2Cl_2:CH_3OH (9:1 and 8:2) to give **11** (44 mg, 4.15%). m.p.: 180 °C (dec.). ^{1}H-NMR (250 MHz, DMSO-d_6, δ ppm): 10.10 (2H, bs), 9.70 (2H, bs), 9.51 (2H, bs), 7.72 (1H, s), 7.46 (1H, d, *J* = 15.5 Hz), 7.15 (1H, d, *J* = 8.8 Hz), 6.84 (1H, d, *J* = 15.4 Hz), 6.64 (3H, m), 5.96 (2H, s), 5.83 (1H, s). ^{13}C-NMR (62.9 MHz, DMSO-d_6, δ ppm): 182.4, 169.7, 160.5, 158.7 (2C), 158.0, 148.7, 141.6, 139.5, 137.6, 134.0, 130.5, 126.9, 125.3, 120.7, 120.6, 114.3, 114.2, 109.9, 103.1, 95.7 (2C). Anal. Calcd. (%) for [$C_{22}H_{16}O_9$]: C, 62.27; H, 3.80; found (%): C, 62.24; H, 3.78.

4.2.9. Synthesis of 7,8-Dihydroxy-3-(7-(2′,4′,5′-trihydroxyphenyl)-3,5-dioxohepta-1,6-dien-1-yl)coumarin (**12**)

Following the procedure described above to obtain compound **11**, reaction of coumarin 5 (0.925 g, 2.5 mmol) and tris(methoxymethoxy)benzaldehyde **10b** [44] (0.715 g, 2.5 mmol) yielded compound **12** as a red solid (40 mg, 3.77%). m.p.: 177 °C (dec.). ^{1}H-NMR (250 MHz, DMSO-d_6, δ ppm): 10.12 (2H, bs), 9.76 (2H, bs), 9.50 (2H, bs), 7.79 (1H, d, *J* = 16.1 Hz), 7.65 (1H, s), 7.07 (1H, d, *J* = 8.8 Hz), 6.78 (1H, d, *J* = 15.4 Hz), 6.65 (1H, d, *J* = 16.1 Hz), 6.59 (2H, m), 6.42 (1H, s), 6.19 (1H, s), 5.83 (1H, s). ^{13}C-NMR (62.9 MHz, DMSO-d_6, δ ppm): 187.4, 179.5, 162.9, 152.5, 148.7, 148.6, 141.6, 140.4, 139.5, 137.6, 134.0, 129.0, 126.9, 124.2, 120.7, 120.6, 115.6, 114.4, 114.3, 114.2, 103.5, 96.9. Anal. Calcd. (%) for [$C_{22}H_{16}O_9$]: C, 62.27; H, 3.80; found (%): C, 62.22; H, 3.77.

4.2.10. Synthesis of 7,8-Dihydroxy-3-(7-(2′,3′,4′-trihydroxyphenyl)-3,5-dioxohepta-1,6-dien-1-yl)coumarin (**13**)

Following the procedure described above to obtain compound **11**, reaction of coumarin **5** (0.925 g, 2.5 mmol) and tris(methoxytrimethoxy)benzaldehyde **10c** [45,46] (0.715 g, 2.5 mmol) yielded compound **13** as a red solid (46 mg, 4.34%). m.p.: 185 °C (dec.). ^{1}H-NMR (250 MHz, DMSO-d_6, δ ppm): 10.10 (2H, bs), 9.74 (2H, bs), 9.52 (2H, bs), 7.79 (1H, d, *J* = 16.0 Hz), 7.72 (1H, s), 7.08 (1H, d, *J* = 8.8 Hz), 6.86 (1H, d, *J* = 15.8 Hz), 6.78 (1H, d, *J* = 16.1 Hz), 6.58 (2H, m), 6.51 (1H, d, *J* = 8.4 Hz), 6.12 (1H, d, *J* = 8.4 Hz), 5.78 (1H, s). ^{13}C-NMR (62.9 MHz, DMSO-d_6, δ ppm): 184.6, 162.4, 158.9, 148.8, 148.7, 147.4, 141.6, 139.5, 137.6, 134.0, 133.8, 128.4, 126.9, 124.4, 120.8, 120.7, 120.6, 114.7, 114.3, 114.2, 107.8, 101.6. Anal. Calcd. (%) for [$C_{22}H_{16}O_9$]: C, 62.27; H, 3.80; found (%): C, 62.26; H, 3.79.

4.2.11. Synthesis of 7,8-Dihydroxy-3-(7-(3′,4′,5′-trihydroxyphenyl)-3,5-dioxohepta-1,6-dien-1-yl)coumarin (**14**)

Following the procedure described above to obtain compound **11**, reaction of coumarin **5** (0.925 g, 2.5 mmol) and tris(methoxymethoxy)benzaldehyde **10d** [47–49] (0.715 g, 2.5 mmol) yielded compound **14** as a red solid (48 mg, 4.53%). m.p.: 183 °C (dec.). ^{1}H-NMR (250 MHz, DMSO-d_6, δ ppm): 10.09 (2H, bs), 9.70 (2H, bs), 9.53 (2H, bs), 7.70 (2H, m), 7.08 (1H, d, *J* = 8.8 Hz), 6.91 (1H, d, *J* = 16.1 Hz), 6.83 (1H, d, *J* = 16.1 Hz), 6.74 (1H, d, *J* = 15.8 Hz), 6.58 (1H, d, *J* = 8.8 Hz), 6.25 (2H, s), 5.85 (1H, s). ^{13}C-NMR (62.9 MHz, DMSO-d_6, δ ppm): 185.8, 180.1, 159.6, 148.7, 146.7 (2C), 141.6, 139.5, 137.6, 135.3, 135.1, 134.0, 128.9, 126.9, 123.0, 120.7, 120.6, 114.3, 114.2, 108.0 (2C), 101.9. Anal. Calcd. (%) for [$C_{22}H_{16}O_9$]: C, 62.27; H, 3.80; found (%): C, 62.21; H, 3.78.

4.2.12. Synthesis of 6,7-Dihydroxy-3-(7-(2′,4′,6′-trihydroxyphenyl)-3,5-dioxohepta-1,6-dien-1-yl)coumarin (**15**)

Following the procedure described above to obtain compound **11**, reaction of coumarin **9** (0.925 g, 2.5 mmol) and tris(methoxymethoxy)benzaldehyde **10a** (0.715 g, 2.5 mmol) yielded compound **15** as a red solid (50 mg, 4.72%). m.p.: 179 °C (dec). ^{1}H-NMR (250 MHz, DMSO-d_6, δ ppm): 10.09 (2H, bs), 9.75 (2H, bs), 9.51 (2H, bs), 7.77 (1H, d, *J* = 15.3 Hz), 7.58

(1H, s), 6.87 (1H, s), 6.75 (1H, d, *J* = 15.3 Hz), 6.60 (2H, m), 6.45 (1H, s), 5.97 (2H, s), 5.84 (1H, s). ^{13}C-NMR (62.9 MHz, DMSO-d_6, δ ppm): 185.2, 180.1, 160.5, 159.4, 158.7 (2C), 148.7, 148.1, 146.3, 139.2, 137.6, 130.5, 126.9, 125.3, 118.8, 113.7, 113.0, 104.9, 103.5, 102.1, 95.7 (2C). Anal. Calcd. (%) for [$C_{22}H_{16}O_9$]: C, 62.27; H, 3.80; found (%): C, 62.25; H, 3.79.

4.2.13. Synthesis of 6,7-Dihydroxy-3-(7-(2′,4′,5′-trihydroxyphenyl)-3,5-dioxohepta-1,6-dien-1-yl)coumarin (**16**)

Following the procedure described above to obtain compound **11**, reaction of coumarin **9** (0.925 g, 2.5 mmol) and tris(methoxytrimethoxy)benzaldehyde **10b** (0.715 g, 2.5 mmol) yielded compound **16** as a red solid (53 mg, 5.00%). m.p.: 188 °C (dec.). ^{1}H-NMR (250 MHz, DMSO-d_6, δ ppm): 10.10 (2H, bs), 9.76 (2H, bs), 9.55 (2H, bs), 7.75 (1H, d, *J* = 16.1 Hz), 7.67 (1H, s), 6.95 (1H, d, *J* = 15.5 Hz), 6.86 (1H, s), 6.68 (1H, d, *J* = 16.1 Hz), 6.53 (1H, d, *J* = 15.5 Hz), 6.38 (1H, s), 6.30 (1H, s), 6.15 (1H, s), 5.84 (1H, s). ^{13}C-NMR (62.9 MHz, DMSO-d_6, δ ppm): 183.8, 181.7, 159.1, 152.5, 148.7, 148.6, 148.1, 146.3, 140.4, 139.7, 137.6, 129.0, 126.9, 124.2, 118.8, 115.6, 114.3, 113.7, 113.0, 103.7, 102.2, 95.4. Anal. Calcd. (%) for [$C_{22}H_{16}O_9$]: C, 62.27; H, 3.80; found (%): C, 62.20; H, 3.77.

4.2.14. Synthesis of 6,7-Dihydroxy-3-(7-(2′,3′,4′-trihydroxyphenyl)-3,5-dioxohepta-1,6-dien-1-yl)coumarin (**17**)

Following the procedure described above to obtain compound **11**, reaction of coumarin **9** (0.925 g, 2.5 mmol) and tris(methoxytrimethoxy)benzaldehyde **10c** (0.715 g, 2.5 mmol) yielded compound **17** as a red solid (45 mg, 4.24%). m.p.: 190 °C (dec.). ^{1}H-NMR (250 MHz, DMSO-d_6, δ ppm): 10.05 (2H, bs), 9.70 (2H, bs), 9.45 (2H, bs), 7.75 (2H, m), 7.04 (1H, s), 6.95 (1H, d, *J* = 15.8 Hz), 6.71 (1H, d, *J* = 15.8 Hz), 6.57 (2H, m), 6.43 (1H, d, *J* = 8.4 Hz), 6.14 (1H, d, *J* = 8.4 Hz), 5.85 (1H, s). ^{13}C-NMR (62.9 MHz, DMSO-d_6, δ ppm): 186.4, 179.3, 159.0, 148.8, 148.7, 148.1, 147.4, 146.3, 139.2, 137.6, 133.8, 128.4, 126.9, 124.4, 120.8, 118.8, 114.7, 113.7, 113.0, 107.8, 102.1, 96.5. Anal. Calcd. (%) for [$C_{22}H_{16}O_9$]: C, 62.27; H, 3.80; found (%): C, 62.24; H, 3.73.

4.2.15. Synthesis of 6,7-Dihydroxy-3-(7-(3′,4′,5′-trihydroxyphenyl)-3,5-dioxohepta-1,6-dien-1-yl)coumarin (**18**)

Following the procedure described above to obtain compound **11**, reaction of coumarin **9** (0.925 g, 2.5 mmol) and tris(methoxytrimethoxy)benzaldehyde **10d** (0.715 g, 2.5 mmol) yielded compound **18** as a red solid (55 mg, 5.19%). m.p.: 185 °C (dec.). ^{1}H-NMR (250 MHz, DMSO-d_6, δ ppm): 10.12 (2H, bs), 9.77 (2H, bs), 9.53 (2H, bs), 7.78 (1H, d, *J* = 15.7 Hz), 7.69 (1H, s), 7.15 (1H, s), 6.94 (1H, d, *J* = 16.0 Hz), 6.75 (1H, d, *J* = 16.0 Hz), 6.55 (1H, d, *J* = 15.7 Hz), 6.45 (1H, s), 6.25 (2H, s), 5.84 (1H, s). ^{13}C-NMR (62.9 MHz, DMSO-d_6, δ ppm): 183.5, 181.8, 158.6, 148.7, 148.1, 146.7 (2C), 146.3, 139.2, 137.6, 135.3, 135.1, 128.9, 126.9, 123.0, 118.8, 113.7, 113.0, 108.0 (2C), 102.0, 94.6. Anal. Calcd. (%) for [$C_{22}H_{16}O_9$]: C, 62.27; H, 3.80; found (%): C, 62.22; H, 3.73.

4.3. Formulation of Biodegradable Nanoparticles

Resomer® RG503H (Evonic) polymer, which is a 50:50 copolymer of polylactic and polyglycolic acid in its acid form following the nanoprecipitation technique, was used to make the biodegradable nanoparticles [50]. Briefly, a 1% aqueous solution of poloxamer 407 (Sigma Aldrich) was used as the aqueous phase. Acetone was used as a solvent for the polymer and the active principles, the concentration of Resomer® and drug being 0.4% and 0.1% (p/v), respectively. The organic phase was added slowly at room temperature, using a syringe, to the aqueous solution, under magnetic stirring. Finally, the acetone was removed by evaporation at 50 °C in a rotary evaporator (Buchi) until a final volume of 25 mL was obtained. Filtration was performed through 0.22 µm diameter polyamide membrane filters to remove the non-incorporated drug.

To determine the content of the active principle and the encapsulation efficiency, an aliquot of the nanosuspensions was diluted in ethanol to dissolve the Resomer® and release the drugs, determining its concentration by spectrophotometry.

Size distribution (mean diameter and polydispersity index) and zeta potential of nanoparticles were determined in purified water at 25 °C using a Malvern Zetasizer Nano ZS instrument (Malvern Instruments Ltd., Malvern, UK).

4.4. Determination of hMAO-A and hMAO-B In Vitro Activity

The in vitro activity of the synthesized curcumin–coumarin hybrid analogues **11–18** or curcumin on hMAO enzymatic activity was evaluated using an Amplex® Red MAO assay kit and following a fluorimetric method previously described by us [11]. Briefly, 50 µL of sodium phosphate buffer (0.05 M, pH 7.4) containing the test molecules (new compounds or reference inhibitors) in different concentrations and adequate amounts of recombinant hMAO-A or hMAO-B (adjusted to obtain in our experimental conditions the same reaction velocity (hMAO-A: 1.1 µg protein; specific activity: 150 nmol of *para*-tyramine oxidized to *para*-hydroxyphenylacetaldehyde/min/mg protein; hMAO-B: 7.5 µg protein; specific activity: 22 nmol of *para*-tyramine transformed/min/mg protein)) were incubated for 10 min at 37 °C in a flat-black bottom 96-well microtest plate, placed in the dark fluorimeter chamber. After this incubation period, the reaction was started by adding 50 µL of the mixture containing (final concentrations) 200 µM of the Amplex® Red reagent, 1 U/mL of horseradish peroxidase and 1 mM of *para*-tyramine. The production of H_2O_2 and, consequently, of resorufin, was quantified at 37 °C in a multidetection microplate fluorescence reader (Fluo-star OptimaTM, BMG LABTECH, Offenburg, Germany) based on the fluorescence generated (excitation, 545 nm, emission, 590 nm) over a 10 min period, in which the fluorescence increased linearly. Control experiments were carried out simultaneously by replacing the tested molecules with appropriate dilutions of the vehicles. In addition, the possible capacity of these molecules to modify the fluorescence generated in the reaction mixture due to non-enzymatic inhibition (i.e., for directly reacting with Amplex® Red reagent) was determined by adding these molecules to solutions containing only the Amplex® Red reagent in sodium phosphate buffer. The specific fluorescence emission (used to obtain the final results) was calculated after subtraction of the background activity, which was determined from wells containing all components except the hMAO isoforms, which were replaced by sodium phosphate buffer solution. The IC_{50} values for each compound were calculated by linear regression representing the logarithm of the concentration (M) of the studied compound (abscissa axis) against the percentage of inhibition of the control MAO activity (ordinate axis). This linear regression was performed with 4–6 concentrations of each evaluated compound capable of inhibiting the control enzymatic activity of the MAO isoenzymes between 20% and 80%.

4.5. Determination of AChE and BuChE In Vitro Activity

Ellman's method [51] was used to determine in vitro ChE activity. 0.01 U/mL human recombinant AChE expressed in HEK 293 cells or 0.0005 U/mL BuChE isolated from human serum were added to a 50 mM phosphate buffer solution (pH 7.2) containing different concentrations of curcumin–coumarin hybrid analogues **11–18** or curcumin. The mixture was preincubated at 37 °C for 5 min followed by the addition of 5 mM acetylthiocholine or butyrylthiocholine and 0.25 mM 5,5′-dithio-bis(2-nitrobenzoic acid) (DNTB). The activity was measured by the absorbance increasing at λ 412 nm at 1 min intervals for 10 min at 37 °C (Fluo-Star OptimaTM, BMG LABTECH, Offenburg, Germany). Control experiments were performed simultaneously by replacing the test drugs with appropriate dilutions of the vehicles. The specific absorbance (used to obtain the final results) was calculated after subtraction of the background activity, which was determined in wells containing all components except the AChE or BuChE, which was replaced by a sodium phosphate buffer solution.

4.6. DPPH Radical Scavenging Assay

The DPPH was dissolved in methanol (50 µM), and 99 µL of the solution was transferred to each well of a 96-well microplate. Curcumin–coumarin hybrid analogues **11–18**, coumarin or reference drug (vitamin C) were added to each well at a final concentration of 100 μM. Solutions were incubated at room temperature for 30 min. Absorbance was determined at λ 517 nm using a microplate reader (Fluo-star OptimaTM, BMG LABTECH, Offenburg, Germany). DPPH radical solution in methanol was used as a control, whereas a mixture of methanol and sample served as blank. The scavenging activity percentage (AA%) was determined according to the equation described by Mensor et al. [52]: AA% = 100 − ((Abssample − Absblank) × 100/Abscontrol)

4.7. Cell Culture

4.7.1. Primary Culture of Rat Motor Cortex Neurons

Embryos were extracted by cesarean section from 18 days pregnant Wistar Kyoto rats which were euthanized by CO_2 inhalation. Brains were carefully dissected out, and after removing meninges, a portion of the motor cortex was isolated. Fragments obtained from several embryos were mechanically digested and cells were resuspended in Neurobasal medium. The Neurobasal medium was supplemented with 2% B-27 to obtain cortex neuronal cultures. Cells were seeded in 96-well plates at a density of 200,000 cells/mL. Cultures were grown for 7–8 days in an incubator (Form Direct Heat CO_2, Thermo Electron Corporation, Madrid, Spain) under saturated humidity at a partial pressure of 5% CO_2 in air at 37 °C until a dense neuronal network could be observed [53].

4.7.2. Human Neuroblastoma SH-SY5Y Cell Culture and Maintaining

The SH-SY5Y cells grew in a culture medium containing Ham's F12 and MEM (mixture 1:1) and supplemented with 15% FBS, 1% L-Glutamine, 1% non-essential amino acids and 1% of penicillin G/streptomycin sulfate (all of them from Sigma-Aldrich S.A.) [54]. The cells were grown in 75 cm^2 flasks in an incubator, under conditions of saturated humidity with a partial pressure of 5% CO_2 in the air, at 37 °C. Cell culture medium was replaced every 2 days, and, at 80–90% of confluence, the cells were sub-cultured. To carry out the viability assays, the cells were seeded in sterile 96-well plates, with a density of 2×10^5 cells/mL and grown distributed in aliquots of 100 μL for 24 h under the conditions described above.

4.7.3. Cell Viability

Cells grown in 96-well plates were treated with H_2O_2 (100 μM) and curcumin or test compounds **11–18** (10 μM). When cells were treated with curcumin or curcumin–coumarin hybrid analogues **14** and **16** formulated in nanoparticles, they were added in the 24 h prior to H_2O_2 treatment. Then, cultures were incubated for 24 h. After this time, cell viability was determined using MTT (5 mg/mL in Hank's). 10 μL of MTT solution was added to each well containing 100 μL of culture medium and the cells were incubated for 2 h as described above. Then, the culture medium was removed, 100 μL DMSO/well was added to solve the formazan crystals formed by the viable cells and the absorbance (λ 540 nm) was quantified in a plate reader. The viability (percentage) was calculated as (Absorbance (treatment)/Absorbance (negative control))100% [55]. Statistical analysis was performed using one way ANOVA test followed by Dunnett's multiple comparison test by using GraphPad software.

5. Conclusions

A new series of curcumin–coumarin hybrid analogues **11–18** were synthesized in low yield starting from 2′,2′-diphenyl-1,3-dioxol[h]coumarin-3-carbaldehyde (**5**), or 2′,2′-diphenyl-1,3-dioxol[g]coumarin-3-carbaldehyde (**9**), and the corresponding tris (methoxymethoxy)benzaldehyde **10a–d**. Synthesized derivatives did not reach a potential as a multitarget drug. In general, they were either better at inhibiting MAO isoforms or AChE and BuChE activity. Only compound **12** inhibited BuChE and MAO isoforms with similar potency. In addition, compounds **14** and **16** resulted in being neuroprotective against H_2O_2 in SH-SY5Y cells. The formulation of these compounds in nanoparticles improves their neuroprotective activity at low concentrations. Results suggest that by modulating the substitution pattern on both coumarin moiety and phenyl ring, ChEs and MAO-targeted derivatives or derivatives with activity in cell-based phenotypic assays can be obtained.

Supplementary Materials: The following are available online. 1H and 13C NMR of compounds **5**, **9**, **10a–d**, **11–18**.

Author Contributions: Conceptualization, D.V., R.L. and F.O.; methodology, E.Q., F.R.-E. and E.C.; validation, E.Q., F.R.-E. and E.C.; formal analysis, D.V. and E.U.; investigation, E.Q., F.R.-E. and E.C.; resources, D.V. and F.O.; data curation, D.V. and F.O.; writing—original draft preparation, D.V. and R.L.; writing—review and editing, D.V. and E.U.; visualization, D.V.; supervision, D.V., F.O.; project administration, D.V.; funding acquisition, D.V. and F.O. All authors have read and agreed to the published version of the manuscript.

Funding: This research was funded by Consellería de Cultura, Educación e Ordenación Universitaria (EM2014/016) and Centro Singular de Investigación de Galicia and the European Regional Development Fund (ERDF) (accreditation 2016–2019, ED431G/05).

Institutional Review Board Statement: The study was conducted according to the European regulations on the protection of animals (Directive 2010/63/UE), the Spanish Real Decreto 53/2013 (1 February) and the Guide for the Care and Use of Laboratory Animals as adopted and promulgated by the United States National Institutes of Health. In this context, the experimental protocol was approved by the Institutional Animal Care and Use Committee of the University of Santiago de Compostela and Xunta de Galicia, Spain (Register Number 15007DE/12/INVMED02/NERV02/B/MCT3).

Informed Consent Statement: Not applicable.

Data Availability Statement: Not applicable.

Conflicts of Interest: The authors declare no conflict of interest. The funders had no role in the design of the study; in the collection, analyses, or interpretation of data; in the writing of the manuscript, or in the decision to publish the results.

Sample Availability: Samples of the compounds **5**, **9**, **10a–d**, **11–18** are available from the authors.

References

1. Gabr, M.T.; Yahiaoui, S. Multitarget therapeutics for neurodegenerative diseases. *BioMed Res. Int.* **2020**, *2020*, 6532827. [CrossRef]
2. Ramsay, R.R.; Majekova, M.; Medina, M.; Valoti, M. Key targets for multi-target ligands designed to combat neurodegeneration. *Front. Neurosci.* **2016**, *10*, 375. [CrossRef] [PubMed]
3. Eghbaliferiz, S.; Farhadi, F.; Barreto, G.E.; Majeed, M.; Sahebkar, A. Effects of cucrcumin on neurological diseases: Focus on astrocytes. *Pharmacol. Rep.* **2020**, *72*, 769–782. [CrossRef] [PubMed]
4. Abass, S.; Latif, M.S.; Shafie, N.S.; Ghazali, M.I.; Kormin, F. Neuroprotective expression of turmeric and curcumin. *Food Res.* **2020**, *4*, 2366–2381. [CrossRef]
5. Nelson, K.M.; Dahlin, J.L.; Bisson, J.; Graham, J.; Pauli, G.F.; Walters, A. The essential medicinal chemistry of curcumin. *J. Med. Chem.* **2017**, *60*, 1620–1637. [CrossRef]
6. Gera, M.; Sharma, N.; Ghosh, M.; Luong, H.D.; Lee, S.L.; Min, T.; Kwon, T.; Jeong, D.K. Nanoformulations of curcumin: An emerging paradigm for improved remedial application. *Oncotarget* **2017**, *8*, 66680–66698. [CrossRef]
7. Salehi, B.; Calina, D.; Docea, A.O.; Koirala, N.; Aryal, S.; Lombardo, D.; Pasqua, L.; Taheri, Y.; Salgado Castillo, C.M.; Martorell, M.; et al. Curcumin's nanomedicine formulations for therapeutic application in neurological diseases. *J. Clin. Med.* **2020**, *9*, 430. [CrossRef] [PubMed]
8. Chainoglou, E.; Hadjipavlou-Litina, D. Curcumin in health and diseases: Alzheimer's disease and curcumin analogues, derivatives, and hybrids. *Int. J. Mol. Sci.* **2020**, *21*, 1975. [CrossRef]
9. Lo Cascio, F.; Marzullo, P.; Kayed, R.; Palumbo Piccionello, A. Curcumin as scaffold for drug discovery against neurodegenerative diseases. *Biomedicines* **2021**, *9*, 173. [CrossRef] [PubMed]
10. Jameel, E.; Umar, T.; Kumar, J.; Hoda, N. Coumarin: A privileged scaffold for the design and development of antineurodegenerative agents. *Chem. Biol. Drug Des.* **2016**, *87*, 21–38. [CrossRef]
11. Matos, M.J.; Terán, C.; Pérez-Castillo, Y.; Uriarte, E.; Santana, L.; Viña, D. Synthesis and study of a series of 3-arylcoumarins as potent and selective monoamine oxidase B inhibitors. *J. Med. Chem.* **2011**, *54*, 7127–7137. [CrossRef]
12. Matos, M.J.; Vilar, S.; Gonzalez-Franco, R.M.; Uriarte, E.; Santana, L.; Friedman, C.; Tatonetti, N.P.; Viña, D.; Fontenla, J.A. Novel (coumarin-3-yl)carbamates as selective MAO-B inhibitors: Synthesis, in vitro and in vivo assays, theoretical evaluation of ADME properties and docking study. *Eur. J. Med. Chem.* **2013**, *63*, 151–161. [CrossRef]

13. Matos, M.J.; Rodríguez-Enríquez, F.; Borges, F.; Santana, L.; Uriarte, E.; Estrada, M.; Rodríguez-Franco, M.I.; Laguna, R.; Viña, D. 3-Amidocoumarins as potential multifunctional agents against neurodegenerative diseases. *ChemMedChem* **2015**, *10*, 2071–2079. [CrossRef]
14. Delogu, G.; Picciau, C.; Ferino, G.; Quezada, E.; Podda, G.; Uriarte, E.; Viña, D. Synthesis, human monoamine oxidase inhibitory activity and molecular docking studies of 3-heteroarylcoumarin derivatives. *Eur. J. Med. Chem.* **2011**, *46*, 1147–1152. [CrossRef]
15. Costas-Lago, M.C.; Besada, P.; Rodríguez-Enríquez, F.; Viña, D.; Vilar, S.; Uriarte, E.; Borges, F.; Terán, C. Synthesis and structure-activity relationship study of novel 3-heteroarylcoumarins based on pyridazine scaffold as selective MAO-B inhibitors. *Eur. J. Med. Chem.* **2017**, *139*, 1–11. [CrossRef]
16. Matos, M.J.; Herrera Ibatá, D.M.; Uriarte, E.; Viña, D. Coumarin-rasagiline hybrids as potent and selective hMAO-B inhibitors, antioxidants, and neuroprotective agents. *ChemMedChem* **2020**, *15*, 532–538. [CrossRef]
17. Rodríguez-Enríquez, F.; Costas-Lago, M.C.; Besada, P.; Alonso-Pena, M.; Torres-Terán, I.; Viña, D.; Fontenla, J.Á.; Sturlese, M.; Moro, S.; Quezada, E.; et al. Novel coumarin-pyridazine hybrids as selective MAO-B inhibitors for the Parkinson's disease therapy. *Bioorg. Chem.* **2020**, *104*, 104203. [CrossRef]
18. Finberg, J.P.M.; Rabey, J.M. Inhibitors of MAO-A and MAO-B in psychiatry and neurology. *Front Pharmacol.* **2016**, *7*, 340. [CrossRef]
19. Zuo, Y.; Huang, J.; Zhou, B.; Wang, S.; Shao, W.; Zhu, C.; Lin, L.; Wen, G.; Wang, H.; Du, J.; et al. Synthesis, cytotoxicity of new 4-arylidene curcumin analogues and their multi-functions in inhibition of both NF-κB and Akt signalling. *Eur. J. Med. Chem.* **2012**, *55*, 346–357. [CrossRef] [PubMed]
20. Lin, L.; Shi, Q.; Nyarko, A.K.; Bastow, K.F.; Wu, C.C.; Su, C.Y.; Shih, C.C.Y.; Lee, K.H. Antitumor Agents. 250. Design and synthesis of new curcumin analogues as potential anti-prostate cancer agents. *J. Med. Chem.* **2006**, *49*, 3963–3972. [CrossRef]
21. Lee, K.H.; Lin, L.; Shih, C.C.Y.; Su, C.Y.; Ishida, J.; Ohtsu, H.; Wang, H.-K.; Itokawa, H.; Chang, C. Curcumin Analogues and uses Thereof. U.S. Patent 7355081 B2 20080408, 2008.
22. Qiu, X.; Du, Y.; Lou, B.; Zuo, Y.; Shao, W.; Huo, Y.; Huang, J.; Yu, Y.; Zhou, B.; Du, J.; et al. Synthesis and identification of new 4-arylidene curcumin analogues as potential anticancer agents targeting nuclear factor-κB signaling pathway. *J. Med. Chem.* **2010**, *53*, 8260–8273. [CrossRef] [PubMed]
23. Reinke, A.A.; Gestwicki, J.E. Structure-activity relationships of amyloid beta-aggregation inhibitors based on curcumin: Influence of linker length and flexibility. *Chem. Biol. Drug Des.* **2007**, *70*, 206–215. [CrossRef]
24. Finberg, J.P.M. Update on the pharmacology of selective inhibitors of MAO-A and MAO-B: Focus on modulation of CNS monoamine neurotransmitter release. *Pharmacol. Therapeut.* **2014**, *143*, 133–152. [CrossRef] [PubMed]
25. Cai, Z. Monoamine oxidase inhibitors: Promising therapeutic agents for Alzheimer's disease. *Mol. Med. Report.* **2014**, *9*, 1533–1541. [CrossRef] [PubMed]
26. Lane, R.M.; Potkin, S.G.; Enz, A. Targeting acetylcholinesterase and butyrylcholinesterase in dementia. *Int. J. Neuropsychopharmacol.* **2006**, *9*, 101–124. [CrossRef]
27. Kandezi, K.; Mohammadi, M.; Ghaffari, M.; Gholami, M.; Motaghinejad, M.; Safari, S. Novel insight to neuroprotective potential of curcumin: A mechanistic review of possible involvement of mitochondrial biogenesis and PI3/Aky/GSK3 or PI3/Akt/CREB/BDNF signaling pathways. *Int. J. Mol. Cell. Med.* **2020**, *9*, 20–32. [CrossRef]
28. Sandoval-Avila, S.; Diaz, N.F.; Gómez-Pinedo, U.; Canales-Aguirre, A.A.; Gutiérrez-Mercado, Y.K.; Padilla-Camberos, E.; Marquez-Aguirre, A.L.; Díaz-Martínez, N.E. Neuroprotective effects of phytochemicals on dopaminergic neuron cultures. *Neurología* **2019**, *34*, 114–124. [CrossRef]
29. Mufti, S.; Bautista, A.; Pino-Figueroa, A. Evaluation of the neuroprotective effects of curcuminoids on B35 and SH-SY5Y neuroblastoma cells. *Med. Aromat. Plants* **2015**, *4*, 1000197. [CrossRef]
30. Yu, X.; Chen, L.; Tang, M.; Yang, Z.; Fu, A.; Wang, Z.; Wang, H. Revealing the effects of curcumin on SH-SY5Y neuronal cells: A combined study from cellular viability, morphology, and biomechanics. *J. Agric. Food Chem.* **2019**, *67*, 4273–4279. [CrossRef]
31. Zhai, K.; Brockmüller, A.; Kubatka, P.; Shakibaei, M.; Büsselberg, D. Curcumin's beneficial effects on neuroblastoma: Mechanisms, challenges, and potential solutions. *Biomolecules* **2020**, *10*, 1469. [CrossRef]
32. Moghaddam, N.S.A.; Oskouie, M.N.; Butler, A.E.; Petit, P.X.; Barreto, G.E.; Sahebkar, A. Hormetic effects of curcumin: What is the evidence? *J. Cell Physiol.* **2019**, *234*, 10060–10071. [CrossRef]
33. Kim, M.K.; Choo, H.; Chong, Y. Water-soluble and cleavable quercetin–amino acid conjugates as safe modulators for P-glycoprotein-based multidrug resistance. *J. Med. Chem.* **2014**, *57*, 7216–7233. [CrossRef]
34. Casagrande, C.; Ferrini, R.; Miragoli, G.; Ferrari, G. β-Adrenergic receptor inhibitors. IV. 1-(Hydroxy and dihydroxyphenoxy)-3-isopropylamine-2-propanols. *Boll. Chim. Farm.* **1973**, *112*, 445–454.
35. Akselsen, O.W.; Skattebøl, L.; Hansen, T.V. ortho-Formylation of oxygenated phenols. *Tetrahedron Lett.* **2009**, *50*, 6339–6341. [CrossRef]
36. Medvedeva, A.S.; Andreev, M.V.; Safronova, L.P.; Afonin, A.V. Synthesis of trimethylsilylpropynoyl chloride. *Russ. J. Org. Chem.* **2005**, *41*, 1463–1466. [CrossRef]
37. Nakagawa-Goto, K.; Lee, K. Anti-AIDS agents 68. The first total synthesis of a unique potent anti-HIV chalcone from genus Desmos. *Tetrahedron Lett.* **2006**, *47*, 8263–8266. [CrossRef]

38. Matsuya, Y.; Hayashi, K.; Nemoto, H. A new protocol for the consecutive a- and b-activation of propiolates towards electrophiles, involving conjugate addition of tertiary amines and intramolecular silyl migration. *Chem. Eur. J.* **2005**, *11*, 5408–5418. [CrossRef] [PubMed]
39. Pereira, A.R. Marine Natural Products: Synthesis and Isolation of Bioactive Analogues. Ph.D. Thesis, University of British Columbia, Vancouver, BC, Canada, December 2007. Available online: http://hdl.handle.net/2429/7526 (accessed on 27 July 2021).
40. Chen, D.; Chen, R.; Wang, R.; Li, J.; Xie, K.; Bian, C.; Sun, L.; Zhang, X.; Liu, J.; Yang, L.; et al. Probing the catalytic promiscuity of a regio- and stereospecific C-glycosyltransferase from Mangifera indica. *Ang. Chem. Int. Ed.* **2015**, *54*, 12678–12682. [CrossRef]
41. Chao, S.W.; Su, M.Y.; Chiou, L.C.; Chen, L.C.; Chang, C.I.; Huang, W.J. Total synthesis of hispidulin and the structural basis for its inhibition of proto-oncogene kinase Pim-1. *J. Nat. Prod.* **2015**, *78*, 1969–1976. [CrossRef]
42. Mazumder, A.; Neamati, N.; Sunder, S.; Schulz, J.; Pertz, H.; Eich, E.; Pommier, Y. Curcumin analogs with altered potencies against HIV-1 integrase as probes for biochemical mechanisms of drug action. *J. Med. Chem.* **1997**, *40*, 3057–3063. [CrossRef]
43. Feldman, K.S.; Lawlor, M.D. Ellagitannin Chemistry. The first total synthesis of a dimeric ellagitannin, coriariin A. *J. Am. Chem. Soc.* **2000**, *122*, 7396–7397. [CrossRef]
44. Xu, J.; Wang, H.; Sim, M.M. First synthesis of (±)-C-3-prenylated flavanones. *Synth. Commun.* **2003**, *33*, 2737–2750. [CrossRef]
45. Teng, Y.; Li, X.; Yang, K.; Li, X.; Zhang, Z.; Wang, L.; Deng, Z.; Song, B.; Yan, Z.; Zhang, Y.; et al. Synthesis and antioxidant evaluation of desmethylxanthohumol analogs and their dimers. *Eur. J. Med. Chem.* **2017**, *125*, 335–345. [CrossRef]
46. Hofmann, E.; Webster, J.; Do, T.; Kline, R.; Snider, L.; Hauser, Q.; Higginbottom, G.; Campbell, A.; Ma, L.; Paula, S. Hydroxylated chalcones with dual properties: Xanthine oxidase inhibitors and radical scavengers. *Bioorg. Med. Chem.* **2016**, *24*, 578–587. [CrossRef] [PubMed]
47. Hirotaka, E.; Naoko, Y. Adhesive composition containing copolymer having gallol group-like side chain and method for producing the copolymer. *Jpn. Kokai Tokkyo Koho* **2019**, JP2019147857 A 2019-09-05.
48. Zhan, K.; Sung, K.; Ejima, H. Tunicate-inspired gallol polymers for underwater adhesive: A comparative study of catechol and gallol. *Biomacromolecules* **2017**, *18*, 2959–2966. [CrossRef]
49. Roschek, B., Jr.; Fink, R.C.; McMichael, M.D.; Li, D.; Alberte, R.S. Elderberry flavonoids bind to and prevent H1N1 infection in vitro. *Phytochemistry* **2009**, *70*, 1255–1261. [CrossRef] [PubMed]
50. Lopalco, A.; Ali, H.; Denora, N.; Rytting, E. Oxcarbazepine-loaded polymeric nanoparticles: Development and permeability studies across in vitro models of the blood-brain barrier and human placental trophoblast. *Int. J. Nanomed.* **2015**, *10*, 1985–1996. [CrossRef]
51. Ellman, G.L.; Counrtney, K.D.; Andres, V.; Featherstone, R.M.A. A new and rapid colorimetric determination of acetylcholinesterase activity. *Biochem. Pharmacol.* **1961**, *7*, 88–90. [CrossRef]
52. Mensor, L.L.; Menezes, F.S.; Leitão, G.G.; Reis, A.S.; dos Santos, T.C.; Coube, C.S.; Leitão, S.G. Screening of Brazilian plant extracts for antioxidant activity by the use of DPPH free radical method. *Phytother. Res.* **2001**, *15*, 127–130. [CrossRef]
53. Yáñez, M.; Matías-Guiu, J.; Arranz-Tagarro, J.-A.; Galán, L.; Viña, D.; Gómez-Pinedo, U.; Vela, A.; Guerrero, A.; Martínez-Vila, E.; García, A.G. The neuroprotection exerted by memantine, minocycline and lithium, against neurotoxicity of CSF from patients with amyotrophic lateral sclerosis, is antagonized by riluzole. *Neurodegen. Dis.* **2014**, *13*, 171–179. [CrossRef] [PubMed]
54. Xicoy, H.; Wiering, B.; Martens, G.J.M. SH-SY5Y cell line in Parkinson's disease research: A systematic review. *Mol. Neurodegener.* **2017**, *12*, 10. [CrossRef] [PubMed]
55. Liu, Y.; Peterson, D.A.; Kimura, H.; Schubert, D. Mechanism of cellular 3-(4,5-dimethylthiazol-2-yl)-2,5-diphenyltetrazolium bromide (MTT) reduction. *J. Neurochem.* **1997**, *69*, 581–593. [CrossRef] [PubMed]

Article

Inhibition of Butyrylcholinesterase and Human Monoamine Oxidase-B by the Coumarin Glycyrol and Liquiritigenin Isolated from *Glycyrrhiza uralensis*

Geum Seok Jeong [1], Myung-Gyun Kang [2], Joon Yeop Lee [3], Sang Ryong Lee [1], Daeui Park [2], MyoungLae Cho [3] and Hoon Kim [1,*]

[1] Department of Pharmacy, and Research Institute of Life Pharmaceutical Sciences, Sunchon National University, Suncheon 57922, Korea; fever41@naver.com (G.S.J.); wholsr@naver.com (S.R.L.)

[2] Department of Predictive Toxicology, Korea Institute of Toxicology, Daejeon 34114, Korea; myung-gyun.kang@kitox.re.kr (M.-G.K.); daeui.park@kitox.re.kr (D.P.)

[3] National Institute for Korean Medicine Development, Gyeongsan 38540, Korea; chool9090@nikom.or.kr (J.Y.L.); meanglae@nikom.or.kr (M.C.)

* Correspondence: hoon@sunchon.ac.kr; Tel.: +82-61-750-3751

Academic Editor: Maria João Matos

Received: 11 August 2020; Accepted: 25 August 2020; Published: 26 August 2020

Abstract: Eight compounds were isolated from the roots of *Glycyrrhiza uralensis* and tested for cholinesterase (ChE) and monoamine oxidase (MAO) inhibitory activities. The coumarin glycyrol (GC) effectively inhibited butyrylcholinesterase (BChE) and acetylcholinesterase (AChE) with IC_{50} values of 7.22 and 14.77 μM, respectively, and also moderately inhibited MAO-B (29.48 μM). Six of the other seven compounds only weakly inhibited AChE and BChE, whereas liquiritin apioside moderately inhibited AChE (IC_{50} = 36.68 μM). Liquiritigenin (LG) potently inhibited MAO-B (IC_{50} = 0.098 μM) and MAO-A (IC_{50} = 0.27 μM), and liquiritin, a glycoside of LG, weakly inhibited MAO-B (>40 μM). GC was a reversible, noncompetitive inhibitor of BChE with a K_i value of 4.47 μM, and LG was a reversible competitive inhibitor of MAO-B with a K_i value of 0.024 μM. Docking simulations showed that the binding affinity of GC for BChE (−7.8 kcal/mol) was greater than its affinity for AChE (−7.1 kcal/mol), and suggested that GC interacted with BChE at Thr284 and Val288 by hydrogen bonds (distances: 2.42 and 1.92 Å, respectively) beyond the ligand binding site of BChE, but that GC did not form hydrogen bond with AChE. The binding affinity of LG for MAO-B (−8.8 kcal/mol) was greater than its affinity for MAO-A (−7.9 kcal/mol). These findings suggest GC and LG should be considered promising compounds for the treatment of Alzheimer's disease with multi-targeting activities.

Keywords: *Glycyrrhiza uralensis*; glycyrol; liquiritigenin; cholinesterases; human monoamine oxidases; kinetics; docking simulation

---

1. Introduction

Acetylcholinesterase (AChE, EC 3.1.1.7) catalyzes the breakdown of acetylcholine (ACh), a neurotransmitter found in synapses of the cerebral cortex [1]. AChE inhibitors reduce AChE activity and maintain or increase ACh levels, which are typically deficient in Alzheimer's disease (AD), and thus, enhance cholinergic transmission in brain [2,3]. AD is a chronic, devastating manifestation of neuronal dysfunction and is characterized by progressive mental failure, disordered cognitive functions, and speech impairment. Various cholinesterase inhibitors (e.g., tacrine, donepezil, galantamine, and rivastigmine), immunotherapies, antisense oligonucleotides, phyto-pharmaceuticals, and nutraceuticals are being used to treat AD [4].

Butyrylcholinesterase (BChE) breaks down butyrylcholine (BCh), and BChE levels are significantly elevated in the AD brain [5,6]. Interestingly, AChE and BChE, which are both related to AD and act independently, are viewed as diagnostic markers and as potential targets for drug development [7].

On the other hand, monoamine oxidases (MAO, EC 1.4.3.4) catalyze the oxidative deamination of monoamine neurotransmitters and are present as two MAO isoforms, that is, MAO-A and MAO-B, in the outer mitochondrial membranes of all tissues [8]. MAO-A is primarily targeted to treat depression and anxiety, whereas MAO-B is targeted to treat AD and Parkinson's disease (PD) [9]. In addition, MAOs are critically related to amyloid plaque formation in AD, and MAO-B is co-expressed at high levels in the AD brain with γ-secretase [10].

Due to the complex etiology of AD, multi-targeting therapeutic agents have been devised to inhibit MAOs and AChE, and thus, elevate monoamine and choline ester levels [11]. Recently, multi-targeting agents such as homoisoflavonoid derivatives [12], donepezil-butylated hydroxytoluene hybrids [13], coumarin-dithiocarbamate hybrids [14], alcohol-bearing dual inhibitors [15], and chalcone oxime ethers [16] have been reported to target MAO-B and AChE. Dual function inhibitors of AChE and BChE have been studied using in silico approaches, such as pharmacophore-based virtual screening and molecular docking [17]. In addition, compounds targeting for MAO-A, MAO-B, AChE, and BChE like TV 3326 [18] and 1,2,3,4-tetrahydrochromeno[3,2-c]pyridin-10-one derivatives [19] have also been described.

Glycyrol (GC), a coumarin derivative, has been reported to have anticancer [20–22], anti-fungal [23], anti-bacterial [24], anti-viral [25,26], anti-inflammatory [27–29], and immunosuppressive activities [30]. However, the ChE inhibitory activity of GC has not been reported to date. Liquiritigenin (LG), a flavonoid, is also known to have many biological activities including MAO inhibitory activity [31,32]. However, these MAO studies were conducted using fractions of rat liver mitochondrial MAO and rat brain MAO-A and MAO-B, respectively.

In the present study, eight compounds were isolated from *Glycyrrhiza uralensis* (also known as Chinese licorice) and investigated for their inhibitory activities against AChE, BChE, and human MAO-A and MAO-B. The two most potent compounds (GC and LG) were subjected to kinetic analysis and their affinities for the enzymes were investigated using molecular docking simulations.

2. Results

2.1. Isolation and Identification of Compounds

Eight natural products were isolated from *Glycyrrhiza uralensis* and identified by comparing NMR data (Supplementary Information) with literature values: GC (**1**) [33], isoliquiritin (**2**) [33], LG (**3**) [33], glycyrrhetinic acid (**4**) [34], liquiritin (**5**) [33], liquiritin apioside (**6**) [35], isoliquiritin apioside (**7**) [36], and glycyrrhizin (**8**) [37]. The structures of the eight compounds are detailed in Figure 1.

2.2. Analysis of Inhibitory Activities

The AChE, BChE, MAO-A, and MAO-B inhibitory activities of the eight compounds were investigated at a concentration of 10 μM. Most of the eight inhibited AChE, BChE, MAO-A, and MAO-B by less than 50%. However, GC and LG potently inhibited AChE and BChE, and MAO-A and MAO-B, respectively (Table 1). GC inhibited BChE and AChE with IC_{50} values of 7.22 and 14.77 μM, respectively, with a selectivity index (SI) of 2.0 for BChE with respect to AChE, and also moderately inhibited MAO-B (29.48 μM). Other compounds showed weak inhibitory activities against AChE or BChE, except liquiritin apioside, which moderately inhibited AChE (IC_{50} = 36.68 μM). LG potently inhibited MAO-B (IC_{50} = 0.098 μM) and MAO-A (IC_{50} = 0.27 μM). The SI value of LG for MAO-B with respect to MAO-A was 2.8 (Table 1). Liquiritin, a LG glycoside, weakly inhibited MAO-A and MAO-B (>40 μM). Thus, GC and LG were found to be effective inhibitors of BChE and MAO-B, respectively.

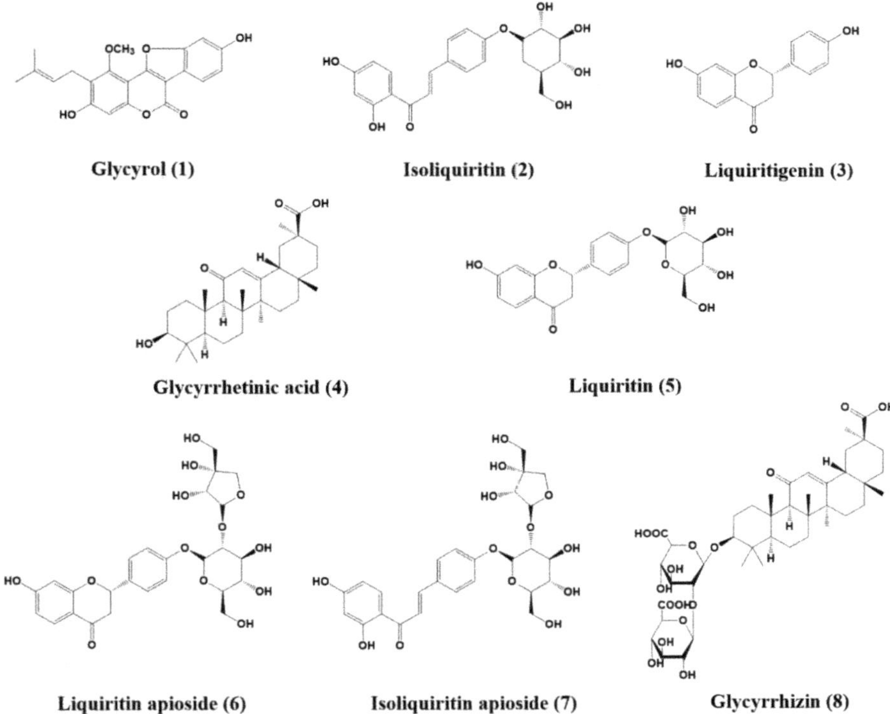

Figure 1. Chemical structures of eight compounds isolated from the roots of *Glycyrrhiza uralensis*.

2.3. Analysis of the Reversibilities of BChE and MAO-B Inhibitions

Reversibilities of BChE inhibition by GC and of MAO-B inhibition by LG were investigated by dialysis and dilution methods. Residual BChE activity after GC inhibition recovered partially from 34.6% (undialyzed activity; A_U) to 58.4% (dialyzed activity; A_D) by dialysis, whereas inhibition by tacrine (a known reversible inhibitor) significantly recovered from 10.3% to 74.1% (Figure 2A). We also confirmed reversibility using the dilution method by measuring and comparing residual BChE activities of a sample preincubated with GC at a concentration of 50 × IC_{50} and then diluted to a concentration of 1 × IC_{50} with a control sample treated at a GC concentration of 1 × IC_{50}. We found that residual activities were similar before and after dilution (51.1% and 40.1%, respectively), and that the activity of the sample at a concentration of 50 × IC_{50} was 10.3% (Figure 2A). These results suggested that GC is a reversible inhibitor of BChE, because if it acted as an irreversible inhibitor, activity would have been reduced by dilution. On the other hand, the relative residual activity of MAO-B after LG inhibition recovered from 38.4% (A_U) to 87.2% (A_D) by dialysis, which was similar to activity recovery observed for the reversible MAO-B inhibitor lazabemide (from 36.1% to 88.0%). On the other hand, values for the irreversible inhibitor pargyline were 17.0% and 8.4%, respectively (Figure 2B). These results showed that GC and LG reversibly inhibited BChE and MAO-B, respectively.

2.4. Analysis of Inhibitory Patterns

Modes of BChE inhibition by GC and of MAO-B inhibition by LG were investigated by analyzing Lineweaver–Burk plots. Plots of BChE inhibition by GC were linear and intersected the *x*-axis (Figure 3A). Secondary plots of the slopes of Lineweaver–Burk plots against inhibitor concentration showed the K_i value of GC for BChE inhibition was 4.47 ± 0.29 μM (Figure 3B).

Table 1. Inhibitions of AChE, BChE, MAO-A, and MAO-B by compounds isolated from the roots of *Glycyrrhiza uralensis*.

Compounds	Residual Activity at 10 μM (%)				IC_{50} (μM)			
	MAO-A	MAO-B	AChE	BChE	MAO-A	MAO-B	AChE	BChE
GC	70.5 ± 1.61	74.2 ± 3.46	46.1 ± 4.40	44.6 ± 5.36	>40	29.48 ± 0.67	14.77 ± 0.19	7.22 ± 0.37
Isoliquiritin	81.8 ± 1.61	75.4 ± 2.27	69.9 ± 2.20	91.9 ± 7.16	>40	>40	>40	>40
LG	0.46 ± 1.60	0.00 ± 3.34	95.3 ± 3.59	82.5 ± 0.26	0.27 ± 0.041	0.098 ± 0.00079	>40	>40
Glycyrrhetinic acid	96.3 ± 2.64	84.0 ± 2.16	97.3 ± 1.61	95.0 ± 4.12	>40	>40	-	-
Liquiritin	93.5 ± 0.00	90.2 ± 0.56	93.5 ± 5.12	95.5 ± 4.67	>40	>40	>40	>40
Liquiritin apioside	86.9 ± 2.41	94.8 ± 0.57	63.5 ± 2.56	97.6 ± 0.93	>40	>40	36.68 ± 1.42	>40
Isoliquiritin apioside	86.6 ± 3.27	80.3 ± 5.57	93.5 ± 3.07	95.6 ± 0.88	>40	>40	-	-
Glycyrrhizin	95.8 ± 3.30	93.1 ± 4.32	97.7 ± 2.14	82.3 ± 7.95	>40	>40	-	-
Toloxatone					1.08 ± 0.025	-	-	-
Lazabemide					-	0.063 ± 0.015	-	-
Clorgyline					0.007 ± 0.00070	-	-	-
Pargyline					-	0.028 ± 0.0043	-	-
Tacrine						-	0.27 ± 0.019	0.014 ± 0.0043

-, not determined. Values above are the means ± SEs of triplicate experiments, and IC_{50} values were graphically determined at three different inhibitor concentrations around its concentration showing 50% of residual activity.

(A)

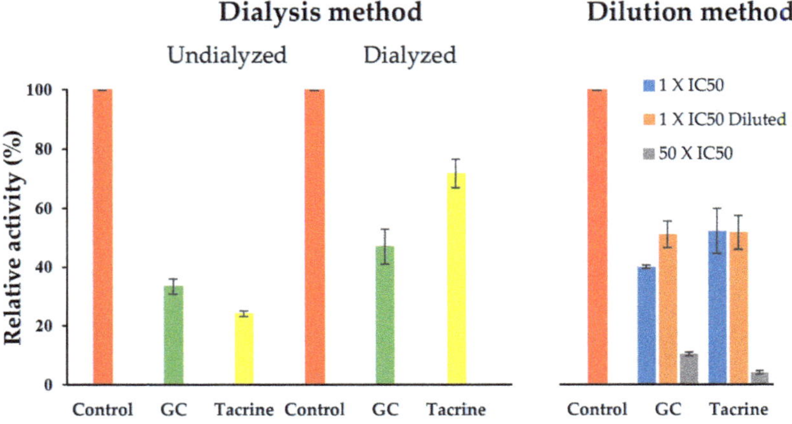

(B)

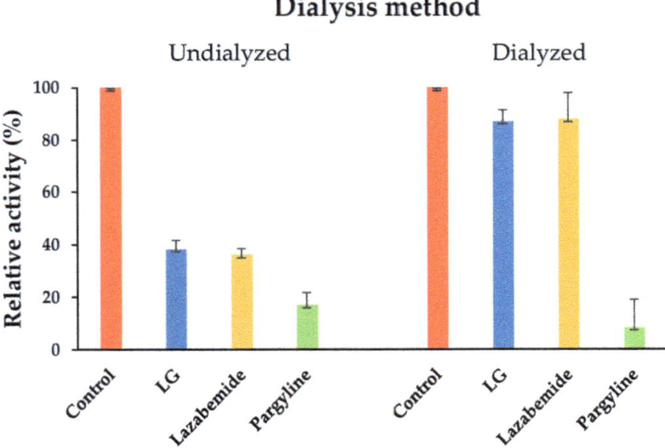

Figure 2. Recoveries of BChE inhibition by GC (**A**) and MAO-B inhibition by LG (**B**). The concentrations of the inhibitors used were: GC, 14.0 μM; LG, 0.2 μM; tacrine, 0.03 μM; lazabemide, 0.12 μM; and pargyline, 0.06 μM. In the dilution experiments, we measured residual activities of BChE at an inhibitor concentration of $1 \times IC_{50}$, at a concentration of $50 \times IC_{50}$ and then diluted to a concentration of $1 \times IC_{50}$ after preincubation, at an inhibitor concentration of $50 \times IC_{50}$. Results are the averages of duplicate or triplicate (GC dialysis) experiments. Tacrine was used as a reversible reference BChE inhibitor. Lazabemide and pargyline were used as reversible and irreversible reference MAO-B inhibitors, respectively.

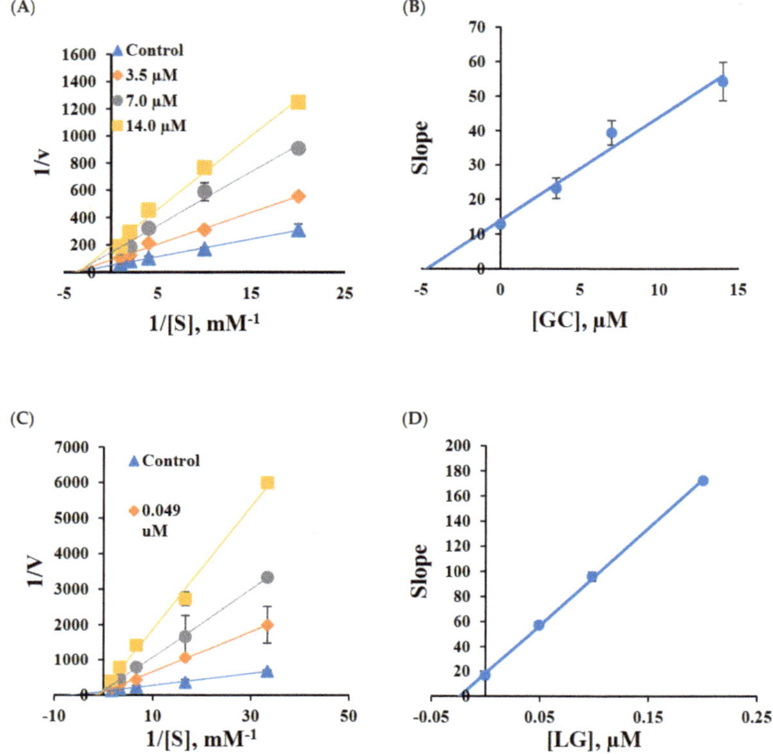

Figure 3. Lineweaver-Burk plots of inhibitions of BChE by GC (**A**) and of MAO-B by LG (**C**), and their respective secondary plots (**B**,**D**) of slopes of Lineweaver-Burk plots versus inhibitor concentrations. Five different substrate concentrations were used; 0.05, 0.1, 0.25, 0.5 or 1.0 mM for BChE, and 0.03, 0.06, 0.15, 0.3, or 0.6 mM for MAO-B. Inhibition studies were carried out at three inhibitor concentrations, that is, at 0.5×, 1.0×, and 2.0× of the IC_{50} values of GC and LG. The errors were determined by duplicate experiments.

These results indicate GC acted as a noncompetitive inhibitor of BChE and bound to a site other than the understood substrate binding site of BChE. On the other hand, plots of MAO-B inhibition by LG were linear and intersected the *y*-axis (Figure 3C) and secondary plots showed the K_i value of LG for MAO-B inhibition was 0.023 ± 0.00061 μM (Figure 3D), indicating LG is a competitive inhibitor of MAO-B.

2.5. Molecular Docking Simulation

Docking simulations showed that GC located at the binding site of 3-[(1*S*)-1-(dimethylamino)ethyl] phenol (SAF) in AChE (PDB: 1GQS) and the binding site of *N*-{[(3*R*)-1-(2,3-dihydro-1H-inden-2-yl) piperidin-3-yl]methyl}-*N*-(2-methoxyethyl)naphthalene-2-carboxamide in BChE (PDB ID: 4TPK). The binding affinity (−7.8 kcal/mol) of GC for BChE was greater than its affinity for AChE (−7.1 kcal/mol) as determined by AutoDock Vina (Table 2), and these binding affinity values concurred with the IC_{50} values (Table 1). Docking simulation results suggested that GC did not form a hydrogen bond with AChE (Figure 4A), but that GC forms two hydrogen bonds with the Thr284 and Val288 residues of BChE (distances: 2.42 and 1.92 Å, respectively) (Figure 4B). These results explain the preference of GC for BChE.

Table 2. Docking scores of GC, LG, and liquiritin with AChE, BChE, MAO-A, and MAO-B.

Compounds	Docking Scores (kcal/mol)			
	AChE	BChE	MAO-A	MAO-B
GC	−7.1	−7.8	-	-
LG	-	-	−7.9	−8.8
Liquiritin	-	-	−2.9	−4.1

The values were obtained using AutoDock Vina.

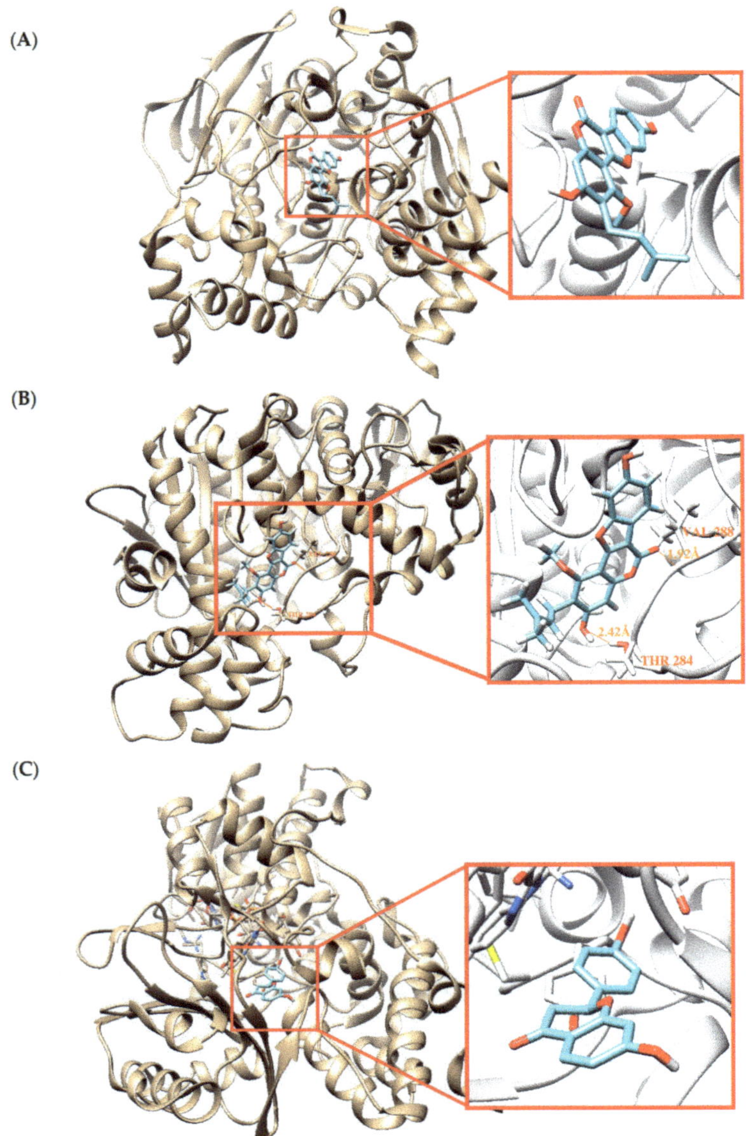

Figure 4. *Cont.*

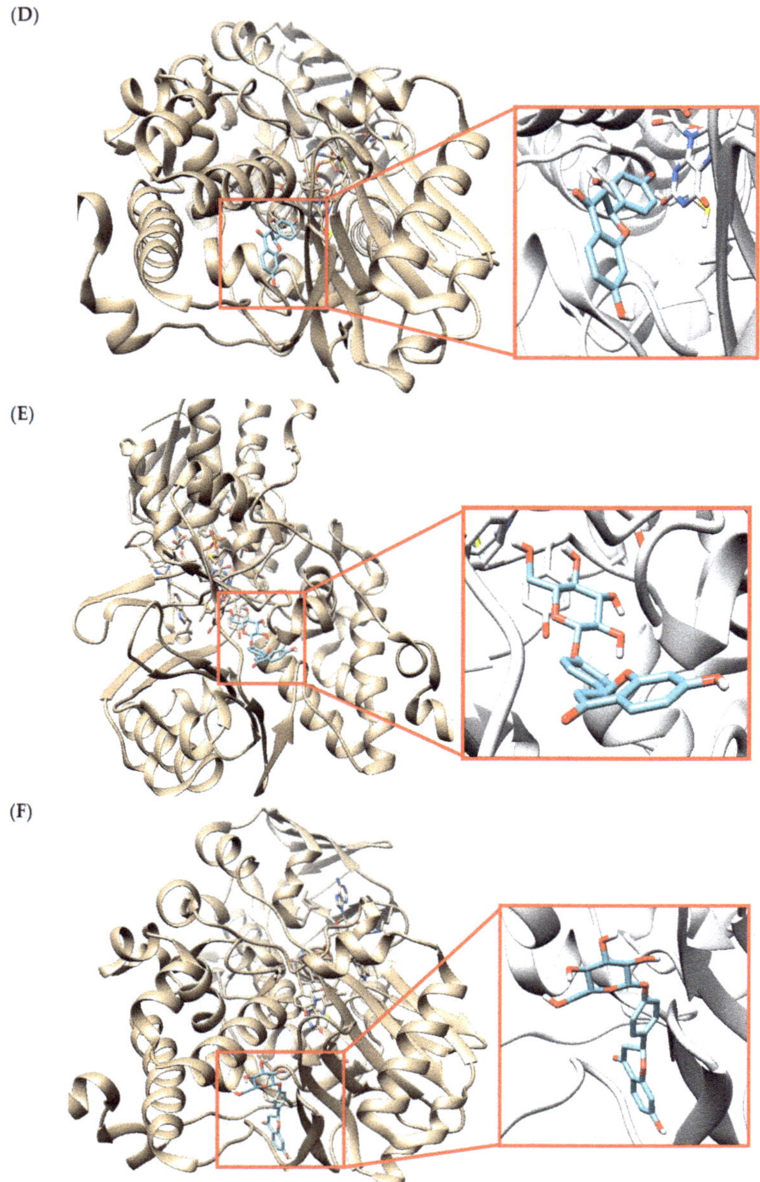

Figure 4. Docking simulations of GC with AChE (**A**) and BChE (**B**), LG with MAO-A (**C**) and MAO-B (**D**), and liquiritin with MAO-A (**E**) and MAO-B (**F**). AChE (1GQS), BChE (4TPK), MAO-A (2Z5X), and MAO-B (4A79) were subjected to docking analysis.

LG and liquiritin located at the binding site of 7-methoxy-1-methyl-9H-beta-carboline complexed with MAO-A (PDB: 2Z5X) and of pioglitazone complexed with MAO-B (PDB: 4A79). The binding affinities of LG and liquiritin with MAO-B were greater than their binding affinities with MAO-A (Table 2), and LG binding affinities were in-line with the IC_{50} values shown in Table 1. However, docking simulations did not predict hydrogen bond formation between LG or liquiritin with MAO-A or MAO-B (Figure 4C–F).

3. Discussion

In the present study, GC (a coumarin) was isolated from *G. uralensis* and its BChE inhibitory activity was evaluated. Coumarins are characterized by the presence of 1,2-benzopyrone or benzopyran-2-one groups, which are the most common oxygen-containing heterocyclic compounds found in Nature. The ChE inhibitory activities by coumarins have been previously reviewed for synthetic and natural compounds [38]. Most of the known coumarins have a lower IC_{50} value for AChE than for BChE, and selectivity for AChE or BChE is dependent on scaffold substituents, as exemplified by 3-(4-aminophenyl)-coumarin derivatives [39]. Furthermore, potencies of natural coumarins for AChE or BChE are much weaker than those of synthetic analogues. Nevertheless, natural coumarins exhibit significant inhibitory activities against AChE, examples include xanthotoxin from *Ferula lutea* (IC_{50} = 0.76 μM) [40] and a 4-phenylcoumarin mesuagenin B from *Mesua elegans* (IC_{50} = 0.70 μM) [41]. Based on the classification of natural AChE inhibitors, those with IC_{50} values ≤ 15 μM are termed high potency inhibitors and those with values ranging from 15 to 50 μM moderate potency inhibitors [42]. According to this classification, GC is a high potency AChE inhibitor (IC_{50} = 14.77 μM), though the value is near the threshold. In a previous study, osthenol, a prenylated coumarin obtained from *Angelica pubescens*, selectively inhibited MAO-A, and exhibited moderate AChE inhibitory activity (IC_{50} = 25.3 μM) [43].

Natural coumarins have been reported to have low BChE inhibitory activities; sphondin and pimpinellin from *Heracleum platytaenium* inhibited BChE by 63.69% and 78.02%, respectively, at a concentration of 25 μg/mL concentration (115.7 and 101.5 μM, respectively) [44], and notably, all these IC_{50} values are higher than that of GC (IC50 = 7.22 μM) as determined in the present study.

As regards other natural compounds, inhibition of BChE by GC was greater than that by boldine (IC_{50} = 321 μM) [45], hyperforin and hyuganin C (IC_{50} = 141.60 and 38.86 μM, respectively) [46], cremaphenanthrene F (14.62) [47], scopoletin (IC_{50} = 9.11 μM) [48], and broussonin A and sagachromanol I (IC_{50} = 7. 50 and 10.79 μM, respectively) [49], but less than those of norditerpenoids isograndifoliol and (1*R*,15*R*)-1-acetoxycryptotanshinone (IC_{50} = 0.9 and 2.4 μM, respectively) [50]. Notably, these IC_{50} values were much greater than those for AChE inhibition by tannic acid (IC_{50} = 0.087 μM) [51], or hesperidin (IC_{50} = 0.00345 μM) [52].

Dual inhibitions of ChE and MAO-B have been investigated in the context of AD [11,16]. In the present study, GC potently inhibited BChE with an IC_{50} value of 7.22 μM, and moderately inhibited AChE and MAO-B, indicating GC should be considered as a multi-function inhibitor of BChE, AChE, and MAO-B.

Pan et al. reported that MAO-B inhibition by LG in rat liver mitochondria was weaker than MAO-A inhibition by a mixed type [32]. However, in our study, LG more potently inhibited human MAO-B (IC_{50} = 0.098 μM) than human MAO-A (IC_{50} = 0.27 μM) and functioned as a competitive inhibitor. The IC_{50} of LG for MAO-B was lower than that of the flavonoid acacetin (IC_{50} = 0.17 μM) [53], which is one of the lowest IC_{50} values reported for a natural compound to date. Liquiritin was less effective than LG, aglycone of liquiritin, likely observed in acacetin and acacetin 7-*O*-(6-*O*-malonylglucoside) [53].

In our docking analysis, GC showed greater binding affinity with BChE than with AChE, and LG and liquiritin were predicted to bind to MAO-B more strongly than to MAO-A, and these results agreed well with determined IC_{50} values. In particular, our kinetic study showed that GC noncompetitively inhibited BChE. Docking simulation was performed to identify BChE binding sites. The docked pose for GC indicated that it interacted with BChE beyond the active site and hydrogen bonded with Thr284 and Val288. The active-site of BChE is composed of 4 subdomains, i.e., a peripheral site, a choline binding pocket, a catalytic site, and an acyl binding pocket [54], and the acyl binding pocket contains Trp231, Leu286, and Val288, which permit binding and hydrolysis of ligands and substrates bulkier than those of AChE [54], which is considered to be largely responsible for the different ligand-binding specificities of AChE and BChE [55]. Jannat et al. reported that (2*S*,3*R*)-pretosin C is a noncompetitive inhibitor of BChE and that it hydrophobically interacts with Val288, Lue286, and Phe357, and hydrogen bonds with Gly283 and Asn397, and docks at a non-ligand binding site [56]. It was also observed that hydrogen bond formation was the main driving force behind BChE–coumarin complex formation, whereas hydrophobic and halogen interactions underpinned AChE interactions with *N*1-(coumarin-7-yl) derivatives [57]. Similarly, we found that GC hydrogen bonded with Thr284 and Val288 located outside the ligand binding site. Such results suggest that GC might bind noncompetitively at the acyl binding pocket of BChE.

4. Materials and Methods

4.1. General

The dried roots of *Glycyrrhiza uralensis* were purchased in April 2011 at a commercial herbal market (Human-herb, Gyeongsan, Gyeongbuk, South Korea). Organic solvents (e.g., methanol (MeOH), chloroform ($CHCl_3$), methylene chloride (MC), ethyl acetate (EtOAc), and *n*-hexane (Hx)) were purchased from Duksan Chemical Co. (Seoul, South Korea). Column chromatography was performed using silica gel 60 (70–230 mesh, 230–400 mesh, ASTM, Merck, Darmstadt, Germany), octadecyl silica gel (ODS-A, 12 nm, S-150 m, YMC, Tokyo, Japan), and Sephadex LH-20 gel (GE Healthcare, Uppsala, Sweden). NMR spectra were recorded on a JEOL ECX-500 spectrometer, operating at 500 MHz for ^{1}H- and 125 MHz for ^{13}C-NMR (JEOL Ltd., Tokyo, Japan). High performance liquid chromatography (HPLC) was performed using an Agilent 1260 series system (Agilent Inc., Palo Alto, CA, USA) equipped with a binary pump, an autosampler, a column oven, a Phenomenex Kinetex C18 column (2.6 μm, 150 × 4.6 mm; Phenomenex, Torrance, CA, USA), a photodiode array detector (DAD), and an evaporative light scattering detector (ELSD). Trifluoroacetic acid (TFA, 0.1%, *v*/*v*) was used in water (solvent A) and in acetonitrile (ACN; solvent B). Gradient was applied to the elution from 95% A/5% B (0–3 min) to 0% A/100% B (3–30 min) at 0.5 mL/min using 3 μL of injection volume.

4.2. Extraction and Isolation of Coumarin Derivatives

The dried roots of *Glycyrrhiza uralensis* (1 kg) were extracted with MeOH at room temperature for 24 h (3 × 10 L) to obtain a crude MEOH extract, and 110 g of this extract was then suspended in 2000 mL of distilled water and partitioned versus the same volume of $CHCl_3$ and EtOAc. The $CHCl_3$ extract (25.2 g) obtained was separated into five fractions (GHC 1–5) by silica gel column chromatography using a Hx and EtOAc gradient (30:1 to 1:1). GHC 3 (5.2 g) was then separated by silica column chromatography using an Hx and EtOAc gradient (1:0 to 0:1) to yield ten subfractions (GHC 3-1–3-10). Subfraction GHC 3-5 (1.1 g) was subjected to reverse-phase column chromatography using ODS-A gel (50% aqueous MeOH, *v*/*v*) to obtain GC (**1**, 261.0 mg, purity: 99%). Subfraction GHC 3-7 was subjected to reverse-phase column chromatography using ODS-A gel (60% aqueous MeOH, *v*/*v*) to obtained isoliquiritin (**2**, 372.0 mg, purity: 98.3%). Pure LG (**3**, 530 mg, purity: 99%) was obtained from fraction GHC 4 using silica gel column chromatography with an Hx and EtOAc gradient (15:1 to 5:1). In addition, the EtOAc soluble extract (16.7 g) was separated into nine fractions (GHE 1–9) by silica gel column chromatography using an MC and MeOH gradient (40:1 to 4:1). Fraction GHE 2 (1.2 g) was separated by silica column chromatography using an Hx and EtOAc gradient (10:1) to obtain glycyrrhetinic acid (**4**, 113 mg, purity: 99%). Fraction GHE 6 (0.9 g) was subjected to silica gel column chromatography using an MC and MeOH gradient (15:1 to 6:1) to obtain liquiritin (**5**, 128.0 mg, purity: 99%). Fraction GHE 9 (3.6 g) was isolated by reverse-phase column chromatography using ODS-A gel (40% aqueous MeOH, *v*/*v*) to yield four fractions (GHE 9-1~9-4). Subsequently, subfraction GHE 9-3 was subjected to silica gel column chromatography using an isocratic MC-EtOAc-MeOH (3.5:3.5:1) mixture to obtain liquiritin apioside (**6**, 70.0 mg, purity: 95.1%). The water-soluble extract (20.2 g) was separated into five fractions (GHD 1–5) by chromatography on an LH-20 gel column using a H_2O and MeOH gradient (0:1 to 1:1). Fraction GHD 4 (12.6 g) was separated by reverse-phase column chromatography using a MeOH and H_2O gradient (5:6 to 3:2, *v*/*v*) to yield seven fractions (GHD 4-1–4-7). Subfraction GHD 4-5 was subjected to silica column chromatography using a $CHCl_3$-MeOH (4:1) as eluent and yielded isoliquiritin apioside (**7**, 35.0 mg, purity: 96.3%). Glycyrrhizin was obtained from hot water extracts. The hot water extracts (28.1 g) was aggregated by reducing its pH to 2.0 with 10% H_2SO_4 and filtering through Whatman No. 1 paper. The precipitate obtained was suspended in distilled water (1000 mL), the pH was adjusted to 7.0 using ammonia water, and glycyrrhizin (**8**, 223.0 mg, purity: 99%) was obtained by subjecting this solution to ODS-A gel column chromatography using 60% aqueous ACN as eluant. HPLC chromatograms of the eight compounds were provided in Supplementary Figure S1.

4.3. Chemicals and Enzyme Assays

Enzymes (recombinant human MAO-A and MAO-B, AChE from *Electrophorus electricus*, and BChE from equine serum), substrates (kynuramine and benzylamine, acetylthiocholine iodide (ATCI), *S*-butyrylthiocholine iodide (BTCI)), inhibitors (toloxatone, lazabemide, and tacrine), and other chemicals including 5,5′-dithiobis(2-nitrobenzoic acid) (DTNB) were purchased from Sigma-Aldrich (St. Louis, MO, USA) [49,58]. The irreversible inhibitors (clorgyline and pargyline) were obtained from BioAssay Systems (Hayward, CA, USA) [59].

MAO-A and MAO-B activities were measured continuously at 316 nm for 20 min, and at 250 nm for 30 min, respectively, as described previously [60,61]. The concentrations used were; kynuramine (0.06 mM) for MAO-A and benzylamine (0.3 mM) for MAO-B. AChE activity was assayed continuously for 10 min at 412 nm using 0.2 U/mL of enzyme in the presence of 0.5 mM DTNB and 0.5 mM ATCI in 0.5 mL of reaction mixture, as previously described [49,58], based on the method developed by Ellman et al. [62]. BChE activity was assayed using the same method as AChE, except using BTCI [49]. Substrate concentrations of BTCI for BChE and benzylamine for MAO-B were 2.3- and 2.1-fold of the respective K_m values (0.22 and 0.14 mM).

4.4. Inhibitory Activities and Enzyme kinetics

Inhibitions of MAO-A, MAO-B, AChE, and BChE were initially observed at an inhibitor concentration of 10 μM. IC_{50} values of compounds exhibiting >50% inhibition were determined. Kinetic parameters, inhibition types, and K_i values were determined for the most potent inhibitors, i.e., GC for BChE and LG for MAO-B, as previously described [49,58]. The kinetics of BChE and MAO-B inhibitions were investigated at five different substrate concentrations; 0.05, 0.1, 0.25, 0.5 or 1.0 mM for BChE, and 0.03, 0.06, 0.15, 0.3, or 0.6 mM for MAO-B. Inhibition studies were conducted in the absence or presence of each inhibitor at about 0.5×, 1.0×, and 2.0× their IC_{50} values [58]. Inhibitory patterns and K_i values were determined using Lineweaver-Burk plots and secondary derivative plots.

4.5. Analysis of Inhibitor Reversibility

The reversibilities of BChE inhibition by GC and of MAO-B inhibition by LG were investigated by dialysis at concentrations of $2 \times IC_{50}$ values, as previously described [63]. After preincubating GC or LG with BChE or MAO-B, respectively, for 30 min, residual activities for undialyzed and 6 h-dialyzed samples were measured; relative values for A_U and A_D were then calculated and compared with each control without inhibitor. Reversibilities were determined by comparing A_U and A_D values of inhibitors with those of references. In addition, the dilution method was used to access BChE activity recovery after inhibition by GC (i.e., after preincubating BChE with GC at $50 \times IC_{50}$ for 15 min) and diluting to a GC concentration of $1 \times IC_{50}$ [60]. Residual activity of the preincubated and then diluted mixture was measured and compared to those of mixtures at 1× or $50 \times IC_{50}$ concentrations.

4.6. Docking Simulations of GC with AChE and BChE and of LG or Liquiritin with MAO-A and MAO-B

To simulate docking of GC with AChE or BChE, we used Autodock Vina [64], which has an automated docking facility. To define enzyme docking pockets, we used a set of predefined active sites defined using a complex of AChE with 3-[(1*S*)-1-(dimethylamino)ethyl]phenol (SAF) (PDB ID: 1GQS) or a complex of BChE with *N*-{[(3*R*)-1-(2,3-dihydro-1H-inden-2-yl)piperidin-3-yl]methyl} -*N*-(2-methoxyethyl) naphthalene-2-carboxamide (3F9) (PDB ID: 4TPK). In addition, to define MAO-A or MAO-B docking sites with LG or liquiritin, we used a set of predefined active sites obtained using MAO-A/7-methoxy-1-methyl-9*H*-β-carboline complex (PDB ID: 2Z5X) or MAO-B/pioglitazone complex (PDB ID: 4A79). To prepare for docking simulations, we performed the following steps: created 2D structures, converted 2D into 3D structures, performed energy minimization using the ChemOffice program (http://www.cambridgesoft.com) and docking simulations using Chimera [65], and checked for possible hydrogen bonding interactions using 0.4 Å and 20.0° constraints using Chimera [66].

5. Conclusions

GC effectively inhibited BChE and AChE (IC_{50} = 7.22 and 14.77 μM, respectively), and also moderately inhibited MAO-B (IC_{50} = 29.48 μM). LG potently inhibited MAO-B (IC_{50} = 0.098 μM) and MAO-A (IC_{50} = 0.27 μM). GC was found to be a noncompetitive inhibitor of BChE and LG to be a competitive inhibitor of MAO-B. The binding affinity of GC for BChE (−7.8 kcal/mol) was higher than its affinity for AChE (−7.1 kcal/mol), and this binding was driven by hydrogen bond formation with Thr284 and Val288 of BChE. These findings regarding the multi-inhibitory effects of GC and LG suggest that they be considered potential candidates for the treatment of Alzheimer's disease.

Supplementary Materials: ^{1}H- and ^{13}C-NMR spectral data are available in the Supplementary Materials.

Author Contributions: Conceptualization: M.C. and H.K.; biological activity: G.S.J. and S.R.L.; kinetics: G.S.J. and S.R.L.; compound isolation and identification: J.Y.L. and M.C.; docking analysis: M.-G.K. and D.P.; original draft writing: G.S.J., M.-G.K. and J.Y.L.; review and editing: D.P., M.C., and H.K.; supervision: H.K.; funding acquisition: H.K. All authors have read and agreed to the published version of the manuscript.

Funding: This study was supported by the Cooperative Research Program for Agriculture Science and Technology Development (#PJ01319104) of the Rural Development Administration, Republic of Korea.

Conflicts of Interest: The authors have no conflict of interest to declare.

References

1. Bierer, L.M.; Haroutunian, V.; Gabriel, S.; Knott, P.J.; Carlin, L.S.; Purohit, D.P.; Perl, D.P.; Schmeidler, J.; Kanof, P.; Davis, K.L. Neurochemical correlates of dementia severity in Alzheimer's disease: Relative importance of the cholinergic deficits. *J. Neurochem.* **1995**, *64*, 749–760. [CrossRef] [PubMed]
2. Colović, M.B.; Krstić, D.Z.; Lazarević-Pašti, T.D.; Bondžić, A.M.; Vasić, V.M. Acetylcholinesterase inhibitors: Pharmacology and toxicology. *Curr. Neuropharmacol.* **2013**, *11*, 315–335. [CrossRef] [PubMed]
3. Anand, P.; Singh, B. A review on cholinesterase inhibitors for Alzheimer's disease. *Arch. Pharm. Res.* **2013**, *36*, 375–399. [CrossRef] [PubMed]
4. Ghai, R.; Nagarajan, K.; Arora, M.; Grover, P.; Ali, N.; Kapoor, G. Current strategies and novel drug approaches for Alzheimer disease. *CNS Neurol. Disord. Drug Targets* **2020**. [CrossRef] [PubMed]
5. Kumar, A.; Pintus, F.; Di Petrillo, A.; Medda, R.; Caria, P.; Matos, M.J.; Vina, D.; Pieroni, E.; Delogu, F.; Era, B.; et al. Novel 2-phenylbenzofuran derivatives as selective butyrylcholinesterase inhibitors for Alzheimer's disease. *Sci. Rep.* **2018**, *8*, 4424. [CrossRef]
6. Korábečný, J.; Nepovimová, E.; Cikánková, T.; Špilovská, K.; Vašková, L.; Mezeiová, E.; Kuča, K.; Hroudová, J. Newly developed drugs for Alzheimer's disease in relation to energy metabolism, cholinergic and monoaminergic neurotransmission. *Neuroscience* **2018**, *370*, 191–206. [CrossRef]
7. Lake, F. BChE reported to be associated with plaque level in Alzheimer's disease. *Biomark. Med.* **2013**, *7*, 197–198.
8. Ramsay, R.R. Monoamine oxidases: The biochemistry of the proteins as targets in medicinal chemistry and drug discovery. *Curr. Top. Med. Chem.* **2012**, *12*, 2189–2209. [CrossRef]
9. Youdim, M.B.; Edmondson, D.; Tipton, K.F. The therapeutic potential of monoamine oxidase inhibitors. *Nat. Rev. Neurosci.* **2006**, *7*, 295–309. [CrossRef]

10. Schedin-Weiss, S.; Inoue, M.; Hromadkova, L.; Teranishi, Y.; Yamamoto, N.G.; Wiehager, B.; Bogdanovic, N.; Winblad, B.; Sandebring-Matton, A.; Frykman, S.; et al. Monoamine oxidase B is elevated in Alzheimer disease neurons, is associated with γ-secretase and regulates neuronal amyloid β-peptide levels. *Alzheimers Res. Ther.* **2017**, *9*, 57. [CrossRef]
11. Ramsay, R.R.; Tipton, K.F. Assessment of enzyme inhibition: A review with examples from the development of monoamine oxidase and cholinesterase inhibitory drugs. *Molecules* **2017**, *22*, 1192. [CrossRef] [PubMed]
12. Li, Y.; Qiang, X.; Luo, L.; Yang, X.; Xiao, G.; Zheng, Y.; Cao, Z.; Sang, Z.; Su, F.; Deng, Y. Multitarget drug design strategy against Alzheimer's disease: Homoisoflavonoid mannich base derivatives serve as acetylcholinesterase and monoamine oxidase B dual inhibitors with multifunctional properties. *Bioorg. Med. Chem.* **2017**, *25*, 714–726. [CrossRef] [PubMed]
13. Cai, P.; Fang, S.Q.; Yang, H.L.; Yang, X.L.; Liu, Q.H.; Kong, L.Y.; Wang, X.B. Donepezil-butylated hydroxytoluene (BHT) hybrids as anti-Alzheimer's disease agents with cholinergic, antioxidant, and neuroprotective properties. *Eur. J. Med. Chem.* **2018**, *157*, 161–176. [CrossRef] [PubMed]
14. He, Q.; Liu, J.; Lan, J.S.; Ding, J.; Sun, Y.; Fang, Y.; Jiang, N.; Yang, Z.; Sun, L.; Jin, Y.; et al. Coumarin-dithiocarbamate hybrids as novel multitarget AChE and MAO-B inhibitors against Alzheimer's disease: Design, synthesis and biological evaluation. *Bioorg. Chem.* **2018**, *81*, 512–528. [CrossRef]
15. Pisani, L.; Iacobazzi, R.M.; Catto, M.; Rullo, M.; Farina, R.; Denora, N.; Cellamare, S.; Altomare, C.D. Investigating alkyl nitrates as nitric oxide releasing precursors of multitarget acetylcholinesterase-monoamine oxidase B inhibitors. *Eur. J. Med. Chem.* **2019**, *161*, 292–309. [CrossRef]
16. Oh, J.M.; Rangarajan, T.M.; Chaudhary, R.; Singh, R.P.; Singh, M.; Singh, R.P.; Tondo, A.R.; Gambacorta, N.; Nicolotti, O.; Mathew, B.; et al. Novel class of chalcone oxime ethers as potent monoamine oxidase-B and acetylcholinesterase inhibitors. *Molecules* **2020**, *25*, 2356. [CrossRef]
17. Mascarenhas, A.M.S.; de Almeida, R.B.M.; de Araujo Neto, M.F.; Mendes, G.O.; da Cruz, J.N.; Dos Santos, C.B.R.; Botura, M.B.; Leite, F.H.A. Pharmacophore-based virtual screening and molecular docking to identify promising dual inhibitors of human acetylcholinesterase and butyrylcholinesterase. *J. Biomol. Struct. Dyn.* **2020**. [CrossRef]
18. Uddin, M.S.; Kabir, M.T.; Rahman, M.M.; Mathew, B.; Shah, M.A.; Ashraf, G.M. TV 3326 for Alzheimer's dementia: A novel multimodal ChE and MAO inhibitors to mitigate Alzheimer's-like neuropathology. *J. Pharm. Pharmcol.* **2020**, *72*, 1001–1012. [CrossRef]
19. Purgatorio, R.; Kulikova, L.; Pisani, L.; Catto, M.; de Candia, M.; Carrieri, A.; Cellamare, S.; De Palma, A.; Beloglazkin, A.; Raesi, G.R.; et al. Scouting around 1,2,3,4-Tetrahydrochromeno[3,2-c]pyridin-10-ones for single- and multi-target ligands directed towards relevant Alzheimer's targets. *ChemMedChem* **2020**. [CrossRef]
20. Lee, S.; Oh, H.M.; Lim, W.B.; Choi, E.J.; Park, Y.N.; Kim, J.A.; Choi, J.Y.; Hong, S.J.; Oh, H.K.; Son, J.K.; et al. Gene induction by glycyrol to apoptosis through endonuclease G in tumor cells and prediction of oncogene function by microarray analysis. *Anticancer Drugs* **2008**, *19*, 503–515. [CrossRef]
21. Xu, M.Y.; Kim, Y.S. Antitumor activity of glycyrol via induction of cell cycle arrest, apoptosis and defective autophagy. *Food Chem. Toxicol.* **2014**, *74*, 311–319. [CrossRef] [PubMed]
22. Lu, S.; Ye, L.; Yin, S.; Zhao, C.; Yan, M.; Liu, X.; Cui, J.; Hu, H. Glycyrol exerts potent therapeutic effect on lung cancer via directly inactivating T-LAK cell-originated protein kinase. *Pharmacol. Res.* **2019**, *147*, 104366. [CrossRef] [PubMed]
23. Rhew, Z.I.; Han, Y. Synergic effect of combination of glycyrol and fluconazole against experimental cutaneous candidiasis due to *Candida albicans*. *Arch. Pharm. Res.* **2016**, *39*, 1482–1489. [CrossRef] [PubMed]
24. Tanaka, Y.; Kikuzaki, H.; Fukuda, S.; Nakatani, N. Antibacterial compounds of licorice against upper airway respiratory tract pathogens. *J. Nutr. Sci. Vitaminol.* **2001**, *47*, 270–273. [CrossRef]
25. Ryu, Y.B.; Kim, J.H.; Park, S.J.; Chang, J.S.; Rho, M.C.; Bae, K.H.; Park, K.H.; Lee, W.S. Inhibition of neuraminidase activity by polyphenol compounds isolated from the roots of Glycyrrhiza uralensis. *Bioorg. Med. Chem. Lett.* **2010**, *20*, 971–974. [CrossRef]
26. Adianti, M.; Aoki, C.; Komoto, M.; Deng, L.; Shoji, I.; Wahyuni, T.S.; Lusida, M.I.; Fuchino, H.; Kawahara, N.; Hotta, H. Anti-hepatitis C virus compounds obtained from *Glycyrrhiza uralensis* and other Glycyrrhiza species. *Microbiol. Immunol.* **2014**, *58*, 180–187. [CrossRef]

27. Shin, E.M.; Zhou, H.Y.; Guo, L.Y.; Kim, J.A.; Lee, S.H.; Merfort, I.; Kang, S.S.; Kim, H.S.; Kim, S.; Kim, Y.S. Anti-inflammatory effects of glycyrol isolated from *Glycyrrhiza uralensis* in LPS-stimulated RAW264.7 macrophages. *Int. Immunopharmacol.* **2008**, *8*, 1524–1532. [CrossRef]
28. Fu, Y.; Zhou, H.; Wang, S.; Wei, Q. Glycyrol suppresses collagen-induced arthritis by regulating autoimmune and inflammatory responses. *PLoS ONE* **2014**, *9*, e98137. [CrossRef]
29. Bai, H.; Bao, F.; Fan, X.; Han, S.; Zheng, W.; Sun, L.; Yan, N.; Du, H.; Zhao, H.; Yang, Z. Metabolomics study of different parts of licorice from different geographical origins and their anti-inflammatory activities. *J. Sep. Sci.* **2020**, *43*, 1593–1602. [CrossRef]
30. Li, J.; Tu, Y.; Tong, L.; Zhang, W.; Zheng, J.; Wei, Q. Immunosuppressive activity on the murine immune responses of glycyrol from *Glycyrrhiza uralensis* via inhibition of calcineurin activity. *Pharm. Biol.* **2010**, *48*, 1177–1184. [CrossRef]
31. Tanaka, S.; Kuwai, Y.; Tabata, M. Isolation of monoamine oxidase inhibitors from *Glycyrrhiza uralensis* roots and the structure-activity relationship. *Planta Med.* **1987**, *53*, 5–8. [CrossRef] [PubMed]
32. Pan, X.; Kong, L.D.; Zhang, Y.; Cheng, C.H.K.; Tan, R.X. In vitro inhibition of rat monoamine oxidase by liquiritigenin and isoliquiritigenin isolated from *Sinofranchetia chinensis*. *Acta Pharmacol. Sin.* **2000**, *21*, 949–953. [PubMed]
33. Lee, J.S.; Kim, J.A.; Cho, S.H.; Son, A.R.; Jang, T.S.; So, M.S.; Chung, S.R.; Lee, S.H. Tyrosinase inhibitors isolated from the roots of *Glycyrrhiza glabra* L. *Korean J. Pharmacogn.* **2003**, *34*, 33–39.
34. Maatooq, G.T.; Marzouk, A.M.; Gray, A.I.; Rosazza, J.P. Bioactive microbial metabolites from glycyrrhetinic acid. *Phytochemistry* **2010**, *71*, 262–270. [CrossRef] [PubMed]
35. Montoro, P.; Maldini, M.; Russo, M.; Postorino, S.; Piacente, S.; Pizza, C. Metabolic profiling of roots of liquorice (*Glycyrrhiza glabra*) from different geographical areas by ESI/MS/MS and determination of major metabolites by LC-ESI/MS and LC-ESI/MS/MS. *J. Pharm. Biomed. Anal.* **2011**, *54*, 535–544. [CrossRef] [PubMed]
36. Wang, H.G.; Liu, Y.N.; Le, X.Y.; Wang, S.F. Study on active constituents from Glycyrrhizae Radix et Rhizoma against NO production in LPS-induced Raw 264.7 macrophages. *Chin. Tradit. Herb. Drugs* **2016**, *47*, 4155–4159.
37. Chaturvedula, V.S.P.; Yu, O.; Mao, G. NMR analysis and hydrolysis studies of glycyrrhizin acid, a major constituent of *Glycyrrhia glabra*. *Eur. Chem. Bull.* **2014**, *3*, 104–107.
38. de Souza, L.G.; Renn, M.N.; Figueroa-Villar, J.D. Coumarins as cholinesterase inhibitors: A review. *Chem. Biol. Interact.* **2016**, *254*, 11e23. [CrossRef]
39. Hu, Y.H.; Yang, J.; Zhang, Y.; Liu, K.C.; Liu, T.; Sun, J.; Wang, X.J. Synthesis and biological evaluation of 3–(4-aminophenyl)-coumarin derivatives as potential anti-Alzheimer's disease agents. *J. Enzym. Inhib. Med. Chem.* **2019**, *34*, 1083–1092. [CrossRef]
40. Salem, S.B.; Jabrane, A.; Harzallah-Skhiri, F.; Jannet, H.B. New bioactive dihydrofuranocoumarins from the roots of the Tunisian *Ferula lutea* (Poir.) maire. *Bioorg. Med. Chem. Lett.* **2013**, *23*, 4248–4252. [CrossRef]
41. Awang, K.; Chan, G.; Litaudon, M.; Ismail, N.H.; Martin, M.T.; Gueritte, F. 4-Phenylcoumarins from *Mesua elegans* with acetylcholinesterase inhibitory activity. *Bioorg. Med. Chem.* **2010**, *18*, 7873–7877. [CrossRef] [PubMed]
42. Dos Santos, T.C.; Gomes, T.; Pinto, B.A.S.; Camara, A.L.; de Andrade Paes, A.M. Naturally occurring acetylcholinesterase inhibitors and their potential use for Alzheimer's disease therapy. *Front. Pharmacol.* **2018**, *9*, 1192. [CrossRef] [PubMed]
43. Baek, S.C.; Kang, M.G.; Park, J.E.; Lee, J.P.; Lee, H.; Ryu, H.W.; Park, C.M.; Park, D.; Cho, M.L.; Oh, S.R.; et al. Osthenol, a prenylated coumarin, as a monoamine oxidase A inhibitor with high selectivity. *Bioorg. Med. Chem. Lett.* **2019**, *29*, 839–843. [CrossRef] [PubMed]
44. Dincel, D.; Hatipoglu, S.D.; Gæren, A.C.; Topçu, G. Anticholinesterase furocoumarins from heracleumplatytaenium, a species endemic to the Ida mountains. *Turk. J. Chem.* **2013**, *37*, 675e683.
45. Kostelnik, A.; Pohanka, M. Inhibition of acetylcholinesterase and butyrylcholinesterase by a plant secondary metabolite boldine. *Biomed. Res. Int.* **2018**, *2018*, 9634349. [CrossRef]
46. Orhan, I.E.; Senol Deniz, F.S.; Traedal-Henden, S.; Cerón-Carrasco, J.P.; den Haan, H.; Peña-García, J.; Pérez-Sánchez, H.; Emerce, E.; Skalicka-Wozniak, K. Profiling auspicious butyrylcholinesterase inhibitory activity of two herbal molecules: Hyperforin and hyuganin C. *Chem. Biodivers.* **2019**, *16*, e1900017. [CrossRef]

47. Liu, L.; Yin, Q.M.; Gao, Q.; Li, J.; Jiang, Y.; Tu, P.F. New biphenanthrenes with butyrylcholinesterase inhibitory activity from *Cremastra appendiculata*. *Nat. Prod. Res.* **2019**, 1–7. [CrossRef]
48. Kashyap, P.; Ram, H.; Shukla, S.D.; Kumar, S. Scopoletin: Antiamyloidogenic, anticholinesterase, and neuroprotective potential of a natural compound present in argyreia speciosa roots by in vitro and in silico study. *Neurosci. Insights* **2020**, *15*, 2633105520937693. [CrossRef]
49. Lee, J.P.; Kang, M.G.; Lee, J.Y.; Oh, J.M.; Baek, S.C.; Leem, H.H.; Park, D.; Cho, M.L.; Kim, H. Potent inhibition of acetylcholinesterase by sargachromanol I from Sargassum siliquastrum and by selected natural compounds. *Bioorg. Chem.* **2019**, *89*, 103043. [CrossRef]
50. Ślusarczyk, S.; Senol Deniz, F.S.; Abel, R.; Pecio, Ł.; Pérez-Sánchez, H.; Cerón-Carrasco, J.P.; den-Haan, H.; Banerjee, P.; Preissner, R.; Krzyżak, E.; et al. Norditerpenoids with selective anti-cholinesterase activity from the roots of *perovskia atriplicifolia* benth. *Int. J. Mol. Sci.* **2020**, *21*, 4475. [CrossRef]
51. Türkan, F.; Taslimi, P.; Saltan, F.Z. Tannic acid as a natural antioxidant compound: Discovery of a potent metabolic enzyme inhibitor for a new therapeutic approach in diabetes and Alzheimer's disease. *J. Biochem. Mol. Toxicol.* **2019**, *33*, e22340. [CrossRef] [PubMed]
52. Taslimi, P.; Caglayan, C.; Gulcin, İ. The impact of some natural phenolic compounds on carbonic anhydrase, acetylcholinesterase, butyrylcholinesterase, and α-glycosidase enzymes: An antidiabetic, anticholinergic, and antiepileptic study. *Biochem. Mol. Toxicol.* **2017**. [CrossRef] [PubMed]
53. Lee, H.W.; Ryu, H.W.; Baek, S.C.; Kang, M.G.; Park, D.; Han, H.Y.; An, J.H.; Oh, S.R.; Kim, H. Potent inhibitions of monoamine oxidase A and B by acacetin and its 7-*O*-(6-*O*-malonylglucoside) derivative from *Agastache rugosa*. *Int. J. Biol. Macromol.* **2017**, *104*, 547–553. [CrossRef]
54. Brus, B.; Košak, U.; Turk, S.; Pišlar, A.; Coquelle, N.; Kos, J.; Stojan, J.; Colletier, J.P.; Gobec, S. Discovery, biological evaluation, and crystal structure of a novel nanomolar selective butyrylcholinesterase inhibitor. *J. Med. Chem.* **2014**, *57*, 8167–8179. [CrossRef]
55. Radic, Z.; Pickering, N.A.; Vellom, D.C.; Camp, S.; Taylor, P. Three distinct domains in the cholinesterase molecule confer selectivity for acetyl- and butyrylcholinesterase inhibitors. *Biochemistry* **1993**, *32*, 12074–12084. [CrossRef] [PubMed]
56. Jannat, S.; Balupuri, A.; Ali, M.Y.; Hong, S.S.; Choi, C.W.; Choi, Y.H.; Ku, J.M.; Kim, W.J.; Leem, J.Y.; Kim, J.E.; et al. Inhibition of β-site amyloid precursor protein cleaving enzyme 1 and cholinesterases by pterosins via a specific structure-activity relationship with a strong BBB permeability. *Exp. Mol. Med.* **2019**, *51*, 12. [CrossRef]
57. Abu-Aisheh, M.N.; Al-Aboudi, A.; Mustafa, M.S.; El-Abadelah, M.M.; Ali, S.Y.; Ul-Haq, Z.; Mubarak, M.S. Coumarin derivatives as acetyl- and butyrylcholinesterase inhibitors: An in vitro, molecular docking, and molecular dynamics simulations study. *Heliyon* **2019**, *5*, e01552. [CrossRef]
58. Baek, S.C.; Park, M.H.; Ryu, H.W.; Lee, J.P.; Kang, M.G.; Park, D.; Park, C.M.; Oh, S.R.; Kim, H. Rhamnocitrin isolated from *Prunus padus* var. seoulensis: A potent and selective reversible inhibitor of human monoamine oxidase A. *Bioorg. Chem.* **2019**, *83*, 317–325. [CrossRef]
59. Baek, S.C.; Lee, H.W.; Ryu, H.W.; Kang, M.G.; Park, D.; Kim, S.H.; Cho, M.L.; Oh, S.R.; Kim, H. Selective inhibition of monoamine oxidase A by hispidol. *Bioorg. Med. Chem. Lett.* **2018**, *28*, 584–588. [CrossRef]
60. Lee, H.W.; Ryu, H.W.; Kang, M.G.; Park, D.; Oh, S.R.; Kim, H. Potent selective monoamine oxidase B inhibition by maackiain, a pterocarpan from the roots of *Sophora flavescens*. *Bioorg. Med. Chem. Lett.* **2016**, *26*, 4714–4719. [CrossRef]
61. Baek, S.C.; Choi, B.; Nam, S.J.; Kim, H. Inhibition of monoamine oxidase A and B by demethoxycurcumin and bisdemethoxycurcumin. *J. Appl. Biol. Chem.* **2018**, *61*, 187–190. [CrossRef]
62. Ellman, G.L.; Courtney, K.D.; Andres, J.V.; Feather-Stone, R.M. A new and rapid colorimetric determination of acetylcholinesterase activity. *Biochem. Pharmacol.* **1961**, *7*, 88–95. [CrossRef]
63. Oh, J.M.; Kang, M.G.; Hong, A.; Park, J.E.; Kim, S.H.; Lee, J.P.; Baek, S.C.; Park, D.; Nam, S.J.; Cho, M.L.; et al. Potent and selective inhibition of human monoamine oxidase-B by 4-dimethylaminochalcone and selected chalcone derivatives. *Int. J. Biol. Macromol.* **2019**, *137*, 426–432. [CrossRef] [PubMed]
64. Trott, O.; Olson, A.J. AutoDock Vina: Improving the speed and accuracy of docking with a new scoring function, efficient optimization, and multithreading. *J. Comput. Chem.* **2010**, *31*, 455–461. [CrossRef] [PubMed]

65. Pettersen, E.F.; Goddard, T.D.; Huang, C.C.; Couch, G.S.; Greenblatt, D.M.; Meng, E.C.; Ferrin, T.E. UCSF Chimera—A visualization system for exploratory research and analysis. *J. Comput. Chem.* **2004**, *25*, 1605–1612. [CrossRef] [PubMed]
66. Mills, J.E.; Dean, P.M. Three-dimensional hydrogen-bond geometry and probability information from a crystal survey. *J. Comput. Aided Mol. Des.* **1996**, *10*, 607–622. [CrossRef] [PubMed]

Sample Availability: Samples of the compounds are available from the authors.

Article

Adenosine Receptor Ligands: Coumarin–Chalcone Hybrids as Modulating Agents on the Activity of *h*ARs

Saleta Vazquez-Rodriguez [1,*,†], Santiago Vilar [1], Sonja Kachler [2], Karl-Norbert Klotz [2], Eugenio Uriarte [1,3], Fernanda Borges [4] and Maria João Matos [1,4,*]

[1] Departamento de Química Orgánica, Facultad de Farmacia, Universidad de Santiago de Compostela, 15782 Santiago de Compostela, Spain; qosanti@yahoo.es (S.V.); eugenio.uriarte@usc.es (E.U.)

[2] Institut für Pharmakologie und Toxikologie, Universität Würzburg, 97078 Würzburg, Germany; sonja.kachler@toxi.uni-wuerzburg.de (S.K.); klotz@toxi.uni-wuerzburg.de (K.-N.K.)

[3] Instituto de Ciencias Químicas Aplicadas, Universidad Autónoma de Chile, 7500912 Santiago, Chile

[4] CIQUP/Department of Chemistry and Biochemistry, Faculty of Sciences, University of Porto, Rua Campo Alegre 687, 4169-007 Porto, Portugal; mfernandamborges@gmail.com

* Correspondence: saleta.vazquezrodriguez@cmd.ox.ac.uk (S.V.-R.); mariajoao.correiapinto@usc.es or maria.matos@fc.up.pt (M.J.M.)

† Current address SVR: Target Discovery Institute, University of Oxford, NDM Research building, Old Road Campus, Oxford OX3 7FX, UK.

Academic Editor: Pascal Richomme
Received: 27 August 2020; Accepted: 18 September 2020; Published: 19 September 2020

Abstract: Adenosine receptors (ARs) play an important role in neurological and psychiatric disorders such as Alzheimer's disease, Parkinson's disease, epilepsy and schizophrenia. The different subtypes of ARs and the knowledge on their densities and status are important for understanding the mechanisms underlying the pathogenesis of diseases and for developing new therapeutics. Looking for new scaffolds for selective AR ligands, coumarin–chalcone hybrids were synthesized (compounds **1–8**) and screened in radioligand binding (hA_1, hA_{2A} and hA_3) and adenylyl cyclase (hA_{2B}) assays in order to evaluate their affinity for the four human AR subtypes (*h*ARs). Coumarin–chalcone hybrid has been established as a new scaffold suitable for the development of potent and selective ligands for hA_1 or hA_3 subtypes. In general, hydroxy-substituted hybrids showed some affinity for the hA_1, while the methoxy counterparts were selective for the hA_3. The most potent hA_1 ligand was compound **7** (K_i = 17.7 μM), whereas compound **4** was the most potent ligand for hA_3 (K_i = 2.49 μM). In addition, docking studies with hA_1 and hA_3 homology models were established to analyze the structure–function relationships. Results showed that the different residues located on the protein binding pocket could play an important role in ligand selectivity.

Keywords: coumarin; chalcone; neurodegenerative diseases; adenosine receptors; binding affinity; docking

1. Introduction

Adenosine receptors (ARs) are cell membrane receptors, belonging to the G protein-coupled receptor (GPCRs) superfamily. ARs comprised of four different subtypes: A_1, A_{2A}, A_{2B} and A_3 [1]. Adenosine is a purine nucleoside and an endogenous modulator of several physiological processes [1–4]. Extracellular adenosine activates the G_i-coupled receptors of the A_1 and A_3 subtypes, depressing the action of the brain, heart, kidneys, and the immune system, amongst other systems, as a consequence of the inhibition of adenylyl cyclase [5]. The A_3 subtype of AR has been cloned [6,7], making it possible to establish its pharmacological [8–11] and regulatory features [12].

Due to their widespread presence in cells, ARs proved to be promising targets in drug discovery. During the last decade, the search for selective ligands has been raised [13–15]. Several AR antagonists appeared as promising drug candidates for different pathological processes such as inflammation (A_3) [14], heart and renal failure (A_1) [16], or neurological disorders including Parkinson [17,18] and Alzheimer's diseases (A_{2A} and/or A_1) [19]. ARs can work as targets for various diseases and can open a new window for new therapeutic approaches.

In particular, A_1 antagonists are effective as diuretic agents [20,21] and also show neuroprotective activity in animal models of in vivo ischemia [22]. On the other hand, A_3 antagonists are being investigated as potential agents against renal injury [23] and also as neuroprotective agents [24,25], while A_3 agonists are also under consideration for treating conditions of the central nervous system (CNS) and peripheral nervous system [26,27].

From the arsenal of molecules presenting high potency and selectivity on ARs, the xanthine scaffold was the first to be used to develop the so-called classical AR antagonists [28,29]. In the search for non-xanthine AR ligands, numerous structures were discovered over the years. Flavones and isoflavones have played a remarkable role. As an example, genistein, was described as a competitive antagonist at A_1 in FRTL (thyroid) cells [30], and galangin was found to bind to the three subtypes of ARs displaying micromolar affinity for the A_3 [31]. The affinity of flavonoids and other phytochemicals to ARs brings about the hypothesis that probably other types of natural substances, namely those present in the diet, can interact with this type of receptor.

Coumarins (chromone isosteres) and chalcones (a flavonoid precursor) are naturally occurring benzopyran-related molecules presenting a variety of pharmacological activities [32–34]. Having in mind that both the coumarin and chalcone nuclei are structurally close to flavonoids, the design of novel AR ligands based on their scaffolds emerged as an interesting idea. Our study was also motivated by the structural similarity between the coumarin and the chromone scaffolds, which were previously described as AR ligands [35,36], and by the similarities with some coumarin derivatives previously described in our group [37–42]. In this context, we focused our attention on the 3-benzoylcoumarin core, considered as a hybrid scaffold in which the chalcone is fixed in a *trans* conformation through the double bond of the pyrone ring of the coumarin skeleton (Figure 1), presenting a more restricted conformation compared to the previously described coumarin–chalcone hybrids [36].

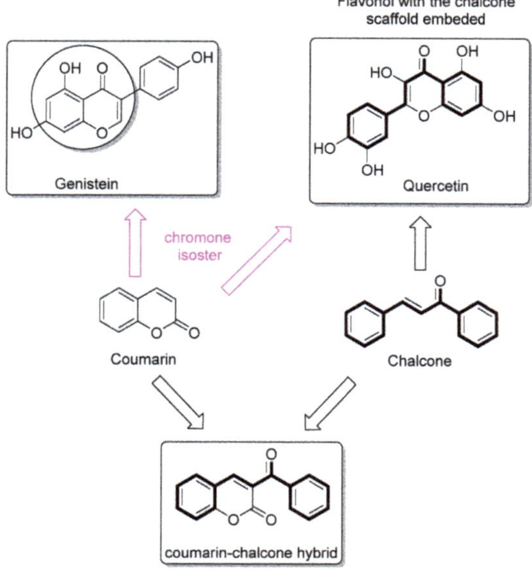

Figure 1. Rational design of coumarin–chalcone hybrids.

Therefore, based on the structural similarities between flavones, chalcones and coumarins, we decided to design and synthesize a novel family of coumarin–chalcone hybrid derivatives and study their activity on the different subtypes of human AR.

2. Results and Discussion

2.1. Chemistry

Two sets of coumarin–chalcone hybrids have been synthesized: one decorated with methoxy substituents (**1–4**) and another with hydroxy substituents (**5–8**). An efficient and versatile Knoevenagel reaction, treating a commercially available salicylaldehyde and the corresponding methoxylated ethyl benzoylacetate with piperidine in ethanol (EtOH) at reflux for 2–6 h, allowed the desired methoxy-3-benzoylcoumarins **1–4** with 85–97% yield. The hydroxy-3-benzoylcoumarins **5–8** were obtained by hydrolysis of the corresponding methoxy derivatives, with 75–94% yield, by employing boron tribromide (BBr_3) as deprotecting reagent in dichloromethane (DCM) at 80 °C in a Schlenk tube for 48 h [43]. The synthetic approach is illustrated in Scheme 1 and described in the methods and materials section.

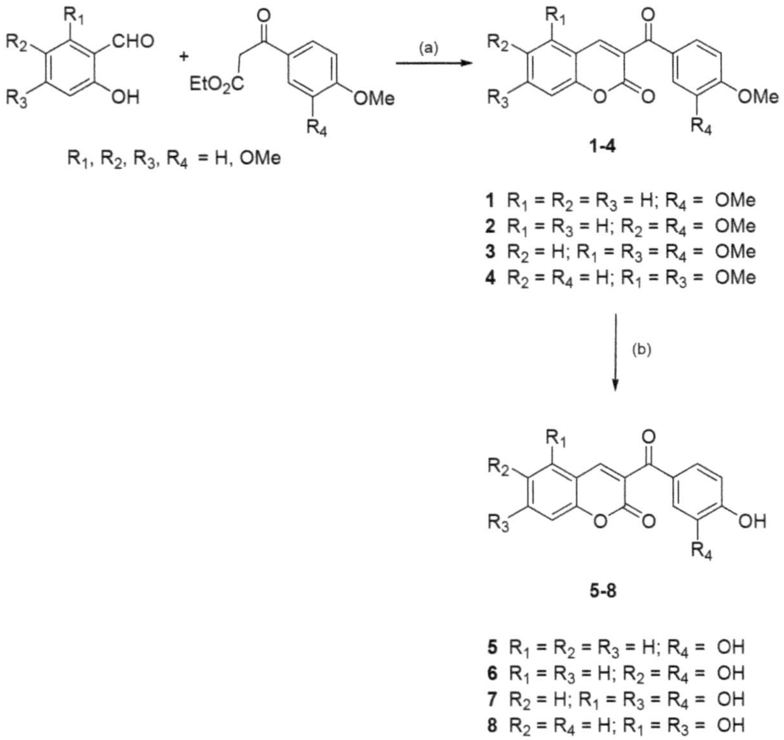

R1
R2
CHO
+
R3
OH
O
EtO2C
OMe
R4
R1, R2, R3, R4 = H, OMe
(a)
R1
O
R2
R3
O
O
OMe
R4
1-4
1 R1 = R2 = R3 = H; R4 = OMe
2 R1 = R3 = H; R2 = R4 = OMe
3 R2 = H; R1 = R3 = R4 = OMe
4 R2 = R4 = H; R1 = R3 = OMe
(b)
R1
O
R2
R3
O
O
OH
R4
5-8
5 R1 = R2 = R3 = H; R4 = OH
6 R1 = R3 = H; R2 = R4 = OH
7 R2 = H; R1 = R3 = R4 = OH
8 R2 = R4 = H; R1 = R3 = OH

Scheme 1. Synthetic route to obtain the coumarin-chalcone hybrids. *Reagents and conditions:* (**a**) piperidine, EtOH, reflux, 2–6 h; (**b**) BBr_3, DCM, 80 °C, 48 h.

2.2. Pharmacology

Adenosine Receptor Binding Affinity Assays

The adenosine binding affinity of derivatives **1–8** for the human AR subtypes hA_1, hA_{2A} and hA_3, expressed in Chinese Hamster Ovary (CHO) cells, was determined in radioligand competition experiments [43,44]. In the binding affinity assay, it is measured the competition of ligands for specific binding of [^{3}H]CCPA (2-chloro-N^6-cyclopentyladenosine) to hA_1; specific binding of [^{3}H]NECA (5′-*N*-ethylcarboxamidoadenosine) to hA_{2A}; and specific binding of [^{3}H]HEMADO (2-(1-hexynyl)-N^6-methyladenosine) to hA_3. The results are expressed as K_i (dissociation constants), which were calculated with the program SCTFIT, and given as geometric means of at least three experiments, including 95% confidence intervals. Due to the lack of a suitable radioligand for the hA_{2B} receptor, the potency of antagonists at the hA_{2B} receptor was determined by inhibition of NECA-stimulated adenylyl cyclase activity with increasing concentrations of antagonist [43,44]. As a result, cAMP (cyclic adenosine monophosphate) production was inhibited in a concentration-dependent fashion, and K_i values were calculated from the measured IC_{50} values [45].

Derivatives **1–8** were efficiently synthesized and their in vitro binding affinity for human AR subtypes hA_1, hA_{2A}, hA_{2B} and hA_3, expressed in CHO cells, was evaluated. In the present communication, the studies were focused on the inspection of the effect on the binding affinity of different number and position of methoxy or hydroxy substituents on the 3-benzoylcoumarin scaffold. Data obtained for the binding affinity for hA_1 and hA_3 is summarized in Table 1. For all the tested compounds, no significant affinity was detected for the hA_{2A} ($Ki > 100$ μM) or hA_{2B} ($K_i > 10$ μM).

Table 1. Binding affinity (K_i) of compounds **1–8** on hA_1 and hA_3 AR.

Compound	hA_1 (K_i μM) [a]	hA_3 (K_i μM) [a]
1	>100	>100
2	>100	>100
3	>30 [b]	9.03 (6.28–13.0)
4	>100	2.49 (2.33–2.66)
5	39.5 (25.3–61.5)	34.5 (29.7–40.1)
6	54.0 (49.8–58.5)	>60 [b]
7	17.7 (16.0–19.5)	>30 [b]
8	29.1 (20.4–41.5)	>60 [b]
Theophylline	6.77 (4.07–11.3)	86.4 (73.6–101)

[a] Results are geometric means of 3 experiments and given with 95% confidence intervals (in parentheses). [b] At higher concentrations, the compounds precipitate.

The binding affinity results show that derivatives **1** and **2**, without substitutions on the coumarin scaffold or with a single methoxy group at the position 6 of the coumarin core, respectively, display no detectable binding affinity for the evaluated receptors ($K_i > 100$ μM). However, the presence of two methoxy groups at positions 5 and 7 (compounds **3** and **4**, respectively) lead to an increment on both the potency and selectivity for the hA_3. Compound **3**, presenting three methoxy groups at positions 5, 7 and 4′ proved to be hA_3 selective, displaying a $K_i = 9.03$ μM, whereas compound **4**, presenting an extra methoxy groups at position 3′ is not only selective for hA_3, but also displays a increase in potency ($K_i = 2.49$ μM). Compared to theophylline, classically used as a reference compound, we would like to highlight that both compounds **3** and **4** are more potent and hA_3 selective molecules.

Based on this data, it can be concluded that both nature and position of the substitution patterns on the coumarin–chalcone scaffold play a key role in the interaction with the hA_3. It can be highlighted that positions 5 and 7 of the studied scaffold seem to be relevant for the observed selectivity and potency. Analyzing the methoxylated derivatives **1–4**, only the molecules presenting substituents at these two positions (compounds **3** and **4**) are hA_3 active and selective ligands.

Interestingly, a similar tendency was observed for hA_1 binding of the hydroxylated derivatives (**5–8**), which bear hydroxy groups instead of methoxy groups at positions 5 and 7 (compounds **7** and **8**). Derivatives **7** and **8** display the highest potency and selectivity of the studied series towards hA_1, but their activity towards this receptor is still low with $K_i = 17.7$ μM and $K_i = 29.1$ μM, respectively.

2.3. Theoretical Evaluation of ADME Properties

In order to explore the drug-like properties of compounds **1–8**, the lipophilicity, expressed as the octanol/water partition coefficient and herein named clogP, as well as other theoretical calculations such as number of hydrogen acceptors and number of hydrogen bond donors, and topological polar surface area (TPSA), were calculated using the Molinspiration software [46]. Theoretical prediction of absorption, distribution, metabolism and excretion (ADME) properties of all derivatives is summarized in Table 2.

Table 2. Theoretical evaluation of the ADME properties of coumarin–chalcone hybrids.[a]

Compound	clogP	TPSA ($Å^2$)	*n*-OH Acceptors	*n*-OHNH Donors	Volume ($Å^3$)
1	3.04	65.75	5	0	270.07
2	3.08	74.98	6	0	295.62
3	3.06	84.22	7	0	321.16
4	3.47	74.98	6	0	295.16
5	2.43	87.74	5	2	235.01
6	1.93	107.97	6	3	243.03
7	1.63	128.20	7	4	251.05
8	2.12	107.97	6	3	243.03

[a] TPSA, topological polar surface area; n-OH, number of hydrogen acceptors; n-OHNH, number of hydrogen bond donors.

Based on this theoretical data, it can be concluded that the study molecules **1–8** do not violate any of Lipinski's rules (namely molecular weight, clogP, number of hydrogen donors and acceptors). In addition, TPSA, described as an indicator of membrane permeability, was favorable for the studied compounds.

2.4. Molecular Modeling

hA_1 and hA_3 homology models were successfully constructed (Materials and methods section). A selection of models obtained from Induce Fit calculations were tested based on their ability to discriminate between known ligands, decoys and between subtype-selective compounds. The models selected for the docking calculations showed excellent results in both tests. A dataset of 200 randomly selected decoys from the ZINC database [47] were mixed up with 22 known ligands of each adenosine receptor subtype [48] Glide SP precision was used to dock the database to the hA_1 and hA_3 models [49]. Table 3 presents the area under the receiver operating characteristic (ROC) curve (AUROC) for both systems. To differentiate between subtype-selective ligands, a second and more challenging test was performed. As in a previous study [48], 66 subtype-selective molecules (22 hA_1, 22 hA_{2A} and 22 hA_3 compounds) were docked to the hA_1 model (22 true positives vs. 44 false positives) and to the hA_3 (22 true positives vs. 44 false positives). Results corroborate those previously published by Katritch et al. [50] and proved that the developed homology models are able to discriminate between subtype-selective compounds (Table 3).

Table 3. Area under the ROC curve (AUROC) for the two homology models.

AUROC	hA_1	hA_3
test 1 [a]	0.91	0.95
test 2 [b]	0.86	0.82

[a] 22 hA_1 or 22 hA_3 ligands as true positives (TP) and 200 random decoys as false positives (FP) were considered.
[b] For hA_1, 22 hA_1 selective compounds as TP and 22 hA_{2A} + 22 hA_3 compounds as FP were considered. For hA_3, 22 hA_3 compounds as TP and 22 hA_{2A} + 22 hA_1 compounds as FP were considered.

Glide SP molecular docking simulations were run with our data using the hA_1 and hA_3 selected homology models as protein structures to detect the hypothetical binding mode of the new synthesized compounds [51]. The Prime module was used to optimize the protein structure for each binding mode [52]. Molecular docking simulations are represented in Figure 2.

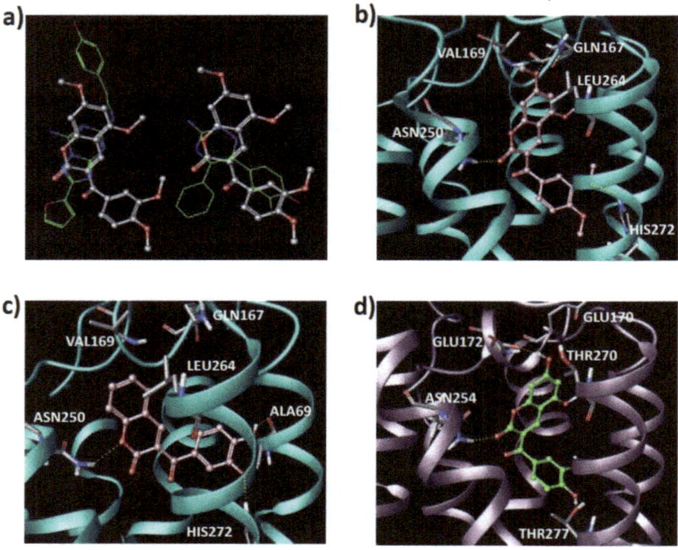

Figure 2. (**a**) Comparative study of the co-crystallized ligands (green carbons) in the hA_{2A} [3EML (left) and 3UZC (right)] with the pose of compound **3** extracted from the hA_3 docking calculations (grey carbons). Binding pockets in hA_{2A} and hA_3 were superposed. (**b**) Pose extracted for compound **3** inside the hA_3 after docking. Hydrogen bonds are represented in yellow color. (**c**) Hypothetical binding mode for compound **5** (pink carbons) in the hA_3. (**d**) Pose obtained through docking simulations for compound **7** (green carbons) in the hA_1 protein pocket.

Docking calculations and the established homology models for the hA_1 and hA_3 identified the hypothetical binding mode and rationalized the interaction of these derivatives with their respective ARs binding sites.

The calculations showed a high level of variability since all the synthetized derivatives yielded different possible binding modes inside the pockets. Selection of the hypothetical binding pose was accomplished considering the number of similar poses extracted from the simulations and geometrical correspondence to crystallized ligands in the hA_{2A} (Figure 2a).

Docking results disclosed important data about the binding mode: the oxygens presented in the benzopyrone system are oriented towards the Asn250 residue and the benzoyl moiety was buried in the hA_3 pocket. This hypothetical binding mode corroborates the conformations shown by the co-crystallized ligands in the hA_{2A} (PDB: 3EML and 3UZC) [48,53] (Figure 2a,b). The pose of compound **3** produced effective hydrogen bonds with Gln167, Asn250 and His272 residues.

Interestingly, when methoxy substituents were demethylated and changed into hydroxy equivalents (compounds **5–8**) a modification in the profile of the studied derivatives was noticed: a loss of affinity for hA_3 and a tendency for interaction with hA_1. The only compound that discloses some affinity for both receptors was compound **5** (hA_1 K_i = 39.5 and hA_3 K_i = 34.5 μM), which presents a catechol at positions 3′ and 4′ and no substitutions in the coumarin fragment. The hypothetical binding mode for compound **5** in the hA_3 pocket is represented in Figure 2c. The compound can establish hydrogen bonds with Ala69, Asn250 and His272 residues. As observed in the hA_{2A} crystallized structure and previously published studies [54,55], the corresponding Asn250 residue seems to play an important role in ligand recognition. The compound **5** pose inside the hA_1 pocket is likewise the described pose in the hA_3 one. However, the position was slightly shifted, and calculations were not able to retrieve a hydrogen bond with the Asn250 residue. The introduction of an additional hydroxy group at position 6 of the coumarin scaffold (compound **6**), resulted in a loss of measurable hA_3 binding affinity. The most noticeable binding affinities were found for derivatives with hydroxy substitutions at positions 5 and 7 of the coumarin core, as stated for methoxy equivalents. Thereby, compound **7**, with the same substitution pattern as quercetin (Figure 1), that is, hydroxy groups at positions 5, 7, 3′ and 4′, displays hA_1 selectivity, and the best binding affinity (K_i = 17.7 μM). Compound **8**, with the same substitution pattern as genistein (Figure 1, hydroxy substituents at positions 5, 7 and 4′) shows a similar hA_1 selectivity (K_i = 29.1 μM). The pose obtained through docking calculations for compound **7** in the hA_1 protein pocket showed the possibility of establishment of hydrogen bonds with Glu172, Asn254 and Thr277 residues (Figure 2d).

Moreover, we calculated the interaction energy contributions of the residues in hA_3 and hA_1 pockets with compounds **3** and **7**, respectively (Figure 3). The sum of different individual contributions, such as Coulomb, *van der Waals* and hydrogen bond energies, was taken into account in the calculation of the interaction energies for each residue.

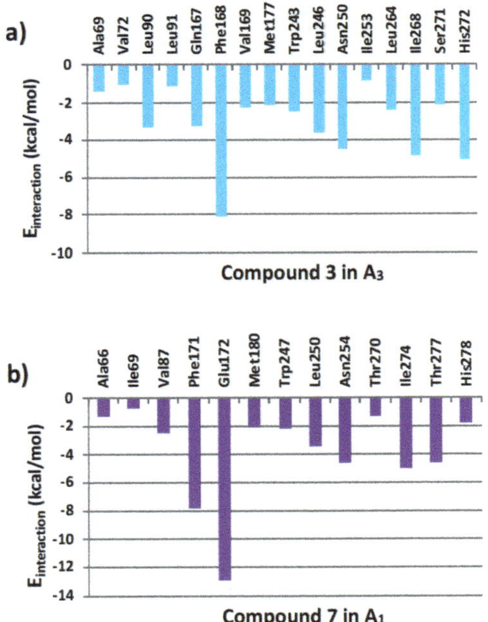

Figure 3. Interaction energy contribution (sum of Coulomb, *van der Waals* and hydrogen bond energies) between the residues in the (**a**) hA_3 and (**b**) hA_1 and the respective derivatives **3** and **7** (residues in a distance of 3 Å from the ligand).

In addition, Figure 4 shows the molecular surface around the two residues in the hA_1 and hA_3 that could be responsible for the observed selectivity.

Regarding the interaction energy contributions (Figure 4), calculations showed that the molecular surface around the two residues in the hA_1 and hA_3 could be responsible for the observed selectivity. Phe168, Asn250, Ile268 and His272 are important residues in the interaction between compound **3** and the hA_3. Residues with important contributions in the stabilization of compound **7** inside the hA_1 are Phe171, Gln172, Asn254, Ile274 and Thr277.

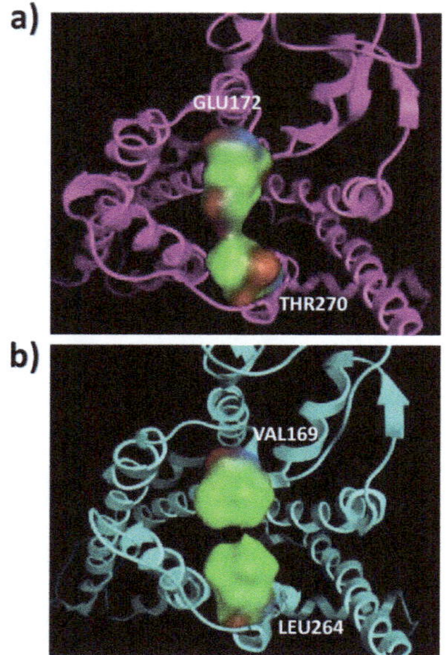

Figure 4. Molecular surface showing favored interaction areas generated in the (**a**) hA_1 and (**b**) hA_3. Red color represents hydrogen-bond areas, green color shows hydrophobic areas, and blue represents mildly polar interfaces. Protein structures are viewed from the extracellular side.

There are different residues in both hA_1 and hA_3 with different hydrophobic/hydrophilic characteristics, which may be important to understand the observed selectivity. Hydrophobic residues in the hA_3, such as Val169 and Leu264, could establish hydrophobic interactions and contribute towards stabilizing the ligand when the derivatives present hydrophobic substituents, like methoxy groups (i.e., **3** and **4**) (Figure 4). However, in the case of hA_1, the corresponding residues are Glu172 and Thr270. They have hydrophilic characteristics and so can stabilize the binding of derivatives with polar substituents, such as the hybrids with hydroxy groups (compounds **6–8**). Yet, compound **5**, with no substituents in the coumarin ring, can be stabilized in the pocket of both proteins.

3. Materials and Methods

3.1. Chemistry

3.1.1. General Methods

Starting materials and reagents were obtained from commercial suppliers and were used without purification. Melting points (mp) were determined using a Reichert Kofler thermopan or in capillary tubes on a Büchi 510 (Flawil, Switzerland) apparatus and were uncorrected. ^{1}H-NMR (300 MHz) and ^{13}C-NMR (75.4 MHz) spectra were recorded with a Bruker AMX spectrometer (Bruker Daltonics Inc., Fremont, CA, USA) using DMSO-d_6 or $CDCl_3$ as solvent. Chemical shifts (δ) are expressed in parts per million (ppm) using TMS as an internal standard. Coupling constants *J* are expressed in hertz (Hz). Spin multiplicities are given as s (singlet), bs (broad singlet), d (doublet), dd (doublet of doublets) and m (multiplet). Mass spectrometry was carried out with a Kratos MS-50 or a Varian MAT-711 spectrometer (Thermo Fisher Scientific, Waltham, MA, USA). Elemental analyses were performed by a Perkin–Elmer 240B microanalyzer (Thermo Fisher Scientific, Waltham, MA, USA) and were within ±0.4% of the calculated values in all cases. The analytical results were ≥95% purity for all compounds. Flash Chromatography (FC) was performed on silica gel (Merck 60, 230–400 mesh,

Kenilworth, NJ, USA) and analytical TLC on precoated silica gel plates (Merck 60 F254, Kenilworth, NJ, USA). Organic solutions were dried over anhydrous sodium sulfate. Concentration and evaporation of the solvent after reaction or extraction was carried out on a Büchi rotavapor (BÜCHI Labortechnik AG, Switzerland) operating at reduced pressure. The purity of compounds was assessed by high performance liquid chromatography (HPLC) coupled at diode array detector (DAD) on a Thermo Quest Spectrasystem (Thermo Fisher Scientific, Waltham, MA, USA) equipped with a P4000 pump, an UV6000 UV-Vis diode array detector, and a SN4000 interface to be operated via a personal computer. The instrument software ChromQuest 5.0 (Thermo Fisher Scientific, Waltham, MA, USA) was used for data acquisition. Different analytical columns and mobile phases (all solvents were HPLC grade) were tested. The mobile phase was H_2O:CH_3CN = 70:30 and an Eclipse xdb C18 column (5 µm particle size, 0.46 mm i.d., 25 cm length; Agilent Technologies, CA, USA) was used. The purity of the compounds was found to be higher than 95%.

3.1.2. Synthetic Protocol to Obtain the Methoxy-3-benzoylcoumarins **1–4**

To a solution of the appropriate β-ketoester (1 mmol) and the corresponding salicylaldehyde (1 mmol) in ethanol (5 mL) piperidine in catalytic amount (0.10 mL) was added. The reaction mixture was refluxed for 2–6 h and, after completion (followed by TLC), the reaction was cooled, and the precipitate was filtered and washed with cold ethanol and ether. The obtained solid was recrystallized in DCM to afford the corresponding methoxy-3-benzoylcoumarin compounds.

3.1.3. Synthetic Protocol to Obtain the Hydroxy-3-benzoylcoumarins **5–8**

In a Schlenk tube, the appropriate methoxy derivative compound **1–4** (1 mmol) was dissolved in DCM (1 mL), and BBr_3 (20 mmol, 1M) was added dropwise. The tube was sealed, and the reaction mixture was heated at 80 °C for 48 h. The resulting crude product was treated with MeOH and rotated to dryness. The obtained crude solid was recrystallized in MeOH or purified by flash chromatography using hexane/ethyl acetate mixtures as eluent, to afford the desired hydroxy derivatives.

3-(3′,4′-Dimethoxybenzoyl)coumarin (**1**): 85% yield; white solid; mp 190–191 °C; ^{1}H-NMR (300 MHz, $CDCl_3$) δ ppm 8.01 (s, 1H, H-4), 7.71–7.52 (m, 3H, 3x Ar-H), 7.50–7.29 (m, 3H, 3x Ar-H), 6.87 (d, *J* = 8.4 Hz, 1H, H-5′), 3.95 (s, 6H, 2x OCH_3); ^{13}C-NMR (75 MHz, $CDCl_3$) δ ppm 190.3, 154.8, 154.4, 149.5, 144.6, 133.6, 129.3, 129.2, 127.8, 125.7, 125.2, 118.5, 117.2, 111.2, 110.2, 56.4, 56.3; EI-MS *m/z* (%): 311 ([M + 1]$^+$, 59), 310 (M$^+$, 100), 173 (41), 166 (25), 165 (99), 79 (22), 77 (22); Anal. Calcd. For $C_{18}H_{14}O_5$: C 69.67, H 4.55. Found: C 69.69, H 4.58.

6-Methoxy-3-(3′,4′-dimethoxybenzoyl)coumarin (**2**): 97% yield; beige solid; mp 202–203 °C; ^{1}H-NMR (300 MHz, $CDCl_3$) δ ppm 7.78 (s, 1H, H-4), 7.38 (d, *J* = 1.9 Hz, 1H, H-2′), 7.26 (dd, *J* = 8.4, 2.0 Hz, 1H, H-6′), 7.16 (d, *J* = 9.1 Hz, 1H, H-8), 7.04 (dd, *J* = 9.1, 2.9 Hz, 1H, H-7), 6.82 (d, *J* = 2.9 Hz, 1H, H-5), 6.70 (d, *J* = 8.4 Hz, 1H, H-5′), 3.78 (s, 6H, 2x OCH_3), 3.69 (s, 3H, OCH_3); ^{13}C-NMR (75 MHz, $CDCl_3$) δ ppm 190.4, 156.6, 154.4, 149.5, 149.3, 144.4, 129.4, 128.0, 125.7, 121.6, 118.8, 118.2, 111.2, 110.8, 110.2, 56.4, 56.3, 56.1; EI-MS *m/z* (%): 341 ([M + 1]$^+$, 58), 340 ([M]$^+$, 94), 165 (100), 77 (22); Anal. Calcd. For $C_{19}H_{16}O_6$: C 67.05, H 4.74. Found: C 67.09, H 4.75.

5,7-Dimethoxy-3-(3′,4′-dimethoxybenzoyl)coumarin (**3**): 91% yield; pale yellow solid; mp 210–211 °C; ^{1}H-NMR (300 MHz, $CDCl_3$) δ ppm 8.19 (s, 1H, H-4), 7.34 (d, *J* = 1.9 Hz, 1H, H-2′), 7.26 (dd, *J* = 8.4, 2.0 Hz, 1H, H-6′), 6.70 (d, *J* = 8.4 Hz, 1H, H-5′), 6.29 (d, *J* = 2.0 Hz, 1H, H-6), 6.14 (d, *J* = 2.0 Hz, 1H, H-8), 3.77 (bs, 6H, 2x OCH_3), 3.72 (bs, 6H, 2x OCH_3); ^{13}C-NMR (75 MHz, $CDCl_3$) δ ppm 190.9, 165.8, 159.4, 158.4, 158.0, 153.9, 149.3, 141.5, 130.0, 125.4, 121.3, 111.5, 110.1, 103.9, 95.4, 93.0, 56.3; EI-MS *m/z* (%): 371 ([M + 1]$^+$, 24), 370 (M$^+$, 100), 339 (21), 233 (30), 165 (63); Anal. Calcd. For $C_{20}H_{18}O_7$: C 64.86, H 4.90. Found: C 64.88, H 4.93.

5,7-Dimethoxy-3-(4′-methoxybenzoyl)coumarin (**4**): 97% yield; pale yellow solid; mp 174–175 °C; ^{1}H-NMR (300 MHz, $CDCl_3$) δ ppm 8.21 (s, 1H, H-4), 7.69 (d, *J* = 8.8 Hz, 2H, H-2′, H-6′), 6.77 (d, *J* = 8.8 Hz,

2H, H-3′, H-5′), 6.29 (d, J = 2.2 Hz, 1H, H-6), 6.13 (d, J = 2.2 Hz, 1H, H-8), 3.72 (2s, 3H + 3H, 2x OCH_3), 3.70 (s, 3H, OCH_3); ^{13}C-NMR (75 MHz, $CDCl_3$) δ ppm 190.7, 165.6, 163.8, 159.1, 158.2, 157.8, 141.5, 132.1, 129.8, 121.2, 113.7, 103.8, 95.2, 92.8, 56.1, 56.0, 55.5; EI-MS m/z (%): 341 ($[M + 1]^+$, 33), 340 (M^+, 88), 325 (28) 312 (30), 309 (45), 297 (20), 233 (48), 135 (100), 92 (27), 77 (38). Anal. Calcd. For $C_{19}H_{16}O_6$: C 67.05, H 4.74. Found: C 67.08, H 4.76.

5,7-Dihydroxy-3-(4′-hydroxybenzoyl)coumarin (**8**): 88% yield; pale green solid; mp 290–292 °C; ^{1}H-NMR (300 MHz, DMSO-d_6) δ ppm 11.10 (s, 1H), 10.85 (s, 1H), 10.53 (s, 1H), 8.09 (d, J = 1.4 Hz, 1H), 7.67 (d, J = 8.6 Hz, 2H), 6.80 (d, J = 8.7 Hz, 2H), 6.25 (d, J = 2.0 Hz, 1H), 6.22 (d, J = 1.8 Hz, 1H); ^{13}C-NMR (75 MHz, DMSO-d_6) δ ppm 190.4, 164.2, 162.4, 158.8, 157.5, 157.2, 141.1, 132.3, 128.3, 119.0, 115.3, 101.5, 98.5, 94.3. EI-MS m/z (%): 299 ($[M + 1]^+$, 9), 298 (M^+, 31), 283 (16), 218 (20), 121 (100), 93 (26), 65 (27). Anal. Calcd. For $C_{16}H_{10}O_6$: C 64.43, H 3.38. Found: C 64.39, H 3.37.

3.2. Biological Assays

3.2.1. Binding Affinity Assays

The binding affinity for hA_1, hA_{2A}, hA_3 of the synthetized compounds was evaluated using radioligand competition experiments in CHO cells that were stably transfected with the individual receptor subtypes [44,45]. The radioligands used were 1 nM [^{3}H]CCPA for hA_1 (K_D = 0.61 nM), 10 nM [^{3}H]NECA for hA_{2A} (K_D = 20.1 nM), and 1 nM [^{3}H]HEMADO for hA_3 (K_D = 1.2 nM) receptors. Due to the lack of a suitable radioligand for the hA_{2B} receptor, the potency of antagonists at the hA_{2B} receptor (expressed on CHO cells) was determined by inhibition of NECA-stimulated adenylyl cyclase activity [44,45]. The IC_{50} for inhibition of cAMP (cyclic adenosine monophosphate) production was determined and converted to K_i values using the Cheng and Prusoff equation [56]. For all the tested compounds, no measurable activity for the hA_{2B} (K_i > 10 μM) was detected.

3.2.2. Statistical Methods

K_i values (dissociation constants) were determined in radioligand competition experiments with 7–8 different concentrations of test compound and each concentration was tested in duplicate. K_i values are given as geometric means of three independent experiments with 95% confidence intervals. The program Prism 6 (GraphPad Software) was used for the analysis of the competition curves.

3.3. Theoretical Evaluation of ADME Properties

cLogP was calculated by the methodology developed by Molinspiration as a sum of fragment-based contributions and correction factors. Topological Polar Surface Area (TPSA) was calculated based on the methodology published by Ertl et al. as a sum of fragment contributions [57]. Oxygen- and nitrogen-centered polar fragments are considered. TPSA has been shown to be a very good descriptor characterizing drug absorption, including intestinal absorption, bioavailability, Caco-2 permeability and blood–brain barrier penetration. The method for calculation of molecule volume developed at Molinspiration is based on group contributions. These have been obtained by fitting the sum of fragment contributions to "real" 3D volume for a training set of about twelve thousand, mostly drug-like molecules. Three-dimensional molecular geometries for a training set were fully optimized by the semiempirical AM1 method.

3.4. Molecular Modeling

Homology modeling was carried out using the Molecular Operating Environment (MOE) suite [49]. Molecular docking simulations were performed with the Schrodinger package [51,52].

3.4.1. Homology Models of hA_1 and hA_3

Homology models of the $hA1$ and hA_3 were constructed. The crystallized structure of the hA_{2A} receptor (PDB: 3EML) was used as a template [48]. Protein sequence alignment of the 3 receptors (hA_1, hA_{2A} and hA_3) used to generate the homology models was performed as previously described by Katritch et al. [50]. The alignment was made considering the highly conserved residues in the different TM helices. MOE software was used to generate the homology models [49]. Protein geometry was evaluated for the models taking into account Phi–Psi plots, rotamers, bond angles, bond lengths, atom clashes, dihedrals and contact energies. The presence of different conserved disulfide bridges was manually checked, such as the bridge between the corresponding Cys77 and Cys166 residues in the hA_{2A}. Induce Fit Docking Workflow in the Schrodinger package was used to optimize the final models [58]. Selective high affinity ligands (compounds coll_11 and jaco_mre3008_f20) [50] were used to adapt the protein pocket for the hA_1 and hA_3, respectively. This procedure involved three steps: 1) Glide-based docking of the ligands using SP mode (standard-precision); 2) Protein pocket optimization using Prime and considering the residues within 5Å from the ligand poses; 3) Glide-based docking of the ligands in the refined pocket using XP mode (Extra-precision). As previously described [50], homology models were tested for their capability to discriminate ligands from decoys and between known subtype-selective compounds. ROC curves were performed, and the best models were selected for further molecular docking studies.

3.4.2. Molecular Docking of hA_1 and hA_3 ARs

Molecular docking studies using the hA_1 and hA_3 homology models, selected in the previous step, were carried out. Compounds were docked using Glide SP mode [52]. Ten poses for each ligand were collected and optimized using MM-GBSA in Prime [53], taking into account a flexible protein region defined by 5 Å from the ligand. Final binding modes were selected, taking into account the number of similar poses extracted from the calculations and geometrical correspondence to co-crystallized ligands in the hA_{2A}.

4. Conclusions

The current study was focused on the synthesis and the evaluation of binding affinity towards the four subtypes of human ARs of a selected series of methoxy and hydroxy coumarin–chalcone hybrids. Structure–activity relationship (SAR) studies of the new molecules highlighted that, in general, methoxy substitutions, as in the example of compounds **3** and **4,** allow a superior hA_3 binding affinity and selectivity, whereas the hydroxy substitutions, as in the example of compounds **5–8**, allow a modest hA_1 binding affinity. Substitutions at positions 5 and 7 of the coumarin scaffold proved to be essential for the potency and selectivity in both series of compounds. Compound **4**, a methoxy derivative, and compound **7**, a hydroxy derivative, proved to be the most potent compounds of the studied series, displaying a hA_3 K_i = 2.49 µM and a hA_1 K_i = 17.7 µM, respectively. Docking calculations allow an understanding the binding preference of the studied molecules. Finally, the theoretical values for the ADME properties show that all the coumarin–chalcone hybrids **1–8** do not break any of Lipinski's rules, being promising scaffolds for further compound optimization.

Author Contributions: Conceptualization, S.V.-R. and M.J.M.; methodology, S.V.-R., S.V. and S.K.; software, S.V.; validation, K.-N.K., F.B. and E.U.; formal analysis, S.V.-R., S.K. and S.V.; investigation, S.V.-R., S.K., S.V. and M.J.M.; resources, S.V.-R., S.K., S.V., K.-N.K., E.U., F.B. and M.J.M.; data curation, S.V.-R., S.K., S.V.; writing—original draft preparation and editing, S.V.-R.; writing—review and editing, M.J.M.; visualization, K.-N.K.; supervision, F.B. and E.U.; project administration, S.V.-R. and M.J.M.; funding acquisition, S.V.-R., S.K., S.V., K.-N.K., E.U., F.B. and M.J.M. All authors have read and agreed to the published version of the manuscript.

Funding: This research was funded by Xunta de Galicia (Galician Plan of Research, Innovation and Growth 2011–2015, Plan I2C, ED481B 2014/027-0, ED481B 2014/086–0 and ED481B 2018/007), Angeles Alvariño program from Xunta de Galicia, European Social Fund (ESF) and Fundação para a Ciência e Tecnologia (FCT, CEECIND/02423/2018 and UIDB/00081/2020).

Acknowledgments: The authors would like to thank Lourdes Santana for her scientific support. The authors would like to thank the use of RIAIDT-USC analytical facilities.

Conflicts of Interest: The authors declare no conflict of interest.

References

1. Fredholm, B.B.; IJzerman, A.P.; Jacobson, K.A.; Linden, J.; Müller, C.E. International union of basic and clinical pharmacology. LXXXI. Nomenclature and classification of adenosine receptors-an update. *Pharmacol. Rev.* **2011**, *63*, 1–34. [CrossRef] [PubMed]
2. Poulsen, S.A.; Quinn, R.J. Adenosine receptors: New opportunities for future drugs. *Bioorg. Med. Chem.* **1998**, *6*, 619–641. [CrossRef]
3. Fredholm, B.B.; IJzerman, A.P.; Jacobson, K.A.; Klotz, K.N.; Linden, J. International union of pharmacology. XXV. Nomenclature and classification of adenosine receptors. *Pharmacol. Rev.* **2001**, *5*, 527–552.
4. Fredholm, B.B.; Arslan, G.; Halldner, L.; Kull, B.; Schulte, G.; Wasserman, W. Structure and function of adenosine receptors and their genes. *Arch. Pharmacol.* **2000**, *362*, 364–374. [CrossRef] [PubMed]
5. Draper-Joyce, C.J.; Khoshouei, M.; Thal, D.M.; Liang, Y.-L.; Nguyen, A.T.N.; Furness, S.G.B.; Venugopal, H.; Baltos, J.-A.; Plitzko, J.M.; Danev, R.; et al. Structure of the adenosine-bound human adenosine A1 receptor-Gi complex. *Nature* **2018**, *558*, 559–563. [CrossRef] [PubMed]
6. Zhou, Q.Y.; Li, C.; Olah, M.E.; Johnson, R.A.; Stiles, G.L.; Civelli, O. Molecular cloning and characterization of an adenosine receptor: The A3 adenosine receptor. *Proc. Natl. Acad. Sci. USA* **1992**, *89*, 7432–7436. [CrossRef]
7. Salvatore, C.A.; Jacobson, M.A.; Taylor, H.E.; Linden, J.; Johnson, R.G. Molecular cloning and characterization of the human A_3 adenosine receptor. *Proc. Natl. Acad. Sci. USA* **1993**, *90*, 10365–10369. [CrossRef]
8. Ramkumar, V.; Stiles, G.L.; Beaven, M.A.; Ali, H. The A_3 adenosine receptor is the unique adenosine receptor which facilitates release of allergic mediators in mast cells. *J. Biol. Chem.* **1993**, *268*, 16887–16890.
9. Linden, J. Cloned adenosine A_3 receptors: Pharmacological properties, species differences and receptor functions. *Trends Pharmacol. Sci.* **1994**, *15*, 298–306. [CrossRef]
10. Hannon, J.P.; Pfannkuche, H.J.; Fozard, J.R. A role for mast cells in adenosine A_3 receptor-mediated hypotension in the rat. *J. Pharmacol.* **1995**, *115*, 945–952. [CrossRef]
11. Jacobson, K.A.; Nikodijević, O.; Shi, D.; Gallo-Rodriguez, C.; Olah, M.E.; Stiles, G.L.; Daly, J.W. A role for central A_3-adenosine receptors. Mediation of behavioral depressant effects. *FEBS Lett.* **1993**, *336*, 57–60. [CrossRef]
12. Vecchio, E.A.; Baltos, J.A.; Nguyen, A.T.N.; Christopoulos, A.; White, P.J.; May, L.T. New paradigms in adenosine receptor pharmacology: Allostery, oligomerization and biased agonism. *Br. J. Pharmacol.* **2018**, *175*, 4036–4046. [CrossRef] [PubMed]
13. Kim, Y.; de Castro, S.; Gao, Z.G.; IJzerman, A.P.; Jacobson, K.A. Novel 2- and 4-substituted 1*H*-imidazo [4,5-*c*]quinolin-4-amine derivatives as allosteric modulators of the A_3 adenosine receptor. *J. Med. Chem.* **2009**, *52*, 2098–2108. [CrossRef] [PubMed]
14. Lopes, L.V.; Sebastião, A.M.; Ribeiro, J.A. Adenosine and related drugs in brain diseases: Present and future in clinical trials. *Curr. Top. Med. Chem.* **2011**, *11*, 1087–1101. [CrossRef]
15. Jespers, W.; Schiedel, A.C.; Heitman, L.H.; Cooke, R.M.; Kleene, L.; van Westen, G.J.P.; Gloriam, D.E.; Müller, C.E.; Sotelo, E.; Gutiérrez-de-Terán, H. Structural mapping of adenosine receptor mutations: Ligand binding and signaling mechanisms. *Trends Pharmacol. Sci.* **2018**, *39*, 75–89. [CrossRef]
16. Shah, R.H.; Frishmanet, W.H. Adenosine1 receptor antagonism: A new therapeutic approach for the treatment of decompensated heart failure. *Cardiol. Rev.* **2009**, *17*, 125–131. [CrossRef]
17. Shook, B.; Rassnick, S.; Wallace, N.; Crooke, J.; Ault, M.; Chakravarty, D.; Barbay, J.K.; Wang, A.; Powell, M.T.; Leonard, K.; et al. Design and characterization of optimized adenosine A_{2A}/A_1 receptor antagonists for the treatment of Parkinson's disease. *J. Med. Chem.* **2012**, *55*, 1402–1417. [CrossRef]
18. Liu, Y.J.; Chen, J.; Li, X.; Zhou, X.; Hu, Y.M.; Chu, S.F.; Peng, Y.; Chen, N.H. Research progress on adenosine in central nervous system diseases. *CNS Neurosci. Ther.* **2019**, *25*, 899–910. [CrossRef]
19. Stone, T.W.; Ceruti, S.; Abbracchio, M.P. Adenosine receptors and neurological disease: Neuroprotection and neurodegeneration. *Handb. Exp. Pharmacol.* **2009**, *193*, 535–587.
20. Wilcox, C.S.; Welch, W.J.; Schreiner, G.F.; Belardinelli, L. Natriuretic and diuretic actions of a highly selective adenosine receptor antagonist. *J. Am. Soc. Nephrol.* **1999**, *10*, 714–720.

21. Gottlieb, S.S.; Skettino, S.L.; Wolff, A.; Beckman, E.; Fisher, M.L.; Freudenberger, R.; Gladwell, T.; Marshall, J.; Cines, M.; Bennett, D.; et al. Effects of BG9719 (CVT-124), an A_1 adenosine receptor antagonist, and Furosemide on glomerular filtration rate and natriuresis in patients with congestive heart failure. *J. Am. Coll. Cardiol.* **2000**, *35*, 56–59. [CrossRef]
22. de Mendonça, A.; Sebastião, A.M.; Ribeiro, J.A. Adenosine: Does it have a neuroprotective role after all? *Brain Res. Rev.* **2000**, *33*, 258–274. [CrossRef]
23. Lee, H.T.; Ota-Setlik, A.; Xu, H.; D'Agati, V.D.; Jacobson, M.A.; Emala, C.W. A_3 adenosine receptor knockout mice are protected against ischemia- and myoglobinuria-induced renal failure. *Am. J. Physiol. Renal Physiol.* **2003**, *284*, 267–273. [CrossRef]
24. Kolahdouzan, M.; Hamadeh, M.J. The neuroprotective effects of caffeine in neurodegenerative diseases. *CNS Neurosci. Ther.* **2017**, *23*, 272–290. [CrossRef]
25. Pugliese, A.M.; Coppi, E.; Spalluto, G.; Corradetti, R.; Pedata, F. A_3 adenosine receptor antagonists delay irreversible synaptic failure caused by oxygen and glucose deprivation in the rat CA_1 hippocampus in vitro. *Br. J. Pharmacol.* **2006**, *147*, 524–532. [CrossRef]
26. Fishman, P.; Bar-Yehuda, S.; Liang, B.T.; Jacobson, K.A. Pharmacological and therapeutic effects of A_3 adenosine receptor (A_3AR) agonists. *Drug Discov. Today* **2012**, *17*, 359–366. [CrossRef]
27. Chen, Z.; Janes, K.; Chen, C.; Doyle, T.; Tosh, D.K.; Jacobson, K.A.; Salvemini, D. Controlling murine and rat chronic pain through A_3 adenosine receptor activation. *FASEB J.* **2012**, *26*, 1855–1865. [CrossRef]
28. Baraldi, P.G.; Tabrizi, M.A.; Gessi, S.; Borea, P.A. Adenosine receptor antagonists: Translating medicinal chemistry and pharmacology into clinical utility. *Chem. Rev.* **2008**, *108*, 238–263. [CrossRef]
29. Moro, S.; Gao, Z.G.; Jacobson, K.A.; Spalluto, G. Progress in the pursuit of therapeutic adenosine receptor antagonists. *Med. Res. Rev.* **2006**, *26*, 131–159. [CrossRef]
30. Okajima, F.; Akbar, M.; Abdul Majid, M.; Sho, K.; Tomura, H.; Kondo, Y. Genistein, an inhibitor of protein tyrosine kinase, is also a competitive antagonist for P1-purinergic (adenosine) receptor in FRTL-5 thyroid cells. *Biochem. Biophys. Res. Commun.* **1994**, *203*, 1488–1495. [CrossRef]
31. Jacobson, K.A.; Moro, S.; Manthey, J.A.; West, P.L.; Ji, X.D. Interactions of flavones and other phytochemicals with adenosine receptors. *Adv. Exp. Med. Biol.* **2002**, *505*, 163–171.
32. Borges, F.; Roleira, F.; Milhazes, N.; Santana, L.; Uriarte, E. Simple coumarins and analogues in medicinal chemistry: Occurrence, synthesis and biological activity. *Curr. Med. Chem.* **2005**, *12*, 887–916. [CrossRef]
33. Go, M.L.; Wu, X.; Liu, X.L. Chalcones: An update on cytotoxic and chemoprotective properties. *Curr. Med. Chem.* **2005**, *12*, 483–499. [CrossRef]
34. Nowakowska, Z. A review of anti-infective and anti-inflammatory chalcones. *Eur. J. Med. Chem.* **2007**, *42*, 125–137. [CrossRef]
35. Langmead, C.J.; Andrews, S.P.; Congreve, M.; Errey, J.C.; Hurrell, E.; Marshall, F.H.; Mason, J.S.; Richardson, C.M.; Robertson, N.; Zhukov, A.; et al. Identification of novel adenosine $A_{(2A)}$ receptor antagonists by virtual screening. *J. Med. Chem.* **2012**, *55*, 1904–1909. [CrossRef]
36. Vazquez-Rodriguez, S.; Matos, M.J.; Santana, L.; Uriarte, E.; Borges, F.; Kachler, S.; Klotz, K.N. Chalcone-based derivatives as new scaffolds for hA_3 adenosine receptor antagonists. *J. Pharm. Pharmacol.* **2013**, *65*, 607–703. [CrossRef]
37. Gaspar, A.; Reis, J.; Matos, M.J.; Uriarte, E.; Borges, F. In search for new chemical entities as adenosine receptor ligands: Development of agents based on benzo-γ-pyrone skeleton. *Eur. J. Med. Chem.* **2012**, *54*, 914–918. [CrossRef]
38. Matos, M.J.; Vilar, S.; Vazquez-Rodriguez, S.; Kachler, S.; Klotz, K.N.; Buccioni, M.; Delogu, G.; Santana, L.; Uriarte, E.; Borges, F. Structure-based optimization of coumarin hA_3 adenosine receptor antagonists. *J. Med. Chem.* **2020**, *63*, 2577–2587. [CrossRef]
39. Matos, M.J.; Vilar, S.; Kachler, S.; Fonseca, A.; Santana, L.; Uriarte, E.; Borges, F.; Tatonetti, N.P.; Klotz, K.N. Insight into the interactions between novel coumarin derivatives and human A_3 adenosine receptors. *ChemMedChem* **2014**, *9*, 2245–2253. [CrossRef]
40. Matos, M.J.; Vilar, S.; Kachler, S.; Celeiro, M.; Vazquez-Rodriguez, S.; Santana, L.; Uriarte, E.; Hripcsak, G.; Borges, F.; Klotz, K.N. Development of novel adenosine receptor ligands based on the 3-amidocoumarin scaffold. *Bioorg. Chem.* **2015**, *61*, 1–6. [CrossRef]

41. Matos, M.J.; Vilar, S.; Kachler, S.; Vazquez-Rodriguez, S.; Varela, C.; Delogu, G.; Hripcsak, G.; Santana, L.; Uriarte, E.; Klotz, K.N.; et al. Progress in the development of small molecules as new human A_3 adenosine receptor ligands based on the 3-thiophenylcoumarin core. *MedChemComm* **2016**, *7*, 845–852. [CrossRef]
42. Fonseca, A.; Matos, M.J.; Vilar, S.; Kachler, S.; Klotz, K.N.; Uriarte, E.; Borges, F. Coumarins and adenosine receptors: New perceptions in. structure–affinity relationships. *Chem. Biol. Drug Des.* **2018**, *91*, 245–256. [CrossRef]
43. Vazquez-Rodriguez, S.; Figueroa-Guiñez, R.; Matos, M.J.; Santana, L.; Uriarte, E.; Lapier, M.; Maya, J.D.; Olea-Azar, C. Synthesis of coumarin–chalcone hybrids and evaluation of their antioxidant and trypanocidal properties. *Med. Chem. Commun.* **2013**, *4*, 993–1000. [CrossRef]
44. Klotz, K.N.; Hessling, J.; Hegler, J.; Owman, C.; Kull, B.; Fredholm, B.B.; Lohse, M.J. Comparative pharmacology of human adenosine receptor subtypes-characterization of stably transfected receptors in CHO cells. *Arch. Pharmacol.* **1997**, *357*, 1–9. [CrossRef]
45. Klotz, K.N.; Falgner, N.; Kachler, S.; Lambertucci, C.; Vittori, S.; Volpini, R.; Cristalli, G. [3H]HEMADO–a novel tritiated agonist selective for the human adenosine A_3 receptor. *Eur. J. Pharmacol.* **2007**, *556*, 14–18. [CrossRef]
46. M cheminformatics, Bratislava, Slovak Republic. 2009. Available online: http://www.molinspiration.com/services/properties.html (accessed on 19 September 2020).
47. Irwin, J.J.; Shoichet, B.K. ZINC–a free database of commercially available compounds for virtual screening. *J. Chem. Inf. Model.* **2005**, *45*, 177–182. [CrossRef]
48. Jaakola, V.P.; Griffith, M.T.; Hanson, M.A.; Cherezov, V.; Chien, E.Y.; Lane, J.R.; Ijzerman, A.P.; Stevens, R.C. The 2.6 angstrom crystal structure of a human A_{2A} adenosine receptor bound to an antagonist. *Science* **2008**, *322*, 1211–1217. [CrossRef]
49. *MOE, Version 2011.10*; Chemical Computing Group, Inc.: Montreal, QC, Canada, 2011.
50. Katritch, V.; Kufareva, I.; Abagyan, R. Structure based prediction of subtype-selectivity for adenosine receptor antagonists. *Neuropharmacology* **2011**, *60*, 108–115. [CrossRef]
51. *Glide, Version 5.7*; Schrödinger, LLC: New York, NY, USA, 2011.
52. *Prime, Version 3.0*; Schrödinger, LLC: New York, NY, USA, 2011.
53. Congreve, M.; Andrews, S.P.; Doré, A.S.; Hollenstein, K.; Hurrell, E.; Langmead, C.J.; Mason, J.S.; Ng, I.W.; Tehan, B.; Zhukov, A.; et al. Discovery of 1,2,4-triazine derivatives as adenosine $A_{(2A)}$ antagonists using structure based drug design. *J. Med. Chem.* **2012**, *55*, 1898–1903. [CrossRef]
54. Lenzi, O.; Colotta, V.; Catarzi, D.; Varano, F.; Poli, D.; Filacchioni, G.; Varani, K.; Vincenzi, F.; Borea, P.A.; Paoletta, S.; et al. 2-Phenylpyrazolo[4,3-*d*]pyrimidin-7-one as a new scaffold to obtain potent and selective human A_3 adenosine receptor antagonists: New insights into the receptor-antagonist recognition. *J. Med. Chem.* **2009**, *52*, 7640–7652. [CrossRef]
55. Gaspar, A.; Reis, J.; Kachler, S.; Paoletta, S.; Uriarte, E.; Klotz, K.N.; Moro, S.; Borges, F. Discovery of novel A_3 adenosine receptor ligands based on chromone scaffold. *Biochem. Pharmacol.* **2012**, *84*, 21–29. [CrossRef] [PubMed]
56. Cheng, Y.; Prusoff, W.H. Relationship between the inhibition constant (K_1) and the concentration of inhibitor which causes 50 per cent inhibition (IC_{50}) of an enzymatic reaction. *Biochem. Pharmacol.* **1973**, *22*, 3099–3108. [PubMed]
57. Ertl, P.; Rohde, B.; Selzer, P. Fast calculation of molecular polar surface area as a sum of fragment-based contributions and its application to the prediction of drug transport properties. *J. Med. Chem.* **2000**, *43*, 3714–3717. [CrossRef] [PubMed]
58. Sherman, W.; Day, T.; Jacobson, M.P.; Friesner, R.A.; Farid, R. Novel procedure for modeling ligand/receptor induced fit effects. *J. Med. Chem.* **2006**, *49*, 534–553. [CrossRef] [PubMed]

Sample Availability: Samples of the compounds are available from the authors.

Article

Coumarin-Chalcone Hybrids as Inhibitors of MAO-B: Biological Activity and In Silico Studies

Guillermo Moya-Alvarado [1,†], Osvaldo Yañez [2,3,†], Nicole Morales [4], Angélica González-González [5], Carlos Areche [6], Marco Tulio Núñez [7], Angélica Fierro [8,*] and Olimpo García-Beltrán [9,10,*]

1 Biology Department, Johns Hopkins University, Baltimore, MD 21218, USA; gmoya@bio.puc.cl
2 Center of New Drugs for Hypertension (CENDHY), Santiago 8330015, Chile; osvyanezosses@gmail.com
3 Computational and Theoretical Chemistry Group, Departamento de Ciencias Químicas, Facultad de Ciencias Exactas, Universidad Andres Bello, República 498, Santiago 7550196, Chile
4 Department of Physiology, Faculty of Biological Sciences, Pontificia Universidad Católica de Chile, Santiago 8331150, Chile; nlmorales@uc.cl
5 Laboratorio de Interacciones Insecto-Planta, Instituto de Ciencias Biológicas, Universidad de Talca, Casilla 747, Talca 3460000, Chile; angelica.gonzalez@utalca.cl
6 Department of Chemistry, Faculty of Sciences, Universidad de Chile, Las Palmeras 3425, Nuñoa, Santiago 7800024, Chile; areche@uchile.cl
7 Biology Department, Faculty of Sciences, Universidad de Chile, Santiago 7800024, Chile; mnunez@uchile.cl
8 Department of Organic Chemistry, Faculty of Chemistry, Pontificia Universidad Católica de Chile, Casilla 306, Santiago 6094411, Chile
9 Centro Integrativo de Biología y Química Aplicada (CIBQA), Universidad Bernardo O'Higgins, General Gana 1702, Santiago 8370854, Chile
10 Facultad de Ciencias Naturales y Matemáticas, Universidad de Ibagué, Carrera 22 Calle 67, Ibagué 730002, Colombia
* Correspondence: afierroh@uc.cl (A.F.); jose.garcia@unibague.edu.co (O.G.-B.)
† These authors contributed equally to this work.

Abstract: Fourteen coumarin-derived compounds modified at the C3 carbon of coumarin with an α,β-unsaturated ketone were synthesized. These compounds may be designated as chalcocoumarins (3-cinnamoyl-2*H*-chromen-2-ones). Both chalcones and coumarins are recognized scaffolds in medicinal chemistry, showing diverse biological and pharmacological properties among which neuroprotective activities and multiple enzyme inhibition, including mitochondrial enzyme systems, stand out. The evaluation of monoamine oxidase B (MAO-B) inhibitors has aroused considerable interest as therapeutic agents for neurodegenerative diseases such as Parkinson's. Of the fourteen chalcocumarins evaluated here against MAO-B, **ChC4** showed the strongest activity in vitro, with IC_{50} = 0.76 ± 0.08 μM. Computational docking, molecular dynamics and MM/GBSA studies, confirm that **ChC4** binds very stably to the active rMAO-B site, explaining the experimental inhibition data.

Keywords: chalcocoumarin; MAO-B; molecular dynamics; in silico studies; neurodegenerative diseases

Citation: Moya-Alvarado, G.; Yañez, O.; Morales, N.; González-González, A.; Areche, C.; Núñez, M.T.; Fierro, A.; García-Beltrán, O. Coumarin-Chalcone Hybrids as Inhibitors of MAO-B: Biological Activity and In Silico Studies. *Molecules* **2021**, *26*, 2430. https://doi.org/10.3390/molecules26092430

Academic Editor: Maria João Matos

Received: 1 April 2021
Accepted: 18 April 2021
Published: 22 April 2021

1. Introduction

Coumarins (α-benzopyrones, 2*H*-chromen-2-ones) are a large family of compounds, of natural and synthetic origin, that show numerous biological [1–6] and medicinal chemistry activities, such as anticoagulant, anticancer, antioxidant, antiviral, anti-diabetic, anti-inflammatory, antibacterial, antifungal and anti-neurodegerative properties [7–9], among which recent studies have paid special attention to enzyme inhibition. With regard to monoamine oxidase (MAO) inhibition, recent findings have revealed that MAO affinity and selectivity can be efficiently modulated by appropriate substitutions on the coumarin ring system [1,10–13].

MAOs (EC 1.4.3.4) are flavoproteins located in the outer mitochondrial membrane and involved in the oxidative deamination of endogenous and exogenous monoamines using oxygen (O_2) as electron acceptor. In humans they exist in two isoforms called MAO-A and MAO-B. The high resolution crystal structures of both human isoforms A and B (hMAO) rat MAO-A (rMAO-A) have made it possible to analyze binding modes of ligands inside these macromolecules [14]. While the active site is formed by the common FAD cofactor and similar amino acid residues in the different forms, these are distinguished by their selectivity for substrates and inhibitors [15]. Thus, serotonin and noradrenaline are substrates of MAO-A which is selectively inhibited by clorgyline, while MAO-B oxidizes β-phenylethylamines and benzylamines and is selectively inhibited by l-deprenyl. MAO genes are expressed in various tissues. However, in the brain, although both isoforms are widely distributed, MAO-B is expressed in high concentrations in the hypothalamus, striatum, globus pallidus and thalamus, and mainly in serotonergic cells while the A isoform is rather evenly distributed, mainly in the cortex, and in nuclei containing preferably catecholaminergic and glial cells [16–21].

Although knowledge about MAO inhibition by compounds containing coumarin scaffolds is scarce, publications of articles describing new inhibitors of this class of compounds are increasing. The variety of substitutions on the coumarin ring provide insight into the influence on the activity-structure relationship. Among the most reported modifications of the coumarin ring with MAO activity are on C3 and the steric effect of the substituent appears to be important in modulating MAO-B inhibitory activity [11]. In addition, it has been reported that the introduction of various substituents at the *para* position of the 3-phenyl ring is a good strategy for improving the desired MAO-B inhibitory activity [22] and when the 3-phenyl skeleton is replaced by a 3-benzoyl group, the activity is strongly diminished [20]. It has also been observed that coumarins substituted with 3-indolyl and 3-thiophenyl shows greatest selective inhibition was against MAO-B [11,23,24].

In this work a merger of the coumarin scaffold and a 3-cinnamyl group led to new hybrid (chalcocoumarin) derivatives (Scheme 1) that preserve structural characteristics of compounds with the ability to interact with MAO. The synthetic strategy chosen allowed a large variety of substituents on the cinnamyl benzene ring to be accessed using different readily accessible benzaldehydes. Thus, the quantity and/or type of interactions with the enzyme were explored involving some bulky groups to determine their possible contribution to the biological activity as MAOIs. Our new compounds were screened versus both MAO isoforms, and in silico studies were carried out to rationalize their main interactions in the MAO active cavity. The computational biochemistry tools were used considering the geometrical restrictions and most probable positions in the formation of the ligand-receptor complex. The chalcocoumarin molecules were subjected to theoretical studies in which binding energies were estimated using docking and MM/GBSA analysis. In addition, physicochemical parameters that are responsible for governing the pharmacokinetic properties of drug molecules were determined.

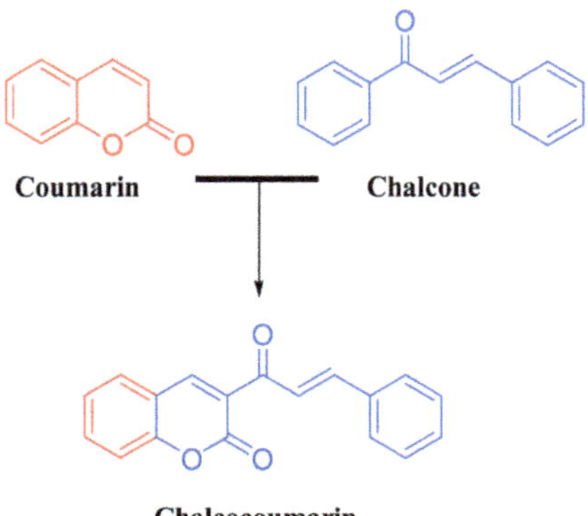

Scheme 1. Hybrids of chalcocoumarin.

In the present study, a series of coumarin-chalcone hybrid compounds were synthesized and tested on the 2 MAO isoforms. The activity shown was selective for MAO-B and in particular, compound **ChC4** showed the highest inhibitory activity on rMAO-B at submicromolar concentrations. The results obtained will be useful to understand the mode of inhibition of chalcocoumarins against rMAO-B, and to help predict the activities of these new inhibitors that could be promising as therapeutics to treat neurodegenerative diseases such as Parkinson's disease.

2. Results and Discussion

2.1. Chemistry

The route employed to synthesize the compounds is summarized in Scheme 1. The compounds were obtained starting from resorcinol (**1**), which was formylated using the Vilsmaier-Haack reaction [25]. Knoevenagel condensation of the aldehyde intermediate with ethyl acetoacetate afforded hydroxycoumarin **2**. The 7-hydroxycoumarin obtained was methylated using a Williamson reaction using methyl sulfate as methylating agent, obtaining the compound **3**, finally the compounds derived the 3-cinnamoyl-2H-chromen-2-one (**ChC1–ChC14**) (Table 1) were prepared in moderate yields (25–47%, unoptimized) by Claisen-Schmidt condensation with the respective aldehyde (Supplementary Materials; Scheme S1) [26]. Coumarin-chalcone hybrids have been studied and are currently still being synthesized for various uses and their spectroscopy is well known, however, we will detail some signals that are key to their identification. The ^{1}H-NMR spectra of the compounds **ChC1–ChC14** present very similar chemical shift patterns with a particular signal that identifies this type of molecules, the neighboring vinyl protons of the α,β-unsaturated ketone appear at very close low field from the aromatic proton region. These protons present signals corresponding to two doublets with variable δ between 8.5 and 7.0 and with J_{ab} = 16 Hz on average. this high constant corresponds to a trans isomer [27–29]. As for the ^{13}C-NMR spectrum, we will mention typical signals such as carbonyl shifts. first of all, we will detail that the carbon of the α,β-unsaturated ketone has a δ 190–180 ppm and carbonyl carbons of α-pyrone δ 165–155 pmm on average [27–29], the compounds were characterized by ^{1}H and ^{13}C NMR (Supplementary Materials; Figures S1–S15).

Table 1. Structures of the synthetized chalcocoumarin hybrids.

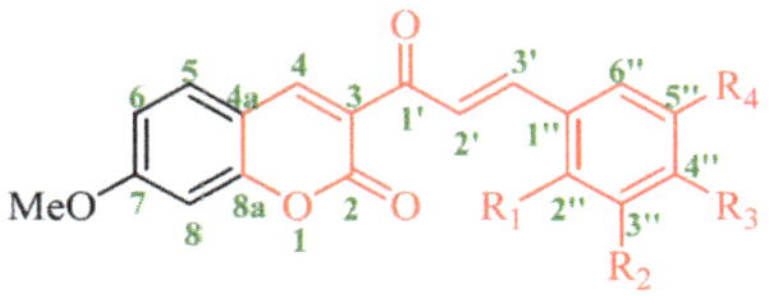

Compounds	R_1	R_2	R_3	R_4
ChC1	H	H	H	H
ChC2	OH	H	H	H
ChC3	OCH_3	H	H	H
ChC4	H	OH	H	H
ChC5	H	OCH_3	H	H
ChC6	H	H	OCH_3	H
ChC7	OCH_3	OCH_3	H	H
ChC8	OCH_3	H	H	OCH_3
ChC10	H	H	OH	OCH_3
ChC11	H	H	SCH_3	H
ChC12	H	Br	OCH_3	Br

Table 1. *Cont.*

Compounds	R_1	R_2	R_3	R_4
ChC13	H	H	$N(CH_3)_2$	H
ChC14	H	H	Br	H

ChC9

2.2. Biological Analysis in Rat MAO

Fourteen derivatives differing in the substitution pattern of the cinnamyl benzene ring were studied, these compounds were tested on rat MAO-A and B to determine their inhibitory activity MAO. A general screening was carried out at 10 μM finding moderate activity for some of the compounds against rMAO-B but none against rMAO-A. Thus, five molecules were identified as possibly selective IMAO-B.

ChC4, **ChC5**, **ChC6**, **ChC9** and **ChC11** in MAO-B exhibited micromolar or submicromolar in vitro potencies, all below 10 μM (Table 2). Out of these **ChC4**, substituted with a hydroxyl group on the meta position of the variable ring, displayed the highest rMAO-B inhibitory activity (IC_{50} = 0.76 μM). Interestingly, changing the position of the hydroxyl group from meta to ortho or para (**ChC2** and **ChC10** respectively) led to loss of the inhibitory activity. An approximately 12-fold lower IMAO activity was observed when the hydroxyl group (in **ChC4**) was methylated (**ChC5**). This might be attributed to steric hindrance and/or to the loss of hydrogen bonding donor quality which could be crucial for some interaction in the binding site. Moving the methoxyl group from the meta to the para position (**ChC6** vs. **ChC5**) slightly increased potency.

Table 2. IC_{50} of the compounds in rMAO-A and rMAO-B.

Compounds	IC_{50} (μM) rMAO A	IC_{50} (μM) rMAO B
ChC1	>10	>10
ChC2	>10	>10
ChC3	>10	>10
ChC4	>10	0.76 ± 0.08
ChC5	>10	9.63 ± 0.90
ChC6	>10	6.96 ± 0.07
ChC7	>10	>10
ChC8	>10	>10
ChC9	>10	3.71 ± 0.68
ChC10	>10	>10
ChC11	>10	5.88 ± 0.57
ChC12	>10	>10
ChC13	>10	>10
ChC14	>10	>10

Each IC_{50} value was obtained from an average of three evaluations (n = 3).

Replacing the methoxy group of **ChC6** with a bulkier, less electronegative and more polarisable methylthio group (**ChC11**) only produced **ChC6** is less potent than **ChC11**. The second most potent molecule was **ChC9**, with a methylenedioxy group bridging the meta and para carbons. The methylenedioxy group increases the rigidity of the molecule, possibly stabilizing the complex protein-ligand interaction. The same effect, extending the rigidity, has been observed, on other derivatives as IMAO [30,31].

2.3. Molecular Docking and Ligand Efficiency Analysis

To analyze the changes in potency of the coumarin-chalcone hybrids, docking studies were carried out using the crystal structure of rMAO-A and the homology model of rMAO-B (Supplementary Materials; Figures S16 and S17), analyzing the possibility that each one of them has to form a stable complex with each of the 14 molecules synthesized by us. Table 3 shows that most of these molecules present better interactions with the rMAO-B binding site, since the corresponding energies are at least 2.5 kcal mol more negative in all but one of the cases. This difference could be due to the substitutes present in the molecules. The results of the molecular docking experiments showed more favorable interactions (more negative $\Delta E_{binding}$) for the complexes in rMAO-B than in rMAO-A, with average values around -9.26 kcal·mol^{-1} vs. -6.57 kcal·mol^{-1} respectively) which are in accord with the experimental data for the whole series. In the rMAO-B, although no major differences were observed in the binding modes of the active compounds, subtle energy differences were identified. The results of this molecular docking study point to strong interactions of the chalcocoumarins in the binding pocket of rMAO-B, but considerably weaker in rMAO-A.

Table 3. Molecular docking results for **ChC1–ChC14** in the rMAO-A/rMAO-B models. Intermolecular docking energy values ($\Delta E_{binding}$), K_d values and calculated Ligand Efficiency (*LE*) for the rMAO-A and rMAO-B complexes.

Compound	Docking Results [a]		Ligand Efficiency			
	rMAO-A	rMAO-B	rMAO-A		rMAO-B	
	$\Delta E_{binding}$ (kcal·mol^{-1})	$\Delta E_{binding}$ (kcal·mol^{-1})	K_d	*LE* (kcal·mol^{-1})	K_d	*LE* (kcal·mol^{-1})
ChC1	−6.4	−9.3	2.03×10^{-5}	0.278	1.52×10^{-7}	0.404
ChC2	−6.7	−9.8	1.22×10^{-5}	0.279	6.57×10^{-8}	0.408
ChC3	−6.7	−9.1	1.22×10^{-5}	0.268	2.14×10^{-7}	0.364
ChC4	**−6.3**	**−9.8**	$\mathbf{2.41\times10^{-5}}$	**0.262**	$\mathbf{6.57\times10^{-8}}$	**0.408**
ChC5	−6.5	−9.6	1.72×10^{-5}	0.260	9.21×10^{-8}	0.384
ChC6	−6.6	−9.3	1.45×10^{-5}	0.264	1.52×10^{-7}	0.372
ChC7	−6.2	−8.7	2.85×10^{-5}	0.229	4.20×10^{-7}	0.322
ChC8	−6.7	−9.6	1.22×10^{-5}	0.248	9.21×10^{-8}	0.355
ChC9	−6.8	−9.9	1.03×10^{-5}	0.261	5.55×10^{-8}	0.380
ChC10	−6.4	−9.2	2.03×10^{-5}	0.246	1.80×10^{-7}	0.353
ChC11	−6.4	−8.9	2.03×10^{-5}	0.256	3.00×10^{-7}	0.356
ChC12	−6.7	−7.2	1.22×10^{-5}	0.248	5.28×10^{-6}	0.266
ChC13	−7.0	−9.9	7.41×10^{-6}	0.269	5.55×10^{-8}	0.380
ChC14	−6.6	−9.4	1.45×10^{-5}	0.275	1.29×10^{-7}	0.391

[a] In each site, the energy was calculated to see which site bound more strongly to the ligand. In bold ChC4 displayed the highest rMAO-B inhibitory activity.

When analyzing the docking results for rMAO-B from the conformational viewpoint, it is necessary to consider the residues that constitute the substrate-binding site of rMAO-B, which is composed of the FAD cofactor, two flanking residues, Tyr398 and Tyr435, that form an "aromatic box", and a number of others, particularly Cys172, Tyr326, Met341, Ser200 Gln206 and Thr314 [32,33]. The results show that all the chalcocoumarins settle in the active site of rMAO-B (Supplementary Materials; Figure S18), with the benzene ring of the coumarin moiety close to the FAD, more specifically the central N-5, at a distance of about 4.0 Å. The benzene ring of the cinnamyl moiety extends into the generally hydrophobic entrance cavity adjoining the substrate-binding site. The mere length of the ChC molecules indicates that to bind in the active site of MAO-A the latter must undergo a rearrangement of the residues separating the entrance and the substrate cavities, which may explain their general preference for MAO-B.

ChC4 was located inside the cavity interacting with Tyr435, Tyr398, Tyr60, Phe343, Asn83, Arg307, Thr314, and Leu328. Two hydrogen bonds where generated with Asn83 and, via its C-3′ hydroxyl group, Thr316. **ChC2** actually when interact with the amino acids into the pocket adopt a planar conformation because the hydrogen bond confirms our discussion that could be responsible for none activity of **ChC2** in rMAO-B. A quantum geometric optimization of **ChC2** and **ChC4** at the M05-2X-D3/6-31G(d,p) level of theory, showed their C-2′ and C-3′ hydroxyl groups pointing in opposite directions, suggesting different preferred intramolecular interactions (Figure 1). Both the different electronic potential distribution and the resulting preferred intermolecular interaction might be responsible for the difference in in IC_{50} values.

The best three ligands obtained from the docking exhibit low K_d values, these ligands are **ChC4**, **ChC9** and **ChC13**, which means that these ligand/rMAO-B complexes are the most stable in the series. These results are consistent with those obtained in the docking experiments in which these complexes were the most stable according to their $\Delta E_{binding}$ values. The proposed tolerable values of *LE* for inhibitor candidates are $LE > 0.3$ kcal·mol^{-1} [34–36]. According to this reference value, **ChC4** is a good prospect for development as an rMAO-B inhibitor with a *LE* value of 0.408. Although **ChC2** and **ChC4** have similar K_d, *LE* and $\Delta E_{binding}$ values, the **ChC2** molecule does not show in vitro activity against rMAO-B, on the other hand, ChC4 has a good inhibitory activity against rMAO-B, since micromolar concentrations are needed to inhibit it, which is consistent with the values obtained for the K_d. ChC1 would appear to be almost as good, with $LE = 0.404$, but again its activity, if any, is worse than our cutoff value. The low micromolar-active **ChC5**, **ChC6**, **ChC9** and **ChC11** have *LE* values of 0.384, 0.372, 0.380 and 0.356, respectively, and the less (or in-) active **ChC13** and **ChC14** have *LE* values of 0.380 and 0.391.

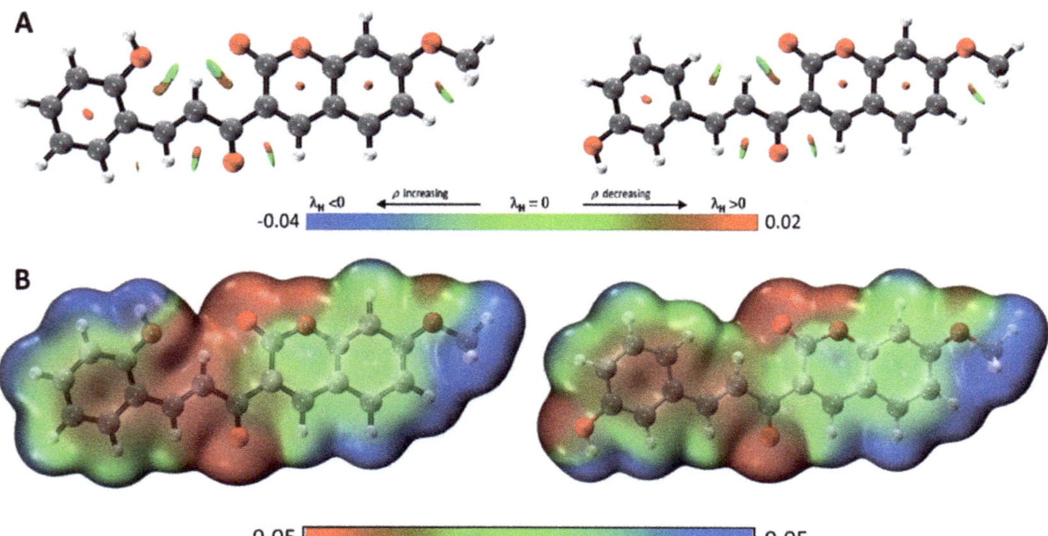

Figure 1. (**A**) NCIplot of the non-covalent interaction regions with isosurface gradient (0.6 au) for **ChC2** (left) and **ChC4** (right). (**B**) Electrostatic potential (in a.u.) of **ChC2** (left) and **ChC4** (right) mapped on the 0.001 a.u. isodensity surface for the selected structure computed at the M05-2X-D3/6-31G(d,p) level of theory.

2.4. *Analysis of Molecular Dynamics Simulations*

Molecular dynamics simulations were performed for 100 ns to analyze the conformational stability of the rMAO-B/**ChC2** and rMAO-B/**ChC4** complexes. The RMSD, a quantitative parameter, was used to estimate the stability of the protein-ligand systems and the apoprotein. The RMSD in Figure 2A shows that the rMAO-B/**ChC2** and rMAO-B/**ChC4** complexes remain highly stable throughout the simulation time. We can see that the structures of the complexes does not change significantly. The RMSD values for the **ChC4** complex are remarkably constant about 1.5 Å, with a very slight instability and increase near the end of the simulation. The **ChC2** complex shows similar, somewhat less stable behavior for almost 40 ns, and then its RMSD value falls abruptly to about 1.0 Å and rises slowly with appreciable fluctuations to about 1.2 Å at 100 ns, indicative of weaker binding in the rMAO-B site. However, a maximum difference of 3.0 Å in the RMSD is taken to indicate that a system is in equilibrium [37], so this condition is fulfilled by both compounds. To complement the analysis carried out calculating the RMSD, the Radius of Gyration (R_{Gyr}) was analyzed for the same runs. From this analysis (Figure 2B), we can conclude that the R_{Gyr} of **ChC2** and **ChC4** oscillate in a narrow interval between 4.3–4.8 Å. These stable values during the 100 ns simulation indicate again that ligand binding does not induce major conformational changes in the protein structure.

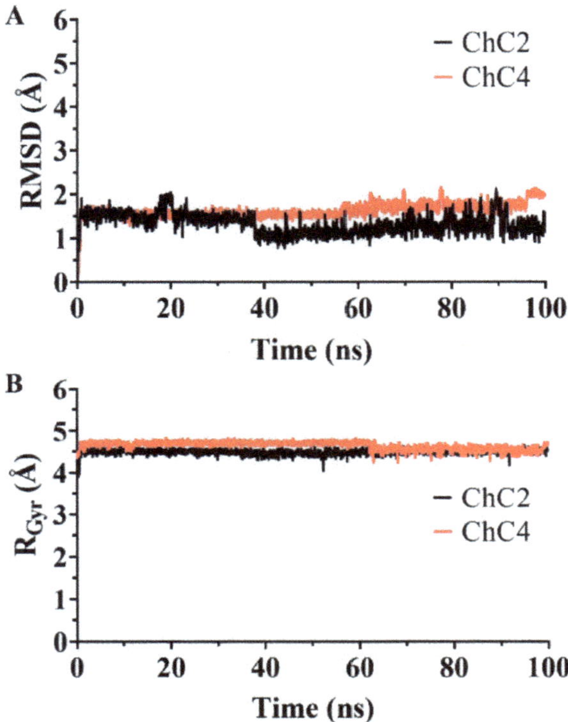

Figure 2. (**A**) Root Mean Square Deviation (RMSD) and (**B**) Radius of gyration (R_{Gyr}) as a function of simulation times for the complexes formed between rMAO-B and **ChC2** and **ChC4**.

Structural studies in MAO have shown that two residues Tyr398 and Tyr435 in MAO-B located in the active site approximately perpendicular to the FAD play a functional role in this enzyme, acting as a cofactor stabilizing the active site, forming an aromatic box whose function is to stabilize the ligand [13]. Molecular simulation results show a difference in the interaction of the compounds **ChC2** and **ChC4** with the FAD cofactor, see Figure 3A. Compound **ChC2** shows a spacing that fluctuates between 17.0 Å and 20.0 Å from its original position, signifying a null interaction with the FAD cofactor. On the other hand, the compound **ChC4** is within the range of interaction with the FAD cofactor. This distance was measured between the nitrogen atom of the alloxazine planar ring of FAD and the center of the benzaldehyde aromatic ring of compounds, Figure 3B.

Molecular dynamics simulations showed of rMAO-B that residues that interact with the ligands ChC2 and ChC4, see Figure 4. The most frequent residues in rMAO-B/**ChC2** were Ile164, Ile199, Leu167, Leu171, Phe168, Pro104, Trp119, Val316, Phe103, Pro102, Tyr115 and Thr196. In contrast, the most frequent residues in rMAO-B/**ChC4** were Ile164, Ile199, Leu171, Phe168, Pro104, Trp119, Tyr326, Val316, Cys172 and Tyr115 with van der Waals and hydrogen bonds interactions. Highlighting residues Cys172 and Tyr326, which are important for the active site of the rMAO-B flavoprotein. Tyr326 and Cys172 are key residues that determines substrate and inhibitor specificity, also exhibits conformational changes on the inhibitor binding and restricts the binding of certain inhibitors (e.g., harmine) to human MAO-B [38]. These results documents that ChC4 is a reversible inhibitor of rMAO-B.

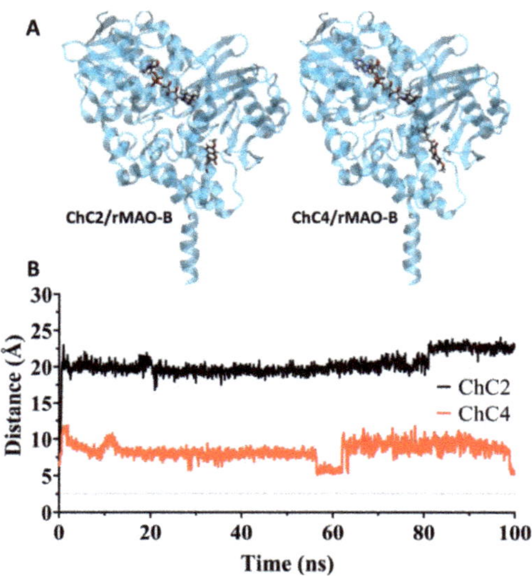

Figure 3. (**A**) Last frame of the molecular simulation showing the positions between the FAD molecule and **ChC2–ChC4** compounds interacting with rMAO-B. (**B**) Distance as a function of simulation time, between the nitrogen atom of the aloxazine planar ring of FAD and the center of the benzaldehyde aromatic ring, for compounds **ChC2** and **ChC4**. Dashed lines represent the position of the nitrogen atom of the aloxazine planar ring.

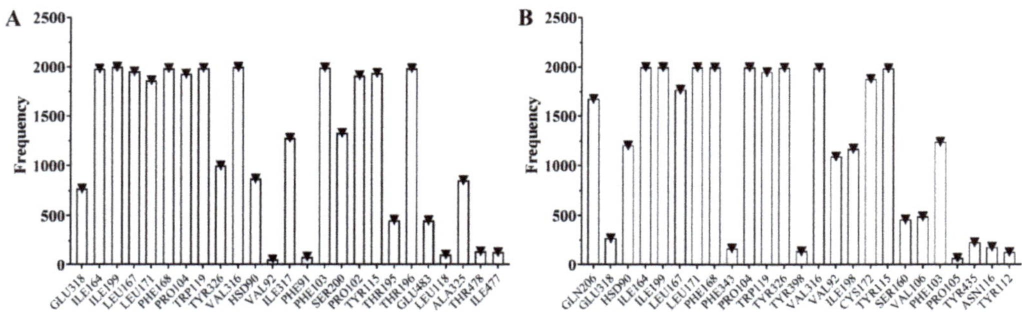

Figure 4. Frequency of the appearance of residues at a distance of 3.0 Å or closer from ligands (**A**) **ChC2** and (**B**) **ChC4** calculated using MD procedures.

The analyses of trajectories indicate that during most of the simulation the ligand **ChC4** maintain hydrogen bonds with residues of the active site of rMAO-B. However, the number of hydrogen bonds formed was different for **ChC2** and **ChC4** (Figure 5). **ChC2** formed two hydrogen bonds between the residues Glu483 and Tyr115, highlighting the participation of the residues Val316, Ala325, Ile164 and Leu167. Finally, **ChC4** formed two hydrogen bonds with the Phe168, Cys172, Ile164 and Tyr115, highlighting the participation of the residues Ile199, Trp119 and Tyr326. These residues, see Figure 6, are consistent with previous theoretical-experimental studies carried out [39,40] where they detail the interaction that some of the synthesized compounds have with the active site of the rMAO-B. This difference in the formation of hydrogen bonds with key residues in rMAO-B could be explained the difference in experimental activity between the **ChC2** and **ChC4** compounds.

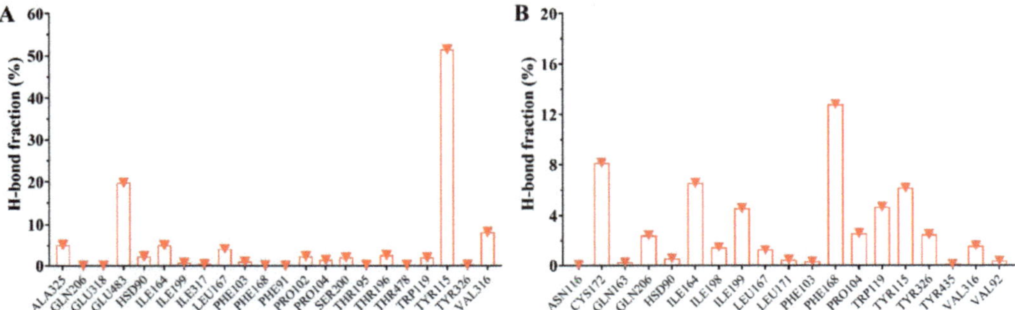

Figure 5. Fraction (in %) of intermolecular hydrogen bonds for rMAO-B interacting with (**A**) **ChC2** and (**B**) **ChC4**. The graph bar shows the most common hydrogen bonds formed between the residues on the pocket and the inhibitors.

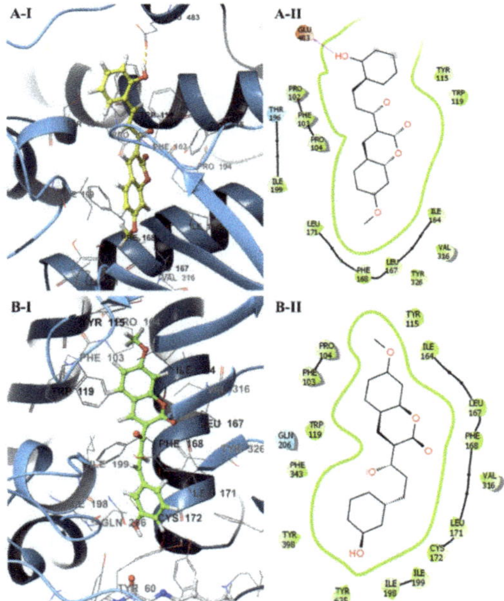

Figure 6. Schematic representations at the end (100 ns) of their respective production runs for ligands (**A**) **ChC2** and (**B**) **ChC4** bound to rMAO-B. (**I**) The surrounding amino acid residues in the binding pocket of rMAO-B within 4Å from ligands. (**II**) Two-dimensional interaction map of **ChC2** and **ChC4** and rMAO-B. The arrows indicate potential interactions between amino acid residues and the ligands.

Finally, the binding free energy (MM/GBSA) was computed after the MD simulation; the last 70 ns for all the complexes and the results are given in Table 4. The compound **ChC2** has a binding free energy of -29.06 kcal·mol^{-1} with rMAO-B enzyme, while the compound **ChC4** showed relatively binding free energy of –25.87 kcal·mol^{-1}. The results obtained from MM/GBSA show a slight difference in their binding free energy between **ChC2** and **ChC4** compounds bound to rMAO-B. This slight difference is due to the R1 to R2 position of the hydroxyl group in benzaldehyde aromatic ring. In particular, the **ChC4** compound has a better activity at the experimental and in silico level.

Table 4. Predicted binding free energies (kcal·mol^{-1}) and individual energy terms calculated from molecular dynamics simulation through the MM-GBSA protocol for rMAO-B complexes.

Compounds	Calculated Free Energy Descomposition (kcal·mol^{-1})			
	$\Delta G_{binding}$	ΔE_{vdW}	ΔE_{elect}	ΔE_{pot}
ChC2	−29.06 ± 0.10	−39.69 ± 0.13	15.03 ± 0.17	−24.65 ± 0.10
ChC4	−25.87 ± 0.09	−44.94 ± 0.09	24.90 ± 0.04	−20.03 ± 0.09

2.5. In Silico Pharmacokinetic Prediction

A good drug candidate is absorbed in required time and well distributed throughout the system for its effective metabolism and action. Toxicity is another very important factor that often overshadows the ADME behaviour. SwissADME explorer online was used for in silico prediction of drug likeness of the synthesized compounds (**ChC1–ChC14**) based on various molecular descriptors and the results are presented in Table 5.

Table 5. In silico predicted physicochemical properties of all compounds **ChC1–ChC14**.

Compounds	Log P	MW (g/mol)	TPSA (Å^2)	HBA	HBD	RB	Log S	Log K_p (cm/s)	N° Violations
ChC1	3.38	307.32	63.6	4	1	4	−4.89	−5.43	0
ChC2	2.9	323.32	83.83	5	2	4	−4.94	−5.79	0
ChC3	3.36	337.35	72.83	5	1	5	−5.06	−5.64	0
ChC4	2.97	323.32	83.83	5	2	4	−4.94	−5.79	0
ChC5	3.27	337.35	72.83	5	1	5	−5.06	−5.64	0
ChC6	3.27	337.35	72.83	5	1	5	−5.06	−5.64	0
ChC7	3.28	367.37	82.06	6	1	6	−5.22	−5.84	0
ChC8	3.27	367.37	82.06	6	1	6	−5.22	−5.84	0
ChC9	3.23	351.33	82.06	6	1	4	−5.08	−5.84	0
ChC10	2.92	353.35	93.06	6	2	5	−5.11	−5.99	0
ChC11	3.82	353.41	88.9	4	1	5	−5.95	−5.35	0
ChC12	4.48	495.14	72.83	5	1	5	−6.49	−5.62	0,1 *
ChC13	4.24	351.42	63.6	4	1	6	−5.86	−5.04	0
ChC14	3.91	386.22	63.6	4	1	4	−5.61	−5.43	0

MW = 150–500 g/mol; *TPSA* = 20 Å^2–130 Å^2; HBA = N° of H-bond acceptors ≤ 10; HBD = N° of H-bond donor ≤ 5; *RB* = 0–9; Log S = Insoluble < −10 < Poorly < −6 < Moderately < −4 < Soluble < −2; Log P ≤ 5; log K_p ≥ −2.5 considered to be permeable; N° Violations of Lipinski, Ghose, Veber, Egan and Muegge rules. * Violation of Ghose and Muegge rules.

The most potent compound ChC4 in biological experiment data having logP value of 2.97, it's clear that it doesn't violate of five Lipinski rules, while the other molecules have logP values in the range of 2.90–4.48 and are expected to be orally active. In addition, the logS values for **ChC4** have a value of −4.94 indicating proper solubility, which is an indication of favorable drug like property, makes compound **ChC4** promising drug candidate for further research and development. Thirteen of fourteen synthesized molecules do not break the rules of Lipinski, Ghose, Veber, Egan and Muegge, since the molecule **ChC12** breaks the rules of Ghose and Muegge.

The Boiled-egg model is proposed as an accurate predictive model that works by computing the lipophilicity and polarity of small molecules. The Boiled-egg analysis of the fourteen molecules (Figure 7) has shown that compounds **ChC1, ChC3, ChC5, ChC6, ChC12, ChC13** and **ChC14** are highly absorbable at the brain barrier, whereas compounds **ChC2, ChC4, ChC7, ChC8, ChC9, ChC10** and ChC11 are highly absorbable in the gastrointestinal tract.

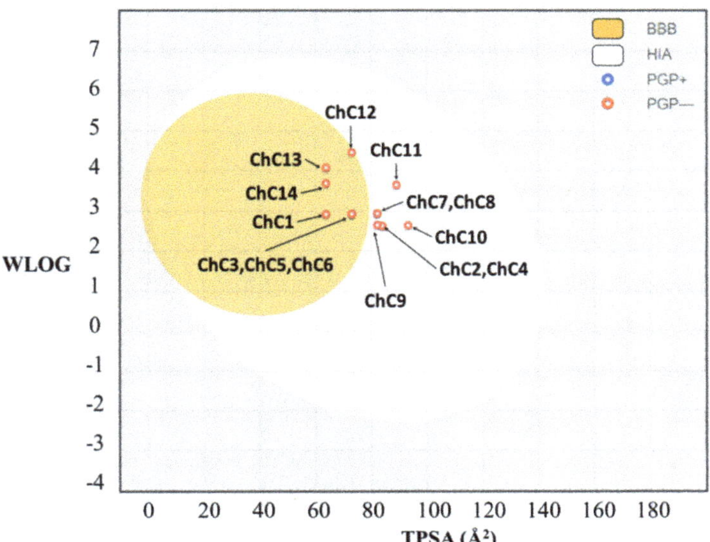

Figure 7. Predictive human intestinal absorption (HIA) model and blood-brain barrier permeation (BBB) method (boiled-egg plot) of the 14 compounds.

The ADMET properties showed much similarity among the thirteen molecules that can be used for advanced clinical trials.

3. Materials and Methods

3.1. Solvents and Reagents

Solvents and reagents (analytical grade and spectroscopic grade) were obtained from Sigma-Aldrich (St. Louis, MO, USA) and Merck (Darmstadt, Germany). Melting points were determined on a Galen III hot-plate microscope (Reichert-Jung, St. Louis, MO, USA) equipped with a thermocouple. ^{1}H- and ^{13}C-NMR spectra were recorded on a 400 MHz multidimensional spectrometer (Bruker Corporation, Billerica, MA, USA) using the solvent or the TMS signal as an internal standard.

3.2. Synthesis

3-Cinnamoyl-7-methoxy-2H-chromen-2-one (**ChC1**). 3-Acetyl-7-methoxy-2H-chromen-2-one (0.44 g, 2.0 mmol) and benzaldehyde (0.21 g, 2.0 mmol) were dissolved in 25 mL of DCM and to this solution 0.5 mL of piperidine were added. The mixture was kept at reflux temperature, monitoring the reaction by TLC for 10 h. The solution was concentrated under reduced pressure and dissolved in a small aliquot of DCM and then MeOH was added in excess to induce precipitation. This procedure was performed twice. The precipitate was finally purified by column chromatography on silica gel eluting with DCM: 0.25 g yellow solid, 40.8%, m.p. 190–192 °C; ^{1}H NMR (CDCl$_3$): δ 8.59 (s, 1H, Ar-H), 8.01 (d, 1H, *J* = 15.8 Hz, Ar-CH), 7.85 (d, 1H, *J* = 15.8 Hz, CO-CH=), 7.67 (s, 2H, Ar-H), 7.56 (d, 1H, *J* = 8.6 Hz, Ar-H), 7.40 (s, 3H, Ar-H,), 6.90 (dd, 1H, *J* = 8.6, 1.0 Hz, Ar-H), 6.85 (s, 1H, Ar-H), 3.91 (s, 3H, OCH$_3$). ^{13}C-NMR (CDCl$_3$): δ 56.1, 100.2, 112.4, 113.8, 124.0, 129.0, 130.6, 131.5, 135.3, 144.3, 148.5, 157.8, 160.0, 165.0, 186.3.

(E)-3-(3-(2-Hydroxyphenyl)acryloyl)-7-methoxy-2H-chromen-2-one (**ChC2**). 3-Acetyl-7-methoxy-2*H*-chromen-2-one (0.44 g, 2.0 mmol) and *o*-hydroxybenzaldehyde (0.24 g, 2.0 mmol) were reacted and worked up according to the previous procedure: 1.75 g, pale white solid, 95.6%, m.p.; 188–190 °C. ^{1}H-NMR (DMSO-d$_6$), δ 10.43 (s, 1H, OH), 8.71 (s, 1H, Ar-H), 8.04 (d, 1H, *J* = 15.9 Hz, Ar-CH=), 7.92 (d, 1H, *J* = 8.8 Hz, Ar-H), 7.89 (d, 1H, *J* = 15.9 Hz), 7.69 (dd, 1H, *J* = 7.7, 1.0 Hz, Ar-H), 7.35 (t, 1H, *J* = 7.0 Hz, Ar-H), 7.13 (d, 1H, *J* = 2.2 Hz, Ar-H),

7.08 (dd, 1H, *J* = 8.6, 2.2 Hz, Ar-H), 7.00 (d, 1H, *J* = 8.0 Hz, Ar-H), 6.94 (t, 1H, *J* = 7.5 Hz, Ar-H), 3.96 (s, 3H, OCH_3), ^{13}C-NMR (DMSO-d_6): δ 56.7, 100.8, 112.5, 113.9, 116.8, 120.0, 121.8, 121.9, 124.3, 129.2, 132.3, 132.6, 139.5, 148.2, 157.4, 157.8, 159.4, 165.1, 187.1.

(E)-7-methoxy-3-(3-(2-methoxyphenyl)acryloyl)-2H-chromen-2-one (**ChC3**). 3-Acetyl-7-methoxy-2*H*-chromen-2-one (0.44 g, 2.0 mmol) and *o*-methoxybenzaldehide (0.27 g, 2.0 mmol), were reacted and worked up according to the previous procedure: 0.330 g, pale white solid, 49%, m.p.. 184–186 °C; ^{1}H-NMR ($CDCl_3$): δ 8.55 (s, 1H, =C-H), 8.21 (d, 1H, Ar-CH=, *J* = 15.8 Hz), 8.04 (d, 1H, CO-CH=, *J* = 15.8 Hz), 7.71 (d, 1H, Ar-H, *J* = 7.6 Hz), 7.56 (d, 1H, Ar-H, *J* = 8.4 Hz), 7.37 (t, 1H, Ar-H, *J* = 7.9), 6.98 (t, 1H, Ar-H, *J* = 7.7 Hz), 6.92 (d, 1H, Ar-H, *J* = 8.2 Hz), 6.90 (d, 1H, Ar-H, *J* = 8.5 Hz), 6.85 (s, 1H, Ar-H), 3.91 (s, 6H, 2 OCH_3); ^{13}C-NMR (DMSO-d_6): δ 55.6, 56.0, 100.4, 111.2, 112.5, 113.7, 120.8, 121.9, 124.1, 124.6, 129.3, 131.3, 132.0, 139.9, 148.2, 157.6, 159.0, 159.8, 165.0, 186.8.

(E)-3-(3-(3-hydroxyphenyl)acryloyl)-7-methoxy-2H-chromen-2-one (**ChC4**). 3-Acetyl-7-methoxy-2*H*-chromen-2-one (0.44 g, 2.0 mmol) and *m*-hydroxybenzaldehyde (0.24 g, 2.0 mmol) were reacted and worked up according to the previous procedure: 0.195 g, white solid, 30%, m.p. 184–186 °C; ^{1}H NMR (DMSO-d_6): δ 9.77 (sbr, 1H), 8.76 (s, 1H, =C-H), 7.94 (d, 1H, Ar-H, *J* = 8.0 Hz), 7.82 (d, 1H, Ar-CH=, *J* = 15.9 Hz), 7.71 (d, 1H, CO-CH=, *J* = 15.9 Hz), 7.62 (d, 1H, Ar-H, *J* = 8.6 Hz), 7.33 (m, 1H, Ar-H), 7.25-7.17 (m, 2H, Ar-H), 7.15 (s, 1H, Ar-H), 7.09 (dd, 1H, Ar-H, *J* = 8.0, 1.0 Hz), 6.94 (dd, 1H, Ar-H, *J* = 8.0, 1.0 Hz), 3.97 (s, 3H, OCH_3); ^{13}C NMR (DMSO-d_6): 56.7, 100.9, 112.5, 114.0, 114.8, 118.5, 120.5, 121.6, 124.9, 130.6, 132.5, 136.3, 143.9, 148.6, 157.5, 158.2, 159.4, 165.3, 186.7.

(E)-7-methoxy-3-(3-(3-methoxyphenyl)acryloyl)-2H-chromen-2-one (**ChC5**). 3-Acetyl-7-methoxy-2*H*-chromen-2-one (0.44 g, 2.0 mmol) and *m*-methoxybenzaldehide (0,27 g, 2.0 mmol), were reacted and worked up according to the previous procedure: 0,280 g, pale white solid, 42%, m.p. 164–166 °C; ^{1}H-NMR ($CDCl_3$): δ 8.58 (s, 1H, =C-H), 7.99 (d, 1H, Ar-CH=, *J* = 15.9 Hz), 7.82 (d, 1H, CO-CH=, *J* = 15.9 Hz), 7.57 (d, 1H, Ar-H, *J* = 8.0 Hz), 7.35-7.25 (m, 2H, Ar-H), 7.18 (s, 1H, Ar-H) 6.96 (dd, 1H, Ar-H, *J* = 8.0, 2.0 Hz), 6.91 (dd, 1H, Ar-H, *J* = 8.8, 2.0 Hz), 6.85 (d, 1H, Ar-H, *J* = 2.0 Hz), 3.83 (s, 3H, OCH_3), 3.76 (s, 3H, OCH_3); ^{13}C-NMR ($CDCl_3$): δ 55.8, 56.5, 100.8, 112.8, 113.9, 114.3, 117.1, 121.7, 122.0, 124.9, 130.3, 131.8, 136.8, 144.8, 149.0, 158.1, 160.2, 160.3, 165.8, 186.8.

(E)-7-methoxy-3-(3-(4-methoxyphenyl)acryloyl)-2H-chromen-2-one (**ChC6**). 3-Acetyl-7-methoxy-2*H*-chromen-2-one (0.44 g, 2.0 mmol) and *p*-methoxybenzaldehide (0.27 g, 2.0 mmol), were reacted and worked up according to the previous procedure: 0.26 g, pale white solid, 39%, m.p. 158–160 °C; ^{1}H-NMR ($CDCl_3$): δ 8.58 (s, 1H, =C-H), 7.91 (d, 1H, Ar-CH=, *J* = 15.9 Hz), 7.84 (d, 1H, CO-CH=, *J* = 15.9 Hz), 7.64 (d, 2H, Ar-H, *J* = 8.0 Hz), 7.56 (d, 1H, Ar-H, *J* = 8.8 Hz), 6.93 (d, 2H, Ar-H, *J* = 8.0 Hz), 6.91 (dd, 1H, Ar-H, *J* = 8.8, 2.0 Hz), 6.85 (d, 1H, Ar-H, *J* = 2.0 Hz), 3.92 (s, 3H, OCH_3), 3.86 (s, 3H, OCH_3); ^{13}C-NMR ($CDCl_3$): δ 57.2, 57.8, 102.0, 114.2, 115.5, 116.1, 123.4, 123.6, 129.6, 132.5, 133.0, 146.3, 150.0, 159.4, 161.6, 163.5, 166.8, 188.0.

(E)-3-(3-(3,4-dimethoxyphenyl)acryloyl)-7-methoxy-2H-chromen-2-one (**ChC7**). 3-Acetyl-7-methoxy-2*H*-chromen-2-one (0.44 g, 2.0 mmol) and 3,4-dimethoxybenzaldehide (0.33 g, 2.0 mmol) were reacted and worked up according to the previous procedure: 0.42 g, bright yellow solid, 57%, m.p. 182–184 °C; ^{1}H-NMR ($CDCl_3$): δ 8.55 (s, 1H, =C-H), 8.16 (d, 1H, Ar-CH=, *J* = 15.9 Hz), 8.00 (d, 1H, CO-CH=, *J* = 15.9 Hz), 7.55 (d, 1H, Ar-H, *J* = 8.6 Hz), 7.34 (d, 1H, Ar-H, *J* = 1.0 Hz), 7.19 (dd, 1H, AR-H, *J* = 8.0, 8.0 Hz) 7.07 (d, 1H, Ar-H, *J* = 8.0 Hz), 7.00 (dd, 1H, Ar-H, *J* = 8.0, 2.0 Hz), 6.94(d, 1H, Ar-H, *J* = 2.0 Hz), 3.90 (s, 3H, OCH_3), 3.89 (s, 3H, OCH_3), 3.87 (s, 3H, OCH_3); ^{13}C-NMR ($CDCl_3$): δ 55.9, 56.0, 61.5, 100.4, 112.4, 113.8, 114.4, 119.9, 121.6, 124.2, 125.5, 129.2, 131.3, 139.1, 148.4, 149.2, 153.2, 157.7, 159.7, 165.1, 186.7.

(E)-3-(3-(2,5-dimethoxyphenyl)acryloyl)-7-methoxy-2H-chromen-2-one (**ChC8**). 3-Acetyl-7-methoxy-2*H*-chromen-2-one (0.44 g, 2.0 mmol) and 2,5-dimethoxybenzaldehide (0.33 g, 2.0 mmol), were reacted and worked up according to the previous procedure: 0,45 g, bright yellow solid, 62%, m.p. 174–176 °C; ^{1}H-NMR ($CDCl_3$): δ 8.55 (s, 1H, =C-H), 8.17 (d, 1H, Ar-CH=, *J* = 15.6 Hz), 8.0 (d, 1H, CO-CH=, *J* = 15.6 Hz), 7.6 (dd, 1H, Ar-H, *J* = 8.0, 9.0 Hz), 7.2 (d, 1H, Ar-H, *J* = 2.0 Hz), 6.92-6.97 (m, 4H, Ar-H), 3.91 (s, 3H, OCH_3), 3.87 (s, 3H, OCH_3),

3.82 (s, 3H, OCH_3); ^{13}C-NMR (CDCl3): δ 56.2, 56.4, 56.6, 100.7, 100.7, 112.8, 112.9, 113.7, 114.1, 114.3, 118.3, 125.1, 131.7, 131.9, 139.9, 148.6, 153.9, 154.0, 158.0, 165.4, 187.1.

(E)-3-(3-(benzo[d][1,3]dioxol-5-yl)acryloyl)-7-methoxy-2H-chromen-2-one (**ChC9**). 3-Acetyl-7-methoxy-2*H*-chromen-2-one (0.44 g, 2.0 mmol) and piperonal (0.30 g, 2.0 mmol) were reacted and worked up according to the previous procedure: 0.29 g, yellow solid, 41%, m.p. 178–180 °C; ^{1}H-NMR (DMSO-d_6): δ 8.62 (s, 1H, =C-H), 7.84 (d, 1H, *J* = 12.0 Hz), 7.63 (d, 1H, Ar-CH=, *J* = 15.7 Hz), 7.56 (d, 1H, CO-CH= *J* = 15.7 Hz), 7.68 (d, 1H, Ar-H, *J* = 8.0 Hz), 7.33 (s, 1H, Ar-H), 7.25 (d, 1H, Ar-H, *J* = 8.0 Hz), 7.06 (s, 1H, Ar-H), 7.03-6.94 (m, 3H, Ar-H), 6.07 (s, 2H, OCH_2O), 3.87 (s, 3H, OCH_3); ^{13}C-NMR ($CDCl_3$): δ 55.8, 100.6, 101.2, 106.7, 108.4, 111.5, 122.8, 125.7, 128.1, 130.8, 134.6, 142.2, 147.3, 147.9, 153.6, 160.9, 163.5, 187.1.

(E)-3-(3-(4-hydroxy-3-methoxyphenylacryloyl)-7-methoxy-2H-chromen-2-one (**ChC10**). 3-Acetyl-7-methoxy-2*H*-chromen-2-one (0.44 g, 2.0 mmol) and vainillin (0.30 g, 2.0 mmol), were reacted and worked up according to the previous procedure: 0,340 g, yellow solid, 48.3%, m.p. 210–212 °C; ^{1}H-NMR ($CDCl_3$): δ 8.65 (s, 1H, =C-H), 8.27 (d, 1H, Ar-CH=, *J* = 16 Hz), 8.10 (d, 1H, CO-CH=, *J* = 16 Hz), 7.65 (d, 1H, Ar-H, *J* = 8.0 Hz), 7.32 (d, 1H, Ar-H, *J* = 2.0 Hz), 7.45 (dd, 1H, Ar-H, *J* = 8.8, 2.0 Hz) 7.00 (dd, 1H, Ar-H, *J* = 8.8, 2.0 Hz), 6.91 (d, 1H, Ar-H, *J* = 8.8 Hz), 6.93 (d, 1H, Ar-H, *J* = 2.0 Hz), 3.97 (s, 3H, OCH_3), 3.92 (s, 3H, OCH_3); ^{13}C-NMR ($CDCl_3$): δ 55.9, 56.0, 100.4, 112.4, 112.5, 113.3, 113.7, 117.9, 121.7, 124.6, 124.8, 131.3, 139.6, 148.2, 153.5, 153.6, 157.6, 159.7, 165.7, 186.7.

(E)-3-(3-(4-mercaptophenyl)acryloyl)-7-methoxy-2H-chromen-2-one (**ChC11**). 3-Acetyl-7-methoxy-2*H*-chromen-2-one (0.44 g, 2.0 mmol) and methyl(4-vinylphenyl)sulfane (0.27g, 2.0 mmol), were reacted and worked up according to the previous procedure: 0,305 g, yellow solid, 43%.m.p. 196–198 °C; ^{1}H-NMR ($CDCl_3$): δ 8.58 (s, 1H, =C-H), 7.99 (d, 1H, Ar-CH=, *J* = 15.7 Hz), 7.82 (d, 1H, CO-CH=, *J* = 15.7 Hz), 7.58 (m, 3H, Ar-H), 7.26 (m, 3H, Ar-H 6.91 (dd, 1H, Ar-H, *J* = (8.0, 2.0), 6.86 (d, 1H, Ar-H, *J* = 2.0 Hz), 3.91 (s, 3H, OCH_3), 2.50 (s, 3H, SCH_3); ^{13}C-NMR ($CDCl_3$): δ 15.6, 56.4, 100.8, 112.9, 114.3, 121.9, 123.6, 126.3, 129.6, 131.7, 131.9, 142.9, 144.4, 148.8, 158.1, 160.2, 165.5, 165.6, 186.7.

(E)-3-(3-(3,5-dibromo-4-methoxyphenyl)acryloyl)-7-methoxy-2H-chromen-2-one (**ChC12**). 3-Acetyl-7-methoxy-2*H*-chromen-2-one (0,44 g, 2.0 mmol) and 3,5-dibromo-4-methoxybenzaldehyde (0.58 g, 2.0 mmol), were reacted and worked up according to the previous procedure: 0.21 g, yellow solid, 21.4%, m.p. 228–230 °C; ^{1}H-NMR ($CDCl_3$): δ 8.57 (s, 1H, =C-H), 8.18. (s, 2H, Ar-H), 7.95 (d, 1H, Ar-CH=, *J* = 16 Hz), 7.81 (d, 1H, CO-CH=, *J* = 16 Hz), 7.63 (d, 1H, Ar-H, *J* = 8.0 Hz), 7.22 (dd, 1H, Ar-H, *J* = 8.0, 2.0 Hz), 6.98 (d, 1H, Ar-H, *J* = 2.0 Hz) 4.01 (s, 3H, OCH_3), 3.95 (s, 3H, OCH_3); ^{13}C-NMR ($CDCl_3$): δ 60.9, 56.5, 100.5, 110.8, 111.7, 117.4, 123.7, 126.9, 132.3, 133.6, 136.1, 141.9, 147.7, 154.0, 154.5, 157.6, 159.4, 163.2, 188.1.

(E)-3-(3-(4-(dimethylamino)phenyl)acryloyl)-7-methoxy-2H-chromen-2-one (**ChC13**). 3-Acetyl-7-methoxy-2*H*-chromen-2-one (0.44 g, 2.0 mmol) and 4-(dimethylamino)benzaldehyde (0.27 g, 2.0 mmol), were reacted and worked up according to the previous procedure: 0.15 g, red solid, 22%, m.p. 220–222 °C; ^{1}H-NMR ($CDCl_3$): δ 8.56 (s, 1H, =C-H), 7.97 (d, 1H, Ar-CH=, *J* = 15.7 Hz), 7.81 (d, 1H, CO-CH=, *J* = 15.7 Hz), 7.58 (d, 1H, Ar-H, *J* = 8.8 Hz), 7.54 (d, 2H, Ar-H, *J* = 8.6 Hz), 6.88 (d, 1H, Ar-H, *J* = 8.0, 2.0 Hz), 6.85 (d, 1H, Ar-H, *J* = 2.0 Hz), 6.68 (d, 2H, Ar-H, *J* = 8.8 Hz), 3.90 (s, 3H, OCH_3), 3.04 (s, 6H, $N(CH_3)_2$); ^{13}C-NMR ($CDCl_3$): δ 40.5, 56.4, 64.1, 100.7, 112.2, 113.0, 114.0, 119.3, 122.6, 123.3, 131.4, 131.5, 146.4, 148.2, 152.6, 158.0, 160.3, 165.2, 186.4.

(E)-3-(3-(4-bromophenyl)acryloyl)-7-methoxy-2H-chromen-2-one (**ChC14**). 3-Acetyl-7-methoxy-2*H*-chromen-2-one (0.44 g, 2.0 mmol) and *p*-methoxybenzaldehide (0.27 g, 2.0 mmol), were reacted and worked up according to the previous procedure: 0.27 g, pale white solid, 35%, m.p. 158–160 °C; ^{1}H-NMR ($CDCl_3$): δ 8.76 (s, 1H, =C-H), 7.94 (d, 1H, Ar-H, *J* = 8.0, 8.0 Hz), 7.82 (d, 1H, Ar-CH=, *J* = 16 Hz), 7.71 (d, 1H, Ar-CH=, *J* = 16 Hz), 7.33 (d, 1, Ar-H, *J* = 8.0, 8.0 Hz), 7.25-7.18 (m, 2H, Ar-H), 7.15 (s, 1H, Ar-H), 7.09 (dd, 1H Ar-H, *J* = 8.0, 2.0 Hz), 6.94 (dd, 1H Ar-H, *J* = 8.0, 2.0 Hz), 3.97 (s, 3H, OCH_3); ^{13}C-NMR ($CDCl_3$): δ 57.2, 60.8, 110.2, 112.4, 118.5, 126.1, 129.6, 133.0, 134.2, 136.5, 142.0, 147.3, 154.1, 154.4, 159.6, 160.5, 186.7.

3.3. Biological Assessment

The effect of coumarin derivatives on MAO-A and MAO-B were measured using a suspension of crude rat brain mitochondria as enzyme source. 4-Dimethylaminophenethylamine (4-DMAPEA, 2.5 μM) and 5-hydroxytryptamine (5-HT, 100 mM) were used as substrates selective of MAO-B or MAO-A, respectively. Evaluation of the test compounds on rMAO activity was executed by measuring their effects on the production of 4-dimethylaminophenylacetic acid (DMAPAA) by rMAO-B and 5-hydroxyindoleacetic acid (5-HIAA) by rMAO-A with O_2 using HPLC-ED (L-7110 LaChrom and amperometric detector L-3500 LaChrom Recipe, Hitachi, (Tokyo, Japan) (for more detail see methodological references [41,42]). The IC_{50} values (average ± SD was measured in two independent experiments each in triplicate) were assessed representing percentage of inhibition in function of the negative logarithm of different inhibitor concentrations (10^{-4} to 10^{-8}) using the GraphPad Prism software [43].

3.4. Computational Analysis

3.4.1. Homology Modeling

Human monoamine oxidase B (hMAO-B) at 1.6Å resolution was used as template (PDB code 1OJ9) to obtain a 3D structure of rat MAO-B (rMAO-B) using homology modeling. The amino acid sequence and crystal structure of the protein was extracted from NCBI and PDB databases [44,45] considering the high level of amino acid identity (around 90%) the target protein and template were aligned through a single alignment using MultAlin interface [46]. MODELLER9v6 program [47] was used and 100 structures were prepared using standard parameters and the outcomes were ranked on the basis of the internal scoring function of the program (DOPE score). The best model was chosen as the target model. The cofactor FAD was placed inside of MAO using the corresponding crystal coordinates. To analyze the rMAO-B model, VMD program [48] was used to evaluate the 3D distribution and general physical chemistry characteristics. Then, stereochemical and energetic quality of the homology models was evaluated using PROSAII server [49], ANOLEA server [50] and Procheck program [51]. The crystal structure of rMAO-A (PDB code 1O5W [52]) and model of rMAO-B isoform were submited to H++ server [53,54] to computes pK values of ionizable groups and adds missing hydrogen atoms according to the specified pH of the environment as is described in H++ server.

3.4.2. Molecular Docking

Coumarin-chalcone hybrids were docked in the binding cavity of rMAO-A (PDB code 1O5W) and homology model for rMAO-B using AutoDock 4.012 suite. In general, the grid maps were calculated using the AutoGrid 4.0 option and were centered on the sites described before. The volume chosen for the grid maps were made up of 60 × 60 × 60 points, with a grid-point spacing of 0.375 Å. The author's option was used to define the rotating bond in the ligand. In the Lamarckian genetic algorithm (LGA) dockings, an initial population of random individuals with a population size of 150 individuals, a maximum number of 2.5×10^7 energy evaluations, a maximum number generation of 27,000, a mutation rate of 0.02 and crossover rate of 0.80 were employed. Each complex was built using the lowest docked-energy binding positions. Van der Waals interaction cutoff distances were set at 12 Å and dielectric constant was 10. The partial charges of each ligand were determined with PM6-D3H4 semi-empirical method [55,56] implemented in the MOPAC2016 [57] software. PM6-D3H4 [56] introduces dispersion and hydrogen-bonded corrections to the PM6 method.

3.4.3. Ligand Efficiency Approach

Ligand efficiency (LE) calculations were performed using one parameter K_d. The K_d parameter corresponds to the dissociation constant between a ligand/protein, and their value indicates the bond strength between the ligand/protein [34–36]. Low values indicate strong binding of the molecule to the protein. K_d calculations were done using the following Equations (1) and (2):

$$\Delta G^0 = -2.303RTlog(K_d) \quad (1)$$

$$K_d = 10^{\frac{\Delta G^0}{2.303RT}} \quad (2)$$

where ΔG^0 is the binding energy (kcal·mol^{-1}) obtained from docking experiments, R is the gas constant, and T is the temperature in Kelvin. In standard conditions of aqueous solution at 298.15 K, neutral pH and remaining concentrations of 1 M. The LE allows us to compare molecules according to their average binding energy [36,58]. Thus, it determined as the ratio of binding energy per non-hydrogen atom, as follows (Equation (3)) [34–36,59]:

$$LLE = -\frac{2.303RT}{HAC}\log(K_d) \quad (3)$$

where K_d is obtained from Equation (2) and HAC denotes the heavy atom count (i.e., number of non-hydrogen atoms) in a ligand.

3.4.4. Molecular Dynamic Simulations

Two complexes were built for each modeled **ChC2**/rMAO-B and **ChC4**/rMAO-B, and each model was confined inside a periodic simulation box. Water model TIP3P [60] with 20.459 molecules was used as solvent. Furthermore, Na^+ and Cl^- ions were added to neutralize the systems and maintain an ionic concentration of 0.15 mol·L^{-1}.The full geometry optimizations of the two molecules were carried out with the density functional theory method by a M05-2X [61]-D3 [62] in conjunction with the 6-31G(d,p) basis set. **ChC2**, **ChC4** and FAD compounds were parametrized using LigParGen web server and implementing the OPLS-AA/1.14*CM1A(-LBCC) force field parameters for organic ligands [63–65]. The partial charges of each ligand were determined with generated by the restrained electrostatic potential (RESP) model [66]. MD simulations were carried out using the modeled CHARMM22 and CHARMM36 force fields [67,68] within the NAMD software [69]. First, each system included 20,000 steps of conjugate-gradient energy minimization followed by 10 ns of simulation with the protein backbone atoms fixed and gradually releasing the backbone over 50,000 ps with 10 to 0.001 kcal·mol^{-1}Å^{-2} restraints. The total duration of simulation was approximately 100 ns for each system. During the MD simulations, motion equations were integrated with a 1 fs time step in the NPT ensemble at a pressure of 1 atm. The SHAKE algorithm was applied to all hydrogen atoms, and the van der Waals cutoff was set to 12 Å. The temperature was maintained at 310 K, employing the Nosée-Hoover thermostat method with a relaxation time of 1 ps. The Nosée-Hoover-Langevin piston was used to control the pressure at 1 atm. Long-range electrostatic forces were taken into account by means of the particle-mesh Ewald approach. Data were collected every 1 ps during the MD runs. Molecular visualization of the systems and MD trajectory analysis were carried out with the VMD software package [48].

3.4.5. Free Energy Calculation

The molecular MM/GBSA method was employed to estimate the binding free energy of the rMAO-B/ligand complexes. For calculations from a total of 100 ns of MD, the last 70 ns were extracted for analysis, and the explicit water molecules and ions were removed. The MM/GBSA analysis was performed on three subsets of each system: the protein alone, the ligand alone, and the complex (protein-ligand). For each of these subsets, the total free energy (ΔG_{tot}) was calculated as follows (Equation (4)):

$$\Delta G_{tot} = E_{MM} + G_{solv} - T\Delta S_{conf} \quad (4)$$

where E_{MM} is the bonded and Lennard–Jones energy terms; G_{solv} is the polar contribution of solvation energy and non-polar contribution to the solvation energy; T is the temperature;

and ΔS_{conf} corresponds to the conformational entropy [70]. Both E_{MM} and G_{solv} were calculated using NAMD software with the generalized Born implicit solvent model [71,72]. ΔG_{tot} was calculated as a linear function of the solvent-accessible surface area, which was calculated with a probe radius of 1.4 Å [73]. The binding free energy of rMAO-B and ligand complexes (ΔG_{bind}) were calculated by the difference where ΔS_{conf} values are the averages over the simulation (Equation (5)):

$$\Delta G_{bind} = \Delta G_{tot}(complex) - \Delta G_{tot}(protein) - \Delta G_{tot}(ligand) \quad (5)$$

3.4.6. ADMET Prediction

The ADMET properties of a compound deal with its absorption, distribution, metabolism, excretion, and toxicity in and through the human body. ADMET, which constitutes the pharmacokinetic profile of a drug molecule, is very essential in evaluating its pharmacodynamic activities. In this study for all molecules, we have used the SwissADME [74] prediction tool, for in silico physicochemical properties such as molecular hydrogen bond acceptor (*HBA*), hydrogen bond donor (*HBD*), weight (*MW*), topological polar surface area (*TPSA*), rotatable bond count (*RB*), octanol/water partition coefficient (*LogP*), water solubility (*LogS*) and skin permeation (*logKp*). Further the ligands were analyzed for Bioavailability property using Boiled Egg analysis [75].

4. Conclusions

Fourteen compounds derived from chalcocoumarins were synthesized and evaluated against monoamine oxidase enzyme isoforms. The experimental results obtained against MAO-A and MAO-B show that the compounds **ChC4**, **ChC5**, **ChC6**, **ChC9** and **ChC11** exhibit MAO-B affinity at micro and sub-micromolar concentrations, in particular **ChC4** which shows an IC_{50} value of 0 0.76 ± 0.08 µM. Where compound **ChC4** is highlighted in molecular modeling, ADMET predictions, docking and MM/GBSA calculations, these results suggest that compound **ChC4** has the appropriate interactions with the active site of rMAO-B. Furthermore, the ADMETox values obtained for the compound **ChC4** indicate adequate solubility in the gastrointestinal tract, which is a favourable indication for it to be a promising drug candidate for further research and development. This compound complies with the interactions described for the active site of rMAO-, fitting into a distance close enough to the nitrogen atom of the aloxazine planar ring of FAD to form an interaction necessary for the inhibition of rMAO-B. These analyses may be important initial steps towards the development of new drugs in the fight against depressive disorder and Parkinson's disease.

Supplementary Materials: The following are available online. **Scheme S1**. Synthetic route to compounds **ChC1–ChC14; Figure S1–S15**: ^{1}H and ^{13}C NMR spectrum of **ChC1–ChC14**: **Figure S16**. Homology modeling analysis; **Figure S17**. Ramachandran plot generated via PROCHECK for the rMAO-B model; **Figure S18**. Alignment of all the ChC ligands docked in complex with rMAO-B.

Author Contributions: O.Y., M.T.N., C.A., A.F. and O.G.-B. contributed to the conception and design of the study; O.Y., A.F. and A.G.-G. preformed the theoretical calculations; O.Y., A.F., and O.G.-B. organized the database; A.F., M.T.N., G.M.-A. and N.M. design and performance of biological assay; O.Y., A.F. and O.G.-B. wrote the first draft of the manuscript. All authors contributed to manuscript revision, read and approved the submitted version. All authors have read and agreed to the published version of the manuscript.

Funding: Ministry of Science, Technology and Innovation, the Ministry of Education, the Ministry of Industry, Commerce and Tourism, and ICETEX, Programme Ecosistema Científico-Colombia Científica, from the Francisco José de Caldas Fund, Grand RC-FP44842-212-2018.

Institutional Review Board Statement: Not applicable.

Informed Consent Statement: Not applicable.

Conflicts of Interest: The authors declare that there is no conflict of interest.

Sample Availability: Samples of compounds **ChC1–ChC14** are available from the authors.

References

1. Rodríguez-Enríquez, F.; Costas-Lago, M.C.; Besada, P.; Alonso-Pena, M.; Torres-Terán, I.; Viña, D.; Fontenla, J.Á.; Sturlese, M.; Moro, S.; Quezada, E.; et al. Novel coumarin-pyridazine hybrids as selective MAO-B inhibitors for the Parkinson's disease therapy. *Bioorg. Chem.* **2020**, *104*, 104203. [CrossRef] [PubMed]
2. Garćia-Beltran, O.; Mena, N.P.; Aguirre, P.; Barriga-Gonzalez, G.; Galdamez, A.; Nagles, E.; Adasme, T.; Hidalgo, C.; Nuñez, M.T. Development of an iron-selective antioxidant probe with protective effects on neuronal function. *PLoS ONE* **2017**, *12*, e0189043. [CrossRef]
3. Aguirre, P.; García-Beltrán, O.; Tapia, V.; Muñoz, Y.; Cassels, B.K.; Núñez, M.T. Neuroprotective Effect of a New 7,8-Dihydroxycoumarin-Based Fe^{2+}/Cu^{2+} Chelator in Cell and Animal Models of Parkinson's Disease. *ACS Chem. Neurosci.* **2017**, *8*, 178–185. [CrossRef]
4. Al-Warhi, T.; Sabt, A.; Elkaeed, E.B.; Eldehna, W.M. Recent advancements of coumarin-based anticancer agents: An up-to-date review. *Bioorg. Chem.* **2020**, *103*, 104163. [CrossRef] [PubMed]
5. Liu, G.L.; Liu, L.; Hu, Y.; Wang, G.X. Evaluation of the antiparasitic activity of coumarin derivatives against Dactylogyrus intermedius in goldfish (*Carassius auratus*). *Aquaculture* **2021**, *533*, 736069. [CrossRef]
6. Sahoo, C.R.; Sahoo, J.; Mahapatra, M.; Lenka, D.; Kumar Sahu, P.; Dehury, B.; Nath Padhy, R.; Kumar Paidesetty, S. Coumarin derivatives as promising antibacterial agent(s). *Arab. J. Chem.* **2021**, *14*, 102922. [CrossRef]
7. Cione, E.; La Torre, C.; Cannataro, R.; Caroleo, M.C.; Plastina, P.; Gallelli, L. Quercetin, Epigallocatechin Gallate, Curcumin, and Resveratrol: From Dietary Sources to Human MicroRNA Modulation. *Molecules* **2020**, *25*, 63. [CrossRef]
8. Oliver, D.M.A.; Reddy, P.H. Small molecules as therapeutic drugs for Alzheimer's disease. *Mol. Cell. Neurosci.* **2019**, *96*, 47–62. [CrossRef] [PubMed]
9. Lv, D.; Hu, Z.; Lu, L.; Lu, H.; Xu, X. Three-dimensional cell culture: A powerful tool in tumor research and drug discovery. *Oncol. Lett.* **2017**, *14*, 6999–7010. [CrossRef]
10. Rodríguez-Enríquez, F.; Viña, D.; Uriarte, E.; Fontenla, J.A.; Matos, M.J. Discovery and optimization of 3-thiophenylcoumarins as novel agents against Parkinson's disease: Synthesis, in vitro and in vivo studies. *Bioorg. Chem.* **2020**, *101*, 103986. [CrossRef]
11. Patil, P.O.; Bari, S.B.; Firke, S.D.; Deshmukh, P.K.; Donda, S.T.; Patil, D.A. A comprehensive review on synthesis and designing aspects of coumarin derivatives as monoamine oxidase inhibitors for depression and Alzheimer's disease. *Bioorg. Med. Chem.* **2013**, *21*, 2434–2450. [CrossRef]
12. Carotti, A.; Altomare, C.; Catto, M.; Gnerre, C.; Summo, L.; De Marco, A.; Rose, S.; Jenner, P.; Testa, B. Lipophilicity plays a major role in modulating the inhibition on monoamine oxidase B by 7-substituted coumarins. *Chem. Biodivers.* **2006**, *3*, 134–149. [CrossRef] [PubMed]
13. Akyüz, M.A.; Erdem, S.S.; Edmondson, D.E. The aromatic cage in the active site of monoamine oxidase B: Effect on the structural and electronic properties of bound benzylamine and p-nitrobenzylamine. *J. Neural Transm.* **2007**, *114*, 693. [CrossRef] [PubMed]
14. Can, N.O.; Osmaniye, D.; Levent, S.; Sa Sağlik, B.N.; Inci, B.; Ilgin, S.; Özkay, Y.; Kaplancikli, Z.A. Synthesis of new hydrazone derivatives for MAO enzymes inhibitory activity. *Molecules* **2017**, *22*, 1381. [CrossRef] [PubMed]
15. Billett, E.E. Monoamine Oxidase (MAO) in Human Peripheral Tissues. *Neurotoxicology* **2004**, *25*, 139–148. [CrossRef]
16. Binda, C.; Wang, J.; Pisani, L.; Caccia, C.; Carotti, A.; Salvati, P.; Edmondson, D.E.; Mattevi, A. Structures of human monoamine oxidase B complexes with selective noncovalent inhibitors: Safinamide and coumarin analogs. *J. Med. Chem.* **2007**, *50*, 5848–5852. [CrossRef]
17. Novaroli, L.; Daina, A.; Favre, E.; Bravo, J.; Carotti, A.; Leonetti, F.; Catto, M.; Carrupt, P.A.; Reist, M. Impact of species-dependent differences on screening, design, and development of MAO B inhibitors. *J. Med. Chem.* **2006**, *49*, 6264–6272. [CrossRef] [PubMed]
18. Reyes-Parada, M.; Fierro, A.; Iturriaga-Vasquez, P.; Cassels, B. Monoamine Oxidase Inhibition in the Light of New Structural Data. *Curr. Enzym. Inhib.* **2006**, *1*, 85–95. [CrossRef]
19. Shih, J.C.; Chen, K.; Ridd, M.J. Monoamine oxidase: From genes to behavior. *Annu. Rev. Neurosci.* **1999**, *22*, 197–217. [CrossRef]
20. Matos, M.J.; Vazquez-Rodriguez, S.; Uriarte, E.; Santana, L.; Viña, D. MAO inhibitory activity modulation: 3-Phenylcoumarins versus 3-benzoylcoumarins. *Bioorg. Med. Chem. Lett.* **2011**, *21*, 4224–4227. [CrossRef]
21. Tong, J.; Meyer, J.H.; Furukawa, Y.; Boileau, I.; Chang, L.J.; Wilson, A.A.; Houle, S.; Kish, S.J. Distribution of monoamine oxidase proteins in human brain: Implications for brain imaging studies. *J. Cereb. Blood Flow Metab.* **2013**, *33*, 863–871. [CrossRef]
22. Viña, D.; Matos, M.J.; Ferino, G.; Cadoni, E.; Laguna, R.; Borges, F.; Uriarte, E.; Santana, L. 8-Substituted 3-Arylcoumarins as Potent and Selective MAO-B Inhibitors: Synthesis, Pharmacological Evaluation, and Docking Studies. *ChemMedChem* **2012**, *7*, 464–470. [CrossRef] [PubMed]
23. Sahoo, A.; Yabanoglu, S.; Sinha, B.N.; Ucar, G.; Basu, A.; Jayaprakash, V. Towards development of selective and reversible pyrazoline based MAO-inhibitors: Synthesis, biological evaluation and docking studies. *Bioorg. Med. Chem. Lett.* **2010**, *20*, 132–136. [CrossRef]
24. Chimenti, F.; Secci, D.; Bolasco, A.; Chimenti, P.; Granese, A.; Carradori, S.; Befani, O.; Turini, P.; Alcaro, S.; Ortuso, F. Synthesis, molecular modeling studies, and selective inhibitory activity against monoamine oxidase of N,N′-bis[2-oxo-2H-benzopyran]-3-carboxamides. *Bioorg. Med. Chem. Lett.* **2006**, *16*, 4135–4140. [CrossRef] [PubMed]

25. García-Beltrán, O.; Cassels, B.K.; Mena, N.; Nuñez, M.T.; Yañez, O.; Caballero, J. A coumarinylaldoxime as a specific sensor for Cu^{2+} and its biological application. *Tetrahedron Lett.* **2014**, *55*, 873–976. [CrossRef]
26. García-Beltrán, O.; Mena, N.; Pérez, E.G.; Cassels, B.K.; Nuñez, M.T.; Werlinger, F.; Zavala, D.; Aliaga, M.E.; Pavez, P. The development of a fluorescence turn-on sensor for cysteine, glutathione and other biothiols. A kinetic study. *Tetrahedron Lett.* **2011**, *52*, 6606–6609. [CrossRef]
27. Konidala, S.K.; Kotra, V.; Danduga, R.C.S.R.; Kola, P.K. Coumarin-chalcone hybrids targeting insulin receptor: Design, synthesis, anti-diabetic activity, and molecular docking. *Bioorg. Chem.* **2020**, *104*, 104207. [CrossRef]
28. Emam, S.H.; Sonousi, A.; Osman, E.O.; Hwang, D.; Do Kim, G.; Hassan, R.A. Design and synthesis of methoxyphenyl- and coumarin-based chalcone derivatives as anti-inflammatory agents by inhibition of NO production and down-regulation of NF-κB in LPS-induced RAW264.7 macrophage cells. *Bioorg. Chem.* **2021**, *107*, 104630. [CrossRef]
29. Kurt, B.Z.; Ozten Kandas, N.; Dag, A.; Sonmez, F.; Kucukislamoglu, M. Synthesis and biological evaluation of novel coumarin-chalcone derivatives containing urea moiety as potential anticancer agents. *Arab. J. Chem.* **2020**, *13*, 1120–1129. [CrossRef]
30. Vilches-Herrera, M.; Miranda-Sepúlveda, J.; Rebolledo-Fuentes, M.; Fierro, A.; Lühr, S.; Iturriaga-Vasquez, P.; Cassels, B.K.; Reyes-Parada, M. Naphthylisopropylamine and N-benzylamphetamine derivatives as monoamine oxidase inhibitors. *Bioorg. Med. Chem.* **2009**, *17*, 2452–2460. [CrossRef] [PubMed]
31. Lühr, S.; Vilches-Herrera, M.; Fierro, A.; Ramsay, R.R.; Edmondson, D.E.; Reyes-Parada, M.; Cassels, B.K.; Iturriaga-Vásquez, P. 2-Arylthiomorpholine derivatives as potent and selective monoamine oxidase B inhibitors. *Bioorg. Med. Chem.* **2010**, *18*, 1388–1395. [CrossRef] [PubMed]
32. Binda, C.; Aldeco, M.; Mattevi, A.; Edmondson, D.E. Interactions of Monoamine Oxidases with the Antiepileptic Drug Zonisamide: Specificity of Inhibition and Structure of the Human Monoamine Oxidase B Complex. *J. Med. Chem.* **2011**, *54*, 909–912. [CrossRef] [PubMed]
33. Binda, C.; Newton-Vinson, P.; Hubálek, F.; Edmondson, D.E.; Mattevi, A. Structure of human monoamine oxidase B, a drug target for the treatment of neurological disorders. *Nat. Struct. Biol.* **2002**, *9*, 22–26. [CrossRef] [PubMed]
34. Abad-Zapatero, C. Ligand efficiency indices for effective drug discovery. *Expert Opin. Drug Discov.* **2007**, *2*, 469–488. [CrossRef]
35. Abad-Zapatero, C.; Perišić, O.; Wass, J.; Bento, A.P.; Overington, J.; Al-Lazikani, B.; Johnson, M.E. Ligand efficiency indices for an effective mapping of chemico-biological space: The concept of an atlas-like representation. *Drug Discov. Today* **2010**, *15*, 804–811. [CrossRef]
36. Abad-Zapatero, C. *Ligand Efficiency Indices for Drug Discovery: Towards an Atlas-Guided Paradigm*; Elsevier: Amsterdam, The Netherlands, 2013; ISBN 978-0-12-404635-1.
37. Carugo, O. How root-mean-square distance (r.m.s.d.) values depend on the resolution of protein structures that are compared. *J. Appl. Crystallogr.* **2003**, *36*, 125–128. [CrossRef]
38. Shalaby, R.; Petzer, J.P.; Petzer, A.; Ashraf, U.M.; Atari, E.; Alasmari, F.; Kumarasamy, S.; Sari, Y.; Khalil, A. SAR and molecular mechanism studies of monoamine oxidase inhibition by selected chalcone analogs. *J. Enzyme Inhib. Med. Chem.* **2019**, *34*, 863–876. [CrossRef]
39. Binda, C.; Hubálek, F.; Li, M.; Edmondson, D.E.; Mattevi, A. Crystal structure of human monoamine oxidase B, a drug target enzyme monotopically inserted into the mitochondrial outer membrane. *FEBS Lett.* **2004**, *564*, 225–228. [CrossRef]
40. Milczek, E.M.; Binda, C.; Rovida, S.; Mattevi, A.; Edmondson, D.E. The 'gating' residues Ile199 and Tyr326 in human monoamine oxidase B function in substrate and inhibitor recognition. *FEBS J.* **2011**, *278*, 4860–4869. [CrossRef]
41. Fierro, A.; Osorio-Olivares, M.; Cassels, B.K.; Edmondson, D.E.; Sepúlveda-Boza, S.; Reyes-Parada, M. Human and rat monoamine oxidase-A are differentially inhibited by (S)-4-alkylthioamphetamine derivatives: Insights from molecular modeling studies. *Bioorg. Med. Chem.* **2007**, *15*, 5198–5206. [CrossRef]
42. Morales-Camilo, N.; Salas, C.O.; Sanhueza, C.; Espinosa-Bustos, C.; Sepúlveda-Boza, S.; Reyes-Parada, M.; Gonzalez-Nilo, F.; Caroli-Rezende, M.; Fierro, A. Synthesis, biological evaluation, and molecular simulation of chalcones and aurones as selective MAO-B inhibitors. *Chem. Biol. Drug Des.* **2015**, *85*, 685–695. [CrossRef]
43. *GraphPad One-Way ANOVA with Dunnett's Post Test was Performed Using GraphPad Prism*, version 9.00 for Windows 2021; GraphPad: San Diego, CA, USA.
44. Berman, H.M.; Westbrook, J.; Feng, Z.; Gilliland, G.; Bhat, T.N.; Weissig, H.; Shindyalov, I.N.; Bourne, P.E. The Protein Data Bank. *Nucleic Acids Res.* **2000**, *28*, 235–242. [CrossRef] [PubMed]
45. Geer, L.Y.; Marchler-Bauer, A.; Geer, R.C.; Han, L.; He, J.; He, S.; Liu, C.; Shi, W.; Bryant, S.H. The NCBI BioSystems database. *Nucleic Acids Res.* **2009**, *38*, D492–D496. [CrossRef] [PubMed]
46. Corpet, F. Multiple sequence alignment with hierarchical clustering. *Nucleic Acids Res.* **1988**, *16*, 10881–10890. [CrossRef]
47. Sali, A.; Blundell, T.L. Comparative protein modelling by satisfaction of spatial restraints. *J. Mol. Biol.* **1993**, *234*, 779–815. [CrossRef] [PubMed]
48. Humphrey, W.; Dalke, A.; Schulten, K. VMD: Visual molecular dynamics. *J. Mol. Graph.* **1996**, *14*, 33–38. [CrossRef]
49. Wiederstein, M.; Sippl, M.J. ProSA-web: Interactive web service for the recognition of errors in three-dimensional structures of proteins. *Nucleic Acids Res.* **2007**, *35*, W407–W410. [CrossRef]
50. Melo, F.; Devos, D.; Depiereux, E.; Feytmans, E. ANOLEA: A www server to assess protein structures. *Proc. Int. Conf. Intell. Syst. Mol. Biol.* **1997**, *5*, 187–190.

51. Laskowski, R.A.; MacArthur, M.W.; Moss, D.S.; Thornton, J.M. PROCHECK: A program to check the stereochemical quality of protein structures. *J. Appl. Crystallogr.* **1993**, *26*, 283–291. [CrossRef]
52. Ma, J.; Yoshimura, M.; Yamashita, E.; Nakagawa, A.; Ito, A.; Tsukihara, T. Structure of Rat Monoamine Oxidase A and Its Specific Recognitions for Substrates and Inhibitors. *J. Mol. Biol.* **2004**, *338*, 103–114. [CrossRef]
53. Gordon, J.C.; Myers, J.B.; Folta, T.; Shoja, V.; Heath, L.S.; Onufriev, A. H++: A server for estimating p Ka s and adding missing hydrogens to macromolecules. *Nucleic Acids Res.* **2005**, *33*, W368–W371. [CrossRef] [PubMed]
54. Anandakrishnan, R.; Onufriev, A. Analysis of Basic Clustering Algorithms for Numerical Estimation of Statistical Averages in Biomolecules. *J. Comput. Biol.* **2008**, *15*, 165–184. [CrossRef] [PubMed]
55. Stewart, J.J.P. Optimization of parameters for semiempirical methods V: Modification of NDDO approximations and application to 70 elements. *J. Mol. Model.* **2007**, *13*, 1173–1213. [CrossRef] [PubMed]
56. Řezáč, J.; Hobza, P. Advanced corrections of hydrogen bonding and dispersion for semiempirical quantum mechanical methods. *J. Chem. Theory Comput.* **2012**, *8*, 141–151. [CrossRef] [PubMed]
57. Stewart, J.J.P. *MOPAC 2016*, Stewart Computational Chemistry: Colorado Springs, CO, USA.
58. Reynolds, C.H.; Tounge, B.A.; Bembenek, S.D. Ligand Binding Efficiency: Trends, Physical Basis, and Implications. *J. Med. Chem.* **2008**, *51*, 2432–2438. [CrossRef]
59. Cavalluzzi, M.M.; Mangiatordi, G.F.; Nicolotti, O.; Lentini, G. Ligand efficiency metrics in drug discovery: The pros and cons from a practical perspective. *Expert Opin. Drug Discov.* **2017**, *12*, 1087–1104. [CrossRef]
60. Neria, E.; Fisher, S.; Karplus, M. Simulation of activation free energies in molecular systems. *J. Chem. Phys.* **1996**, *105*, 1902–1921. [CrossRef]
61. Zhao, Y.; Schultz, N.E.; Truhlar, D.G. Design of Density Functionals by Combining the Method of Constraint Satisfaction with Parametrization for Thermochemistry, Thermochemical Kinetics, and Noncovalent Interactions. *J. Chem. Theory Comput.* **2006**, *2*, 364–382. [CrossRef]
62. Grimme, S.; Ehrlich, S.; Goerigk, L. Effect of the damping function in dispersion corrected density functional theory. *J. Comput. Chem.* **2011**, *32*, 1456–1465. [CrossRef]
63. Dodda, L.S.; Vilseck, J.Z.; Tirado-Rives, J.; Jorgensen, W.L. 1.14*CM1A-LBCC: Localized Bond-Charge Corrected CM1A Charges for Condensed-Phase Simulations. *J. Phys. Chem. B* **2017**, *121*, 3864–3870. [CrossRef]
64. Dodda, L.S.; Cabeza de Vaca, I.; Tirado-Rives, J.; Jorgensen, W.L. LigParGen web server: An automatic OPLS-AA parameter generator for organic ligands. *Nucleic Acids Res.* **2017**, *45*, W331–W336. [CrossRef]
65. Jorgensen, W.L.; Tirado-Rives, J. Potential energy functions for atomic-level simulations of water and organic and biomolecular systems. *Proc. Natl. Acad. Sci. USA* **2005**, *102*, 6665–6670. [CrossRef]
66. Bayly, C.I.; Cieplak, P.; Cornell, W.D.; Kollman, P.A. A well-behaved electrostatic potential based method using charge restraints for deriving atomic charges: The RESP model. *J. Phys. Chem.* **1993**, *97*, 10269–10280. [CrossRef]
67. MacKerell, A.D., Jr.; Bashford, D.; Bellott, M.L.D.R.; Dunbrack, R.L., Jr.; Evanseck, J.D.; Field, M.J.; Fischer, S.; Gao, J.; Guo, H.; Ha, S.; et al. All-Atom Empirical Potential for Molecular Modeling and Dynamics Studies of Proteins. *J. Phys. Chem. B* **1998**, *102*, 3586–3616. [CrossRef]
68. Huang, J.; MacKerell, A.D., Jr. CHARMM36 all-atom additive protein force field: Validation based on comparison to NMR data. *J. Comput. Chem.* **2013**, *34*, 2135–2145. [CrossRef] [PubMed]
69. Kalé, L.; Skeel, R.; Bhandarkar, M.; Brunner, R.; Gursoy, A.; Krawetz, N.; Phillips, J.; Shinozaki, A.; Varadarajan, K.; Schulten, K. NAMD2: Greater scalability for parallel molecular dynamics. *J. Comput. Phys.* **1999**, *151*, 283–312. [CrossRef]
70. Hayes, J.M.; Archontis, G. MM-GB(PB)SA Calculations of Protein-Ligand Binding Free Energies. In *Molecular Dynamics*; Wang, L., Ed.; IntechOpen: Rijeka, Croatia, 2012.
71. Song, L.; Lee, T.-S.; Zhu, C.; York, D.M.; Merz, K.M., Jr. Using AMBER18 for Relative Free Energy Calculations. *J. Chem. Inf. Model.* **2019**, *59*, 3128–3135. [CrossRef] [PubMed]
72. Götz, A.W.; Williamson, M.J.; Xu, D.; Poole, D.; Le Grand, S.; Walker, R.C. Routine Microsecond Molecular Dynamics Simulations with AMBER on GPUs. 1. Generalized Born. *J. Chem. Theory Comput.* **2012**, *8*, 1542–1555. [CrossRef] [PubMed]
73. Abroshan, H.; Akbarzadeh, H.; Parsafar, G.A. Molecular dynamics simulation and MM-PBSA calculations of sickle cell hemoglobin in dimer form with Val, Trp, or Phe at the lateral contact. *J. Phys. Org. Chem.* **2010**, *23*, 866–877. [CrossRef]
74. Daina, A.; Michielin, O.; Zoete, V. SwissADME: A free web tool to evaluate pharmacokinetics, drug-likeness and medicinal chemistry friendliness of small molecules. *Sci. Rep.* **2017**, *7*, 1–13. [CrossRef]
75. Daina, A.; Zoete, V. A BOILED-Egg To Predict Gastrointestinal Absorption and Brain Penetration of Small Molecules. *ChemMedChem* **2016**, *11*, 1117–1121. [CrossRef]

Article

Identification and Quantification of Coumarins by UHPLC-MS in *Arabidopsis thaliana* Natural Populations

Izabela Perkowska [1], Joanna Siwinska [1], Alexandre Olry [2], Jérémy Grosjean [2], Alain Hehn [2], Frédéric Bourgaud [3], Ewa Lojkowska [1] and Anna Ihnatowicz [1,*]

[1] Intercollegiate Faculty of Biotechnology of University of Gdansk and Medical University of Gdansk, University of Gdansk, Abrahama 58, 80-307 Gdansk, Poland; izabela.perkowska@phdstud.ug.edu.pl (I.P.); siwinskajoanna@gmail.com (J.S.); ewa.lojkowska@biotech.ug.edu.pl (E.L.)

[2] Université de Lorraine-INRAE, LAE, 54000 Nancy, France; alexandre.olry@univ-lorraine.fr (A.O.); jeremy.grosjean@univ-lorraine.fr (J.G.); alain.hehn@univ-lorraine.fr (A.H.)

[3] Plant Advanced Technologies, 54500 Vandœuvre-lès-Nancy, France; frederic.bourgaud@plantadvanced.com

* Correspondence: anna.ihnatowicz@biotech.ug.edu.pl; Tel.: +48-58-5236330

Abstract: Coumarins are phytochemicals occurring in the plant kingdom, which biosynthesis is induced under various stress factors. They belong to the wide class of specialized metabolites well known for their beneficial properties. Due to their high and wide biological activities, coumarins are important not only for the survival of plants in changing environmental conditions, but are of great importance in the pharmaceutical industry and are an active source for drug development. The identification of coumarins from natural sources has been reported for different plant species including a model plant *Arabidopsis thaliana*. In our previous work, we demonstrated a presence of naturally occurring intraspecies variation in the concentrations of scopoletin and its glycoside, scopolin, the major coumarins accumulating in Arabidopsis roots. Here, we expanded this work by examining a larger group of 28 Arabidopsis natural populations (called accessions) and by extracting and analysing coumarins from two different types of tissues–roots and leaves. In the current work, by quantifying the coumarin content in plant extracts with ultra-high-performance liquid chromatography coupled with a mass spectrometry analysis (UHPLC-MS), we detected a significant natural variation in the content of simple coumarins like scopoletin, umbelliferone and esculetin together with their glycosides: scopolin, skimmin and esculin, respectively. Increasing our knowledge of coumarin accumulation in Arabidopsis natural populations, might be beneficial for the future discovery of physiological mechanisms of action of various alleles involved in their biosynthesis. A better understanding of biosynthetic pathways of biologically active compounds is the prerequisite step in undertaking a metabolic engineering research.

Keywords: analytical methods; model plant; natural genetic variation; natural products; simple coumarins

Citation: Perkowska, I.; Siwinska, J.; Olry, A.; Grosjean, J.; Hehn, A.; Bourgaud, F.; Lojkowska, E.; Ihnatowicz, A. Identification and Quantification of Coumarins by UHPLC-MS in *Arabidopsis thaliana* Natural Populations. *Molecules* **2021**, 26, 1804. https://doi.org/10.3390/molecules26061804

Academic Editor: Maria João Matos

Received: 27 February 2021
Accepted: 18 March 2021
Published: 23 March 2021

1. Introduction

Coumarins are secondary metabolites widely distributed throughout the plant kingdom. They are synthetized via the phenylpropanoid biosynthesis pathway. We can distinguish several simple coumarins like coumarin, scopoletin (7-hydroxy-6-methoxycoumarin), esculetin (6,7-dihydroxycoumarin), umbelliferone (7-hydroxycoumarin), fraxetin (7,8-dihydroxy-6-methoxycoumarin), sideretin (5,7,8-trihydroxy-6-methoxycoumarin) and their respective glycosylated forms–scopolin, esculetin, skimmin, fraxin and sideretin-glycoside, respectively [1]. Figure 1 presents the semi-developed formula of simple coumarins and their glycosides derivatives identified in this research.

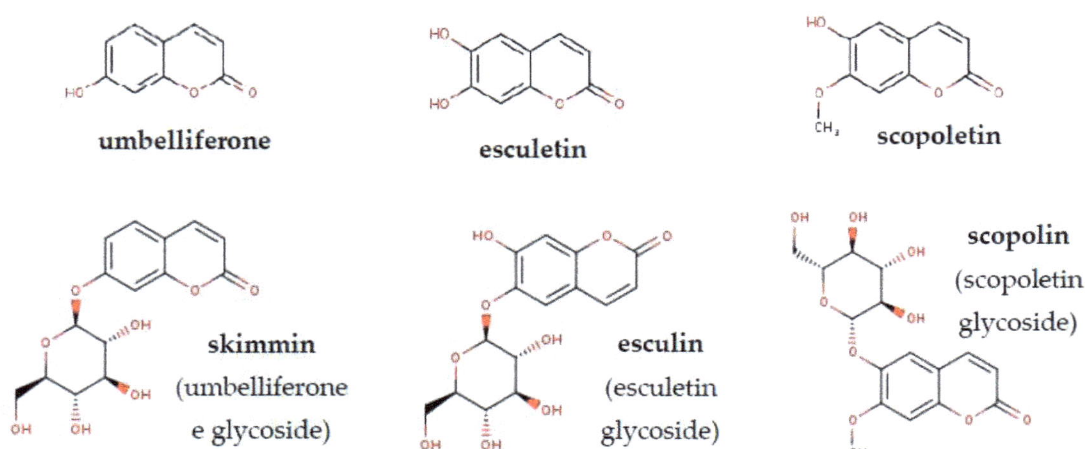

Figure 1. Chemical structures of simple coumarins and their glycosides analysed in this work (www.chem-space.com (accessed on 20 January 2021)).

Coumarins have been recognized for many years as an important class of pharmacologically active compounds. They have anticoagulant, anticancer, antiviral and anti-inflammatory properties [2]. In addition to the listed medicinal benefits, it was shown recently in numerous studies that coumarins play an important role in iron (Fe) homeostasis, oxidative stress response, plant-microbe interactions and that they can act as signalling molecules in plants [3–8]. In the last few years, an increasing number of reports concern the analysis of root extracts and root exudates that are rich in phenolic compounds, such as simple coumarins, which mediate multiple interactions in the rhizosphere. Coumarins were shown to have a strong impact on the plant interactions with microorganisms and play a crucial role in nutrient acquisition [6,9–18]. Moreover, the root-secreted scopoletin was proved to exert a selective antimicrobial action in the rhizosphere [8]. These numerous reports examining the biochemical and physiological functions of coumarins, make this class of specialized metabolites extremely interesting from a scientific and commercial point of view. The vast majority of these studies were performed using a reference accession, Col-0, of the model plant *Arabidopsis thaliana* (hereinafter Arabidopsis), and its mutants defective at various steps of coumarin biosynthesis.

Here, we conducted the qualitative and quantitative assessment of coumarin content in leaf and root tissue of a set of Arabidopsis natural populations (accessions). Numerous studies on primary and specialized metabolites profiling were conducted using the Arabidopsis model system [19–28]. Previously, due to the importance of coumarins for human health, most research on their metabolic profiling were carried out on plants of economic importance, such as e.g., sweet potato (*Ipomoea batatas* L.), rue (*Ruta graveolens* L.) or lettuce (*Lactuca sativa* L.) [29–34]. One of the first metabolic profiling of root exudates using Arabidopsis natural populations (Col-0, C24, Cvi-0, Ler) was made by Micallef et al. [35] who attempted to correlate them with the compositions of rhizobacterial communities. However, the authors of this work did not undertake the qualitative and quantitative evaluation of the isolated compounds. Consequently, they could only conclude that there are differences between Arabidopsis accessions in terms of the quality and quantity of released substances, which may have an impact on the composition of the rhizobacterial communities.

So far, only a few more studies focusing on the accumulation of coumarins in Arabidopsis natural populations have been published. As shown by our group [36], a significant natural variation in the accumulation of coumarins is present among the roots of Arabidopsis accessions. Using HPLC and GC-MS analytical methods, we identified and quantified coumarins in the roots of selected seven accessions-Antwerpen (An-1, Belgium),

Columbia (Col-0, Germany), Estland (Est-1, Estonia), Kashmir (Kas-2, India), Kondara (Kond, Tajikistan), Landsberg *erecta* (Ler, Poland) and Tsu (Tsu-1, Japan). Subsequently, we conducted a QTL mapping and identified new loci possibly underlying the observed variation in scopoletin and scopolin accumulation. Thereby, we demonstrated that Arabidopsis natural variation is an attractive tool for elucidating the basis of coumarin biosynthesis. Other studies focusing on differential accumulation of coumarins between Arabidopsis accessions were conducted by Mönchgesang et al. [14]. A non-targeted metabolite profiling of root exudates revealed the existence of distinct metabolic phenotypes for 19 Arabidopsis accessions. Scopoletin and its glycosides were among phenylpropanoids that differed in the exudates of tested accessions. This research group also focused on the plant-to-plant variability in root metabolite profiles of 19 Arabidopsis accessions [15]. In the current study, a larger set of 28 accessions was chosen, that represent a wide genetic variation existing in Arabidopsis. To increase the scope of this work, we extracted and quantified coumarins from two different types of tissue–roots and leaves. The latter one, in the light of our best knowledge, have never been tested for coumarin content using Arabidopsis natural variation. We believe our results will be beneficial for further studies focusing on a better understanding of coumarin physiological functions and the exact role of enzymes involved in their biosynthesis.

2. Results

2.1. UHPLC-MS Targeted Metabolite Profiling of Root and Leaf Tissues Reveals Distinct Metabolic Phenotypes for 28 Arabidopsis Accessions

The average content of each tested coumarins (Table 1) grouped by the 28 Arabidopsis accessions (Table 2), type of tissue (extracts from roots and leaves) and method of preparing extracts for analysis (without and after hydrolysis) were depicted through a general heatmap. For each compound, we quantified both the non-glycosylated coumarins—scopoletin (Figure 2A), umbelliferone (Figure 3A), esculetin (Figure 4A), and their respective glycosylated forms—scopolin (Figure 2B), skimmin (Figure 3B), and esculin (Figure 4B), respectively. The concentration μM was based on the fresh weight (FW).

Table 1. Coumarins and their glycosides identified in this study.

Peak Number	Retention Time tR (min)	Compound	Mass (*m/z* Ratio)	LOQ
(1)	14.5	Umbelliferone	163 (M+H$^+$)	0.2 μM
(2)	11.8	Esculetin	179 (M+H$^+$)	0.5 μM
(3)	14.8	Scopoletin	193 (M+H$^+$)	0.2 μM
(5)	10.1	Skimmin (glycosylated umbelliferone)	325 (M+H$^+$)	0.1 μM
(5)	11.8	Esculin (glycosylated esculetin)	341 (M+H$^+$)	0.1 μM
(6)	11	Scopolin (glycosylated scopoletin)	355 (M+H$^+$)	0.1 μM

Our analyses made evidence a significant variation in accumulation of all tested compounds between Arabidopsis accessions, both in roots and leaves. In accordance with the current state of knowledge [6,8,36–39], we identified the coumarin scopoletin and its glycoside, scopolin, to be the major metabolites that accumulate in Arabidopsis roots (Figure 2A,B).

Table 2. Basic information on the Arabidopsis accessions used in this study.

No.	Full Name	Abbreviation	Country of Origin
1	Antwerpen	An-1	Belgium
2	Bayreuth	Bay-0	Germany
3	Brunn	Br-0	Czech Republic
4	Coimbra	C24	Portugal
5	Canary Islands	Can-0	Spain
6	Columbia	Col-0	USA
7	Cape Verdi	Cvi-1	Cape Verde Islands
8	Eilenburg	Eil-0	Germany
9	Eringsboda	Eri-1	Sweden
10	Estland	Est-1	Russia
11	St. Maria d. Feiria	Fei-0	Portugal
12	Fukuyama	Fuk-1	Japan
13	Gabelstein	Ga-0	Germany
14	Hodja-Obi-Garm	Hog	Tajikistan
15	Kashmir	Kas-2	India
16	Kondara	Kondara	Tajikistan
17	Kyoto	Kyo-1	Japan
18	Lebjasche	Leb 3/4	Russia
19	Landsberg *erecta*	Ler-1	Germany
20	Nossen	No-0	Germany
21	Richmond	Ri-0	Canada
22	San Feliu	Sf-2	Spain
23	Shakdara	Sha-1	Tajikistan
24	Sorbo	Sorbo	Tajikistan
25	Tossa del Mar	Ts-5	Spain
26	Tsushima	Tsu-1	Japan
27	Vancouver	Van-0	Canada
28	Wassilewskija	Ws-0	Belarus

Scopoletin was the most abundant compound in each of the 28 Arabidopsis accessions studied, especially in the roots (from 2.61 to 151.90 μM), but interestingly this phytochemical was also detected in the leaf extracts (Table S1). The highest amount of scopoletin was detected in Bay-0, Br-0 and Kondara, respectively (Figure 2A), in samples prepared from the roots and subjected to hydrolysis. In non-hydrolyzed root samples, the highest content of scopoletin was detected for the same accessions–Kondara, Bay-0 and Br-0. As expected, scopoletin content in the leaf extracts was several dozen times smaller (from 0.03 to 2.6 μM) when compared with extracts prepared from the root tissue. Amount of scopoletin in the leaf sample was the highest in Br-0, Est-1 and Bay-0 when subjected to hydrolysis and in Est-1, Br-0, Col-0 and Bay-0 when not hydrolyzed.

Relatively large amounts of scopolin (from 2.94 to 67.26 μM) were found in almost all root extracts that were not subjected to hydrolysis (Figure 2B), the highest in Br-0, Fei-0, Ga-0, Leb 3/4, Ri-0, C24 and Bay-0 accessions (Table S1). We also identified some accessions (Fuk-1, Bay-0, Ri-0, Sha-1 and Eri-1) with relatively high content of scopolin in root extracts after hydrolysis (from 2.04 to 8.09 μM), most probably due to non-effective enzymatic reaction. Interestingly, another set of accessions (Est-1, Sorbo, Bay-0, Kyo-0, No-0, Ga-0, Fei-0, Ws-0) with relatively high scopolin concentration (from 2.58 to 6.48 μM) was also detected in leaf extracts not subjected to hydrolysis. As could be expected, in hydrolysed leaf samples in which sugar residues were cut off and most of scopolin was transformed into scopoletin, the amounts of scopolin were quite low or close to the LOQ.

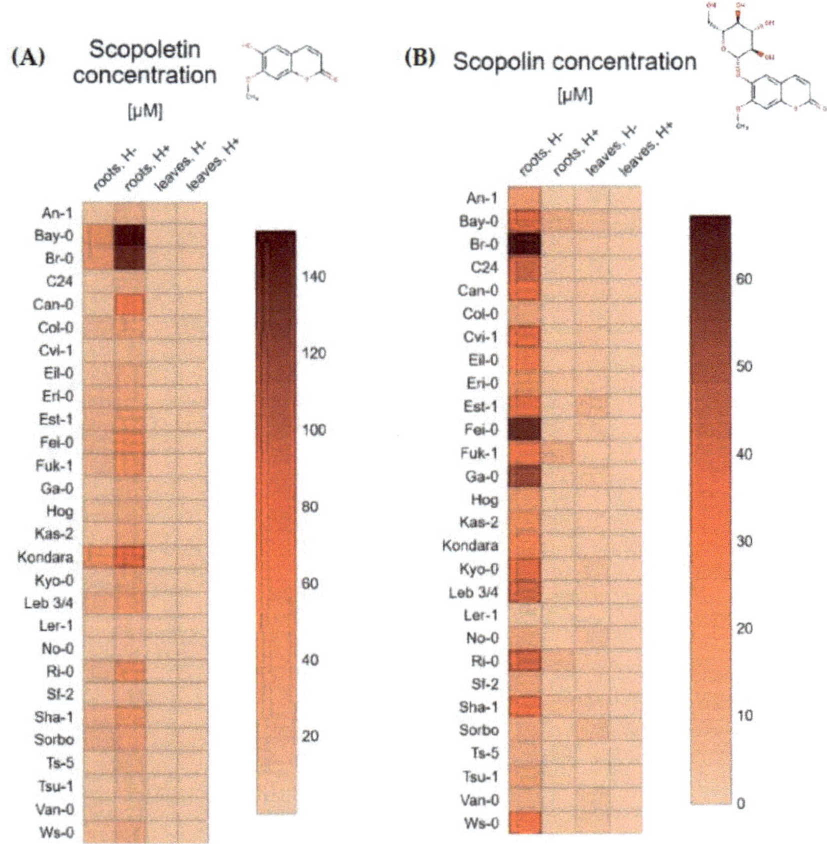

Figure 2. Heat maps based on the average (**A**) scopoletin and (**B**) scopolin concentration (µM/FW) in Arabidopsis tissue extracts from roots and leaves, without and after hydrolysis. The values used in the plots (https://app.displayr.com (accessed on 20 January 2021)) are the mean of 3 biological replicates. The mean values and standard deviations (±SD) are gathered in the supplementary materials (Table S1).

Interestingly, in this study, we identified small amounts of umbelliferone (from 0.02 to 1.64 µM) in Arabidopsis plants (Figure 3A). Importantly, we detected this phytochemical in all of the hydrolyzed root extracts (Table S2). The highest levels of umbelliferone were found in Bay-0, Ri-0, Est-1, Col-0, Br-0, C24, Sorbo, Fuk-1 and Leb 3/4, respectively. The quantity of umbelliferone in all leaf extract samples was below LOQ.

Skimmin, which is a glucoside of umbelliferone, was detected and quantified (from 0.69 to 19.80 µM) mostly in root samples of Arabidopsis accessions that were not subjected to enzymatic hydrolysis (Figure 3B). The highest levels were detected in extracts originating from Ga-0, C24, Ws-0, Ri-0, Est-1, Kyo-0 and Eri-0 accessions. It should be noted that skimmin could also be quantified (concentration from 0.04 to 18.76 µM) in all hydrolyzed root extract samples (Table S2), which needs further investigation.

Most of the results obtained for the leaf tissues were very low and near the LOQ, however in Eri-1, An-1, C24, Col-0, Van-0, Kondara, Ws-0, Ga-0, Fuk-1, Can-0 and Tsu-1 accessions, we observed values slightly above the limit.

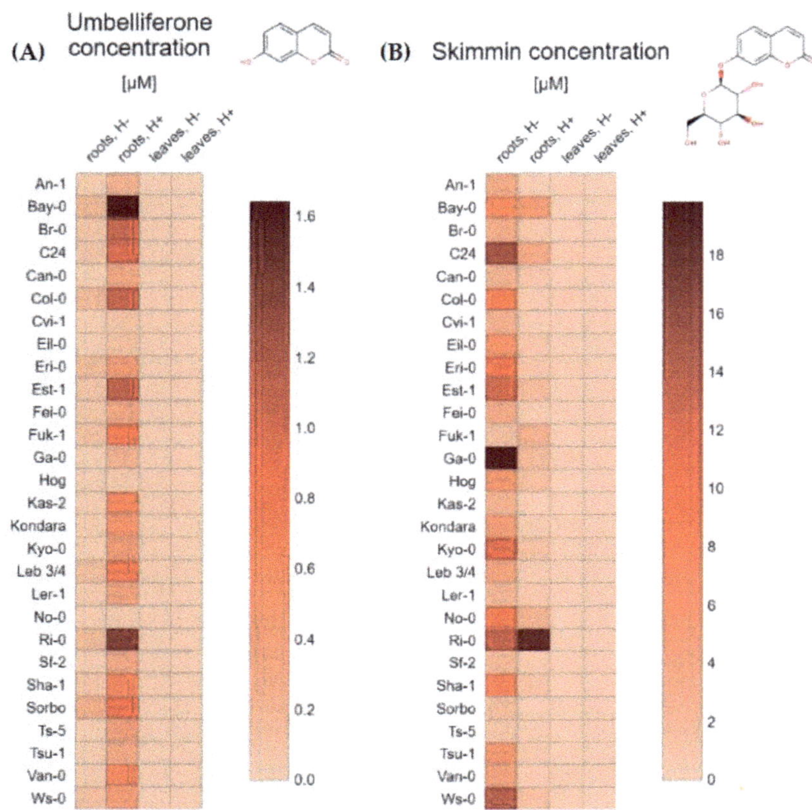

Figure 3. Heat maps based on the average (**A**) umbelliferone and (**B**) skimmin concentration (μM/FW) in Arabidopsis tissue extracts from roots and leaves, without and after hydrolysis. The values used in the plots (https://app.displayr.com (accessed on 20 January 2021)) are the mean of 3 biological replicates. The mean values and standard deviations (±SD) are gathered in the supplementary materials (Table S2).

Small amounts of esculetin were detected only in a few of root extracts (max. concentration 0.29 μM) and leaf samples (max. concentration 0.16 μM) (Figure 4A). In root non-hydrolysed samples, esculetin was present in Bay-0, Br-0 and Can-0 accessions, while in hydrolysed extracts it was detected in Can-0, Bay-0, Col-0, Ri-0 and Tsu-1 (Table S1). It may be puzzling that in some accessions, esculetin was only detected in samples which were not subjected to hydrolysis but not in the hydrolysed ones. This is the case for the root extract of Br-0 (0.07 μM), and leaf samples of C24, Br-0, An-1, Col-0, No-0, Ws-0 and Ri-0 accessions (from 0.01 to 0.16 μM, Table S3). In leaf samples after hydrolysis, only trace amount of esculetin was detected in Ws-0.

Esculin, which is a glycoside form of esculetin, was not found in any root extract (Figure 4B), except Col-0 sample with quantity near to LOQ (0.01 μM). Trace amounts of esculin were detected in some leaf extracts without hydrolysis (from 0.01 to 0.15 μM) with the highest content in Col-0 accession, and in the leaf samples subjected to hydrolysis (from 0.01 to 0.36 uM). Here, the highest esculin content was detected in Ws-0 accession (Table S3).

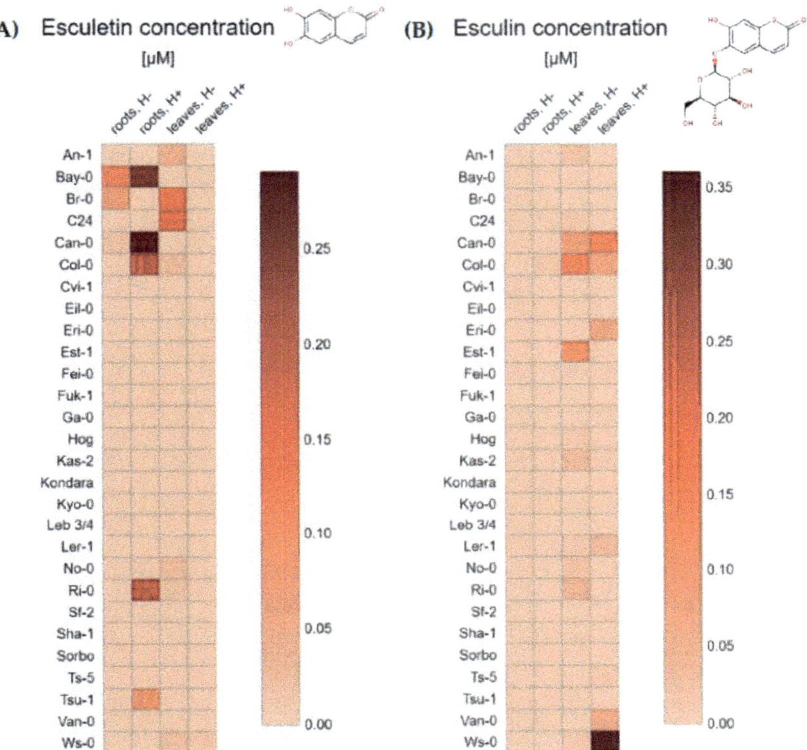

Figure 4. Heat maps based on the average (**A**) esculetin and (**B**) esculin concentration (μM/ FW) in Arabidopsis tissue extracts from roots and leaves, without and after hydrolysis. The values used in the plots (https://app.displayr.com (accessed on 20 January 2021)) are the mean of 3 biological replicates. The mean values and standard deviations (±SD) are gathered in the supplementary materials (Table S3).

2.2. Principal Component Analysis (PCA) for 28 Arabidopsis Accessions Using Coumarin Quantification by UHPLC-MS in Selected Geographic and under Diverse Climatic Factors

In order to compare and visualize the possible relationship between coumarin content variability present among 28 Arabidopsis accessions in selected geographic and in various climatic factors (maximal altitude [m], average winter minimal temperature [°C], average summer maximal temperature [°C] and average annual precipitation [mm]), we performed Principal Component Analysis (PCA). About half of the variance of used dataset was covered by the first two principal components, explaining 49% of the overall data variance (27.1% and 21.9% for PC1 and PC2, respectively) (Figure 5A).

According to the results presented on the Variables-PCA plot (Figure 5B), we assumed that there is a positive correlation between scopoletin, umbelliferone and scopolin concentration in root samples before hydrolysis. Despite the fact that scopolin content has relatively small contribution in explaining the variability between tested accessions, it can be also positively correlated with skimmin and umbelliferone concentration. Skimmin content is positively correlated with annual precipitation data.

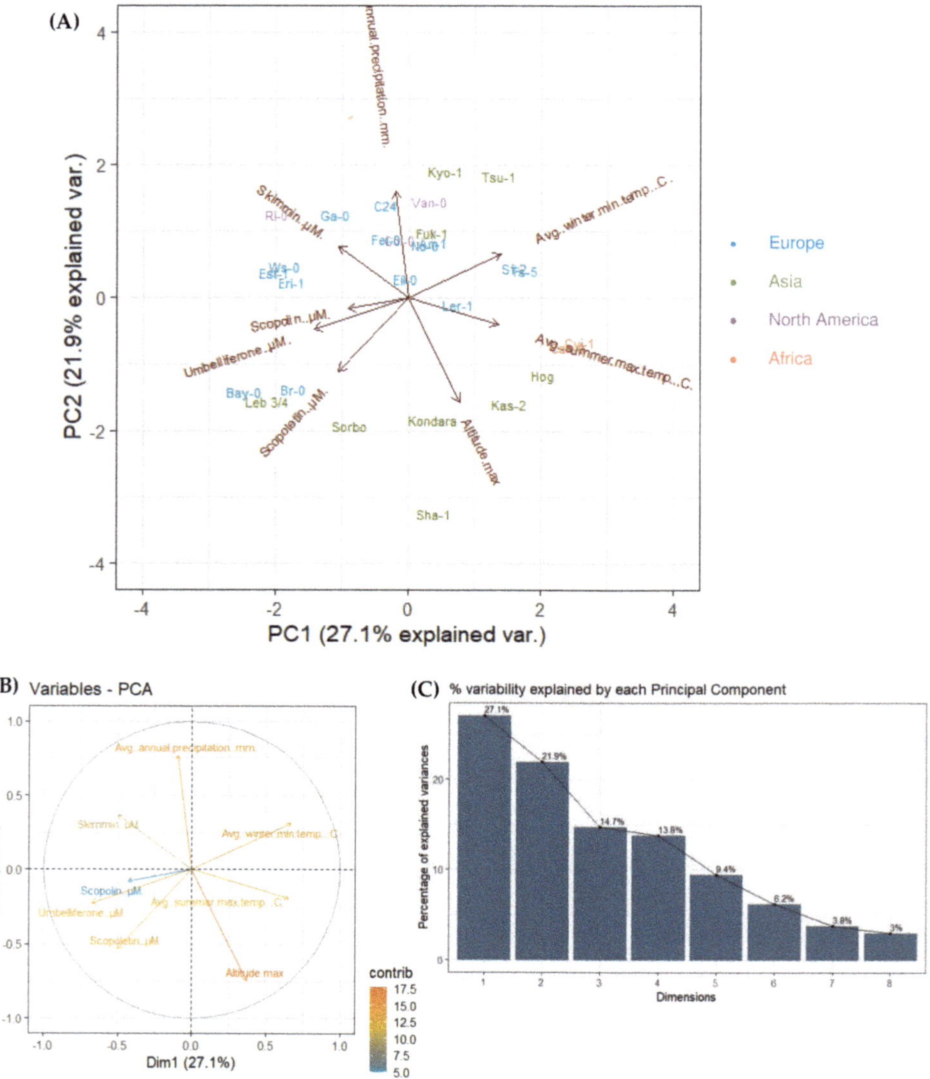

Figure 5. (**A**) Principal component analysis (PCA) for 28 Arabidopsis accessions using the concentration of umbelliferone, scopoletin and their corresponding glycosides (skimmin and scopolin, respectively) in root samples without hydrolysis, and four geographic and climatic factors (maximal altitude [m], average winter minimal temperature [°C], average summer maximal temperature [°C] and average annual precipitation [mm]; Table S4). Factor coordinates are marked with arrows. Observations indicated by blue accession names represent European locations (n = 14), green represent Asian locations (n = 9), violet represent North American locations (n = 3) and red represent African locations (n = 2). The abbreviations indicate the accessions according to Table 2. Component one and two explain 49% of the point variability. (**B**) The Variables-PCA contribution plot shows the correlation of the variables used in PCA with the respective contribution of each factor (contrib) indicated with a colour gradient. (**C**) The scree plot/graph of variables demonstrate the percentage of variability explained by each dimension (PC). Principal Component 1 and 2 explain 27.1% and 21.9% of the variance respectively.

A negative correlation is highlighted between the following variables: (1) umbelliferone concentration and temperatures (average winter minimal temperature and average summer maximal temperature); (2) scopolin concentration and temperatures (average winter minimal temperature and average summer maximal temperature); (3) skimmin concentration and average summer maximal temperature, as well as skimmin concentration and maximal altitude; (4) scopoletin concentration and annual precipitation, as well as scopoletin concentration and average winter minimal temperature (Figure 5B).

On the Figure 5C we demonstrate the graph of variables (scree plot) which indicates the percentage of variability explained by each dimension (PC). Principal Component 1 and 2 explain 27.1% and 21.9% of the variance respectively, while the other 6 dimensions account for the total remaining variability between each accession (PC3 = 14.7%, PC4 = 13.8%, PC5 = 9.4%, PC6 = 6.2%, PC7 = 3.8% and PC8 = 3%).

3. Discussion

Our previous study strongly suggest that Arabidopsis is an excellent model for elucidating the basis of natural variation in coumarin accumulation [36]. Here, we identified and quantified a set of coumarin compounds in the root and leaf methanol extracts prepared from 28 Arabidopsis accessions. In the light of our best knowledge, it is the largest set of Arabidopsis natural populations used in the coumarin profiling analysis that should well represent a wide genetic variation existing in this model plant. It is assumed that these accessions reflect genetic adaptation to local environmental factors [40]. As a result of evolutionary pressure differentially acting on the studied accessions originating from various geographical locations, a large number of genetic polymorphisms is present that have led to different levels of expression of genes involved in the biosynthesis, transport and metabolism of coumarins, and ultimately to different levels of their accumulation. In the current work, we detected a significant natural variation in the content of simple coumarins present in the root and leaf extracts of 28 Arabidopsis accessions. Among tested compounds, scopoletin and its glycosylated form, scopolin, were the most abundant, which is in line with the current state of knowledge [6,14,15,36–38].

The previous study on differential accumulation of coumarins between 19 Arabidopsis accessions belonging to the MAGIC lines characterized by a high genetic variability [14], confirm the hypothesis that the composition of Arabidopsis accessions root exudates is genetically determined. They revealed the existence of distinct root metabolic phenotypes among tested natural populations, including variation in the accumulation of scopoletin and its glycoside. Another study focused on extensive profiling of specialized metabolites in root exudates of Arabidopsis reference accession, Col-0, by non-targeted metabolite profiling using reversed-phase UPLC/ESI-QTOFMS [16]. As many as 103 compounds were detected in exudates of hydroponically grown Col-0 plants. Among them, 42 were identified by authenticated standards, including the following coumarins: esculetin, scopoletin, and their glucosides esculin and scopolin. In addition to these coumarins, further esculetin and scopoletin conjugates were initially identified in the root exudates based on their mass spectral fragmentation pattern [16].

It has to be noted that among other coumarin compounds identified in our study, we detected umbelliferone for the first time in Arabidopsis model plant. Authors of the first publication on the accumulation of coumarins in Arabidopsis [37], in which various type of tissues (roots, shoots and callus) were tested, detected trace amounts of skimmin (umbelliferone glucoside) in the wild type plants and slightly increased skimmin level in mutants of CYP98A3. No umbelliferone was detected in that study, or in any other work with Arabidopsis to date, in the light of our best knowledge. It cannot be excluded that we were able to detect umbelliferone due to the sample types tested. We conducted coumarin profiling of extracts prepared from the plant tissues grown in in vitro liquid cultures. Moreover, umbelliferone was detected in root methanol extracts additionally subjected to enzymatic hydrolysis prior to quantification done by UHPLC-MS in order to hydrolyze the glycoside forms of coumarins, while its glycosylated form, skimmin, was detected in samples without enzymatic treatment. It should be highlighted that low amounts of umbelliferone were also detected in non-hydrolysed extracts.

Coumarins have become important players both in optimizing Fe uptake and shaping the root microbiome, thus affecting plant health [5,41]. The link between plant specialized metabolites, in particular coumarins, nutrient deficiencies and microbiome composition that was discovered in recent studies [7,8,42,43], could provide a new set of tools for rationally manipulating the plant microbiome [44]. The selection of underground tissue was an obvious choice in such analyses, considering that coumarins are essentially synthesized in roots where optimization of Fe uptake is coordinated with plant requirements and interaction with soil microorganisms. Therefore, most of the previous coumarin metabolic profiling analysis, including functional characterization of Arabidopsis mutants defective in genes encoding enzymes involved in coumarin biosynthesis or transport, were performed using the root exudates and root tissue [3,6,12,17,38,44,45]. It was also the case in research conducted on the effects of Fe, phosphorus (Pi) or both deficiencies on coumarin profiles in the root tissue of several T-DNA insertional mutants defective in genes involved in Pi or Fe homeostasis [11]. Importantly, in the current study we detected variation in accumulation of esculetin and esculin. The latter one was identified in Arabidopsis leaf extracts, both with and without enzymatic hydrolysis. This requires further research and is of particular interest in the light of recent research findings on coumarin cellular localization, trafficking and signalling [5,7]. Coumarins were found to be involved in the plant response to pathogens in aerial tissues [41,46] and proposed to play an important signalling role in bidirectional chemical communication along the microbiome-root-shoot axis [7].

The study of natural variation in coumarin content present among Arabidopsis accessions is a starting point in elucidating direct links between metabolic phenotypes and genotypes. In the presented research, we also checked whether the climatic and geographic data on the regions from which particular accessions originate, are correlated with the concentrations of tested coumarins. The conducted PCA showed a number of positive and negative correlations between climatic factors and coumarin content. Further investigation is needed to draw a more precise conclusion about possible relationship between the accumulation of coumarins and habitat data. Taking into account, the recent studies showing an important role of coumarins in plant interactions with soil microorganisms and nutrient acquisition, a more in-depth analysis, including data on soil parameters at the origin sites of a given accession, would explain the greater variance and give us more information on the potential correlations. It will be beneficial for the future discovery of physiological mechanisms of action of various alleles involved in the coumarin biosynthesis and can help to select biosynthetic enzymes for further metabolic engineering research.

4. Materials and Methods

4.1. Chemicals and Reagents

The coumarins standards umbelliferone (purity $\geq$ 99%), coumarin (>99% purity), esculin ($\geq$98% purity) were purchased from Sigma-Aldrich (St. Louis, MO, USA), scopoletin (>95% purity) and esculetin (>98% purity) from Extrasynthese (Genay, France), skimmin (98% purity) from Biopurify Phytochemicals (Chengdu, China), scopolin (>98% purity) from Chemicals Aktin Inc. (Chengdu, China). Stock solutions of each standard at a concentration of 10 mmol/L were prepared by diluting the powder in dimethyl sulfoxide (Fisher scientific, Illkirch, France) and kept at −18 °C until use. HPLC-grade methanol was purchased from CarloErba Reagents (Val de Reuil, France), formic acid was purchased from Fisher Scientific (Illkirch, France). Water was purified by a PURELAB Ultra system (Veolia Water S.T.I., Antony, France).

4.2. Plant Material

All seeds of the 28 Arabidopsis accessions (Table 2) from various habitats which were used in this study were obtained courtesy of prof. Maarten Koornneef.

4.3. In Vitro Plant Culture

All the Arabidopsis accessions seeds were surface sterilized with 70% ethanol for 2 min, 5% calcium hypochlorite solution for 8 min and then washed 3 times with sterile ultrapure water. The seeds were placed in Petri dishes with $\frac{1}{2}$ Murashige-Skoog (MS) medium solidified with agar (Sigma-Aldrich) for in vitro plant culture and incubated for 72 h in the dark at 4 °C. Then the plates were transferred to a growth chamber (daily cycle: 16 h light 35 μmol m^{-2} s^{-1} temperature 20 °C and 8 h dark temperature 18 °C) for 10 days. After that time, seedlings were transferred from agar plates into 200 mL flasks (three individuals per flask) containing 5 mL of $\frac{1}{2}$ MS liquid medium containing 1% sucrose, MS salts, 100 mg/L myo-inositol, 1 mg/L thiamine hydrochloride, 0.5 mg/L pyridoxine hydrochloride and 0.5 mg/L nicotinic acid (Sigma-Aldrich/Merck KGaA, Darmstadt, Germany). Plants were grown in the growth chamber on a rotary platform with shaking 120 rpm. After one week, 3 mL of fresh $\frac{1}{2}$ MS medium was added. Plants were grown for 17 supplementary days and after that time were rinsed with demineralized water, dried on paper towels. Roots and leaves samples were weighted (50 ± 2 mg fresh weight (FW)) and frozen in liquid nitrogen. The plant material was stored in a freezer at −80 °C until extraction process. All accessions were grown in three biological replicates (in three independent flasks, three seedlings per flask).

4.4. Metabolites Extraction

For the metabolites extraction, plant tissue frozen in liquid nitrogen was grinded by the usage of 5 mm diameter stainless steel beads (Qiagen, Hilden, Germany). To the 2 mL microtubes, 2 clean beads were added and samples were frozen in liquid nitrogen. Then, using vortex, samples were mixed. For the better performance, the freezing and vortexing procedure was repeated several times until all tissue was powdered. To the powdered tissues 0.5 mL of 80% methanol containing 5 μM 4-methylumbelliferone as an internal standard was added. After that, samples were sonicated for 30 min with ultrasonic cleaner (Proclean 3.0DSP, Ulsonix, Expondo, Berlin, Germany) (70% frequency, sweep function) and incubated in 4 °C in darkness for 24 h. Next day, all samples were vortexed, centrifuged at 13,000× *g* for 10 min and the supernatant was transferred into new microtubes. Centrifugation was repeated in order to get rid of any sediment. The extracts were firstly dried for 2 h in an incubator at 45 °C and then, for the next 2 h in a vacuum centrifuge (Savant SpeedVac vacuum concentrator, Thermo Fisher Scientific, Waltham, MA, USA). To the dried extracts 100 μL of 80% methanol was added to dissolve samples during the night at 4 °C. Then the extracts were vortexed for 10 min and separated by 50 μL. One of the replicates was subjected to enzymatic hydrolysis, and the second one was stored at −20 °C until UHPLC-MS analysis (Shimadzu Corp., Kyoto, Japan).

4.5. Enzymatic Hydrolysis

The enzymatic hydrolysis was performed according to Nguyen et al. [47]. Methanolic extracts were subjected to enzymatic hydrolysis with a β-glucosidase (Fluka Chemie GmbH, Buchs, Switzerland) in 0.1 M acetate buffer at a concentration of 0.5 mg/mL in order to determine the amounts of glycosylated compounds (o-glycosides). 50 μL of acetate buffer with β-glucosidase at pH 5.0 (0.1 M sodium acetate, 0.1 M acetic acid and 0.5 mg/mL β-glucosidase buffer) was added to 50 μL of the prepared extract and incubated for 22 h at 37 °C. The reaction was stopped by adding 100 μL of 96% ethanol to the reaction mixture. The extracts were dried in an incubator at 45 °C for 2 h and then for about 1 h in a vacuum centrifuge (Savant SpeedVac vacuum concentrator). The obtained extract was dissolved in 50 μL of 80% methanol overnight and stored at −20 °C until UHPLC-MS analysis.

4.6. UHPLC Separation

The coumarins analyses were performed using a NEXERA UPLC-MS system (Shimadzu Corp., Kyoto, Japan) equipped with two UHPLC pumps (LC-30AD), an automatic sampler (SIL-30AC), a photodiode array detector (PDA, SPDM-20A) and combined with a mass spectrometer (single quadrupole, LCMS-2020). Coumarins separation was done on a C18 reversed phase column (ZORBAX Eclipse Plus), 150 × 2.1 mm, 1.8 μm (Agilent Technologies, Santa Clara, CA, USA) protected with an Agilent Technologies 1290 Infinity filter. The column was kept at 40 °C in a column oven (Shimadzu CTO-20AC). Mobile phase consisted of 0.1% formic acid in ultrapure water (buffer A) and 0.1% formic acid in methanol (buffer B) at a constant flow rate of 200 μL/min. The linear gradient solvent system was set as follows: 0 min, 10% B; 16 min, 70% B; 18 min, 99% B; 18.01 min, 10% B; 20 min, 10% B. The total analysis duration was 20 min. The injection volume was 5 μL.

4.7. MS Detection

The UHPLC system was connected to the MS by an electrospray ionization source (ESI), operating in positive mode (ESI+) and scanning in single ion monitoring mode (SIM). The inlet, desolvation line and heating block temperatures were set at 350 °C, 250 °C, and 400 °C, respectively. The capillary voltage was set at 4.5 kV. Dry gas flow was set at 15 L/min and nebulizing gas at 1.5 L/min. The instrument was operated and data were processed using LabSolution software version 5.52 sp2 (Shimadzu Corp., Kyoto, Japan).

4.8. Peak Identification and Quantitation

Each standard molecule was individually injected in the UHPLC-MS in full scan mode to determine retention time and *m/z* ratio for the analysis. The quantitation of each molecule (Table 2) was based on the signal obtained from the MS detection, using angelicin, as an analytical internal standard. Angelicin was added at the same concentration (5 μM) in all the samples before injection as well as in 7 calibration solutions. The calibration solutions contained all of the standard molecules at the same concentrations ranging from 0.1 to 10 μM (0.1, 0.2, 0.5, 1, 2, 5 and 10 μM). Calibration curves were drawn for each compound by linking its relative peak area (compound area divided by the angelicin area) and its concentration. Each curve fit type was linear. The limit of quantitation (LOQ) was calculated as the analyte concentration giving signal to signal to noise ratios (S/N) of 10. Three measurements were assessed per accession.

4.9. Principal Component Analysis (PCA)

Principal Component Analysis (PCA) were performed using *prcomp()* package and visualize with the *factoextra* 1.0.7 version package in the R 4.0.4 environment developed by the R Core Team (2020). R: A language and environment for statistical computing. R Foundation for Statistical Computing, Vienna, Austria (www.R-project.org (accessed on 20 January 2021)) and the RStudio Team (2019). RStudio: Integrated Development for R. RStudio, Inc., Boston, MA, USA (www.rstudio.com (accessed on 20 January 2021)). All the variables were standardized before analysis. Data used for the analysis are presented in Table S4.

5. Conclusions

Multi-pharmacological properties of coumarins that are widely used in medical applications, make the study of coumarin biosynthesis attractive from the commercial point of view. Considering that all medicinal plants currently used in studying the biosynthesis of coumarins are non-model organisms and many approaches are not available in those species, makes a model plant Arabidopsis, with its extensive genetic variation and numerous publicly accessible web-based databases, an excellent model to study accumulation of coumarins in natural populations. The presented results focusing on qualitative and quantitative characterization of natural resources provide a basis for further research on identification of genetic variants involved in coumarin biosynthesis in plants, which is the first step in metabolic engineering for the production of natural compounds. We identified scopoletin, and its glycosylated form, scopolin, to be the most abundant coumarins in Arabidopsis tissues. It should be emphasized that among other coumarin compounds identified in this study, we detected umbelliferone for the first time in Arabidopsis. In view of the considerable importance of umbelliferone in synthesis and its pharmacological properties, this is a significant step in the study of biosynthesis of coumarins using this model plant.

Supplementary Materials: The following are available online, Table S1: The mean values (scopoletin and scopolin concentrations) and standard deviations (±SD) that were used in making heat maps shown in Figure 2, Table S2: The mean values (scopoletin and scopolin concentrations) and standard deviations (±SD) that were used in making heat maps shown in Figure 3, Table S3: The mean values (scopoletin and scopolin concentrations) and standard deviations (±SD) that were used in making heat maps shown in Figure 4, Table S4. Coumarin concentrations in root samples before hydrolysis and four geographic and climatic factors used in principal component analysis (PCA).

Author Contributions: Conceptualization, A.O., A.H. and A.I.; methodology, A.O., A.H. and A.I.; validation, I.P., J.S. and A.O.; investigation, I.P., J.S., A.O. and J.G.; writing—original draft preparation, I.P., E.L. and A.I.; writing—review and editing, E.L., A.O., A.H. and A.I.; supervision, E.L., F.B. and A.I.; funding acquisition, A.I. and A.O. All authors have read and agreed to the published version of the manuscript.

Funding: This research was funded by Narodowe Centrum Nauki (Polish National Science Centre) grant number 2014/15/B/NZ2/01073 and Narodowa Agencja Wymiany Akademickiej (Polish National Agency for Academic Exchange) grant number PPN/BFR/2019/1/00050 to A.I.; and by Campus France (PHC POLONIUM) grant number 45026ZE to A.O.

Institutional Review Board Statement: Not applicable.

Informed Consent Statement: Not applicable.

Data Availability Statement: The data presented in this study are available within the article and supplementary materials [Tables S1–S4].

Acknowledgments: We thank Maarten Koornneef for providing all Arabidopsis seeds used in this study and Thibaut Duval for technical support.

Conflicts of Interest: The authors declare no conflict of interest.

Sample Availability: Coumarin standards used in the study and seeds corresponding to each accessions are available from the authors.

References

1. Matos, M.J.; Santana, L.; Uriarte, E.; Abreu, O.A.; Molina, E.; Yordi, E.G. Coumarins—An Important Class of Phytochemicals. *Phytochem. Isol. Charact. Role Hum. Health* **2015**, 113–140. [CrossRef]
2. Riveiro, M.E.; De Kimpe, N.; Moglioni, A.; Vazquez, R.; Monczor, F.; Shayo, C.; Davio, C. Coumarins: Old Compounds with Novel Promising Therapeutic Perspectives. *Curr. Med. Chem.* **2010**, *17*, 1325–1338. [CrossRef]
3. Rajniak, J.; Giehl, R.F.H.; Chang, E.; Murgia, I.; Von Wirén, N.; Sattely, E.S. Biosynthesis of redox-active metabolites in response to iron deficiency in plants. *Nat. Chem. Biol.* **2018**, *14*, 442–450. [CrossRef]
4. Reen, F.J.; Gutiérrez-Barranquero, J.A.; Parages, M.L.; O´gara, F. Coumarin: A novel player in microbial quorum sensing and biofilm formation inhibition. *Appl. Microbiol. Biotechnol.* **2018**, *102*, 2063–2073. [CrossRef] [PubMed]
5. Robe, K.; Conejero, G.; Gao, F.; Lefebvre-Legendre, L.; Sylvestre-Gonon, E.; Rofidal, V.; Hem, S.; Rouhier, N.; Barberon, M.; Hecker, A.; et al. Coumarin accumulation and trafficking in *Arabidopsis thaliana*: A complex and dynamic process. *New Phytol.* **2021**, *229*, 2062–2079. [CrossRef]
6. Siwinska, J.; Siatkowska, K.; Olry, A.; Grosjean, J.; Hehn, A.; Bourgaud, F.; A Meharg, A.; Carey, M.; Lojkowska, E.; Ihnatowicz, A. Scopoletin 8-hydroxylase: A novel enzyme involved in coumarin biosynthesis and iron-deficiency responses in Arabidopsis. *J. Exp. Bot.* **2018**, *69*, 1735–1748. [CrossRef]
7. Stassen, M.J.; Hsu, S.-H.; Pieterse, C.M.; Stringlis, I.A. Coumarin Communication Along the Microbiome–Root–Shoot Axis. *Trends Plant Sci.* **2021**, *26*, 169–183. [CrossRef] [PubMed]
8. A Stringlis, I.; De Jonge, R. The Age of Coumarins in Plant–Microbe Interactions. *Plant Cell Physiol.* **2019**, *60*, 1405–1419. [CrossRef] [PubMed]
9. Ziegler, J.; Schmidt, S.; Chutia, R.; Müller, J.; Böttcher, C.; Strehmel, N.; Scheel, D.; Abel, S. Non-targeted profiling of semi-polar metabolites in Arabidopsis root exudates uncovers a role for coumarin secretion and lignification during the local response to phosphate limitation. *J. Exp. Bot.* **2016**, *67*, 1421–1432. [CrossRef] [PubMed]
10. Badri, D.V.; Loyola-Vargas, V.M.; Broeckling, C.D.; Vivanco, J.M. Root Secretion of Phytochemicals in Arabidopsis Is Predominantly Not Influenced by Diurnal Rhythms. *Mol. Plant* **2010**, *3*, 491–498. [CrossRef] [PubMed]

11. Chutia, R.; Abel, S.; Ziegler, J. Iron and Phosphate Deficiency Regulators Concertedly Control Coumarin Profiles in *Arabidopsis thaliana* Roots During Iron, Phosphate, and Combined Deficiencies. *Front. Plant Sci.* **2019**, *10*, 113. [CrossRef] [PubMed]
12. Fourcroy, P.; Sisó-Terraza, P.; Sudre, D.; Savirón, M.; Reyt, G.; Gaymard, F.; Abadía, A.; Abadia, J.; Álvarez-Fernández, A.; Briat, J.-F. Involvement of the ABCG37 transporter in secretion of scopoletin and derivatives by Arabidopsis roots in response to iron deficiency. *New Phytol.* **2014**, *201*, 155–167. [CrossRef] [PubMed]
13. Hildreth, S.B.; Foley, E.E.; Muday, G.K.; Helm, R.F.; Winkel, B.S.J. The dynamic response of the Arabidopsis root metabolome to auxin and ethylene is not predicted by changes in the transcriptome. *Sci. Rep.* **2020**, *10*, 1–15. [CrossRef]
14. Mönchgesang, S.; Strehmel, N.; Schmidt, S.; Westphal, L.; Taruttis, F.; Müller, E.; Herklotz, S.; Neumann, S.; Scheel, D. Natural variation of root exudates in *Arabidopsis thaliana*-linking metabolomic and genomic data. *Sci. Rep.* **2016**, *6*, 29033. [CrossRef]
15. Mönchgesang, S.; Strehmel, N.; Trutschel, D.; Westphal, L.; Neumann, S.; Scheel, D. Plant-to-Plant Variability in Root Metabolite Profiles of 19 *Arabidopsis thaliana* Accessions Is Substance-Class-Dependent. *Int. J. Mol. Sci.* **2016**, *17*, 1565. [CrossRef]
16. Strehmel, N.; Böttcher, C.; Schmidt, S.; Scheel, D. Profiling of secondary metabolites in root exudates of *Arabidopsis thaliana*. *Phytochemistry* **2014**, *108*, 35–46. [CrossRef] [PubMed]
17. Ziegler, J.; Schmidt, S.; Strehmel, N.; Scheel, D.; Abel, S. Arabidopsis Transporter ABCG37/PDR9 contributes primarily highly oxygenated Coumarins to Root Exudation. *Sci. Rep.* **2017**, *7*, 1–11. [CrossRef] [PubMed]
18. Clemens, S.; Weber, M. The essential role of coumarin secretion for Fe acquisition from alkaline soil. *Plant Signal. Behav.* **2016**, *11*, e1114197. [CrossRef]
19. Mitchell-Olds, T.; Pedersen, D. The molecular basis of quantitative genetic variation in central and secondary metabolism in Arabidopsis. *Genetics* **1998**, *149*, 739–747. [PubMed]
20. Hirai, M.Y.; Yano, M.; Goodenowe, D.B.; Kanaya, S.; Kimura, T.; Awazuhara, M.; Arita, M.; Fujiwara, T.; Saito, K. From the Cover: Integration of transcriptomics and metabolomics for understanding of global responses to nutritional stresses in *Arabidopsis thaliana*. *Proc. Natl. Acad. Sci. USA* **2004**, *101*, 10205–10210. [CrossRef] [PubMed]
21. Von Roepenack-Lahaye, E.; Degenkolb, T.; Zerjeski, M.; Franz, M.; Roth, U.; Wessjohann, L.; Schmidt, J.; Scheel, D.; Clemens, S. Profiling of Arabidopsis Secondary Metabolites by Capillary Liquid Chromatography Coupled to Electrospray Ionization Quadrupole Time-of-Flight Mass Spectrometry. *Plant Physiol.* **2004**, *134*, 548–559. [CrossRef]
22. Keurentjes, J.J.; Fu, J.; De Vos, C.R.; Lommen, A.; Hall, R.D.; Bino, R.J.; Van Der Plas, L.H.; Jansen, R.C.; Vreugdenhil, D.; Koornneef, M. The genetics of plant metabolism. *Nat. Genet.* **2006**, *38*, 842–849. [CrossRef]
23. Fiehn, O. Metabolite Profiling in Arabidopsis. In *Arabidopsis Protocols*; Springer Science and Business Media LLC: Berlin, Germany, 2006; Volume 323, pp. 439–448.
24. Ebert, B.; Zöller, D.; Erban, A.; Fehrle, I.; Hartmann, J.; Niehl, A.; Kopka, J.; Fisahn, J. Metabolic profiling of *Arabidopsis thaliana* epidermal cells. *J. Exp. Bot.* **2010**, *61*, 1321–1335. [CrossRef] [PubMed]
25. Hannah, M.A.; Caldana, C.; Steinhauser, D.; Balbo, I.; Fernie, A.R.; Willmitzer, L. Combined Transcript and Metabolite Profiling of Arabidopsis Grown under Widely Variant Growth Conditions Facilitates the Identification of Novel Metabolite-Mediated Regulation of Gene Expression. *Plant Physiol.* **2010**, *152*, 2120–2129. [CrossRef] [PubMed]
26. Routaboul, J.-M.; Dubos, C.; Beck, G.; Marquis, C.; Bidzinski, P.; Loudet, O.; Lepiniec, L. Metabolite profiling and quantitative genetics of natural variation for flavonoids in Arabidopsis. *J. Exp. Bot.* **2012**, *63*, 3749–3764. [CrossRef] [PubMed]
27. Watanabe, M.; Balazadeh, S.; Tohge, T.; Erban, A.; Giavalisco, P.; Kopka, J.; Mueller-Roeber, B.; Fernie, A.R.; Hoefgen, R. Comprehensive Dissection of Spatiotemporal Metabolic Shifts in Primary, Secondary, and Lipid Metabolism during Developmental Senescence in Arabidopsis. *Plant Physiol.* **2013**, *162*, 1290–1310. [CrossRef]
28. Wu, S.; Alseekh, S.; Cuadros-Inostroza, Á.; Fusari, C.M.; Mutwil, M.; Kooke, R.; Keurentjes, J.B.; Fernie, A.R.; Willmitzer, L.; Brotman, Y. Combined Use of Genome-Wide Association Data and Correlation Networks Unravels Key Regulators of Primary Metabolism in *Arabidopsis thaliana*. *PLoS Genet.* **2016**, *12*, e1006363. [CrossRef]
29. Orlita, A.; Sidwa-Gorycka, M.; Paszkiewicz, M.; Malinski, E.; Kumirska, J.; Siedlecka, E.M.; Łojkowska, E.; Stepnowski, P. Application of chitin and chitosan as elicitors of coumarins and furoquinolone alkaloids in *Ruta graveolens* L. (common rue). *Biotechnol. Appl. Biochem.* **2008**, *51*, 91–96. [CrossRef]
30. Sidwa-Gorycka, M.; Krolicka, A.; Orlita, A.; Maliński, E.; Golebiowski, M.; Kumirska, J.; Chromik, A.; Biskup, E.; Stepnowski, P.; Lojkowska, E. Genetic transformation of *Ruta graveolens* L. by *Agrobacterium rhizogenes*: Hairy root cultures a promising approach for production of coumarins and furanocoumarins. *Plant Cell Tissue Organ Cult. (PCTOC)* **2009**, *97*, 59–69. [CrossRef]
31. Sidwa-Gorycka, M.; Królicka, A.; Kozyra, M.; Głowniak, K.; Bourgaud, F.; Łojkowska, E. Establishment of a co-culture of *Ammi majus* L. and *Ruta graveolens* L. for the synthesis of furanocoumarins. *Plant Sci.* **2003**, *165*, 1315–1319. [CrossRef]
32. Matsumoto, S.; Mizutani, M.; Sakata, K.; Shimizu, B.-I. Molecular cloning and functional analysis of the ortho-hydroxylases of p-coumaroyl coenzyme A/feruloyl coenzyme A involved in formation of umbelliferone and scopoletin in sweet potato, *Ipomoea batatas* (L.) Lam. *Phytochemistry* **2012**, *74*, 49–57. [CrossRef]
33. Vialart, G.; Hehn, A.; Olry, A.; Ito, K.; Krieger, C.; Larbat, R.; Paris, C.; Shimizu, B.-I.; Sugimoto, Y.; Mizutani, M.; et al. A 2-oxoglutarate-dependent dioxygenase from *Ruta graveolens* L. exhibits p-coumaroyl CoA 2′-hydroxylase activity (C2′H): A missing step in the synthesis of umbelliferone in plants. *Plant J.* **2012**, *70*, 460–470. [CrossRef]
34. Yang, X.; Wei, S.; Liu, B.; Guo, D.; Zheng, B.; Feng, L.; Liu, Y.; Tomás-Barberán, F.A.; Luo, L.; Huang, D. A novel integrated non-targeted metabolomic analysis reveals significant metabolite variations between different lettuce (*Lactuca sativa.* L.) varieties. *Hortic. Res.* **2018**, *5*, 1–14. [CrossRef] [PubMed]

35. Micallef, S.A.; Shiaris, M.P.; Colón-Carmona, A. Influence of *Arabidopsis thaliana* accessions on rhizobacterial communities and natural variation in root exudates. *J. Exp. Bot.* **2009**, *60*, 1729–1742. [CrossRef] [PubMed]
36. Siwinska, J.; Kadzinski, L.; Banasiuk, R.; Gwizdek-Wisniewska, A.; Olry, A.; Banecki, B.; Lojkowska, E.; Ihnatowicz, A. Identification of QTLs affecting scopolin and scopoletin biosynthesis in *Arabidopsis thaliana*. *BMC Plant Biol.* **2014**, *14*, 280. [CrossRef]
37. Kai, K.; Shimizu, B.-I.; Mizutani, M.; Watanabe, K.; Sakata, K. Accumulation of coumarins in *Arabidopsis thaliana*. *Phytochemistry* **2006**, *67*, 379–386. [CrossRef]
38. Kai, K.; Mizutani, M.; Kawamura, N.; Yamamoto, R.; Tamai, M.; Yamaguchi, H.; Sakata, K.; Shimizu, B.-I. Scopoletin is biosynthesized via ortho-hydroxylation of feruloyl CoA by a 2-oxoglutarate-dependent dioxygenase in *Arabidopsis thaliana*. *Plant J.* **2008**, *55*, 989–999. [CrossRef] [PubMed]
39. Schmid, N.B.; Giehl, R.F.; Döll, S.; Mock, H.-P.; Strehmel, N.; Scheel, D.; Kong, X.; Hider, R.C.; Von Wirén, N. Feruloyl-CoA 6′-Hydroxylase1-Dependent Coumarins Mediate Iron Acquisition from Alkaline Substrates in Arabidopsis. *Plant Physiol.* **2014**, *164*, 160–172. [CrossRef] [PubMed]
40. Koornneef, M.; Alonso-Blanco, C.; Vreugdenhil, D. Naturally occurring genetic variation in *Arabidopsis thaliana*. *Annu. Rev. Plant Biol.* **2004**, *55*, 141–172. [CrossRef]
41. Stringlis, I.A.; Yu, K.; Feussner, K.; De Jonge, R.; Van Bentum, S.; Van Verk, M.C.; Berendsen, R.L.; Bakker, P.A.H.M.; Feussner, I.; Pieterse, C.M.J. MYB72-dependent coumarin exudation shapes root microbiome assembly to promote plant health. *Proc. Natl. Acad. Sci. USA* **2018**, *115*, E5213–E5222. [CrossRef]
42. Harbort, C.J.; Hashimoto, M.; Inoue, H.; Niu, Y.; Guan, R.; Rombolà, A.D.; Kopriva, S.; Voges, M.J.; Sattely, E.S.; Garrido-Oter, R.; et al. Root-Secreted Coumarins and the Microbiota Interact to Improve Iron Nutrition in Arabidopsis. *Cell Host Microbe* **2020**, *28*, 825–837.e6. [CrossRef] [PubMed]
43. Voges, M.J.E.E.E.; Bai, Y.; Schulze-Lefert, P.; Sattely, E.S. Plant-derived coumarins shape the composition of an Arabidopsis synthetic root microbiome. *Proc. Natl. Acad. Sci. USA* **2019**, *116*, 12558–12565. [CrossRef] [PubMed]
44. Jacoby, R.P.; Koprivova, A.; Kopriva, S. Pinpointing secondary metabolites that shape the composition and function of the plant microbiome. *J. Exp. Bot.* **2021**, *72*, 57–69. [CrossRef]
45. Schmidt, H.; Günther, C.; Weber, M.; Spörlein, C.; Loscher, S.; Böttcher, C.; Schobert, R.; Clemens, S. Metabolome Analysis of *Arabidopsis thaliana* Roots Identifies a Key Metabolic Pathway for Iron Acquisition. *PLoS ONE* **2014**, *9*, e102444. [CrossRef]
46. Döll, S.; Kuhlmann, M.; Rutten, T.; Mette, M.F.; Scharfenberg, S.; Petridis, A.; Berreth, D.-C.; Mock, H.-P. Accumulation of the coumarin scopolin under abiotic stress conditions is mediated by the *Arabidopsis thaliana* THO/TREX complex. *Plant J.* **2018**, *93*, 431–444. [CrossRef]
47. Nguyen, C.; Bouque, V.; Bourgaud, F.; Guckert, A. Quantification of Daidzein and Furanocoumarin Conjugates of *Psoralea cinerea* L. (Leguminosae). *Phytochem. Anal.* **1997**, *8*, 27–31. [CrossRef]

Article

3-Carboxylic Acid and Formyl-Derived Coumarins as Photoinitiators in Photo-Oxidation or Photo-Reduction Processes for Photopolymerization upon Visible Light: Photocomposite Synthesis and 3D Printing Applications

Mahmoud Rahal [1,2,3], Bernadette Graff [1,2], Joumana Toufaily [3], Tayssir Hamieh [3,4], Guillaume Noirbent [5], Didier Gigmes [5], Frédéric Dumur [5,*] and Jacques Lalevée [1,2,*]

1 Université de Haute-Alsace, CNRS, IS2M UMR 7361, F-68100 Mulhouse, France; mahmoud-rahal@outlook.com (M.R.); bernadette.graff@uha.fr (B.G.)
2 Université de Strasbourg, 67081 Strasbourg, France
3 Laboratory of Materials, Catalysis, Environment and Analytical Methods (MCEMA) and LEADDER Laboratory, Faculty of Sciences, Doctoral School of Sciences and Technology (EDST), Lebanese University, Beirut 6573-14, Lebanon; joumana.toufaily@ul.edu.lb (J.T.); tayssir.hamieh@ul.edu.lb (T.H.)
4 SATIE-IFSTTAR, Université Gustave Eiffel, Campus de Marne-La-Vallée, 25, allée des Marronniers, F-78000 Versailles, France
5 Aix Marseille Univ, CNRS, ICR UMR 7273, F-13397 Marseille, France; guillaume.noirbent@outlook.fr (G.N.); didier.gigmes@univ-amu.fr (D.G.)
* Correspondence: frederic.dumur@univ-amu.fr (F.D.); jacques.lalevee@uha.fr (J.L.)

Citation: Rahal, M.; Graff, B.; Toufaily, J.; Hamieh, T.; Noirbent, G.; Gigmes, D.; Dumur, F.; Lalevée, J. 3-Carboxylic Acid and Formyl-Derived Coumarins as Photoinitiators in Photo-Oxidation or Photo-Reduction Processes for Photopolymerization upon Visible Light: Photocomposite Synthesis and 3D Printing Applications. *Molecules* **2021**, *26*, 1753. https://doi.org/10.3390/molecules26061753

Academic Editor: Maria João Matos

Received: 22 February 2021
Accepted: 18 March 2021
Published: 21 March 2021

Abstract: In this paper, nine organic compounds based on the coumarin scaffold and different substituents were synthesized and used as high-performance photoinitiators for free radical photopolymerization (FRP) of meth(acrylate) functions under visible light irradiation using LED at 405 nm. In fact, these compounds showed a very high initiation capacity and very good polymerization profiles (both high rate of polymerization (Rp) and final conversion (FC)) using two and three-component photoinitiating systems based on coum/iodonium salt (0.1%/1% *w*/*w*) and coum/iodonium salt/amine (0.1%/1%/1% *w*/*w*/*w*), respectively. To demonstrate the efficiency of the initiation of photopolymerization, several techniques were used to study the photophysical and photochemical properties of coumarins, such as: UV-visible absorption spectroscopy, steady-state photolysis, real-time FTIR, and cyclic voltammetry. On the other hand, these compounds were also tested in direct laser write experiments (3D printing). The synthesis of photocomposites based on glass fiber or carbon fiber using an LED conveyor at 385 nm (0.7 W/cm^2) was also examined.

Keywords: coumarin; free radical polymerization; LED; photocomposites; direct laser write

1. Introduction

The development of new low-cost, environmentally friendly, and energy-efficient polymer synthesis remains more than ever at the heart of academic and industrial concerns and the subject of many new research strategies. In fact, thanks to technological development, light sources which are at the same time inexpensive, efficient, and with low energy consumption have been developed recently to induce photopolymerization reactions [1–4]. Nowadays, photopolymers are present in several fields such as coatings [5], dentistry [6], automotive [7], cosmetics [8], 3D printing, and holography [9], etc. For most of these industrial fields, photochemical polymerization uses ultraviolet radiation, a technique widely known as UV curing. However, this pathway based on UV lamps (Hg lamps) remains energy-consuming. Moreover, the ultraviolet light is harmful to human health (carcinogenic) and characterized by particularly low light penetration, which is a challenge for the photopolymerization of thick and filled samples [10]. Therefore, alternatives to UV lamps and the use of longer wavelengths (near UV or visible) can be advantageous.

The use of light-emitting diodes (LEDs) perfectly fit this requirement for safer/cheaper, and more efficient irradiation devices than UV lamps or UV lasers [11–14]. In parallel, it is important to develop new photoinitiating systems able to absorb in the near UV or the visible range where their absorption spectrum overlaps that of the LED emission. To obtain this type of system, it is necessary to develop new organic molecules carrying chromophore groups capable of shifting their absorption spectrum towards the near-UV-visible range. These molecules will be called photoinitiator (PI), which can absorb the light and generate reactive species (in combination with additives) able to initiate the photopolymerization process.

In this paper, nine coumarin derivatives (noted Coum in Scheme 1) varying by the substitution pattern at the 3- and 7-positions of the coumarin core were synthesized and evaluated as photoinitiators for the FRP of acrylate and methacrylate monomers. In fact, coumarin derivatives have already been tested as photoinitiators of FRP and they have shown good polymerization profile (Rp and FC) as well as good photochemical and photophysical properties [15–19].

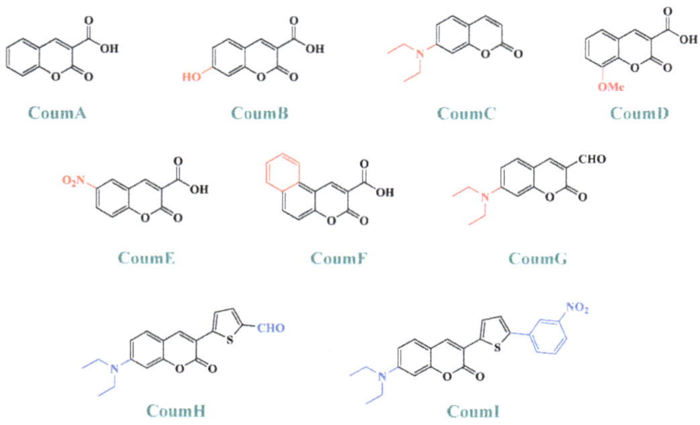

Scheme 1. The new series of coumarins (CoumA–CoumI) examined as photoinitiators of polymerization.

However, in the present work, coumarin-3-carboxylic acids, coumarin-3-aldehydes varying by the substitution pattern of the coumarin core and a coumarin of extended aromaticity have been studied as photoinitiators. Comparisons of the three families of coumarins have revealed that the substitution of the 3-position by electron-withdrawing groups such as a formyl group could improve the reactivity. The presence of a strong electron-donating group at the 7-position, such as diethylamine or a naphthalene group, could reinforce the electronic delocalization and the photoinitiating ability of the different systems. An optimum situation was found when electron-donating and electron-accepting groups were attached at both extremities of the coumarin core. Considering that the nitro group is among the most electron-withdrawing group, a coumarin bearing this electron acceptor was also designed and synthesized.

In fact, coumarin derivatives are usually characterized by very high fluorescence emission and can be used as fluorescent chromophores for several applications [20]. They are also characterized by high molar extinction coefficients in the near-UV and the visible range [19]. These novel coumarin-based photoinitiators were tested in photopolymerization of acrylate functions (TMPTA or TA) in both Thick (1.4 mm) and Thin sample (25 µm) using two and three-component photoinitiating systems PISs based on Coum/Iodonium salt (0.1%/1% *w*/*w*) and Coum/Iodonium salt/amine (NPG) (0.1%/1%/1% *w*/*w*/*w*). These systems were also used in 3D printing and photocomposite synthesis. These dyes are characterized by very high extinction coefficients with a broad absorption extending over the near UV/visible and high quantum yields were determined by fluorescence quenching.

It is important to note that coumarin shows a dual photo-oxidation and photo-reduction character.

2. Results

Photoinitiation ability, the performance of photopolymerization, photophysical and photochemical properties as well as chemical mechanisms associated with the photopolymerization processes will be discussed in detail.

2.1. Synthesis of the Different Dyes

As mentioned in the introduction section, three families of coumarins have been examined as photoinitiators of polymerization. The first family concerned coumarin-3-carboxylic acids. The five dyes were prepared in solution by condensation of diethyl malonate with *ortho*-hydroxyarylaldehydes [21]. After hydrolysis of esters in acidic conditions (a mixture of hydrochloric acid and acetic acid), the solution was neutralized to provide the different dyes with reaction yields ranging from 75% for CoumA to 86% yield for CoumE. A similar procedure was used for CoumC except that the hydrolysis of the intermediate ester coumarin resulted in a decarboxylation reaction, providing CoumC in 72% yield. The presence of the dimethylamino group in CoumC is essential to activate the decarboxylation reaction since this reaction was not observed for the other coumarins, maintaining the acidic function on the coumarins (See Scheme 2) [22]. Using the Vilsmeier Haack reaction, CoumC could be converted as CoumG in a 77% yield.

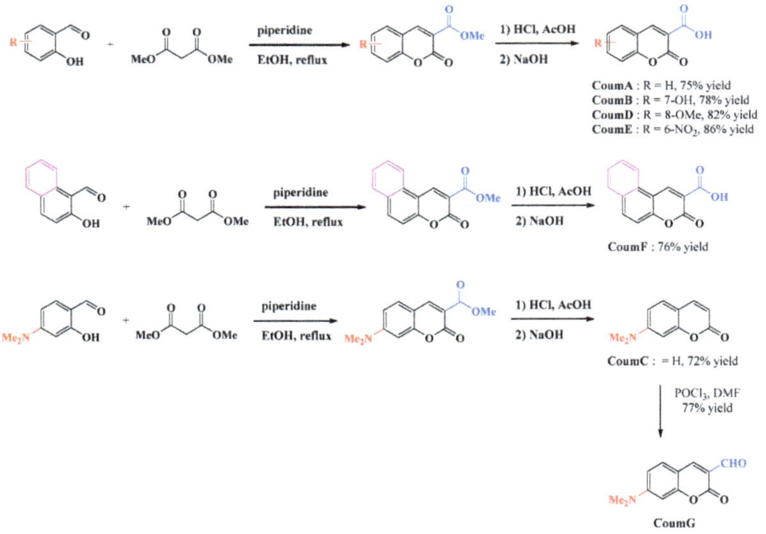

piperidine, EtOH, reflux; 1) HCl, AcOH 2) NaOH

CoumA : R = H, 75% yield
CoumB : R = 7-OH, 78% yield
CoumD : R = 8-OMe, 82% yield
CoumE : R = 6-NO_2, 86% yield

CoumF : 76% yield

CoumC : = H, 72% yield

$POCl_3$, DMF 77% yield

CoumG

Scheme 2. Synthetic routes to CoumA–CoumG.

Finally, CoumH and CoumI could be prepared starting from 2-thiopheneacetic acid and 4-diethylamino-2-hydroxybenzaldehyde. By Knoevenagel reaction, 7-(diethylamino)-3-(thiophen-2-yl)-2*H*-chromen-2-one could be obtained in 58% yield and by means of a Vilsmeier Haack reaction, CoumH was isolated in pure form in 88% yield. Conversely, CoumI was prepared in two steps, first by bromination of 7-(diethylamino)-3-(thiophen-2-yl)-2*H*-chromen-2-one in 86% yield, followed by a Suzuki cross-coupling reaction with 3-nitrophenylboronic acid. Using this procedure, CoumI was obtained in 67% yield (See Scheme 3).

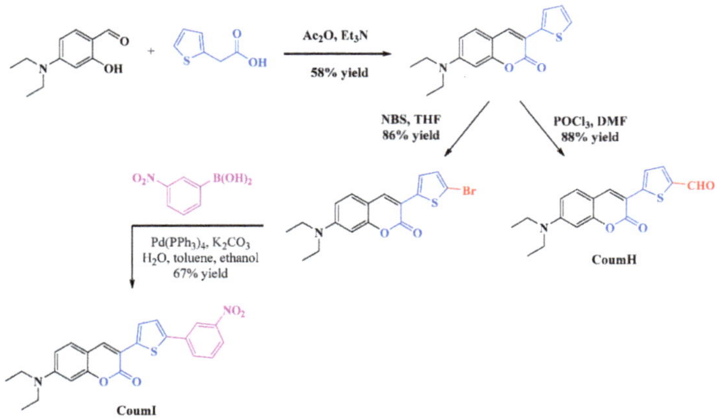

Scheme 3. Synthetic routes to CoumH and CoumI.

2.2. Light Absorption Properties

UV-visible absorption spectra of the different coumarins in acetonitrile are depicted in Figure 1 (See also Table 1). These organic compounds are characterized by a high molar extinction coefficient in both near-UV and visible range (e.g., CoumC ε ~ 18000 $M^{-1}cm^{-1}$ at 374 nm and 3500 $M^{-1}cm^{-1}$ at 405 nm, and CoumF ε ~ 10200 $M^{-1}cm^{-1}$ at 376 nm and 4800 $M^{-1}cm^{-1}$ at 405 nm). So, these absorption properties afford a good overlap with the emission spectrum of the LEDs used in this work (LED at 405 nm for FRP, LED at 375 nm for the photolysis experiments and LED at 385 nm for the photocomposites synthesis).

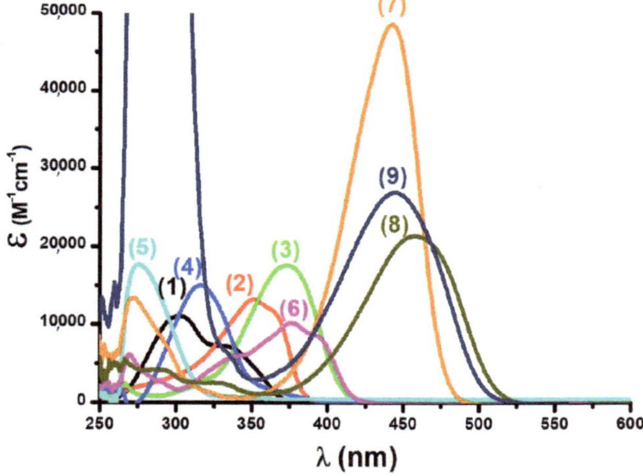

Figure 1. UV-visible absorption spectra of the investigated compounds based on coumarin derivatives in ACN: (**1**) CoumA, (**2**) CoumB, (**3**) CoumC, (**4**) CoumD, (**5**) CoumE, (**6**) CoumF, (**7**) CoumG, (**8**) CoumH, and (**9**) CoumI.

In fact, the presence of different substituents on the coumarin scaffold can affect the absorption properties (Figure 2) of these compounds and their molar extinction coefficients can be affected. For example, taking CoumA as a standard structure among these 10 compounds, we observed a shift towards higher absorption range (e.g., CoumB, CoumD, and CoumF are strongly shifted), and towards lower absorption range (e.g., CoumE), so a bathochromic effect is observed by introduction of electron donor group (such as OH,

OMe, and NR_2) and a hypsochromic effect is observed by introduction of electron acceptor group (e.g., NO_2 in case of CoumE).

Table 1. Light absorption properties of coumarins at 405 nm and at λ_{max}; singlet state energy (E_{S1}) determined from the crossing point of absorption and fluorescence spectra.

	λ_{max} (nm)	ε_{max} (M^{-1} cm^{-1})	ε_{405nm} (M^{-1} cm^{-1})	E_{S1} (eV)
CoumA	302	11,000	40	3.35
CoumB	351	13,000	80	3.24
CoumC	374	18,000	3500	3.01
CoumD	322	14,000	100	3.12
CoumE	275	18,000	60	3.91
CoumF	376	10,000	4800	2.98
CoumG	442	48,000	19,000	2.62
CoumH	458	21,000	6500	2.48
CoumI	445	27,000	1400	2.55

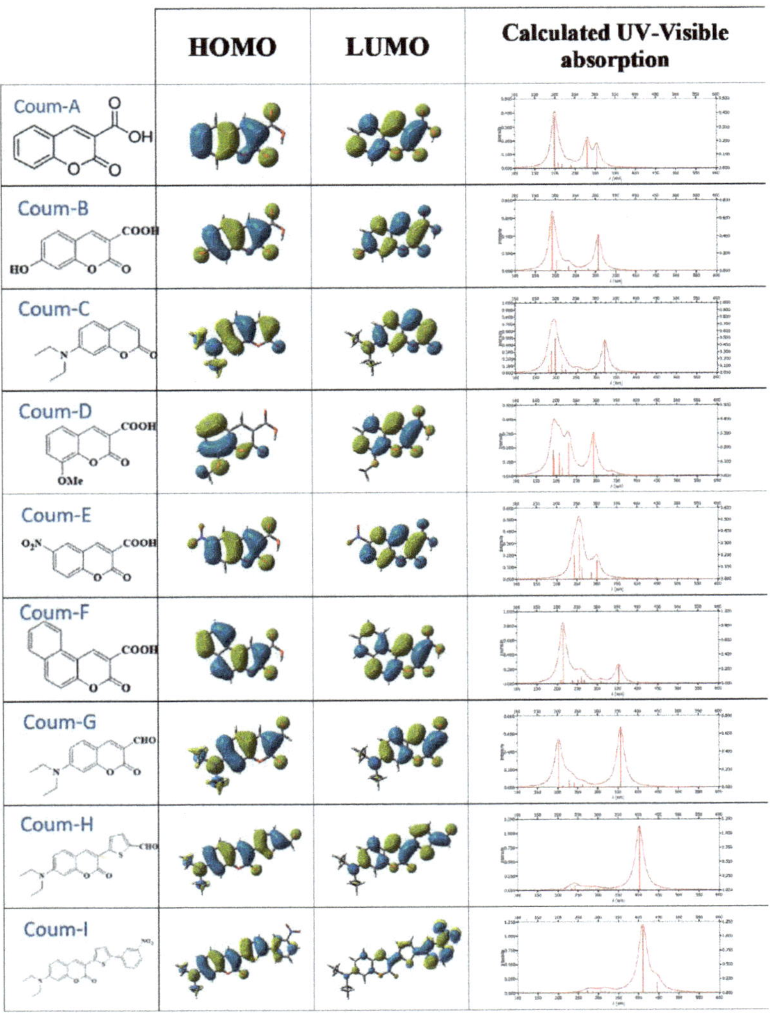

Figure 2. HOMO and LUMO frontier orbitals and their respective calculated UV-visible absorption spectra for the different investigated compounds at the UB3LYP/6-31G* level.

The electron-donating effect of these substituents is presented by ascending order: CoumH > CoumI > CoumG > CoumF > CoumC > CoumD > CoumB.

2.3. Free Radical Photopolymerization

2.3.1. Photopolymerization of Methacrylate Function of Mix-MA

The FRP profiles of methacrylate functions using Mix-MA as the benchmark monomer was performed in thick sample and in the presence of two or three-component PISs based on Coum/Iod (or NPG) (0.1%/1% *w*/*w*) or Coum/Iod/NPG (0.1%/1%/1% *w*/*w*/*w*) respectively, upon visible light irradiation with a LED at 405nm are given in Figure 3 (See also Table 2).

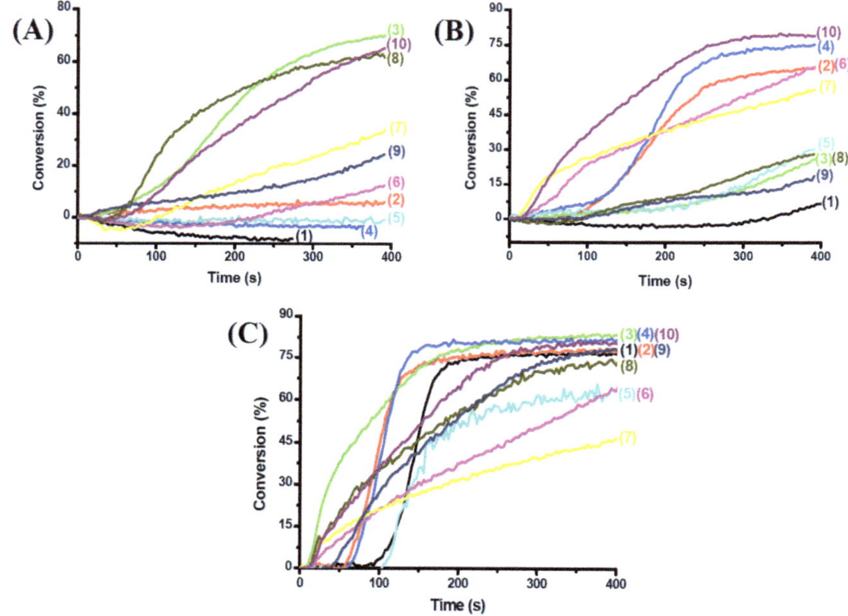

Figure 3. Photopolymerization profiles of methacrylate functions (conversion vs. irradiation time) using MIX-MA in thick sample (1.4 mm) under visible light irradiation using a LED at 405 nm: (**A**) Coum/Iod (0.1%/1% *w*/*w*), (**B**) Coum/NPG (0.1%/1% *w*/*w*) and (**C**) Coum/Iod/NPG (0.1%/1%/1% *w*/*w*/*w*): (1) CoumA, (2) CoumB, (3) CoumC, (4) CoumD, (5) CoumE, (6) CoumF, (7) CoumG, (8) CoumH, (9) CoumI and (10) Iod/NPG (1%/1% *w*/*w*). Irradiation starts at t = 10 s.

The obtained results show that Coum/Iod (or NPG) is less reactive than three-component PISs (Coum/Iod/NPG), this result can be explained by a higher yield of reactive species (radicals) in the presence of Iod/NPG which is not able, alone, to initiate the FRP (e.g., FC = 24% for CoumI/Iod vs. 76% for CoumI/Iod/NPG, and FC = 0% for CoumA/Iod vs. 76% for CoumA/Iod/NPG; show Figure 3A,C curve 1). In fact, CoumB, CoumD, CoumF and CoumG showed a photoreduction process rather than a photo-oxidation process (e.g., FC = 0% for CoumB/Iod vs. 67% for CoumB/NPG, and FC = 0% for CoumD/Iod vs. 75% for CoumD/NPG), but CoumC and Coum8 show an opposite behavior with a photo-oxidation process probably more favorable than the photoreduction (FC = 70% for CoumC/Iod vs. 28% for CoumC/NPG; Figure 3A,B curve 3). The FRP profiles also show a low rate of polymerization, this can be due to the high oxygen inhibition effect.

Table 2. Final reactive functions conversion (FC%) for different monomers and different PISs upon visible light irradiation using a LED at 405 nm (400 s of irradiation and thickness = 1.4 mm).

	Two-Component PISs Coum/Additives (0.1%/1% *w*/*w*)			Three-Component PISs Coum/Iod/NPG (0.1%/1%/1% *w*/*w*/*w*)		
	TMPTA	TA	Mix-MA	TMPTA	TA	Mix-MA
CoumA	n.p [a] 61% [b]	n.p [a] 30% [b]	n.p [a] n.p [b]	80%	85%	76%
CoumB	n.p [a] 78% [b]	n.p [a] 81% [b]	n.p [a] 67% [b]	78%	88%	76%
CoumC	60% [a] 42% [b]	80% [a] 56% [b]	70% [a] 28% [b]	80%	88%	80%
CoumD	33% [a] 83% [b]	74% [a] 90% [b]	n.p [a] 75% [b]	86%	92%	79%
CoumE	n.p [a] 36% [b]	n.p [a] 25% [b]	n.p [a] 30% [b]	73%	75%	64%
CoumF	60% [a] 81% [b]	86% [a] 88% [b]	13% [a] 63% [b]	80%	88%	64%
CoumG	70% [a] 75% [b]	80% [a] 82% [b]	35% [a] 55% [b]	50%	40%	46%
CoumH	65% [a] 53% [b]	84% [a] 59% [b]	62% [a] 29% [b]	81%	87%	75%
CoumI	58% [a] 60% [b]	78% [a] 31% [b]	24% [a] 18% [b]	70%	75%	76%

[a] Coum/Iod (0.1%/1% *w*/*w*); [b] Coum/NPG (0.1%/1% *w*/*w*).

2.3.2. Photopolymerization of Acrylates (TMPTA or TA)

In fact, iodonium salt or NPG alone cannot initiate the FRP of acrylate at 405 nm due to their absorption in the UV range [17,23]. Therefore, the coumarins derivatives are introduced in order to improve the absorption properties of photosensitive formulations.

Firstly, the most of Coumarin derivatives show high extinction coefficients at 405 nm. The photopolymerization profiles of acrylate functions in thick (1.4 mm) or thin (25 µm) samples (conversion vs. irradiation time) using TMPTA (or TA) as benchmark monomers are depicted in Figure 4 (see also Tables 2 and 3). The obtained results show that the two-component PISs based on Coum/Iod (0.1%/1% *w*/*w*) (or Coum/NPG) are able to strongly initiate the FRP, but a very higher performance [Final conversion (FC) and polymerization rate (Rp)] was acquired using the three-component PISs based on Coum/Iod/NPG which is quite efficient in the FRP of acrylate functions upon LED at 405 nm (e.g., FC = 60% for CoumC/Iod (0.1%/1% *w*/*w*) vs. 80% for CoumC/Iod/NPG (0.1%/1%/1% *w*/*w*/*w*), Figure 4A and B curve 3).

Moreover, the Iod/NPG (1%/1% *w*/*w*) couple weakly initiates the FRP (FC = 47%). This is ascribed to the formation of a charge transfer complex (CTC) between Iod and NPG [24] which is able to generate reactive species when it absorbs light. Clearly, the presence of Coumarin as photoinitiator is improving the performance of the photopolymerization processes.

Some of the coumarins can show both photoreduction (electron transfer from NPG to Coumarin) and photoxidation (electron transfer from Coumarin to Iod) processes, while other derivatives show only photoreduction process, such as CoumA, CoumB and CoumE (e.g., FC = 0% for CoumB/Iod (0.1%/1% *w*/*w*) vs. FC = 78% for CoumB/NPG (0.1%/1% *w*/*w*)).

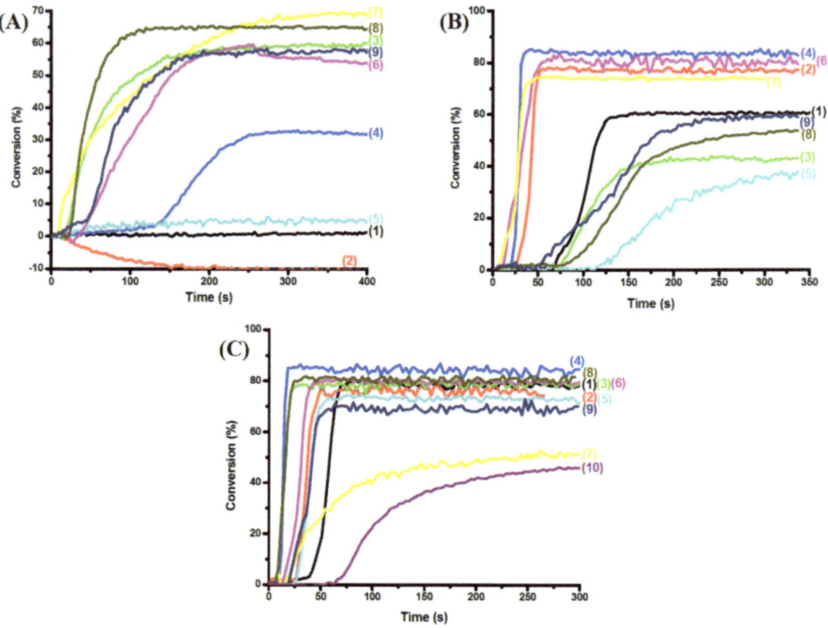

Figure 4. Photopolymerization profiles of acrylate functions (conversion vs irradiation time) using TMPTA in thick sample (1.4 mm) under visible light irradiation using a LED at 405 nm: (**A**) Coum/Iod (0.1%/1% *w/w*), (**B**) Coum/NPG (0.1%/1% *w/w*) and (**C**) Coum/Iod/NPG (0.1%/1%/1% *w/w/w*): (1) CoumA, (2) CoumB, (3) CoumC, (4) CoumD, (5) CoumE, (6) CoumF, (7) CoumG, (8) CoumH, (9) CoumI and (10) Iod/NPG (1%/1% *w/w*). The molar concentrations for 0.1 % *w/w* are 0.0055, 0.0051, 0.0049, 0.0048, 0.0045, 0.0044, 0.0043, 0.0032, and 0.0025 M for CoumA, CoumB, CoumC, CoumD, CoumE, CoumF, CoumG, CoumH, CoumI, respectelievly. The irradiation starts at t = 10 s.

Table 3. Final reactive functions conversion (FC%) for different monomers and different PISs upon visible light irradiation using a LED at 405 nm (150 s of irradiation and thickness = 25 μm).

	Two-Component PISs Coum/Additives (0.1%/1% *w/w*)			**Three-Component PISs Coum/Iod/NPG (0.1%/1%/1% *w/w/w*)**		
	TMPTA	TA	Mix-MA	TMPTA	TA	Mix-MA
CoumA	n.p [a] 32% [b]	n.p [a] n.p [b]	n.p [a] 22% [b]	25%	38%	36%
CoumB	49% [a] 13% [b]	32% [a] 24% [b]	14% [a] 15% [b]	28%	44%	52%
CoumC	25% [a] 25% [b]	53% [a] 15% [b]	n.p [a] 17% [b]	50%	75%	72%
CoumD	42% [a] 34% [b]	42% [a] 45% [b]	17% [a] 22% [b]	42%	70%	57%
CoumE	n.p [a] n.p [b]	n.p [a] n.p [b]	n.p [a] n.p [b]	30%	40%	n.p
CoumF	11% [a] 47% [b]	42% [a] 65% [b]	18% [a] 70% [b]	51%	73%	79%
CoumG	38% [a] 48% [b]	68% [a] 67% [b]	43% [a] 64% [b]	55%	81%	74%
CoumH	19% [a] 14% [b]	43% [a] 45% [b]	36% [a] 48% [b]	46%	68%	74%
CoumI	15% [a] 25% [b]	37% [a] 45% [b]	12% [a] 29% [b]	42%	58%	66%

[a] Coum/Iod (0.1%/1% *w/w*); [b] Coum/NPG (0.1%/1% *w/w*).

2.4. D Printing Experiments Using Coum/Iod/amine PISs and Optical Microscopy Characterization

New 3D patterns were obtained by direct laser write experiments of Coum/Iod/amine PISs using a laser diode at 405 nm and characterized by optical microscopy. These 3D patterns were obtained under air using different PISs based on Coum/Iod/TMA (0.05%/0.5%/0.235% *w*/*w*/*w*) in TA or TMPTA (See Figure 5). In fact, the high photosensitivity of this resin allowed an efficient polymerization process in the irradiated area so a high spatial resolution is observed. Markedly, a great thickness is obtained (~2090 µm) and these patterns were carried out in a very short irradiation time (~2–3 min). Using a well-established Type I photoinitiator (diphenyl(2,4,6-trimethylbenzoyl)phosphine oxide—TPO) in similar direct laser write conditions; similar performances can be reached but requiring a higher content (0.5% *w*/*w*). This latter result demonstrates the interest in using Coum derivatives. It is important to note that the 3D patterns based on CoumC exhibit a blue fluorescence when these structures are characterized by the light of the microscope.

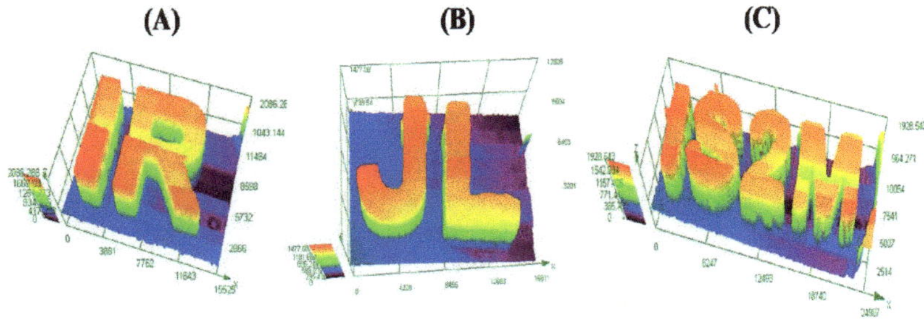

Figure 5. Characterization of 3D patterns by numerical optical microscopy obtained by free radical photopolymerization experiment (using TA or TMPTA as benchmark monomer) using a diode laser at 405 nm: (**A**) CoumC/Iod/4,N-N-TMA (0.05%/0.5%/0.235% *w*/*w*/*w*) in TA, (**B**) CoumH/Iod/TMA (0.05%/0.5%/0.19% *w*/*w*/*w*) in TMPTA and (**C**) CoumD/Iod/4,N-N-TMA (0.05%/0.5%/0.275% *w*/*w*/*w*) in TMPTA.

2.5. Near-UV Conveyor Experiments for the Synthesis of Photocomposites Using Coum/Iod/NPG (0.1%/1%/1% w/w/w)

Generally, photocomposites are materials composed of at least two components: matrix and reinforcement. The mixture of these two components leads to new interesting properties that the two components separately do not have. The production of composites in the last decades and until today represents a very dynamic market in different fields such as aeronautics, automotive, wind power, and buildings. So, due to their very high mechanical resistance and chemical resistance, the glass fibers are used in this work as a matrix for the photocomposite synthesis.

In this work, the proposed coumarin derivatives were tested for access to photocomposites upon near-UV light using a LED conveyor at 385 nm (0.7 W/cm^2). The curing results obtained are summarized in Figure 6. Firstly, photocomposites were prepared by impregnation of glass fibers with an acrylic resin (TMPTA) (50% glass fibers/50% acrylic resin) and irradiated upon a LED at 385 nm. Remarkably, a very fast polymerization was observed using Coum/Iod/NPG (0.1%/1%/1% *w*/*w*/*w*), where both the surface and the bottom are tack-free after some passes.

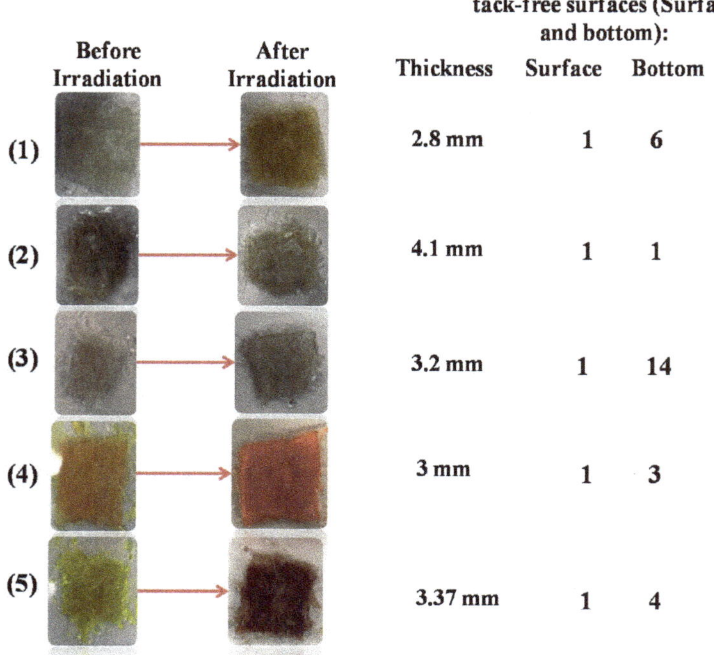

Figure 6. Photocomposites manufactured upon near-UV irradiation at 385 nm (0.7 W/cm^2) using glass fiber/resin (50%/50% *w*/*w*) in the presence of three-component PISs based on Coum/Iod/NPG (0.1%/1%/1% *w*/*w*/*w*): **(1)** CoumC, **(2)** CoumD, **(3)** CoumF, **(4)** CoumG, and **(5)** CoumH.

3. Discussion

For a better understanding of the photoinitiation ability, the photochemical properties of the studied coumarins were investigated. More particularly, their photolysis behaviors, fluorescence quenchings, and redox properties were investigated in the presence of additives (amine/iodonium salt), allowing to establish the photochemical mechanisms (see Scheme 4 below).

$\text{Coumarin } (h\nu) \rightarrow {}^{1,3}\text{Coumarin}$	r1
${}^{1,3}\text{Coumarin} + \text{ArI}^{+} \rightarrow \text{Ar}^{\bullet} + \text{ArI} + \text{Coumarin}^{\bullet+}$	r2
$\text{NPG} + \text{Iod} \rightarrow [\text{NPG-Iod}]_{CTC}$	r3
$[\text{NPG-Iod}]_{CTC} (h\nu) \rightarrow \text{Ar}^{\bullet}$	r4
${}^{1,3}\text{Coumarin} + \text{NPG} \rightarrow \text{Coumarin-H}^{\bullet} + \text{NPG}_{(-H)}^{\bullet}$	r5
$\text{NPG}_{(-H)}^{\bullet} \rightarrow \text{NPG}_{(-H;-CO_2)}^{\bullet} + CO_2$	r6
$\text{NPG}_{(-H;-CO_2)}^{\bullet} + \text{Ar}_2\text{I}^{+} \rightarrow \text{NPG}_{(-H;-CO_2)}^{+} + \text{Ar}^{\bullet} + \text{ArI}$	r7
$\text{Coumarin}^{\bullet+} + \text{NPG} \rightarrow \text{Coumarin} + \text{NPG}^{\bullet+}$	r8
$\text{Coumarin-H}^{\bullet} + \text{Ar}_2\text{I}^{+} \rightarrow \text{Coumarin} + \text{Ar}^{\bullet} + \text{ArI} + \text{H}^{+}$	r9

Scheme 4. Proposed chemical mechanisms.

3.1. *Steady-State Photolysis of Coumarins*

Steady-state photolysis of coumarins derivatives in ACN and under irradiation light using a LED at 375 nm have been performed to explain the obtained results in FRP. So, the photolysis of one of these compounds (CoumC) is presented in Figure 7. First of all, the photolysis of CoumC alone upon irradiation at 375 nm is very slow compared to that obtained with Iod, which is very fast. In fact, the appearance of a weak peak between 425 and 500 nm and the evolution of the absorption peak of CoumC shows that a high interaction between CoumC and Iod took place by an electron transfer process, this process induced, during the irradiation, a photolysis of the CoumC and generation of new photoproducts. On the other hand, the photolysis of CoumC with Iod/NPG couple was very slow (Figure 7D curve 3) and poor consumption was obtained; these results can be explained by a high regeneration of CoumC in three-component PISs.

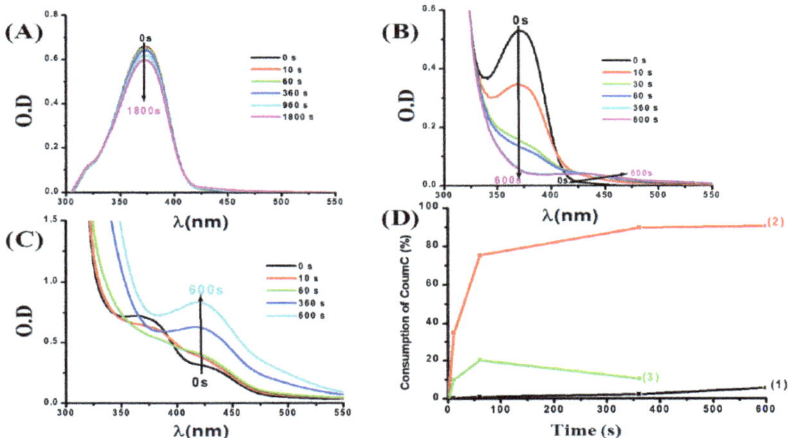

Figure 7. (**A**) Photolysis of CoumC alone in ACN, (**B**) photolysis of CoumC with Iod (10^{-2} M) in ACN, (**C**) photolysis of CoumC with Iod/NPG (10^{-2}M) couple, (**D**) percentage of consumption of CoumC (1) without Iod, and with (2) with Iod (10^{-2} M), and (3) with Iod/NPG vs. irradiation time—upon exposure to the LED@375 nm in ACN.

3.2. Fluorescence Quenching and Cyclic Voltammetry Experiments for the Coumarins

Fluorescence quenching and emission spectra of the different coumarins (e.g., CoumC) have been carried out in ACN and reported in Figure 8. Firstly, where the emission intensity of CoumC decreases when we added Iod or NPG, so an interaction between ^{1}Coum-C and Iod (or NPG) occurs, this result is in full agreement with FRP and photolysis experiments shown above. To compare the reactivity of different coumarin with Iod or NPG, the Stern-Volmer coefficient (Ksv) have been calculated according to Equation (1). For example, a very high quenching of CoumF with NPG and poor quenching of CoumC with NPG were observed, so Ksv for CoumF is higher than that of CoumC (Ksv = 44 M^{-1} for CoumC and 400 M^{-1} for CoumF), therefore a high electron transfer quantum yield is obtained for CoumF ϕ = 0.9) compared to that obtained for CoumC (ϕ = 0.6) (Table 4)

$$\phi_{S1} = K_{SV}[\mathrm{Iod}]/(1 + K_{SV}[\mathrm{Iod}]) \quad (1)$$

The free energy change (ΔG) for the electron transfer between coumarins and Iod or NPG is an important parameter to evaluate the feasibility of this process. ΔG can be extracted from the E_{S1} and the electrochemical properties (E_{ox} and E_{red}) (using Equation (1)) e.g., $\Delta G = -2.39$ eV for CoumF/Iod which is more reactive in FRP of acrylate functions (TA monomer) (FC = 86%). All these data are gathered in Table 4.

Finally, the FRP results of acrylate functions can be explained by a global mechanism based on the different results obtained by the characterization techniques (steady-state photolysis, Fluorescence quenching and cyclic voltammetry). First of all, the photoinitiator (Coumarin) goes to its excited state once it absorbs suitable light energy, and as it is not able to give reactive species alone, the Iod salt (or NPG), therefore, interacts with its excited state and will be able to dissociate and give reactive species responsible to initiate the FRP (r1–r2). The addition of NPG to the photosensitive formulation is very important because of the formation of a charge-transfer complex between Iod salt and NPG [Iod-NPG]$_{CTC}$ able to generate reactive species (r3–r4). Moreover, a hydrogen transfer process from NPG to Coumarins can occur which generates two types of radicals (Coum-H$^•$, NPG$_{(-H)}{}^•$) (r5). In fact, a decarboxylation of NPG$_{(-H)}{}^•$ can take place and leads to the radical formation (NPG$_{(-H, -CO2)}{}^•$), which react with Iod salt to produce reactive species (Ar$^•$ and NPG$_{(-H, CO2)}{}^+$) (r6–r7). Ar$^•$ and NPG$_{(-H,-CO2)}{}^•$) (r1–r9) radicals are assumed as the reactive species responsible to the FRP of the (meth)acrylate functions. The coumarins consumption is reduced in three-component PIS (Figure 6); this can be explained by a regeneration of the photoinitiator, which is in agreement on r8-r9 (See Scheme 4).

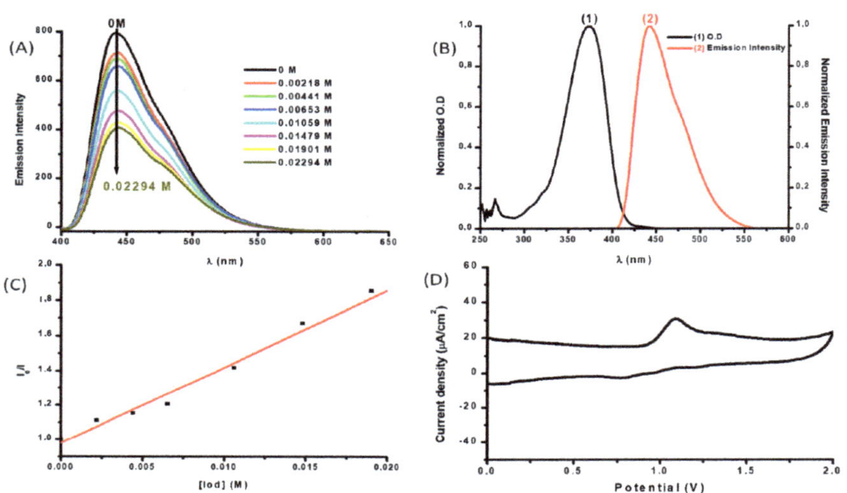

Figure 8. (**A**) Fluorescence quenching of CoumC by Iod, (**B**) E_{S1} determination, (**C**) determination of K_{SV} (Stern–Volmer coefficient), and (**D**) oxidation potential (E_{ox}) determination of CoumC.

Table 4. Parameters characterizing the chemical mechanisms associated with ^{1}Coum/Iod or ^{1}Coum/NPG interaction in acetonitrile. For Iod and NPG, a reduction and oxidation potential of −0.2 and 1.03 eV were used respectively for the ΔGet calculation.

	E_{ox} (V)	E_{red} (V)	ΔG (eV) (Coum/Iod)	ΔG (eV) (Coum/NPG)	K_{SV} M^{-1} (Coum/Iod)	K_{SV} M^{-1} (Coum/NPG)	Φ (Coum/Iod)	Φ (Coum/NPG)
CoumA	-	−0.61	-	−1.71	7.25	104	0.35	0.78
CoumB	-	-	-	-	21	174	0.39	0.82
CoumC	1	-	−1.81	-	44	44	0.5	0.6
CoumD	0.6	−0.98	−2.32	−1.11	14	46.5	0.3	0.5
CoumE	-	-	-	-	5297	773	0.972	0.9
CoumF	0.39	−1.01	−2.39	−0.94	9	400	0.3	0.9
CoumG	0.46	−1.35	−1.96	−0.24	6	16	0.2	0.5
CoumH	0.96	−1.40	−1.32	−0.05	48	22	0.6	0.6
CoumI	-	−1.06	-	−0.460	4	-	0.25	-

The photoinitiation ability is a strong interplay between these different reactions (r1-r9), but their light absorption properties and intersystem crossing behavior (singlet vs. triplet state pathways, lifetimes) must also be taken into account. Therefore, a deeper characterization of their structure/reactivity/efficiency relationship is beyond the scope of the present work.

4. Materials and Methods

4.1. Synthesis of the Coumarins

All reagents and solvents were purchased from Aldrich, Alfa Aesar, or TCI Europe and used as received without further purification. Mass spectroscopy was performed by the Spectropole of Aix-Marseille University. ESI mass spectral analyses were recorded with a 3200 QTRAP (Applied Biosystems SCIEX) mass spectrometer. The HRMS mass spectral analysis was performed with a QStar Elite (Applied Biosystems SCIEX) mass spectrometer. Elemental analyses were recorded with a Thermo Finnigan EA 1112 elemental analysis apparatus driven by the Eager 300 software. ^{1}H and ^{13}C NMR spectra were determined at room temperature in 5 mm o.d. tubes on a Bruker Avance 400 spectrometer and on a Bruker Avance 300 spectrometer of the Spectropole: The ^{1}H chemical shifts were referenced to the solvent peak $CDCl_3$ (7.26 ppm), and the ^{13}C chemical shifts were referenced to the solvent peak $CDCl_3$ (77 ppm). 7-(Diethylamino)-3-(thiophen-2-yl)-2*H*-chromen-2-one and 3-(5-bromothiophen-2-yl)-7-(diethylamino)-2*H*-chromen-2-one used as intermediates of reaction have been synthesized according to procedures previously reported in the literature, without modifications and in similar yields [18].

Synthesis of 2-oxo-2*H*-chromene-3-carboxylic acid (CoumA)

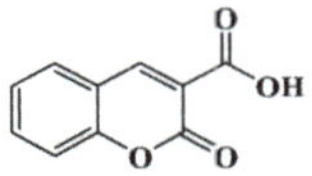

Dimethyl malonate (2.64 g, 20 mmol, M = 132.11 g/mol) and piperidine (1 mL, 10 mmol) were mixed and added to a solution of salicylaldehyde (1.22 g, 10 mmol, M = 122.12 g/mol) dissolved in absolute ethanol (30 mL). After stirring and heating to reflux for 6 h, the solvent was removed under reduced pressure. Then, concentrated HCl (20 mL) and acetic acid (20 mL) were added for hydrolysis and the solution was refluxed overnight. After cooling, the solution was poured onto ice and aq. 40% NaOH was added until pH = 5. A pale precipitate formed. After stirring for another 30 min, the mixture was filtered, washed with water, pentane, and dried under vacuum (1.43 g, 75% yield). ^{1}H NMR (400 MHz, DMSO-d_6) δ 8.75 (s, 1H), 7.91 (dd, *J* = 7.7, 1.2 Hz, 1H), 7.80–7.68 (m, 1H), 7.42 (dd, *J* = 15.4, 7.9 Hz, 2H). Analyses were consistent with those previously reported in the literature [25].

Synthesis of 7-hydroxy-2-oxo-2*H*-chromene-3-carboxylic acid (CoumB),

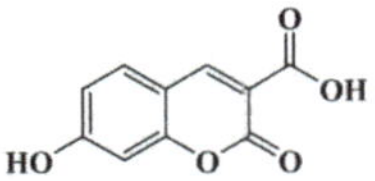

Dimethyl malonate (2.64 g, 20 mmol, M = 132.11 g/mol) and piperidine (1 mL, 10 mmol) were mixed and added to the solution of 2,4-dihydroxybenzaldehyde (1.38 g, 10 mmol, M = 138.12 g/mol) dissolved in absolute ethanol (30 mL). After stirring and heating to reflux for 6 h, the solvent was removed under reduced pressure. Then, concentrated HCl (20 mL) and acetic acid (20 mL) were added for hydrolysis, and the solution was refluxed overnight. After cooling, the solution was poured onto ice and aq. 40% NaOH was added until pH = 5. A pale precipitate formed. After stirring for another 30 min, the mixture was filtered, washed with water, pentane, and dried under vacuum (1.61 g, 78% yield). ^{1}H NMR (400 MHz, DMSO-d_6) δ 8.68 (s, 1H), 7.75 (d, *J* = 8.6 Hz, 1H), 6.85 (dd, *J* = 8.6, 2.3 Hz, 1H), 6.74 (d, *J* = 2.1 Hz, 1H). Analyses were consistent with those previously reported in the literature [25].

Synthesis of 7-(diethylamino)-2*H*-chromen-2-one (CoumC)

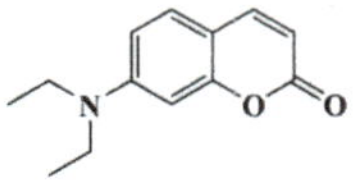

Dimethyl malonate (2.64 g, 20 mmol, M = 132.11 g/mol) and piperidine (1 mL, 10 mmol) were mixed and added to the solution of 4-(Diethylamino)-salicylaldehyde (1.93 g, 10 mmol) dissolved in absolute ethanol (30 mL). After stirring and heating to reflux for 6 h, the solvent was removed under reduced pressure. Then, concentrated HCl (20 mL) and acetic acid (20 mL) were added for hydrolysis and the solution was refluxed overnight. After cooling, the solution was poured onto ice and aq. 40% NaOH was added until pH = 5. A pale precipitate formed. After stirring for another 30 min, the mixture was filtered, washed with water, pentane, and dried under vacuum (1.56 g, 72% yield). ^{1}H NMR (400 MHz, $CDCl_3$) δ 7.53 (d, J = 9.3 Hz, 1H), 7.24 (d, J = 8.8 Hz, 1H), 6.60–6.44 (m, 2H), 6.03 (d, J = 9.3 Hz, 1H), 3.41 (q, J = 7.1 Hz, 4H), 1.21 (t, J = 7.1 Hz, 6H). Analyses were consistent with those previously reported in the literature [26].

Synthesis of 8-methoxy-2-oxo-2*H*-chromene-3-carboxylic acid (CoumD)

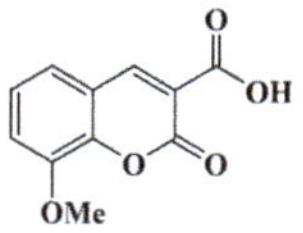

Dimethyl malonate (2.64 g, 20 mmol, M = 132.11 g/mol) and piperidine (1 mL, 10 mmol) were mixed and added to the solution of 3-methoxysalicylaldehyde (1.52 g, 10 mmol, M = 152.15 g/mol) dissolved in absolute ethanol (30 mL). After stirring and heating to reflux for 6 h, the solvent was removed under reduced pressure. Then, concentrated HCl (20 mL) and acetic acid (20 mL) were added for hydrolysis and the solution was refluxed overnight. After cooling, the solution was poured onto ice and aq. 40% NaOH was added until pH = 5. A pale precipitate formed. After stirring for another 30 min, the mixture was filtered, washed with water, pentane, and dried under vacuum (1.80 g, 82% yield). ^{1}H NMR (400 MHz, DMSO-d_6) δ 8.50 (s, 1H), 7.42–7.26 (m, 3H), 3.91 (s, 3H). Analyses were consistent with those previously reported in the literature [27].

Synthesis of 6-nitro-2-oxo-2*H*-chromene-3-carboxylic acid (CoumE)

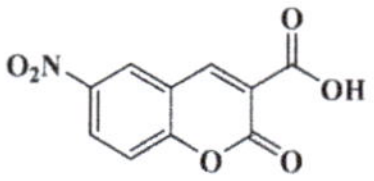

Dimethyl malonate (2.64 g, 20 mmol, M = 132.11 g/mol) and piperidine (1 mL, 10 mmol) were mixed and added to the solution of 2-hydroxy-5-nitrobenzaldehyde (1.67 g, 10 mmol, M = 167.12 g/mol) dissolved in absolute ethanol (30 mL). After stirring and heating to reflux for 6 h, the solvent was removed under reduced pressure. Then, concentrated HCl (20 mL) and acetic acid (20 mL) were added for hydrolysis and the solution was refluxed overnight. After cooling, the solution was poured onto ice and aq. 40% NaOH was added until pH = 5. A pale precipitate formed. After stirring for another 30 min, the mixture was filtered, washed with water, pentane, and dried under vacuum (2.02 g, 86% yield). ^{1}H NMR (400 MHz, DMSO-d$_6$) δ 8.81 (d, J = 2.8 Hz, 1H), 8.51–8.33 (m, 2H), 7.59 (d, J = 9.1 Hz, 1H) Analyses were consistent with those previously reported in the literature [24].

Synthesis of 3-oxo-3*H*-benzo[f]chromene-2-carboxylic acid (CoumF)

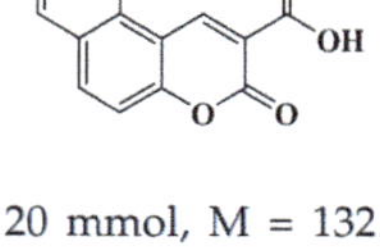

Dimethyl malonate (2.64 g, 20 mmol, M = 132.11 g/mol) and piperidine (1 mL, 10 mmol) were mixed and added to the solution of 2-hydroxy-1-naphthaldehyde (1.72 g, 10 mmol, M = 172.18 g/mol) dissolved in absolute ethanol (30 mL). After stirring and heating to reflux for 6 h, the solvent was removed under reduced pressure. Then, concentrated HCl (20 mL) and acetic acid (20 mL) were added for hydrolysis and the solution was refluxed overnight. After cooling, the solution was poured onto ice and aq. 40% NaOH was added until pH = 5. A pale precipitate formed. After stirring for another 30 min, the mixture was filtered, washed with water, pentane, and dried under vacuum (1.82 g, 76% yield). ^{1}H NMR (400 MHz, DMSO-d_6) δ 9.37 (s, 1H), 8.60 (d, *J* = 8.3 Hz, 1H), 8.31 (d, *J* = 9.0 Hz, 1H), 8.09 (d, *J* = 7.8 Hz, 1H), 7.77 (ddd, *J* = 8.3, 7.0, 1.3 Hz, 1H), 7.65 (dt, *J* = 12.9, 2.9 Hz, 1H), 7.60 (d, *J* = 9.1 Hz, 1H). Analyses were consistent with those previously reported in the literature [28].

Synthesis of 7-(diethylamino)-2-oxo-2*H*-chromene-3-carbaldehyde (CoumG)

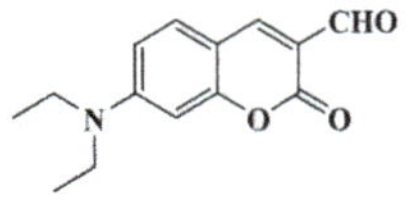

CHO
N O O

20 mL of $POCl_3$ was added dropwise to 20mL of dry DMF at 0 °C under Argon and stirred during 30 minutes at 50 °C. Then 7-(diethylamino)-2H-chromen-2-one (15.0 g, 69.1 mmol, M = 217.27g/mol) in 100 mL of DMF was added to the mixture and the mixture was heated to 60 °C overnight. Afterward, the mixture was poured into 500 mL of ice water and a solution of NaOH 20% was added. The precipitate was filtered and washed with water. (13.12 g, 77% yield). ^{1}H NMR (300 MHz, $CDCl_3$) δ 10.08 (s, 1H), 8.21 (s, 1H), 7.46–7.32 (m, 1H), 6.61 (dd, *J* = 9.0, 2.5 Hz, 1H), 6.45 (d, *J* = 2.3 Hz, 1H), 3.46 (q, *J* = 7.1 Hz, 4H), 1.23 (t, *J* = 7.1 Hz, 6H). Analyses were consistent with those previously reported in the literature [26].

Synthesis of 5-(7-(diethylamino)-2-oxo-2*H*-chromen-3-yl)thiophene-2-carbaldehyde (CoumH)

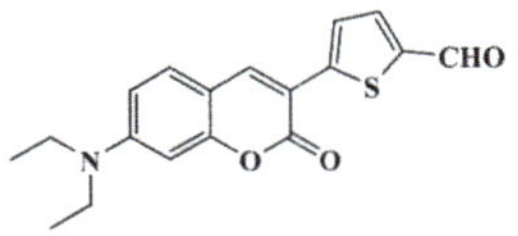

CHO
S
N O O

7-(Diethylamino)-3-(thiophen-2-yl)-2*H*-chromen-2-one (3.00 g, 10 mmol, M = 299.39 g/mol) was dissolved in DMF (7 mL) and $POCl_3$ (1.8 mL, 20 mmol) was slowly added at 0 °C. The mixture was heated up to 80 °C overnight. After cooling, the solution was quenched with water. The mixture was extracted with DCM several times. The organic phases were combined, dried over magnesium sulfate and the solvent removed under reduced pressure. It was used without any further purification (2.88 g, 88% yield). ^{1}H NMR (400 MHz, $CDCl_3$) δ 9.90 (s, 1H), 8.02 (s, 1H), 7.77 (d, *J* = 4.1 Hz, 1H), 7.73 (d, *J* = 4.1 Hz, 1H), 7.38 (d, *J* = 8.9 Hz, 1H), 6.69 (dd, *J* = 8.9, 2.5 Hz, 1H), 6.56 (d, *J* = 2.4 Hz, 1H), 3.46 (q, *J* = 7.1 Hz, 4H), 1.25 (t, *J* = 7.1 Hz, 6H). Analyses were consistent with those previously reported in the literature [29].

Synthesis of 7-(diethylamino)-3-(5-(3-nitrophenyl)thiophen-2-yl)-2*H*-chromen-2-one (CoumI)

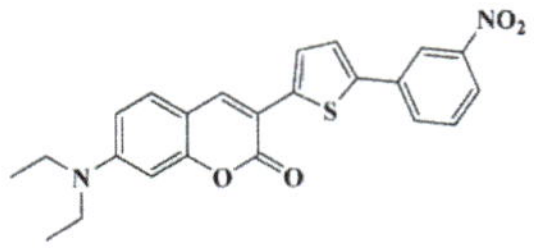

NO_2
S
N O O

Tetra*kis*(triphenylphosphine)palladium (0) (0.46 g, 0.744 mmol, M = 1155.56 g.mol^{-1}) was added to a mixture of 3-(5-bromothiophen-2-yl)-7-(diethylamino)-2*H*-chromen-2-one (2.31 g, 6.11 mmol, M = 378.28 g.mol^{-1}), 3-nitrophenylboronic acid (1.53 g, 9.16 mmol, M = 166.93 g.mol^{-1}), toluene (54 mL), ethanol (26 mL) and an aqueous potassium carbonate solution (2 M, 6.91 g in 25 mL water, 26 mL) under vigorous stirring. The mixture was stirred at 80 °C for 48 h under a nitrogen atmosphere. After cooling to room temperature, the reaction mixture was poured into water and extracted with ethyl acetate. The organic layer was washed with brine several times, and the solvent was then evaporated. The residue was purified by filtration on a plug of silicagel using a mixture of DCM/ethanol as the eluent (67% yield, 1.72 g). ^{1}H NMR (400 MHz, $CDCl_3$) δ 8.49 (t, *J* = 1.9 Hz, 1H), 8.10 (ddd, *J* = 8.2, 2.2, 1.0 Hz, 1H), 7.98–7.92 (m, 2H), 7.64 (d, *J* = 4.0 Hz, 1H), 7.56 (d, *J* = 8.0 Hz, 1H), 7.43 (d, *J* = 4.0 Hz, 1H), 7.35 (d, *J* = 8.9 Hz, 1H), 6.64 (dd, *J* = 8.8, 2.5 Hz, 1H), 6.55 (d, *J* = 2.4 Hz, 1H), 3.45 (q, *J* = 7.1 Hz, 4H), 1.24 (t, *J* = 7.1 Hz, 6H); HRMS (ESI MS) *m/z*: theor: 420.1144 found: 420.1147 ($M^{+\cdot}$ detected); Anal. calc. for $C_{23}H_{20}N_2O_4S$: C, 65.7, H, 4.8, O, 15.2; found: C 65.5, H 4.7, O 15.4.

4.2. Other Chemical Compounds

All the other chemicals (Figure 9) were selected with the highest purity available and used as received. Di-*tert*-butyl-diphenyliodonium hexafluorophosphate (Iod) and TMA (4,*N*,*N*-Trimethylaniline) were obtained from Lambson Ltd. (Wetherby, UK). Trimethylolpropane triacrylate (TMPTA), di(trimethylolpropane) tetraacrylate (TA), Mix-MA, *N*-Phenylglycine (NPG) were obtained from Allnex or Sigma Aldrich (Darmstadt, Germany). TMPTA, TA, and Mix-MA were selected as benchmark monomers for the radical polymerizations.

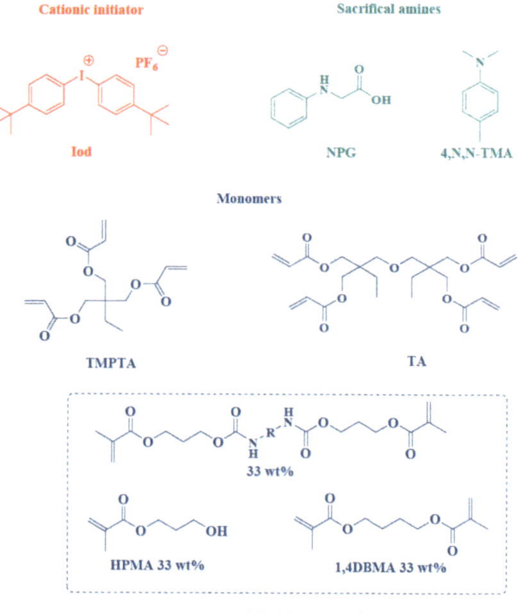

Figure 9. Other chemical compounds used in this paper.

4.3. Irradiation Light Sources

We used different light-emitting diodes (LEDs) as light sources: (1) at 405 nm (I = 110 mW/cm^2) for the photopolymerization experiments, (2) at 375 nm (I = 40 mW/cm^2) for the photolysis of Coumarins and (3) at 385 nm (I = 0.7 W/cm^2) for the photocomposite synthesis.

4.4. Real-Time Fourier Transform Infrared Spectroscopy (RT-FTIR): Kinetic Followed and Final Conversion (FC) Determination for the Photopolymerisation

In this research, the ability of coumarins to initiate the photopolymerization of (meth)acrylate functions (FRP) was studied using two and three-component photoinitiating systems based on Coum/Iod salt (or NPG) (0.1%/1% *w/w*) and Coum/Iod/NPG (0.1%/1%/1% *w/w/w*). The percentage of the different chemical compounds is calculated according to the monomer weight. Kinetic study, as well as the reactive function conversion, were monitored by the evolution of the double bond vs. time. In fact, the polymerization experiments were performed in both thick (1.4 mm) and thin (25 μm) samples, they were obtained by deposition of the formulation into the mold (1.4 mm) or between two propylene films in order to reduce O_2 inhibition, respectively. In addition, excellent solubility of all coumarin derivatives (excluding the CoumE) were observed. For the thick and thin samples, the evolution of the (meth)acrylate functions for TMPTA or Mix-MA were followed by RT-FTIR spectroscopy (JASCO FTIR 6600) at about 6150 cm^{-1} and 1630 cm^{-1}, respectively. The procedure used to monitor the photopolymerization profile was described in detail in [30,31].

4.5. Redox Potentials

The reduction (E_{red}) or oxidation (E_{ox}) potentials for the different coumarin derivatives were determined by cyclic voltammetry in ACN using tetrabutylammonium hexafluorophosphate as the supporting electrolyte (potential vs. saturated calomel electrode–SCE). The free energy change (ΔG_{et}) for an electron transfer reaction was calculated using equation (2) [27], where E_{ox}, E_{red}, E*, and C represent the oxidation potential of the electron donor, the reduction potential of the electron acceptor, the excited state energy level (determined from luminescence experiments) and the coulombic term for the initially formed ion pair, respectively. Here, C was neglectedm as usually done for polar solvents.

$$\Delta G_{et} = E_{ox} - E_{red} - E^* + C \quad (2)$$

4.6. UV-Vis Absorption and Photolysis Experiments

The absorption properties (UV-visible absorption spectrum and molar extinction coefficient) as well as the steady state photolysis of the investigated Coumarin derivatives (CoumA–CoumI) in acetonitrile have been investigated using a JASCO V730 spectrometer.

4.7. Fluorescence Experiments

The fluorescence properties of these organic compounds in ACN were studied using a JASCO FP-6200 spectrofluorimeter. The fluorescence quenching of ^{1}coumarin by Iod or NPG were examinated from the classical Stern-Volmer treatment [32] ($I_0/I = 1 + kq\ \tau_0[Q]$, where I_0 and I stand for the fluorescent intensity of coumarin in the absence and the presence of Iod or NPG, respectively; τ_0 stands for the lifetime of coumarin in the absence of Iod and [Q] stand for the concentration of quencher, in our study Iod or NPG).

4.8. Computational Procedure

Molecular orbital calculations were carried out with the Gaussian 03 suite of programs [33,34]. Electronic absorption spectra for the different compounds were calculated with the time-dependent density functional theory at the MPW1PW91-FC/6-31G* level of theory on the relaxed geometries calculated at the UB3LYP/6-31G* level of theory.

4.9. Near-UV Conveyor for Photocomposite Synthesis

Photocomposites have been achieved using glass fibers for the reinforcement and an organic resin based on acrylates (50%/50% *w/w*). The photosensitive resin has been deposited on glass fibers, then the mixture was irradiated using a LED conveyor at 385 nm (0.7 W/cm^2). A Dymax-UV conveyor was used, the distance between the belt and the LED was fixed to 15 mm, and the belt speed was fixed to 2 m/min.

4.10. Laser Writing and 3D patterns Characterization

For the direct laser write experiments, a computer-controlled laser diode at 405 nm (spot size = 50 µm) was used, and the 3D patterns obtained were characterized by a numerical optical microscope (DSX-HRSU from OLYMPUS Corporation) [35].

5. Conclusions

Nine coumarins varying by the substitution pattern at the 3- and 7-positions of the coumarin core have been tested and proposed as highly efficient photoinitiators for the FRP of meth(acrylates) functions under visible light irradiation using a LED at 405 nm. Remarkably, these photoinitiators can be used in 3D printing experiments and these dyes showed a very high efficiency in the photocomposite synthesis (significant curing of the surface and the bottom) using a LED conveyor at 385 nm. The challenge remains, therefore, to develop new coumarins absorbing at longer wavelengths e.g., in the near-infrared range for a better penetration of light into thick/filled samples.

Author Contributions: Conceptualization, J.L. and F.D.; methodology, J.L. and F.D.; software, B.G.; validation, all authors; formal analysis, J.L.; investigation, M.R. and G.N.; resources, J.L., F.D., J.T., T.H., and D.G.; data curation, J.L.; writing—original draft preparation, M.R., B.G., F.D., and J.L.; writing—review and editing, all authors; visualization, J.L.; supervision, J.L.; project administration, J.L.; funding acquisition, J.L. All authors have read and agreed to the published version of the manuscript.

Funding: This research was funded by The Association of Specialization and Scientific Guidance, the Centre National de la Recherche Scientifique, Aix Marseille Université and the Université de Haute Alsace. This research was also funded by the Agence Nationale de la Recherche (ANR agency) through the PhD grant of Guillaume Noirbent (ANR-17-CE08-0054 VISICAT project).

Institutional Review Board Statement: Not applicable.

Informed Consent Statement: Not applicable.

Data Availability Statement: The data presented in this study are available on request from the corresponding author.

Acknowledgments: The Lebanese group would like to thank "The Association of Specialization and Scientific Guidance" (Beirut, Lebanon) for funding and supporting this scientific work.

Conflicts of Interest: The authors declare no conflict of interest. The funders had no role in the design of the study; in the collection, analyses, or interpretation of data; in the writing of the manuscript, or in the decision to publish the results.

Sample Availability: Samples of the compounds are not available from the authors.

References

1. Fouassier, J.-P.; Lalevée, J. *Photoinitiators for Polymer Synthesis, Scope, Reactivity, and Efficiency*; Wiley-VCH Verlag GmbH &Co. KGaA: Weinheim, Germany, 2012.
2. Fouassier, J.-P. *Photoinitiator, Photopolymerization and Photo-curing: Fundamentals and Applications*; Gardner Publications: New York, NY, USA, 1995.
3. Dietliker, K.A. *Compilation of Photoinitiators Commercially Available for UV Today*; Sita Technology Ltd.: Edinbergh, London, 2002.
4. Davidson, S. *Exploring the Science, Technology and Application of UV and EB Curing*; Sita Technology Ltd.: Edinbergh, London, 1999.
5. Wu, L.; Baghdachi, J. *Functional Polymer Coatings: Principles, Methods, and Applications*; Wiley Series on Polymer Engineering and Technology: New York, NY, USA, 2015.
6. Bouzrati-Zerelli, M.; Maier, M.; Dietlin, C.; Morlet-Savary, F.; Fouassier, J.-P.; Klee, J.-E.; Lalevée, J. A novel photoinitiating system producing germyl radicals for the polymerization of representative methacrylate resins: Camphorquinone/R3GeH/iodonium salt. *Dent. Mater.* **2016**, *32*, 1226–1234. [CrossRef]
7. Garcès, J.M.; Moll, D.J.; Bicerano, J.; Fibiger, R.; McLeod, D.G. Polymeric Nanocomposites for Automotive Applications. *Adv. Mater.* **2000**, *12*, 1835–1839. [CrossRef]
8. Morgan, S.E.; Havelka, K.O.; Lochhead, R.Y. *Cosmetic Nanotechnology: Polymers and Colloids in Cosmetics in Person Care*; ACS Symposium Series: Washington, DC, USA, 2007; 961p.
9. Lawrence, J.R.; O'Neill, F.T.; Sheridan, J.T. Photopolymer holographic recording material. *Optik* **2001**, *112*, 449–463. [CrossRef]

10. Lee, J.H.; Prud'Homme, R.K.; Aksay, I.A. Cure Depth in Photopolymerization: Experiments and Theory. *J. Mater. Res.* **2001**, *16*, 3536–3544. [CrossRef]
11. Fouassier, J.-P.; Lalevée, J. Recent Advances in Photoinduced Polymerization Reactions under 400−700 nm Light. *Photochemistry* **2014**, *42*, 215–232.
12. Dumur, F.; Gigmes, D.; Fouassier, J.-P.; Lalevée, J. Organic Electronics: An El Dorado in the Quest of New Photocatalysts for Polymerization Reactions. *Acc. Chem. Res.* **2016**, *49*, 1980–1989. [CrossRef]
13. Dietlin, C.; Schweizer, S.; Xiao, P.; Zhang, J.; Morlet-Savary, F.; Graff, B.; Fouassier, J.-P.; Lalevée, J. Photopolymerization upon LEDs: New Photoinitiating Systems and Strategies. *Polym. Chem.* **2015**, *6*, 3895–3912. [CrossRef]
14. Zivic, N.; Bouzrati-Zerelli, M.; Kermagoret, A.; Dumur, F.; Fouassier, J.-P.; Gigmes, D.; Lalevée, J. Photocatalysts in Polymerization Reactions. *Chem. Cat. Chem.* **2016**, *8*, 1617–1631. [CrossRef]
15. Abdallah, M.; Hijazi, A.; Graff, B.; Fouassier, J.-P.; Rodeghiero, G.; Gualandi, A.; Dumur, F.; Cozzi, P.G.; Lalevée, J. Coumarin derivatives as versatile photoinitiators for 3D printing, polymerization in water and photocomposite synthesis. *Polym. Chem.* **2019**, *10*, 872–884. [CrossRef]
16. Abdallah, M.; Hijazi, A.; Dumur, F.; Lalevée, J. Coumarins as Powerful Photosensitizers for the Cationic Polymerization of Epoxy-Silicones under Near-UV and Visible Light and Applications for 3D Printing Technology. *Molecules* **2020**, *25*, 2063. [CrossRef]
17. Abdallah, M.; Hijazi, A.; Graff, B.; Fouassier, J.-P.; Dumur, F.; Lalevée. In-silico based development of photoinitiators for 3D printing and composites: Search on the coumarin scaffold. *J. Photochem. Photobiol. A Chem.* **2020**, *400*, 112698. [CrossRef]
18. Abdallah, M.; Hijazi, A.; Lin, J.-T.; Graff, B.; Dumur, F.; Lalevée. Coumarin Derivatives as Photoinitiators in Photo-Oxidation and Photo-Reduction Processes and a Kinetic Model for Simulations of the Associated Polymerization Profiles. *ACS Appl. Polym. Mater.* **2020**, *2*, 2769–2780. [CrossRef]
19. Rahal, M.; Mokbel, H.; Graff, B.; Toufaily, J.; Hamieh, T.; Dumur, F.; Lalevée, J. Mono vs. Difunctional Coumarin as Photoinitiators in Photocomposite Synthesis and 3D Printing. *Catalyst* **2020**, *10*, 1202. [CrossRef]
20. Jiang, Z.-J.; Lv, H.-S.; Zhu, J.; Zha, B.-X. New fluorescent chemosensor based on quinoline and coumarin for Cu^{2+}. *Synth. Met.* **2012**, *162*, 2112–2116. [CrossRef]
21. Loncaric, M.; Gašo-Sokac, D.; Jokic, S.; Molnar, M. Recent Advances in the Synthesis of Coumarin Derivatives from Different Starting Materials. *Biomolecules* **2020**, *10*, 151. [CrossRef] [PubMed]
22. Zeydi, M.M.; Kalantarian, S.J.; Kazeminejad, Z. Overview on developed synthesis procedures of coumarin heterocycles. *J. Iran. Chem. Soc.* **2020**, *17*, 3031–3094. [CrossRef]
23. Mousawi, A.A.; Dumur, F.; Garra, P.; Toufaily, J.; Hamieh, T.; Graff, B.; Gigmes, D.; Fouassier, J.-P.; Lalevée, J. Carbazole Scaffold Based Photoinitiator/Photoredox Catalysts: Toward New High-Performance Photoinitiating Systems and Application in LED Projector 3D Printing Resins. *Macromolecules* **2017**, *50*, 2747–2758. [CrossRef]
24. Qiu, W.; Zhu, J.; Dietliker, K.; Li, Z. Polymerizable Oxime Esters: An Efficient Photoinitiator with Low Migration Ability for 3D Printing to Fabricate Luminescent Devices. *Chem. Photo. Chem.* **2020**, *4*, 5296–5303.
25. Jiang, X.; Guo, J.; Lv, Y.; Yao, C.; Zhang, C.; Mi, Z.; Shi, Y.; Gu, J.; Zhou, T.; Bai, R.; et al. Rational design, synthesis and biological evaluation of novel multitargeting anti-AD iron chelators with potent MAO-B inhibitory and antioxidant activity, Bioorg. *Med. Chem.* **2020**, *28*, 115550. [CrossRef]
26. Markad, D.; Khullar, S.; Mandal, S.K. A Primary Amide-Functionalized Heterogeneous Catalyst for the Synthesis of Coumarin-3-carboxylic Acids via a Tandem Reaction. *Inorg. Chem.* **2020**, *59*, 11407–11416. [CrossRef]
27. Wang, X.; Bai, X.; Su, D.; Zhang, Y.; Li, P.; Lu, S.; Gong, Y.; Zhang, W.; Tang, B. Simultaneous Fluorescence Imaging Reveals N-Methyl-D-aspartic Acid Receptor Dependent Zn^{2+}/H^{+} Flux in the Brains of Mice with Depression. *Anal. Chem.* **2020**, *92*, 4101–4107.
28. Liu, H.; Xia, D.-G.; Chu, Z.-W.; Hu, R.; Cheng, X.; Lv, X.-H. Novel coumarin-thiazolyl ester derivatives as potential DNA gyrase Inhibitors: Design, synthesis, and antibacterial activity. *Bioorg. Chem.* **2020**, *100*, 103907. [CrossRef]
29. Bhagwat, A.A.; Avhad, K.C.; Patil, D.S.; Sekar, N. Design and Synthesis of Coumarin–Imidazole Hybrid Chromophores: Solvatochromism, Acidochromism and Nonlinear Optical Properties. *Photochem. Photobiol.* **2019**, *95*, 740–754. [CrossRef] [PubMed]
30. Lalevée, J.; Blanchard, N.; Tehfe, M.-A.; Peter, M.; Morlet-Savary, F.; Gigmes, D.; Fouassier, J.P. Efficient Dual Radical/Cationic Photoinitiator under Visible Light: A New Concept. *Polym. Chem.* **2011**, *2*, 1986–1991. [CrossRef]
31. Lalevée, J.; Blanchard, N.; Tehfe, M.-A.; Peter, M.; Morlet-Savary, F.; Fouassier, J.P. A Novel Photopolymerization Initiating System Based on an Iridium Complex Photocatalyst. Macromol. *Rapid Commun.* **2011**, *32*, 917–920. [CrossRef] [PubMed]
32. Rehm, D.; Weller, A. Kinetics of Fluorescence Quenching by Electron and H-Atom Transfer. *Isr. J. Chem.* **1970**, *8*, 259–271. [CrossRef]
33. Foresman, J.B.; Frisch, A. Exploring Chemistry with Electronic Structure Methods. In *Exploring Chemistry with Electronic Structure Methods*, 2nd ed.; Gaussian Inc.: Pittsburgh, PA, USA, 1996.

34. Frisch, M.J.; Trucks, G.W.; Schlegel, H.B.; Scuseria, G.E.; Robb, M.A.; Cheeseman, J.R.; Zakrzewski, V.G.; Montgomery, J.A.; Stratmann, J.R.E.; Burant, J.C.; et al. *Gaussian 03, Revision B-2*; Gaussian Inc.: Pittsburgh, PA, USA, 2003.
35. Zhang, J.; Dumur, F.; Xiao, P.; Graff, B.; Bardelang, D.; Gigmes, D.; Fouassier, J.-P.; Lalevée, J. Structure Design of Naphthalimide Derivatives: Toward Versatile Photoinitiators for Near-UV/Visible LEDs, 3D Printing, and Water-Soluble Photoinitiating Systems. *Macromolecules* **2015**, *48*, 2054–2063. [CrossRef]

Article

Synthesis and Biological Activity Evaluation of Coumarin-3-Carboxamide Derivatives

Weerachai Phutdhawong [1], Apiwat Chuenchid [2], Thongchai Taechowisan [3], Jitnapa Sirirak [2] and Waya S. Phutdhawong [2,*]

[1] Department of Science, Faculty of Liberal Arts and Science, Kamphaeng Sean Campus, Kasetsart University, Nakhon Pathom 73140, Thailand; phutdhawong@gmail.com

[2] Department of Chemistry, Faculty of Science, Silpakorn University, Nakhon Pathom 73000, Thailand; chuenchid.a@gmail.com (A.C.); jitnapasirirak@gmail.com (J.S.)

[3] Department of Microbiology, Faculty of Science, Silpakorn University, Nakhon Pathom 73000, Thailand; tewson84@hotmail.com

* Correspondence: Phutdhawong_w@su.ac.th; Tel.: +66-34-255797

Abstract: A series of novel coumarin-3-carboxamide derivatives were designed and synthesized to evaluate their biological activities. The compounds showed little to no activity against gram-positive and gram-negative bacteria but specifically showed potential to inhibit the growth of cancer cells. In particular, among the tested compounds, 4-fluoro and 2,5-difluoro benzamide derivatives (**14b** and **14e**, respectively) were found to be the most potent derivatives against HepG2 cancer cell lines (IC_{50} = 2.62–4.85 μM) and HeLa cancer cell lines (IC_{50} = 0.39–0.75 μM). The activities of these two compounds were comparable to that of the positive control doxorubicin; especially, 4-flurobenzamide derivative (**14b**) exhibited low cytotoxic activity against LLC-MK2 normal cell lines, with IC_{50} more than 100 μM. The molecular docking study of the synthesized compounds revealed the binding to the active site of the CK2 enzyme, indicating that the presence of the benzamide functionality is an important feature for anticancer activity.

Keywords: coumarin3-carboxamides; coumarins; pyranocoumarins; anticancer activity; antibacterial activity

Citation: Phutdhawong, W.; Chuenchid, A.; Taechowisan, T.; Sirirak, J.; Phutdhawong, W.S. Synthesis and Biological Activity Evaluation of Coumarin-3-Carboxamide Derivatives. *Molecules* **2021**, *26*, 1653. https://doi.org/10.3390/molecules26061653

Academic Editor: Maria João Matos

Received: 26 February 2021
Accepted: 12 March 2021
Published: 16 March 2021

Publisher's Note: MDPI stays neutral with regard to jurisdictional claims in published maps and institutional affiliations.

1. Introduction

Coumarin is one of the potent secondary metabolites in plants [1,2] and fungi [3], and it is characterized by several pharmacological properties [4]. Like decursin **1** and decursinol **2**, these coumarins have pyranocoumarin moiety (Figure 1), having been isolated from the medicinal plant Angelica genus and shown potential for treating inflammatory diseases such as cancer, hepatic fibrosis, diabetic retinopathy, and neurological disorders [5]. The dehydrated derivative of decursinol, xanthyletin **3**, has also been shown to possess several biological properties, such as anti-tumor and antibacterial activities [6]. With a benzopyrone skeleton, coumarin is versatile and can be easily transformed into a large variety of functionalized skeletons. As a result, many coumarin derivatives have been designed, synthesized, and evaluated to address broad biological activities [7] such as antibacterial [8], antifungal [9], antioxidant [10], anti-inflammatory [11], anticancer [12], anticoagulant [13], and antiviral activities [14]. The synthetic *N*-phenyl coumarin carboxamide **4a** has been designed and shown to possess high antibacterial activity against *Helicobacter pylori* (*H. pylori*), with the minimum inhibitory concentration (MIC) = 1 μg/mL [15], while the benzyl substitution of coumarin carboxamides **4b–d** has been shown to arrest breast cancer cell (BT474 and SKBR-3) growth by inhibiting ErbB-2 and ERK1 MAP kinase. Moreover, compounds **4b–d** are specific to cancer cell lines, with no cytotoxicity against normal human fibroblasts [16]. In our ongoing search for novel compounds to overcome drug resistance, the diverse biological activities of coumarins have been interesting. In the current study, we designed novel pyranocoumarin-3-carboxamide derivatives with the expectation that the carboxamide part could possess active pharmacological properties **4a–d** and that the pyran ring moiety could also show specific biological proteins, as in the case of xanthyletin **3**. The synthesized compounds were examined to evaluate their antibacterial activity and cytotoxicity against HepG2 and HeLa cell lines. As the coumarins were attractive casein kinase 2 (CK2) inhibitors [17], molecular docking was used to study the possible interactions of novel coumarin-3-carboxamides with the CK2 enzymes.

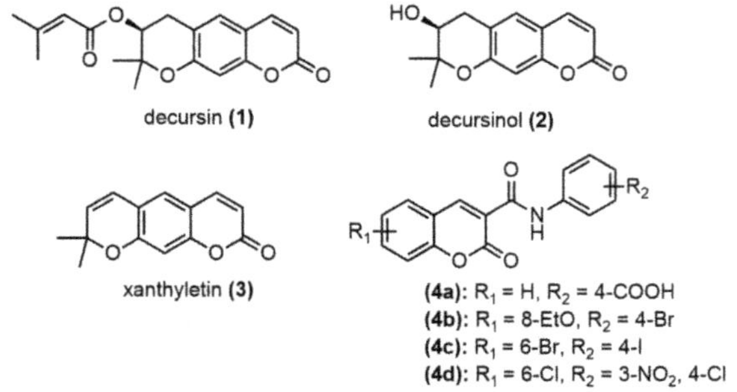

decursin **(1)**

decursinol **(2)**

xanthyletin **(3)**

(4a): R_1 = H, R_2 = 4-COOH
(4b): R_1 = 8-EtO, R_2 = 4-Br
(4c): R_1 = 6-Br, R_2 = 4-I
(4d): R_1 = 6-Cl, R_2 = 3-NO_2, 4-Cl

Figure 1. Synthetic and natural-occurring coumarins with biological activities.

2. Results and Discussion

2.1. Chemistry

The preparation of pyranocoumarin-3-carboxamide was applied from the previous synthetic strategies reported by Faulgues and colleagues [18] and was described in Scheme 1. Commercially available 2,4-dihydroxy benzyldehyde **5** and 3-hydroxy-3-methyl-1-butene **6** were used as the starting materials and were subjected to Lewis acid–promoted Friedel-Crafts alkylation reaction in dioxane with BF_3-diethylate to obtain the aldehyde **7** in 53% yield as a major product together with many unidentified products [19]. The aldehyde **7** was cyclized to form a pyrano ring using 2,3-dichloro-5,6-dicyano-l,4-benzoquinone (DDQ), and the benzopyran **8** was obtained in a very good yield. The cyclization of benzopyran **8** with the malonic anhydride **9** in pyridine and aniline at room temperature for 24 h according to a previous report [20] gave a poor yield of the pyranocoumarin-3-carboxylic acid **10**, due to the difficulty of purification. The amidation of the acid **10** with aniline using *N,N′*-dicyclohexyl carbodiimide (DCC) and 4-dimethylaminopyridine (DMAP) gave the amide **12** in 11% yield after recrystallization.

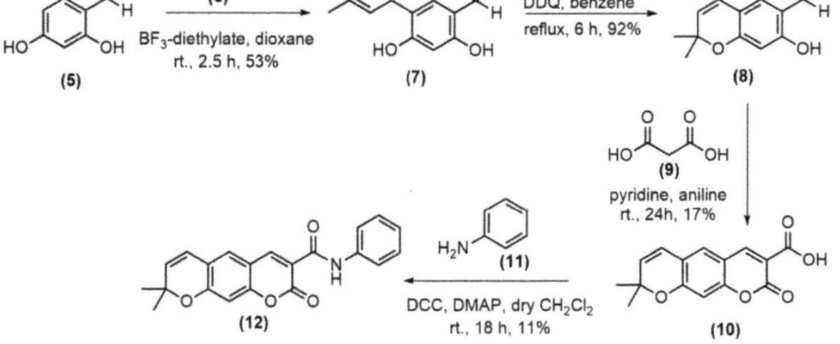

(6)
BF_3-diethylate, dioxane
rt., 2.5 h, 53%
(5)
(7)
DDQ, benzene
reflux, 6 h, 92%
(8)
(9)
pyridine, aniline
rt., 24h, 17%
(10)
(11)
DCC, DMAP, dry CH_2Cl_2
rt., 18 h, 11%
(12)

Scheme 1. Synthesis of pyranocoumarin-3-carboxamide **12**.

To improve the yield of pyranocoumarin-3-carboxamide **12**, the coumarin-3-carboxylic acid **13** was prepared in good yield prior to amidation with appropriate anilines using hexafluorophosphate azabenzotriazole tetramethyl Uronium (HATU) and Et_3N to obtain amide **14a–g** in 43%–51% yields (Scheme 2). Then, the cyclization of **14a** by refluxing with DDQ in benzene gave pyranocoumarin-3-carboxamide **12** in 66% yield (Scheme 3). To study the effect of the substituent at C3 of coumarin ring, the carboxyl group was decarboxylated using Cu powder to give coumarin **15** in 53% yield (Scheme 4). Then, the synthesized coumarins were further evaluated for their antibacterial and anticancer activities.

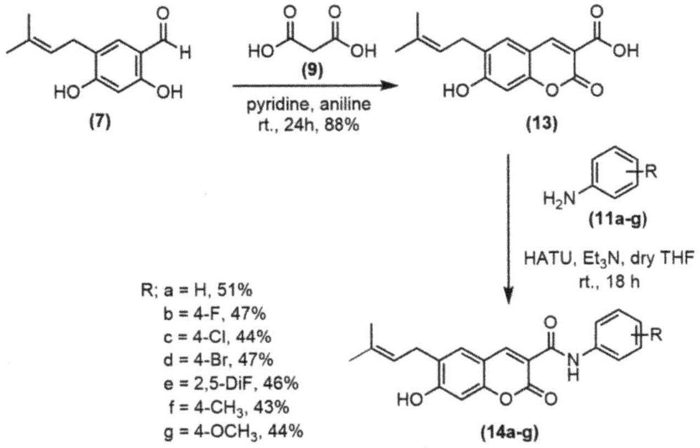

R; a = H, 51%
b = 4-F, 47%
c = 4-Cl, 44%
d = 4-Br, 47%
e = 2,5-DiF, 46%
f = 4-CH_3, 43%
g = 4-OCH_3, 44%

(7) → (13): (9), pyridine, aniline, rt., 24h, 88%
(13) → (14a-g): (11a-g), HATU, Et_3N, dry THF, rt., 18 h

Scheme 2. Synthesis of coumarin-3-carboxamides **14a–g**.

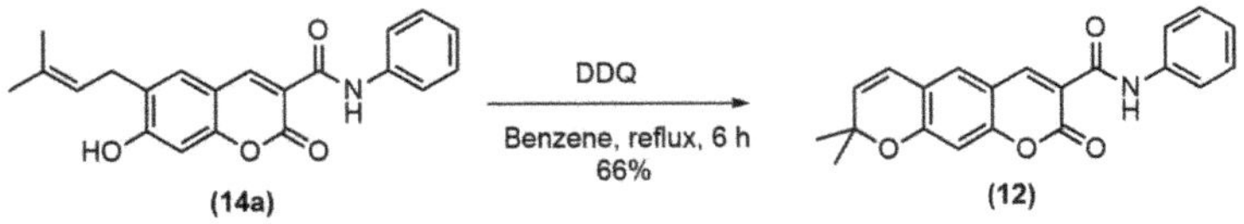

(14a) → (12): DDQ, Benzene, reflux, 6 h, 66%

Scheme 3. Cyclization of pyranocoumarin-3-carboxamide **12**.

(13) → (15): Cu Powder, heat, 53%

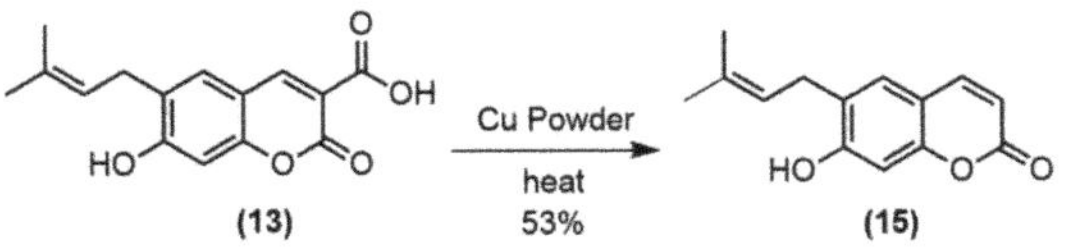

Scheme 4. Decarboxylation of coumarin-3-carboxylic acid.

2.2. Antibacterial Activity

Coumarin derivatives **10**, **12**, **13**, **14a–g**, and **15** were evaluated for their antibacterial activity against *Bacillus cereus*, *Bacillus subtilis*, *Staphylococcus aureus*, *Escherichia coli*, *Salmonella typhimuriumthrough* using the microbroth dilution method. Penicillin G and solvent were used as positive and negative controls, respectively, and the MIC (μg/mL) values were obtained (Table 1). The results show that only compounds **10** and **13** exhibited moderate antibacterial activities against gram-positive bacteria, while the other tested compounds displayed MIC values of more than 128 μg/mL. This may be because the carboxylic acid at the C3 position played an essential role in the antibacterial activity. Compound **15**, without the carboxyl group, showed no antibacterial activity, and the carboxamides **14a–g**, displayed no activity. Meanwhile, coumarin-3-carboxylic acid **13** was the most active among the tested compounds, with an MIC value of 32 μg/mL against *B. cereus*; however, it was less active than the reference drug penicillin G. Moreover, all the tested compounds showed no activity against any gram-negative bacteria.

Table 1. MIC of coumarin derivatives **10**, **12a**, **13**, **14a–g**, **15** and penicillin G against *B. cereus*, *B. subtilis*, *S. aureus*, *E. coli*, *S. typhimurium*.

Compounds	R	MIC (μg/mL)				
		Gram-Positive			Gram-Negative	
		B. cereus	*B. subtilis*	*S. aureus*	*E. coli*	*S. typhimurium*
10	-	128	128	64	>128	> 128
12	4-H	> 128	> 128	> 128	>128	> 128
13	-	32	128	64	>128	> 128
14a	4-H	> 128	> 128	> 128	>128	> 128
14b	4-F	> 128	> 128	> 128	>128	> 128
14c	4-Cl	> 128	> 128	> 128	>128	> 128
14d	4-Br	> 128	> 128	> 128	>128	> 128
14e	2,5-diF	> 128	> 128	> 128	>128	> 128
14f	4-CH_3	> 128	> 128	> 128	>128	> 128
14g	4-OCH_3	> 128	> 128	> 128	>128	> 128
15	-	> 128	> 128	> 128	>128	> 128
penicillin G	-	16	2	16	32	64

2.3. *Anticancer Activity*

All synthesized compounds were evaluated for in vitro cytotoxic activity against two cancer cell lines (HepG2 and HeLa cell lines) and normal cell lines (LLC-MK2) through an MTT assay, and the results are presented in Table 2. Most of the tested compounds displayed potent anticancer activity. The *N*-phenyl coumarin-3-carboxamides **12** and **14a** showed significantly more potency than the parent acids **10** and **13**, respectively, against HepG2 cell lines. Moreover, compound **15**, with no substituent at C3, showed better activity than the parent acid **13**. The effect of substituents on the phenyl ring was compared with the effect of substituents on the carboxamide **14a** and it was found that the phenyl bearing fluorine atoms **14b** and **14e** possessed similar effects on the potency, while, the phenyl bearing 4-chlorine and 4-bromine atoms showed less activity against both cancer cell lines. Moreover, the phenyl bearing 4-methyl and 4-methoxyl groups displayed weak activity against the test cancer cell lines. From these results, the size and electron-donating group of the *para*-substituted benzene ring may affect anticancer properties. From these tested compounds, the amide **14e** displayed the most potent anticancer activity; however, it exhibited high cytotoxic activity against the normal cell line, with IC_{50} = 1.33 μM. Interestingly, amide **14b** displayed slightly lower activity than **14e**, but it showed low cytotoxicity against the normal cell line. This compound possessed anticancer activity comparable to those of the tested anticancer drugs doxorubicin and acridine orange.

2.4. *Molecular Docking*

Casein Kinase 2 enzyme is a key player in the pathophysiology of cancer [21,22]. Using the iGEMDOCK v2.1 software [23], molecular docking was performed to investigate binding positions and intermolecular interactions between coumarin **10**, **12**, **13**, **14a–f**, and **15** and the binding site of CK2. Coumarins **10**, **12**, **13**, **14a–f**, and **15** were docked to the active site of CK2 co-crystallized with **G12** (PDB ID: 2QC6). Moreover, **G12** was also redocked to CK2, and its total energy and hydrogen bond length were compared with those of coumarins **10**, **12**, **13**, **14a–f**, and **15**. The molecular docking results show that the binding position of redocked **G12** was roughly the same as that of co-crystallized **G12** in CK2 (Figure 2a). Moreover, all synthesized compounds were bound to the active site of CK2, and the binding positions were similar to that of **G12** (Figure 2b,c), and their binding energies (−118.99 to −89.19 kcal/mol) were lower than that of **G12** (−79.10 kcal/mol) (Table 3). Figure 3 also illustrates the hydrogen bond interactions in the binding of coumarin **12**, **14a**, **14b**, **14d**, and **14e** in the cavity of CK-2 compared with that of **G12**. Coumarins **14a–f** had much lower binding energy (−118.99 to −105.53 kcal/mol) than **G12**, and their *N*-phenyl ring was also bound to similar positions of the **G12** phenolic ring (Figure 2c). Key amino acid residues, including LYS68, ASN118, ASN117, and ASP175, formed a hydrogen bond with **14a–f**, where ASN117 and ASN118 interacted with the hydroxy group of **14a–f**, while *N*-phenyl ring interacted with LYS68 and ASP175. The substitution group on the *N*-phenyl ring of **14a–f** also influenced the number of hydrogen bonds in the CK2 active site. Moreover, a comparison of the halogen substitution groups on the *N*-phenyl ring showed that the Cl and Br substitution groups on the *N*-phenyl ring formed no hydrogen bond, while the F substitution group **14b** and **14e** could form hydrogen bonds with LYS68 and ASP175 (Figure 3d,f), which may be because Cl and Br are larger than F in size. Additionally, the experimental results show that the coumarins **14b** and **14e** had good anticancer activities.

Table 2. In vitro anticancer activities of synthesized coumarins compared with doxorubicin and acridine orange.

Compounds	R	IC_{50} (μM)		
		Cancer Cell Line		Normal Cell Line
		HepG2	HeLa	LLC-MK2
10	-	> 100	80.38	> 100
12	4-H	2.35	> 100	> 100
13	-	81.73	26.42	37.95
14a	4-H	4.33	10.40	0.48
14b	4-F	4.85	0.75	>100
14c	4-Cl	30.28	14.04	66.80
14d	4-Br	45.76	37.36	> 100
14e	2,5-diF	2.62	0.39	1.33
14f	4-CH_3	> 100	> 100	85.31
14g	4-OCH_3	> 100	> 100	> 100
15	-	9.13	13.47	6.87
Doxorubicin	-	1.91	1.18	63.47
Acridine orange	-	5.72	6.59	82.30

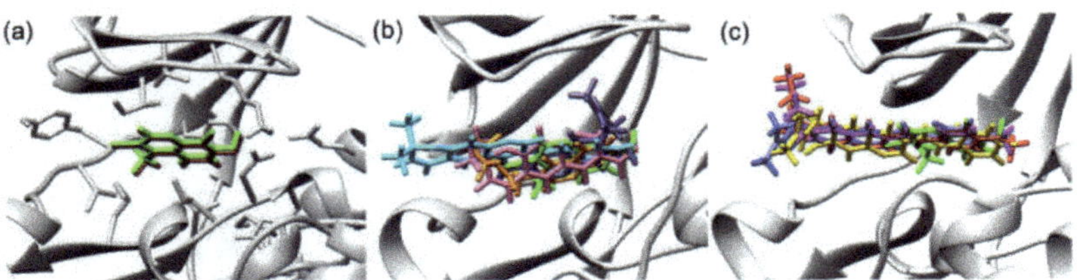

Figure 2. (**a**) Redocked **G12** (brown) and **G12** (green) in the cavity of CK-2 (PDB ID: 2QC6). (**b**) Comparison of the bindings of **10** (pink), **12** (sky blue), **13** (orange), 15 (violet), and **G12** (green) in the CK-2 cavity. (**c**) Comparison of the bindings of **14a** (yellow), **14e** (blue), **14f** (purple), **14g** (red), and **G12** (green) in the CK-2 cavity (PDB ID: 2QC6).

Table 3. Summary of binding energy, amino acid interactions, and hydrogen bond length of coumarin derivatives in CK2 binding site.

Compounds	Total Energy (kcal/mol)	Amino Acid Residue	Hydrogen Bond Length (Å)
10	−89.19	LYS68, ASP175	2.44, 2.34
12	−100.88	ASP175	2.91
13	−91.13	LYS68	1.92
14a	−108.21	ASN118, ASN117, ASP175	2.19, 2.99,2.51
14b	−107.08	ASN117, LYS68, ASP175	2.93, 2.42, 2.74
14c	−106.07	ASN117	2.94
14d	−105.91	ASN117	2.95
14e	−111.19	ASN118, LYS68	2.27, 2.44
14f	−105.53	ASN117	2.93
14g	−118.99	ASN117, LYS68, GLU81	2.74, 1.90, 2.72
15	−86.27	LYS68, GLU114, ASP175,	2.08, 1.99, 2.42
G12	−79.10	ASP175	2.43

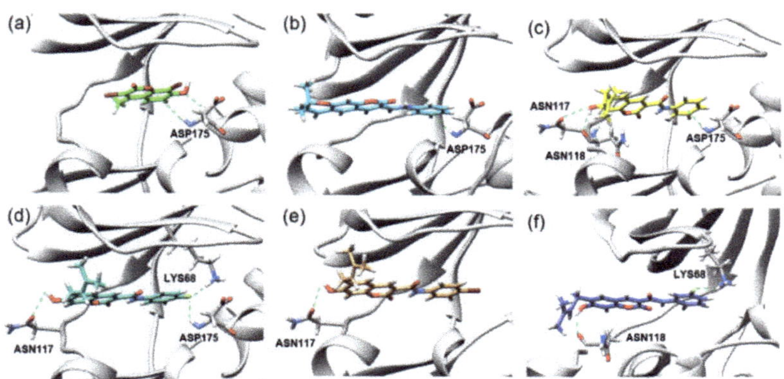

Figure 3. Hydrogen bond interactions in the bindings of (**a**) **G12**,(**b**) **12**,(**c**) **14a**, (**d**) **14b**,(**e**) **14d**, and (**f**) **14e** in the CK-2 cavity.

3. Materials and Methods

3.1. Chemistry

General information: Solvents and reagents were purchased from commercial suppliers TCI Chemicals (Tokyo, Japan), Sigma-Aldrich (Bangalore, India), and Fluka (Dorset, UK). Structure determination was conducted by analyzing the ^{1}H, ^{13}C, and ^{19}F NMR spectra (Bruker 300 apparatus) and the infrared (IR) spectrum was determined using PerkinElmer Frontier Fourier-transform infrared spectrometer. Melting point was conducted using Stuart SMP2 melting point apparatus and high-resolution mass spectroscopy was analyzed by Thermo scientific, Orbitrap Q Exactive Focus.

3.1.1. Synthesis of 2,4-Dihydroxy-5-(3-methylbut-2-en-1-yl)-benzaldehyde **7**

First, 2,4-dihydroxy benzaldehyde **5** (1.5 g, 10.8 mmol) in dioxane (5 mL) was added to a stirred solution of 3-hydroxy-3-methyl-1-butene **6** (1.5 mL, 14.3 mmol) and boron trifluoride diethyl etherate (BF_3-OEt_2, 1.5 mL) in dioxane (3 mL), and stirring was continued for 2.5 h at room temperature. Dichloromethane (50 mL) was added, and the resulting solution was extracted with water (3 × 50 mL). The combined organic layer was dried over Na_2SO_4 before evaporation to dryness and then purified via column chromatography (silica gel, 4:1 hexane:EtOAc) to obtain a white solid of 2,4-dihydroxy-5-(3-methylbut-2-en-1-yl) benzaldehyde **7** (0.48 g, 53% yield): m.p. 124–125 °C (lit. [19] 121–123 °C), ^{1}H-NMR (300 MHz, $CDCl_3$): δ 11.27 (s, 1H), 9.69 (s, 1H, OH), 7.26 (s, 1H), 6.37 (s, 1H), 6.10 (s, 1H, OH), 5.30 (tt, J = 4.53, 1.35 Hz, 1H), 3.30 (d, J = 7.2 Hz, 2H), 1.77 (s, 3H), 1.61 (s, 3H) ppm.

3.1.2. Synthesis of 7-Hydroxy-2,2-dimethyl-2H-chromene-6-carbaldehyde **8**

A mixture of compound **7** (1.0 g, 4.85 mmol) and DDQ (1.2 g, 5.28 mmol) in benzene (10 mL) was refluxed for 6 h, and the precipitate was filtered off. The filtrate was evaporated to dryness to afford the crude product, which was purified via column chromatography (silica gel, 10:1 hexane:EtOAc) to obtain a white solid of 7-hydroxy-2,2-dimethyl-2H-chromene-6-carbaldehyde **8** (0.91 g, 92% yield): m.p. 82–83 °C (lit. [20] 95–96 °C), ^{1}H NMR (300 MHz, $CDCl_3$): δ 11.41 (br, OH), 9.68 (s, 1H), 7.16 (s, 1H), 6.35 (s, 1H), 6.30 (d, J = 7.86 Hz, 1H), 5.59 (d, J = 9.93 Hz, 1H), 1.59 (s, 3H) 1.48 (s, 3H) ppm.

3.1.3. Synthesis of 8,8-Dimethyl-2-oxo-2H,8H-pyrano[3,2-g]chromene-3-carboxylic acid **10**

Compound **8** (1.0 g, 4.90 mmol) and malonic acid **9** (1.0 g, 9.60 mmol) were dissolved in pyridine (5.5 mL) containing aniline (0.5 mL) and stirred for 24 h at room temperature. Afterward, the rection mixture was poured into ice-cold 10% HCl (80 mL). The yellow precipitate was washed with cold water to remove mineral acid and then air-dried to yield a yellow solid (recrystallization by 2:1:2; EtOAc:EtOH:hexane) of 8,8-dimethyl-2-oxo-2H,8H-pyrano[3,2-g]chromene-3-carboxylic acid **10** (0.23 g, 17% yield): m.p. 187–188 °C, ^{1}H NMR (300 MHz, $CDCl_3$): δ 8.80 (s, 1H), 7.29 (s, 1H), 6.85 (s, 1H), 6.40 (d, J = 10.02 Hz, 1H), 5.81 (d, J = 10.02 Hz, 1H), 1.52 (s, 6H) ppm., ^{13}C NMR (75 MHz, $CDCl_3$): δ 164.54 (C), 163.28 (C), 160.87 (C), 156.60 (C), 150.99 (CH), 132.34 (C), 127.13 (CH), 120.17 (CH), 120.06 (CH), 112.50 (C), 110.59 (C), 104.38 (CH), 79.33 (C), 28.75 ($2CH_3$) ppm. IR: 3051.36 (C-H, aromatic), 2922.46 (C-H, aliphatic), 1743.12 (C=O, acid) cm^{-1}; HREI-MS (m/z) calculated for $C_{15}H_{13}O_5$ $(M)^+$ 273.0758, found 273.0757.

3.1.4. Synthesis of 8,8-Dimethyl-2-oxo-*N*-phenyl-2H,8H-pyrano[3,2-g]chromene-3-carboxamide **12**

A mixture of compound **10** (0.10 g, 0.36 mmol), aniline **11** (0.040 mL, 0.43 mmol), DCC (0.10 g, 0.44 mmol), and DMAP (8 mg, 0.065 mmol) in dry CH_2Cl_2 (5 mL) was stirred at room temperature for 18 h. Afterward, the reaction mixture was filtered, and the filtrate was evaporated under vacuum. The residue was recrystallized using EtOH to obtain a yellow solid of 8,8-dimethyl-2-oxo-*N*-phenyl-2H,8H-pyrano[3,2-g]chromene-3-carboxamide **12** (14 mg, 11% yield): m.p. 187–188 °C, ^{1}H NMR (300 MHz, $CDCl_3$): δ 10.81(s, 1H), 8.89 (s, 1H), 7.75 (d, J = 1.14 Hz, 2H), 7.39 (t, J = 6.54 Hz, 2H), 7.30 (s, 1H), 7.18 (t, J = 1.14 Hz, 1H), 6.80 (s, 1H), 6.40 (d, J = 9.90 Hz, 1H), 5.78 (d, J = 9.96 Hz, 1H), 1.55 (s, 6H) ppm., ^{13}C NMR (75 MHz, $CDCl_3$): δ 162.20 (C), 160.03 (C), 159.55 (C), 156.35 (C), 148.67 (CH), 137.30 (C), 131.82 (CH), 129.02 (2CH), 126.72 (CH), 124.57 (CH), 120.53 (2CH), 120.43 (CH), 119.58 (C), 114.71 (C), 112.73 (C), 75.72 (C), 28.64 ($2CH_3$) ppm. IR: 3198.94 (N-H), 3059.35 (C-H, aromatic), 2969.10 (C-H, aliphatic), 1699.85 (C=O, amide) cm^{-1}; HREI-MS (m/z) calculated for $C_{21}H_{18}O_4N$ $(M)^+$ 348.1230, found 348.1230.

3.1.5. Synthesis of 3-Carboxy-6-(3-methyl-2-butenyl)-7-hydroxy-coumarin **13**

Compound **7** (1.0 g, 4.85 mmol) and malonic acid **9** (1.0 g, 9.60 mmol) were dissolved in pyridine (5.5 mL) containing aniline (0.5 mL), and stirred for 24 h at room temperature. Afterward, the reaction mixture was poured into ice-cold 10% HCl (80 mL), and the yellow precipitate obtained was washed with cold water to remove mineral acid and then air-dried to yield a yellow solid of 3-carboxy-6-(3-methyl-2-butenyl)-7-hydroxy-coumarin **13** (1.10g, 88% yield): m.p. 237–238 °C (lit. [20] 218–224 °C), ^{1}H NMR (300 MHz, $CDCl_3$+methanol-d_4) δ 8.79 (s. 1H), 7.40 (s, 1H), 6.85 (s, 1H), 5.32 (tt, J = 7.4, 1.3 Hz, 1H), 3.34 (d, J = 7.29 Hz, 2H), 1.79 (s, 3H), 1.70 (s, 3H) ppm., ^{13}C-NMR (75 MHz, $CDCl_3$+methanol-d_4): δ 163.58 (C), 162.79 (C), 162.37 (C), 154.78 (C), 150.43 (CH), 133.49 (C), 129.32 (CH), 127.92 (C), 119.16 (CH), 110.24 (C), 107.85 (C), 100.76 (CH), 26.34 (CH_2), 24.42 (CH_3), 16.50 (CH_3) ppm., IR: 3303.61 (O-H), 3049.81 (C-H, aromatic), 2911.92 (C-H, aliphatic), 1733.10 (C=O, acid), 1718.83 (C=O, lactone) cm^{-1}.

3.1.6. General Procedure of Coumarin-3-carboxamides Preparation (**14a–g**)

Triethanolamine (TEA) (0.1 mL, 1.36 mmol) was added to a solution of compound **13** (70 mg, 0.26 mmol) and HATU (0.12 g, 0.36 mmol) in dry THF (5 mL), and the mixture was stirred at room temperature for 30 min. The obtained dark clear mixture was treated with aniline derivatives (**11a–g**) (1.2 eq.). The resulting mixture was stirred at room temperature for 18 h. Dichloromethane (50 mL) was added, and the resulting solution was extracted with sat. NaCl (3 × 30 mL). The remaining organic layer was dried over Na_2SO_4 before evaporation to dryness. After evaporation of the solvent in vacuo, the crude product was purified via preparative thin-layer chromatography (silica gel, 4:1 hexane:EtOAc) to give yellow solids of coumarin-3-carboxamide (**14a–g**)

7-Hydroxy-6-(3-methylbut-2-en-1-yl)-2-oxo-*N*-phenyl-2H-chromene-3-carboxamide **14a** (47 mg, 51% yield): m.p. 258–259 °C, ^{1}H NMR (300 MHz, $CDCl_3$+pyridine-d_5): δ 10.86 (s, NH), 8.90 (s, 1H), 7.73 (d, *J* = 7.62 Hz, 2H), 7.42 (s, 1H), 7.35 (t, *J* = 7.62 Hz, 2H), 7.12 (t, *J* = 7.38 Hz, 1H), 6.84 (s, 1H), 5.40 (tt, *J* = 7.26, 1.35 Hz, 1H), 3.43 (d, *J* = 7.26 Hz, 2H), 1.81 (s, 3H), 1.74 (s, 1H) ppm., ^{13}C NMR (75 MHz, $CDCl_3$+pyridine-d_5): δ 163.38 (C), 162.81 (C), 160.56 (C), 155.49 (C), 149.36 (CH), 138.08 (CH), 134.13(C), 130.04 (CH), 128.97 (2CH), 128.68 (C), 124.35 (CH), 121.14 (CH), 120.49 (2CH), 112.94 (C), 111.33 (C), 101.68 (CH), 27.85 (CH_2), 25.84 (CH_3), 17.85 (CH_3) ppm., IR: 3195.27 (O-H), 2917.31 (C-H, aliphatic), 1695.54 (C=O, amide) cm^{-1}; HREI-MS (m/z) calculated for $C_{21}H_{20}O_4N$ $(M)^+$ 350.1387, found 350.1386.

N-(4-Fluorophenyl)-7-hydroxy-6-(3-methylbut-2-en-1-yl)-2-oxo-2H-chromene-3-carboxamide **14b** (45 mg, 47% yield): m.p. 259–261 °C, ^{1}H NMR (300 MHz, $CDCl_3$): δ 10.56 (br, NH), 8.89 (s, 1H), 7.68 (dd, *J* = 8.97, 4.92 Hz, 2H), 7.42 (s, 1H), 7.07 (dd, *J* = 9.12, 8.79 Hz 2H), 6.83 (s, 1H), 5.34 (t, *J* = 7.23 Hz, 1H), 3.35 (d, *J* = 7.29 Hz, 2H), 1.80 (s, 3H), 1.72 (s, 1H) ppm., ^{13}C NMR (75 MHz, CDC13): δ 163.04 (C), 162.57 (C), 160.72 (C), 159.64 (d, *J* = 242.5 Hz, C), 155.46 (C), 149.71 (CH), 134.43 (C), 133.84 (d, *J* = 3.0 Hz, C), 130.23 (CH), 128.80 (C), 120.91 (CH), 115.72 (d, *J* = 22.5 Hz, 2CH), 112.86 (C), 112.31 (d, *J* = 7.5 Hz, 2CH), 111.68 (C), 101.70 (CH), 27.74 (CH_2), 25.87 (CH_3), 17.82 (CH_3) ppm., ^{19}F NMR (282 MHz, $CDCl_3$, std. TFA): −118.08 (s, 1F) ppm., IR: 3208.66 (O-H), 3155.21 (N-H), 3048.54 (C-H, aromatic), 2913.86 (C-H, aliphatic), 1698.12 (C=O, amide) cm^{-1}; HREI-MS (m/z) calculated for $C_{21}H_{19}O_4NF$ $(M)^+$ 368.1293, found 368.1293.

N-(4-Chlorophenyl)-7-hydroxy-6-(3-methylbut-2-en-1-yl)-2-oxo-2H-chromene-3-carboxamide **14c** (44 mg, 44% yield): m.p. 282–283 °C, ^{1}H NMR (300 MHz, $CDCl_3$+pyridine-d_5): δ 10.92 (s, NH), 8.89 (s, 1H), 7.69 (d, *J* = 8.88 Hz, 2H), 7.37 (s, 1H), 7.30 (dd, *J* = 8.89, 2.01 Hz, 2H), 6.83 (s, 1H), 5.41 (t, *J* = 7.23 Hz, 1H), 3.43 (d, *J* = 7.23 Hz, 2H), 1.81 (s, 3H), 1.74 (s, 1H) ppm., ^{13}C NMR (75 MHz, $CDCl_3$+pyridine-d_5): δ 163.65 (C), 162.81 (C), 160.64 (C), 155.57 (C), 149.50 (CH), 136.76 (C), 134.06 (C), 130.09 (CH), 129.15 (C), 128.95 (2CH), 128.83 (C), 121.64 (2CH), 121.17 (CH), 112.56 (C), 111.27 (C), 101.68 (CH), 27.85 (CH_2), 25.83 (CH_3), 17.84 (CH_3) ppm. IR: 3196.03 (O-H), 3122.34 (N-H), 3073.18 (C-H, aromatic), 2911.85 (C-H, aliphatic), 1698.47 (C=O, amide) cm^{-1}; HREI-MS (m/z) calculated for $C_{21}H_{19}O_4N^{35}Cl$ $(M)^+$ 384.0997, found 384.0996.

N-(4-Bromophenyl)-7-hydroxy-6-(3-methylbut-2-en-1-yl)-2-oxo-2H-chromene-3-carboxamide **14d** (52 mg, 47% yield)): m.p. 276–277 °C, ^{1}H NMR (300 MHz, $CDCl_3$+pyridine-d_5): δ 10.92 (s, NH), 8.89 (s, 1H), 7.69 (d, *J* = 8.88 Hz, 2H), 7.44 (d, *J* = 9.66 Hz, 2H), 7.39 (s, 1H), 6.83 (s, 1H), 5.40 (t, *J* = 7.14 Hz, 1H), 3.43 (d, *J* = 7.26 Hz, 2H), 1.81 (s, 3H), 1.73 (s, 1H) ppm., ^{13}C NMR (75 MHz, $CDCl_3$+pyridine-d_5): δ 163.67 (C), 162.80 (C), 160.65 (C), 155.57 (C), 149.51 (CH), 137.25 (C), 134.04 (C), 131.89 (2CH), 130.09 (CH), 128.83 (C), 121.96 (2CH), 121.16 (CH), 116.80 (C), 112.52 (C), 111.25 (C), 101.67 (CH), 27.85 (CH_2), 25.83 (CH_3), 17.84 (CH_3) ppm. IR: 3192.94 (O-H), 3070.45 (C-H, aromatic), 2915.14 (C-H, aliphatic), 1697.23 (C=O, amide) cm^{-1}; HREI-MS (m/z) calculated for $C_{21}H_{19}O_4N^{79}Br$ $(M)^+$ 428.0492, found 428.0492.

N-(2,5-Difluorophenyl)-7-hydroxy-6-(3-methylbut-2-en-1-yl)-2-oxo-2H-chromene-3-carboxamide **14e** (46 mg, 46% yield): m.p. 248–250 °C, ^{1}H NMR (300 MHz, $CDCl_3$+pyridine-d_5): δ 12.27 (s, NH), 8.90 (s, 1H), 8.40 (ddd, *J* = 10.46, 6.66, 3.15 Hz, 1H), 7.43 (s, 1H), 7.04 (ddd, *J* = 9.57, 9.18, 4.89 Hz, 1H), 6.86 (s, 1H), 6.65–6.75 (m,1H), 5.41 (t, *J* = 7.35 Hz, 1H), 3.43

(d, J = 7.23 Hz, 2H), 1.81 (s, 3H), 1.74 (s, 1H) ppm., ^{13}C NMR (75 MHz, $CDCl_3$+pyridine-d_5): δ 163.90 (C), 162.63 (C), 160.90 (C), 158.57, (d, J = 238.5 Hz, C), 158.55, (d, J = 239.3 Hz, C), 155.75 (C), 149.73 (C), 134.03 (C), 130.18 (CH), 128.88 (C), 127.65 (d, J = 11.3 Hz, CH), 121.19 (CH), 115.20 (dd, J = 27.4, 9 Hz, CH), 112.28 (C), 111.18 (C), 110.03 (dd, J = 24.0, 7.5 Hz, CH), 109.09 (d, J = 30 Hz, CH), 101.74 (CH), 27.83 (CH_2), 25.82 (CH_3), 17.83 (CH_3) ppm. ^{19}F NMR (282 MHz, $CDCl_3$+pyridine-d_5, std. TFA): −117.72 (d, J = 14.10 Hz, 1F), -136.03 (d, J = 14.10 Hz, 1F) ppm., IR: 3252.42 (O-H), 3130.54 (N-H), 2073.1 (C-H, aromatic), 2915.45 (C-H, aliphatic), 1701.64 (C=O, amide) cm^{-1}; HREI-MS (m/z) calculated for $C_{21}H_{17}O_4NF_2$ $(M+Na)^+$ 408.1018, found 408.1013.

7-Hydroxy-6-(3-methylbut-2-en-1-yl)-2-oxo-N-(p-tolyl)-2H-chromene-3-carboxamide **14f** (41 mg, 43% yield): m.p. 276–277 °C, ^{1}H NMR (300 MHz, $CDCl_3$+pyridine-d_5): δ 10.80 (s, NH), 8.90 (s, 1H), 7.65 (d, J = 7.44 Hz, 2H), 7.41 (s, 1H), 7.15 (d, J = 8.31 Hz, 2H), 6.85 (s, 1H), 5.41 (tt, J = 7.32, 1.35 Hz, 1H), 3.43 (d, J = 7.20 Hz, 2H), 2.31 (s, 3H), 1.81 (s, 3H), 1.74 (s, 1H) ppm., ^{13}C NMR (75 MHz, $CDCl_3$+pyridine-d_5): δ 163.33 (C), 162.79 (C), 160.40 (C), 155.45 (C), 149.19 (CH), 135.56 (C), 134.01 (C), 133.91 (C), 130.00 (CH), 129.47 (2CH), 121.22 (CH), 120.45 (2CH), 113.03 (C), 111.31 (C), 101.66 (CH), 27.86 (CH_2), 25.83 (CH_3), 20.91 (CH_3), 17.84 (CH_3) ppm. IR: 3187.07 (O-H), 3130.54 (N-H), 3073.13 (C-H, aromatic), 2913.70 (C-H, aliphatic), 1698.29 (C=O, amide) cm^{-1}; HREI-MS (m/z) calculated for $C_{22}H_{22}O_4N$ $(M)^+$ 364.1543, found 364.1540.

7-Hydroxy-*N*-(4-methoxyphenyl)-6-(3-methylbut-2-en-1-yl)-2-oxo-2H-chromene-3-carboxamide **14g** (43 mg, 44% yield): m.p. 259–261 °C, ^{1}H NMR (300 MHz, $CDCl_3$+pyridine-d_5): δ 10.74 (s, NH), 8.88 (s, 1H), 7.63 (d, J = 9.03 Hz, 2H), 7.41 (s, 1H), 6.91 (d, J = 9.06 Hz, 2H), 6.82 (s, 1H), 5.34 (tt, J = 7.35, 1.47 Hz, 1H), 3.82 (s, 3H), 3.36 (d, J = 7.20 Hz, 2H), 1.80 (s, 3H), 1.72 (s, 1H) ppm., ^{13}C NMR (75 MHz, $CDCl_3$+pyridine-d_5): δ 163.00 (C), 162.28 (C), 160.44 (C), 156.65 (C), 156.65 (C), 155.34 (C), 149.37 (CH), 134.40 (C), 131.01 (C), 130.16 (CH), 128.33 (C), 122.29 (2CH), 120.94 (CH), 113.20 (C), 111.74 (C), 101.71 (CH), 55.56 (CH_3), 27.76 (CH_2), 25.81 (CH_3), 17.82 (CH_3) ppm. IR: 3182.53 (O-H), 3111.40 (N-H), 3073.14 (C-H, aromatic), 2911.91 (C-H, aliphatic), 1695.83 (C=O, amide) cm^{-1}; HREI-MS (m/z) calculated for $C_{22}H_{22}O_5N$ $(M)^+$ 380.1493, found 380.1490.

3.1.7. Synthesis of 8,8-Dimethyl-2-oxo-*N*-phenyl-2H,8H-pyrano[3,2-g]chromene-3-carboxamide **12**

A mixture of compound **14a** (1.7 g, 4.85 mmol) and DDQ (1.2 g, 5.28 mmol) in benzene (10 mL) was refluxed for 6 h, and the precipitate was filtered off. The filtrate was evaporated to dryness to afford the crude product, which was purified via column chromatography (silica gel, 10:1 hexane:EtOAc) to obtain a white solid of 8,8-dimethyl-2-oxo-N-phenyl-2H,8H-pyrano[3,2-g]chromene-3-carboxamide **12** (1.11 g, 66% yield): m.p. 187–188 °C, ^{1}H NMR (300 MHz, $CDCl_3$): δ 10.81(s, 1H), 8.89 (s, 1H), 7.75 (d, J = 1.14 Hz, 2H), 7.39 (t, J = 6.54 Hz, 2H), 7.30 (s, 1H), 7.18 (t, J = 1.14 Hz, 1H), 6.80 (s, 1H), 6.40 (d, J = 9.90 Hz, 1H), 5.78 (d, J = 9.96 Hz, 1H), 1.55 (s, 6H) ppm.

3.1.8. Synthesis of 6-(3-Methyl-2-buteny1)-7-hydroxycoumarin **15**

Compound **13** (0.20 g, 0.73 mmol) in 2 mL quinoline containing 0.3 g Cu powder was heated for 3 min at 215–220 °C in an oil bath. The mixture was cooled to room temperature and diluted with CH_2Cl_2 (30 mL) prior to extraction with 10% HCI (2 × 30 mL) and then with water (30 mL). The solvent was evaporated, leaving a tacky orange solid, which was purified via column chromatography (silica gel, 1:1 hexane:EtOAc) to yield a cream solid of 6-(3-Methyl-2-buteny1)-7-hydroxycoumarin **15** (0.09 g, 53% yield), m.p. 134–136 °C (lit. 133 °C [24]), ^{1}H NMR (300 MHz, $CDCl_3$): δ 7.67 (d, J = 9.42 Hz, 1H), 7.20 (s, 1H), 7.06 (s, 1H), 6.24 (d, J = 9.42 Hz, 1H), 5.33 (tt, J = 8.73, 1.41 Hz, 1H), 3.38 (d, J = 7.23 Hz, 2H), 1.78 (s, 3H), 1.75 (s, 3H).

3.2. Determination of Antibacterial Activity

The antibacterial activities of coumarin derivatives **10**, **12**, **13**, **14a–g**, and **15** were evaluated against five reference standard bacteria, both gram-positive and gram-negative:

B. cereus TISTR 2372, *B. subtilis* TISTR 001, *S. aureus* TISTR 2392, *E. coli* TISTR 073, and *S. typhimurium* TISTR 2519, using a standard microbroth dilution method [25].

The MIC values of coumarin derivatives **10**, **12**, **13**, **14a–g**, and **15** were determined through the microbroth dilution method in 96-well microtitre plates. The bacterial cultures were prepared from overnight cultures on nutrient broth (NB) at 37 °C for 24 h by diluting in NB compared with 0.5 McFarland. Coumarin derivatives **10**, **12**, **13**, **14a–g**, and **15** (5000 μg/mL) were prepared in EtOH, and 128 μg/mL of these were added to the first wells. Two-fold serial dilutions were prepared, and final concentrations of 128 to 1 μg/mL were achieved. The positive controls for penicillin G were determined, with the final concentrations from 128 to 1 μg/mL. In addition, an extra row of EtOH was used as a vehicle control to determine its possible inhibitory activity. Finally, 10 μL of bacterial suspension was added to each well. After the bacteria were incubated at 37 °C for 24 h, the microtitre plates were visually examined for bacterial growth; the growth rate was monitored at the optical density at 600 nm with a microplate reader. In each row, the well containing the lowest concentration that showed no visible growth was considered the MIC.

3.3. Cell Viability Assay

The cell viability assays of coumarin derivatives **10**, **12**, **13**, **14a–g**, and **15** were conducted against three cancer cell lines (HepG2, HeLa, and MDA-MB-231) and one normal cell line (LLC-MK2) using an MTT assay [26].

Stock solutions of coumarin derivatives **10**, **12a**, **13**, **14a–g**, and **15** were prepared in EtOH at a concentration of 5000 μg/mL. Prior to use, the stock solutions were further diluted to 128 μg/mL and added to the first wells. Two-fold serial dilutions were prepared, and final concentrations of 128 to 1 μg/mL in culture medium were achieved. Cells were seeded at a density of 5 × 104 cells/well in a 96-well plate and incubated for 16 h, followed by treatment with the test compounds. The control culture contained the carrier solvent of 2.5% dimethyl sulfoxide (DMSO). After 24 h, HepG2, HeLa, MDA-MB-231, and LLC-MK2 cells were then incubated with MTT (500 μg/mL) for 4 h. Then, DMSO was added to dissolve the blue formazan crystals formed, which were formed as a result of the action of cellular oxidoreductase enzymes on the MTT dye. Finally, the optical density at 450 nm was determined using a microplate reader.

4. Conclusions

We designed and synthesized a series of coumarin-3-carboxamides and evaluated their antibacterial and anticancer activities. The carboxylic acid at the C3 position of coumarins was necessary for the antibacterial activity, as seen for compounds 10 and 13, which showed moderate antibacterial activities against the tested gram-positive bacteria. Meanwhile, most of the tested compounds showed potent anticancer activity, and the 4-fluorophenyl coumarin-3′-carboxazine 4b was by far the most active anticancer, with activity comparable to that of the anticancer drug doxorubicin, and it had low cytotoxicity against a normal cell line. The molecular docking study revealed the binding to the active site of the CK2 enzyme, indicating that the presence of the phenyl carboxamide is important for anticancer activity.

Author Contributions: Conceptualization, project administration, supervision, W.P.; Methodology, investigation, validation, writing—original draft, A.C.; conceptualization, methodology, microbiology testing supervision, T.T.; molecular modeling supervision, J.S.; methodology, conceptualization, project administration, writing—review and editing, supervision, W.S.P. All authors have read and agreed to the published version of the manuscript.

Funding: This research received no external funding.

Institutional Review Board Statement: Not applicable.

Informed Consent Statement: Not applicable.

Data Availability Statement: The data presented in this study are available on request from the corresponding author.

Acknowledgments: We gratefully acknowledge the Department of Chemistry and the Department of Microbiology, Faculty of Science, Silpakorn University, for the financial support and antibacterial and anticancer assay. We also thank the Chulabhorn Research Institute for the measurements of the HR-ESI mass spectroscopy investigation and the Rice Department.

Conflicts of Interest: The authors declare no conflict of interest.

References

1. Murray, R.D.H. Coumarins. *Nat. Prod. Rep.* **1995**, *12*, 477–505. [CrossRef]
2. Xu, J.; Kjer, J.; Sendker, J.; Wray, V.; Guan, H.; Edrada, R.; Muller, W.E.; Bayer, M.; Lin, W.; Wu, J.; et al. Cytosporones, coumarins, and an alkaloid from the endophytic fungus Pestalotiopsis sp. isolated from the Chinese mangrove plant Rhizophoramucronata. *Bioorg. Med. Chem.* **2009**, *17*, 7362–7367. [CrossRef]
3. Do Nascimento, J.S.; Conceição, J.C.S.; de Oliveira Silva, E. Biotransformation of Coumarins by Filamentous Fungi: An Alternative Way for Achievement of Bioactive Analogs. *Mini. Rev. Org. Chem.* **2019**, *16*, 568–577. [CrossRef]
4. Penta, S. Introduction to Coumarin and SAR. In *Advances in Structure and Activity Relationship of Coumarin Derivatives*; Academic Press: Cambridge, MA, USA, 2016; pp. 1–8. [CrossRef]
5. Shehzad, A.; Parveen, S.; Qureshi, M.; Subhan, F.; Lee, Y.S. Decursin and decursinol angelate: Molecular mechanism and therapeutic potential in inflammatory diseases. *Inflamm. Res.* **2018**, *67*, 209–218. [CrossRef]
6. Chun-Ching, C.; Ming-Jen, C.; Peng, C.F.; Huang, H.Y.; Chen, I.S. A Novel Dimeric Coumarin Analog and Antimycobacterial Constituents from *Fatoua pilosa*. *Chem. Biodivers.* **2010**, *7*, 1728–1736. [CrossRef] [PubMed]
7. Stefanachi, A.; Leonetti, F.; Pisani, L.; Catto, M.; Carotti, A. Coumarin: A Natural, Privileged and Versatile Scaffold for Bioactive Compounds. *Molecules* **2018**, *23*, 250. [CrossRef]
8. Al-Majedy, Y.K.; Kadhum, A.A.H.; Al-Amiery, A.A.; Mohamad, A.B. Coumarins: The Antimicrobial agents. *Sys. Rev. Pharm.* **2017**, *8*, 62–70. [CrossRef]
9. Montagner, C.; de Souza, S.M.; Groposoa, C.; Delle Monache, F.; Smania, E.F.; Smania, A., Jr. Antifungal activity of coumarins. *Z. Naturforsch. C J. Biosci.* **2008**, *63*, 21–28. [CrossRef]
10. Kadhum, A.A.; Al-Amiery, A.A.; Musa, A.Y.; Mohamad, A.B. The antioxidant activity of new coumarin derivatives. *Int. J. Mol. Sci.* **2011**, *12*, 5747–5761. [CrossRef] [PubMed]
11. Kirsch, G.; Abdelwahab, A.B.; Chaimbault, P. Natural and Synthetic Coumarins with Effects on Inflammation. *Molecules* **2016**, *21*, 1322. [CrossRef] [PubMed]
12. Song, X.F.; Fan, J.; Liu, L.; Liu, X.F.; Gao, F. Coumarin derivatives with anticancer activities: An update. *Arch. Pharm. (Weinheim)* **2020**, *353*, e2000025. [CrossRef]
13. Golfakhrabadi, F.; Abdollahi, M.; Ardakani, M.R.; Saeidnia, S.; Akbarzadeh, T.; Ahmadabadi, A.N.; Ebrahimi, A.; Yousefbeyk, F.; Hassanzadeh, A.; Khanavi, M. Anticoagulant activity of isolated coumarins (suberosin and suberenol) and toxicity evaluation of Ferulago carduchorum in rats. *Pharm. Biol.* **2014**, *52*, 1335–1340. [CrossRef]
14. Neyts, J.; Cleucq, E.; Singha, R.; Chang, Y.H.; Das, A.R.; Chakraborty, S.K.; Hong, S.C.; Tsay, S.C.; Hsu, M.H.; Hwu, J.R. Structure-Activity Relationship of New Anti-Hepatitis C Virus Agents: Heterobicycle-Coumarin Conjugates. *J. Med. Chem.* **2009**, *52*, 1486–1490. [CrossRef]
15. Chimenti, F.; Bizzarri, B.; Bolasco, A.; Secci, D.; Chimenti, P.; Carradori, S.; Granese, A.; Rivanera, D.; Lilli, D.; Scaltrito, M.M.; et al. Synthesis and in vitro selective anti-Helicobacter pylori activity of N-substituted-2-oxo-2H-1-benzopyran-3-carboxamides. *Eur. J. Med. Chem.* **2006**, *41*, 208–212. [CrossRef]
16. Reddy, N.S.; Gumireddy, K.; Mallireddigari, M.R.; Cosenza, S.C.; Venkatapuram, P.; Bell, S.C.; Reddy, E.P.; Reddy, M.V. Novel coumarin-3-(N-aryl)carboxamides arrest breast cancer cell growth by inhibiting ErbB-2 and ERK1. *Bioorg. Med. Chem.* **2005**, *13*, 3141–3147. [CrossRef] [PubMed]
17. Xiao, Y.H.; Zhi, J.S.; Hong, L.Z.; Li, S.; Xiao, Y.Z. Study on the Anticancer Activity of Coumarin Derivatives by Molecular Modeling. *Chem. Biol. Drug Des.* **2011**, *78*, 651–658. [CrossRef]
18. Faulques, M.; Rene, L.; Royer, R.; Averbeck, D.; Moradi, M. Synthesis and photo induced biological properties of demethyl derivatives of natural pyranocoumarins xanthyletin or seselin. *Eur. J. Med. Chem.* **1983**, *18*, 9–14. [CrossRef]
19. Wang, Z.; Cao, Y.; Paudel, S.; Yoon, G.; Cheon, S.H. Concise synthesis of licochalcone C and its regioisomer, licochalcone H. *Arch. Pharm. Res.* **2013**, *36*, 1432–1436. [CrossRef]
20. Steck, W. New Syntheses of Demethylsuberosin, Xanthyletin, (±)-Decursinol, (+)-Marmesin, (-)-Nodakenetin, (L-)-Decursin, and (+)-Prantschimginl. *Can. J. Chem.* **1971**, *49*, 2297–2301. [CrossRef]
21. Trembley, J.H.; Wang, G.; Unger, G.; Slaton, J.; Ahmed, K. Protein kinase CK2 in health and disease: CK2: A key player in cancer biology. *Cell Mol. Life Sci.* **2009**, *66*, 1858–1867. [CrossRef]
22. Chilin, A.; Battistutta, R.; Bortolato, A.; Cozza, G.; Zanatta, S.; Poletto, G.; Mazzorana, M.; Zagotto, G.; Uriarte, E.; Guiotto, A.; et al. Coumarin as Attractive Casein Kinase 2 (CK2) Inhibitor Scaffold: An Integrate Approach To Elucidate the Putative Binding Motif and Explain Structure–Activity Relationships. *J. Med. Chem.* **2008**, *51*, 752–759. [CrossRef] [PubMed]

23. Hsu, K.-C.; Chen, Y.-F.; Lin, S.-R.; Yang, J.-M. iGEMDOCK: A graphical environment of enh ancing GEMDOCK using pharmacological interactions and post-screening analysis. *BMC Bioinform.* **2011**, *12*, S33. [CrossRef] [PubMed]
24. Trumble, J.T.; Millar, J.G. Biological Activity of Marmesin and Demethylsuberosin against a Generalist Herbivore, *Spodoptera exigua* (Lepidoptera: Noctuidae). *J. Agric. Food Chem.* **1996**, *44*, 2859–2864. [CrossRef]
25. Mbaveng, A.T.; Ngameni, B.; Kuete, V.; Simo, I.K.; Ambassa, P.; Roy, R.; Bezabih, M.; Etoa, F.X.; Ngadjui, B.T.; Abegaz, B.M.; et al. Antimicrobial activity of the crude extracts and five flavonoids from the twigs of Dorstenia barteri (Moraceae). *J. Ethnopharmacol.* **2008**, *116*, 483–489. [CrossRef]
26. Tim, M. Rapid Colorimetric Assay for Cellular Growth and Survival: Application to Proliferation and Cytotoxicity Assays. *J. Immunol. Methods* **1983**, *65*, 55–63. [CrossRef]

Article

Studies of Coumarin Derivatives for Constitutive Androstane Receptor (CAR) Activation

Shin-Hun Juang [1], Min-Tsang Hsieh [1,2], Pei-Ling Hsu [1,†], Ju-Ling Chen [1,3,†], Hui-Kang Liu [4], Fong-Pin Liang [1], Sheng-Chu Kuo [1], Chen-Yuan Chiu [5], Shing-Hwa Liu [5], Chen-Hsi Chou [3], Tian-Shung Wu [3] and Hsin-Yi Hung [3,*]

1 School of Pharmacy, China Medical University, Taichung 404, Taiwan; Paul.juang@gmail.com (S.-H.J.); t21917@mail.cmu.edu.tw (M.-T.H.); Peiling96@livemail.tw (P.-L.H.); ericababa@msn.com (J.-L.C.); cellpharmacology@gmail.com (F.-P.L.); sckuo@mail.cmu.edu.tw (S.-C.K.)

2 Chinese Medicine Research and Development Center, China Medical University Hospital, 2 Yude Road, Taichung 404, Taiwan

3 School of Pharmacy, College of Medicine, National Cheng Kung University, Tainan 701, Taiwan; chenhsi@mail.ncku.edu.tw (C.-H.C.); tswu@mail.ncku.edu.tw (T.-S.W.)

4 National Research Institute of Chinese Medicine, Ministry of Health and Welfare, Taipei 112, Taiwan; hk.liu@nricm.edu.tw

5 Graduate Institute of Toxicology, College of Medicine, National Taiwan University, Taipei 100, Taiwan; kidchiou@gmail.com (C.-Y.C.); shinghwaliu@ntu.edu.tw (S.-H.L.)

* Correspondence: z10308005@email.ncku.edu.tw; Tel.: +886-6-2353535 (ext. 6803)

† Authors contribute equally to this work.

Abstract: Constitutive androstane receptor (CAR) activation has found to ameliorate diabetes in animal models. However, no CAR agonists are available clinically. Therefore, a safe and effective CAR activator would be an alternative option. In this study, sixty courmarin derivatives either synthesized or purified from *Artemisia capillaris* were screened for CAR activation activity. Chemical modifications were on position 5,6,7,8 with mono-, di-, tri-, or tetra-substitutions. Among all the compounds subjected for in vitro CAR activation screening, 6,7-diprenoxycoumarin was the most effective and was selected for further preclinical studies. Chemical modification on the 6 position and unsaturated chains were generally beneficial. Electron-withdrawn groups as well as long unsaturated chains were hazardous to the activity. Mechanism of action studies showed that CAR activation of 6,7-diprenoxycoumarin might be through the inhibition of EGFR signaling and upregulating PP2Ac methylation. To sum up, modification mimicking natural occurring coumarins shed light on CAR studies and the established screening system provides a rapid method for the discovery and development of CAR activators. In addition, one CAR activator, scoparone, did showed anti-diabetes effect in *db/db* mice without elevation of insulin levels.

Keywords: Yin Chen Hao; constitutive androstane receptor; coumarin; scoparone

Citation: Juang, S.-H.; Hsieh, M.-T.; Hsu, P.-L.; Chen, J.-L.; Liu, H.-K.; Liang, F.-P.; Kuo, S.-C.; Chiu, C.-Y.; Liu, S.-H.; Chou, C.-H.; et al. Studies of Coumarin Derivatives for Constitutive Androstane Receptor (CAR) Activation. *Molecules* **2021**, *26*, 164. https://doi.org/10.3390/molecules26010164

Received: 16 November 2020
Accepted: 28 December 2020
Published: 31 December 2020

1. Introduction

Constitutive androstane receptor (CAR, NR1I3), a member of the superfamily of nuclear receptors, is a xenobiotic receptor responsible for the regulation of drug metabolism as well as the pathological involvement of various diseases such as cancer, diabetes, inflammatory disease, metabolic disease and liver disease, suggesting a potential target for drug discovery [1]. Predominantly expressed in the liver, once CAR becomes activated, it disassociates from the cytoplasmic protein complex and with retinoid X receptor (RXR) binds to specific DNA region and influence target gene expression [2].

CAR has been proven to involve in various energy pathways as well as metabolic pathways [3]. The well-established function of CAR is to regulate bile acid detoxification and bilirubin clearance via the induction of metabolizing enzymes and transporters, such as Mrp2–4, Oatp 2, Cyp3a11, Sult2a1 and Ugt1a1 [4–6]. Moreover, CAR knock out (KO) mice led to high incidence of hepatic necrosis and high level of alanine aminotransferase (ALT) and bilirubin, suggesting an important function of CAR activation for hepatoprotective effect and a potential role in treating cholestasis [7]. A direct CAR agonist, TCPOBOP, was found to decrease liver injury and hepatocyte apoptosis in the treated mice as well as to improve insulin sensitivity in ob/ob mice model [8]. In addition, CAR activation ameliorates hyperglycemia as well as fatty liver by suppressing glucose production, stimulating glucose uptake and usage in the liver and inhibiting hepatic lipogenesis and induction of β–oxidation [9]. Recent metabolomic research revealed CAR activation significantly lower mRNA expression that involved in gluconeogenic pathway, up-regulate glucose utilization pathways, enhance specific fatty acid synthesis and impair β–oxidation [10].

Since CAR is a potential target for various diseases, neither CAR agonists nor antagonists are seen clinically. 6-(4-Chlorophenyl) imidazo [2,1-b] [1,3] thiazole-5-carbaldehyde *O*-(3,4-dichlorobenzyl) oxime (CITCO), a human CAR agonist, is used extensively in biological studies, but failed to be further developed due to toxicity. Another commonly used agonist, TCPOBOP (Figure 1), is a mouse CAR agonist, which also points out the difficulty of developing CAR activators—species specificity [11]. Luckily, Yin Chen Hao (*Artemisia capillaris*) in the Asteraceae family has long been used to treat jaundice in Traditional Chinese Medicine (TCM). The extract of Yin Chen Hao Tang (YCHT) was found to activate CAR and then accelerated bilirubin clearance in animal models [12,13]. In 2015, a meta-analysis studied 15 randomized controlled trials (RCT) including 1405 cases ranging from 2000 to 2014 and found YCHT reduced the elevated levels of cholestasis serum markers significantly either in short or long curative time periods and also improved the clinical efficiency of other anti-cholestasis medications [14]. Further, 6,7-dimethylesculetin (scoparone), a major active principal isolated from Yin Chin Hao, also exerted CAR activation activity both in vitro and in vivo [12]. Scoparone exerted plenty of pharmacological activites, such as hepatoprotective effect, antioxidation, anti-inflammation, and anti-cholestasis [15]. In addition, scoparone (10 μM) potentiated chenodeoxycholic acid (10 μM, a potent FXR agonist) effect to increase the expression of bile salt export pump. However, scoparone alone (up to 100 μM) had no effect on FXR activation [16].

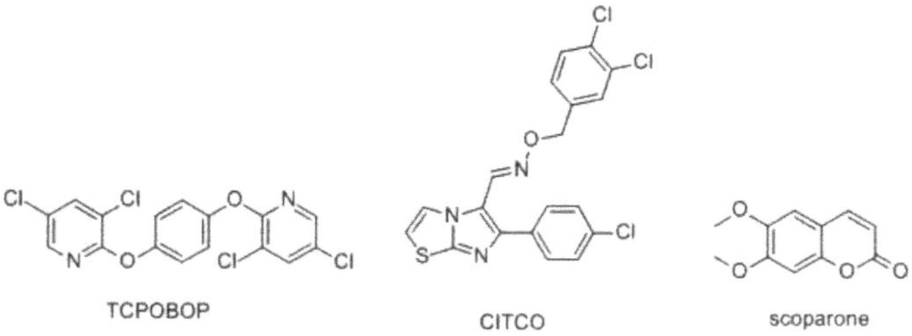

Figure 1. Structures of CAR direct activators. TCPOBOP is a mouse CAR direct activator. CITCO is a human CAR direct activator.

Inspired by the findings above, this study was designed to mimic coumarins from Yin Chen Hao, to systemically synthesize coumarin derivatives to study structure-activity relationship of coumarin derivatives on CAR activation and to establish in vitro CAR activation screening method for discovering and developing therapeutic CAR activation agents in the future. Further, the mechanism of action of the most active compound will be discussed and a possible indication, anti-diabetes, of a CAR activator will be evaluated in vivo.

2. Results and Discussion

Chemistry

Although coumarin derivatives have been studied extensively, their effects on CAR activation are largely underinvestigated. In coumarins isolated from Yin Chen Hao, all the substitutions are on C5, C6, C7, C8 positions [17]. Since the exact binding site of scoparone is unclear, ligand-based drug design is applied. The first strategy is to evaluate the substitution effect on C5, C6, C7, C8 positions alone and together to establish SAR (Tables 1–3). Various alkoxy groups including different chain length, branch, and unsaturation were synthesized and studied in these four positions. These modifications included the structures of natural coumarins, such as 5-hydroxycoumarin, 5-methoxycoumarin, umbeliferone, methylumbeliferone, aurapten, 8-hydroxycoumarin, 8-methoxycoumarin, esculetin, isoscopoletin, scopoletin, scoparone, ayapin, isoscopolin, scopolin, fraxinol, and leptodactylone. Their natural occurrence is provided in the reference column of the tables. From the references, we can find these coumarins can be found in many clinically used TCMs or herbs. Moreover, we also synthesized eight new derivatives (compounds **11**, **20**, **44**, **46–49**, **50**) for comparison of SAR. The syntheses of the coumarin derivatives were mainly via Perkin reaction [18] or Knoevenagel reaction [19] or Pechmann reaction [20] or Wittig reactions [21,22] or others [23,24] and the spectral data for the known compounds were identical with the literature values.

Table 1. Mono-substituted coumarin structures and their CAR activation.

Compound	R_1	R_2	R_3	R_4	CAR Activation Fold [a]	Ref
1 **5-hydroxycoumarin**	OH	H	H	H	6% ± 22%	[25]
2 **5-methoxycoumarin**	*O*-methyl	H	H	H	−20% ± 6%	[26,27]
3	*O*-acetyl	H	H	H	−32% ± 12%	[28]
4	*O*-allyl	H	H	H	−15% ± 6%	[29]
5	*O*-*n*-butyl	H	H	H	−9% ± 14%	[29]
6	*O*-prenyl	H	H	H	−5% ± 9%	[30]
7	*O*-geranyl	H	H	H	−18% ± 11%	[30]
8	*O*-farnasyl	H	H	H	2% ± 14%	[30]
9 **6-hydroxycoumarin**	H	OH	H	H	−8% ± 11 %	[31]
10 **6-methoxycoumarin**	H	*O*-methyl	H	H	32% ± 5%	[32]
11	H	*O*-trifluoromethyl	H	H	−34% ± 1%	New
12	H	*O*-acetyl	H	H	−22% ± 3%	[33]
13	H	*O*-allyl	H	H	−17% ± 6%	[34]
14	H	*O*-n-butyl	H	H	37% ± 6%	[35]
15	H	*O*-prenyl	H	H	22% ± 2%	[30]
16	H	*O*-geranyl	H	H	18% ± 7%	[30]
17	H	*O*-farnasyl	H	H	−41% ± 9%	[30]
18 **umbelliferone**	H	H	OH	H	−1% ± 18%	[31,36]
19 **7-methoxycoumarin**	H	H	*O*-methyl	H	−9% ± 6%	[25,27,36]
20	H	H	*O*-trifluoromethyl	H	18% ± 2%	New
21	H	H	*O*-acetyl	H	−37% ± 12%	[37]
22	H	H	*O*-allyl	H	−46% ± 11%	[37]
23	H	H	*O*-n-butyl	H	−51% ± 7%	[38]
24	H	H	*O*-prenyl	H	−35% ± 13%	[30,39]
25 **aurapten**	H	H	*O*-geranyl	H	−13% ± 4%	[30]

Table 1. *Cont.*

Compound	R_1	R_2	R_3	R_4	CAR Activation Fold [a]	Ref
26	H	H	*O*-farnasyl	H	−29% ± 10%	[30]
27 **8-hydroxycoumarin**	H	H	H	OH	40% ± 38%	[40]
28 **8-methoxycoumarin**	H	H	H	*O*-methyl	−8% ± 10%	[41]
29	H	H	H	*O*-acetyl	−33% ± 13%	[28]
30	H	H	H	*O*-allyl	−34% ± 12%	[42]
31	H	H	H	*O*-n-butyl	4% ± 9%	[43]
32	H	H	H	*O*-prenyl	2% ± 23%	[30]
33	H	H	H	*O*-geranyl	−30% ± 4%	[30]
34	H	H	H	*O*-farnasyl	19% ± 25%	[30]

[a] CAR activation fold was calculated: (the luminescence value of the tested compound- the luminescence value of scoparone)/the luminescence value of scoparone (as positive control) × 100%.

Table 2. Di-substituted coumarins and their CAR activation.

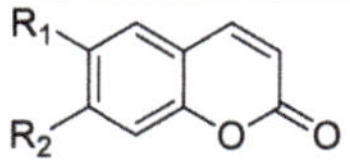

Compound	R_1	R_2	CAR Activation Fold [a]	Ref
35 **esculetin**	OH	OH	−22% ± 12%	[36,44]
36	methyl	methyl	−17% ± 5%	[24]
37 **isoscopoletin**	OH	*O*-methyl	−10% ± 6%	[45]
38 **scopoletin**	*O*-methyl	OH	−1% ± 5%	[44]
39 **scoparone**	*O*-methyl	*O*-methyl	-	[15,17]
40 **ayapin**	$-OCH_2O-$		7% ± 2%	[46]
41	$-OCH_2CH_2O-$		−50% ± 9%	[47]
42	*O*-ethyl	*O*-ethyl	−26% ± 8%	[37]
43	*O*-acetyl	*O*-acetyl	−11% ± 2%	[37]
44	*O*-allyl	*O*-allyl	−18% ± 15%	New
45	*O*-propyl	*O*-propyl	−37% ± 5%	[37]
46	*O*-*n*-butyl	*O*-*n*-butyl	−39% ± 13%	New
47	*O*-pentyl	*O*-pentyl	−10% ± 18%	New
48	*O*-isopentyl	*O*-isopentyl	−61% ± 2%	New
49	*O*-hexyl	*O*-hexyl	−5% ± 9%	New
50	*O*-prenyl	*O*-prenyl	50% ± 1%	[48–50]
51	*O*-geranyl	*O*-geranyl	−32% ± 11%	New
52 **isoscopolin**	*O*-Glc	*O*-methyl	35% ± 5%	[17,51]
53 **scopolin**	*O*-methyl	*O*-Glc	17% ± 26%	[17]

[a] CAR activation fold was calculated: (the luminescence value of the tested compound- the luminescence value of scoparone)/the luminescence value of scoparone (as positive control) × 100%.

Table 3. Tri- and tetra-substituted coumarins and their CAR activation.

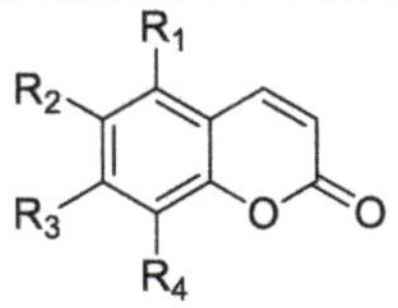

Compound	R_1	R_2	R_3	R_4	CAR Activation Fold [a]	Ref
54	$-OCH_3$	$-OCH_3$	$-OCH_3$	H	−19% ± 3%	[39,44]
55 **fraxinol**	$-OCH_3$	-OH	$-OCH_3$	H	−17% ± 2%	[39]
56	$-OCH_3$	H	$-OCH_3$	$-OCH_3$	−14% ± 2%	[39]
57 **leptodactylone**	$-OCH_3$	H	$-OCH_3$	-OH	2% ± 4%	[39]
58	H	$-OCH_3$	$-OCH_3$	$-OCH_3$	0% ± 2%	[52]
59	H	$-OCH_3$	$-OCH_2O-$		36% ± 23%	[17,53]
60	$-OCH_3$	$-OCH_3$	$-OCH_2O-$		12% ± 1%	[17,53]

[a] CAR activation fold was calculated: (the luminescence value of the tested compound- the luminescence value of scoparone)/the luminescence value of scoparone (as positive control) × 100%.

Moreover, syntheses of coumarins disubstituted mainly on the C6 and C7 positions were carried out (Table 2) The starting material, 6,7-dihydroxycoumarin, was commercially available. Various alkoxyl substitutions were synthesized by treating 6,7-dihydroxycoumarin in dimethylformamide with the corresponding alkyl bromide. Tri-and tetrasubstituted coumarins were also synthesized and are summarized in Table 3. Scopolin (**53**), isoscopolin (**52**), compound **59** and **60** were isolated from Yin Chen Hao [17].

3. Establishment of In Vitro CAR Activation Screening Assay

Until now, there is no commercially available high throughput screening method for CAR activation. Traditionally, CAR activation can be detected in western blots by its downstream signals such as UGT1A1, MRP2 or CYP2B6. In Figure 2, we confirmed that the protein expressions of downstream genes of CAR, UGT1A and MRP2, were upregulated after scoparone and CITCO treatment. However, this method was slow and pricy. In order to rapidly screen all the compounds for CAR activation at less cost, an in vitro CAR activation screening assay was developed: DNA was extracted from Hep3B cells and UGT1A1 promoter was obtained using PCR amplification. Ligation between UGT1A1 promoter and pGL4 vector was performed and the plasmid was amplified via *E. coli* replication. The desired plasmid was extracted from E. coli and transfected into Hep3B. The desired colony bearing UGT1A1 promoter-pGL4 plasmid was selected by neomycin. CAR activation was measured by adding tested compounds to Hep3B for several hours and luminescence was measured. Higher activation will get higher luminescence readout. This assay system was validated with two different CAR agonists: CITCO (direct agonist) and scoparone (an indirect agonist). CITCO could induce about 50% elevation of luciferase activity while scoparone as a positive control could enhance luciferase activity too. Different concentration of scoparone at 1, 5, 10 and 20 μM had been tested either for luciferase assay or downstream UGT1A1 and MRP2 expression (data not shown) and 10 μM was selected to be the best condition for the following work.

After successfully establishing the in vitro assay, all of the coumarin compounds were subjected to the screening (at 10 μM) and their SAR was discussed. Although the reporter system can successfully identify CAR-activating compounds under 8 h and display better efficiency and sensitivity compared with western blots, it only presented a medium luciferase response. According to previous studies, CITCO was reported to only moderately enhance human CAR (less than 2-fold), compared with the enhance provided by TCPOBOP for murine CAR (5- to 10-fold) in luciferase reporter assays [11,54]. These results suggested that human CAR might have a weaker response to agonists compared to murine CAR, which might be the same situation as in our reporter system. Another possibility for the medium response in our reporter system might be the contribution of the whole gtPBREM element which might include response-suppressive elements that was used and the single nucleotide variation of DR3 element which was reported to slightly reduce the original UGT1A1 activity [55].

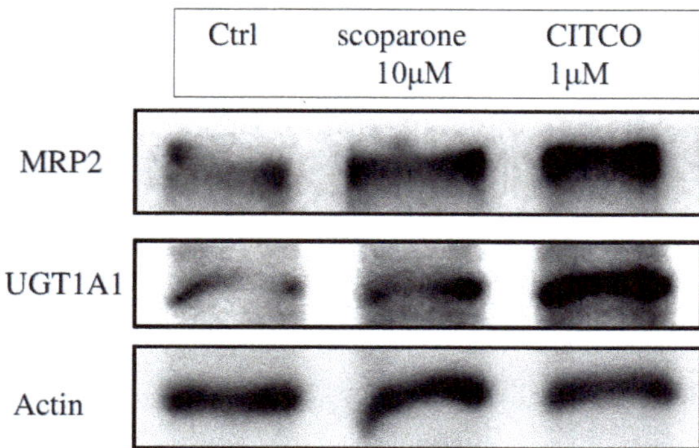

Figure 2. CAR downstream signal protein expression after treatment of CITCO or scoparone.

4. Structure-Activity Relationships

All the compounds were compared to scoparone (**39**) and the resulting CAR activation fold results (%) are shown in Tables 1–3. This is the readout of the tested compound minus the readout of scoparone and then divided by the scoparone readout. Among all sixty compounds, twenty compounds exhibited better or similar CAR activation ability than scoparone and 6,7-diprenoxycoumarin (**50**) was the best compound with a 50% increased CAR activation fold. In addition, none of the compounds have any cytotoxicity at this concentration (IC_{50} > 50 μM) except for **47** (IC_{50} = 31 μM). As for a detailed SAR discussion of the mono- and disubstituted coumarins, hydroxyl groups except on position 8 exhibited similar or less activity than scoparone. Methoxy substitution is better on position 6 than positions 5, 7 or 8. Electron withdrawing groups, such as an acetyl group or a trifluoromethyl group decrease the activity, suggesting the importance of enough electrons. A medium carbon chain length such as a *n*-butyl group on the 6 position is good. Prenyl groups are general good, except in the 7 position. Longer unsaturated chains such as geranyl or farnesyl reduce the activity. To sum up, electron withdrawing groups would be detrimental to the activity. Modification on the 6 position generally increases the activity. Longer unsaturated chains are not good for the activity. Among tri- and tetra-substituted coumarins, 5,7-dimethoxy-8-hydroxycoumarin (**57**) and 6,7,8-trimethoxycoumarin (**58**) showed comparable CAR activation effects to scoparone. 6-Methoxy-7,8-methylenedioxycoumarin (**59**) exhibited a good activation effect (36% increase) and 5,6-dimethoxy-7,8- methylenedioxycoumarin (**60**) also had good activity.

Moreover, if we examined coumarins reported to exist in Yin Chen Hao, such as esculetin (**35**), isoscopoletin (**37**), scopoletin (**38**), scoparone (**39**), scopolin (**53**), isoscopolin (**52**), 5,6,7-trimethoxycoumarin (**54**), leptodactylone (**57**), compound **59** and **60**, we can find that compounds **52**, **53**, **57**, **59** and **60** had superior CAR activating activity than scoparone, implying that the anti-cholestasis effect of YCHT might be an added effect of all the coumarins inside.

5. Mechanistic Studies of 6,7-Diprenoxycoumarin (50)

First, the translocation of CAR protein was confirmed after treatment with **50** (Figure 3A), and the translocation activity could be inhibited by okadaic acid (OA), the phosphatase inhibitor which inhibits CAR from translocating into the nucleus. In addition, the protein expression of nuclear CAR increased after **50** treatment as shown in Figure 3B.

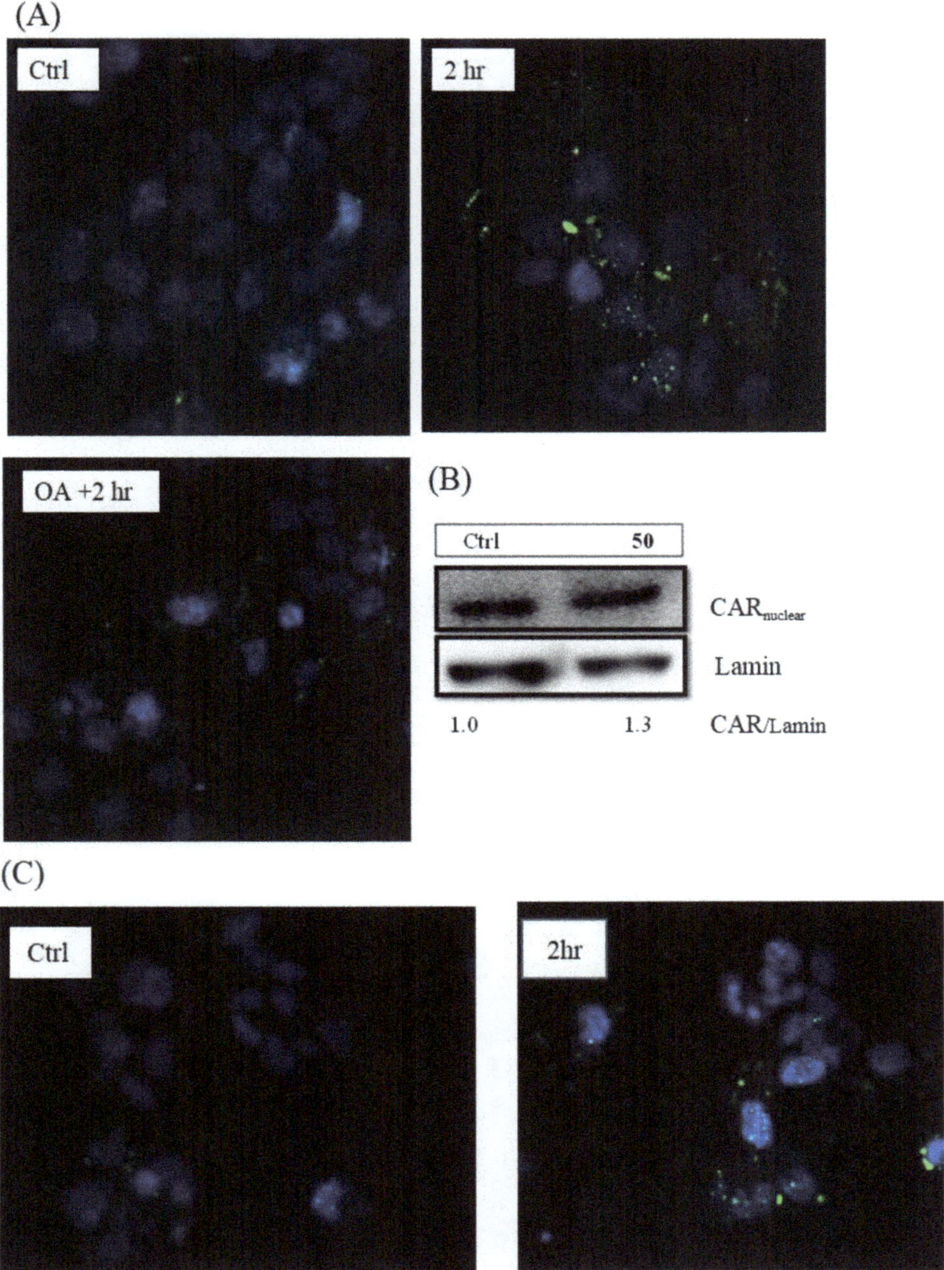

Figure 3. Image of immunofluorescence staining for CAR translocation in Hep3B cells after 6,7-diprenoxycoumarin (**50**) treatment (**A**) and scoparone treatment (**C**). Hep3B cells were treated with **50** (10 μM) for 2 h. The negative control group was pre-treated with 10 nM okadaic acid (OA). The blue and green fluorescence stands for signals of nuclei and CAR protein, respectively. The protein expression of nuclear CAR after 6,7-diprenoxycoumarin (**50**) treatment for 2 h (**B**).

Like phenobarbital-mediated CAR activation, **50** activated CAR through inhibiting EGFR phosphorylation [56]. Western blot results indicated that **50** could reduce EGF-induced phosphorylation of EGFR in Hep3B (Figure 4). The up-stream protein, PP2Ac, which is responsible for CAR dephosphorylation was increased with its active form, methyl-PP2Ac (Figure 5).

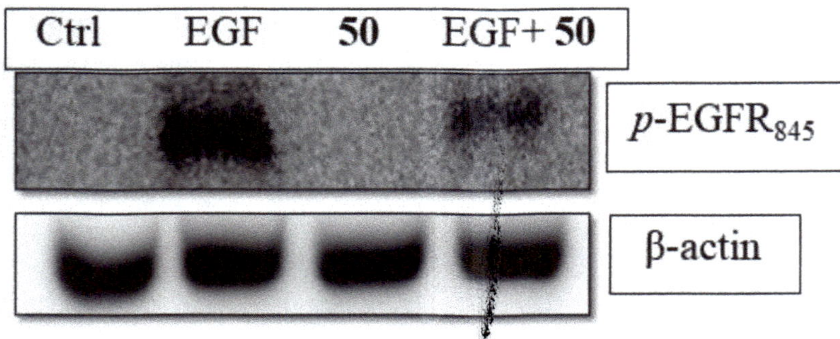

Figure 4. 6,7-Diprenoxycoumarin (**50**) antagonized EGF-induced phosphorylation EGFR in Hep3B. The Hep3B cells (1 × 10^6) were treated without (Ctrl) or with EGF in the absent **50** or present (EGF + **50**) of 10 μM **50** for 15 min and phosphorylation status of EGFR was measured.

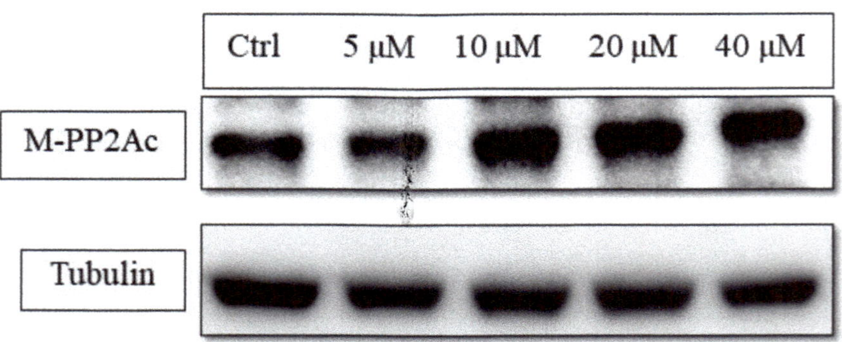

Figure 5. The concentration-dependent study of PP2Ac methylation status under 6,7-diprenoxycoumarin (**50**) treatment. Hep3B cells (1 × 10^6) were treated with **50** (5, 10, 20 and 40 μM) for 24 h. Treated cell lysates were prepared as described in the Methods section and methyl-PP2Ac expression was analyzed by western blot.

The mRNA and protein expression level of CAR-related down-stream genes, MRP2 and UGT1A1, were also enhanced by **50** (UGT1A1 protein level was not significantly increased in our study) (Figures 6 and 7). These data strongly supported that the CAR activation of coumarin derivatives might through the inhibiting the EGFR signaling and upregulating PP2Ac methylation.

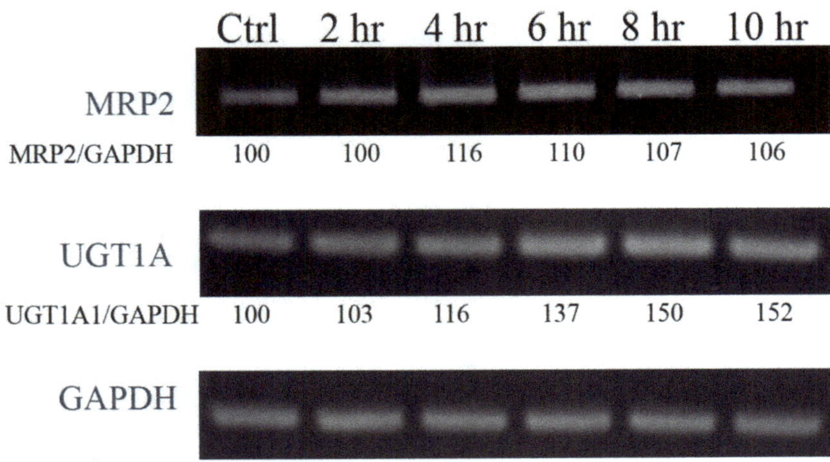

Figure 6. The time course study of mRNA expression of MRP2 and UGT1A1 after 10 µM 6,7-diprenoxycoumarin treatment. Hep3B cells were treated with 10 µM **50** for 2, 4, 6, 8 and 10 h. The total mRNA was obtained and the expression of downstream genes were analyzed with PCR. The ratio of MRP2 to GAPDH was calculated using Image J and normalization.

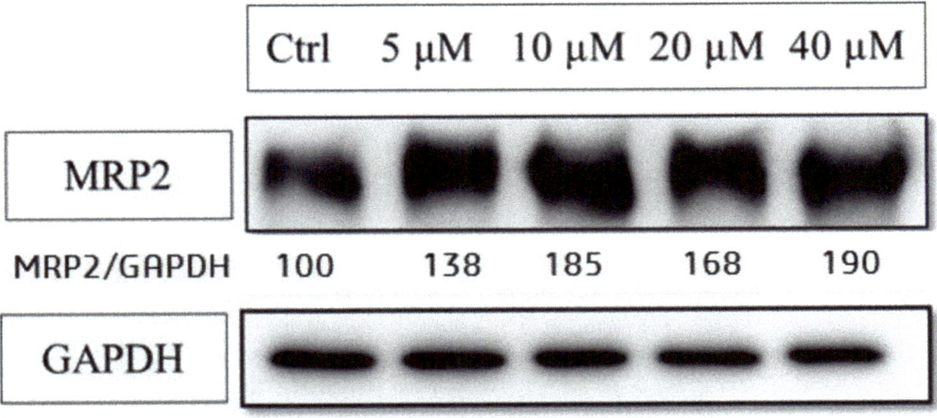

Figure 7. The protein expressions of MRP2 after 6,7-diprenoxycoumarin (**50**) treatment. Hep3B cell (1×10^6) were treated with **50** (5, 10, 20 and 40 µM) for 24 h. Treated cell lysates were prepared as described in the Methods section and the expression of MRP2 was analyzed by western blot.

6. Further In Vivo Hypoglycemic Effect and Pharmacokinetic Studies

From previous literature, CAR activator has been found to regulate glucose metabolizing gene expression [9]. Therefore, in vivo hypoglycemic effect was evaluated using scoparone and compound **50** (Figure 8). Db/db mice were treated with scoparone (100 mg/kg, P.O.) or vehicle (CMC) for two weeks. The scoparone-treated group showed a significant improvement in oral glucose tolerance test (OGTT), which was not observed in vehicle group (A). Further analysis showed the level of fructosamine, a glycation end product, also decreased after treatment, indicating blood glucose levels became lower after treatment (B). However, the insulin level (C) or HOMA-IR (D) did not change, indicating that glucose-lowering effect was not due to an insulinotropic effect. Compound **50** didn't show any in vivo efficacy even by intraperitoneal injection, which led us to perform a pharmacokinetic study of scoparone and **50** (Supplementary Figure S21). After analytical method validation, C57BL/6 mice were given scoparone (i.p.) or **50** and the resulting plasma concentration-time are presented in Tables S1 and S2. In the pharmacological assay, the desired scoparone concentration is ~2 μg/mL (10 μM), which can be achieved when the dose given is 0.5 mg/mouse (Table S1). However, the desired concentration of **50** (~3 μg/mL, 10 μM) cannot be obtained even if the amount of drug is increased to 0.5 mg/mice. Thus, the CAR activating effect of **50** may not be seen in vivo, which can account for the fact no blood sugar lowering effects of **50** were observed in vivo.

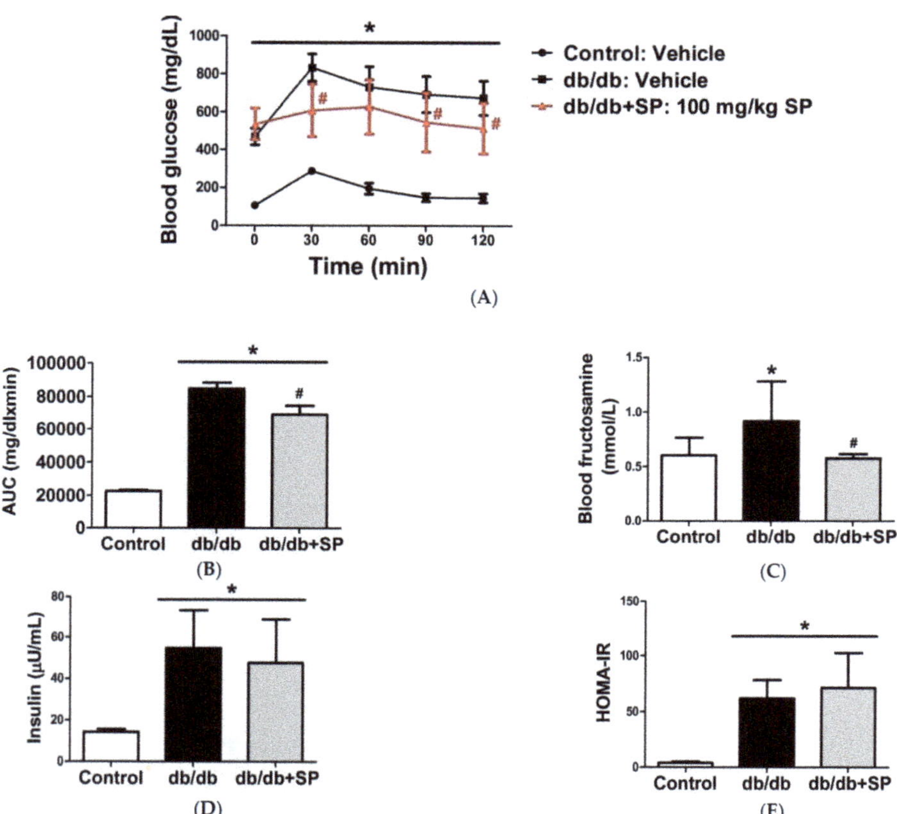

Figure 8. Treatment of scoparone on db/db mice for two weeks improves glucose tolerance and insulin sensitivity. Treatment of scoparone on db/db mice for two weeks improves glucose tolerance. (**A**) db/db mice were treated with scoparone or vehicle (P.O.) for 2 weeks. After 12 h of fasting, glucose tolerance tests were performed and blood glucose levels were assessed (n = 6) (**B**) Area under curve of blood glucose level were calculated. Fructosamine (**C**) and insulin (**D**) levels were measured. (**E**) Homeostasis Model Assessment (HOMA) was calculated as HOMA-IR index = insulin(μU/mL) × glucose(mmol/L)/22.5. (* statistically significant compared with Control group; # statistically significant compared with db/db group; Probability of < 0.05 was considered to be significant)

7. Conclusions

CAR activation has already been associated with amelioration of various diseases such as diabetes. However, clinical CAR agonists are still lacking, hterefore, a safe and effective CAR activator would be an interesting alternative option for these diseases. In this study, sixty coumarin derivatives either synthesized or purified from Yin Chen were screened for CAR activation activity. Among all the compounds, **50** is the most effective for CAR activation and was selected for further studies. Modification on the 6 position is generally more beneficial than at other positions. Electron-withdrawn groups are detrimental to the activity. Mechanism of action studies showed that CAR activation of **50** might be through the inhibition of EGFR signaling and upregulated PP2Ac methylation. In vivo OGTT, scoparone can lower blood sugar and fructosamine levels without insulinotropic effects. In addition, a preliminary pharmacokinetic study also indicated the desired scoparone blood concentration can be achieved. This study established also a screening system providing a rapid method for CAR activator discovery and development and provided scientific evidence for the effects of coumarins on CAR activation.

8. Materials and Methods

8.1. General Information

All of the solvents and reagents were obtained commercially and used without further purification. The progress of all the reactions was monitored by TLC on 2 × 6 cm pre-coated silica gel 60 F_{254} plates of thickness 0.25 mm (Merck). The chromatograms were visualized under 254 or 366 nm UV light. Silica gel 60 (Merck, particle size 0.063–0.200 mm) was used as the adsorbant for column chromatography.

NMR spectra were obtained on an AV400 NMR spectrometer (Bruker). MS spectra were measured either at the instruments center of National Chung Hsing University (JMS-700 spectrometer, JEOL, Japan) or National Tsing Hua University (MAT-95XL HRMS, Finnigan, San Jose, CA). CITCO was purchased from Sigma-Aldrich (Oakville, ON, Canada).

8.2. Chemistry

6-Trifluoromethoxycoumarin (**11**). 2-Hydroxy-4-trifluoromethoxylbenzaldehyde (0.5 g, 2.4 mmol) and methoxycarbonylmethylene-triphenylphosphorane (0.9 g, 2.8 mmol) were dissolved in *N*, *N*-diethylaniline (5.0 mL) and refluxed for 4 h. The reaction mixture was purified by silica gel column chromatography (*n*-hexane/ethyl acetate (8:2)) to give the intermediate. Then the intermediate was dissolved in diphenyl ether (5.0 mL), heated to 200 °C for 4 h. The mixture was partitioned with water and dichloromethane and dried over $MgSO_4$. The desired compound was obtained by silica gel column chromatography (*n*-hexane/ethyl acetate (8:2)) to give 6-trifluoromethoxycoumarin (**11**, 0.3 g, 1.3 mmol). Yellow needle-like crystals. Yield: 55.0%. M.P.: 83–85 °C. ^{1}H-NMR (400 MHz, $CDCl_3$): δ 7.71 (1H, d, J = 9.6 Hz), 7.43–7.38 (2H, m), 7.28 (1H, s), 6.53 (1H, d, J = 9.6 Hz). ^{13}C-NMR (100 MHz, $CDCl_3$): δ 159.8, 152.1, 145.0, 142.3, 124.7, 120.3 (q, J = 258 Hz), 119.7, 119.4, 118.3, 117.9. ^{19}F-NMR (470 MHz, $CDCl_3$): δ = −58.31. HRMS [ESI]$^+$ calculated for $C_{10}H_5F_3O_3$: 230.0191; found [M]$^+$ 230.0192.

7-Trifluoromethoxycoumarin (**20**). 3-Trifluoromethoxylphenol (1.0 g, 5.6 mmol), palladium (II) acetate (5.0 mol%) and trifluoroacetic acid (8.4 mL) were mixed in dichloromethane (5.0 mL) under nitrogen and stirred at room temperature for 7 days. Brine solution was added and the misture extracted with ethyl acetate. The ethyl acetate layer was washed with brine and dried over anhydrous $MgSO_4$. The solvent was removed in vacuo. The residue was purified by column chromatography (*n*-hexane/ethyl acetate (7:3)) to give 7-trifluoromethoxycoumarin (**20**, 0.06 mg, 0.3 mmol) and recovered starting material. White crystals. Yield: 4.0%. M.P.: 69–71 °C. ^{1}H-NMR (400 MHz, $CDCl_3$): δ 7.68 (1H, d, J = 9.6 Hz), 7.50 (1H, d, J = 8.4 Hz), 7.15 (1H, s), 7.11 (1H, d, J = 8.4 Hz), 6.24 (1H, d, J = 9.6 Hz). ^{13}C-NMR (100 MHz, $CDCl_3$): δ 159.7, 154.6, 151.2, 142.4, 129.0, 120.1 (q, J = 258 Hz), 117.2, 116.7, 116.6, 109.1. ^{19}F-NMR (470 MHz, $CDCl_3$): δ = −57.48. HRMS [ESI]$^+$ calculated for $C_{10}H_5F_3O_3$: 230.0191; found [M]$^+$ 230.0190.

8.3. General Procedure for the Synthesis of Various Alkoxyl Courmarin Derivatives

Varied alkoxy-substituted coumarins were prepared by using the corresponding alkyl bromide. The 8-hydroxycoumarin or 6,7-dihydroxycoumarin, alkyl bromide and potassium carbonate were mixed in DMF. The reaction mixture was heated to 200 °C for 1–2 h and then cooled to room temperature. The reaction mixture was diluted with

H_2O and extracted with dichloromethane. The organic layers were combined, dried over anhydrous $MgSO_4$ and concentrated under vacuum. The residue was purified by silica gel column chromatography (*n*-hexane/rthyl acetate (7:3)) to give the various alkoxy-substituted coumarins.

8-n-Butoxycoumarin (**31**). Yield: 22.2%. White needle-like crystals. M.P.: 79–81°C. ^{1}H-NMR (400 MHz, $CDCl_3$): δ 7.65 (1H, d, *J* = 9.6), 7.17–7.13 (1H, m), 7.06–7.00 (2H, m), 6.40 (1H, d, *J* = 9.6 Hz), 4.09–4.06 (2H, m), 1.86–1.79 (2H, m), 1.54–1.52 (2H, m), 0.98–0.94 (3H, m). ^{13}C-NMR (100 MHz, $CDCl_3$): δ 160.4, 146.8, 143.9, 143.6, 124.2, 119.5, 119.1, 116.7, 115.0, 69.1, 31.1, 19.1, 13.7; HRMS [ESI]$^+$ calculated for $C_{13}H_{15}O_3$ [M + H]$^+$ 218.0943; found 218.0944.

6,7-Dialloxycoumarin (**44**). Yield: 74.6%. White needle-like crystals. M.P.: 84–85 °C. ^{1}H-NMR (400 MHz, $CDCl_3$): δ 7.56 (1H, d, *J* = 9.6 Hz), 6.86 (1H, s), 6.82 (1H, s), 6.24 (1H, d, *J* = 9.6 Hz), 6.10–5.99 (2H, m), 5.46–5.28 (4H, m), 4.65–4.59 (4H, m). ^{13}C-NMR (100 MHz, $CDCl_3$): δ 161.3, 152.3, 149.9, 145.3, 143.2, 132.7, 131.9, 118.5, 118.0, 113.4, 111.5, 110.9, 101.5, 70.4, 69.8; HRMS [ESI]$^+$ calculated for $C_{15}H_{14}O_4$: 258.0892; found [M + H]$^+$ 259.0893.

6,7-Di-n-butoxycoumarin (**46**). Yield: 89.9%. Yellow needle-like crystals. M.P.: 77–79 °C. ^{1}H- NMR (400 MHz, $CDCl_3$): δ 7.56 (1H, d, *J* = 9.6 Hz), 6.84 (1H, s), 6.78 (1H, s), 6.22 (1H, d, *J* = 9.6 Hz), 4.04–3.96 (4H, m), 1.85–1.75 (4H, m), 1.51–1.46 (4H, m), 0.98–0.96 (6H, m). ^{13}C-NMR (100 MHz, $CDCl_3$): δ 161.5, 153.1, 150.0, 146.6, 143.3, 113.1, 111.2, 110.5, 110.9, 69.6, 68.9, 31.1, 30.7, 19.1, 19.1, 13.7, 13.7; HRMS [ESI]$^+$ calculated for $C_{17}H_{22}O_4$: 290.1518; found [M]$^+$ 290.1519.

6,7-Dipentoxycoumarin (**47**). Yield: 48.6%. Yellow solid. M.P.: 57–59 °C. ^{1}H-NMR (400 MHz, $CDCl_3$): δ 7.56 (1H, d, *J* = 9.6 Hz), 6.83 (1H, s), 6.78 (1H, s), 6.23 (1H, d, *J* = 9.6 Hz), 4.03–3.96 (4H, m), 1.88–1.78 (4H, m), 1.46–1.35 (4H, m), 0.93–0.89 (6H, m). ^{13}C-NMR (100 MHz, $CDCl_3$): δ 161.5, 153.1, 150.0, 146.0, 143.3, 113.1, 111.2, 110.4, 100.9, 69.8, 69.2, 28.8, 28.4, 28.1, 28.0, 22.3, 22.3, 13.9, 13.9; HRMS [ESI]$^+$ calculated for $C_{19}H_{26}O_4$: 318.1831; found [M]$^+$ 318.1834.

6,7-Di-isopentoxycoumarin (**48**). Yield: 84.2%. White solid. ^{1}H-NMR (400 MHz, $CDCl_3$): δ 7.57 (1H, d, *J* = 9.6 Hz), 6.84 (1H, s), 6.79 (1H, s), 6.23(1H, d, *J* = 9.2 Hz), 4.07–3.99 (4H, m), 1.87–1.80 (2H, m), 1.76–1.55 (4H, m), 0.96–0.94 (12H, m). ^{13}C-NMR (100 MHz, $CDCl_3$): δ 161.5, 153.1, 150.0, 146.0, 143.3, 113.1, 111.2, 110.4, 100.9, 68.3, 67.7, 37.8, 37.4, 25.1, 22.5, 22.5; HRMS [ESI]$^+$ calculated for $C_{19}H_{26}O_4$: 318.1831; found [M]$^+$ 318.1832.

6,7-Dihexoxycoumarin (**49**). Yield: 51.0%. Yellow solid. ^{1}H-NMR (400 MHz, $CDCl_3$): δ 7.56 (1H, d, *J* = 9.5 Hz), 6.83 (1H, s), 6.78 (1H, s), 6.22 (1H, d, *J* = 9.5 Hz), 4.03–3.96 (4H, m), 1.86–1.77 (4H, m), 1.46–1.32 (12H, m), 0.89–0.87 (6H, m). ^{13}C-NMR (100 MHz, $CDCl_3$): δ 161.5, 153.1, 150.0, 146.0, 143.3, 113.0, 111.2, 110.4, 100.8, 69.8, 69.2, 31.4, 31.4, 29.8, 28.7, 25.6, 25.5, 22.5 (2C), 13.9 (2C); HRMS [ESI]$^+$ calculated for $C_{21}H_{31}O_4$ [M + H]$^+$347.2215; found [M]$^+$ 346.4605.

6,7-Digeranoxycoumarin (**51**). Yield: 55.2%. Yellow solid. ^{1}H-NMR (400MHz, $CDCl_3$): δ 7.54 (1H, d, *J* = 9.6 Hz), 6.82 (1H, s), 6.75 (1H, s), 6.19 (1H, d, *J* = 9.2 Hz), 5.46–5.40 (2H, m), 5.01 (2H, s), 4.46–4.51 (4H, m), 2.89 (2H, s), 2.82 (2H, s), 2.06–1.91 (8H, m), 1.73–1.53 (18H, m). ^{13}C-NMR (100 MHz, $CDCl_3$): δ = 161.5, 152.2, 150.0, 145.6, 143.3, 141.5, 141.2, 131.8, 131.8, 123.6, 123.5 119.3, 118.8, 113.1, 111.2, 110.8, 101.3, 66.6, 66.3, 39.4, 39.4, 26.2, 26.1, 25.6, 25.5, 17.4 (2C), 16.7, 16.6; HRMS [ESI]$^+$ calculated for $C_{29}H_{38}O_4$: 450.2770; found 450.2772.

8.4. Cell Lines

Human hepatocellular carcinoma cell line (Hep 3B) was purchased from the Bioresource Collection and Research Center (Hsinchu, Taiwan).

8.5. Cell Culture

Cells were maintained in minimum essential medium (MEM, Invitrogen) containing 10% fetal bovine serum (HyClone®, Tianjin, China), 100 units/mL penicillin, 100 μg/mL streptomycin, 2 mM L-glutamine (penicillin-streptomycin-glutamine 100X from GIBCO, Invitrogen) and 1 mM sodium pyruvate at 37 °C humidified incubator with 5% CO_2.

8.6. Cell Cytotoxicity Assay of Coumarin Derivatives

The colorimetric assay for cellular growth and survivals described by Hansen et al. with modifications [57]. The MTT (3-(4,5-cimethylthiazol-2yl)-2,5-diphenyltetrazolium bromide, Sigma-Aldrich Inc., Germany) assay was utilized to determine the IC_{50} value of each compound. Cells (5000 cells/well) were seeded into 96-well plates for 24 h and subsequently the cells were treated with vehicle or tested compounds at pre-determined concentration. After 72 h treatment, MTT solution was added to the final concentration of 0.5 mg/mL and continue cultured for 2 h at 37 °C. Afterward, the cells were lysed with lysis buffer (40% DMF and 20% SDS in H_2O) overnight at 37 °C to lysate the cells. The absorbance at 570 nm was then detected by a microplate reader and the IC_{50} value was calculated.

8.7. Reverse Transcription Polymerase Chain Reaction (RT-PCR)

Total RNA was isolated from control and treated cells using a TRIzol reagent (Invitrogen) according to the manufacturer's instruction. The RNA concentration was measured by a spectrophotometer and equal amounts of total RNA (2 μg) were reverse—transcribed using a RevertAid First Strand cDNA Synthesis Kit (K1622, Fermentas, Hanover, MD, USA). Amplification of cDNA was performed in the PCR reaction buffer (0.2 mM dNTP, 1.5 mM $MgCl_2$, and 0.5 μM of each primer) containing 2.5 units of *Taq* DNA polymerase. The PCR products were resolved by agarose electrophoresis, visualized by ethidium bromide straining and quantified by Image J®. The PCR cycling procedures and sequences of primers are listed below. PCR conditions: (1) UGT1A1: 94 °C, 30 s; 62 °C, 45 s; 72 °C, 30 s; 27 cycles; (2) MRP2: 94 °C, 30 s; 62 °C, 45 s; 72 °C, 30 s; 27 cycles; (3) CAR: 94 °C, 30 s; 59 °C, 45 s; 72 °C, 30 s; 35 cycles; (4) GADPH: 94 °C, 30 s; 62 °C, 45 s; 72 °C, 30 s; 27 cycles; (4) gtPBREM: 94 °C, 30 s; 63 °C, 30 s; 68 °C, 60 s; 30 cycles.

8.8. Primer Sequences

Primer	Sequence
UGT1A1	Forward: 5′-TGCAAAGCGCATGGAGACTA-3′ Reverse: 5′-GAGGCGCATGATGTTCTCCT-3′
MRP2	Forward: 5′-AGGTGAGGATTGACACCAACCA-3′ Reverse: 5′-AGGCAGTTTGTGAGGGATGACT-3′
CAR	Forward: 5′-GTGCTGCCTCTGGTCACACACT-3′ Reverse: 5′-GAGGCCCGCAGAGGAAGTTT-3′
GAPDH	Forward: 5′-GTCTCCTCTGACTTCAACAGCG-3′ Reverse: 5′-ACCACCCTGTTGCTGTAGCCAA-3′
gtPBREM	Forward: 5′-GTTTCCGCTAGCTACACTAGTAAAGGTCACTC-3′ Reverse: 5′-GTTTAACTCGAGCCCTCTAGCCATTCTGGATC-3′

8.9. Cloning and Transfection

The genomic DNA was isolated from Hep3B cells using Wizard® Genomic DNA Purification Kit (Promega, Fitchburg, WI, USA). The promoter region of *UGT1A1* gene, gtPBREM element, was amplified by AccuPrime™ Taq DNA Polymerase High Fidelity Kit (Invitrogen, MA, USA). The sequence of the amplified element was confirmed by DNA sequencing company and cloned into pGL4 luciferase reporter vector. The pGL4-gtPBREM plasmid was transfected into Hep3B cells. The stable pGL4-gtPBREM transfected Hep3B cells, Hep3B-gtPBREM cells, were established after G418 selection. The luciferase activity represented of CAR activation ability was measured using individual established clone. Further, CAR-shRNA was obtained from the Genomic Research Center and transfected into Hep3B-gtPBREM cells. The transient transfected shCAR-Hep3B-gtPBREM cells were used for further luciferase assays.

8.10. Luciferase Assays

The Hep3B-gtPBREM cells and shCAR-Hep3B-gtPBREM cells (1.5 × 10^5 cells/well) were seeded into 12-well plates for overnight then pre-determined concentration compounds were added for desired time. After treatment, transfected cells were washed twice in PBS, lysated with 1× passive lysis buffer and luciferase intensity of equal amount of cell lysate was measured by a Synergy HT Multi-Mode Microplate Reader (BioTek, Winooski, VT, USA).

8.11. Nuclear and Cytoplasmic Extraction

Cells were harvest by centrifugation after trypsin-EDTA treatment. The ice-cold CER I (10 μL/1 × 10^6 cells) was added to the cell pellet following the manufacturer's instruction. The tube was vigorously vortexed on the highest setting for 15 s to fully suspend the cell pellet, incubated on ice for 10 min, then ice-cold CER II was added to the tube and vortexed for 5 s on the highest setting. After incubating in ice for 1 min, tube was vortexed for 5 s on the highest setting then subjected to centrifuge for 5 min at 16,000× g. The supernatant (cytoplasmic fraction) was transfected to a clean pre-chilled tube and the nuclear fraction (pellet) was re-suspended in ice-cold NER. After 40 min incubation on ice, the nuclear fraction was subjected to centrifuge for 10 min at 16,000× g and the supernatant was transferred to a pre-chilled tube for further investigation.

8.12. Protein Extraction

1 × 10^6 cells were seeded in a 10 cm plate, incubated at 37 °C for overnight and treated with various concentrations of compounds for the indicated times. After incubation, the culture medium was removed and washed twice with 2 mL ice-cold 1× PBS. Three hundred μL of protein lysis buffer (150 mM NaCl, 50 mM Tris-HCl (pH 7.4), 1% SDS, 1% Triton X-100, 1 mM PMSF) was added into each plate, scraped with a rubber policeman, collected into a premarked microtube and the incubated in ice for 1 h, sonicated by Ultrasonic Cell Disruptor for three time at 5 s each. The cell lysated was cleared by centrifugation at 12,000× g for 15 min at 4 °C. Supernatants were transferred into a fresh marked microtube and stored at −20 °C for future use.

8.13. Western Blots

Protein concentration of each sample was determined using a BCA Protein Assay Kit (Pirece®, Thermo Scientific). Different percentages of SDS-PAGE gels were prepared according to the molecular size of proteins of interest. After complete polymerization of stacking gels, each sample with equal amount of total protein was separated in polyacrylamide gels (stacking gel: 50 v, 60 min; separating gel: 100 v, 60 min). After separating, wet-transfer method was utilized to transfer proteins onto the PVDF membrane. The gels were assembled in the transfer sandwich and put in a transfer tank filled with transfer buffer (25 mM glycine, 0.15% ethanolamine, 20% methanol). The electro-transfer was performed at the voltage of 100 volt for 80 min at 4 °C. After transfer, PVDF-membrane was blocked with 5% BSA in 1× TBS-Tween at room temperature for 1 h. Membrane was washed with 10 mL 1× TBS-Tween three times and incubated with primary antibodies overnight at 4 °C with gentle agitation. After primary antibody was removed, the membrane was washed three times with 10 mL 1× TBS-Tween for 10 min at room temperature with constant shaking. The horseradish peroxide (HRP)-conjugated secondary antibody was added to react with the membrane for 1 h at room temperature with gentle agitation. After incubation, membrane was washed with 1× TBS-Tween 3 times for 10 min. For detection, the membrane was immersed completely in the mixture of 0.5 mL luminosol reagent and 0.5 mL peroxidase for 1 min. The emitted fluorescence from the membrane was caught on a LAS-4000 biomolecule imager (FUJI.).

8.14. Immunofluorescence

Hep3B cells (4×10^5) were grown on coverslips for overnight then washed twice with 1× PBS and fixed with 4% paraformaldehyde in 1× PBS for 30 min at room temperature. Cells were then permeabilized with 1× PBS containing 0.1% (*v/v*) Triton X-100 for 10–15 min and blocked in 1× PBS containing 5% BSA for 1 h at room temperature. CAR antibody (1:200 dilutions) was added and reacted at room temperature for 2 h in 1× PBS containing 5% BSA. FITC-conjugated goat anti-rabbit IgG antibody was added for another 1 h at room temperature. Coverslips were mounted and images were acquired under a fluorescence microscope.

8.15. Animals

Six-week old male C57/BL6 mice (n = 6) were acquired from the Animal Center of the College of Medicine, National Taiwan University (Taipei, Taiwan) and 6-week-old male *db/db* mice (BKS.Cg-Dock7m +/+ Leprdb/JNarl) (n = 12) were acquired from the National Laboratory Animal Center (Tainan, Taiwan). Experimental animals were allowed to acclimate to the controlled photoperiod (a cycle of 12-h light/12-h dark), humidity (40–60% relative humidity) and temperature (22 ± 2 °C) with ad libitum supply of standard chow diets and drinking water for one week prior to the experimental treatment in accordance with the Guide for the Care and Use of Laboratory Animals (Institute of Laboratory Animal Resources Commission on Life Sciences, 2011) [58]. Acclimated animals were randomly separated into three cohorts: control group (C57/BL6 mice treated with vehicle for two weeks, n = 6), db/db group (*db/db* mice treated with vehicle for two weeks, n = 6), and db/db+SP group (*db/db* mice treated with 100 mg/kg scoparone for two weeks, n = 6). (IACUC Approval NO.: 20140532; Duration: 2015.01.01–2015.12.31; Ethical Committee: National Taiwan University College of Medicine and College of Public Health Institutional Animal Care and Use Committee (IACUC).)

8.16. Oral Glucose Tolerance Test (OGTT)

The control mice or diabetic mice with or without drugs treatment for 2 weeks received an oral glucose challenge (2 g/kg). Mice were under slight anesthesia by an intraperitoneal injection of pentobarbital (50 mg/kg) and blood samples (20–50 µL via orbital sinus at each time point) were collected following: 0, 15, 30, 60, 90, and 120 min after delivery of the glucose load. Blood glucose levels were determined using SURESTEP blood glucose meter (Lifescan). Glucose tolerance was determined and performed as the curve (AUC) and delta AUC (ΔAUC) using Prism 5 software (GraphPad).

8.17. Determination of Blood Insulin and Fructosamine

To determine the amount of insulin and fructosamine after fasting overnight, blood samples (0.5–0.8 mL) of euthanasia animals were collected. After centrifugation, the serum was analyzed by insulin and fructosamine immunoassay kits according to the respective instructions of the manufacturer (Mercodia AB Inc., Uppsala, Sweden; Hospitex Diagnostics Lp, League City, TX, USA). The homeostasis model assessment of insulin resistance (HOMA-IR) index is calculated with the formula of the values of fasting glucose and insulin divided by 22.5 as previously described [59].

8.18. Statistical Analysis

The values in the graphs are given as mean ± S.E.M. The significance of difference was evaluated by the paired Student's t-test. When more than one group was compared with one control, significance was evaluated according to one-way analysis of variance (ANOVA). Probability values of <0.05 were considered to be significant.

Supplementary Materials: The online version of this article contains Supplementary Materials.

Author Contributions: H.-Y.H., S.-H.J. suggested the work protocol, interpreted the results, prepared and revised the manuscript; M.-T.H. and F.-P.L. curative data; P.-L.H., J.-L.C., and C.-Y.C. performed the laboratory experiments; H.-Y.H., S.-H.J. and S.-H.L. for funding acquisition, resources and data analysis. H.-K.L. performed PEPCK assay; C.-H.C. provided support for PK study; S.-C.K. and T.-S.W. provided natural coumarins and interpretated data. All authors have read and agreed to the published version of the manuscript.

Funding: This research was funded by Ministry of Science and Technology, Taiwan, granted to H.-Y. Hung. And the APC was funded by Hsin-Yi Hung.

Data Availability Statement: The data presented in this study are available on request from the corresponding author.

Acknowledgments: This work was supported by Ministry of Science and Technology Taiwan granted to H.-Y. Hung.

Conflicts of Interest: The authors declare no conflict of interest.

References

1. Banerjee, M.; Robbins, D.; Chen, T. Targeting xenobiotic receptors PXR and CAR in human diseases. *Drug Discov. Today* **2015**, *20*, 618–628. [CrossRef]
2. Trauner, M.; Wagner, M.; Fickert, P.; Zollner, G. Molecular regulation of hepatobiliary transport systems: Clinical implications for understanding and treating cholestasis. *J. Clin. Gastroenterol.* **2005**, *39*, S111–S124. [CrossRef]
3. Kachaylo, E.M.; Pustylnyak, V.O.; Lyakhovich, V.V.; Gulyaeva, L.F. Constitutive androstane receptor (CAR) is a xenosensor and target for therapy. *Biochemistry* **2011**, *76*, 1087–1097. [CrossRef]
4. Goodwin, B.; Hodgson, E.; D'Costa, D.J.; Robertson, G.R.; Liddle, C. Transcriptional regulation of the human CYP3A4 gene by the constitutive androstane receptor. *Mol. Pharmacol.* **2002**, *62*, 359–365. [CrossRef]
5. Echchgadda, I.; Song, C.S.; Oh, T.; Ahmed, M.; De La Cruz, I.J.; Chatterjee, B. The xenobiotic-sensing nuclear receptors pregnane X receptor, constitutive androstane receptor, and orphan nuclear receptor hepatocyte nuclear factor 4alpha in the regulation of human steroid-/bile acid-sulfotransferase. *Mol. Endocrinol.* **2007**, *21*, 2099–2111. [CrossRef]
6. Wagner, M.; Halilbasic, E.; Marschall, H.U.; Zollner, G.; Fickert, P.; Langner, C.; Zatloukal, K.; Denk, H.; Trauner, M. CAR and PXR agonists stimulate hepatic bile acid and bilirubin detoxification and elimination pathways in mice. *Hepatology* **2005**, *42*, 420–430. [CrossRef] [PubMed]
7. Zhang, J.; Huang, W.; Qatanani, M.; Evans, R.M.; Moore, D.D. The constitutive androstane receptor and pregnane X receptor function coordinately to prevent bile acid-induced hepatotoxicity. *J. Biol. Chem.* **2004**, *279*, 49517–49522. [CrossRef] [PubMed]
8. Gao, J.; He, J.; Zhai, Y.; Wada, T.; Xie, W. The constitutive androstane receptor is an anti-obesity nuclear receptor that improves insulin sensitivity. *J. Biol. Chem.* **2009**, *284*, 25984–25992. [CrossRef] [PubMed]
9. Dong, B.; Saha, P.K.; Huang, W.; Chen, W.; Abu-Elheiga, L.A.; Wakil, S.J.; Stevens, R.D.; Ilkayeva, O.; Newgard, C.B.; Chan, L.; et al. Activation of nuclear receptor CAR ameliorates diabetes and fatty liver disease. *Proc. Natl. Acad. Sci. USA* **2009**, *106*, 18831–18836. [CrossRef]
10. Chen, F.; Coslo, D.M.; Chen, T.; Zhang, L.; Tian, Y.; Smith, P.B.; Patterson, A.D.; Omiecinski, C.J. Metabolomic Approaches Reveal the Role of CAR in Energy Metabolism. *J. Proteome Res.* **2019**, *18*, 239–251. [CrossRef]
11. Tzameli, I.; Pissios, P.; Schuetz, E.G.; Moore, D.D. The xenobiotic compound 1,4-bis [2-(3,5-dichloropyridyloxy)] benzene is an agonist ligand for the nuclear receptor CAR. *Mol. Cell. Biol.* **2000**, *20*, 2951–2958. [CrossRef] [PubMed]
12. Huang, W.; Zhang, J.; Moore, D.D. A traditional herbal medicine enhances bilirubin clearance by activating the nuclear receptor CAR. *J. Clin. Investig.* **2004**, *113*, 137–143. [CrossRef] [PubMed]
13. Elferink, R.O. Yin Zhi Huang and other plant-derived preparations: Where herbal and molecular medicine meet. *J. Hepatol.* **2004**, *41*, 691–693. [CrossRef]
14. Chen, Z.; Ma, X.; Zhao, Y.; Wang, J.; Zhang, Y.; Li, J.; Wang, R.; Zhu, Y.; Wang, L.; Xiao, X. Yinchenhao decoction in the treatment of cholestasis: A systematic review and meta-analysis. *J. Ethnopharm.* **2015**, *168*, 208–216. [CrossRef] [PubMed]
15. Cai, Y.; Zhang, Q.; Sun, R.; Wu, J.; Li, X.; Liu, R. Recent progress in the study of *Artemisiae Scopariae* Herba (Yin Chen), a promising medicinal herb for liver diseases. *Biomed. Pharmacother.* **2020**, *130*, 11053. [CrossRef]
16. Yang, D.; Yang, J.; Shi, D.; Deng, R.; Yan, B. Scoparone potentiates transactivation of the bile salt export pump gene and this effect is enhanced by cytochrome P450 metabolism but abolished by a PKC inhibitor. *Br. J. Pharmacol.* **2011**, *164*, 154–157. [CrossRef]
17. Wu, T.S.; Tsang, Z.J.; Wu, P.L.; Lin, F.W.; Li, C.Y.; Teng, C.M.; Lee, K.H. New constituents and antiplatelet aggregation and anti-HIV principles of *Artemisia capillaris*. *Bioorg. Med. Chem.* **2001**, *9*, 77–83. [CrossRef]
18. Perkin, W.H. On the artificial production of coumarin and formation of its homologues. *J. Chem. Soc.* **1868**, *21*, 53–63. [CrossRef]
19. Bogdal, D. Coumarins: Fast synthesis by Knoevenagel condensation under microwave irradiation. *J. Chem. Res.* **1998**, *8*, 468–469. [CrossRef]
20. Joule, J.A.; Mills, K. *Heterocyclic Chemistry*; Wiley: Hoboken, NJ, USA, 2013.
21. Desai, V.G.; Shet, J.B.; Tilve, S.G.; Mali, R.S. Intramolecular Wittig reactions. A new synthesis of coumarins and 2-quinolones. *J. Chem. Res.* **2003**, *2003*, 628–629. [CrossRef]

22. Boeck, F.; Blazejak, M.; Anneser, M.R.; Hintermann, L. Cyclization of ortho-hydroxycinnamates to coumarins under mild conditions: A nucleophilic organocatalysis approach. *Beilstein J. Org. Chem.* **2012**, *8*, 1630–1636. [CrossRef] [PubMed]
23. Rao, H.S.P.; Sivakumar, S. Condensation of α-Aroylketene dithioacetals and 2-hydroxyarylaldehydes results in facile synthesis of a combinatorial library of 3-aroylcoumarins. *J. Org. Chem.* **2006**, *71*, 8715–8723. [CrossRef] [PubMed]
24. Oyamada, J.; Kitamura, T. Synthesis of coumarins by Pt-catalyzed hydroarylation of propiolic acids with phenols. *Tetrahedron* **2006**, *62*, 6918–6925. [CrossRef]
25. Patnam, R.; Chang, F.-R.; Chen, C.-Y.; Kuo, R.-Y.; Lee, Y.-H.; Wu, Y.-C. Hydrachine A, a novel alkaloid from the roots of Hydrangea chinensis. *J. Nat. Prod.* **2001**, *64*, 948–949. [CrossRef]
26. Borba, I.C.G.; Bridi, H.; Soares, K.D.; Apel, M.A.; Poser, G.L.; Ritter, M.R. New natural coumarins from *Trichocline macrocephala* (Asteraceae). *Phytochem. Lett.* **2019**, *32*, 129–133. [CrossRef]
27. Bratulescu, G. A Quick and advantageous synthesis of 2*H*-1-benzopyran-2-ones unsubstituted on the pyranic nucleus. *Synthesis* **2008**, *2008*, 2871–2873. [CrossRef]
28. Kaighen, M.; Williams, R.T. The metabolism of [3-^{14}C] coumarin. *J. Med. Pharm. Chem.* **1961**, *3*, 25–43. [CrossRef]
29. Takaishi, K.; Izumi, M.; Baba, N.; Kawazu, K.; Nakajima, S. Synthesis and biological evaluation of alkoxycoumarins as novel nematicidal constituents. *Bioorg. Med. Chem. Lett.* **2008**, *18*, 5614–5617. [CrossRef]
30. Iranshahi, M.; Jabbari, A.; Orafaie, A.; Mehri, R.; Zeraatkar, S.; Ahmadi, T.; Alimardani, M.; Sadeghian, H. Synthesis and SAR studies of mono *O*-prenylated coumarins as potent 15-lipoxygenase inhibitors. *Eur. J. Med. Chem.* **2012**, *57*, 134–142. [CrossRef]
31. Farooq, S.; Shakeel-u-Rehman Dangroo, N.A.; Priya, D.; Banday, J.A.; Sangwan, P.L.; Qurishi, M.A.; Koul, S.; Saxena, A.K. Isolation, cytotoxicity evaluation and HPLC-quantification of the chemical constituents from *Prangos pabularia*. *PLoS ONE* **2014**, *9*, e108713. [CrossRef]
32. Ding, C.; Zhang, W.; Li, J.; Lei, J.; Yu, J. Cytotoxic constituents of ethyl acetate fraction from *Dianthus superbus*. *Nat. Prod. Res.* **2013**, *27*, 1691–1694. [CrossRef] [PubMed]
33. Garro, H.A.; Reta, G.F.; Donadel, O.J.; Pungitore, C.R. Cytotoxic and antitumor activity of some coumarin derivatives. *Nat. Prod. Commun.* **2016**, *11*, 1289–1292. [CrossRef] [PubMed]
34. Chattopadhyay, S.K.; Mondal, P.; Ghosh, D. A new route to pyranocoumarins and their benzannulated derivatives. *Synthesis* **2014**, *46*, 3331–3340. [CrossRef]
35. Adfa, M.; Itoh, T.; Hattori, Y.; Koketsu, M. Inhibitory effects of 6-alkoxycoumarin and 7-alkoxycoumarin derivatives on lipopolysaccharide/interferon γ-stimulated nitric oxide production in RAW264 cells. *Biol. Pharm. Bull.* **2012**, *35*, 963–966. [CrossRef]
36. Sánchez-Recillas, A.; Navarrete-Vázquez, G.; Hidalgo-Figueroa, S.; Rios, M.Y.; Ibarra-Barajas, M.; Estrada-Soto, S. Semisynthesis, ex vivo evaluation, and SAR studies of coumarin derivatives as potential antiasthmatic drugs. *Eur. J. Med. Chem.* **2014**, *77*, 400–408. [CrossRef]
37. Liu, G.-L.; Hao, B.; Liu, S.-P.; Wang, G.-X. Synthesis and anthelmintic activity of osthol analogs against Dactylogyrus intermedius in goldfish. *Eur. J. Med. Chem.* **2012**, *54*, 582–590. [CrossRef]
38. Huang, W.-J.; Wang, Y.-C.; Chao, S.-W.; Yang, C.-Y.; Chen, L.-C.; Lin, M.-H.; Hou, W.-C.; Chen, M.-Y.; Lee, T.-L.; Yang, P.; et al. Synthesis and biological evaluation of *ortho*-aryl *N*-hydroxycinnamides as potent histone deacetylase (HDAC) 8 isoform-selective inhibitors. *Chem. Med. Chem.* **2012**, *7*, 1815–1824. [CrossRef]
39. Riveiro, M.E.; Maes, D.; Vazquez, R.; Vermeulen, M.; Mangelinckx, S.; Jacobs, J.; Debenedetti, S.; Shayo, C.; De Kimpe, N.; Davio, C. Toward establishing structure-activity relationships for oxygenated coumarins as differentiation inducers of promonocytic leukemic cells. *Bioorg. Med. Chem.* **2009**, *17*, 6547–6559. [CrossRef]
40. Zhu, L.J.; Hou, Y.L.; Shen, X.Y.; Pan, X.D.; Zhang, X.; Yao, X.S. Monoterpene pyridine alkaloids and phenolics from *Scrophularia ningpoensis* and their cardioprotective effect. *Fitoterapia* **2013**, *88*, 44–49. [CrossRef]
41. Nagarajan, G.R.; Rani, U.; Parmar, V.S. Coumarins from *Fraxinus floribunda* leaves. *Phytochemistry* **1980**, *19*, 2494–2495. [CrossRef]
42. Litinas, K.E.; Mangos, A.; Nikkou, T.E.; Hadjipavlou-Litina, D.J. Synthesis and biological evaluation of fused oxepinocoumarins as free radicals scavengers. *J. Enzyme Inhib. Med. Chem.* **2011**, *26*, 805–812. [CrossRef]
43. Sellers, E.M.; Tyndale, R.F. NICOGEN INC., CYP2a Enzymes and Their Use in Therapeutic and Diagnostic Methods. WO1999027919A2, 6 October 1999.
44. Garg, S.S.; Gupta, J.; Sharma, S.; Sahu, D. An insight into the therapeutic applications of coumarin compounds and their mechanisms of action. *Eur. J. Pharm. Sci.* **2020**, *152*, 105424. [CrossRef] [PubMed]
45. Maqua, M.P.; Vines, A.C.G.; Caballero, E.; Grande, M.C.; Medarde, M.; Bellido, I.S. Components from *Santolina rosmarinifolia*, subspecies *rosmarinifolia* and *canescens*. *Phytochemistry* **1988**, *27*, 3664–3667. [CrossRef]
46. Bohlmann, F.; Suwita, A.; King, R.M.; Robinson, H. Neue ent-labdan-derivate aus *Austroeupatorium chaparense*. *Phytochemistry* **1980**, *19*, 111–114. [CrossRef]
47. Guillaumet, G.; Hretani, M.; Coudert, G.; Averbeck, D.; Averbeck, S. Synthesis and photoinduced biological properties of linear dioxinocoumarins. *Eur. J. Med. Chem.* **1990**, *25*, 45–51. [CrossRef]
48. Ballantyne, M.M.; McCabe, P.H.; Murray, R.D.H. Claisen rearrangements—II: Synthesis of six natural coumarins. *Tetrahedron* **1971**, *27*, 871–877. [CrossRef]
49. Genovese, S.; Epifano, F.; Medina, P.; Caron, N.; Rives, A.; Poirot, M.; Silvente-Poirot, S.; Fiorito, S. Natural and semisynthetic oxyprenylated aromatic compounds as stimulators or inhibitors of melanogenesis. *Bioorg. Chem.* **2019**, *87*, 181–190. [CrossRef] [PubMed]

50. Orhan, I.E.; Deniz, F.S.S.; Salmas, R.E.; Durdagi, S.; Epifano, F.; Genovese, S.; Fiorito, S. Combined molecular modeling and cholinesterase inhibition studies on some natural and semisynthetic *O*-alkylcoumarin derivatives. *Bioorg. Chem.* **2019**, *84*, 355–362. [CrossRef]
51. Blagbrough, I.S.; Bayoumi, S.A.; Rowan, M.G.; Beeching, J.R. Cassava: An appraisal of its phytochemistry and its biotechnological prospects. *Phytochemistry* **2010**, *71*, 1940–1951. [CrossRef]
52. Medeiros-Neves, B.; Teixeira, H.F.; Poser, G.L. The genus Pterocaulon (Asteraceae)—A review on traditional medicinal uses, chemical constituents and biological properties. *J. Ethnopharm.* **2018**, *224*, 451–464. [CrossRef]
53. Kim, K.S.; Lee, S.; Shin, J.S.; Shim, S.H.; Kim, B. Arteminin, a new coumarin from *Artemisia apiacea*. *Fitoterapia* **2002**, *73*, 266–268. [CrossRef]
54. Maglich, J.M.; Parks, D.J.; Moore, L.B.; Collins, J.L.; Goodwin, B.; Billin, A.N.; Stoltz, C.A.; Kliewer, S.A.; Lambert, M.H.; Willson, T.M.; et al. Identification of a novel human constitutive androstane receptor (CAR) agonist and its use in the identification of CAR target genes. *J. Biol. Chem.* **2003**, *278*, 17277–17283. [CrossRef] [PubMed]
55. Sugatani, J. Function, genetic polymorphism, and transcriptional regulation of human UDP-glucuronosyltransferase (UGT) 1A1. *Drug Metab. Pharmacokinet.* **2013**, *28*, 83–92. [CrossRef] [PubMed]
56. Mutoh, S.; Sobhany, M.; Moore, R.; Perera, L.; Pedersen, L.; Sueyoshi, T.; Negishi, M. Phenobarbital indirectly activates the constitutive active androstane receptor (CAR) by inhibition of epidermal growth factor receptor signaling. *Sci. Signal.* **2013**, *6*, ra31. [CrossRef] [PubMed]
57. Hansen, M.B.; Nielsen, S.E.; Berg, K. Re-examination and further development of a precise and rapid dye method for measuring cell growth/cell kill. *J. Immunol. Methods* **1989**, *119*, 203–210. [CrossRef]
58. Institute of Laboratory Animal Resources Commission on Life Sciences. National Research Council. In *Guide for the Care and Use of Laboratory Animals*, 8th ed.; National Academy Press: Washington, DC, USA, 2011.
59. Wallace, T.M.; Levy, J.C.; Matthews, D.R. Use and abuse of HOMA modeling. *Diabetes Care* **2004**, *27*, 1487–1495. [CrossRef]

MDPI

Article

Synthesis and Biochemical Evaluation of Warhead-Decorated Psoralens as (Immuno)Proteasome Inhibitors

Eva Shannon Schiffrer [1], Matic Proj [1], Martina Gobec [1], Luka Rejc [2], Andrej Šterman [1], Janez Mravljak [1], Stanislav Gobec [1] and Izidor Sosič [1,*]

[1] Faculty of Pharmacy, University of Ljubljana, Aškerčeva cesta 7, SI-1000 Ljubljana, Slovenia; eva.shannon.schiffrer@ffa.uni-lj.si (E.S.S.); matic.proj@ffa.uni-lj.si (M.P.); martina.gobec@ffa.uni-lj.si (M.G.); andrej.sterman@ffa.uni-lj.si (A.Š.); janez.mravljak@ffa.uni-lj.si (J.M.); stanislav.gobec@ffa.uni-lj.si (S.G.)

[2] Faculty of Chemistry and Chemical Technology, University of Ljubljana, Večna pot 113, 1000 Ljubljana, Slovenia; luka.rejc@fkkt.uni-lj.si

* Correspondence: izidor.sosic@ffa.uni-lj.si; Tel.: +386-1-4769-569

Abstract: The immunoproteasome is a multicatalytic protease that is predominantly expressed in cells of hematopoietic origin. Its elevated expression has been associated with autoimmune diseases, various types of cancer, and inflammatory diseases. Selective inhibition of its catalytic activities is therefore a viable approach for the treatment of these diseases. However, the development of immunoproteasome-selective inhibitors with non-peptidic scaffolds remains a challenging task. We previously reported 7*H*-furo[3,2-*g*]chromen-7-one (psoralen)-based compounds with an oxathiazolone warhead as selective inhibitors of the chymotrypsin-like (β5i) subunit of immunoproteasome. Here, we describe the influence of the electrophilic warhead variations at position 3 of the psoralen core on the inhibitory potencies. Despite mapping the chemical space with different warheads, all compounds showed decreased inhibition of the β5i subunit of immunoproteasome in comparison to the parent oxathiazolone-based compound. Although suboptimal, these results provide crucial information about structure–activity relationships that will serve as guidance for the further design of (immuno)proteasome inhibitors.

Keywords: immunoproteasome; psoralen core; non-peptidic; electrophilic compounds; warhead scan

Citation: Schiffrer, E.S.; Proj, M.; Gobec, M.; Rejc, L.; Šterman, A.; Mravljak, J.; Gobec, S.; Sosič, I. Synthesis and Biochemical Evaluation of Warhead-Decorated Psoralens as (Immuno)Proteasome Inhibitors. *Molecules* **2021**, *26*, 356. https://doi.org/10.3390/molecules26020356

Academic Editor: Maria João Matos
Received: 13 December 2020
Accepted: 9 January 2021
Published: 12 January 2021

1. Introduction

In mammals, most intracellular proteins are destined for degradation, which involves the proteasome, a multiprotease complex [1–3]. The 26S proteasome represents the heart of the ubiquitin-proteasome system that is responsible for the maintenance of protein homeostasis and the regulation of various cellular processes [4–6]. It is a nucleophilic hydrolase with *N*-terminal Thr1 acting as a nucleophile to cleave the peptide bond of proteins [7]. The 26S proteasome is comprised of a 20S core particle (CP) and 19S regulatory units. The 20S core is a 720 kDa large barrel-shaped structure assembled of four stacked rings, each consisting of seven subunits. The two outer α rings provide structural integrity and act like "gates" allowing the entry of unfolded proteins to the two inner β rings, which contain three catalytically active subunits responsible for proteolysis of substrates [8]. Subunit β1 shows caspase-like activity, subunit β2 trypsin-like activity, whereas subunit β5 exhibits chymotrypsin-like activity [9,10]. There are three individual CP types: the constitutive proteasome (cCP), which is expressed in all eukaryotic cells, the thymoproteasome (tCP) [11], which is exclusive to cortical thymic epithelial cells, and the immunoproteasome (iCP) [12], which is expressed in cells of hematopoietic origin, but can also be induced in other tissues. Namely, the induction of iCP in other cell types is possible during acute immune and inflammatory responses [13–15]. Exposure to inflammatory factors, such as tumor necrosis factor α and interferon-γ causes the expression of the iCP active subunits β (designated as β1i, β2i, β5i), which replace their constitutive counterparts [12,16].

Increased expression of cCP and iCP can lead to a number of diseases. These include many types of cancer, infections, inflammatory and autoimmune diseases (Crohn's disease, ulcerative colitis, hepatitis, and rheumatoid arthritis), as well as neurological disorders [17–23]. The cCP and the iCP therefore represent validated targets for the design of new pharmacologically active compounds [24–27]. The druggability of both CPs is clearly represented by the clinically used covalent inhibitors bortezomib, carfilzomib, and ixazomib, which are used for the treatment of multiple myeloma and mantle-cell-lymphoma [27]. Selective inhibition of the iCP's β5i [28] subunit or simultaneously acting on β1i and β5i catalytic activities [29,30] are both approaches that are being investigated in the treatment of autoimmune and inflammatory diseases. In addition, such strategy should cause fewer adverse effects, as the expression of iCP is induced during the course of disease processes [31,32]. By avoiding cCP inhibition, the protein degradation would thus not be inhibited in most eukaryotic cells.

The most advanced iCP inhibitors that are frequently utilized in functional studies of iCP inhibition are represented in Figure 1. Please note that only a selected number of derivatives is depicted; namely, the most studied β5i-selective inhibitor PR-957 [28], β1i and β5i dual inhibitors KZR-616 [29] and 'compound 22′ [33], as well as the most selective β5i inhibitor DPLG-3 [34]. Structurally, these compounds all possess a peptidic backbone. Moreover, the former three compounds are all endowed with an electrophilic warhead, which reacts with the catalytic Thr1 of the proteasome subunits to form a covalent bond and to confer improved inhibition [24].

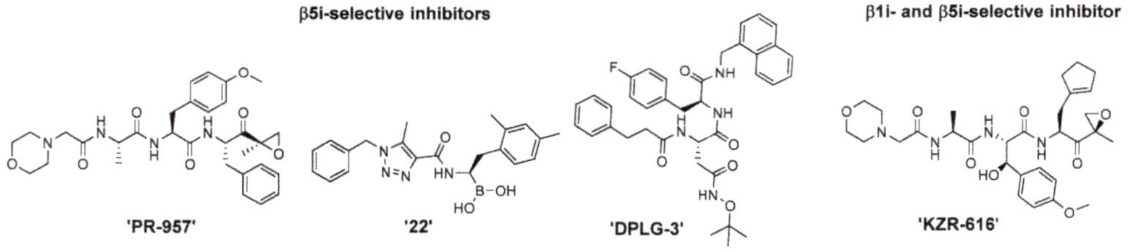

β5i-selective inhibitors

β1i- and β5i-selective inhibitor

'PR-957' '22' 'DPLG-3' 'KZR-616'

Figure 1. Structures of the most studied iCP-selective peptidic inhibitors. For a more thorough overview on subunit-selective iCP inhibitors, the reader is referred to recent reviews [32,35].

Because peptidic compounds, such as bortezomib and carfilzomib, are prone to poor metabolic stability and low bioavailability due to the unfavorable physico-chemical characteristics [36–38], there is a need to develop inhibitors with non-peptidic scaffolds. Despite being significantly less represented, there were some recent reports on non-peptidic inhibitors of the iCP (mostly inhibiting the β5i subunit) and the representative compounds are shown in Figure 2 [39–44]. As with peptidic compounds, irreversible inhibitors of non-peptidic nature can be obtained through structure-guided optimization, whereby an electrophilic warhead is properly positioned onto the structure of the non-covalently binding scaffold [45]. An essential prerequisite for this strategy to work is that the position of the electrophilic moiety allows the formation of the covalent bond between the inhibitor and the catalytic Thr1.

Recently, we discovered non-peptidic and β5i-selective inhibitors with a central psoralen core [39]. The most potent non-covalent inhibitor obtained during structure-activity relationship (SAR) studies possessed a phenyl substituent at position 4′ (see Figure 3 for psoralen atom numbering). This compound was also transformed into two potent irreversible covalent inhibitors by adding electrophilic warheads at position 3, i.e., succinimidyl ester and oxathiazolone. Of these two compounds, the oxathiazolone-based inhibitor showed the most promising inhibitory characteristics (Figure 2, 'compound 42′) as it was a potent and selective iCP inhibitor [39]. It was demonstrated previously that oxathiazolones inhibit iCP via cyclocarbonylation of the β-OH and α-NH_2 of the active site Thr1 [41]. Nevertheless, this structural fragment is deemed hydrolytically unstable making it less suitable for further development [41]. This fact prompted us to investigate other possible warheads that could be attached at the same position of the psoralen core. Previously, we already determined that acrylamides and nitrile-based warheads led to worse inhibition of the iCP [39]. However, to further map the warhead chemical space attached onto the psoralen core, we prepared a new focused set of compounds with different electrophilic fragments attached at position 3 (Figure 3), and evaluated their influence on the inhibition of all six catalytic subunits of both CPs. The selection of warheads in this study was based both on previously well described Thr targeting warheads (e.g., vinyl sulfones, α',β'-epoxyketones) [24], as well as on biologically less represented electrophilic moieties. In addition, to minimize the influence of non-covalently binding portion of the molecule on overall inhibitory potency, we used the same core compound with a phenyl substituent at position 4′.

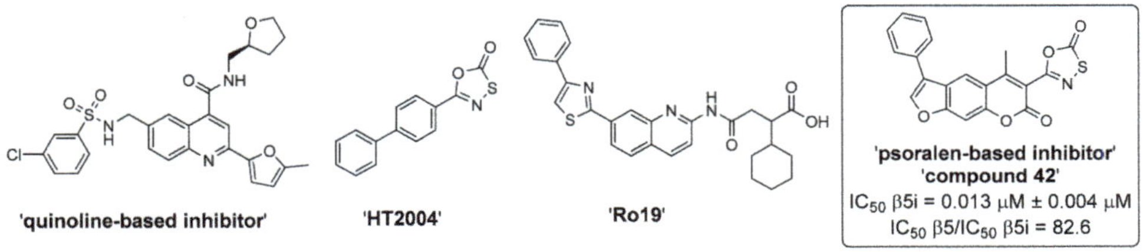

Figure 2. A selection of non-peptidic iCP inhibitors. 'Compound 42' [39] was the most selective irreversible β5i subunit inhibitor from the initial series of psoralen-based inhibitors. It represents the parent compound for studies in this manuscript.

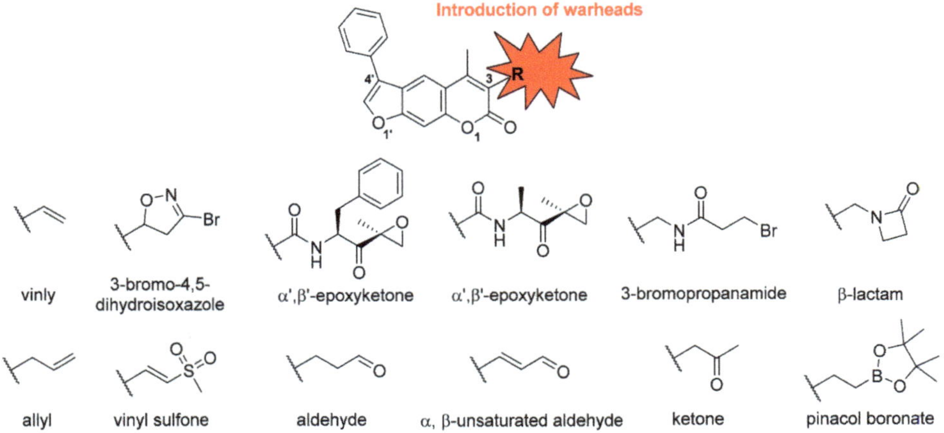

Figure 3. Schematic representation of the work described in this study. The numbering system for the psoralen ring is shown for clarity, as well as general nomenclature for the warhead moieties used.

2. Results and Discussion

2.1. *Syntheses of 3-Substituted Psoralens*

To prepare 3-allyl-substituted psoralen, ethyl acetoacetate was used as a starting material (Scheme 1). It was first alkylated using NaH as a base to obtain compound **1**, which was subjected to Pechmann reaction conditions to yield 7-hydroxycoumarin derivative **2**. After OH group alkylation with 2-bromoacetophenone, the final allyl-substituted compound **4** was obtained by base-catalyzed condensation of the coumarin derivative **3** into psoralen ring (Scheme 1). A compound with 3-vinyl-based warhead attached at position 3 (compound **7**) was obtained via a similar route. The crucial intermediate

7-hydroxy-4-methyl-4-vinyl-2*H*-chromen-2-one (**5**) was obtained in high yield by heating resorcinol derivative and crotonyl chloride at 60 °C in acetone. This was followed by a 2-bromoacetophenone-mediated alkylation and cyclization into psoralen yielding compounds **6** and **7**, respectively.

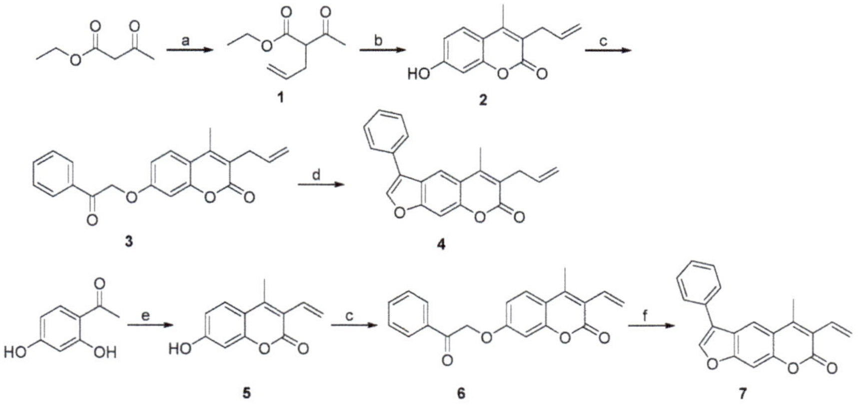

Scheme 1. Synthesis of compounds with allyl (**4**) and vinyl (**7**) warheads attached at position 3 of the psoralen ring. Reagents and conditions: (a) allyl bromide, NaH (60%), THF, 0 °C to rt, overnight; (b) resorcinol, 98% H_2SO_4, dioxane, 0 °C to rt, overnight; (c) 2-bromoacetophenone, K_2CO_3, KI, dioxane, 100 °C, 24 h; (d) 1 M NaOH, propan-2-ol, 80 °C, 40 min; (e) crotonyl chloride, K_2CO_3, acetone, 60 °C, 24 h; (f) 1 M KOH, EtOH, 85 °C, 2 h.

Compounds **4** and **7** were further used as synthons to prepare derivatives with other electrophilic moieties at position 3 (Scheme 2). The former was used in a Wacker-type oxidation of the terminal olefin by the combination of $Pd(OAc)_2$ and Dess-Martin periodinane to prepare the compound with ethyl methyl ketone moiety, i.e., compound **8**. The vinyl-substituted derivative **7** was a starting point for three different warhead-decorated psoralens, namely vinyl sulfone **9** (via NH_4I-induced sulfonylation of vinyl at position 3 with DMSO), 3-bromo-4,5-dihydroisoxazole **10** [46] (via cycloaddition of the alkene with 1,1,-dibromoformaldoxime), and pinacolate ester **11** (via transition-metal-free synthesis of alkylboronate from vinyl and bis(pinacolato)diboron) (Scheme 2).

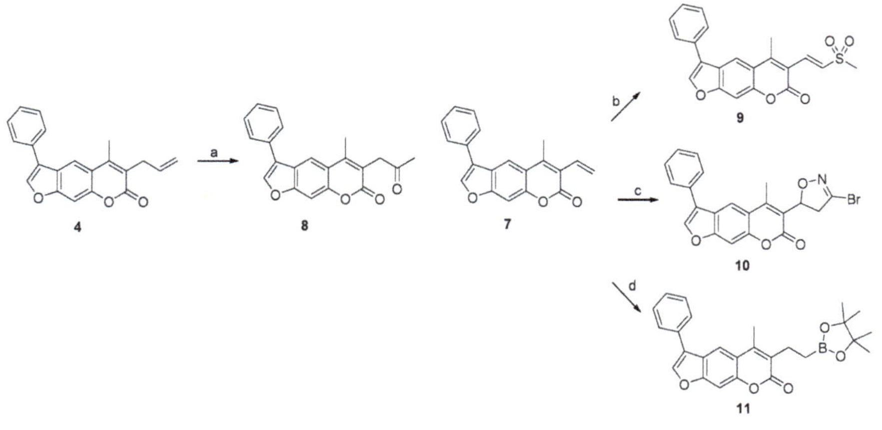

Scheme 2. Synthesis of compounds **8**, **9**, **10**, and **11** with ketone, vinyl sulfone, 3-bromo-4,5-dihydroisoxazole, and pinacolate ester, respectively, as warheads. Reagents and conditions: (a) Dess–Martin periodinane, $Pd(OAc)_2$, CH_3CN, H_2O, 50 °C, overnight; (b) DMSO, H_2O, NH_4I, 130 °C, 36 h; (c) 1,1-dibromoformaldoxime, DMF, $NaHCO_3$, −15 °C to rt, 5 h; (d) bis(pinacolato)diboron, CsF, 1,4-dioxane, MeOH, 100 °C, 12 h.

The synthesis of 3-propanal-substituted psoralen **15** was initiated by a coumarin derivative **12** possessing ethyl propionate moiety at position 3 (Scheme 3). The acidic hydrolysis yielded propanoic acid **13**, which was transformed into aldehyde derivative **14** by first forming an acid chloride, followed by in situ reduction with hydrogen gas using Pd/$BaSO_4$ as a catalyst. Interestingly, an attempt to prepare α-ketoaldehyde (which is a known Thr-targeting warhead [24]) from compound **15** by Riley oxidation with SeO_2 resulted in the formation of α,β-unsaturated aldehyde derivative **16** (Scheme 3, Figures 4 and 5).

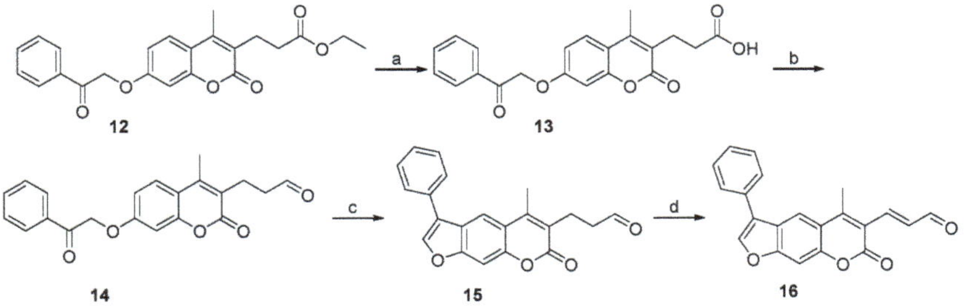

Scheme 3. Synthesis of compounds with aldehyde- (**15**) and α,β-unsaturated aldehyde-based (**16**) warheads attached at position 3 of the psoralen ring. Reagents and conditions: (a) 1 M HCl, dioxane, reflux, 2 h; (b) i. $SOCl_2$, DMF, toluene, rt, 17 h; ii. H_2, Pd/$BaSO_4$, toluene, 100 °C, 2 h; (c) 1 M NaOH, propan-2-ol, 60 °C, 15 min; (d) SeO_2, dioxane, H_2O, MW, 150 °C, 1 h. Synthesis of compound **12** was described previously [39].

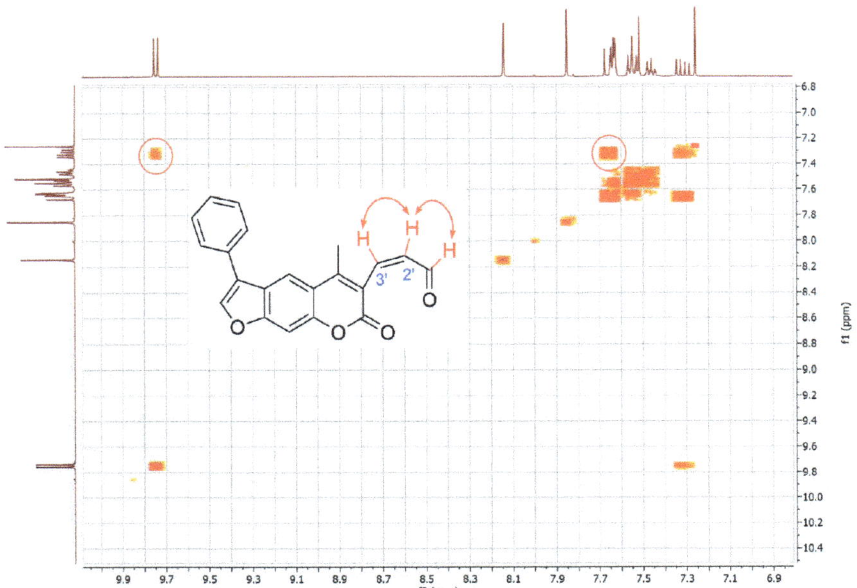

Figure 4. COSY experiment for **16**. Circled cross-peaks indicate coupling between aldehyde proton C<u>H</u>O and the adjacent C2′-<u>H</u>, and between C2′-<u>H</u> and C3′-<u>H</u>.

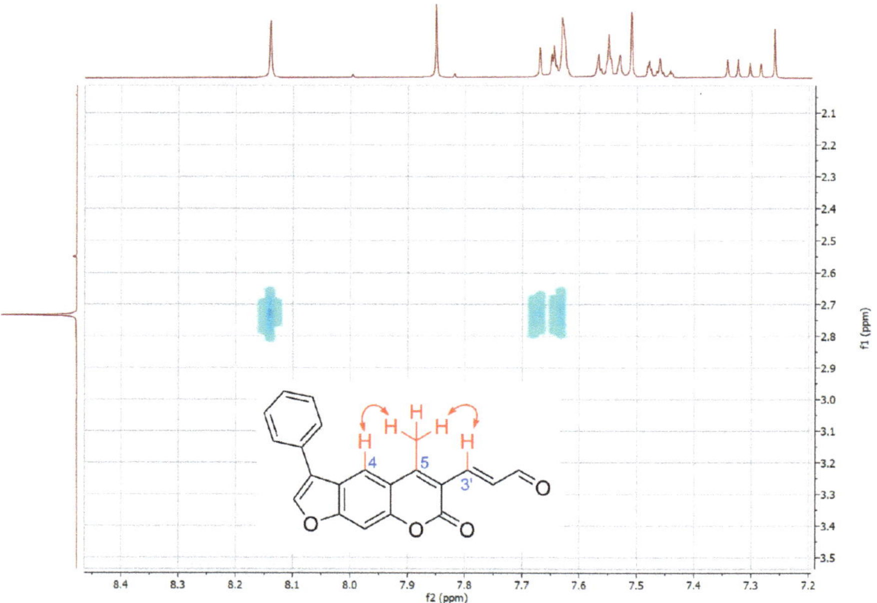

Figure 5. NOESY experiment for **16**. Only cross-peaks that indicate coupling between CH_3 protons and C4-H and C3′-H are shown.

To further confirm the structure of α,β-unsaturated aldehyde **16**, two-dimensional NMR experiments correlation spectroscopy (COSY) and Nuclear Overhauser effect spectroscopy (NOESY) were recorded. In the COSY spectrum (Figure 4), a clear correlation between the aldehyde proton C<u>H</u>O and the adjacent C2′-<u>H</u> was observed. In addition, the NOESY experiment showed a coupling between the C<u>H</u>$_3$ protons and C4-<u>H</u> and C3′-<u>H</u> (Figure 5).

The fact that the most advanced selective iCP inhibitors and also carfilzomib, which is a marketed cCP and iCP inhibitor, possess an α′,β′-epoxyketone fragment as the Thr-targeting warhead, encouraged us to prepare two such psoralen-based compounds (Scheme 4). Both **20** and **21** were synthesized from the corresponding precursors **17**, **18**, and **19** by a HATU-mediated amide bond formation (Scheme 4).

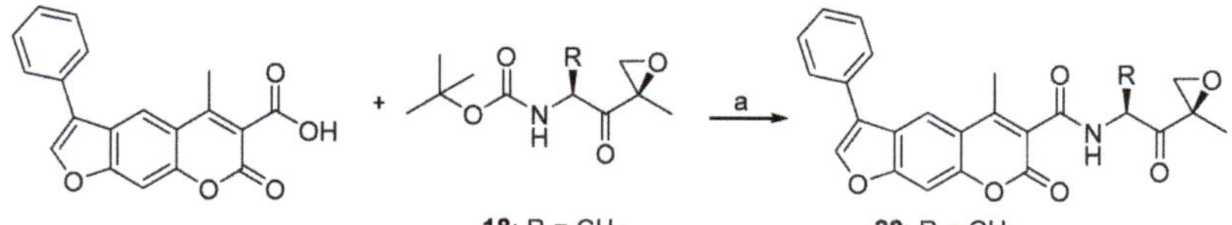

17 + **18**: R = CH_3; **19**: R = CH_2Ph →(a) **20**: R = CH_3; **21**: R = CH_2Ph

Scheme 4. Synthesis of epoxyketone-based compounds **20** and **21**. Reagents and conditions: (a) i. TFA, CH_2Cl_2, 0 °C, 30 min; ii. HATU, HOBt×H_2O, DIPEA, DMF (**20**) or CH_2Cl_2 (for **21**), rt, 24 h. Syntheses of compounds **17**, **18**, and **19** were based on previously described procedures [39], compound **17**; [47], compound **18**; [48], compound **19**. All spectral data (^{1}H-NMR, HRMS) corresponded well to the original reports.

To prepare 3-azetidin-2-one-substituted psoralen **24**, a previously synthesized compound **22** was used as a crucial intermediate. It was first *N*-acylated with 3-bromopropanoyl chloride to yield 3-bromopropanamide **23**, and then cyclized into the β-lactam ring by using NaO*t*Bu as a base (Scheme 5).

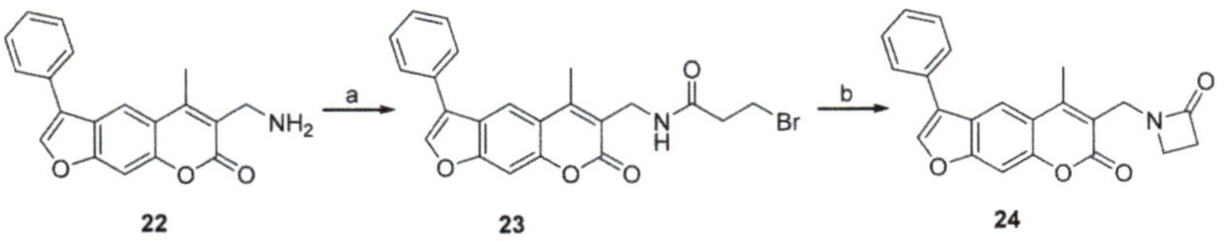

Scheme 5. Synthesis of alkyl bromide-based psoralen **23** and psoralen **24** with azetidin-2-one as a warhead. Reagents and conditions: (a) 3-bromopropanoyl chloride, K_2CO_3, CH_2Cl_2, 0 °C to rt, 3 h; (b) NaO*t*Bu, DMF, 0 °C to rt, 24 h. Synthesis of compound **22** was described previously [39].

2.2. Biochemical Evaluation

The target compounds were evaluated for their inhibitory potencies on both CPs (Table 1) using subunit selective fluorogenic substrates (for details, see Materials and Methods Section). The data were calculated as residual activities (RAs) of individual subunits of CPs in the presence of 1 μM of each compound. This concentration was used due to poor solubility of all final compounds at higher concentrations, emphasizing the need for development of inhibitors with improved solubility. The previously described oxathiazolone derivative 'compound 42′ and carfilzomib were used as positive control using the same concentration (1 μM) to enable a better comparison between compounds.

Table 1. Inhibitory potencies of compounds against all catalytically active subunits (β5i, β2i, and β1i) of the iCP and against all catalytically active subunits (β5, β2, β1) of the human cCP. In the assays, the following substrates were used: Suc-LLVY-AMC for β5i and β5; Boc-LRR-AMC for β2i and β2; Ac-PAL-AMC for β1i; Ac-nLPnLD-AMC for β1.

Cpd	β5i (RA [%]) [1]	β2i (RA [%]) [1]	β1i (RA [%]) [1]	β5 (RA [%]) [1]	β2 (RA [%]) [1]	β1 (RA [%]) [1]
4	78 ± 5	100 ± 0	95 ± 24	80 ± 21	82 ± 7	88 ± 0
7	76 ± 3	100 ± 0	87 ± 15	81 ± 18	86 ± 7	88 ± 5
8	70 ± 0	100 ± 0	90 ± 21	78 ± 21	87 ± 7	89 ± 2
9	69 ± 13	109 ± 3	76 ± 7	72 ± 18	90 ± 2	99 ± 5
10	62 ± 5	102 ± 2	87 ± 19	79 ± 20	86 ± 2	90 ± 3
11	76 ± 12	109 ± 14	72 ± 8	66 ± 21	89 ± 4	87 ± 2
15	71 ± 1	103 ± 4	94 ± 16	76 ± 20	87 ± 3	97 ± 2
16	65 ± 3	107 ± 3	92 ± 21	77 ± 19	83 ± 4	83 ± 2
20	78 ± 0	88 ± 0	83 ± 18	76 ± 17	81 ± 6	79 ± 5
21	76 ± 1	90 ± 0	81 ± 10	79 ± 14	80 ± 4	86 ± 4
23	74 ± 11	113 ± 7	72 ± 1	72 ± 12	89 ± 6	92 ± 4
24	77 ± 7	109 ± 7	74 ± 5	63 ± 26	88 ± 5	89 ± 7
carf.	3 ± 1	1 ± 1	1 ± 1	0 ± 0	16 ± 6	2 ± 2
'42′	5 ± 2	102 ± 5	97 ± 8	52 ± 4	99 ± 2	99 ± 8

[1] RA values are means from at least three independent determinations. Ac-PAL-AMC, acetyl-Pro-Ala-Leu-7-amino-4-methylcoumarin; Ac-nLPnLD-AMC, acetyl-Nle-Pro-Nle-Asp-AMC; Boc-LRR-AMC, *t*-butyloxycarbonyl-Leu-Arg-Arg-7-amino-4-methylcoumarin; Suc-LLVY-AMC, succinyl-Leu-Leu-Val-Tyr-7-amino-4-methylcoumarin. carf.: carfilzomib.

Given the fact that all assayed compounds possessed the same non-covalently binding portion, we were able to thoroughly assess the contributions of attached warheads to the inhibition of all catalytically active subunits of iCP and cCP. The assay results showed that all new psoralens were worse inhibitors of β5i subunit of iCP in comparison to the parent oxathiazolone-based 'compound 42′ (Table 1, Figure 6). This is most probably due to the mispositioning of the electrophilic carbons of all compounds and the catalytic Thr1O$^{\gamma}$ in the β5i active site. Interestingly, all compounds inhibited β5i activity with a similar potency at 1 μM with RA values ranging from 62 to 78%. Of the 12 prepared compounds, 3-bromo-4,5-dihydroisoxazole-substituted psoralen **10** and compound **16** with an α,β-unsaturated aldehyde as the warhead were the most promising. The former showed RA value of

62 ± 5%, whereas for the latter RA was determined at 65 ± 3% (see also postulated binding modes for 10 and 16 in Figure 7). It was not surprising to see that all 12 compounds also exhibited worse inhibition of the β5 subunit of cCP, albeit these differences were much less pronounced as for the β5i subunit. Of note, compounds **9, 11, 23**, and **24** were slightly better inhibitors of β1i subunit of iCP in comparison to the 'compound 42′'. All psoralen-based compounds (with oxathiazolone included) did not inhibit other subunits of both CPs (i.e., β2i, β2, and β1), whereas carfilzomib completely abolished activity of all subunits at 1 μM (Table 1, Figure 6).

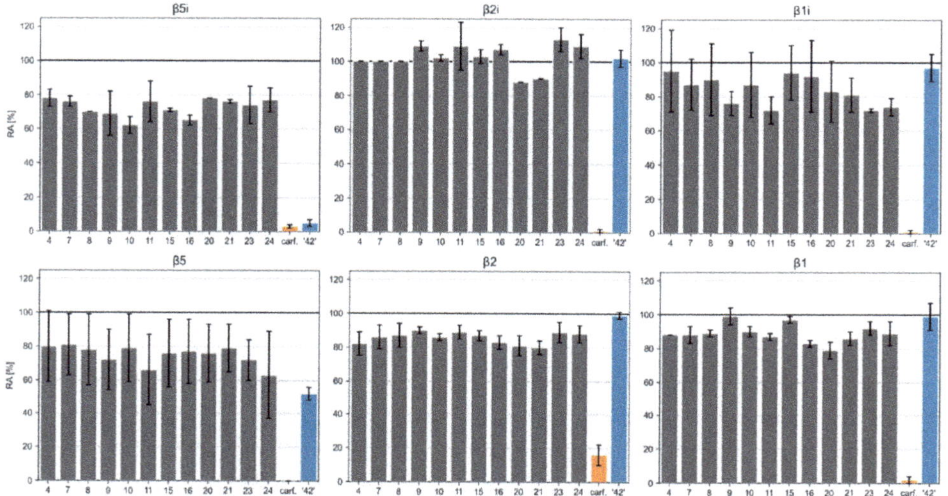

Figure 6. Inhibition results represented as bar charts of inhibition percentage. carf.: carfilzomib.

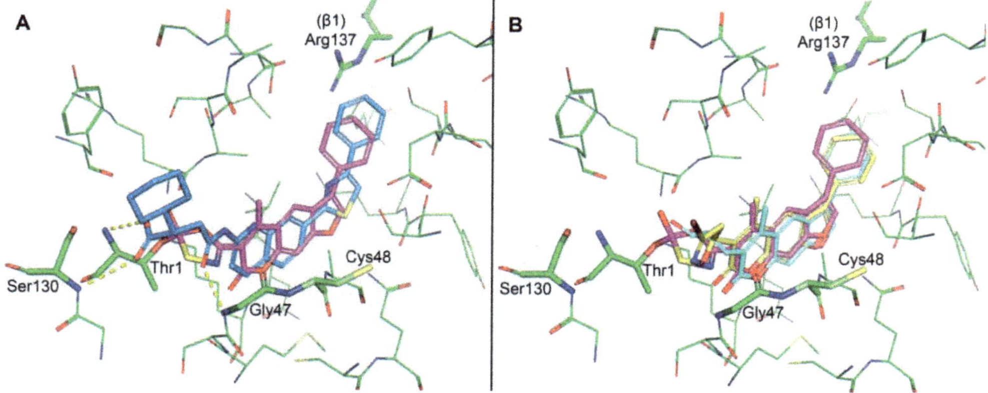

Figure 7. Molecular modelling. Binding site residues are presented as green sticks with labels for some of the key residues. (**A**) Covalent docking of '42′' (magenta) into the β5i subunit (PDB: 5M2B). Please note that only the initial intermediate formed after the nucleophilic attack of OH group of Thr1 onto the carbonyl group of the oxathiazolone is represented. Co-crystalized ligand Ro19 is presented with blue sticks and dashed yellow lines for hydrogen bonds. (**B**) Noncovalent docking of **10** (cyan) and **16** (yellow) reveals good alignment of the psoralen core with the proposed pose of '42′' (magenta). However, the distance from the electrophilic carbons of **10** and **16** to the catalytic Thr1O$^{\gamma}$ is too large to form a covalent bond.

3. Materials and Methods

3.1. General Chemistry Methods

Reagents and solvents were obtained from commercial sources (Acros Organics (Thermo Fisher Scientific, Waltham, MA, USA), Sigma-Aldrich (Merck KGaA, Darmstadt, Germany), TCI Europe (Tokyo Chemical Industry, Tokyo, Japan), Alfa Aesar (Thermo Fisher Scientific, Waltham, MA, USA), Fluorochem (Fluorochem Ltd., Derbyshire, UK) and were used as received. Carfilzomib were purchased from MedChemExpress. For reactions involving air or moisture sensitive reagents, solvents were distilled before use and these reactions were carried out under nitrogen or argon atmosphere. Reactions using microwaves were performed on a standard monomode microwave reactor MONOWAVE 200 (Anton Paar, Graz, Austria). Reactions were monitored using analytical thin-layer chromatography plates (Merck 60 F254, 0.20 mm), and the components were visualized under UV light and/or through staining with the relevant reagent. Normal phase flash column chromatography was performed on Merck Silica Gel 60 (particle size 0.040–0.063 mm; Merck, Germany). ^{1}H and ^{13}C-NMR spectra were recorded at 295 K on a Bruker Avance III 400 MHz spectrometer (Bruker, Billerica, MA, USA) operating at frequencies for ^{1}H-NMR at 400 MHz and for ^{13}C-NMR at 101 MHz. The chemical shifts (δ) are reported in parts per million (ppm) and are referenced to the deuterated solvent used. The coupling constants (J) are given in Hz, and the splitting patterns are designated as follows: s, singlet; br s, broad singlet; d, doublet; app d, apparent doublet; dd, doublet of doublets; ddd, doublet of doublets of doublets; t, triplet; dt, doublet of triplets; td, triplet of doublets; m, multiplet. All ^{13}C-NMR spectra were proton decoupled. Mass spectra data and high-resolution mass measurements were performed on a Thermo Scientific Q Exactive Plus Hybrid Quadrupole-Orbitrap mass spectrometer (Thermo Fisher Scientific, Waltham, MA, USA). The purity of the compounds used in biochemical assays was determined with analytical normal-phase HPLC on an Agilent 1100 LC modular system (Agilent, Santa Clara, CA, USA) that was equipped with a photodiode array detector set to 254 nm. A Kromasil 3-CelluCoat column (150 mm × 4.6 mm; 3 μm particle size) was used, with a flow rate of 1.0 mL/min and a sample injection volume of 5–20 μL. An isocratic eluent system of A (hexane) and B (isopropanol) was used; the ratio used is described for each compound below. The purities of the test compounds used for the biological evaluations were ≥95%, unless stated otherwise.

3.2. Syntheses

Synthesis of ethyl 2-acetylpent-4-enoate (**1**):

To a solution of ethyl acetoacetate (7.28 mL, 7.50 g, 57.60 mmol, 1 equiv.) in 50 mL of anhydrous THF, NaH in mineral oil (60%, 2.30 g, 57.60 mmol, 1 equiv.) was added and the resulting suspension stirred under argon at 0 °C. After 20 min, a solution of allyl bromide (4.99 mL, 6.97 g, 57.60 mmol, 1 equiv.) in 25 mL of anhydrous THF was added dropwise. The reaction mixture was stirred at room temperature overnight. Next, cold H_2O (25 mL) was added and THF was evaporated under reduced pressure. The resulting suspension was extracted with Et_2O (3 × 25 mL), the organic layer separated, dried over anhydrous Na_2SO_4, and evaporated. The product was purified by column chromatography (Et_2O/petroleum ether, 1/5, v/v). Yield: 71%, clear liquid. ^{1}H-NMR (400 MHz, DMSO-d_6) δ 1.17 (t, J = 7.1 Hz, 3H, $\underline{CH_3}CH_2$), 2.18 (s, 3H, $CO\underline{CH_3}$), 2.42–2.47 (m, 2H, $CH_2CH\underline{CH_2}CH$), 3.73 (dd, J = 7.8, 6.8 Hz, 1H, $\underline{CH}$), 4.07–4.15 (qd, 2H, J = 7.1, 1.6 Hz, $CH_3\underline{CH_2}$), 4.98–5.10 (m, 2H, $\underline{CH_2}CHCH_2CH$), 5.67–5.77 (m, 1H, $CH_2\underline{CH}CH_2CH$).

Synthesis of 3-allyl-7-hydroxy-4-methyl-2*H*-chromen-2-one (**2**):

This compound was prepared using Pechmann condensation as follows. A solution of resorcinol (4.06 g, 36.90 mmol, 1 equiv.) and ethyl 2-acetylpent-4-enoate (**1**) (6.90 g, 40.50 mmol, 1.1 equiv.) in dioxane (80 mL) was cooled to 0 °C, followed by drop-wise addition of concentrated H_2SO_4 (98%, 19.60 mL, 405 mmol, 10 equiv.). The reaction mixture was stirred at room temperature overnight. Dioxane was then evaporated under reduced pressure and the semi-solid mixture was added portion-wise to an ice-cold solution of KOH

(40 g) in H_2O (100 mL). The pH was adjusted to 13 with KOH and the resulting white solid was filtered off. The filtrate was extracted with EtOAc (3 × 25 mL) and the combined organic extracts were dried over anhydrous Na_2SO_4 and evaporated under reduced pressure. The compound was purified by column chromatography (EtOAc/*n*-hexane, 1/1.5, *v*/*v*, dry loading) yielding a pale-yellow solid. Yield: 13%. ^{1}H-NMR (400 MHz, DMSO-d_6) δ 2.34 (s, 3H, C$\underline{H}_3$), 3.30 (d, *J* = 6.0 Hz, 2H, Ar-C$\underline{H}_2$CHCH$_2$), 4.98–5.06 (m, 2H, Ar-CH$_2$CHC$\underline{H}_2$), 5.84 (ddt, *J* = 16.3, 10.3, 6.0 Hz, 1H, Ar-CH$_2$C$\underline{H}$CH$_2$), 6.70 (d, *J* = 2.4 Hz, 1H, Ar-$\underline{H}$), 6.80 (dd, *J* = 8.7, 2.4 Hz, 1H, Ar-$\underline{H}$), 7.63 (d, *J* = 8.7 Hz, 1H, Ar-$\underline{H}$), 10.44 (s, 1H, O$\underline{H}$); HRMS (ESI) *m/z* calculated for $C_{13}H_{11}O_3$ $[M - H]^-$ 215.0714, found 215.0707.

Synthesis of 3-allyl-4-methyl-7-(2-oxo-2-phenylethoxy)-2*H*-chromen-2-one (**3**):

This compound was synthesized following a previously described procedure [39]. Briefly, to a solution of 3-allyl-7-hydroxy-4-methyl-2*H*-chromen-2-one (**2**) (0.99 g, 4.56 mmol, 1 equiv.) in dioxane (70 mL), K_2CO_3 (2.52 g, 18.22 mmol, 4 equiv.) and KI (76 mg, 0.46 mmol, 0.1 equiv.) were added. After 10 min of stirring at 100 °C, 2-bromoacetophenone (1.36 g, 6.83 mmol, 1.5 equiv.) was added and the mixture was further stirred at 100 °C for 24 h. The solvent was then removed under reduced pressure, followed by addition of H_2O (30 mL) to the residue. The aqueous phase was extracted with EtOAc (3 × 30 mL), the combined organic extracts were washed with brine (20 mL), dried over anhydrous Na_2SO_4, filtered, and evaporated under reduced pressure. The compound was purified by crystallization from MeOH yielding pale-yellow crystalline solid. Yield: 67%. ^{1}H-NMR (400 MHz, DMSO-d_6) δ 2.38 (s, 3H, C$\underline{H}_3$), 3.32 (d, *J* = 6.0 Hz, 2H, Ar-C$\underline{H}_2$CHCH$_2$), 5.03 (ddd, *J* = 8.5, 3.0, 1.3 Hz, 2H, Ar-CH$_2$CHC$\underline{H}_2$), 5.75 (s, 2H, C$\underline{H}_2$), 5.86 (ddt, *J* = 16.2, 10.2, 6.0 Hz, 1H, Ar-CH$_2$C$\underline{H}$CH$_2$), 7.03 (dd, *J* = 8.9, 2.6 Hz, 1H, Ar-$\underline{H}$), 7.07 (d, *J* = 2.5 Hz, 1H, Ar-$\underline{H}$), 7.59 (app dd, *J* = 10.6, 4.8 Hz, 2H, 2 × Ar-$\underline{H}$), 7.68–7.77 (m, 2H, 2 × Ar-$\underline{H}$), 8.04 (app dd, *J* = 8.4, 1.2 Hz, 2H, 2 × Ar-$\underline{H}$); HRMS (ESI) *m/z* calculated for $C_{21}H_{19}O_4$ $[M + H]^+$ 335.1280, found 335.1272.

Synthesis of 6-allyl-5-methyl-3-phenyl-7*H*-furo[3,2-*g*]chromen-7-one (**4**):

This compound was synthesized following a previously described procedure [39]. Namely, to a heated (80 °C) and stirred solution of **3** (0.25 g, 0.75 mmol, 1 equiv.) in propan-2-ol (25 mL), an aqueous solution of NaOH (7.5 mL, 1 M, 10 equiv.) was added. The reaction mixture was stirred at 80 °C for 40 min. After the reaction was complete (monitored by TLC), propan-2-ol was evaporated under reduced pressure. The aqueous residue was acidified with HCl (6 mL, 1 M) to pH 5, then H_2O (20 mL) was added, the aqueous layer was extracted with CH_2Cl_2 (3 × 25 mL), and the combined organic extracts were evaporated under reduced pressure. The compound was purified by column chromatography (Et_2O/petroleum ether, 1/3, *v*/*v*). White solid, yield: 70%. ^{1}H-NMR (400 MHz, DMSO-d_6) δ 2.54 (s, 3H, C$\underline{H}_3$), 3.40 (d, *J* = 6.0 Hz, 2H, Ar-C$\underline{H}_2$CHCH$_2$), 5.01–5.11 (m, 2H, Ar-CH$_2$CHC$\underline{H}_2$), 5.89 (ddt, *J* = 16.1, 10.2, 6.0 Hz, 1H, Ar-CH$_2$C$\underline{H}$CH$_2$), 7.41–7.48 (m, 1H, Ar-$\underline{H}$), 7.51–7.60 (m, 2H, 2 × Ar-$\underline{H}$), 7.79 (s, 1H, Ar-$\underline{H}$), 7.80–7.82 (m, 1H, Ar-$\underline{H}$), 7.83 (t, *J* = 1.6 Hz, 1H, Ar-$\underline{H}$), 8.18 (s, 1H, Ar-$\underline{H}$), 8.48 (s, 1H, Ar-$\underline{H}$); ^{1}H-NMR (400 MHz, $CDCl_3$) δ 2.49 (s, 3H, C$\underline{H}_3$), 3.49 (d, *J* = 6.0 Hz, 2H, Ar-C$\underline{H}_2$CHCH$_2$), 5.01–5.16 (m, 2H, Ar-CH$_2$CHC$\underline{H}_2$), 5.94 (ddt, *J* = 16.2, 10.1, 6.0 Hz, 1H, Ar-CH$_2$C$\underline{H}$CH$_2$), 7.44 (ddd, *J* = 7.4, 4.0, 1.3 Hz, 1H, Ar-$\underline{H}$), 7.49–7.51 (m, 1H, Ar-$\underline{H}$), 7.51–7.57 (m, 2H, 2 × Ar-$\underline{H}$), 7.61–7.64 (m, 1H, Ar-$\underline{H}$), 7.65 (t, *J* = 1.7 Hz, 1H, Ar-$\underline{H}$), 7.82 (s, 1H, Ar-$\underline{H}$), 8.00 (s, 1H, Ar-$\underline{H}$); ^{13}C-NMR (101 MHz, DMSO-d_6) δ 15.19, 30.94, 99.48, 115.69, 116.46, 116.99, 121.18, 121.29, 122.81, 127.23, 127.84, 129.22, 130.72, 134.50, 144.29, 148.39, 149.94, 155.92, 160.58; HRMS (ESI) *m/z* calculated for $C_{21}H_{17}O_3$ $[M + H]^+$ 317.1172, found 317.1166. Purity by HPLC (0–18 min; 70% *n*-hexane/isopropanol): 99%.

Synthesis of 7-hydroxy-4-methyl-4-vinyl-2*H*-chromen-2-one (**5**):

A suspension of 1-(2,4-dihydroxyphenyl)ethan-1-one (502 mg, 3.3 mmol, 1 equiv.), crotonyl chloride (395 μL, 429 mg, 4.1 mmol, 1.25 equiv.) and K_2CO_3 (1.47 g, 10.6 mmol, 3.2 equiv.) in acetone (25 mL) was heated at 60 °C for 24 h. The solvent was then evaporated under reduced pressure, followed by the addition of EtOAc (100 mL). The organic phase was extracted with H_2O (100 mL), and the aqueous phase acidified with 2 M HCl and further extracted with EtOAc (2 × 100 mL). The combined organic extracts were dried over anhydrous Na_2SO_4, filtered, and evaporated under reduced pressure. The compound was purified by column chromatography (EtOAc/*n*-hexane, 1/4, *v*/*v*). White solid, yield: 71%. ^{1}H-NMR (400 MHz, DMSO-d_6) δ 2.45 (s, 3H, C<u>H</u>$_3$), 5.51 (dd, *J* = 12.0 Hz, 2.4 Hz, 1H, Ar-CHC<u>H</u>$_2$), 6.02 (dd, *J* = 17.4 Hz, 2.4 Hz, 1H, Ar-CHC<u>H</u>$_2$), 6.67 (d, *J* = 2.4 Hz, 1H, Ar-<u>H</u>), 6.72 (dd, *J* = 17.4, 12.0 Hz, 1H, Ar-C<u>H</u>CH$_2$), 6.79 (d, *J* = 8.9, 2.4 Hz, 1H, Ar-<u>H</u>), 7.66 (d, *J* = 8.9 Hz, 1H, Ar-<u>H</u>), 10.51 (br s, 1H, O<u>H</u>). HRMS (ESI) *m/z* calculated for $C_{12}H_9O_3$ $[M - H]^-$ 201.0557, found 201.0549.

Synthesis of 4-methyl-7-(2-oxo-2-phenylethoxy)-3-vinyl-2*H*-chromen-2-one (**6**):

This compound was synthesized following a previously described procedure [39]; using the procedure as for **5**. The compound was purified by crystallization from EtOH yielding off-white crystalline solid. Yield: 79%. ^{1}H-NMR (400 MHz, $CDCl_3$) δ 2.48 (s, 3H, C<u>H</u>$_3$), 5.38 (s, 2H, OC<u>H</u>$_2$), 5.63 (dd, *J* = 11.8, 1.9 Hz, 1H, Ar-CH=C<u>H</u>$_2$), 6.04 (dd, *J* = 17.6, 1.9 Hz, 1H, Ar-CH=C<u>H</u>$_2$), 6.71 (dd, *J* = 17.6, 11.8 Hz, 1H, Ar-C<u>H</u>=CH$_2$), 6.77 (d, *J* = 2.6 Hz, 1H, Ar-<u>H</u>), 6.96 (dd, *J* = 9.0, 2.6 Hz, 1H, Ar-<u>H</u>), 7.49–7.57 (m, 2H, Ar-<u>H</u>), 7.59 (d, *J* = 8.9 Hz, 1H, Ar-<u>H</u>), 7.70–7.63 (m, 1H, Ar-<u>H</u>), 7.95–8.04 (m, 2H, 2 × Ar-<u>H</u>); ^{13}C-NMR (101 MHz, $CDCl_3$) δ 15.37, 70.55, 101.43, 112.79, 114.88, 120.10, 122.26, 126.34, 127.99, 128.99, 129.03, 134.14, 134.26, 146.69, 153.55, 160.18, 160.39, 193.18; HRMS (ESI) *m/z* calculated for $C_{20}H_{17}O_4$ $[M + H]^+$ 321.1121, found 321.1123.

Synthesis of 5-methyl-3-phenyl-6-vinyl-7*H*-furo[3,2-*g*]chromen-7-one (**7**):

To a solution of 4-methyl-7-(2-oxo-2-phenylethoxy)-3-vinyl-2*H*-chromen-2-one (**6**) (132 mg, 0.4 mmol, 1 equiv.) in EtOH (5 mL), KOH (1.2 mL, 1 M, 1.2 mmol, 3 equiv.) was added and the reaction mixture stirred at 85 °C for 2 h. The solvent was then evaporated, followed by the addition of H_2O (20 mL). The suspension was acidified with concentrated HCl to pH = 1 and extracted with CH_2Cl_2 (2 × 50 mL). The combined organic extracts were dried over Na_2SO_4, filtered, and evaporated under reduced pressure. The product was purified by column chromatography (EtOAc/*n*-hexane, 1/1, *v*/*v*). Yellow solid; yield: 78 %; ^{1}H-NMR (400 MHz, $CDCl_3$) δ 2.60 (s, 3H, C<u>H</u>$_3$), 5.69 (dd, *J* = 11.8 Hz, 1.8 Hz, 1H, Ar-CH=C<u>H</u>$_2$), 6.05 (dd, *J* = 17.7 Hz, 1.8 Hz, 1H, Ar-CH=C<u>H</u>$_2$), 6.77 (dd, *J* = 17.7, 11.8, 1H, Ar-C<u>H</u>=CH$_2$), 7.42–7.47 (m, 1H, Ar-<u>H</u>), 7.49 (s, 1H, Ar-<u>H</u>), 7.51–7.56 (m, 2H, Ar-<u>H</u>), 7.62–7.66 (m, 2H, Ar-<u>H</u>), 7.82 (s, 1H, Ar-<u>H</u>), 8.04 (s, 1H, Ar-<u>H</u>); ^{13}C-NMR (101 MHz, $CDCl_3$) δ 15.92, 99.74, 116.23, 117.38, 121.23, 122.31, 122.82, 123.98, 127.59, 128.04, 129.19, 129.26, 131.11, 142.79, 146.88, 150.39, 156.67, 160.21; HRMS (ESI) *m/z* calculated for $C_{20}H_{15}O_3$ $[M + H]^+$ 303.1016, found 303.1019. Purity by HPLC (0–18 min; 70% *n*-hexane/isopropanol): 98%.

Synthesis of 5-methyl-6-(2-oxopropyl)-3-phenyl-7*H*-furo[3,2-*g*]chromen-7-one (**8**):

This compound was synthesized following a previously described procedure [49]. Briefly, to a stirred solution of olefin **4** (158 mg, 0.5 mmol, 1 equiv.) in CH_3CN (3.5 mL) and H_2O (0.5 mL), $Pd(OAc)_2$ (5.6 mg, 0.025 mmol, 5 mol %) and Dess-Martin periodinane (254 mg, 0.6 mmol, 1.2 equiv.) were added. The reaction mixture was warmed to 50 °C and stirred under an argon atmosphere overnight. The reaction mixture was then filtered through a small pad of Celite and washed with EtOAc, and the filtrate was concentrated. The residue was purified by column chromatography (EtOAc/*n*-hexane = 1/2, *v*/*v*, dry loading). White solid, yield: 40%. ^{1}H-NMR (400 MHz, $CDCl_3$) δ 2.32 (s, 3H, CH$_2$COC<u>H</u>$_3$), 2.44 (s, 3H, C<u>H</u>$_3$), 3.88 (s, 2H, C<u>H</u>$_2$COCH$_3$), 7.44 (t, *J* = 7.4 Hz, 1H, Ar-<u>H</u>), 7.48–7.57 (m, 3H, 3 × Ar-<u>H</u>), 7.63 (app dd, *J* = 8.0, 1.0 Hz, 2H, 2 × Ar-<u>H</u>), 7.83 (s, 1H, Ar-<u>H</u>), 8.01 (s, 1H, Ar-<u>H</u>); ^{13}C-NMR (101 MHz, $CDCl_3$) δ 16.19, 30.15, 42.42, 100.13, 116.08, 117.27, 118.41, 122.51, 124.19, 127.77, 128.19, 129.41, 131.26, 143.01, 149.65, 150.81, 156.85, 161.93, 204.58; HRMS (ESI) *m/z* calculated for $C_{21}H_{17}O_4$ $[M + H]^+$ 333.1121, found 333.1127. Purity by HPLC (0–18 min; 70% *n*-hexane/isopropanol): 98%.

Synthesis of (*E*)-5-methyl-6-(2-(methylsulfonyl)vinyl)-3-phenyl-7*H*-furo[3,2-*g*]chromen-7-one (**9**):

To a solution of 5-methyl-3-phenyl-6-vinyl-7*H*-furo[3,2-*g*]chromen-7-one (**7**) (100 mg, 0.33 mmol, 1 equiv.) in DMSO (1 mL), H_2O (0.5 mL) and NH_4I (191 mg, 1.32 mmol, 4 equiv.) were added. The reaction mixture was stirred at 130 °C for 36 h. Then, it was cooled to room temperature, followed by slow addition of $Na_2S_2O_3 \times 5H_2O$ until the discoloration of mixture. Subsequently, H_2O (20 mL) was added and the aqueous phase extracted with EtOAc (3 × 20 mL). The combined organic extracts were dried over Na_2SO_4, filtered, and evaporated under reduced pressure. The product was purified by column chromatography (EtOAc/*n*-hexane, 1/2, *v*/*v*). Yellow solid; yield: 64%; ^{1}H-NMR (400 MHz, $CDCl_3$) δ 2.63 (s, 3H, Ar-C$\underline{H}_3$), 3.14 (s, 3H, $SO_2C\underline{H}_3$), 7.44–7.48 (m, 1H, Ar-CHC$\underline{H}$$SO_2CH_3$), 7.52–7.57 (m, 4H, Ar-C$\underline{H}$CHSO_2CH_3 and 3 × Ar-$\underline{H}$), 7.61–7.65 (m, 3H, Ar-$\underline{H}$), 7.87 (s, 1H, Ar-$\underline{H}$), 8.11 (s, 1H, Ar-$\underline{H}$); ^{13}C-NMR (100 MHz, $CDCl_3$): δ 15.08, 42.96, 60.48, 65.58, 100.25, 116.60, 116.89, 121.68, 122.48, 124.96, 127.66 (2C), 128.31, 129.38 (2C), 130.69, 143.40, 149.56, 150.40, 157.12, 161.99; HRMS (ESI) *m/z* calculated for $C_{21}H_{17}O_5S$ $[M + H]^+$ 381.0791, found 381.0795.; Purity by HPLC (0–18 min; 70% *n*-hexane/isopropanol): 99%.

Synthesis of 6-(3-bromo-4,5-dihydroisoxazol-5-yl)-5-methyl-3-phenyl-7*H*-furo[3,2-*g*]chromen-7-one (**10**)

To a cooled (−15 °C) solution of 5-methyl-3-phenyl-6-vinyl-7*H*-furo[3,2-*g*]chromen-7-one (**7**) (145 mg, 0.48 mmol, 1 equiv.) and 1,1-dibromoformaldoxime (148 mg, 0.73 mmol, 1.5 equiv.) in DMF (10 mL), an aqueous solution of $NaHCO_3$ (1 mL, 112 mg, 1.3 mmol, 2.7 equiv.) was added. The reaction was then warmed to room temperature and stirred for 5 h. Then, the solution was diluted with CH_2Cl_2 (20 mL) and washed with brine (20 mL). The organic extract was dried over Na_2SO_4, filtered, and the solvents removed under reduced pressure. The product was purified by column chromatography (EtOAc/*n*-hexane, 1/4, *v*/*v*) to yield pale yellow solid. Yield: 76%. ^{1}H-NMR (400 MHz, $CDCl_3$) δ 2.60 (s, 3H, Ar-C$\underline{H}_3$), 3.47 (dd, *J* = 17.1, 11.8 Hz, 1H, one H of C$\underline{H}_2$), 3.62 (dd, *J* = 17.1, 10.4 Hz, 1H, one H of C$\underline{H}_2$), 6.09 (dd, *J* = 11.8, 10.4 Hz, 1H, $CH_2C\underline{H}O$), 7.42–7.48 (m, 2H, Ar-$\underline{H}$), 7.51–7.57 (m, 2H, Ar-$\underline{H}$), 7.60–7.64 (m, 2H, Ar-$\underline{H}$), 7.83 (s, 1H, Ar-$\underline{H}$), 8.06 (s, 1H, Ar-$\underline{H}$); ^{13}C-NMR (101 MHz, $CDCl_3$) δ 15.27, 46.05, 77.81, 100.01, 116.61, 116.75, 119.73, 122.36, 124.29, 127.60, 128.19, 129.33, 130.84, 137.27, 143.12, 150.88, 151.48, 157.17, 159.50; HRMS (ESI) *m/z* calculated for $C_{21}H_{15}O_4NBr$ $[M + H]^+$ 424.0179, found 424.0179. Purity by HPLC (0–18 min; 70% *n*-hexane/isopropanol): 99%.

Synthesis of 5-methyl-3-phenyl-6-(2-(4,4,5,5-tetramethyl-1,3,2-dioxaborolan-2-yl)ethyl)-7*H*-furo[3,2-*g*]chromen-7-one (**11**):

5-Methyl-3-phenyl-6-vinyl-7*H*-furo[3,2-*g*]chromen-7-one (**7**) (0.09 mmol, 1 equiv.) was dissolved in 1,4-dioxane (2 mL) and then bis(pinacolato)diboron (1,5 equiv.), cesium fluoride (2,5 equiv.) and MeOH (5 equiv.) were added. The reaction proceeded at 100 °C for 12 h. The reaction mixture was then diluted with EtOAc (15 mL) and filtrated over silica. The filtrate was evaporated under reduced pressure and the product purified from the crude mixture by column chromatography (EtOAc/*n*-hexane, 1/9, gradient to 1/1, *v*/*v*) to yield yellow solid. Yield: 33%. ^{1}H-NMR (400 MHz, CD_3OD): δ 0.94 (t, *J* = 8.5 Hz, 2H, C$\underline{H}_2$), 1.13 (s, 12H, C(C$\underline{H}_3$)$_2$), 2.48 (s, 3H, Ar-C$\underline{H}_3$), 2.68 (t, *J* = 8.5 Hz, 2H, C$\underline{H}_2$), 7.30–7.35 (m, 1H, Ar-$\underline{H}$), 7.41–7.47 (m, 3H, Ar-$\underline{H}$), 7.62–7.62 (m, 2H, Ar-$\underline{H}$), 7.99 (s, 1H, Ar-$\underline{H}$), 8.05 (s, 1H, Ar-$\underline{H}$); ^{13}C-NMR (100 MHz, $CDCl_3$): δ 14.12, 17.77, 23.59 (4C), 29.40, 83.06, 85.22 (2C), 87.21, 102.00, 102.50, 105.99, 127.09 (2C), 127.58, 127.64, 128.75 (2C), 136.06, 141.54, 146.52, 152.51, 157.66, 161.81; HRMS (ESI) *m/z* calculated for $C_{26}H_{28}O_5B$ $[M + H]^+$ 431.2024, found 431.2023.; Purity by HPLC (0–18 min; 95% *n*-hexane/isopropanol): 97%.

Synthesis of 3-(4-methyl-2-oxo-7-(2-oxo-2-phenylethoxy)-2*H*-chromen-3-yl)propanoic acid (**13**):

To a stirred solution of **12** (788 mg, 2 mmol, 1 equiv.) in dioxane (20 mL), HCl (1 M, 20 mL, 10 equiv.) was added. The reaction mixture was heated at reflux temperature for 2 h. Dioxane was then evaporated under reduced pressure, the precipitate that formed filtered off and washed with H_2O. Yield: 94%. ^{1}H-NMR (400 MHz, $CDCl_3$) δ 2.44 (s, 3H, C$\underline{H}_3$), 2.67 (t, *J* = 7.6 Hz, 2H, C$\underline{H}_2$$CH_2COOH$), 2.96 (t, *J* = 7.6 Hz, 2H, $CH_2C\underline{H}_2COOH$), 5.38 (s, 2H, C$\underline{H}_2$), 6.78 (d, *J* = 2.6 Hz, 1H, Ar-$\underline{H}$), 6.95 (dd, *J* = 8.9, 2.6 Hz, 1H, Ar-$\underline{H}$), 7.49–7.60 (m, 3H, 3 × Ar-$\underline{H}$), 7.66 (t, *J* = 7.4 Hz, 1H, Ar-$\underline{H}$), 7.82–8.07 (m, 2H, 2 × Ar-$\underline{H}$); HRMS (ESI) *m/z* calculated for $C_{21}H_{17}O_6$ $[M - H]^-$ 365.1031, found 365.1030.

Synthesis of 3-(4-methyl-2-oxo-7-(2-oxo-2-phenylethoxy)-2*H*-chromen-3-yl)propanal (**14**):

To a suspension of **13** (366 mg, 1 mmol, 1 equiv.) in toluene (20 mL), dried over 3 Å molecular sieves, a catalytic amount of anhydrous DMF (5 drops) and $SOCl_2$ (218 μL, 357 mg, 3 mmol, 3 equiv.) were added under argon. The reaction mixture was stirred at room temperature for 17 h and then the volatiles were evaporated to obtain a white solid that was dried under vacuum for 15 min to remove $SOCl_2$. Toluene (20 mL), dried over 3 Å molecular sieves, was added to the dried solid (under argon), followed by the addition of 10% Pd/$BaSO_4$ (72 mg, 30% [*w*/*w*]). The reaction mixture was heated to 100 °C and stirred under a stream of hydrogen (1 atm) for 2 h. The reaction mixture was then evaporated to dryness and the compound was purified by column chromatography (EtOAc/*n*-hexane, 1/1, *v*/*v*, dry loading). White solid, yield: 59%. ^{1}H-NMR (400 MHz, $CDCl_3$) δ 2.43 (s, 3H, $C\underline{H}_3$), 2.77 (t, *J* = 7.4 Hz, 2H, $C\underline{H}_2CH_2CHO$), 2.94 (t, *J* = 7.4 Hz, 2H, $CH_2C\underline{H}_2CHO$), 5.38 (s, 2H, $C\underline{H}_2$), 6.78 (d, *J* = 2.6 Hz, 1H, Ar-H), 6.95 (dd, *J* = 8.9, 2.6 Hz, 1H, Ar-H), 7.50–7.59 (m, 3H, 3 × Ar-H), 7.66 (t, *J* = 7.4 Hz, 1H, Ar-H), 7.04–8.06 (m, 2H, 2 × Ar-H), 9.83 (t, *J* = 1.0 Hz, 1H, $CH_2CH_2C\underline{H}O$); HRMS (ESI) *m/z* calculated for $C_{21}H_{19}O_5$ $[M + H]^+$ 351.1227, found 351.1221.

Synthesis of 3-(5-methyl-7-oxo-3-phenyl-7*H*-furo[3,2-*g*]chromen-6-yl)propanal (**15**):

To a suspension of **14** (519 mg, 1.48 mmol, 1 equiv.) in propan-2-ol (35 mL), an aqueous solution of NaOH (14.8 mL, 1 M, 10 equiv.) was added. The reaction mixture was stirred at 60 °C for 15 min. After the reaction was complete (monitored by TLC), the resulting red solution was acidified with 1 M HCl (15 mL) to get a yellow precipitate. The reaction mixture was evaporated under reduced pressure. H_2O (50 mL) was added to the dry residue, which was extracted with CH_2Cl_2 (2 × 50 mL), the combined organic extracts were washed with brine (100 mL), dried over anhydrous Na_2SO_4, filtered, and evaporated under reduced pressure. The compound was purified by column chromatography (EtOAc/*n*-hexane, 1/2, *v*/*v*, dry loading). White solid, yield: 61%. ^{1}H-NMR (400 MHz, $CDCl_3$) δ 2.55 (s, 3H, $C\underline{H}_3$), 2.82 (t, *J* = 7.4 Hz, 2H, $C\underline{H}_2CH_2CHO$), 3.01 (t, *J* = 7.4 Hz, 2H, $CH_2C\underline{H}_2CHO$), 7.47–7.41 (m, 1H, Ar-H), 7.49 (s, 1H, Ar-H), 7.50–7.57 (m, 2H, 2 × Ar-H), 7.60–7.67 (m, 2H, 2 × Ar-H), 7.83 (s, 1H, Ar-H), 8.00 (s, 1H, Ar-H), 9.86 (t, *J* = 1.0 Hz, 1H, $CH_2CH_2C\underline{H}O$); ^{13}C-NMR (101 MHz, $CDCl_3$) δ 15.67, 20.85, 42.53, 99.97, 115.99, 117.34, 122.46, 122.98, 124.09, 127.72, 128.18, 129.39, 131.24, 142.94, 147.70, 150.56, 156.58, 161.72, 201.40; HRMS (ESI) *m/z* calculated for $C_{21}H_{17}O_4$ $[M + H]^+$ 333.1121, found 333.1116. Purity by HPLC (0–18 min; 70% *n*-hexane/isopropanol): 87%.

Synthesis of 3-(5-methyl-7-oxo-3-phenyl-7*H*-furo[3,2-*g*]chromen-6-yl)acrylaldehyde (**16**):

To a solution of the aldehyde **15** (53 mg, 0.16 mmol, 1 equiv.) in a mixture of dioxane (1.5 mL) and H_2O (20 μL), SeO_2 (35 mg, 0.32 mmol, 2 equiv.) was added and the reaction mixture was irradiated in a microwave reactor at 150 °C (250 W) for 1 h. The reaction mixture was then evaporated to dryness and the compound was purified by column chromatography (EtOAc/*n*-hexane, 1/2, *v/v*, dry loading). White solid, yield: 10%. ^{1}H-NMR (400 MHz, $CDCl_3$) δ 2.74 (s, 3H, $C\underline{H}_3$), 7.32 (dd, *J* = 15.8, 7.5 Hz, 1H, $CHC\underline{H}CHO$), 7.46 (t, *J* = 7.4 Hz, 1H, Ar-H), 7.51 (s, 1H, Ar-H), 7.52–7.59 (m, 2H, 2 × Ar-H), 7.61–7.69 (m, 3H, 2 × Ar-H and $C\underline{H}CHCHO$), 7.85 (s, 1H, Ar-H), 8.14 (s, 1H, Ar-H), 9.74 (d, *J* = 7.5 Hz, 1H, $CHCHC\underline{H}O$); ^{13}C-NMR (101 MHz, $CDCl_3$) δ 16.21, 100.29, 116.93, 117.39, 118.27, 122.59, 124.84, 127.79, 128.42, 129.49, 130.86, 134.87, 143.45, 143.55, 151.20, 152.79, 157.88, 158.95, 194.53; HRMS (ESI) *m/z* calculated for $C_{21}H_{15}O_4$ $[M + H]^+$ 331.0965, found 331.0979. Purity by HPLC (0–18 min; 70% *n*-hexane/isopropanol): 97%.

Synthesis of 5-methyl-*N*-((*S*)-1-((*R*)-2-methyloxiran-2-yl)-L-oxopropan-2-yl)-7-oxo-3-phenyl-7*H*-furo[3,2-*g*]chromene-6-carboxamide (**20**):

To a cooled (0 °C) solution of compound **17** (160 mg, 0.50 mmol, 1 equiv.) in DMF (4 mL), HATU (285 mg, 0.75 mmol, 1.5 equiv.) and HOBt hydrate (115 mg, 0.75 mmol, 1.5 equiv.) were added. In a separate round-bottom flask, compound **18** (115 mg, 0.50 mmol, 1 equiv.) was dissolved in CH_2Cl_2 (3 mL) at 0 °C, followed by the addition of TFA (3 mL). After 30 min of stirring at 0 °C, the volatiles were evaporated under reduced pressure thoroughly, the residue was dissolved in CH_2Cl_2 and slowly added to the mixture containing compound **17** at 0 °C. After 5 min, DIPEA (348 μL, 285 mg, 2.0 mmol, 4 equiv.)

was added and the reaction mixture stirred at room temperature for 24 h. Then, the solvent was evaporated and the product purified by column chromatography (EtOAc/*n*-hexane, 1/1, *v*/*v*, dry loading) without additional work-up. Off-white solid, yield: 16%. ^{1}H-NMR (400 MHz, $CDCl_3$) δ 1.44 (d, *J* = 7.0 Hz, 3H, CHC<u>H</u>$_3$), 1.57 (s, 3H, C<u>H</u>$_3$), 2.75 (s, 3H, Ar-C<u>H</u>$_3$), 2.96 (d, *J* = 5.0 Hz, 1H, one H of oxirane C<u>H</u>$_2$), 3.39 (app d, *J* = 5.0 Hz, 1H, one H of oxirane C<u>H</u>$_2$), 4.72 (p, *J* = 6.7 Hz, 1H, C<u>H</u>CH$_3$), 7.42–7.48 (m, 1H, Ar-<u>H</u>), 7.51 (s, 1H, Ar-<u>H</u>), 7.52–7.57 (m, 2H, Ar-<u>H</u>), 7.59–7.65 (m, 3H, CON<u>H</u> + Ar-<u>H</u>), 7.85 (s, 1H, Ar-<u>H</u>), 8.13 (s, 1H, Ar-<u>H</u>), ^{13}C-NMR (101 MHz, $CDCl_3$) δ 16.87, 16.92 (2C), 48.68, 52.70, 59.15, 100.04, 116.76, 117.32, 118.53, 122.43, 124.69, 127.63, 128.23, 129.33, 130.71, 143.27, 150.83, 155.64, 157.64, 160.02, 163.86, 208.01; HRMS (ESI) *m/z* calculated for $C_{25}H_{22}O_6N$ [M + H]$^+$ 432.1442, found 432.1438. Purity by HPLC (0–18 min; 70% *n*-hexane/isopropanol): 96%.

Synthesis of 5-methyl-*N*-((*S*)-1-((*R*)-2-methyloxiran-2-yl)-L-oxo-3-phenylpropan-2-yl)-7-oxo-3-phenyl-7*H*-furo[3,2-*g*]chromene-6-carboxamide (**21**):

To a cooled (0 °C) solution of compound **17** (28 mg, 0.087 mmol, 1 equiv.) in CH_2Cl_2 (4 mL), HATU (40 mg, 0.11 mmol, 1.2 equiv.) and HOBt hydrate (17 mg, 0.11 mmol, 1.2 equiv.) were added. In a separate round-bottom flask, compound **19** (27 mg, 0.087 mmol, 1 equiv.) was dissolved in CH_2Cl_2 (2 mL) at 0 °C, followed by the addition of TFA (2 mL). After 30 min of stirring at 0 °C, the volatiles were evaporated under reduced pressure thoroughly, the residue was dissolved in CH_2Cl_2 and slowly added to the mixture containing compound **17** at 0 °C. After 5 min, DIPEA (58 μL, 45 mg, 0.35 mmol, 4 equiv.) was added and the reaction mixture stirred at room temperature for 24 h. Then, the solvent was evaporated and the product purified by column chromatography (EtOAc/*n*-hexane, 1/2, *v*/*v*, dry loading) without additional work-up. Off-white solid, yield: 11%. ^{1}H-NMR (400 MHz, $CDCl_3$) δ 1.55 (s, 3H, C<u>H</u>$_3$), 2.65 (s, 3H, Ar-C<u>H</u>$_3$), 2.88 (dd, *J* = 13.7, 8.7 Hz, 1H, one H of oxirane C<u>H</u>$_2$), 2.98 (d, *J* = 5.0 Hz, 1H, one H of oxirane C<u>H</u>$_2$), 3.27 (dd, *J* = 13.7, 4.8 Hz, 1H, one H of CHC<u>H</u>$_2$Ph), 3.47 (dd, *J* = 5.0, 0.5 Hz, 1H, one H of CHC<u>H</u>$_2$Ph), 4.99 (symm m, 1H, C<u>H</u>CH$_3$), 7.27–7.35 (m, 5H, Ar-<u>H</u>), 7.42–7.47 (m, 1H, Ar-<u>H</u>), 7.50 (d, *J* = 0.4 Hz, 1H, Ar-<u>H</u>), 7.52–7.56 (m, 2H, Ar-<u>H</u>), 7.59–7.63 (m, 2H, Ar-<u>H</u>), 7.84 (s, 1H, Ar-<u>H</u>), 7.85 (br d, *J* = 6.7 Hz, 1H, CON<u>H</u>), 8.11 (s, 1H, Ar-<u>H</u>); ^{13}C-NMR was not recorded due to insufficient amount of the final product; HRMS (ESI) *m/z* calculated for $C_{31}H_{26}O_6N$ [M + H]$^+$ 508.1755, found 508.1755. Purity by HPLC (0–18 min; 95% *n*-hexane/isopropanol): 97%.

Synthesis of 3-bromo-*N*-((5-methyl-7-oxo-3-phenyl-7*H*-furo[3,2-*g*]chromen-6-yl)methyl) propanamide (**23**):

To a cooled (0 °C) solution of compound **22** (92 mg, 0.3 mmol, 1 equiv.) in CH_2Cl_2 (10 mL), K_2CO_3 (50 mg, 0.36 mmol, 1.2 equiv.) was added. After 5 min, 3-bromopropanoyl chloride (36 μL, 62 mg, 0.36 mmol, 1.2 equiv.) was added drop-wise at 0 °C. The reaction mixture was then stirred at room temperature for 3 h. The reaction was quenched by the addition of H_2O (20 mL) and the mixture was extracted with EtOAc (3 × 50 mL). The combined organic extracts were dried over Na_2SO_4, filtered, and evaporated under reduced pressure. The product was purified by column chromatography (EtOAc/*n*-hexane, 1/2, *v*/*v*). White solid, yield: 90%. ^{1}H-NMR (400 MHz, $CDCl_3$) δ 2.73 (s, 3H, C<u>H</u>$_3$), 2,74 (t, *J* = 6.7 Hz, 2H, COC<u>H</u>$_2$), 3.61 (t, *J* = 6.7 Hz, 2H, C<u>H</u>$_2$Br), 4.52 (d, *J* = 6.2 Hz, 2H, Ar-C<u>H</u>$_2$NH), 7.42–7.47 (m, 1H, Ar-<u>H</u>), 7.51–7.56 (m, 3H, Ar-<u>H</u>), 7.61–7.64 (m, 2H, Ar-<u>H</u>), 7.83 (s, 1H, Ar-<u>H</u>), 8.07 (s, 1H, Ar-<u>H</u>); ^{13}C-NMR (101 MHz, $CDCl_3$) δ 15.59, 27.13, 36.69, 39.51, 100.01, 116.75, 117.07, 120.68, 122.43, 124.31, 127.61 (2C), 128.15, 129.32 (2C), 130.96, 142.99, 149.61, 150.66, 156.81, 162.37, 169.52; HRMS (*m/z*) (ESI): calculated for $C_{22}H_{19}O_4NBr$ [M + H]$^+$ 440.0492, found: 440.0490; Purity by HPLC (0–18 min; 70% *n*-hexane/isopropanol): 96%.

Synthesis of 1-((5-methyl-7-oxo-3-phenyl-7*H*-furo[3,2-*g*]chromen-6-yl)methyl)azetidin-2-one (**24**):

To a cooled (0 °C) solution of compound **23** (66 mg, 0.15 mmol, 1 equiv.) in DMF (15 mL), NaO*t*Bu (16 mg, 0.17 mmol, 1.1 equiv.) was added. The reaction mixture was stirred at room temperature for 24 h. The reaction was quenched by the addition of H_2O (20 mL) and the mixture was extracted with EtOAc (3 × 50 mL). The combined organic extracts were dried over Na_2SO_4, filtered, and evaporated under reduced pressure. The product was purified by column chromatography (EtOAc/*n*-hexane, 1/2, *v*/*v*). White solid, yield: 87%. ^{1}H-NMR (400 MHz, $CDCl_3$) δ 2.73 (s, 3H, C$\underline{H}_3$), 2.74 (t, *J* = 6.4 Hz, 2H, azetidin-2-one-C$\underline{H}_2$), 3.61 (t, *J* = 6.4 Hz, 2H, azetidin-2-one-C$\underline{H}_2$), 4.52 (s, 2H, Ar-C$\underline{H}_2$N), 7.42–7.47 (m, 1H, Ar-$\underline{H}$), 7.52 (s, 1H, Ar-$\underline{H}$), 7.53–7.56 (m, 2H, Ar-$\underline{H}$), 7.61–7.65 (m, 2H, Ar-$\underline{H}$), 7.84 (s, 1H, Ar-$\underline{H}$), 8.07 (s, 1H, Ar-$\underline{H}$); ^{13}C-NMR (101 MHz, $CDCl_3$) δ 29.72, 31.95, 37.04, 38.55, 102.72, 112.28, 115.88, 116.54, 116.80, 122.43, 122.51, 127.62 (2C), 128.16, 128.74, 129.33 (2C), 132.78, 143.00, 156.80, 157.68, 178.20; HRMS (*m/z*) (ESI): calculated for $C_{22}H_{18}O_4N$ $[M + H]^+$ 360.1230, found: 360.1222; Purity by HPLC (0–18 min; 70% *n*-hexane/isopropanol): 87%.

3.3. Residual Activity Measurements

The screening of compounds was performed at 1 μM final concentrations in the assay buffer (0.01% SDS, 50 mM Tris-HCl, 0.5 mM EDTA, pH 7.4). Stock solutions of compounds were prepared in DMSO. To 50 μL of each compound, 25 μL of 0.8 nM human iCP or human cCP (both from Boston Biochem, Inc., Cambridge, MA, USA) was added. After 30 min incubation at 37 °C, the reaction was initiated by the addition of 25 μL of 100 μM relevant fluorogenic substrate: acetyl-Nle-Pro-Nle-Asp-AMC (Ac-nLPnLD-AMC, [Bachem, Bubendorf, Switzerland]) for β1, acetyl-Pro-Ala-Leu-7-amino-4-methylcoumarin (Ac-PAL-AMC, [Boston Biochem, Inc., Cambridge, MA, USA]) for β1i, *t*-butyloxycarbonyl-Leu-Arg-Arg-7-amino-4-methylcoumarin (Boc-LRR-AMC, [Bachem, Bubendorf, Switzerland]) for β2 and β2i, succinyl-Leu-Leu-Val-Tyr-7-amino-4-methylcoumarin (Suc-LLVY-AMC [Bachem, Bubendorf, Switzerland]) for β5 and β5i. The reaction progress was recorded on the BioTek Synergy HT microplate reader by monitoring fluorescence at 460 nm (λ_{ex} = 360 nm) for 90 min at 37 °C. The initial linear ranges were used to calculate the velocity and to determine the residual activity.

In the case of the β1, β1i, β2, and β2i activity inhibition determination, the assay buffer was modified; SDS was replaced with the proteasomal activator PA28α (Boston Biochem, Inc., Cambridge, MA, USA).

3.4. Molecular Modelling

Compounds were prepared for docking using LigPrep (Schrödinger Suite 2020-2, Schrödinger, LLC, New York, NY, USA, 2020) to account for all possible tautomers and ionization states at pH 7.0 ± 2.0. The X-ray structure (PDB: 5M2B, [43]) of yeast 20S proteasome with human β5i and β1 subunits in complex with noncovalent inhibitor Ro19 was used for docking. The binding site is defined by the chain K (β5i) and neighbouring L (β1), so all other chains were removed. Protein Preparation Wizard [50] was used to add hydrogen atoms, protonate residues at pH 7, refine the H-bond network and to perform a restrained minimization. The receptor's grid box required for docking calculations was centred on the corresponding co-crystallized ligand. Noncovalent docking was performed using Glide [51], with the following parameters: XP (extra precision), flexible ligand sampling, perform postdocking minimization. Covalent docking was performed with CovDock program [52] using the pose prediction mode with default setup and Thr1 defined as the reactive residue. Nucleophilic addition to a double bond (oxathiazolone) was selected as the reaction.

4. Conclusions

Here, we showed that the introduction of 12 new electrophilic warheads at position 3 of the psoralen ring led to compounds with abrogated inhibition of the iCP (especially β5i subunit). As already described in the Introduction, it is imperative that the initial non-covalent binding of a given compound is followed by the positioning of the electrophilic 'warhead' near the desired nucleophilic amino-acid residue of the protein to achieve covalent interaction. Poor inhibition results were in our cases most probably due to the mispositioning of the electrophilic carbon and the catalytic Thr1O$^\gamma$ (Figure 7). The oxathiazolone thus remains the optimal electrophilic moiety for this compound class. Despite somewhat disappointing results, the obtained data will help steer our future research in the field of psoralen-based iCP inhibitors, e.g., when designing inhibitors which simultaneously inhibit two iCP subunits as it was established that simultaneous inhibition of β1i and β5i is necessary to achieve significant anti-inflammatory effects.

Author Contributions: Conceptualization, J.M., S.G., and I.S.; formal analysis, M.G. and I.S.; funding acquisition, S.G.; investigation, E.S.S., M.P., L.R., and A.Š.; methodology, E.S.S., M.P., M.G., and I.S.; project administration, S.G.; supervision, J.M. and I.S.; writing—original draft, E.S.S.; writing—review & editing, S.G. and I.S. All authors have read and agreed to the published version of the manuscript.

Funding: This research was funded by the Slovenian Research Agency, research core funding No. P1-0208, Grant number N1-0068 to S.G., and Grant number J3-1745 to M.G.

Data Availability Statement: The data presented in this study are available on request from the corresponding author.

Acknowledgments: The authors acknowledge Maja Frelih for HRMS measurements.

Conflicts of Interest: The authors declare no conflict of interest.

Sample Availability: Samples of all compounds, except compound **21** are available from the authors.

References

1. Rousseau, A.; Bertolotti, A. Regulation of proteasome assembly and activity in health and disease. *Nat. Rev. Mol. Cell Biol.* **2018**, *19*, 697–712. [CrossRef]
2. Hershko, A.; Ciechanover, A. The ubiquitin system. *Annu. Rev. Biochem.* **1998**, *67*, 425–479. [CrossRef]
3. DeMartino, G.N.; Gillette, T.G. Proteasomes: Machines for all reasons. *Cell* **2007**, *129*, 659–662. [CrossRef]
4. Collins, G.A.; Goldberg, A.L. The logic of the 26S proteasome. *Cell* **2017**, *169*, 792–806. [CrossRef]
5. Goldberg, A.L. Functions of the proteasome: From protein degradation and immune surveillance to cancer therapy. *Biochem. Soc. Trans.* **2007**, *35*, 12–17. [CrossRef]
6. Gallastegui, N.; Groll, M. The 26S proteasome: Assembly and function of a destructive machine. *Trends Biochem. Sci.* **2010**, *35*, 634–642. [CrossRef]
7. Thibaudeau, T.A.; Smith, D.M. A practical review of proteasome pharmacology. *Pharmacol. Rev.* **2019**, *71*, 170–197. [CrossRef]
8. Groll, M.; Ditzel, L.; Lowe, J.; Stock, D.; Bochtler, M.; Bartunikt, H.D.; Huber, R. Structure of 20S proteasome from yeast at 2.4 A resolution. *Nature* **1997**, *386*, 463–471. [CrossRef]
9. Arendt, C.S.; Hochstrasser, M. Identification of the Yeast 20S Proteasome Catalytic Centers and Subunit Interactions Required for Active-Site Formation. *Proc. Natl. Acad. Sci. USA* **1997**, *94*, 7156–7161. [CrossRef]
10. Budenholzer, L.; Cheng, C.L.; Li, Y.; Hochstrasser, M. Proteasome structure and assembly. *J. Mol. Biol.* **2017**, *429*, 3500–3524. [CrossRef]
11. Murata, S.; Sasaki, K.; Kishimoto, T.; Niwa, S.-I.; Hayashi, H.; Takahama, Y.; Tanaka, K. Regulation of $CD8^+$ T cell development by thymus-specific proteasomes. *Science* **2007**, *316*, 1349–1353. [CrossRef] [PubMed]
12. Groettrup, M.; Kirk, C.J.; Basler, M. Proteasomes in immune cells: More than peptide producers? *Nat. Rev. Immunol.* **2010**, *10*, 73–78. [CrossRef] [PubMed]
13. Tanaka, K. Role of proteasomes modified by interferon-γ in antigen processing. *J. Leukoc. Biol.* **1994**, *56*, 571–575. [CrossRef]
14. Aki, M.; Shimbara, N.; Takashina, M.; Akiyama, K.; Kagawa, S.; Tamura, T.; Tanahashi, N.; Yoshimura, T.; Tanaka, K.; Ichihara, A. Interferon-γ induces different subunit organizations and functional diversity of proteasomes. *J. Biochem.* **1994**, *115*, 257–269. [CrossRef]
15. Nandi, D.; Jiang, H.; Monaco, J.J. Identification of MECL-1 (LMP-10) as the third IFN-gamma-inducible proteasome subunit. *J. Immunol.* **1996**, *156*, 2361–2364.
16. Basler, M.; Kirk, C.J.; Groettrup, M. The immunoproteasome in antigen processing and other immunological functions. *Curr. Opin. Immunol.* **2013**, *25*, 74–80. [CrossRef]
17. Almond, J.B.; Cohen, G.M. The proteasome: A novel target for cancer chemotherapy. *Leukemia* **2002**, *16*, 433–443. [CrossRef]
18. Kuhn, D.J.; Orlowski, R.Z. The immunoproteasome as a target in hematologic malignancies. *Semin. Hematol.* **2012**, *49*, 258–262. [CrossRef]
19. Ito, S. Proteasome inhibitors for the treatment of multiple myeloma. *Cancers* **2020**, *12*, 265. [CrossRef]
20. Verbrugge, E.E.; Scheper, R.J.; Lems, W.F.; de Gruijl, T.D.; Jansen, G. Proteasome inhibitors as experimental therapeutics of autoimmune diseases. *Arthritis Res. Ther.* **2015**, *17*, 17. [CrossRef]
21. Basler, M.; Mundt, S.; Bitzer, A.; Schmidt, C.; Groettrup, M. The immunoproteasome: A novel drug target for autoimmune diseases. *Clin. Exp. Rheumatol.* **2015**, *33*, 74–79.

22. Zhang, C.; Zhu, H.; Shao, J.; He, R.; Xi, J.; Zhuang, R.; Zhang, J. Immunoproteasome-selective inhibitors: The future of autoimmune diseases? *Future Med. Chem.* **2020**, *12*, 269–272. [CrossRef] [PubMed]
23. Limanaqi, F.; Biagioni, F.; Gaglione, A.; Busceti, C.L.; Fornai, F. A sentinel in the crosstalk between the nervous and immune system: The (immuno)-proteasome. *Front. Immunol.* **2019**, *10*, 628. [CrossRef]
24. Huber, E.M.; Groll, M. Inhibitors for the immuno- and constitutive proteasome: Current and future trends in drug development. *Angew. Chem. Int. Ed.* **2012**, *51*, 8708–8720. [CrossRef]
25. Cromm, P.M.; Crews, C.M. The proteasome in modern drug discovery: Second life of a highly valuable drug target. *ACS Cent. Sci.* **2017**, *3*, 830–838. [CrossRef]
26. Richy, N.; Sarraf, D.; Maréchal, X.; Janmamode, N.; Le Guével, R.; Genin, E.; Reboud-Ravaux, M.; Vidal, J. Structure-based design of human immuno- and constitutive proteasomes inhibitors. *Eur. J. Med. Chem.* **2018**, *145*, 570–587. [CrossRef]
27. Sherman, D.J.; Li, J. Proteasome inhibitors: Harnessing proteostasis to combat disease. *Molecules* **2020**, *25*, 671. [CrossRef]
28. Muchamuel, T.; Basler, M.; Aujay, M.A.; Suzuki, E.; Kalim, K.W.; Lauer, C.; Sylvain, C.; Ring, E.R.; Shields, J.; Jiang, J.; et al. A selective inhibitor of the immunoproteasome subunit LMP7 blocks cytokine production and attenuates progression of experimental arthritis. *Nat. Med.* **2009**, *7*, 781–788. [CrossRef]
29. Johnson, H.W.B.; Lowe, E.; Anderl, J.L.; Fan, A.; Muchamuel, T.; Bowers, S.; Moebius, D.C.; Kirk, C.; McMinn, D.L. Required immunoproteasome subunit inhibition profile for anti-inflammatory efficacy and clinical candidate KZR-616 ((2*S*,3*R*)-*N*-((*S*)-3-(cyclopent-1-en-1-yl)-1-((*R*)-2-methyloxiran-2-yl)-1-oxopropan-2-yl)-3-hydroxy-3-(4-methoxyphenyl)-2-((*S*)-2-(2-morpholinoacetamido)propanamido)propenamide). *J. Med. Chem.* **2018**, *61*, 11127–11143. [CrossRef] [PubMed]
30. Basler, M.; Lindstrom, M.M.; LaStant, J.J.; Bradshaw, J.M.; Owens, T.D.; Schmidt, C.; Meurits, E.; Tsu, C.; Overkleeft, H.S.; Kirk, C.J.; et al. Co-inhibition of immunoproteasome subunits LMP2 and LMP7 is required to block autoimmunity. *EMBO Rep.* **2018**, *19*, e46512. [CrossRef] [PubMed]
31. Ettari, R.; Zappalà, M.; Grasso, S.; Musolino, C.; Innao, V.; Allegra, A. Immunoproteasome-selective and non-selective inhibitors: A promising approach for the treatment of multiple myeloma. *Pharmacol. Ther.* **2018**, *182*, 176–192. [CrossRef] [PubMed]
32. Xi, J.; Zhuang, R.; Kong, L.; He, R.; Zhu, H.; Zhang, J. Immunoproteasome-selective inhibitors: An overview of recent developments as potential drugs for hematologic malignancies and autoimmune diseases. *Eur. J. Med. Chem.* **2019**, *182*, 111646. [CrossRef]
33. Ladi, E.; Everett, C.; Stivala, C.E.; Daniels, B.E.; Durk, M.R.; Harris, S.F.; Huestis, M.P.; Purkey, H.E.; Staben, S.T.; Augustin, M.; et al. Design and evaluation of highly selective human immunoproteasome inhibitors reveal a compensatory process that preserves immune cell viability. *J. Med. Chem.* **2019**, *62*, 7032–7041. [CrossRef] [PubMed]
34. Karreci, E.S.; Fan, H.; Uehara, M.; Mihali, A.B.; Singh, P.K.; Kurdi, A.T.; Solhjou, Z.; Riella, L.V.; Ghobrial, I.; Laragione, T.; et al. Brief treatment with a highly selective immunoproteasome inhibitor promotes long-term cardiac allograft acceptance in mice. *Proc. Natl. Acad. Sci. USA* **2016**, *113*, E8425–E8432. [CrossRef] [PubMed]
35. Ogorevc, E.; Schiffrer, E.S.; Sosič, I.; Gobec, S. A patent review of immunoproteasome inhibitors. *Expert Opin. Ther. Pat.* **2018**, *28*, 517–540. [CrossRef]
36. Tan, C.R.C.; Abdul-Majeed, S.; Cael, B.; Barta, S.K. Clinical pharmacokinetics and pharmacodynamics of bortezomib. *Clin. Pharmacokinet.* **2019**, *58*, 157–168. [CrossRef]
37. Yang, J.; Wang, Z.; Fang, Y.; Jiang, J.; Zhao, F.; Wong, H.; Bennett, M.K.; Molineaux, C.J.; Kirk, C.J. Pharmacokinetics, pharmacodynamics, metabolism, distribution, and excretion of carfilzomib in rats. *Drug Metab. Dispos.* **2011**, *39*, 1873–1882. [CrossRef]
38. Wang, Z.; Yang, J.; Kirk, C.; Fang, Y.; Alsina, M.; Badros, A.; Papadopoulos, K.; Wong, A.; Woo, T.; Bomba, D.; et al. Clinical pharmacokinetics, metabolism, and drug-drug interaction of carfilzomib. *Drug Metab. Dispos.* **2013**, *41*, 230–237. [CrossRef]
39. Sosič, I.; Gobec, M.; Brus, B.; Knez, D.; Živec, M.; Konc, J.; Lešnik, S.; Ogrizek, M.; Obreza, A.; Žigon, D.; et al. Nonpeptidic selective inhibitors of the chymotrypsin-like (β5i) subunit of the immunoproteasome. *Angew. Chem. Int. Ed.* **2016**, *55*, 5745–5748. [CrossRef]
40. Schiffrer, E.S.; Sosič, I.; Šterman, A.; Mravljak, J.; Raščan, I.M.; Gobec, S.; Gobec, M. A focused structure-activity relationship study of psoralen-based immunoproteasome inhibitors. *Med. Chem. Commun.* **2019**, *10*, 1958–1965. [CrossRef]
41. Fan, H.; Angelo, N.G.; Warren, J.D.; Nathan, C.F.; Lin, G. Oxathiazolones selectively inhibit the human immunoproteasome over the constitutive proteasome. *ACS Med. Chem. Lett.* **2014**, *5*, 405–410. [CrossRef] [PubMed]
42. Kasam, V.; Lee, N.-R.; Kim, K.-B.; Zhan, C.-G. Selective immunoproteasome inhibitors with non-peptide scaffolds identified from structure-based virtual screening. *Bioorg. Med. Chem. Lett.* **2014**, *24*, 3614–3617. [CrossRef] [PubMed]
43. Cui, H.; Baur, R.; Le Chapelain, C.; Dubiella, C.; Heinemeyer, W.; Huber, E.M.; Groll, M. Structural elucidation of a nonpeptidic inhibitor specific for the human immunoproteasome. *ChemBioChem* **2017**, *18*, 523–526. [CrossRef] [PubMed]
44. Scarpino, A.; Bajusz, D.; Proj, M.; Gobec, M.; Sosič, I.; Gobec, S.; Ferenczy, G.G.; Keseru, G.M. Discovery of immunoproteasome inhibitors using large-scale covalent virtual screening. *Molecules* **2019**, *24*, 2590. [CrossRef] [PubMed]
45. Singh, J.; Petter, R.C.; Baillie, T.A.; Whitty, A. The resurgence of covalent drugs. *Nat. Rev. Drug Discov.* **2011**, *10*, 307–317. [CrossRef] [PubMed]
46. Pinto, A.; Tamborini, L.; Cullia, G.; Conti, P.; De Micheli, C. Inspired by nature: The 3-halo-4,5-dihydroisoxazole moiety as a novel molecular warhead for the design of covalent inhibitors. *ChemMedChem* **2016**, *11*, 10–14. [CrossRef]
47. Huber, E.M.; de Bruin, G.; Heinemeyer, W.; Soriano, G.P.; Overkleeft, H.S.; Groll, M. Systematic analyses of substrate preferences of 20S proteasomes using peptidic epoxyketone inhibitors. *J. Am. Chem. Soc.* **2015**, *137*, 7835–7842. [CrossRef]

48. Zhou, H.J.; Aujay, M.A.; Bennett, M.K.; Dajee, M.; Demo, S.D.; Fang, Y.; Ho, M.N.; Jiang, J.; Kirk, C.J.; Laidig, G.J.; et al. Design and synthesis of an orally bioavailable and selective peptide epoxyketone proteasome inhibitor (PR-047). *J. Med. Chem.* **2009**, *52*, 3028–3038. [CrossRef]
49. Chaudhari, D.A.; Fernandes, R.A. Hypervalent iodine as a terminal oxidant in wacker-type oxidation of terminal olefins to methyl ketones. *J. Org. Chem.* **2016**, *81*, 2113–2121. [CrossRef]
50. Madhavi Sastry, G.; Adzhigirey, M.; Day, T.; Annabhimoju, R.; Sherman, W. Protein and ligand preparation: Parameters, protocols, and influence on virtual screening enrichments. *J. Comput. Aided. Mol. Des.* **2013**, *27*, 221–234. [CrossRef]
51. Friesner, R.A.; Murphy, R.B.; Repasky, M.P.; Frye, L.L.; Greenwood, J.R.; Halgren, T.A.; Sanschagrin, P.C.; Mainz, D.T. Extra precision glide: Docking and scoring incorporating a model of hydrophobic enclosure for protein-ligand complexes. *J. Med. Chem.* **2006**, *49*, 6177–6196. [CrossRef] [PubMed]
52. Zhu, K.; Borrelli, K.W.; Greenwood, J.R.; Day, T.; Abel, R.; Farid, R.S.; Harder, E. Docking covalent inhibitors: A parameter free approach to pose prediction and scoring. *J. Chem. Inf. Model.* **2014**, *54*, 1932–1940. [CrossRef] [PubMed]

Communication

4-Hydroxy-7-Methoxycoumarin Inhibits Inflammation in LPS-activated RAW264.7 Macrophages by Suppressing NF-κB and MAPK Activation

Jin Kyu Kang and Chang-Gu Hyun *

Jeju Inside Agency and Cosmetic Science Center, Department of Chemistry and Cosmetics, Jeju National University, Jeju 63243, Korea; wlsrbtjsrb@naver.com

* Correspondence: cghyun@jejunu.ac.kr; Tel.: +82-64-754-3542

Academic Editor: Maria João Matos

Received: 14 August 2020; Accepted: 24 September 2020; Published: 26 September 2020

Abstract: Coumarins are natural products with promising pharmacological activities owing to their anti-inflammatory, antioxidant, antiviral, anti-diabetic, and antimicrobial effects. Coumarins are present in many plants and microorganisms and have been widely used as complementary and alternative medicines. To date, the pharmacological efficacy of 4-hydroxy-7-methoxycoumarin (4H-7MTC) has not been reported yet. Therefore, in this study, we investigated the anti-inflammatory effects of 4H-7MTC in LPS-stimulated RAW264.7 cells as well as its mechanisms of action. Cells were treated with various concentrations of 4H-7MTC (0.3, 0.6, 0.9, and 1.2 mM) and 40 μM L-N^6-(1-iminoethyl)-L-lysine (L-NIL) were used as controls. LPS-stimulated RAW264.7 cells showed that 4H-7MTC significantly reduced nitric oxide (NO) and prostaglandin E_2 (PGE_2) production without cytotoxic effects. In addition, 4H-7MTC strongly decreased the expression of inducible nitric oxide synthase (iNOS) and cyclooxygenase (COX-2). Furthermore, 4H-7MTC reduced the production of proinflammatory cytokines such as tumor necrosis factor (TNF)-α, interleukin (IL)-1β, and IL-6. We also found that 4H-7MTC strongly exerted its anti-inflammatory actions by downregulating nuclear factor kappa B (NF-κB) activation by suppressing inhibitor of nuclear factor kappa B alpha (IκBα) degradation in macrophages. Moreover, 4H-7MTC decreased phosphorylation of extracellular signal-regulated kinase (ERK1/2) and c-Jun N-terminal kinase/stress-activated protein kinase (JNK), but not that of p38 MAPK. These results suggest that 4H-7MTC may be a good candidate for the treatment or prevention of inflammatory diseases such as dermatitis, psoriasis, and arthritis. Ultimately, this is the first report describing the effective anti-inflammatory activity of 4H-7MTC.

Keywords: 4-hydroxy-7-methoxycoumarin; macrophage; inflammation; NF-κB; MAPK

1. Introduction

Coumarins (benzo-α-pyrones) are oxygen heterocycles that are naturally occurring benzopyrene derivatives which have been identified in plants, bacteria, and fungi [1]. Coumarins represent a broad family of secondary metabolites that are found naturally in over 1300 plant species. The main pathway of coumarin biosynthesis occurs through the shikimic acid pathway, which involves cinnamic acid and phenylalanine metabolism [2,3]. Natural coumarins are subdivided into several classes according to their chemical diversity and complexity, namely, simple coumarins, isocoumarins, furanocoumarins, pyranocoumarins (both angular and linear), biscoumarins, and phenylcoumarins [4].

Coumarins have several desirable features. First, they have a low molecular weight owing to their simple structures. Second, they have high solubility in most organic solvents. Third, they have high bioavailability and low toxicity. Fourth, they have various pharmacological effects such as anticoagulant, antimicrobial, anti-inflammatory, neuroprotective, antidiabetic, anticonvulsant, and antiproliferative activities [4–6]. These characteristics and advantages support their roles as lead compounds in drug research and development [7]. Coumarins have diverse structures owing to the different types of substitutions in their underlying structures, which can affect biological activity. Thus, the structure-system-activity-relationship of coumarin must be carefully studied [1].

During our ongoing screening program designed to identify modulators of skin inflammation and melanogenesis from coumarin and its derivatives, we reported that 8-methoxycoumarin increased melanogenesis via the MAPK signaling pathway [8]. In addition, we identified that auraptene, the most abundant naturally occurring geranyloxycoumarin, possesses anti-melanogenic activity through ERK-mediated MITF downregulation [9]. Furthermore, we reported that 7,8-dimethoxycoumarin stimulates melanogenesis via MAPK-mediated MITF upregulation and attenuates the expression of IL-6, IL-8, and CCL2/MCP-1 in TNF-α-treated HaCaT cells [10,11].

As an extension of this study, we investigated the anti-inflammatory effects of 4-hydroxy-7-methoxycoumarin (4H-7MTC, Figure 1). 4H-7MTC belongs to a class of organic compounds known as hydroxycoumarins. These are coumarins that contain one or more hydroxyl groups attached to the coumarin skeleton. 4H-7MTC can be found in plants such as coriander, artichoke, Tibetan hulless barley, and eggplant [12,13]. To the best of our knowledge, no studies have reported the pharmacological and biochemical properties and therapeutic applications of 4H-7MTC. Therefore, in this study, we investigated whether 4H-7MTC has anti-inflammatory effects; an initial step in the development of 4H-7MTC as a functional compound for use in human health applications.

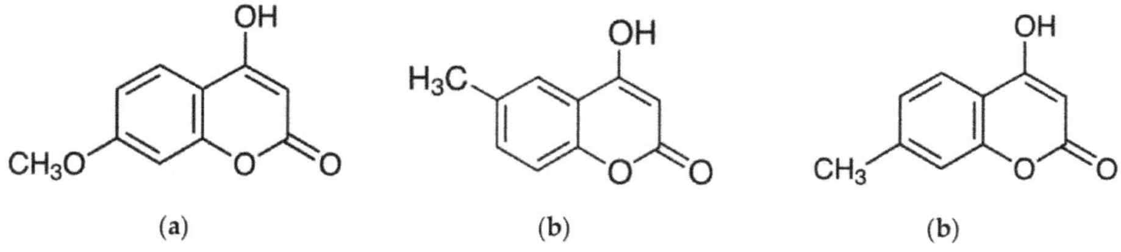

(a) (b) (b)

Figure 1. Structures of 4-hydroxycoumarins: 4-hydroxy-7-methoxycoumarin (**a**), 4-hydroxy-6-methylcoumarin (**b**), and 4-hydroxy-7-methylcoumarin (**c**).

2. Results and Discussion

Macrophages, the main cells responsible for innate immunity, are activated by the invasion of foreign pathogens such as parasites, bacteria, and viruses, or by stimulation with external signals. In particular, lipopolysaccharide (LPS), an endotoxin produced by Gram-negative bacteria, stimulates macrophages, which in turn promotes secretion of proinflammatory cytokines, including tumor necrosis factor (TNF)-α, interleukin (IL)-1β, and IL-6, and induces the expression of inflammatory response factors such as nitric oxide (NO) and prostaglandin E_2 (PGE_2) [14,15]. As such, regulation of the production of NO and proinflammatory cytokines in macrophages is a current research topic for the development of new anti-inflammatory agents, and there have been many attempts to derive new anti-inflammatory agents from natural compounds [16–18].

To demonstrate the anti-inflammatory activity of the three types of 4-hydroxycoumarin, including 4H-7MTC, we first assessed its ability to inhibit NO production in LPS-stimulated macrophage RAW264.7 cells (Figure 1). RAW264.7 cells were treated with various concentrations of 4-hydroxycoumarins, and cell viability was measured using the MTT assay. As shown in Figure 2, NO production increased by 3.43- to 15-fold in LPS-activated macrophages relative to untreated macrophages. Moreover, 4-hydroxycoumarins reduced LPS-induced NO production in a concentration-dependent manner. At 0.6 mM concentration of 4H-7MTC, the production of NO by LPS-treated macrophages decreased by 23.10%. At 0.5 mM concentration of 4H-6MC and 4H-7MC, the production of NO by LPS-treated macrophages decreased by 21.27% and 17.61%, respectively.

These results show that the 4-hydroxy structure of coumarin influences the degree of inhibition of NO production, and the substituents on carbon 6 and 7 of the B-ring structure had little effect on the inhibition of NO production. No concentration of 4-hydroxycoumarins displayed significant cytotoxicity, indicating that the anti-inflammatory effects of 4-hydroxycoumarins were not attributable to cytotoxicity. Among them, we found that 4H-7MTC is a safe substance that does not induce cytotoxicity even at concentrations as high as 1.2 mM.

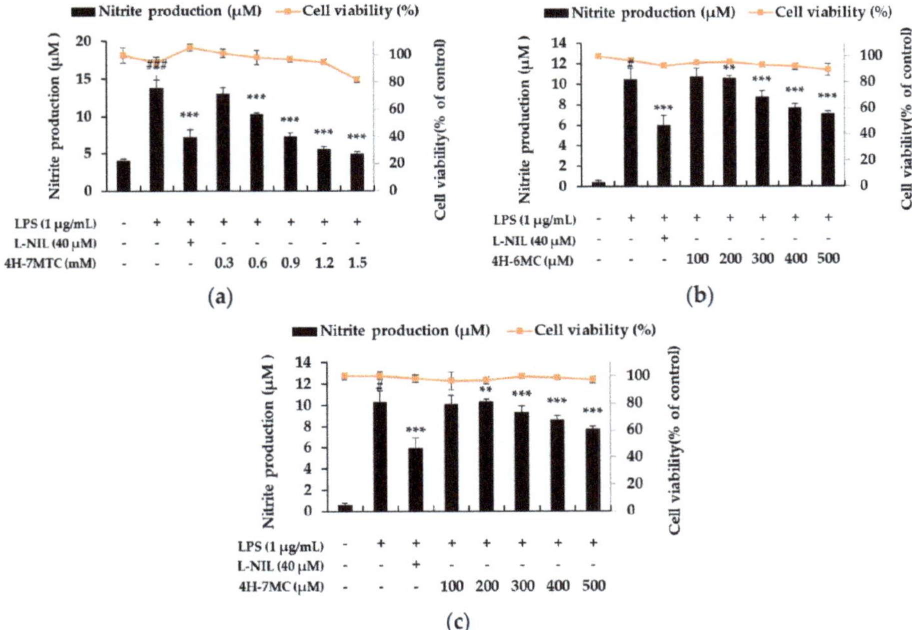

Figure 2. Effect of 4H-7MTC (**a**), 4H-6MC (**b**), and 4H-7MC (**c**), on nitric oxide production in LPS-stimulated RAW264.7 cells. The cells were plated in 24-well plates (1.5×10^5 cells/well), incubated for 24 h, and then pretreated with 4H-7MTC (0.3, 0.6, 0.9, 1.2, and 1.5 mM), 4H-6MC (100, 200, 300, 400, and 500 μM), and 4H-7MC (100, 200, 300, 400, and 500 μM) for 1 h, followed by LPS stimulation for 24 h. Cytotoxicity of 4H-7MTC, 4H-6MC, and 4H-7MC were evaluated using MTT assay. Nitric oxide production was determined by the Griess reagent method. L-N6-(1-Iminoethyl) lysine dihydrochloride (L-NIL) was used as a positive control. The data are presented as mean ± SD. Statistical significance was assessed by one-way analysis of variance (ANOVA), followed by Tukey's post-hoc test and represented as follows: $^{\#}p < 0.05$, $^{\#\#\#}p < 0.005$, $^{**}p < 0.01$, $^{***}p < 0.001$ vs. LPS alone.

To investigate the additional functionalities of 4H-7MTC, which was confirmed to be safe at high concentrations, we aimed to evaluate its potential activity as an anticancer agent or as a preventive of gray hair. As shown in Figure 3a, 4H-7MTC upregulated melanin production in a concentration-dependent manner over a wide concentration range (25–200 μM), without any observed cytotoxicity. Additionally, 4H-7MTC showed no cytotoxicity up to 1.2 mM in normal macrophages, whereas it exhibited a cytotoxic effect on B16F10 melanoma cells at a low concentration of 0.3 mM (Figure 3b). This suggests that 4H-7MTC could be a potential anticancer agent.

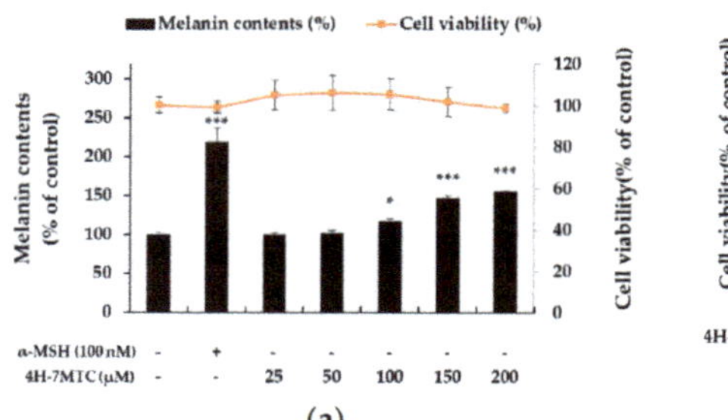

(a)

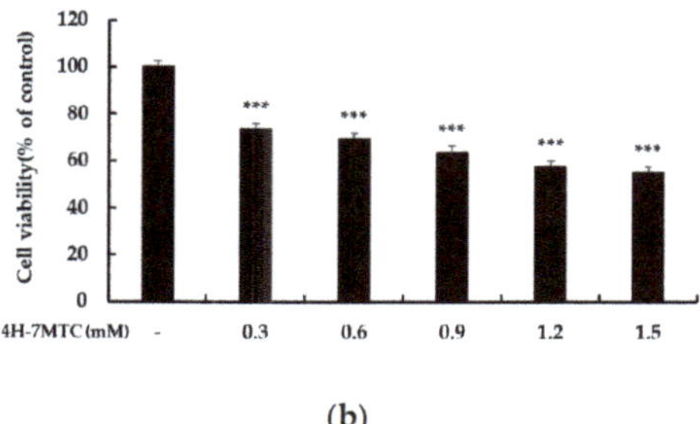

(b)

Figure 3. Effect of 4H-7MTC on the production of melanin (**a**) in α-MSH-stimulated B16F10 cells and Cytotoxicity of 4H-7MTC in B16F10 cells (**b**). Cells were plated in 60 mm cell culture dish (6.0×10^4 cells/dish), incubated for 24 h, and then treated with 4H-7MTC (25, 50, 100, 150 and 200 μM) for 72 h in the presence of α-MSH (100 nM). α-MSH was used as the negative control. Cytotoxicity of 4H-7MTC was evaluated using MTT assay. Cells were plated in 24-well plates (1.5×10^4 cells/well) for 24 h, and then treated with 4H-7MTC (0.3, 0.6, 0.9, 1.2, and 1.5 mM) for 72 h. The data are presented as mean ± SD. Statistical significance was assessed by one-way analysis of variance (ANOVA) followed by Tukey's post-hoc test and represented as follows: * $p < 0.05$, *** $p < 0.001$ vs. LPS alone.

To further elucidate the anti-inflammatory mechanisms of 4H-7MTC, we measured the levels of PGE_2, IL-6, IL-1β, and TNF-α in culture supernatants using ELISA. Treatment of RAW264.7 cells with LPS alone resulted in a significant increase in cytokine production compared to that in the drug groups (Figure 4). However, NO, PGE_2, IL-6, IL-1β, and TNF-α levels in the supernatants of LPS-stimulated cells pretreated with 0.3, 0.6, 0.9, and 1.2 mM 4H-7MTC were significantly reduced compared to those in the LPS group in a concentration-dependent manner (Figure 4).

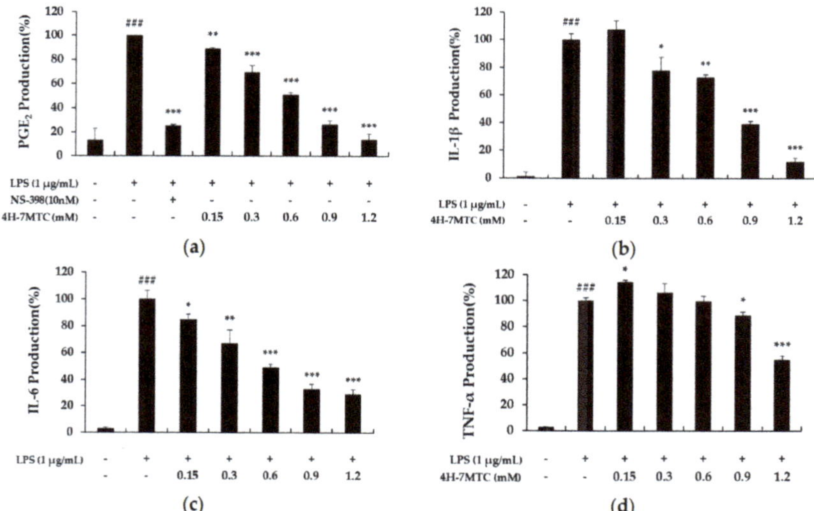

Figure 4. The effect of 4-hydroxy-7-methoxycoumarin (4H-7MTC) on the LPS-induced production of proinflammatory cytokines in RAW264.7 cells. Cells were pretreated with 4H-7MTC (0.15, 0.3, 0.6, 0.9, and 1.2 mM) for 1 h and then stimulated for 20 h with LPS. The production of PGE_2 (**a**), IL-1β (**b**), IL-6 (**c**), and TNF-α (**d**) were determined using ELISA. The data are presented as the mean ± SD. Statistical significance was assessed by one-way analysis of variance (ANOVA) followed by Tukey's post-hoc test and represented as follows: Values are representative of three independent experiments. ### $p < 0.005$ vs. control cells. * $p < 0.05$, ** $p < 0.01$, *** $p < 0.001$ vs. LPS alone.

To further elucidate the mechanisms by which 4H-7MTC inhibited NO and PGE_2 production in LPS-activated macrophages, we analyzed the effects of 4H-7MTC on LPS-induced iNOS and COX-2 gene expression in macrophages. Under normal conditions, RAW264.7 cells expressed non-detectable levels of COX-2 expression, but iNOS and COX-2 protein levels markedly increased after 18 h of LPS stimulation (Figure 5). With the addition of 4H-7MTC (0.3, 0.6, 0.9, and 1.2 mM), concentration-dependent inhibition of iNOS and COX-2 expression was observed, indicating that 4H-7MTC modulates iNOS and COX-2 expression.

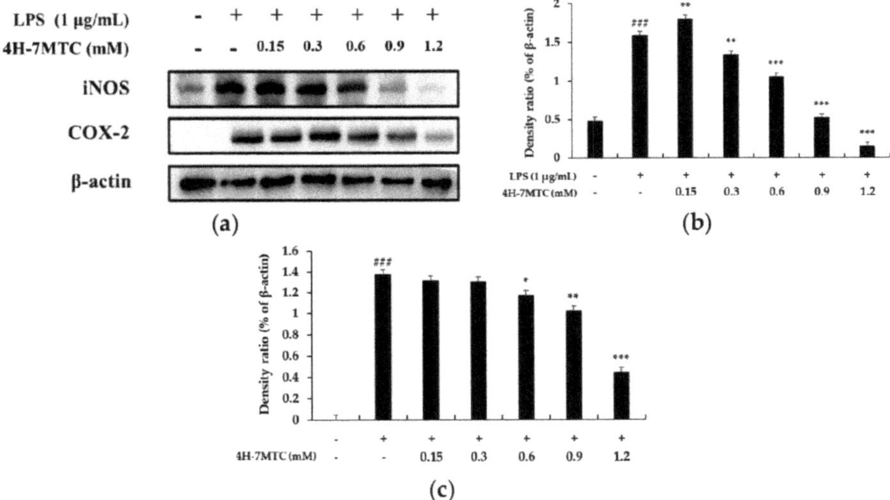

Figure 5. Effect of 4-hydroxy-7-methoxycoumarin (4H-7MTC) on the level of iNOS in LPS-induced RAW264.7 cells. Lysates were prepared from cells pretreated with 4H-7MTC (0.15, 0.3, 0.6, 0.9, and 1.2 mM) for 1 h and treated with LPS (1 μg/mL) for 18 h. β -actin was used as a loading control. Total cellular proteins were separated using SDS-PAGE, transferred to PVDF membranes, and detected using specific antibodies against iNOS and β-actin (**a**). Results are presented as representative of three independent experiments and summarized in the bar graphs (**b**,**c**). ### $p < 0.005$ vs. control cells. * $p < 0.05$, ** $p < 0.01$, *** $p < 0.005$ vs. LPS-induced cells.

A previous study revealed that NF-κB activation in response to pro-inflammatory stimuli involves the rapid phosphorylation of IκBs by the IKK signalosome complex. Free NF-κB produced by this process translocates to the nucleus where it binds to κB-binding sites in the promoter regions of target genes. It then induces the transcription of pro-inflammatory mediators such as iNOS and COX-2. Several studies have shown that anti-inflammatory agents inhibit NF-κB activation by preventing IκB degradation [19–21]. Thus, we attempted to determine whether 4H-7MTC inhibits IκB phosphorylation and degradation. Accordingly, RAW264.7 cells were pretreated for 1 h with 4H-7MTC, and IκB-α protein levels were determined after 20 min of LPS exposure (1 μg/mL). As shown in Figure 6, 4H-7MTC significantly suppressed LPS-induced phosphorylation and degradation of IκB-α. These results show that 4H-7MTC inhibits LPS-induced NF-κB activation by preventing the degradation of IκB-α phosphorylation.

MAPK plays a critical role in regulating cell growth and differentiation and controls cellular responses to cytokines and stress. In addition, three MAP kinases (JNK, p38 MAPK, and ERK 1/2) have been reported to be adjustable in LPS-induced pro-inflammatory cytokine production [22–25].

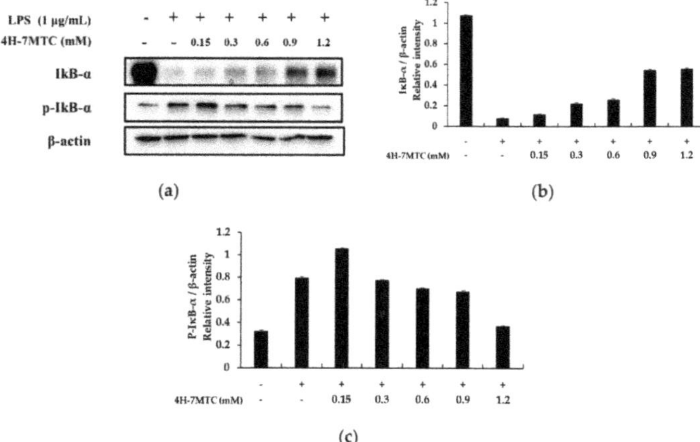

Figure 6. Effect of 4-hydroxy-7-methoxycoumarin (4H-7MTC) on the level of phospho-IκB-α and IκB-α in LPS-induced RAW264.7 cells. Lysates were prepared from cells pretreated with 4H-7MTC (0.15, 0.3, 0.6, 0.9, and 1.2 mM) for 1 h and then treated with LPS (1 μg/mL) for 20 min. Western blotting was performed to detect the expression of IκBα and p-IκBα. β-actin was used as a loading control (**a**). Quantification of immunoreactive protein bands is shown via bar graphs (**b**,**c**).

To investigate the molecular mechanism of MAPK signaling by 4H-7MTC in LPS-stimulated RAW264.7 cells, we studied the inhibition of phosphorylation of ERK1/2, p-38, and JNK. RAW264.7 cells were pretreated with 4H-7MTC at the indicated concentrations for 1 h and then stimulated with 1 μg/mL LPS for 1 h. The total cell lysates were then probed with phosphospecific antibodies for ERK1/2 and JNK. Phosphorylation of ERK1/2 and JNK increased in cells treated with LPS alone. Pretreatment with 4H-7MTC inhibited the LPS-induced phosphorylation of JNK and ERK 1/2 in a concentration-dependent manner, but not that of p38 MAPK. The amount of non-phosphorylated MAPKs was not affected by either LPS or 4H-7MTC treatment (Figure 7).

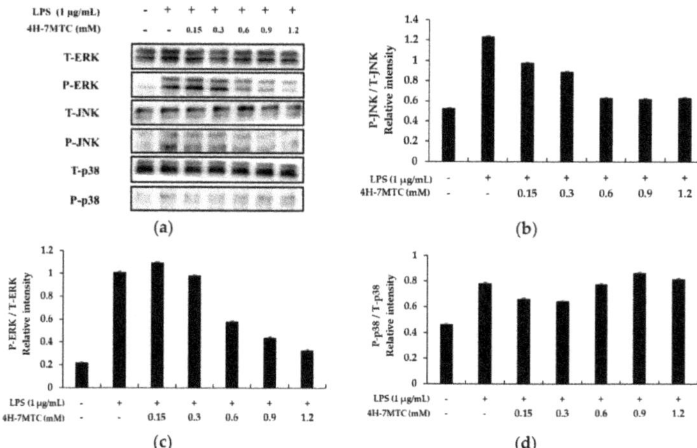

Figure 7. Effect of 4-hydroxy-7-methoxycoumarin (4H-7MTC) on LPS-induced MAPK in RAW264.7 cells. Lysates were prepared from cells pretreated with 4H-7MTC (0.15, 0.3, 0.6, 0.9, and 1.2 mM) for 1 h and treated with LPS (1 μg/mL) for 15 min. Western blotting was performed to detect the expression of phospho-ERK, T-ERK, phospho-JNK, T-JKN, phospho-p38, and T-p38 (**a**). β-actin was used as a loading control. Quantification of immunoreactive protein bands is shown via bar graphs (**b**–**d**).

These results suggest that suppression of MAPK phosphorylation may be involved in the inhibitory effect of 4H-7MTC on LPS-stimulated inflammatory response factors and inflammatory cytokines via NF-κB signaling in RAW264.7 cells.

3. Materials and Methods

3.1. Chemicals and Reagents

4-Hydroxy-7-methoxycoumarin (4H-7MTC), 4-Hydroxy-6-methylcoumarin (4H-6MC), and 4-Hydroxy-7-methylcoumarin (4H-7MC) were obtained from Tokyo Chemical Industry (Tokyo, Kita-ku, Japan). Lipopolysaccharide (LPS) from *Escherichia coli*, 3-(4,5-dimethylthiazol-2-yl)-2,5-diphenyltetrazolium bromide (MTT), α-melanocyte-stimulating hormone (α-MSH), dimethyl sulfoxide (DMSO), Griess reagent, sodium nitrite, and protease inhibitor cocktail were obtained from Sigma-Aldrich (St Louis, MO, USA). Dulbecco's Modified Eagle Medium (DMEM), fetal bovine serum, and penicillin/ streptomycin were obtained from Thermo Fisher Scientific (Waltham, MA, USA). Radioimmunoprecipitation assay buffer, phosphate-buffered saline (PBS), enhanced chemiluminescence (ECL) kit, and tris-buffered saline (TBS) were obtained from Biosesang (Seongnam, Gyeonggi-do, Korea). *N*-[2-(Cyclohexyloxy)-4-nitrophenyl] methanesulfonamide (NS-398), and L-N^6-(1-iminoethyl) lysine dihydrochloride (L-NIL) were obtained from Cayman Chemical Company (Ann Arbor, MI, USA). Prostaglandin E_2 (PGE_2) ELISA kit, interleukin-1β (IL-1β) kit, IL-6 ELISA kit, and tumor necrosis factor (TNF-α) ELISA kits were obtained from R&D System Inc. (St. Louis, MO, USA). The following antibodies were used in this study: β-actin, anti-iNOS, anti-inhibitor of NF-κB (IκBα), Akt, p-Akt, p38, p-p38, JNK, p-JNK, ERK, and p-ERK were obtained from Cell Signaling Technology (Beverly, MA, USA). Anti-COX-2 was obtained from BD Biosciences (San Diego, CA, USA). All reagents used were of analytical grade.

3.2. Cell Culture

RAW264.7 mouse macrophages and B16F10 melanoma cells were obtained from the Korean Cell Line Bank (Seoul, Korea). RAW264.7 cells were subcultured at intervals of 2–3 days. The B16F10 melanoma cells were subcultured at 4-day intervals using DMEM with 10% FBS, 100 U/mL penicillin, and 100 μg/mL streptomycin at 37 °C in a humidified 5% CO_2 atmosphere.

3.3. Cell Viability

Cytotoxicity was determined using the MTT assay. RAW264.7 cells were cultured at a density of 1.5×10^5 cells/well in 24-well plates for 24 h. Cells were treated with various concentrations of 4H-7MTC (0.3, 0.6, 0.9, and 1.2 mM). RAW264.7 cells were incubated for 24 h and MTT solution (0.2 mg/mL) was added to the medium and incubated for 4 h. Next, the medium was removed and formazan crystals in each well were dissolved in DMSO for 20 min. Optical density (OD) was measured at 570 nm, and the percentage of cells showing cell viability relative to the control was determined.

3.4. NO Production

NO production in the cell culture was assayed by measuring the accumulated nitrite using Griess reagent. RAW264.7 cells were plated at a density of 1.5×10^5 cells/well in 24-well plates. Cells were pretreated with various concentrations of 4H-7MTC (0.3, 0.6, 0.9, 1.2 mM) for 1 h and treated with LPS (1 μg/mL) for 24 h. Then, the treated cell culture solution was mixed with the Griess reagent in a 1:1 ratio, reacted for 15 min, and the absorbance measured at 540 nm using a spectrophotometer. NO production in the sample was quantified from a standard curve constructed using sodium nitrite.

3.5. Measurement of Cytokines

RAW264.7 mouse cells were plated at a density of 1.5×10^5 cells/well in 24-well plates. Cells were pretreated with various concentrations of 4H-7MTC (0.3, 0.6, 0.9, and 1.2 mM) for 1 h and treated with

LPS (1 μg/mL) for 24 h. Supernatants were harvested, and PGE_2, IL-1β, IL-6, and TNF-α levels were measured using ELISA kits according to the manufacturer's protocols.

3.6. *Measurement of Melanin Content*

B16F10 melanoma cells were plated in 60 mm cell culture dishes (6.0×10^4 cells/dish), incubated for 24 h, and then treated with 4H-7MTC (25, 50, 100, 150, and 200 μM) for 72 h in the presence of α-MSH (100 nM). After incubation, the cells were washed with 1 × PBS and the pellets were solubilized in 1 N NaOH containing 10% DMSO at 70 °C for 1 h. Absorbance was measured at 405 nm with a spectrophotometer. The protein concentration was determined using a BCA protein analysis kit.

3.7. *Western Blot Analysis*

RAW264.7 mouse cells were plated at a density of 6.0×10^5 cells/dish in 60-mm cell culture dishes for 24 h. Cells were pretreated with various concentrations of 4H-7MTC (0.3, 0.6, 0.9, 1.2 mM) for 1 h and treated with LPS (1 μg/mL) for the indicated times. After incubation, cells were washed with 1 × PBS and lysed on ice with RIPA lysis buffer (150 mM NaCl, 50 mM Tris-HCl [pH 7.5], 2 mM EDTA, 1% Triton X-100, 0.1% SDS, and 1% protease inhibitor cocktail) for 30 min. The harvested cell lysates were centrifuged at −8 °C and 15,000 rpm for 20 min. A standard assay curve of bovine serum albumin (BSA) was prepared using the BCA Protein Assay Kit, and the protein contents of the extracted cell lysates were quantitatively determined. The protein concentration was determined using a BCA protein analysis kit. Whole-cell lysates (30 μg) were separated by SDS-polyacrylamide gel electrophoresis on a 10% gel (SDS-PAGE) and electroblotted onto polyvinylidene fluoride (PVDF) membranes. The membranes were then blocked with 5% skim milk and incubated for 2 h. The membrane was washed 6 times with TBS buffer containing 0.1% Tween 20 (TTBS) and then incubated with specific primary antibodies (1:2500) at 4 °C for 6 h. The membrane was washed 6 times with TTBS buffer and incubated with a peroxidase-conjugated secondary antibody (1:2000) at room temperature for 2 h. The membrane was then washed six times with TTBS buffer and the protein was detected using an ECL kit.

3.8. *Statistical Analysis*

All results are expressed as mean ± standard deviation (SD). Each value represents the mean of three independent experiments. Statistical analysis was performed using a one-way analysis of variance (ANOVA) followed by Tukey's post-hoc test, and survival rates between multiple groups were analyzed using the log-rank test. The significant difference was set at * $p < 0.05$, ** $p < 0.01$, and *** $p < 0.001$.

4. Conclusions

This study is, to the best of our knowledge, the first to elucidate the anti-inflammatory properties of 4H-7MTC, which was mediated through the suppression of NO, PGE_2, IL-6, IL-1β, and TNF-α production in LPS-stimulated RAW264.7 cells via the NF-κB and MAPK signaling pathways. Our findings indicate that 4H-7MTC may be a promising agent for the clinical prevention and treatment of inflammation-associated diseases in the future. Additionally, 4H-7MTC has also been shown to enhance melanin production and has a potential application as an anticancer agent.

Author Contributions: Conceptualization, C.-G.H.; validation and formal analysis, J.K.K.; C.-G.H.; writing—original draft preparation, review, and editing; C.-G.H.; C.-G.H.; funding acquisition, C.-G.H. All authors have read and agreed to the published version of the manuscript.

Funding: This research was supported by the Ministry of Trade, Industry & Energy (MOTIE), Korea Institute for Advancement of Technology (KIAT) through the Encouragement Program for the Industries of Economic Cooperation Region (P0006063).

Acknowledgments: The authors thank all the students in our research group for their helpful cooperation and discussions. English proofreading of this paper was supported by R&D Program of the Establishment Project of Industry-University Fusion District (N0002327).

Conflicts of Interest: The authors declare no conflict of interest.

References

1. Kostova, I.; Raleva, S.; Genova, P.; Argirova, R. Structure-activity relationships of synthetic coumarins as HIV-1 inhibitors. *Bioinorg. Chem. Appl.* **2006**, *2006*, 68274. [CrossRef]
2. Kummerle, A.E.; Vitorio, F.; Franco, D.P.; Pereira, T.M. Coumarin compounds in medicinal chemistry: Some important examples from the last year. *Curr. Top. Med. Chem.* **2018**, *18*, 124–128.
3. Stefanachi, A.; Leonetti, F.; Pisani, L.; Catto, M.; Carotti, A. Coumarin: A natural, privileged and versatile scaffold for bioactive compounds. *Molecules* **2018**, *23*, 250. [CrossRef]
4. Annunziata, F.; Pinna, C.; Dallavalle, S.; Tamborini, L.; Pinto, A. An overview of coumarin as a versatile and readily accessible scaffold with broad-ranging biological activities. *Int. J. Mol. Sci.* **2020**, *21*, 4618. [CrossRef] [PubMed]
5. Hoult, J.R.S.; Payá, M. Pharmacological and biochemical actions of simple coumarins: Natural products with therapeutic potential. *Gen. Pharmacol.* **1996**, *27*, 713–722. [CrossRef]
6. Zhu, J.J.; Jiang, J.G. Pharmacological and nutritional effects of natural coumarins and their structure–activity relationships. *Mol. Nutr. Food Res.* **2018**, *62*. [CrossRef]
7. Srikrishna, D.; Godugu, C.; Dubey, P.K. A review on pharmacological properties of coumarins. *Mini Rev. Med. Chem.* **2016**, *18*, 113–141. [CrossRef]
8. Chung, Y.C.; Kim, S.Y.; Hyun, C.G. 8-Methoxycoumarin enhances melanogenesis via the mapkase signaling pathway. *Pharmazie* **2019**, *74*, 529–535.
9. Kim, M.-J.; Kim, S.S.; Park, K.-J.; An, H.J.; Choi, Y.H.; Lee, N.H.; Hyun, C.-G. Anti-melanogenic activity of auraptene via ERK-mediated MITF downregulation. *Cosmetics* **2017**, *4*, 34. [CrossRef]
10. Lee, N.; Chung, Y.C.; Kim, Y.B.; Park, S.M.; Kim, B.S.; Hyun, C.G. 7,8-Dimethoxycoumarin stimulates melanogenesis via MAPKs mediated MITF upregulation. *Pharmazie* **2020**, *75*, 107–111.
11. Lee, N.; Chung, Y.C.; Kang, C.I.; Park, S.-M.; Hyun, C.-G. 7,8-dimethoxycoumarin attenuates the expression of IL-6, IL-8, and CCL2/MCP-1 in TNF-α-treated HaCaT cells by potentially targeting the NF-κB and MAPK pathways. *Cosmetics* **2019**, *6*, 41. [CrossRef]
12. Yuan, H.; Zeng, X.; Shi, J.; Xu, Q.; Wang, Y.; Jabu, D.; Sang, Z.; Nyima, T. Time-course comparative metabolite profiling under osmotic stress in tolerant and sensitive tibetan hulless barley. *BioMed Res. Int.* **2018**, 9415409. [CrossRef] [PubMed]
13. Abd El-Ghany, Z.M. Evaluation of antibacterial activity, gas chromatography analysis and antioxidant efficacy of artichoke (*Cynara scolymus* L.). *J. Agric. Chem. Biotechnol. Mansoura Univ.* **2017**, *8*, 265–280. [CrossRef]
14. Haque, M.A.; Jantan, I.; Harikrishnan, H.; Ahmad, W. Standardized ethanol extract of *Tinospora crispa* upregulates pro-inflammatory mediators release in LPS-primed U937 human macrophages through stimulation of MAPK, NF-κB and PI3K-Akt signaling networks. *BMC Complement. Med. Ther.* **2020**, *20*, 245. [CrossRef]
15. Kang, H.-K.; Hyun, C.-G. Anti-inflammatory effect of d-(+)-cycloserine through inhibition of NF-κB and MAPK signaling pathways in LPS-induced RAW 264.7 macrophages. *Nat. Prod. Commun.* **2020**, *15*, 1–11. [CrossRef]
16. Kim, M.-S.; Park, J.-S.; Chung, Y.C.; Jang, S.; Hyun, C.-G.; Kim, S.-Y. Anti-inflammatory effects of formononetin 7-O-phosphate, a novel biorenovation product, on LPS-Stimulated RAW 264.7 macrophage cells. *Molecules* **2019**, *24*, 3910. [CrossRef]
17. Ham, Y.M.; Yoon, W.J.; Lee, W.J.; Kim, S.C.; Baik, J.S.; Kim, J.W.; Lee, G.S.; Lee, N.H.; Hyun, C.-G. Anti-inflammatory effects of isoketocharbroic acid from brown alga, *Sargassum micracanthum*. *EXCLI J.* **2015**, *14*, 1116–1121.
18. Kim, M.J.; Kim, S.J.; Kim, S.S.; Lee, N.H.; Hyun, C.G. *Hypochoeris radicata* attenuates LPS-induced inflammation by suppressing p38, ERK, and JNK phosphorylation in RAW 264.7 macrophages. *EXCLI J.* **2014**, *13*, 123–136.
19. Ha, Y.; Lee, W.-H.; Jeong, J.; Park, M.; Ko, J.-Y.; Kwon, O.W.; Lee, J.; Kim, Y.-J. Pyropia yezoensis Extract suppresses IFN-Gamma- and TNF-alpha-induced proinflammatory chemokine production in HaCaT cells via the down-regulation of NF-κB. *Nutrients* **2020**, *12*, 1238. [CrossRef]

20. Lee, S.G.; Brownmiller, C.R.; Lee, S.-O.; Kang, H.W. Anti-inflammatory and antioxidant effects of anthocyanins of *trifolium pratense* (Red Clover) in lipopolysaccharide-stimulated RAW-267.4 macrophages. *Nutrients* **2020**, *12*, 1089. [CrossRef]
21. Wu, M.S.; Aquino, L.B.B.; Barbaza, M.Y.U.; Hsieh, C.L.; Castro-Cruz, K.A.; Yang, L.L.; Tsai, P.W. Anti-inflammatory and anticancer properties of bioactive compounds from *Sesamum indicum* L.—A review. *Molecules* **2019**, *24*, 4426. [CrossRef] [PubMed]
22. Nguyen, T.Q.C.; Duy Binh, T.; Pham, T.L.A.; Nguyen, Y.D.H.; Thi Xuan Trang, D.; Nguyen, T.T.; Kanaori, K.; Kamei, K. Anti-inflammatory effects of *Lasia spinosa* leaf extract in lipopolysaccharide-induced RAW 264.7 macrophages. *Int. J. Mol. Sci.* **2020**, *21*, 3439. [CrossRef] [PubMed]
23. Henamayee, S.; Banik, K.; Sailo, B.L.; Shabnam, B.; Harsha, C.; Srilakshmi, S.; VGM, N.; Baek, S.H.; Ahn, K.S.; Kunnumakkara, A.B. Therapeutic emergence of rhein as a potential anticancer drug: A review of its molecular targets and anticancer properties. *Molecules* **2020**, *25*, 2278. [CrossRef]
24. Abdallah, B.M.; Ali, E.M. Butein promotes lineage commitment of bone marrow-derived stem cells into osteoblasts via modulating ERK1/2 signaling pathways. *Molecules* **2020**, *25*, 1885. [CrossRef] [PubMed]
25. Hu, Y.; Zhou, Q.; Liu, T.; Liu, Z. Coixol suppresses NF-κB, MAPK pathways and NLRP3 inflammasome activation in lipopolysaccharide-induced RAW 264.7 cells. *Molecules* **2020**, *25*, 894. [CrossRef]

Sample Availability: Samples of the compounds are not available from the authors.

Article

Comparison of Anticoagulation Quality between Acenocoumarol and Warfarin in Patients with Mechanical Prosthetic Heart Valves: Insights from the Nationwide PLECTRUM Study

Danilo Menichelli [1], Daniela Poli [2], Emilia Antonucci [3], Vittoria Cammisotto [4], Sophie Testa [5], Pasquale Pignatelli [1], Gualtiero Palareti [3], Daniele Pastori [1,*] and the Italian Federation of Anticoagulation Clinics (FCSA) †

1 I Clinica Medica, Atherothrombosis Centre, Department of Clinical, Internal, Anesthesiological and Cardiovascular Sciences, Sapienza University of Rome, 00185 Rome, Italy; danilo.menichelli@uniroma1.it (D.M.); pasquale.pignatelli@uniroma1.it (P.P.)
2 Thrombosis Centre, Azienda Ospedaliero-Universitaria Careggi, 50134 Florence, Italy; polida@aou-careggi.toscana.it
3 Arianna Anticoagulazione Foundation, 40138 Bologna, Italy; e.antonucci@fondazionearianna.org (E.A.); gualtiero.palareti@unibo.it (G.P.)
4 Department of General Surgery and Surgical Specialties "Paride Stefanini", Sapienza, University of Rome, 00185 Rome, Italy; vittoria.cammisotto@uniroma1.it
5 Department of Laboratory Medicine, Haemostasis and Thrombosis Center, AO Istituti Ospitalieri, 26100 Cremona, Italy; s.testa@ospedale.cremona.it
* Correspondence: daniele.pastori@uniroma1.it
† Members of the FCSA Study Group is provided in the Appendix A.

Citation: Menichelli, D.; Poli, D.; Antonucci, E.; Cammisotto, V.; Testa, S.; Pignatelli, P.; Palareti, G.; Pastori, D.; the Italian Federation of Anticoagulation Clinics (FCSA). Comparison of Anticoagulation Quality between Acenocoumarol and Warfarin in Patients with Mechanical Prosthetic Heart Valves: Insights from the Nationwide PLECTRUM Study. *Molecules* **2021**, *26*, 1425. https://doi.org/10.3390/molecules26051425

Academic Editor: Mee Young Hong

Received: 8 February 2021
Accepted: 3 March 2021
Published: 6 March 2021

Abstract: Vitamin K antagonists are indicated for the thromboprophylaxis in patients with mechanical prosthetic heart valves (MPHV). However, it is unclear whether some differences between acenocoumarol and warfarin in terms of anticoagulation quality do exist. We included 2111 MPHV patients included in the nationwide PLECTRUM registry. We evaluated anticoagulation quality by the time in therapeutic range (TiTR). Factors associated with acenocoumarol use and with low TiTR were investigated by multivariable logistic regression analysis. Mean age was 56.8 ± 12.3 years; 44.6% of patients were women and 395 patients were on acenocoumarol. A multivariable logistic regression analysis showed that patients on acenocoumarol had more comorbidities (i.e., ≥3, odds ratio (OR) 1.443, 95% confidence interval (CI) 1.081–1.927, $p = 0.013$). The mean TiTR was lower in the acenocoumarol than in the warfarin group (56.1 ± 19.2% vs. 61.6 ± 19.4%, $p < 0.001$). A higher prevalence of TiTR (<60%, <65%, or <70%) was found in acenocoumarol users than in warfarin ones ($p < 0.001$ for all comparisons). Acenocoumarol use was associated with low TiTR regardless of the cutoff used at multivariable analysis. A lower TiTR on acenocoumarol was found in all subgroups of patients analyzed according to sex, hypertension, diabetes, age, valve site, atrial fibrillation, and INR range. In conclusion, anticoagulation quality was consistently lower in MPHV patients on acenocoumarol compared to those on warfarin.

Keywords: warfarin; acenocoumarol; mechanical valve; time in therapeutic range; anticoagulation

1. Introduction

The burden of valvular heart disease (VHD) is still rising worldwide due to degenerative valve diseases. Although valve rheumatic disease is decreasing [1]. Implantation of mechanical prosthetic heart valves (MPHV) is associated with a reduction in valve-related morbidity compared to biological valves [2]. In MPHV, long-term antithrombotic treatment with only vitamin K antagonists (VKAs) is needed [3]. Consolidated evidence from studies including patients with atrial fibrillation (AF) showed that during VKA treatment, a poor anticoagulation quality, expressed as low time in therapeutic range (TiTR) (<65%–70%), was associated with an increased risk of thromboembolism [4], cardiovascular events [5], and mortality [6].

However, no specific indication regarding the type of VKAs, such as warfarin or acenocoumarol, is given by international guidelines or expert consensus documents. As a consequence, the use of different VKAs is highly variable among countries, with warfarin being more commonly used in the United Kingdom and Italy, acenocoumarol in Spain, phenprocoumon in Germany, and fluindione in France [7].

Few previous studies investigated potential differences in patients treated with different VKAs. A study including 498 patients with various indications of anticoagulant therapy (AF in 70% of cases) showed that acenocoumarol may be associated with a lower TiTR and a higher international normalized ratio (INR) variability, which improved after switching to phenprocoumon [8].

Previous evidence showed a generally low quality of anticoagulation with VKAs in patients implanted with MPHV [9], but the difference between warfarin and acenocoumarol in terms of clinical characteristics of patients and anticoagulation quality was not investigated in these patients.

The aims of our study were (1) to investigate the clinical characteristics of patients treated with acenocoumarol compared to those treated with warfarin, (2) to describe clinical determinants associated with acenocoumarol use, and (3) to report the proportion of suboptimal anticoagulation quality in acenocoumarol and warfarin use in patients enrolled in the multicenter PLECTRUM registry.

2. Results

The study enrolled 2111 patients with MPHV, of which 1716 (81.3%) were treated with warfarin and 395 (18.7%) with acenocoumarol. The mean age was 56.8 years and 44.6% of patients were women (Table 1).

Table 1. Characteristics of patients according to vitamin K antagonists.

	Whole Cohort (n = 2111)	Warfarin (n = 1716)	Acenocoumarol (n = 395)	p-Value
Age (years)	56.8 ± 12.3	56.8 ± 12.3	56.9 ± 12.2	0.869
Age ≥ 65 years (%)	29.1	28.6	31.1	0.319
Age ≥ 75 years (%)	4.0	4.0	4.1	0.978
Women (%)	44.6	44.4	45.3	0.743
Arterial hypertension (%)	65.9	64.7	70.9	0.020
Diabetes (%)	13.5	13.2	14.7	0.445
Heart failure (%)	14.9	14.2	18.2	0.041
Previous thromboembolism * (%)	7.8	7.5	9.4	0.203
Previous hemorrhage	3.8	4.2	2.3	0.074
Previous ischemic heart disease (%)	12.9	12.6	14.2	0.413
Previous clinical outcomes ^	5.8	6.1	4.3	0.163
Peripheral artery disease ** (%)	9.0	8.2	12.4	0.008
Atrial fibrillation (%)	38.4	38.3	39.0	0.796
Comorbidities §	1.4 ± 1.0	1.4 ± 1.0	1.6 ± 1.1	0.004
Comorbidities ≥ 3	14.5	13.6	18.5	0.013
Concomitant antiplatelet (%)	17.3	16.6	20.5	0.061
Amiodarone users (%)	13.2	13.3	12.4	0.618

Table 1. *Cont.*

	Whole Cohort (n = 2111)	Warfarin (n = 1716)	Acenocoumarol (n = 395)	*p*-Value
MPHV site				
Aortic (%)	60.7	60.6	61.0	0.379
Mitral (%)	28.1	28.6	26.1	
Mitroaortic (%)	11.2	10.8	12.9	
INR ranges				
2.0–3.0 (%)	27.2	27.4	26.3	0.898
2.5–3.5 (%)	63.6	63.4	64.5	
3.0–4.0 (%)	9.2	9.2	9.2	
TiTR (%)	60.6 ± 19.5	61.6 ± 19.4	56.1 ± 19.2	<0.001
Low-quality anticoagulation				
TiTR < 60% (%)	48.5	46.3	58.0	<0.001
TiTR < 65% (%)	60.1	57.8	70.4	<0.001
TiTR < 70% (%)	66.9	64.8	76.2	<0.001

INR: international normalized ratio. MPHV: mechanical prosthetic heart valve. TiTR: time in therapeutic range. * Includes previous stroke/TIA/systemic embolism. ** Includes lower limb and carotid disease. ^ Previous thromboembolism, previous ischemic heart disease, previous hemorrhage. § Includes hypertension, diabetes, heart failure, peripheral artery disease, atrial fibrillation.

The MPHV site most represented in the whole cohort was aortic (60.7%) and 38.4% of patients had concomitant AF. Patients on acenocoumarol were more frequently affected by arterial hypertension, heart failure (HF), and peripheral artery disease (PAD) and had more comorbidities compared to those on warfarin (Table 1). There was no difference between anticoagulant treatment groups concerning age, sex, MPHV site, INR range, diabetes, previous ischemic heart disease, or thromboembolism at baseline (Table 1). Patients treated with acenocoumarol were affected by a higher number of comorbidities at baseline compared to those treated with warfarin (26.6% vs. 20.0%, respectively, $p = 0.004$).

At univariable logistic regression analysis (Table 2), factors associated with acenocoumarol use were the number of comorbidities, in particular arterial hypertension, PAD, and HF. At multivariable logistic regression analysis, only the presence of three or more comorbidities (OR 1.443, 95%CI 1.081–1.927, $p = 0.013$) were associated with acenocoumarol use. In a second model using single comorbidities, we found that PAD (OR 1.536, 95%CI 1.085–2.174, $p = 0.015$) and arterial hypertension (OR 1.292, 95%CI 1.016–1.642, $p = 0.036$) were associated with acenocoumarol use.

Anticoagulation Quality According to Treatment

In the whole cohort, the mean TiTR was 60.6 ± 19.5%; anticoagulation quality was lower in patients treated with acenocoumarol compared to those on warfarin (61.6 ± 19.4% vs. 56.1 ± 19.2%, $p < 0.001$, Table 1).

Analyzing the proportion of suboptimal anticoagulation using different cutoffs of TiTR, we found that acenocoumarol users had a higher prevalence of TiTR < 60%, <65%, or <70% ($p < 0.001$ for all comparisons, Table 1).

Furthermore, after performing a multivariable regression analysis of factors associated with poor anticoagulation quality, acenocoumarol use was found to be associated with low TiTR regardless of the cutoff used: TTR < 60% (OR 1.590, 95%CI 1.262–2.002, $p < 0.001$), TTR < 65% (OR 1.747, 95%CI 1.368–2.232. $p < 0.001$), and TTR < 70% (OR 1.747, 95%CI 1.347–2.266, $p < 0.001$) (Figure 1).

Table 2. Univariable logistic regression analysis of clinical factors associated with acenocoumarol use.

	Odds Ratio	95% Confidence Interval	*p*-Value
Female sex	0.964	0.774–1.201	0.743
Age ≥ 65 years	0.886	0.669–1.124	0.319
Atrial fibrillation	1.030	0.823–1.289	0.796
Hypertension	1.326	1.044–1.683	0.021
Diabetes	1.129	0.827–1.542	0.446
PAD	1.594	1.128–2.252	0.008
Heart failure	1.351	1.012–1.804	0.041
Previous TE	1.282	0.874–1.881	0.204
Previous ischemic heart disease	1.141	0.831–1.566	0.414
Previous hemorrhage	0.532	0.264–1.074	0.078
Comorbidities ≥ 3	1.443	1.081–1.927	0.013
Previous clinical outcomes	0.690	0.408–1.166	0.166
Mitral vs. Aortic	0.907	0.703–1.170	0.453
Mitroaortic vs. Aortic	1.183	0.842–1.662	0.332
Concomitant antiplatelet	1.301	0.988–1.713	0.061
Amiodarone	0.920	0.661–1.279	0.619

PAD: peripheral artery disease. TE: Thromboembolism.

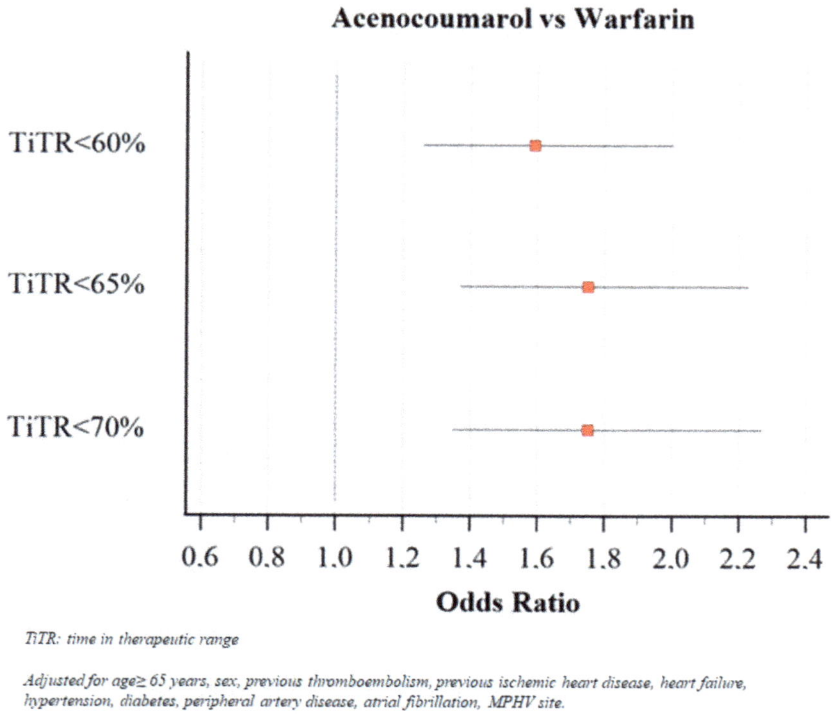

TiTR: time in therapeutic range

Adjusted for age≥ 65 years, sex, previous thromboembolism, previous ischemic heart disease, heart failure, hypertension, diabetes, peripheral artery disease, atrial fibrillation, MPHV site.

Figure 1. Association among acenocoumarol and low time in therapeutic range using different cutoff values in multivariable regression analysis.

To better characterize the association between acenocoumarol and low TiTR, we performed a subgroup analysis according to sex, hypertension, diabetes, age (≥65 years), MPHV site, AF, and INR range (Table 3). A lower anticoagulation quality on acenocoumarol was found in all subgroups of patients analyzed (Table 3).

Table 3. Subgroup analysis of time in therapeutic range according to acenocoumarol or warfarin use.

	OAC Type	Mean TiTR	*p*	TiTR < 60% (%)	*p*	TiTR < 65% (%)	*p*	TiTR < 70% (%)	*p*
Women	*Warfarin*	58.9 ± 19.1	<0.001	52.4	0.001	63.4	<0.001	70.6	0.003
	Acenocoumarol	51.9 ± 18.6		65.9		77.7		81.6	
Men	*Warfarin*	63.8 ± 19.4	0.004	41.5	0.008	53.2	0.003	60.2	0.002
	Acenocoumarol	59.6 ± 19.0		51.4		64.4		71.8	
Arterial hypertension	*Warfarin*	60.6 ± 19.8	<0.001	49.7	<0.001	60.6	<0.001	67.2	<0.001
	Acenocoumarol	54.8 ± 19.1		62.9		73.9		79.3	
Diabetes	*Warfarin*	57.8 ± 19.3	0.019	56.8	0.016	66.1	0.052	70.5	0.060
	Acenocoumarol	51.1 ± 19.9		74.1		79.3		82.8	
Age (≥65 years)	*Warfarin*	60.2 ± 18.9	0.001	48.9	0.006	61.3	0.001	68.2	0.004
	Acenocoumarol	53.6 ± 18.4		62.6		77.2		81.3	
Aortic MPHV	*Warfarin*	65.0 ± 19.3	<0.001	37.8	<0.001	49.6	<0.001	57.2	<0.001
	Acenocoumarol	59.5 ± 18.9		50.2		63.9		69.7	
Mitral/Mitroaortic MPHV	*Warfarin*	56.4 ± 18.5	0.001	59.5	0.014	70.3	0.010	76.5	0.007
	Acenocoumarol	50.8 ± 18.4		70.1		80.5		86.4	
INR range 2.0–3.0	*Warfarin*	70.3 ± 19,0	0.026	26.5	0.125	37.4	0.014	43.6	0.012
	Acenocoumarol	65.7 ± 19.1		34.0		50.5		57.3	
INR range above 2.0–3.0	*Warfarin*	58.4 ± 18.6	<0.001	53.8	<0.001	65.4	<0.001	72.8	<0.001
	Acenocoumarol	52.7 ± 18.1		66.4		77.4		82.9	
Atrial fibrillation	*Warfarin*	58.5 ± 19.4	0.002	41.9	0.002	53.4	<0.001	61.1	0.001
	Acenocoumarol	53.1 ± 19.0		53.1		67.2		72.6	

MPHV: mechanical prosthetic heart valve; OAC: oral anticoagulant; TiTR: time in therapeutic range.

We repeated the analysis, excluding patients treated with antiplatelets, and found similar results in 1746 patients as follows: 56.2 ± 18.8 in acenocoumarol-treated vs. 61.7 ± 19.2 in warfarin-treated patients ($p < 0.001$).

3. Material and Methods

The FCSA-START Valve Study (PLECTRUM) is a retrospective multicenter observational study conducted within the Italian Survey on Anticoagulation Patient Records (START register) and conducted among 33 centers affiliated with the Italian Federation of Thrombosis Diagnosis Centers and Surveillance of Antithrombotic Therapies (FCSA) [10]. The centers were asked to select from their databases patients with a mechanical heart valve prosthesis that was implanted after 1990 and who were followed up on for the management of oral anticoagulant therapy. Patients with MPHV were treated with warfarin or acenocoumarol to prevent thromboembolic event according to European Society of Cardiology guidelines [11]. Each physician prescribed warfarin or acenocoumarol after individualized clinical evaluation. The patients followed by the FCSA centers for the management of oral anticoagulation received an adequate education on the purpose of the treatment, the risk of complications, the INR values, and the management of the dosage of the drugs. The centers performed periodic INR measurements based on INR value, prescribe daily VKA, dosage and scheduled the date for subsequent visits; they also monitored and recorded changes in patient habits, diet, co-medications, intercurrent illness, bleeding, and thrombotic complications during regular follow-up visits through patient interviews. All centers participated in the specially designed external laboratory quality control program, which is performed 3 times a year and uses lyophilized plasma samples obtained from anticoagulated patients. For this reason, to standardize the quality of INR measurements, none of the patients were monitored at home.

Demographic information and clinical data were collected. Patients were classified as having high blood pressure if they were taking medicines to lower their blood pressure. Diabetes mellitus was defined according to the criteria of the American Diabetes Association. Coronary artery disease was defined on the basis of a history of myocardial infarction or stable and unstable angina. Heart failure was defined as the presence of signs and symptoms of right or left ventricular failure or both and confirmed by non-invasive or invasive measurements that demonstrated objective evidence of cardiac dysfunction. The quality of the anticoagulant control, calculated as TiTR using the linear interpolation method of Rosendaal et al. [12], was analyzed considering the INRs recorded in the last year of follow-up. The study protocol complied with the ethical guidelines of the 1975 Helsinki Declaration, as evidenced by the approval of the institution's human research committee, and informed consent was obtained from each patient. Authorization to set up the registry was obtained from the Ethical Committee of the University Hospital "S. Orsola-Malpighi," Bologna, Italy, in October 2011 (N = 142/2010/0/0ss"). The same institution is charged with deploying and upkeeping the registry central database.

Statistical Analysis

Continuous variables were reported as mean and standard deviation and compared by the Student t-test. Categorical variables were reported as count and percentage and compared by Pearson chi-squared test. A first descriptive analysis of clinical characteristics according to acenocoumarol or warfarin use was performed. Univariable and multivariable logistic regression analysis was used to calculate the relative odds ratio (OR) with a 95% confidence interval (95%CI) of factors associated with acenocoumarol use and low TiTR. Significance was set at a p-value < 0.05. All tests were two-tailed and analyses were performed using computer software packages (SPSS-25.0, SPSS Inc. IBM Corp, Armonk, NY, USA).

4. Discussion

The difference between acenocoumarol and warfarin effectiveness in terms of anticoagulation stability was never investigated in a large cohort of patients with MPHV. Findings from our study show that 18.7% of patients implanted with MPHV were treated with acenocoumarol in specialized outpatients' clinics for the management of antithrombotic therapies. Acenocoumarol prescription was more common in complex patients, as indicated by the higher number of comorbidities. Patients treated with acenocoumarol showed lower anticoagulation quality compared to those on warfarin. This difference was consistent in all thresholds of TiTR used and in all subgroups of patients regardless of sex, age, valve site, or INR range.

Acenocoumarol presents some pharmacokinetic and pharmacodynamic differences from warfarin that may turn useful in some patients, such a more rapid onset of action, a shorter half-life, and lower renal excretion. In our study, patients with a higher number of comorbidities and use of antiplatelet agents were more frequently prescribed acenocoumarol instead of warfarin. In this last context, the shorter half-life of acenocoumarol may be an advantage in the case of a major or life-threatening bleeding event in patients treated with dual therapy needing a rapid offset of action of the drug.

We found a generally lower anticoagulation quality in patients treated with acenocoumarol, which persisted after adjustment for confounders. Suboptimal anticoagulation with acenocoumarol compared to warfarin was also consistent in all subgroups of patients analyzed, such as sex, hypertension, diabetes (mostly for TiTR < 60%), AF, MPHV site, and INR range. This finding adds to previous evidence that female sex is associated with lower overall anticoagulation quality in the PLECTRUM registry [13].

In a study performed in Poland including 430 patients with mixed indications for VKAs therapy (65.8% AF, 22.6% venous thromboembolism, and 11.6% MPHV) and treated in most cases with acenocoumarol (78.8%), the mean TiTR was as low as 55%, with acenocoumarol use associated with a nearly threefold higher chance of having INR outside the therapeutic range [14].

A previous small study including patients with various indications of anticoagulation showed a significant improvement of TiTR in patients switched from acenocoumarol to warfarin (from 40.2% to 60.2%) [15].

Furthermore, in a population with similar age affected by venous thromboembolism enrolled within the EINSTEIN-DVT and EINSTEIN-PE studies, the use of acenocoumarol was a risk factor for long-term low TiTR (OR 1.81, 95%CI 1.49–2.20, $p < 0.01$) [16].

As patients treated with acenocoumarol were more frequently prescribed antiplatelet agents, which may lead to an increased risk of bleeding episodes and subsequently to a lower adherence to anticoagulant prescription and to a higher discontinuation rate [17], we also performed a subgroup analysis excluding patients on antiplatelets. In this group of patients, we found similar results than the overall cohort, suggesting that anticoagulation quality in MPHV patients is not affected by concomitant administration of antiplatelet drugs.

Our results may have implications for clinical practice. Prescribing acenocoumarol or switching from warfarin should be considered only in select patients in whom warfarin therapy is not successful, such as those with low TiTR or those with recurrent thrombotic events; in patients with a known or suspected warfarin resistance [18], such as antiphospholipid syndrome [19]; and in patients taking drugs interacting with warfarin metabolism.

Our study has limitations to be acknowledged. First, its retrospective observational design is an intrinsic limitation to establishing any inference on our observation. Second, some additional factors not considered in this study may affect both the choice of acenocoumarol use and TiTR; for instance, use of different VKAs may be affected by national guidelines in different countries. Furthermore, some drugs interacting with VKAs that were not considered in this study may influence the TiTR. Finally, we do not have data on concomitant hospitalizations and interruptions for diagnostic/therapeutic procedures that may lead to low anticoagulation quality. In addition, we included only Caucasian patients with a mean age of 60 years, and the reproducibility of our findings in elderly patients and in patients with different ethnic origins needs to be explored. Indeed, ethnic differences such as environmental factors and genetic variants of isoenzymes may affect pharmacokinetic features, hepatic metabolism, and renal elimination of warfarin [20]. Finally, the difference between acenocoumarol and warfarin in other settings such as AF and venous thrombosis needs to be explored, even if in these contexts the use of direct oral anticoagulation is replacing VKAs in many countries.

In conclusion, warfarin would be the first-choice treatment for thromboprophylaxis in patients with MPHV regardless of valve site and INR range. Switching from acenocoumarol to warfarin may improve TiTR in patients with unstable anticoagulation.

Author Contributions: Conceptualization, D.M. and D.P. (Daniela Poli); Formal analysis, D.P. (Daniela Poli); Investigation, D.M., D.P. (Daniela Poli), V.C., E.A., G.P., S.T., P.P., D.P. (Daniele Pastori); Data Curation, E.A., G.P.; Writing—Original Draft Preparation, D.M., D.P. (Daniela Poli), V.C., E.A., G.P., S.T., P.P., D.P. (Daniele Pastori); Writing—Review and Editing, D.M., D.P. (Daniela Poli), V.C., E.A., G.P., S.T., P.P., D.P. (Daniele Pastori). All authors have read and agreed to the published version of the manuscript.

Funding: This research received no external funding.

Institutional Review Board Statement: The study was conducted according to the guidelines of the Declaration of Helsinki, and approved by the Institutional Review Board of the Coordinating Center at the University Hospital "S. Orsola-Malpighi," Bologna, Italy, in October 2011 (N = 142/2010/0/0ss") and by all participating centers.

Informed Consent Statement: Informed consent was obtained from all subjects involved in the study.

Data Availability Statement: The data presented in this study are available in this article.

Conflicts of Interest: The authors declare no conflict of interest.

Sample Availability: Not available.

Appendix A

Italian Federation of Anticoagulation Clinics (FCSA).

Sophie Testa, Oriana Paoletti; Dipartimento di Medicina di Laboratorio, Centro Emostasi e Trombosi ASST Cremona.

Corrado Lodigiani; Paola Ferrazzi; Ilaria Quaglia; Centro Trombosi e Malattie Emorragiche, Humanitas Research Hospital, IRCCS Rozzano-Milano.

Daniela Poli Centro Trombosi SOD Malattie Aterotrombotiche Azienda Ospedaliero Universitaria Careggi; Firenze.

Nadia Coffetti; Rosa Marotta; Varusca Brusegan, Orazio Bergamelli, Servizio di Immunoematologia e Medicina Trasfusionale Azienda Ospedaliera Bolognini, ASST Bergamo Est, Seriate e Ambulatorio TAO SIMT Ospedale Fernaroli, Alzano Lombardo.

Roberto Facchinetti Laboratorio Analisi Azienda Ospedaliera Universitaria Integrata Ospedale Civile Maggiore Di Borgo Trento; Verona.

Giuseppina Serricchio; Silvia Sarpau, Francesca Brevi; ASST Lariana Como.

Pietro Falco; Guarino Silverio; Poliambulatorio Specialistico MEDICAL PONTINO, Latina.

Catello Mangione; Giacomo Bellomo, Servizio Immunotrasfusionale Ospedale "Santa Caterina Novella" Galatina (Lecce).

Serena Masottini; Alessandra Cosenza; Centro per la prevenzione, diagnosi e trattamento dellemalattie tromboemboliche- Asl 8- Cagliari.

Lucia Ruocco; Paolo Chiarugi, Monica Casini; Ambulatorio Antitrombosi CAT-TAO AOU Pisana, Pisa.

Arturo Cafolla; Antonietta Ferretti; Ematologia, UOS Emostasi e Trombosi, Policlinico Umberto I°, Roma.

Giorgia Micucci; Serena Rupoli; Lucia Canafoglia; Azienda Ospedaliero-Universitaria Ospedali Riuniti Di Ancona.

Paolo Pedico; Rita Galasso; Rosa Rotunno; U.O. MedicinaTrasfusionale Ospedale Mons. Raffaele Dimiccoli Barletta.

Antonio Insana; Paolo Valesella; Servizio Di Patologia Clinica Dipartimento Dei Servizi—Ospedale S. Croce Moncalieri, Moncalieri, Torino.

Angelo Santoro; U.O.C. Patologia Clinica e Centro Trombosi, Presidio Ospedaliero "A. Perrino," ASL Brindisi.

Francesco Marongiu, Doris Barcellona; Centro Di Fisiopatologia dell'emostasi e Terapia Anticoagulante, Azienda Ospedaliera Universitaria di Monserrato, Cagliari.

Vittorio Pengo, Gentian Denas; Istituto di Cardiologia, Policlinico Universitario, Università di Padova.

Carmelo Paparo; Centro Anti Trombosi Ospedale Maggiore ASL TO5, Chieri (Torino).

Eugenio Bucherini; Flavia Tani; Enrico Carioli, Centro di Sorveglianza per la Terapia Anticoagulante, Angiolgia- Medicina Vascolare U.O. Cardiologia O.C. Per gli infermi—Faenza AUSL Romagna, (Ravenna).

Francesco Violi, Pasquale Pignatelli, Daniele Pastori, Mirella Saliola; Dipartimento di Scienze Cliniche, Internistiche, Anestesiologiche e Cardiovascolari, Sapienza Università di Roma.

Lucilla Masciocco, Pasquale Saracino, Angelo Benvenuto, UOC Medicina Interna, Centro Controllo Coagulazione, Ospedale Lastaria, Lucera (Foggia).

Anna Turrini; Stefano Ciaffone; Ospedale "SACRO CUORE" Laboratorio Analisi Cliniche E Medicina Trasfusionale Negrar, Verona.

Andrea Toma; Pietro Barbera UOC di Patologia Clinica Arzignano, Vicenza.

Paolo Gresele; Tiziana Fierro; Stefano Pasquino Department of Medicine, Section of Internal and Cardiovascular Medicine, University of Perugia.

Lucia La Rosa; Rino Morales Centro Trasfusionale e ambulatorio emostasi e trombosi; ASST Vimercate.

Francesco Ronchi, Giuseppe Isu Centro Tao Servizio Di Patologia Clinica Ospedale Ns Signora Di Bonaria Asl 6 Sanluri.

Teresa Lerede, Luca Barcella Centro Trombosi e Emostasi Immunoematologia eMedicina Trasfusionale ASST Papa Giovanni XXIII, Bergamo.

Luigi Ria, Centro Trombosi ed. Emostasi, U.O.C. di Medicina Interna, Ospedale "Sacro Cuore di Gesù" Gallipoli; ASL Lecce, Lecce.

Rosanna Crisantemo; Luciano Suriano, Luciano Lorusso; Mario De Sarlo Servizio di Immunoematologia e Medicina Trasfusionale Ospedale L.Bonomo, Andria.

Pasquale Carrato Istituto Polidiagnostico Santa Chiara, Agropoli, Salerno.

Carmine Oricchio U.O.S. Centro Trasfusionale del P.O. "Luigi Curto" di Polla—ASL Salerno.

Elvira Grandone, Donatella Colaizzo, Centro Trombosi, I.R.C.C.S. Casa Sollievo della Sofferenza, S. Giovanni Rotondo, Foggia.

Maurizio Molinatti Unità Funzionale di Ematologia Centro TAO Humanitas Cell, Torino.

References

1. Iung, B.; Vahanian, A. Epidemiology of valvular heart disease in the adult. *Nat. Rev. Cardiol.* **2011**, *8*, 162–172. [CrossRef] [PubMed]
2. Roumieh, M.; Ius, F.; Tudorache, I.; Ismail, I.; Fleissner, F.; Haverich, A.; Cebotari, S. Comparison between biological and mechanical aortic valve prostheses in middle-aged patients matched through propensity score analysis: Long-term results. *Eur. J. Cardio-Thorac. Surg.* **2014**, *48*, 129–136. [CrossRef] [PubMed]
3. Menichelli, D.; Ettorre, E.; Pani, A.; Violi, F.; Pignatelli, P.; Pastori, D. Update and Unmet Needs on the Use of Nonvitamin K Oral Anticoagulants for Stroke Prevention in Patients with Atrial Fibrillation. *Curr. Probl. Cardiol.* **2021**, *46*, 100410. [CrossRef] [PubMed]
4. Dentali, F.; Pignatelli, P.; Malato, A.; Poli, D.; Di Minno, M.N.D.; Di Gennaro, L.; Rancan, E.; Pastori, D.; Grifoni, E.; Squizzato, A.; et al. Incidence of thromboembolic complications in patients with atrial fibrillation or mechanical heart valves with a subtherapeutic international normalized ratio: A prospective multicenter cohort study. *Am. J. Hematol.* **2012**, *87*, 384–387. [CrossRef] [PubMed]
5. Pastori, D.; Pignatelli, P.; Saliola, M.; Carnevale, R.; Vicario, T.; Del Ben, M.; Cangemi, R.; Barillà, F.; Lip, G.Y.; Violi, F. Inadequate anticoagulation by Vitamin K Antagonists is associated with Major Adverse Cardiovascular Events in patients with atrial fibrillation. *Int. J. Cardiol.* **2015**, *201*, 513–516. [CrossRef] [PubMed]
6. Rivera-Caravaca, J.M.; Roldán, V.; Esteve-Pastor, M.A.; Valdés, M.; Vicente, V.; Marín, F.; Lip, G.Y. Reduced Time in Therapeutic Range and Higher Mortality in Atrial Fibrillation Patients Taking Acenocoumarol. *Clin. Ther.* **2018**, *40*, 114–122. [CrossRef] [PubMed]
7. Le Heuzey, J.; Ammentorp, B.; Darius, H.; de Caterina, R.; Schilling, R.J.; Schmitt, J.; Zamorano, J.L.; Kirchhof, P. Differences among western European countries in anticoagulation management of atrial fibrillation. Data from the PREFER IN AF registry. *Thromb. Haemost.* **2014**, *111*, 833–841. [PubMed]
8. Van Miert, J.H.A.; Veeger, N.; Ten Cate-Hoek, A.J.; Piersma-Wichers, M.; Meijer, K. Effect of switching from acenocoumarol to phenprocoumon on time in therapeutic range and INR variability: A cohort study. *PLoS ONE* **2020**, *15*, e0235639.
9. Pastori, D.; Lip, G.Y.H.; Poli, D.; Antonucci, E.; Rubino, L.; Menichelli, D.; Saliola, M.; Violi, F.; Palareti, G.; Pignatelli, P.; et al. Determinants of low-quality warfarin anticoagulation in patients with mechanical prosthetic heart valves. The nationwide PLECTRUM study. *Br. J. Haematol.* **2020**, *190*, 588–593. [CrossRef] [PubMed]
10. Poli, D.; Antonucci, E.; Pengo, V.; Migliaccio, L.; Testa, S.; Lodigiani, C.; Coffetti, N.; Facchinetti, R.; Serricchio, G.; Falco, P.; et al. Mechanical prosthetic heart valves: Quality of anticoagulation and thromboembolic risk. The observational multicenter PLECTRUM study. *Int. J. Cardiol.* **2018**, *267*, 68–73. [CrossRef] [PubMed]
11. Falk, V.; Baumgartner, H.; Bax, J.J.; de Bonis, M.; Hamm, C.; Holm, P.J.; Iung, B.; Lancellotti, P.; Lansac, E.; Muñoz, D.R.; et al. 2017 ESC/EACTS Guidelines for the management of valvular heart disease. *Eur. J. Cardio-Thorac. Surg.* **2017**, *52*, 616–664. [CrossRef] [PubMed]
12. Rosendaal, F.R.; Cannegieter, S.C.; Van Der Meer, F.J.M.; Briët, E. A Method to Determine the Optimal Intensity of Oral Anticoagulant Therapy. *Thromb. Haemost.* **1993**, *69*, 236–239. [CrossRef] [PubMed]
13. Pastori, D.; Poli, D.; Antonucci, E.; Menichelli, D.; Violi, F.; Palareti, G.; Pignatelli, P.; Testa, S.; Paoletti, O.; Lodigiani, C.; et al. Sex-based difference in anticoagulated patients with mechanical prosthetic heart valves and long-term mortality risk. *Int. J. Clin. Pr.* **2021**, *3*, e14064. [CrossRef]

14. Sawicka-Powierza, J.; Buczkowski, K.; Chlabicz, S.; Gugnowski, Z.; Powierza, K.; Ołtarzewska, A.M. Quality control of oral anticoagulation with vitamin K antagonists in primary care patients in Poland: A multi-centre study. *Kardiol. Pol.* **2018**, *76*, 764–769. [CrossRef] [PubMed]
15. Undas, A.; Cieśla-Dul, M.; Żółciński, M.; Tracz, W. Switching from acenocoumarol to warfarin in patients with unstable anticoagulation and its effect on anticoagulation control. *Pol. Arch. Intern. Med.* **2009**, *119*, 360–365. [CrossRef]
16. Gebel, M.; Sahin, K.; Lensing, A.W.A.; Meijer, K.; Kooistra, H.A.M. Independent predictors of poor vitamin K antagonist control in venous thromboembolism patients. *Thromb. Haemost.* **2015**, *114*, 1136–1143. [CrossRef] [PubMed]
17. O'Brien, E.C.; Simon, D.N.; Allen, L.A.; Singer, D.E.; Fonarow, G.C.; Kowey, P.R.; Thomas, L.E.; Ezekowitz, M.D.; Mahaffey, K.W.; Chang, P.; et al. Reasons for warfarin discontinuation in the Outcomes Registry for Better Informed Treatment of Atrial Fibrillation (ORBIT-AF). *Am. Heart J.* **2014**, *168*, 487–494. [CrossRef] [PubMed]
18. Marusic, S.; Gojo-Tomic, N.; Franić, M.; Božina, N. Therapeutic efficacy of acenocoumarol in a warfarin-resistant patient with deep venous thrombosis: A case report. *Eur. J. Clin. Pharmacol.* **2009**, *65*, 1265–1266. [CrossRef] [PubMed]
19. Pastori, D.; Parrotto, S.; Vicario, T.; Saliola, M.; Mezzaroma, I.; Violi, F.; Pignatelli, P. Antiphospholipid syndrome and anticoagulation quality: A clinical challenge. *Atherosclerosis* **2016**, *244*, 48–50. [CrossRef] [PubMed]
20. El Rouby, S.; Mestres, C.A.; LaDuca, F.M.; Zucker, M.L. Racial and ethnic differences in warfarin response. *J. Heart Valve Dis.* **2004**, *13*, 15–21. [PubMed]

Communication

Chiral Tertiary Amine Catalyzed Asymmetric [4 + 2] Cyclization of 3-Aroylcoumarines with 2,3-Butadienoate

Jun-Lin Li [1], Xiao-Hui Wang [1], Jun-Chao Sun [1], Yi-Yuan Peng [1], Cong-Bin Ji [2] and Xing-Ping Zeng [1,*]

1 Key Laboratory of Small Functional Organic Molecule, Ministry of Education, College of Chemistry and Chemical Engineering, Jiangxi Normal University, Nanchang 330022, China; ljl18870471828@163.com (J.-L.L.); 15879027560@163.com (X.-H.W.); jason96chao@163.com (J.-C.S.); yypeng@jxnu.edu.cn (Y.-Y.P.)

2 Jiangxi Provincial Research of Targeting Pharmaceutical Engineering Technology, Shangrao Normal University, Shangrao 334001, China; jcbpxh521@163.com

* Correspondence: 005173@jxnu.edu.cn

Abstract: Coumarins and 2*H*-pyran derivatives are among the most commonly found structural units in natural products. Therefore, the introduction of 2*H*-pyran moiety into the coumarin structural unit, i.e., dihydrocoumarin-fused dihydropyranones, is a potentially successful route for the identification of novel bioactive structures, and the synthesis of these structures has attracted continuing research interest. Herein, a chiral tertiary amine catalyzed [4 + 2] cyclization of 3-aroylcoumarines with benzyl 2,3-butadienoate was reported. In the presence of Kumar's 6′-(4-biphenyl)-β-iso-cinchonine, the desired dihydrocoumarin-fused dihydropyranone products could be obtained in up to 97% yield and 90% ee values.

Keywords: coumarins; dihydrocoumarin-fused dihydropyranones; 3-aroylcoumarines; benzyl 2,3-butadienoate; 6′-(4-biphenyl)-β-iso-cinchonine

Citation: Li, J.-L.; Wang, X.-H.; Sun, J.-C.; Peng, Y.-Y.; Ji, C.-B.; Zeng, X.-P. Chiral Tertiary Amine Catalyzed Asymmetric [4 + 2] Cyclization of 3-Aroylcoumarines with 2,3-Butadienoate. *Molecules* **2021**, *26*, 489. https://doi.org/10.3390/molecules26020489

Received: 28 November 2020
Accepted: 14 January 2021
Published: 18 January 2021

1. Introduction

Coumarin derivatives are among the most commonly found structural units in natural products, pharmaceuticals, and functional materials [1–5]. Therefore, numerous endeavors have been devoted to develop effective methods for the synthesis of coumarin based compounds [6–11]. On the other hand, 2*H*-pyran moieties also play a vital role in natural and unnatural bioactive compounds. Therefore, the introduction of 2*H*-pyran moiety into coumarin structural unit is a highly potential route for the identification of novel bioactive structures and the synthesis of these structures, i.e., dihydrocoumarin-fused dihydropyranones, have attracted continuing research interest. Among the developed methods, the [4 + 2] reaction of 3-aroylcoumarins are the most commonly used [12–14].

Early in 2012, Shi and co-worker described the first [4 + 2] cyclization of 3-aroylcoumarines (**1**) with ethyl 2,3-butadienoate (**2a**) to construct racemic dihydrocoumarin-fused dihydropyranones **3** in 79–95% yield using DABCO as the Lewis base catalyst (Figure 1a) [15]. This [4 + 2] process was initiated by the necleophilic attack of tertiary amine to 2,3-butadienoate to generate zwitterionic **I**. The γ-carbanion of **I** then attacks the β-carbon of enones **1** to give **II** with Z configuration to avoid the interaction of the ester group with the 3-position substituent. In the following, an intramolecular nucleophilic substitution of **II** could give cycloadduct **3** and regenerate the tertiary amine catalyst. Soon after that, the Ye group reported that chiral dihydrocoumarin-fused dihydropyranones could be accessed in a highly enantioselective manner via chiral NHC catalyzed [4 + 2] cycloaddition of ketenes and 3-aroylcoumarins [16]. In 2016, Lu et al. achieved a phosphine-catalyzed [4 + 2] annulation of allenones with 3-aroylcoumarins to afford chiral dihydrocoumarin-fused dihydropyrans [17]. Chen and co-workers reported the synthesis of chiral dihydrocoumarin-fused dihydropyrans through dienamine catalysis, but only two examples were explored [18]. Despite the above achievements, the identification of new protocols using easily available starting materials and chiral catalysts for the enantioselective construction of dihydrocoumarin-fused dihydropyrans are still highly desirable.

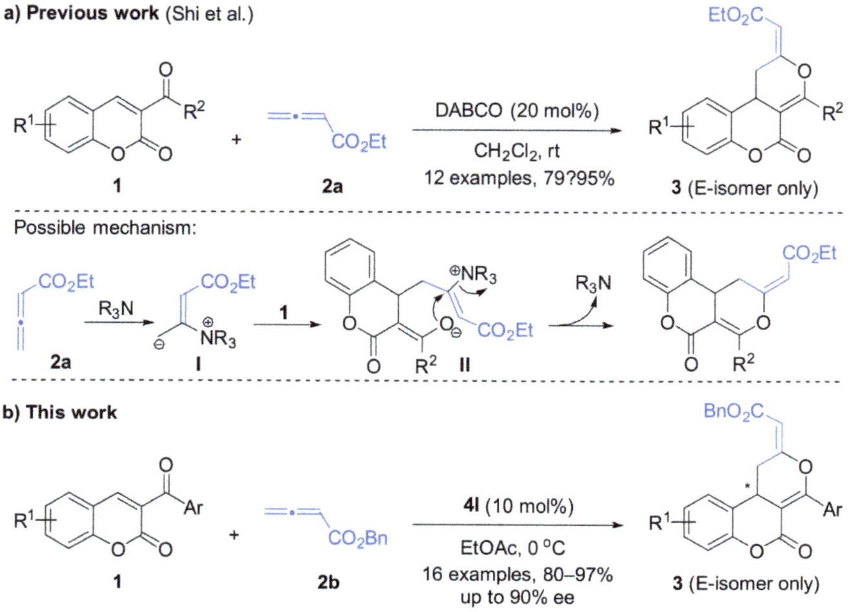

Figure 1. (**a**) DABCO catalyzed [4 + 2] cyclization of 3-aroylcoumarines with 2,3-butadienoate; (**b**) Chiral tertiary amine catalyzed [4 + 2] cyclization of 3-aroylcoumarines with 2,3-butadienoate

Inspired by Shi's pioneering work and based on our interest in the synthesis of chiral coumarin derivatives, we envision that the replacement of DABCO with a suitable chiral tertiary amine catalyst to mediate the [4 + 2] cyclization of 3-aroylcoumarines with 2,3-butadienoate might offer a new method for the synthesis of chiral dihydrocoumarin-fused dihydropyrans (Figure 1b).

2. Results and Discussion

We started our investigations by carrying out the reaction between 3-benzoylcoumarine (**1a**) and benzyl 2,3-butadienoate (**2b**) in $ClCH_2CH_2Cl$ at 25 °C (Table 1). Initially, a series of cinchona alkaloids, including quinine (**4a**), cinchonine (**4b**), and C_2-symmetric (bis)cinchona alkaloid (**4c–f**) were screened (entries 1–6), and the corresponding chiral dihydrocoumarin-fused dihydropyran product (**3a**) was obtained in up to only 45% ee values when **4c** was used, but the yield of **3a** was pretty low even after 48 h (entry 3). To our delight, when the bifunctional β-isocupreidine (**4g**) was tested in our reaction, the reaction was greatly accelerated to complete with 8 h and delivered **3a** in almost quantitative yield with promising 52% ee value (entry 7) [19,20]. Based on this result, we turn our attention to modify β-isocupreidine, so as to improve the enantiocontrol of the reaction. According the methods reported by Kumar and co-workers, a series of 6′-aryl-β-iso-cinchonine (**4h–l**) were prepared and examined in the current reaction [21]. It was observed that 6′-phenyl-β-iso-cinchonine **4h** could facilitate the model reaction to give 82% yield for **3a** with improved 60% ee (entry 8). Further variation of the phenyl into more steric aryl groups turned out to be ineffective, as is demonstrated by the 29–35% ee values obtained from catalyst **4i–k**. A slightly improved 63% ee was achieved when β-iso-cinchonine (**4l**) bearing a longer 4-biphenyl group at the 6′ position was tried (entry 12), but no more improvement was obtained when further increase the length the substituent (entry 13).

Table 1. Condition optimization for the catalytic asymmetric [4 + 2] cyclization.

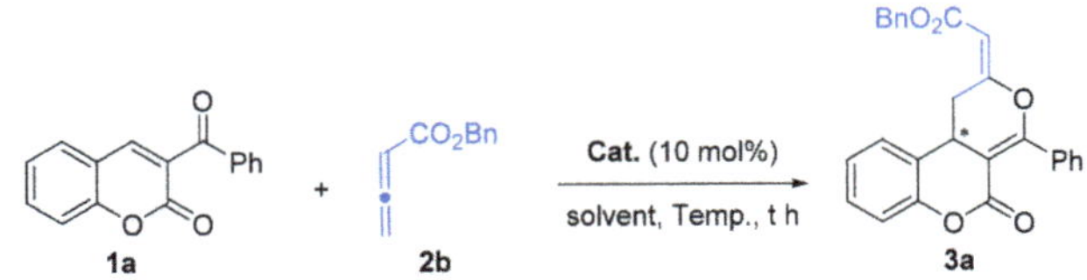

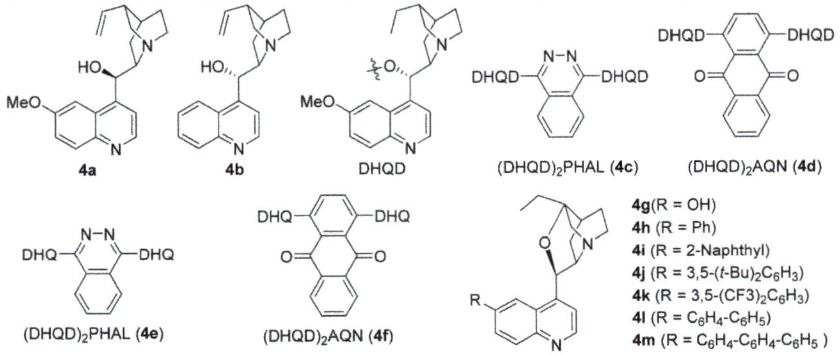

Entry	Cat.	Solvent	Temp. (°C)	t (h)	Yield (%)	Ee (%)
1	Quinine (**4a**)	DCE	25	48	6	39
2	Cinchonine (**4b**)	DCE	25	48	33	16
3	**4c**	DCE	25	48	8	45
4	**4d**	DCE	25	48	83	27
5	**4e**	DCE	25	48	trace	
6	**4f**	DCE	25	48	23	22
7	**4g**	DCE	25	8	96	52
8	**4h**	DCE	25	11	82	60
9	**4i**	DCE	25	23	78	35
10	**4j**	DCE	25	23	60	34
11	**4k**	DCE	25	23	67	29
12	**4l**	DCE	25	11	87	63
13	**4m**	DCE	25	16	78	47
14	**4l**	n-C_6H_{12}	25	48	trace	
15	**4l**	toluene	25	48	68	79
16	**4l**	THF	25	48	77	66
17	**4l**	Acetone	25	2.5	79	58
18	**4l**	EtOAc	25	7.5	87	79
19	**4l**	CH_3CN	25	2.5	95	56
20	**4l**	DMF	25	2.5	88	53
21	**4l**	MeOH	25	48	trace	-
22	**4l**	EtOAc	0	1	93	79

In the following, the solvent effects were examined using **4l** as the catalyst. The reaction was found to be more effective in solvent with moderate polarity (entries 14–21) and EtOAc was found to be the best, which afforded the desired product **3a** in 87% yield and 80% ee (entry 18). In the following, we tried to lower the reaction temperature to 0 °C to improve the ee value of **3a** (entry 22). To our surprise, the reaction finished within 1 h and gave an improved 93% yield, but no improvement of the ee value was observed. The observed higher reactivity at 0 °C than at 25 °C might be attributed to the competitive nucleophilic addition of quinoline nitrogen atoms of the catalyst to 2,3-butadienoate at higher temperature, which deactivate the catalyst and retard the reaction catalytic cycle.

Based on the above optimization, the scope of this tertiary amine catalyzed enantioselective [4 + 2] cyclization of 3-aroylcoumarines with benzyl 2,3-butadienoate (**2b**); we then evaluated this using 10 mol% of **4l** as the catalyst in EtOAc at 0 °C (Figure 2). It was observed that the reaction outcome was significantly affected by the electronic properties of the substituents on the coumarin benzene ring. In general, substrates bearing electron-donating groups (EDG, such as Me, OMe) were relatively less reactive and afforded slightly higher ee values. As is shown by the 79–98% yields and 80–88% ee values for products **3b–e**. The more steric **1f** substrate could give the corresponding product **3f** in highest 90% ee under standard conditions. In contrast, the reactions of electron-withdrawing group (EWG) substituted substrates are found to be more reactive but delivered relatively lower enantioselectivities. For example, products **3g–i** were obtained in excellent yield but with only around 70% ee. Additionally, coumarins bearing both EDG and EWG on the benzyl ring of the aroyl group were also well tolerated under standard conditions, but the enantiomeric excess was significantly affected by the steric effect. The *para*-substituted products (**3k,n**) could be obtained in much higher ee values than the *meta*-substituted products (**3l,o**). Moreover, the current reaction is also suitable for the reaction of (1,1′-biphenyl)-4-carbonyl and thiophene-2-carbonyl substituted coumarins, which afforded the desired products **3o** and **3p** in 79% and 81% ee values, respectively.

The *Z*/*E* configuration of the products was determined by the converting product **3m** into the known compound **5** and comparing their NMR spectrum (Figure 3). Under the above optimized conditions, the reaction of 3-benzoyl coumarin **1m** with ethyl 2,3-butadienoate **2a** afforded product **5** with 45% ee value. The same product **5** could also be obtained via a 3-step sequence from **3m** and **2b** in 79% ee value. The NMR spectrum of these newly synthesized products **5** were identical with the previous report by Shi and co-workers. Thus, the configuration of the products **3** were assigned to be *E*. This process also highlighted the synthetic potential of product **1** to be elaborated into other dihydrocoumarin-fused dihydropyran derivatives. We also tried to recrystallize products **3** and determined their absolute configuration by X-ray crystallography analysis, but turned out to be unsuccessful.

In order to demonstrate the practicability of the current method, we conducted a gram-scale reaction of 3-benzoyl coumarin **1f** with benzyl 2,3-butadienoate **2b** (Figure 4). In the presence of only 2.5 mol% of **4l** as the catalyst, the reaction of 2.5 mmol of **1f** with 1.5 equivalents of **2b** could give rise to the desired dihydrocoumarin-fused dihydropyran **3f** in 91% yield (1.215 g) with slightly improved 93% ee.

In summary, a series of chiral dihydrocoumarin-fused dihydropyranones were synthesized via the enantioselective [4 + 2] cyclization of 3-aroylcoumarines with benzyl 2,3-butadienoate. In the presence of 10 mol% of Kumar's 6′-(4-biphenyl)-β-iso-cinchonine as the chiral tertiary amine catalyst, the desired products could be obtained in up to 97% yield and 90% ee values under mild conditions. The current method used an easily available chiral catalyst and starting materials and could be conducted on gram-scale without loss of enantiomeric excess. The thus obtained products are potential in the construction of other dihydrocoumarin-fused dihydropyran derivatives. Considering the wide existence of coumarins and 2*H*-pyran moieties in natural products and pharmaceuticals, the thus obtained optically active dihydrocoumarin-fused dihydropyranones should be of interest to medicinal chemists.

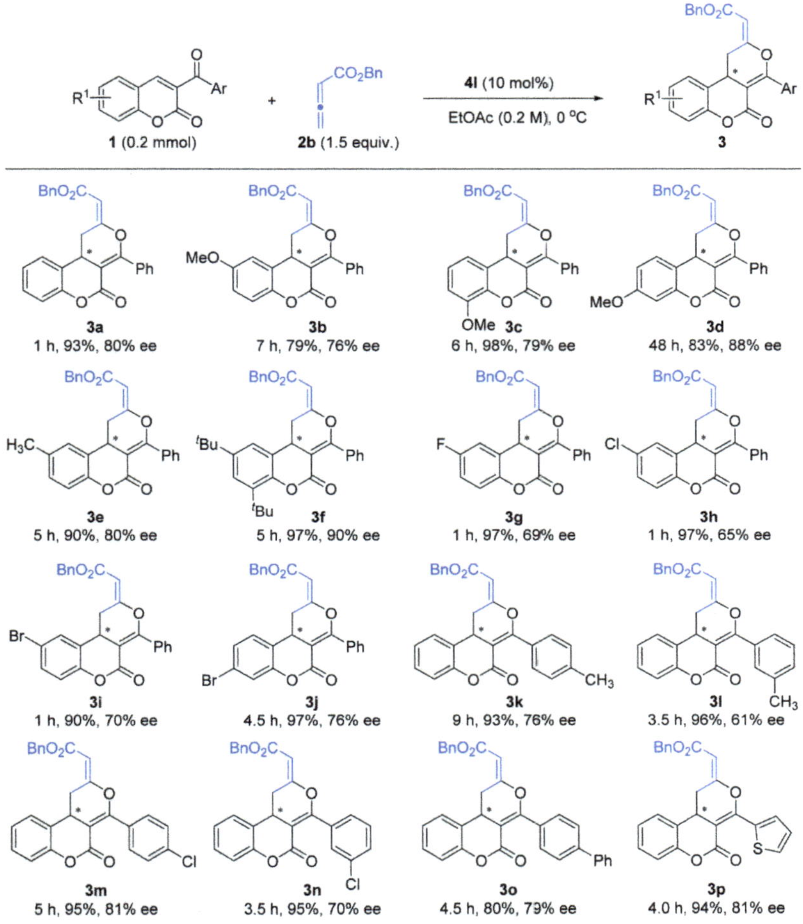

Figure 2. Substrate scope of tertiary amine catalyzed asymmetric [4 + 2] cyclization.

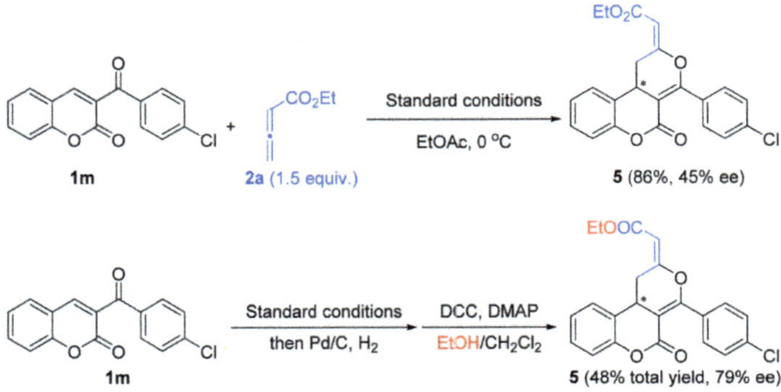

Figure 3. *Z*/*E* Configuration determination.

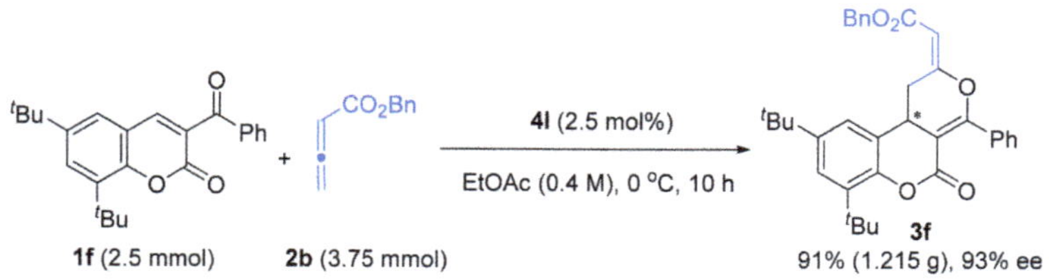

Figure 4. Gram-scale synthesis and product elaboration.

3. Materials and Methods

3.1. General Information

Reactions were monitored by thin layer chromatography using UV light or $KMnO_4$ to visualize the course of reaction. Purification of reaction products was carried out by flash chromatography on silica gel. Chemical yields refer to pure isolated substances. The $[\alpha]_D$ was recorded using PolAAr 3005 High Accuracy Polarimeter (Optical Activity Ltd., Huntingdon, England). Infrared (IR) spectra were obtained using a Bruker tensor 27 infrared spectrometer (Bruker, Borken, Germany). ^1H, ^{13}C and ^{19}F NMR spectra were obtained using Bruker DPX-400 spectrometer (Bruker UK Limited, Coventry, UK). Chiral HPLC analyses were obtained using Agilent Technologies 1260 Infinity series (Agilent Technologies, Inc., Waldbronn, Germany) and DAICEL CHIRALPAK columns (CPI Company, Tokyo, Japan). Chemical shifts were reported in ppm from tetramethylsilane with the solvent resonance as the internal standard. The following abbreviations were used to designate chemical shift multiplicities: s = singlet, d = doublet, t = triplet, q = quartet, h = heptet, m = multiplet, br = broad.

3.2. Tertiary Amine Catalyzed Asymmetric [4 + 2] Cyclization

General procedure: To a 4 mL vial was sequentially added 3-aroylcoumarines **1** (0.2 mmol), catalyst **4l** (0.02 mmol, 10 mol%), and EtOAc (1.0 mL); the mixture was stirred at 0 °C for 15 min before benzyl buta-2,3-dienoate **2b** (0.3 mmol, 1.5 equiv.) was charged. The reaction was monitored by TLC analysis. After completion of the reaction, the solvent was removed by rotary evaporation and the residue was directly subjected to column chromatography using PE/EtOAc (20:1–15:1) as the eluent to afford product **3**.

Benzyl (*E*)-2-(5-oxo-4-phenyl-1,10b-dihydro-2H,5H-pyrano[3,4-c]chromen-2-ylidene) acetate (**3a**).

White solid (m.p. 121.4–122.1 °C); ^1H NMR (400 MHz, $CDCl_3$) δ 7.51–7.31 (m, 12H), 7.20 (t, *J* = 7.2 Hz, 1H), 7.10 (d, *J* = 8.0 Hz, 1H), 5.90 (d, *J* = 1.6 Hz, 1H), 5.23 (s, 2H), 4.87 (dd, *J* = 14.8, 6.0 Hz, 1H), 4.01 (dd, *J* = 12.2, 5.6 Hz, 1H), 2.56–2.49 (m, 1H); ^{13}C NMR (101 MHz, $CDCl_3$) δ 166.6, 164.2, 162.3, 161.5, 150.8, 136.0, 133.1, 130.9, 129.2, 129.1, 128.8, 128.5, 128.3, 128.3, 125.8, 124.9, 122.7, 117.3, 102.2, 101.5, 66.4, 30.3, 26.1; $[\alpha]_D^{26.0}$ = + 44.3 (c = 0.26, $CHCl_3$); The enantiomeric purity of **3a** was determined by HPLC analysis (DAICEL CHIRALPAK AD-H (0.46 cmφ × 25 cm), hexane:2-propanol = 60:40, flow rate = 0.75 mL/min, retention time: 11.2 min (major) and 14.5 min (minor)); HRMS (ESI): Exact mass calcd for $C_{27}H_{21}O_5$ $[M+H]^+$: 425.1389, found: 425.1392.

Benzyl (*E*)-2-(9-methoxy-5-oxo-4-phenyl-1,10b-dihydro-2H,5H-pyrano[3,4-c]chromen-2-ylidene) acetate (**3b**).

White solid (m.p. 131.1–132.8 °C); ^1H NMR (400 MHz, $CDCl_3$) δ 7.51–7.36 (m, 10H), 7.03 (d, *J* = 8.8 Hz, 1H), 6.91 (d, *J* = 2.0 Hz, 1H), 6.85 (dd, *J* = 8.8, 2.8 Hz, 1H), 5.91 (d, *J* = 1.6 Hz, 1H), 5.23 (s, 2H), 4.80 (dd, *J* = 14.6, 5.6 Hz, 1H), 3.98 (dd, *J* = 12.0, 5.6 Hz, 1H), 3.83 (s, 3H), 2.58–2.51(m,1H); ^{13}C NMR (101 MHz, $CDCl_3$) δ 166.6, 164.1, 162.1, 161.9, 156.8, 144.7, 136.0, 133.0, 130.8, 129.1, 128.8, 128.5, 128.3, 128.3, 123.7, 118.0, 114.2, 111.1, 102.2, 101.4, 66.4 56.0, 30.5, 25.9; $[\alpha]_D^{26.0}$ = +242.2 (c = 0.49, $CHCl_3$); The enantiomeric purity of **3b** was determined by HPLC analysis (DAICEL CHIRALPAK AD-H (0.46 cmφ × 25 cm), hexane:2-propanol = 60:40, flow rate = 0.75 mL/min, retention time: 15.2 min (major) and 21.0 min (minor)); HRMS (ESI): Exact mass calcd for $C_{28}H_{23}O_6$ $[M+H]^+$: 455.1495, found: 455.1496.

Benzyl (*E*)-2-(7-methoxy-5-oxo-4-phenyl-1,10b-dihydro-2H,5H-pyrano[3,4-c]chromen-2-ylidene)acetate (**3c**).

White solid (m.p. 108.7–110.2 °C); ^{1}H NMR (400 MHz, $CDCl_3$) δ 7.53–7.34 (m, 10H), 7.14 (t, *J* = 8.4 Hz, 1H), 6.95 (dd, *J* = 13.0, 8.4 Hz, 2H), 5.91 (d, *J* = 1.2 Hz, 1H), 5.24 (s, 2H), 4.83 (dd, *J* = 14.2, 6.0 Hz, 1H), 4.00 (dd, *J* = 12.4, 6.0 Hz, 1H), 3.90 (s, 3H), 2.57–2.50 (m, 1H); ^{13}C NMR (101 MHz, $CDCl_3$) δ 166.5, 164.2, 162.1, 161.0, 147.8, 140.2, 136.0, 132.9, 130.8, 129.1, 128.7, 128.4, 128.3, 128.2, 124.7, 123.8, 117.0, 111.7, 102.1, 101.2, 66.3, 56.2, 30.5, 26.0; $[\alpha]_D^{26.0}$ = +86.5 (c = 0.5, $CHCl_3$); The enantiomeric purity of **3c** was determined by HPLC analysis (DAICEL CHIRALPAK AD-H (0.46 cmφ × 25 cm), hexane:2-propanol = 60:40, flow rate = 0.75 mL/min, retention time: 13.0 min (major) and 17.0 min (minor)). HRMS (ESI): Exact mass calcd for $C_{28}H_{23}O_6$ $[M+H]^+$: 455.1495, found: 455.1493.

Benzyl (*E*)-2-(8-methoxy-5-oxo-4-phenyl-1,10b-dihydro-2H,5H-pyrano[3,4-c]chromen-2-ylidene)acetate (**3d**).

White solid (m.p. 101.9–102.5 °C); ^{1}H NMR (400 MHz, $CDCl_3$) δ 7.50–7.34 (m, 10H), 7.27–7.24 (m, 1H), 6.74 (dd, *J* = 8.6, 2.4 Hz, 1H), 6.63 (d, *J* = 2.4 Hz, 1H), 5.87 (d, *J* = 1.6 Hz, 1H), 5.21 (s, 2H), 4.81 (dd, *J* = 14.8, 5.6 Hz, 1H), 3.92 (dd, *J* = 12.4, 5.6 Hz, 1H), 3.80 (s, 3H), 2.47–2.44 (m, 1H); ^{13}C NMR (101 MHz, $CDCl_3$) δ 166.6, 164.3, 162.2, 161.5, 160.3, 151.5, 136.0, 133.1, 130.8, 129.0, 128.8, 128.4, 128.3, 126.5, 114.5, 111.1, 102.5, 102.0, 101.7, 66.3, 55.7, 29.7, 26.4; $[\alpha]_D^{26.0}$ = −44.3 (c = 0.26, $CHCl_3$); The enantiomeric purity of **3d** was determined by HPLC analysis (DAICEL CHIRALPAK AD-H (0.46 cmφ × 25 cm), hexane:2-propanol = 60:40, flow rate = 0.75 mL/min, retention time: 12.2 min (major) and 17.7 min (minor)). HRMS (ESI): Exact mass calcd for $C_{28}H_{23}O_6$ $[M+H]^+$: 455.1495, found: 455.1498.

Benzyl (*E*)-2-(9-methyl-5-oxo-4-phenyl-1,10b-dihydro-2H,5H-pyrano[3,4-c]chromen-2-ylidene)acetate (**3e**).

White solid (m.p. 127.6–128.9 °C); ^{1}H NMR (400 MHz, $CDCl_3$) δ 7.51–7.35 (m, 10H), 7.18 (s, 1H), 7.12 (d, *J* = 8.4 Hz, 1H), 6.99 (d, *J* = 8.0 Hz, 1H), 5.91 (d, *J* = 1.6 Hz, 1H), 5.24 (s, 2H), 4.87 (dd, *J* = 14.8, 5.6 Hz, 1H), 3.97 (dd, *J* = 12.4, 5.6 Hz, 1H), 2.53–2.46 (m, 1H), 2.37 (s,3H); ^{13}CNMR (101 MHz, $CDCl_3$) δ 166.7, 164.4, 162.1, 161.7, 148.7, 136.0, 134.6, 133.1, 130.8, 129.7, 129.0, 128.8, 128.5, 128.3, 128.2, 126.2, 122.2, 117.0, 102.0, 101.6, 66.4, 30.2, 26.1, 21.0; $[\alpha]_D^{26.0}$ = +155.3 (c = 0.50, $CHCl_3$); The enantiomeric purity of **3e** was determined by HPLC analysis (DAICEL CHIRALPAK AD-H (0.46 cmφ × 25 cm), hexane:2-propanol = 60:40, flow rate = 0.75 mL/min, retention time: 12.7 min (major) and 18.5 min (minor)). HRMS (ESI): Exact mass calcd for $C_{28}H_{23}O_5$ $[M+H]^+$: 439.1545, found: 439.1548.

Benzyl (*E*)-2-(7,9-di-tert-butyl-5-oxo-4-phenyl-1,10b-dihydro-2H,5H-pyrano[3,4-c] chromen-2-ylidene)acetate (**3f**).

Yellow solid (m.p. 83.7–84.9 °C); ^{1}H NMR (400 MHz, $CDCl_3$) δ 7.54–7.37 (m, 11H), 7.28 (s, 1H), 5.95 (s, 1H), 5.28 (s, 2H), 4.76 (dd, *J* = 14.8, 6.0 Hz, 1H), 4.01 (dd, *J* = 11.6, 6.0 Hz, 1H), 2.77 (t, *J* = 13.2 Hz, 1H), 1.49 (s, 9H), 1.39 (s, 9H); ^{13}C NMR (101 MHz, $CDCl_3$) δ 166.6, 164.1, 162.1, 161.9, 156.8, 144.7, 136.0, 133.0, 130.9, 129.1, 128.8, 128.5, 128.3, 128.3 123.7, 118.0, 114.2, 111.1, 102.2, 101.4, 66.4, 56.0, 30.5, 25.9; $[\alpha]_D^{26.0}$ = +58.6 (c = 0.52, $CHCl_3$); The enantiomeric purity of **3f** was determined by HPLC analysis (DAICEL CHIRALPAK AD-H (0.46 cmφ × 25 cm), hexane:2-propanol = 60:40, flow rate = 0.75 mL/min, retention time: 6.0 min (major) and 5.1 min (minor)). HRMS (ESI): Exact mass calcd for $C_{35}H_{37}O_5$ $[M+H]^+$: 537.2641, found: 537.2643.

Benzyl (*E*)-2-(9-fluoro-5-oxo-4-phenyl-1,10b-dihydro-2H,5H-pyrano[3,4-c]chromen-2-ylidene)acetate (**3g**).

Yellow solid (m.p. 118.4–119.7 °C); ^{1}H NMR (400 MHz, $CDCl_3$) δ 7.53–7.39 (m, 10H), 7.11–7.00 (m, 3H), 5.93 (d, *J* = 1.6 Hz, 1H), 5.25 (s, 2H), 4.80 (dd, *J* = 14.6, 5.6 Hz, 1H), 3.97 (dd, *J* = 12.4, 5.6 Hz, 1H), 2.56–2.49 (m, 1H); ^{13}C NMR (101 MHz, $CDCl_3$) δ 166.3, 163.5, 162.6, 161.1, 159.4 (d, ^1J = 243 Hz), 146.7, 135.9, 132.8, 130.9, 129.0, 128.7, 128.4, 128.2, 128.2, 124.3, 124.2, 118.5, 118.4, 115.8 (d, ^2J = 23 Hz), 112.6 (d, ^3J = 25 Hz), 102.4, 100.4, 66.4, 30.3, 25.7; $[\alpha]_D^{26.0}$ = −22.8 (c = 0.47, $CHCl_3$); The enantiomeric purity of **3g** was determined by HPLC analysis (DAICEL CHIRALPAK AD-H (0.46 cmφ × 25 cm), hexane:2-propanol = 60:40, flow rate = 0.75 mL/min, retention time: 11.4 min (major) and 14.1 min (minor)). HRMS (ESI): Exact mass calcd for $C_{27}H_{20}FO_5$ [M+H]$^+$: 443.1295, found: 443.1297.

Benzyl (*E*)-2-(9-chloro-5-oxo-4-phenyl-1,10b-dihydro-2H,5H-pyrano[3,4-c]chromen-2-ylidene)acetate (**3h**).

White solid (m.p. 127.4–128.9 °C); ^{1}H NMR (400 MHz, $CDCl_3$) δ 7.50–7.47 (m, 3H), 7.43–7.40 (m, 8H), 7.39–7.28 (m, 1H), 7.03 (d, *J* = 8.4 Hz, 1H), 5.92 (d, *J* = 2.0 Hz, 1H), 5.24 (s, 2H), 4.83 (dd, *J* = 14.6, 5.6 Hz, 1H), 3.98 (dd, *J* = 12.4, 5.6 Hz, 1H), 2.55–2.47 (m, 1H); ^{13}C NMR (101 MHz, $CDCl_3$) δ 166.4, 163.5, 162.9, 161.0, 149.3, 135.9, 132.8, 131.0, 130.1, 129.2, 129.0, 128.8, 128.5, 128.3, 128.3, 125.9, 124.3, 118.6, 102.6, 100.4, 66.5, 30.3, 25.8; $[\alpha]_D^{26.0}$ = −56.8 (c = 0.19, $CHCl_3$); The enantiomeric purity of **3h** was determined by HPLC analysis (DAICEL CHIRALPAK AD-H (0.46 cmφ × 25 cm), hexane:2-propanol = 60:40, flow rate = 0.75 mL/min, retention time: 13.7 min (major) and 16.4 min (minor)). HRMS (ESI): Exact mass calcd for $C_{27}H_{20}ClO_5$ [M+H]$^+$: 459.0999, found: 459.0997.

Benzyl (*E*)-2-(9-bromo-5-oxo-4-phenyl-1,10b-dihydro-2H,5H-pyrano[3,4-c]chromen-2-ylidene)acetate (**3i**).

White solid (m.p. 131.4–132.7 °C); ^{1}H NMR (400 MHz, $CDCl_3$) δ 7.51–7.35 (m, 12H), 6.98 (d, J = 8.8 Hz, 1H), 5.92 (d, J = 1.2 Hz, 1H), 5.24 (s, 2H), 4.83 (dd, J = 14.6, 6.0 Hz, 1H), 3.99 (dd, J = 12.2, 6.0 Hz, 1H), 2.55–2.48 (m, 1H); ^{13}C NMR (101 MHz, $CDCl_3$) δ 166.4, 163.5, 162.9, 160.9, 149.9, 135.9, 132.8, 132.2, 131.0, 129.0, 128.8, 128.8, 128.5, 128.3, 128.3, 124.8, 119.0, 117.6, 102.6, 100.4, 66.5, 30.3, 25.8; $[\alpha]_D^{26.0}$ = −17.8 (c = 0.22, $CHCl_3$); The enantiomeric purity of **3i** was determined by HPLC analysis (DAICEL CHIRALPAK AD-H (0.46 cmφ × 25 cm), hexane:2-propanol = 60:40, flow rate = 0.75 mL/min, retention time: 17.1 min (major) and 14.2 min (minor)). HRMS (ESI): Exact mass calcd for $C_{27}H_{20}BrO_5$ [M+H]$^+$: 503.0494, found: 503.0490.

Benzyl (*E*)-2-(8-bromo-5-oxo-4-phenyl-1,10b-dihydro-2H,5H-pyrano[3,4-c]chromen-2-ylidene)acetate (**3j**).

White solid (m.p. 119.2–120.7 °C); ^{1}H NMR (400 MHz, $CDCl_3$) δ 7.49–7.46 (m, 3H), 7.45–7.29 (m, 8H),7.24–7.20 (m, 2H), 5.90 (d, *J* = 2.0 Hz, 1H), 5.21 (s, 2H), 4.80 (dd, *J* = 14.8, 5.6 Hz, 1H), 3.90 (dd, *J* = 12.2, 5.6 Hz, 1H), 2.51–2.44 (m, 1H); ^{13}C NMR (101 MHz, $CDCl_3$) δ 166.4, 163.6, 162.8, 160.7, 151.2, 135.9, 132.7, 130.9, 129.0, 128.7, 128.4, 128.3, 127.8, 127.1, 122.0, 121.8, 120.4, 102.4, 100.5, 66.4, 30.0, 25.8; $[\alpha]_D^{26.0}$ = +67.4 (c = 0.52, $CHCl_3$); The enantiomeric purity of **3j** was determined by HPLC analysis (DAICEL CHIRALPAK AD-H (0.46 cmφ × 25 cm), hexane:2-propanol = 60:40, flow rate = 0.75 mL/min, retention time: 14.7 min (major) and 18.0 min (minor)). HRMS (ESI): Exact mass calcd for $C_{27}H_{20}BrO_5$ [M+H]$^+$: 503.0494, found: 503.0497.

Benzyl (*E*)-2-(5-oxo-4-(p-tolyl)-1,10b-dihydro-2H,5H-pyrano[3,4-c]chromen-2-ylidene) acetate (**3k**).

Yellow solid (m.p. 133.2–133.9 °C); ^{1}H NMR (400 MHz, $CDCl_3$) δ 7.44–7.32 (m, 9H), 7.24–7.19 (m, 3H), 7.11 (d, *J* = 8.0 Hz, 1H), 5.91 (d, *J* = 1.2 Hz, 1H), 5.24 (s, 2H), 4.87 (dd, *J* = 14.8, 5.6 Hz, 1H), 3.99 (dd, *J* = 12.2, 6.0 Hz, 1H), 2.55–2.48 (m, 1H), 2.41 (s, 3H); ^{13}C NMR (101 MHz, $CDCl_3$) δ 166.5, 164.3, 162.4, 161.7, 150.8, 141.3, 136.0, 130.0, 129.1, 129.0, 128.7, 128.4, 128.3, 125.7, 124.8, 122.8, 117.2, 102.0, 100.9, 66.3, 30.3, 26.0, 21.7; $[\alpha]_D^{26.0}$ = +32.2 (c = 0.54, $CHCl_3$); The enantiomeric purity of **3k** was determined by HPLC analysis (DAICEL CHIRALPAK AD-H (0.46 cmφ × 25 cm), hexane:2-propanol = 60:40, flow rate = 0.75 mL/min, retention time: 12.1 min (major) and 18.0 min (minor)). HRMS (ESI): Exact mass calcd for $C_{28}H_{23}O_5$ [M+H]$^+$: 439.1545, found: 439.1543.

Benzyl (*E*)-2-(5-oxo-4-(m-tolyl)-1,10b-dihydro-2H,5H-pyrano[3,4-c]chromen-2-ylidene) acetate (**3l**).

Yellow solid (m.p. 135.6–136.8 °C); ^{1}H NMR (400 MHz, $CDCl_3$) δ 7.43–7.31 (m, 9H), 7.24–7.19 (m, 3H), 7.11 (d, *J* = 8.4 Hz, 1H), 5.91 (d, *J* = 1.2 Hz, 1H), 5.24 (s, 2H), 4.87 (dd, J = 14.8, 6.0 Hz, 1H), 3.99 (dd, *J* = 12.4, 5.6 Hz, 1H). 2.55–2.48 (m, 1H), 2.41 (s, 3H); ^{13}C NMR (101 MHz, $CDCl_3$) δ 166.5, 164.3, 162.4, 161.7, 150.8, 141.3, 136.0, 130.0, 129.1, 129.0, 128.7, 128.4, 128.3, 125.7, 124.8, 122.8, 117.2, 102.0, 100.9, 66.3, 30.3, 26.0, 21.7; $[\alpha]_D^{26.0}$ = +42.2 (c = 0.52, $CHCl_3$); The enantiomeric purity of **3l** was determined by HPLC analysis (DAICEL CHIRALPAK AD-H (0.46 cmφ × 25 cm), hexane:2-propanol = 60:40, flow rate = 0.75 mL/min, retention time: 9.9 min (major) and 13.2 min (minor)). HRMS (ESI): Exact mass calcd for $C_{28}H_{23}O_5$ $[M+H]^+$: 439.1545, found: 439.1544.

Benzyl (*E*)-2-(4-(4-chlorophenyl)-5-oxo-1,10b-dihydro-2H,5H-pyrano[3,4-c]chromen-2-ylidene)acetate (**3m**).

Yellow solid (m.p. 122.5–123.7 °C); ^1H NMR (400 MHz, $CDCl_3$) δ 7.46–7.32 (m, 11H), 7.23–7.19 (m, 1H), 7.10 (d, *J* = 8.0 Hz, 1H), 5.90 (d, *J* = 1.6 Hz, 1H), 5.23 (s, 2H), 4.86 (dd, *J* = 14.8, 6.0 Hz, 1H), 4.00 (dd, *J* = 12.4, 5.6 Hz, 1H), 2.55–2.48 (m, 1H); ^{13}C NMR (101 MHz, $CDCl_3$) δ 166.4, 163.9, 161.4, 161.1, 150.7, 137.0, 136.0, 130.5, 129.3, 128.8, 128.6, 128.5, 128.3, 125.8, 125.0, 122.5, 117.3, 102.3, 101.9, 66.4, 30.3, 26.0; $[\alpha]_D^{26.0}$ = −148.3 (c = 0.49, $CHCl_3$); The enantiomeric purity of **3m** was determined by HPLC analysis (DAICEL CHIRALPAK AD-H (0.46 cmφ × 25 cm), hexane:2-propanol = 60:40, flow rate = 0.75 mL/min, retention time: 13.9 min (major) and 21.5 min (minor)). HRMS (ESI): Exact mass calcd for $C_{27}H_{20}ClO_5$ $[M+H]^+$: 459.0999, found: 459.0999.

Benzyl (*E*)-2-(4-(3-chlorophenyl)-5-oxo-1,10b-dihydro-2H,5H-pyrano[3,4-c]chromen-2-ylidene)acetate (**3n**).

Yellow solid (m.p. 124.5–125.3 °C); ^1H NMR (400 MHz, $CDCl_3$) δ 7.49–7.34 (m, 11H), 7.23–7.21 (m, 1H), 7.11–7.09 (m, 1H), 5.92 (d, *J* = 2.0 Hz, 1H), 5.24 (d, *J* = 2.0 Hz, 2H), 4.90–4.85 (m, 1H), 4.01–3.98 (m, 1H), 2.52 (t, *J* = 12.8, 1H); ^{13}C NMR (101 MHz, $CDCl_3$) δ 166.3, 163.8, 161.1, 160.6, 150.6, 135.9, 134.7, 134.2, 131.4, 130.8, 129.5, 129.2, 129.0, 128.7, 128.4, 128.3, 127.4, 125.8, 125.0, 122.3, 117.2, 102.4, 102.2, 66.4, 30.2, 25.9; $[\alpha]_D^{26.0}$ = +100.7 (c = 0.55, $CHCl_3$); The enantiomeric purity of **3n** was determined by HPLC analysis (DAICEL CHIRALPAK AD-H (0.46 cmφ × 25 cm), hexane:2-propanol = 60:40, flow rate = 0.75 mL/min, retention time: 12.4 min (major) and 13.9 min (minor)). HRMS (ESI): Exact mass calcd for $C_{27}H_{20}ClO_5$ $[M+H]^+$: 459.0999, found: 459.1003.

Benzyl (*E*)-2-(4-([1,1′-biphenyl]-4-yl)-5-oxo-1,10b-dihydro-2H,5H-pyrano[3,4-c]chromen-2-ylidene)acetate (**3o**).

White solid (m.p. 130.7–131.4 °C); ^1H NMR (400 MHz, $CDCl_3$) δ 7.66–7.60(m, 6H), 7.49–7.33 (m, 10H), 7.22 (t, *J* = 7.6 Hz, 1H), 7.13 (d, *J* = 8.0 Hz, 1H), 5.95 (s, 1H), 5.26 (s, 2H), 4.89 (dd, *J* = 14.8, 5.6 Hz, 1H), 4.03 (dd, J = 12.0, 5.6 Hz, 1H), 2.55 (t, *J* = 14.0 Hz, 1H); ^{13}C NMR (101 MHz, $CDCl_3$) δ 166.5, 164.2, 162.0, 161.6, 150.8, 143.7, 140.3, 136.0, 129.6, 129.1, 128.9, 128.7, 128.4, 128.3, 127.3, 126.9, 125.8, 117.2, 102.1, 101.4, 66.3, 30.3, 26.0; $[\alpha]_D^{26.0}$ = +91.1 (c = 0,34, $CHCl_3$); The enantiomeric purity of **3o** was determined by HPLC analysis (DAICEL CHIRALPAK AD-H (0.46 cmφ × 25 cm), hexane:2-propanol = 60:40, flow rate = 0.75 mL/min, retention time: 17.9 min (major) and 31.4 min (minor)). HRMS (ESI): Exact mass calcd for $C_{33}H_{25}O_5$ $[M+H]^+$: 501.1702, found: 501.1708.

Benzyl (*E*)-2-(5-oxo-4-(thiophen-3-yl)-1,10b-dihydro-2H,5H-pyrano[3,4-c]chromen-2-ylidene)acetate (**3p**).

White solid (m.p. 135.7–136.2 °C); ^1H NMR (400 MHz, $CDCl_3$) δ 7.83 (d, *J* = 3.6 Hz, 1H), 7.53 (d, *J* = 5.2 Hz, 1H), 7.41–7.30 (m, 7H), 7.18 (t, *J* = 7.6 Hz, 1H), 7.09 (dd, J = 8.4, 4.8 Hz, 2H), 5.92 (d, *J* = 1.2 Hz, 1H), 5.27–5.20 (m,2H), 4.70 (dd, *J* = 15.0, 6.0 Hz, 1H), 4.02 (dd, J = 12.6, 5.6 Hz, 1H), 2.69–2.62 (m,1H); ^{13}C NMR (101 MHz, $CDCl_3$) δ 166.5, 164.0, 163.0, 161.6, 155.2, 150.6, 136.0, 133.6, 132.8, 130.7, 129.1, 128.8, 128.5, 128.3, 127.3, 125.7, 124.9, 122.6, 117.1, 101.8, 100.3, 66.4, 30.9, 29.8, 26.4; $[\alpha]_D^{26.0}$ = +123.7 (c = 0.22, $CHCl_3$); The enantiomeric purity of **3p** was determined by HPLC analysis (DAICEL CHIRALPAK AD-H (0.46 cmφ × 25 cm), hexane:2-propanol = 60:40, flow rate = 0.75 mL/min, retention time: 12.4 min (major) and 14.6 min (minor)). HRMS (ESI): Exact mass calcd for $C_{25}H_{19}O_5S$ $[M+H]^+$: 459.0999, found: 459.1003.

3.3. *Synthesis of* **5** *from* **1m** *and* **2b**

To the reaction mixture obtained under standard condition using **1m** (1 mmol) and **2b** (1.25 mmol) was added Pd/C (wt. 10%), then the mixture was stirred under H_2 atmosphere (H_2 balloon) at room temperature for 48 h. The resulting reaction mixture was filtered through a pad of Celite and eiluted with EtOAc. The filtration was concentrated under reduced pressure and the residue was purified by silica gel column chromatography (PE:EtOAc = 5:1–2:1) to afford the free acid intermediate, which was then dissolved in CH_2Cl_2 (5 mL). This solution was cooled to 0 °C before DCC (2.0 equiv.), DMAP (2.0 equiv.) and EtOH (2 mL) were added. After that, the reaction mixture was moved to rt and stirred overnight. After completion of the reaction by TLC analysis, the solvent was removed by rotary evaporation and the residue was directly subjected to column chromatography using PE/EtOAc (15:1-9:1) as the eluent to afford product **5**.

Ethyl (E)-2-(4-(4-chlorophenyl)-5-oxo-1,10b-dihydro-2H,5H-pyrano[3,4-c]chromen-2-ylidene)acetate (**5**) [4].

^{1}H NMR (400 MHz, $CDCl_3$) δ 7.46–7.31 (m, 6H), 7.21 (t, *J* = 7.6 Hz, 1H), 7.09 (d, *J* = 8.0 Hz, 1H), 5.83 (s, 1H), 4.87 (dd, *J* = 14.8, 5.6 Hz, 1H), 4.24 (q, *J* = 7.2 Hz, 2H), 4.00 (dd, *J* = 12.0, 5.6 Hz, 1H), 2.50 (t, *J* = 14.4 Hz, 1H), 1.33 (t, *J* = 7.2 Hz, 3H); ^{13}C NMR (101 MHz, $CDCl_3$) δ 166.6, 163.4, 161.5, 161.1, 150.6, 136.9, 131.4, 130.5, 129.2, 128.6, 125.8, 125.0, 122.5, 117.2, 102.6, 101.7, 60.6, 30.3, 25.8, 14.4; $[\alpha]_D^{26.0}$ = +78.6 (c = 0.38, $CHCl_3$); The enantiomeric purity of **5** was determined by HPLC analysis (DAICEL CHIRALPAK OD-3 (4.6 mmφ × 150 mml), hexane:2-propanol = 80:20, flow rate = 0.75 mL/min, retention time: 6.1 min (major) and 7.6 min (minor)).

Supplementary Materials: The following are available online: ^{1}H and ^{13}C NMR spectra and HPLC data of compounds **3a–p** and **5**.

Author Contributions: X.-P.Z. conceived and designed the project and wrote the paper after discussing with Y.-Y.P. and C.-B.J.; J.-L.L., X.-H.W. and J.-C.S. performed the experiments and analyzed the data. All authors have read and agreed to the published version of the manuscript.

Funding: This research was funded by National Natural Science Foundation of China (21702080), Natural Science Foundation of Jiangxi Province of China (20181BAB213003), the Open Project Program of Key Laboratory of Functional Small Organic Molecule, Ministry of Education, Jiangxi Normal University (KLFS-KF-201603) and the foundation of Jiangxi Educational Committee (170223).

Institutional Review Board Statement: Not applicable.

Informed Consent Statement: Not applicable.

Data Availability Statement: Samples of the compounds are not available from the authors.

Conflicts of Interest: The authors declare no conflict of interest.

Sample Availability: Samples of the compounds are not available from the authors.

References

1. Hu, X.-L.; Gao, C.; Xu, Z.; Liu, M.-L.; Feng, L.-S.; Zhang, G.-D. Recent Development of Coumarin Derivatives as Potential Antiplasmodial and Antimalarial Agents. *Med. Chem.* **2018**, *18*, 114–123. [CrossRef]
2. Srikrishna, D.; Godugu, C.; Dubey, P.K. A Review on Pharmacological Properties of Coumarins. Mini-Rev. *Med. Chem.* **2018**, *18*, 113–141.
3. Cao, D.; Liu, Z.; Verwilst, P.; Koo, S.; Jangjili, P.; Kim, J.S.; Lin, W. Coumarinbased Small-molecule Fluorescent Chemosensors. *Chem. Rev.* **2019**, *119*, 10403–10519. [CrossRef] [PubMed]
4. Kasperkiewicz, K.; Ponczek, M.B.; Owczarek, J.; Guga, P.; Budzisz, E. Antagonists of Vitamin K-Popular Coumarin Drugs and New Synthetic and Natural Coumarin Derivatives. *Molecules* **2020**, *25*, 1465. [CrossRef] [PubMed]
5. Stefanachi, A.; Leonetti, F.; Pisani, L.; Catto, M.; Carotti, A. Coumarin: A Natural, Privileged and Versatile Scaffold for Bioactive Compounds. *Molecules* **2018**, *23*, 250. [CrossRef] [PubMed]
6. Wang, Z.-H.; Zhu, M.-W.; Chen, Q.-A.; Zhou, Y.-G. Catalytic Biomimetic Asymmetric Reduction of Alkenes and Imines Enabled by Chiral and Regenerable NAD(P)H Models. *Angew. Chem. Int. Ed.* **2019**, *58*, 1813–1817. [CrossRef]
7. Zhang, H.; Luo, Y.; Li, D.-W.; Yao, Q.; Dong, S.-X.; Liu, X.H.; Feng, X.M. Enantioselective Synthesis of 4-Hydroxy-dihydrocoumarins via Catalytic Ring Opening/Cycloaddition of Cyclobutenones. *Org. Lett.* **2019**, *21*, 2388–2392. [CrossRef]
8. Chen, Y.-R.; Ganapuram, M.R.; Hsieh, K.-H.; Chen, K.-H.; Karanam, P.; Vagh, S.S.; Liou, Y.-C.; Lin, W.-W. 3-Homoacyl Coumarin: An All Carbon 1,3-Dipole for Enantioselective Concerted (3 + 2) Cycloaddition. *Chem. Commun.* **2018**, *54*, 12702–12705. [CrossRef]
9. Zhang, X.-Z.; Gan, K.-J.; Liu, X.-X.; Deng, Y.-H.; Wang, F.-X.; Yu, K.-Y.; Zhang, J.; Fan, C.-A. Enantioselective Synthesis of Functionalized 4-Aryl Hydrocoumarins and 4-Aryl Hydroquinolin-2-ones via Intramolecular Vinylogous Rauhut-currier Reaction of *para*-Quinone Methides. *Org. Lett.* **2017**, *19*, 3207–3210. [CrossRef]

10. Jin, J.-H.; Li, X.-Y.; Luo, X.-Y.; Fossey, J.S.; Deng, W.-P. Asymmetric Synthesis of cis-3,4-Dihydrocoumarins via [4+ 2] Cycloadditions Catalyzed by Amidine Derivatives. *J. Org. Chem.* **2017**, *82*, 5424–5432. [CrossRef]
11. Zhao, Y.-L.; Lou, Q.-X.; Wang, L.-S.; Hu, W.-H.; Zhao, J.-L. Organocatalytic Friedel-crafts Alkylation/Lactonization Reaction of Naphthols with 3-Trifluoroe-thylidene Oxindoles: The Asymmetric Synthesis of Dihydrocoumarins. *Angew. Chem. Int. Ed.* **2017**, *56*, 338–342. [CrossRef] [PubMed]
12. Matthias, H.; Geoffrey, A.; Cordell, N.-R.; Payom, T. Candenatone, a Novel Purple Pigment from Dalbergia candenatensis. *J. Org. Chem.* **1988**, *53*, 4162–4165.
13. Wang, W.-J.; Xue, J.-J.; Tian, T.; Zhang, J.; Wei, L.-P.; Shao, J.-D.; Xie, Z.-X.; Li, Y. Total Synthesis of (±)-δ-Rubromycin. *Org. Lett.* **2013**, *15*, 2402–2405. [CrossRef] [PubMed]
14. Li, J.; Zhang, W.-H.; Zhang, F.; Chen, Y.; Li, A. Total Synthesis of Longeracinphyllin A. *J. Am. Chem. Soc.* **2017**, *139*, 14893–14896. [CrossRef] [PubMed]
15. Wang, Y.; Yu, H.-Z.; Zheng, H.-F.; Shi, D.-Q. DABCO and Bu_3P catalyzed [4 + 2] and [3 + 2] cycloadditions of 3-acyl-2Hchromen-ones and ethyl 2,3-butadienoate. *Org. Biomol. Chem.* **2012**, *10*, 7739–7746. [CrossRef]
16. Jian, T.-Y.; Chen, X.-Y.; Sun, L.-H.; Ye, S. *N*-heterocyclic carbene-catalyzed [4 + 2] cycloaddition of ketenes and 3-aroylcoumarins: Highly enantioselective synthesis of dihydrocoumarin-fused dihydropyranones. *Org. Biomol. Chem.* **2013**, *11*, 158–163. [CrossRef]
17. Han, X.-Y.; Ni, H.-Z.; Chan, W.-L.; Gai, X.-K.; Wang, Y.-J.; Lu, Y.-X. Highly enantioselective synthesis of dihydrocoumarin-fused dihydropyrans via the phosphine-catalyzed [4 + 2] annulation of allenones with 3-aroylcoumarins. *Org. Biomol. Chem.* **2016**, *14*, 5059–5064. [CrossRef]
18. Shi, M.-L.; Zhan, G.; Zhou, S.-L.; Du, W.; Chen, Y.-C. Asymmetric Inverse-Electron-Demand Oxa-Diels—Alder Reaction of Allylic Ketones through Dienamine Catalysis. *Org. Lett.* **2016**, *18*, 6480–6483. [CrossRef]
19. Wang, F.; Li, Z.; Wang, J.; Li, X.; Cheng, J.-P. Enantioselective Synthesis of Dihydropyran-Fused Indoles through [4 + 2] Cycloaddition between Allenoates and 3-Olefinic Oxindoles. *J. Org. Chem.* **2015**, *80*, 5279–5286. [CrossRef]
20. Wang, F.; Yang, C.; Xue, X.-S.; Li, X.; Cheng, J.-P. A Highly Efficient Chirality Switchable Synthesis of Dihydropyran-Fused Benzofurans by Fine-Tuning the Phenolic Proton of β-Isocupreidine (β-ICD) Catalyst with Methyl. *Chem. Eur. J.* **2015**, *21*, 10443–10449. [CrossRef]
21. Heiko, D.; Vivek, K.; Herbert, W.; Kamal, K. Lewis Base Catalyzed [4 + 2] Annulation of Electron-Deficient Chromone-Derived Heterodienes and Acetylenes. *Chem. Eur. J.* **2011**, *17*, 5130–5137.

Article

Coumarin's Anti-Quorum Sensing Activity Can Be Enhanced When Combined with Other Plant-Derived Small Molecules

Dmitry Deryabin, Kseniya Inchagova, Elena Rusakova * and Galimzhan Duskaev

Federal Research Centre of Biological Systems and Agro-technologies of the Russian Academy of Sciences, Orenburg 460000, Russia; dgderyabin@yandex.ru (D.D.); ksenia.inchagova@mail.ru (K.I.); gduskaev@mail.ru (G.D.)

* Correspondence: elenka_rs@mail.ru; Tel.: +7-919-860-24-78

Abstract: Coumarins are class of natural aromatic compounds based on benzopyrones (2H-1-benzopyran-2-ones). They are identified as secondary metabolites in about 150 different plant species. The ability of coumarins to inhibit cell-to-cell communication in bacterial communities (quorum sensing; QS) has been previously described. Coumarin and its derivatives in plant extracts are often found together with other small molecules that show anti-QS properties too. The aim of this study was to find the most effective combinations of coumarins and small plant-derived molecules identified in various plants extracts that inhibit QS in *Chromobacterium violaceum* ATCC 31532 violacein production bioassay. The coumarin and its derivatives: 7-hydroxycoumarin, 7.8-dihydroxy-4-methylcoumarin, were included in the study. Combinations of coumarins with gamma-octalactone, 4-hexyl-1.3-benzenediol, 3.4.5-trimethoxyphenol and vanillin, previously identified in oak bark (*Quercus cortex*), and eucalyptus leaves (*Eucalyptus viminalis*) extracts, were analyzed in a bioassay. When testing two-component compositions, it was shown that 7.8-dihydroxy-4-methylcoumarin, 4-hexyl-1.3-benzendiol, and gamma-octalactone showed a supra-additive anti-QS effect. Combinations of all three molecules resulted in a three- to five-fold reduction in the concentration of each compound needed to achieve EC_{50} (half maximal effective concentration) against QS in *C. violaceum* ATCC 31532.

Keywords: coumarins; quorum sensing; QS inhibitors; plant-derived molecules; *Chromobacterium violaceum*

Citation: Deryabin, D.; Inchagova, K.; Rusakova, E.; Duskaev, G. Coumarin's Anti-Quorum Sensing Activity Can Be Enhanced When Combined with Other Plant-Derived Small Molecules. *Molecules* **2021**, *26*, 208. https://doi.org/10.3390/molecules26010208

Academic Editor: Maria João Matos
Received: 30 November 2020
Accepted: 18 December 2020
Published: 3 January 2021

1. Introduction

Coumarins are a class of natural compounds based on benzopyrones (2H-1-benzopyran-2-ones) [1]. These compounds can be classified depending on the core's structure and the presence of substituents. There are the "simplest" coumarins (e.g., coumarin and dihydrocoumarin), followed by oxy-, meth-oxy-, and methylenedioxycoumarins with various substitutions in benzene/pyron rings (e.g., umbelliferon, 3-hydroxycoumarin, and scopoletin). The furancoumarins (e.g., bergamotin) contain an additional condensed furan core. Other, more structurally complex compounds are the result of coumarin condensation with pyran, benzene, and benzofuran rings. Most of the compounds of this class in plants are found in the free state, and only a small number are found in glycosides with D-glucose attached to the C6, C7, or C8 atoms of the coumarin nucleus [2].

Currently, coumarins are identified as secondary metabolites in about 150 different plant species distributed in almost 30 families, of which the most important are *Rutaceae, Umbelliferae, Clusiaceae, Guttiferae, Caprifoliaceae, Oleaceae, Nyctaginaceae,* and *Apiaceae* [3]. These substances are synthesized from phenylalanine via the shikimic acid formation pathway (hydroxylation, glycolysis, and cyclization of cinnamic acid) [4], and often, several different coumarins are found in the same plant.

Coumarins proposed for medical use due to their proven biological activity. They are showed anti-ulcerogenic [5], antiparasitic [6], anti-inflammatory [7–11], and other properties [12–14]. They are also antioxidant [15], and anticoagulant compounds [16–19]. As such, they can be defined as new pharmaceutical candidates [20].

The ability of coumarins to inhibit cell-to-cell communication in bacterial communities—better known as "quorum sensing" (QS)—has been discovered relatively recently. Briefly: QS is a special type of regulator of bacterial gene expression that functions at a high microbial population density. Depending on the chemical nature of the autoinducer, QS can be divided into several types: 1) LuxI/LuxR type (autoinducers–acylated homoserin lactones); 2) type II QS systems (autoinducers–furanone derivatives); 3) QS systems with Gram-positive bacteria (autoinducers–short oligopeptides); 4) QS systems with autoinducers of various natures (e.g., epinephrine, norepinephrine). The first of the described and most common QS systems is a two-component system of the LuxI/LuxR type inherent in many bacterial pathogens [21] where it activates the synthesis of virulence factors and the biofilms formation.

Because the search for plant-derived molecules with anti-QS activity is very actual, the coumarins are interesting object for this screening. Experimental observations of the anti-QS activity of coumarin are mainly related to Gram-negative bacteria that use a LuxI/LuxR type communication system, e.g., *Pseudomonas. aeruginosa* (in which coumarin suppress of phenazine biosynthesis, and motility) and *Aliivibrio fischeri* (coumarin inhibits the bioluminescence) [22]. Another simple coumarin, i.e., dihydrocoumarin, effectively inhibited QS-dependent biosynthesis of violacein in *Chromobacterium violaceum* [23]. The subsequent comparative analysis of seven hydroxycoumarin derivatives in relation to the violacein biosynthesis in *C. violaceum* showed that the promising anti-QS effect is characteristic of 3-hydroxycoumarin [24]. The identification of other functional substitutions of coumarin core, which led to disruption of the bacterial biofilm formation and, at the same time, inhibits QS development, was done by Reen [4]. This variant of bioactivity was also characteristic of a larger group of compounds, including furanocumarins: bergamottin and 6.7-dihydroxybergamottin. Interestingly, the last two compounds found in citrus fruits (*Citrus bergamia, Citrus maxima*, and *Citrus* × *paradisi*) showed their activity against bacteria that use both LuxI/LuxR-type and type II QS system [25].

Significantly, coumarin and its derivatives are often found together with other plant-derived small molecules that also have anti-QS properties. At the same time, our previous studies have shown that such molecules can act synergistically in a single plant [26,27]; however, until now, the possible combination of coumarins and small plant-derived molecules that are part of various plant extracts remains open.

The aim of this study was to find the most effective combinations of coumarins with small plant-derived molecules previously identified in extracts of oak bark (*Quercus cortex*), and eucalyptus leaves (*Eucalyptus viminalis*) to inhibit LuxI/LuxR-type quorum sensing in *C. violaceum* ATCC (American Type Culture Collection) 31532.

2. Results

2.1. Effect of Coumarin and Its Derivatives in Chromobacterium violaceum ATCC 31532 Violacein Production Bioassay

Cultivation of *C. violaceum* ATCC 31532 with coumarin, 7-hydroxycoumarin, and 7.8-dihydroxy-4-methylcoumarin followed registration of the optical density values (OP_{450}) of bacterial biomass and violacein production (OP_{600}), allowed us to evaluate their effect on the growth and QS-dependent biosynthesis in the bioassay. All tested components showed antibacterial activity as follows: minimum inhibitory concentration (MIC_{50}) = 2.689 mg/mL and MIC_{100} = 3.650 mg/mL for coumarin; MIC_{50} = 0.497 mg/mL and MIC_{100} = 1.267 mg/mL for 7-hydroxycoumarin and MIC_{50} = 0.325 mg/mL and MIC_{100} = 2.400 mg/mL for 7.8-dihydroxy-4-methylcoumarin. Simultaneously, sub-inhibitory concentrations of these compounds provided an anti-QS effect evaluated by inhibition on pigment violacein production, which was expressed as follows: effective concentration (EC_{50}) = 1.105 mg/mL and EC_{100} = 3.650 mg/mL for coumarin; EC_{50} = 0.199 mg/mL and EC_{100} = 0.633 mg/mL for 7-hydroxycoumarin and EC_{50} = 0.150 mg/mL and EC_{100} = 1.200 mg/mL for 7.8-dihydroxy-4-methylcoumarin (Table 1). Thus, the highest anti-QS activity was demonstrated by 7.8-dihydroxy-4-methylcoumarin, and this coumarin derivative was taken for further studies on the combination of small plant-derived molecules.

Table 1. Effects of coumarin, 7-hydroxycoumarin and 7.8-dihydroxy-4-methylcoumarin (mg/mL) on growth and QS-controlled violacein pigment biosynthesis in *C. violaceum* ATCC 31532.

Tested Compound	Characteristics of Antibacterial Activity, mg/mL		Characteristics of Anti-QS Activity, mg/mL	
	MIC_{100}	MIC_{50}	EC_{100}	EC_{50}
Coumarin	3.650	2.689	3.650	1.105
7-hydroxycoumarin	1.267	0.497	0.633	0.199
7.8-dihydroxy-4-methylcoumarin	2.400	0.325	1.200	0.150

2.2. *Analysis of the Combined Use of Coumarin Derivatives (7.8-Dihydroxy-4-methylcumarin) with Other Small Plant-Derived Molecules in C. violaceum ATCC 31532*

Results were obtained for combinations of 7.8-dihydroxy-4-methylcumarin (identified in *Baikal skullcap* extract) with small plant-derived molecules from oak bark (4-hexyl-1.3-benzenediol, 3.4.5-trimethoxyphenol, vanillin) or in eucalyptus leaves (gamma-octalactone), which have own anti-QS activity preliminary in *C. violaceum* ATCC 3153 bioassay. While most coumarins combinations showed simple additivity, some two-component mixtures led to pronounced mutual strengthening of anti-QS activity, which was evaluated as a synergetic (supra-additive) effect in 2D isobolographic analysis. The supra-additivity was revealed in combination of 7.8-dihydroxy-4-methylcoumarin with 4-hexyl-1.3-benzenediol (Figure 1a) as well as in 7.8-dihydroxy-4-methylcoumarin and gamma-octalactone mixtures (Figure 1b), where cultivation of *C. violaceum* ATCC 31532 in media enriched these paired molecular compositions showed a two-to four-fold decrease in the concentrations of each compound to achieve 50% inhibition of QS-controlled violacein biosynthesis. In turn, some combination of small molecules from oak bark and eucalyptus leaves showed the antagonistic (infra-additive) effect achieved in the compositions gamma-octalactone and vanillin, gamma-octalactone and 3.4.5-trimethoxyphenol, while the combinations gamma-octalactone and 4-hexyl-1.3-benzenediol have a supra-additive effect.

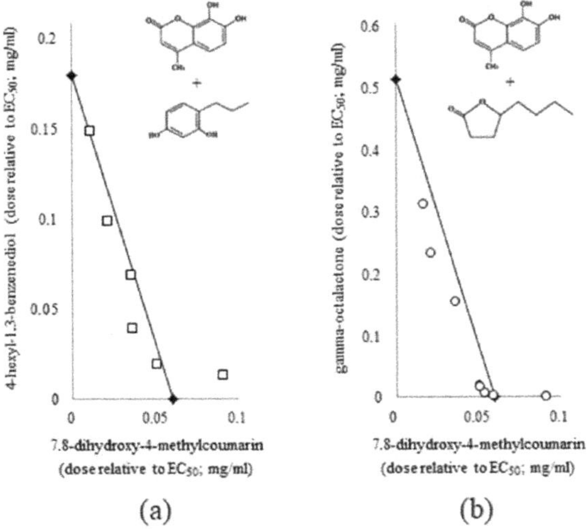

Figure 1. 2D isobolographic analysis of the combined use of 7.8-dihydroxy-4-methylcoumarin and 4-hexyl-1.3-benzenediol (**a**), 7.8-dihydroxy-4-methylcoumarin and gamma-octalactone (**b**) on the QS-controlled violacein biosynthesis in *C. violaceum* ATCC 31532. Isoboles are represented as straight lines connecting the EC_{50} concentrations of each compounds. The points under the isoboles correspond to the supra-additive effect.

On this basis the further research on the combined use of the small molecules from Baikal skullcap, oak bark, and eucalyptus leaves included 7.8-dihydroxy-4-methylcoumarin, 4-hexyl-1.3-benzenediol, and gamma-octalactone, because all pairwise combination of these compounds showed supra-additive anti-QS effect in *C. violaceum* ATCC 31532 bioassay.

2.3. *Evaluation of the Effect of a Three-Component Composition of Small Plant-Derived Molecules on the Quorum Sensing in C. violaceum ATCC 31532*

The bioassay of small plant-derived molecules composition, which included various ratios of 7.8-dihydroxy-4-methylcoumarin, 4-hexyl-1.3-benzenediol and gamma-octalactone, confirmed supra-additive effect of two-component composition, and first showed synergy of three-component composition which manifested in the location of majority of the experimental points below the 3D isobole plane (Figure 2).

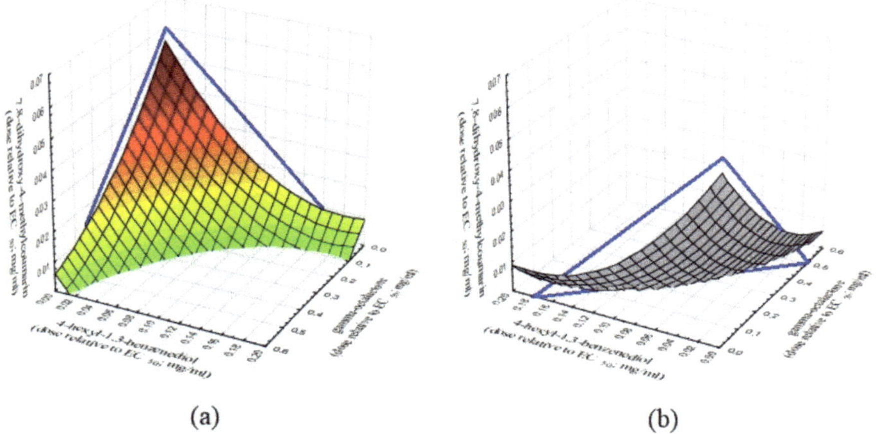

Figure 2. 3D isobolographic analysis of the combined use of 7.8-dihydroxy-4-methylcoumarin, 4-hexyl-1.3-benzenediol, and gamma-octalactone against the QS-controlled violacein biosynthesis in *C. violaceum* ATCC 31532: (**a**) front view; (**b**) back view. The 3D isobole is represented as a blue triangle, the vertices of which correspond to 50% violacein biosynthesis inhibition (EC_{50}) for each compound; the "sail" plane show the supra-additive effect of the experimental samples under the 3D isobole plane.

Figure 2 shows a 3D isobole in the form of a triangle, the vertices of which connect the concentrations of each compounds that cause the same biological effect EC_{50} (50% inhibition of violacein biosynthesis in *C. violaceum* ATCC 31532 bioassay). At the point of maximum supra-additive effect, with the ratio of tested compounds set to 0.6:1:0.8, the concentrations of each compound were three- to five- lower than the concentrations of individual compounds required to achieve EC_{50}. Importantly, the supra-additive effect was detected in at least 85% of samples with various rations of 7.8-dihydroxy-4-methylcoumarin, 4-hexyl-1.3-benzenediol and gamma-octalactone.

Thus, the results of our study described original compositions of various in structure small plant-derived molecules of different origins: 7.8-dihydroxy-4-methylcoumarin from Baikal skullcap, 4-hexyl-1.3-benzenediol from oak bark, and gamma-octalactone from eucalyptus leaves, which enhance each other's anti-quorum activity. In such composition, the content of coumarin's derivative can be significantly decreased while maintaining the anti-QS effect, which makes it possible to avoid unfavorable manifestations of the bioactivity of this group of compounds.

3. Discussion

Coumarins have a wide range of biological properties including antiviral, antimicrobial, anti-inflammatory, and other bioactivities. Some coumarins are approved for use in the treatment of various diseases [28–31]. The most important are vitamin K antagonists, such as warfarin, phenprocumone, or acenocumarol, which are used as anticoagulants [32,33]. Numerous studies have also shown that these compounds do not exhibit significant toxicity to humans and animals (LD_{50} = 275 mg/kg), and are only moderately toxic to the liver and kidneys [34]. In this study we used coumarin and its derivatives at concentrations significantly lower than their LD_{50} for mammals [35–37].

The novel variant of coumarins bioactivity is anti-QS effect that disrupt cell-to-cell chemical communication in bacteria. In this study we continued this direction and followed the path of analyzing the coumarins compositions with other plant-derived molecules in order to enhance the anti-QS effect.

Using *C. violaceum* ATCC 31532 bioassay we found anti-QS effect at sub-inhibitory concentrations of coumarin, 7-hydroxycoumarin, and 7.8-dihydroxy-4-methylcoumarin, that is in good agreement with the same activity of other coumarin derivatives: esculetin (6.7-dihydroxycoumarin) [38,39], scopoletin (7-hydroxy-5-methoxycoumarin) [40], furanocoumarin [25], nodakenetin, fraxin [41], and fizetin [42]. This allows us to state the universality of this bioactivity variant for compounds in this group.

Important, that 7.8-dihydroxy-4-methylcoumarin was characterized as the most effective anti-QS compound in this study. This data has not been previously reported anywhere, whereas the described 7.8-dihydroxy-4-methylcoumarin bioactivity comprised its antioxidant properties only [43,44]. At the same time, its structural features, particularly the hydroxy groups positions, well-corresponded to all known anti-QS active coumarins [23–25]. Our results were consistent with those of Yang et al. which noted a significant increase in the antibacterial effect upon the hydroxylation of coumarins at positions 6, 7, or 8 [45]. Our data also partially agreed with the studies of Lee et al., who showed that hydroxylation at position 7 increased anti-QS activity, while dihydroxylation of coumarin at positions 6 and 7 decreased this activity in comparison to conventional coumarin [46].

The next step was to combine 7.8-dihydroxy-4-methylcoumarin with other small plant-derived molecules in order to get mutual potentiation of the final anti-QS effect. The originality of the proposed approach was that we combined molecules from different plant sources. When testing two-component compositions, it was shown for the first time that 7.8-dihydroxy-4-methylcoumarin, 4-hexyl-1.3-benzendiol, and gamma-octalactone demonstrated a synergetic (supra-additive) anti-Qs effect, and combining all three molecules together decreased the concentration of each compound required to achieve EC_{50} in the composition by three-to-five-fold.

Discussing the mechanism of revealed super-additive effect we assumed that it based on the complementary bioactivity mechanisms for each compound (Figure 3). In this concept, gamma-octolactone is structurally close to LuxI/LuxR quorum sensing autoinducers (acylated homoserine lactones) and probably interferes with him for receptor binding. The 4-hexyl-1.3-benzenediol have a not fully identified mechanism, shown in one of our previous study [47,48], which repress the sensitivity of bacterial cells to autoinducers. Coumarins are characterized by a special mechanism through inhibition of the metabolism of cyclic 3′,5′-diguanilate (c-di-GMP), an intracellular intermediate that is involved in the regulation of bacterial exopolysaccharide synthesis, biofilm formation, adhesion, and virulence [24]. Doing together, these three compounds block the quorum sensing development at different stages, which is manifested in their super-additive anti-QS effect (Figure 3).

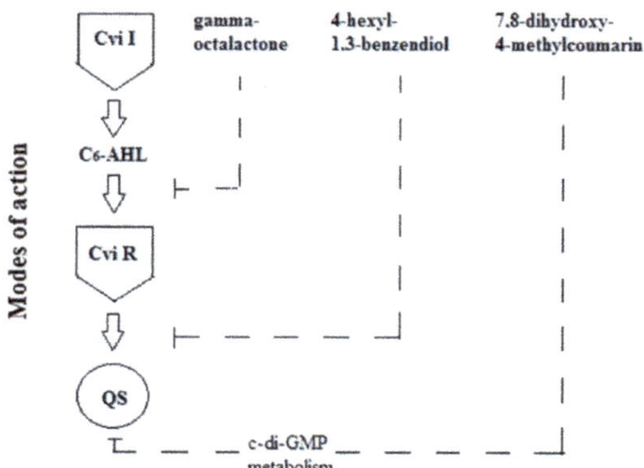

Figure 3. Proposed mechanism of supra-additive anti-QS effect of molecular composition consists of 7.8-dihydroxy-4-methylcoumarin, 4-hexyl-1.3-benzendiol, and gamma-octalactone.

The practical aspect of these results assumes the combined use of coumarin derivatives and other small plant-derived molecules to combat bacterial pathogens of plants, animals, and humans that use quorum sensing systems for the induction of virulence factors and biofilm formation. The implementation of this approach is to use an artificial molecular composition consists of 7.8-dihydroxy-4-methylcoumarin, 4-hexyl-1.3-benzendiol, and gamma-octalactone or the plant materials mixtures with high content of these compounds: Baikal skullcap (*Scutellaria baicalensis*), oak bark (*Quercus cortex*), and eucalyptus leaves (*Eucalyptus viminalis*). Due to the high biological activity of these compositions, they can become a substitute for antibiotics in feeding of farm animals, and should also be considered as candidate pharmaceuticals for further preclinical and clinical studies.

4. Materials and Methods

4.1. Chemical Compounds

Coumarin and its derivatives were used to inhibit QS in *Chromobacterium violaceum* ATCC 31532: coumarin (2H-chromene-2OH; CAS 91-64-5) (Figure 4A), 7-hydroxycoumarin (7-hydroxy-2H-1-benzopyran-2-one; CAS 93-35-6) (Figure 4B), and 7.8-dihydroxy-4-methylcoumarin (4-methyldafnetin; CAS 2107-77-9) (Figure 4C).

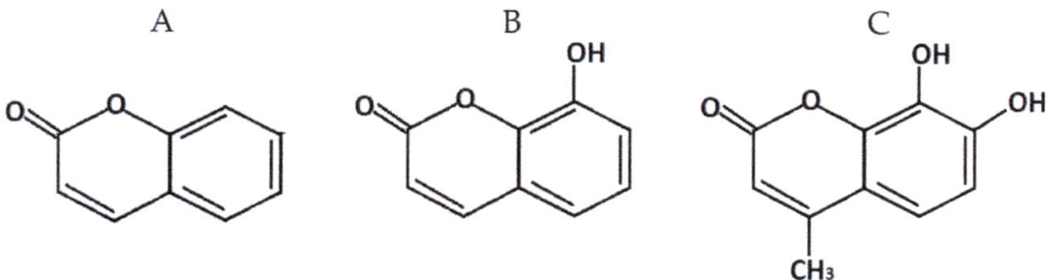

Figure 4. Structural formula of coumarin (**A**), 7-hydroxycoumarin (**B**), and 7.8-dihydroxy-4-methylcoumarin (**C**).

Small plant-derived molecules with previously reported anti-QS activity that were identified in extracts of oak bark (*Quercus cortex*), and eucalyptus leaves (*Eucalyptus viminalis*), were tested in combinations with coumarin derivatives. The analysis included gamma-octalactone (2(3H)-furanone; CAS 147852-83-3), 4-hexyl-1.3-benzenediol (4-n-propylresorcinol; CAS 13331-19-6), 3.4.5-trimethoxyphenol (antiarol; CAS 642-71-7), vanillin (4-hydroxy-3-methoxy benzaldehyde; CAS 121-33-5).

Each of these compounds had a purity at least 99% and was purchased from Sigma-Aldrich (St. Louis, MO, USA).

4.2. Bacterial Strain

The wild strain of *C. violaceum* ATCC 31532 that possessed a two-component LuxI/LuxR-type QS system, was used in bioassay. In this strain CviI synthase (LuxI analog) produce autoinducer N-hexanoyl-L-homoserin lactone (C6-AHL) which bond CviR receptor protein (LuxR analog) and activate QS-controlled transcription of several target genes including *vioABEDC* operon [49]. The encoded VioA, VioB, VioE, VioD, and VioC proteins form a biosynthetic pathway for blue-violet pigment violacein with a maximum absorption at 585 nm. The amount of pigment in the bacterial culture allowed us to directly assess the QS activity.

4.3. Methods for Investigating Anti-QS Activity of Coumarin Derivatives in C. violaceum ATCC 31532 Bioassay

To determine the anti-QS activity of each compound in Luria-Bertani (LB) broth, double dilutions ($n \times 2$) were prepared. The similar samples of LB-broth that did not contain tested compounds were used as positive (growth of the test strain) and negative (sterile) controls. Glass vessels containing 2 mL of experimental dilutions or control samples were inoculated 20 µl of a one-day *C. violaceum* ATCC 31532 culture and cultivated in a static mode at 27 °C. The results were evaluated using a multifunctional microplate reader Infinite 200 PRO (Tecan, Männedorf, Switzerland). The optical density at 450 ± 5 nm (OP_{450}) measured the bacterial biomass and evaluated the effect of the studied compounds on bacterial growth, while the violacein pigment after its ethanol extraction was determined at 600 ± 5 nm (OP_{600}), which was an indicator of the effect on the QS system. The absorption values of the negative control were subtracted. The antibacterial effect of the studied compounds was presented by the MIC_{100} and MIC_{50} values, which were minimal inhibitory concentrations that caused 100% and 50% growth suppression for the test strain relative to the positive control. The inhibition of quorum sensing was expressed as EC_{100} and EC_{50} values, which were equal to 100% and 50% inhibition of violacein pigment biosynthesis in grown culture, respectively.

4.4. Evaluation of the Combined Use of Coumarins and Small Plant-Derived Molecules against in C. violaceum ATCC 31532

To examine the combined effect, double dilutions of test compounds were introduced into plastic 96-well plates in perpendicular directions (connection X: connection Y), so that each well contained their individual ratio. Comparison samples were a series of dilutions containing only one of the tested compounds, as well as positive and negative controls. Further inoculation of *C. violaceum* ATCC 31532, cultivation, and recording of the study results were performed as described above. The effect of the paired compositions was evaluated using isobolographic analysis [49], which is based on the construction of 2D isoboles (i.e., lines connecting the EC_{50} values for the studied compounds X and Y, on the abscissa and ordinate axes) followed by drawing points on this graph which corresponding to the combined effect of compounds X and Y at different concentration ratios. The location of such points on the isobole line corresponded to an additivity (summation effect), their placement above the isobole line described an infra-additive effect (antagonism), and below the isobole showed a supra-additive (synergetic) effect.

Studies of a three-component composition were beginning from the formation of a series of 96-well plates containing the double dilutions of compounds X and Y, as described above. On the next step the wells in each plate were filled with a certain concentration of compound Z (i.e., the total number of used 96-well plates was equal to the number of tested concentrations of compound Z). The control samples were a dilution series containing only compound X, only compound Y, or only compound Z, as well as positive and negative controls. Wells were inoculated with *C. violaceum* ATCC 31532, cultivated and analyzed as described above. The effect of the three-component compositions was evaluated using three-dimensional (3D) isoboles plotted based on the EC_{50} values for each compound.

4.5. Statistical Analysis

All values presented a mean of the 5 experiments. The obtained results were processed using methods of statistical variance in Excel for Windows 10.

5. Conclusions

In this study, coumarin and its derivatives were tested against quorum sensing in *Chromobacterium violaceum* ATCC 31532, and promising activity was shown in 7.8-dihydroxy-4-methylcoumarin. This compound previously detected in Baikal skullcap (*Scutellaria baicalensis*) was combined with other small plant-derived molecules identified in extracts of oak bark (*Quericus cortex*), and eucalyptus leaves (*Eucalyptus viminalis*). It has been shown that 7.8-dihydroxy-4-methylcoumarin, 4-hexyl-1.3-benzenediol and gamma-octalactone exhibit a supra-additive anti-QS effect in two-component combinations, and the combination of all three molecules reduces the concentration of each of them required for reaching EC_{50} against QS, by three- to five- times. It was proposed that the super-additive effect is based on various bioactivity mechanisms of tested molecules, which disrupt the QS development at different stages. The results provide a use for small plant-derived molecule compositions plant materials in the feeding of farm animals, replacing the similar use of prohibited feed antibiotics [50], and also determines the prospects for their testing against human pathogens that use QS to induce virulence factors and biofilm development.

Author Contributions: Conceptualization, D.D.; methodology, D.D. and K.I.; software, K.I.; formal analysis, K.I.; writing—original draft preparation, E.R.; writing—review and editing, D.D.; project administration, E.R. and G.D. All authors have read and agreed to the published version of the manuscript.

Funding: The research of the activity of coumarin and its derivatives was conducted with financial support from the research plan for 2019–2021 of the Federal state budgetary Research Center BST RAS within the framework of the thematic plan for state task No. 0526-2019-0002. The research of supra-additive effects of coumarins and plant-derived molecules was conducted with financial support from the Russian Science Foundation (Grant #21-16-00112).

Informed Consent Statement: Not applicable.

Data Availability Statement: The data that support the findings of this study are available from the corresponding author, upon reasonable request.

Conflicts of Interest: The authors declare no conflict of interest.

Sample Availability: The materials used in the study are commercially available and can be purchased from the relevant firms.

References

1. Venugopala, K.N.; Rashmi, V.; Odhav, B. Review on Natural Coumarin Lead Compounds for Their Pharmacological Activity. *BioMed Res. Int.* **2013**, *3*, 1–14. [CrossRef] [PubMed]
2. Chua, S.L.; Liu, Y.; Li, Y.; Ting, H.J.; Kohli, G.S.; Cai, Z.; Suwanchaikasem, P.; Goh, K.K.K.; Ng, S.P.; Tolker-Nielsen, T. Reduced Intracellular c-di-GMP Content Increases Expression of Quorum Sensing-Regulated Genes in Pseudomonas aeruginosa. *Front. Cell. Infect. Microbiol.* **2017**, *7*, 451. [CrossRef] [PubMed]
3. Matos, M.J.; Santana, L.; Uriarte, E.; Abreu, O.A.; Molina, E.; Yordi, E.G. Coumarins—An Important Class of Phytochemicals. In *Phytochemicals-Isolation, Characterisation and Role in Human Health*; Rao, A.V., Rao, L.G., Eds.; InTech: Rijeka, Croatia, 2015; ISBN 978-953-51-2170-1.
4. Reen, F.J.; Gutierrez-Barranquero, J.A.; Parages, M.L.; O Gara, F. Coumarin: A novel player in microbial Quorum sensing and biofilm formation inhibition. *Appl. Microbiol. Biotechnol.* **2018**, *102*, 2063–2073. [CrossRef] [PubMed]
5. Cruz, L.F.; Figueiredo, G.F.; Pedro, L.P.; Amorin, Y.M.; Andrade, J.T.; Passos, T.F.; Rodrigues, F.F.; Souza, I.L.A.; Gonçalves, T.P.R.; Dos Santos Lima, L.A.R.; et al. Umbelliferone (7-hydroxycoumarin): A non-toxic antidiarrheal and antiulcerogenic coumarin. *Biomed. Pharmacother.* **2020**, *129*, 1–8. [CrossRef] [PubMed]

6. Olanlokun, J.O.; Bodede, O.; Prinsloo, G.; Olorunsogo, O.O. Comparative antimalarial, toxicity and mito-protective effects of *Diospyros mespiliformis* Hochst. ex A. DC. and *Mondia whitei* (Hook.f.) Skeels on Plasmodium berghei infection in mice. *J. Ethnopharmacol.* **2020**, in press. [CrossRef] [PubMed]
7. Williams, K.J.; Gieling, R.G. Preclinical Evaluation of Ureidosulfamate Carbonic Anhydrase IX/XII Inhibitors in the Treatment of Cancers. *Int. J. Mol. Sci.* **2019**, *23*, 6080. [CrossRef]
8. Akkol, E.K.; Genç, Y.; Karpuz, B.; Sobarzo-Sánchez, E.; Capasso, R. Coumarins and Coumarin-Related Compounds in Pharmacotherapy of Cancer. *Cancers* **2020**, *7*, 1959.
9. Shahzadi, I.; Ali, Z.; Baek, S.H.; Mirza, B.; Ahn, K.S. Assessment of the Antitumor Potential of Umbelliprenin, a Naturally Occurring Sesquiterpene Coumarin. *Biomedicines* **2020**, *5*, 126. [CrossRef]
10. Nasser, M.I.; Zhu, S.; Hu, H.; Huang, H.; Guo, M.; Zhu, P. Effects of imperatorin in the cardiovascular system and cancer. *Biomed. Pharmacother.* **2019**, *120*, 109401. [CrossRef]
11. Duan, J.; Shi, J.; Ma, X.; Xuan, Y.; Li, P.; Wang, H.; Fan, Y.; Gong, H.; Wang, L.; Pang, Y.; et al. Esculetin inhibits proliferation, migration, and invasion of clear cell renal cell carcinoma cells. *Biomed. Pharmacother.* **2020**, *125*, 110031. [CrossRef]
12. Usman, H.; Ullah, M.A.; Jan, H.; Siddiquah, A.; Drouet, S.; Anjum, S.; Giglioli-Guviarc'h, N.; Hano, C.; Abbasi, B.H. Interactive Effects of Wide-Spectrum Monochromatic Lights on Phytochemical Production, Antioxidant and Biological Activities of *Solanum Xanthocarpum Callus* Cultures. *Molecules* **2020**, *9*, 2201. [CrossRef] [PubMed]
13. Urbagarova, B.M.; Shults, E.E.; Taraskin, V.V.; Radnaeva, L.D.; Petrova, T.N.; Rybalova, T.V.; Frolova, T.S.; Pokrovskii, A.G.; Ganbaatar, J. Chromones and coumarins from *Saposhnikovia divaricata* (Turcz.) Schischk. Growing in Buryatia and Mongolia and their cytotoxicity. *J. Ethnopharmacol.* **2020**, *261*, 112517. [CrossRef] [PubMed]
14. Bihani, T. *Plumeria rubra* L.—A review on its ethnopharmacological, morphological, phytochemical, pharmacological and toxicological studies. *J. Ethnopharmacol.* **2021**, *264*, 113291. [CrossRef] [PubMed]
15. Starzak, K.; Świergosz, T.; Matwijczuk, A.; Creaven, B.; Podleśny, J.; Karcz, D. Anti-Hypochlorite, Antioxidant, and Catalytic Activity of Three Polyphenol-Rich Super-Foods Investigated with the Use of Coumarin-Based Sensors. *Biomolecules* **2020**, *10*, 723. [CrossRef] [PubMed]
16. Zeng, Z.; Qian, L.; Cao, L.; Tan, H.; Huang, Y.; Xue, X.; Shen, Y.; Zhou, S. Virtual screening for novel quorum sensing inhibitors to eradicate biofilm formation of Pseudomonas aeruginosa. *Appl. Microbiol. Biotechnol.* **2008**, *79*, 119–126. [CrossRef]
17. Takomthong, P.; Waiwut, P.; Yenjai, C.; Sripanidkulchai, B.; Reubroycharoen, P.; Lai, R.; Kamau, P.; Boonyarat, C. Structure-Activity Analysis and Molecular Docking Studies of Coumarins from Toddalia asiatica as Multifunctional Agents for Alzheimer's Disease. *Biomedicines* **2020**, *8*, 107. [CrossRef]
18. Janus, Ł.; Radwan-Pragłowska, J.; Piątkowski, M.; Bogdał, D. Coumarin-Modified CQDs for Biomedical Applications-Two-Step Synthesis and Characterization. *Int. J. Mol. Sci.* **2020**, *21*, 8073. [CrossRef]
19. Lee, E.J.; Kang, M.K.; Kim, Y.H.; Kim, D.Y.; Oh, H.; Kim, S.I.; Oh, S.Y.; Na, W.; Kang, Y.H. Coumarin Ameliorates Impaired Bone Turnover by Inhibiting the Formation of Advanced Glycation End Products in Diabetic Osteoblasts and Osteoclasts. *Biomolecules* **2020**, *10*, 1052. [CrossRef]
20. Koga, H.; Negishi, M.; Kinoshita, M.; Fujii, S.; Mori, S.; Ishigami-Yuasa, M.; Kawachi, E.; Kagechika, H.; Tanatani, A. Development of Androgen-Antagonistic Coumarinamides with a Unique Aromatic Folded Pharmacophore. *Int. J. Mol. Sci.* **2020**, *21*, 5584. [CrossRef]
21. Zhang, S.; Liu, N.; Liang, W.; Han, Q.; Zhang, W.; Li, C. Quorum sensing-disrupting coumarin suppressing virulence phenotypes in *Vibrio Splendidus*. *Appl. Microbiol. Biotechnol.* **2016**, *101*, 3371–3378. [CrossRef]
22. Gutiérrez-Barranquero, J.A.; Reen, F.J.; McCarthy, R.R.; O'Gara, F. Deciphering the role of coumarin as a novel quorum sensing inhibitor suppressing virulence phenotypes in bacterial pathogens. *Appl. Microbiol. Biotechnol.* **2015**, *99*, 3303–3316. [CrossRef] [PubMed]
23. Hou, H.M.; Jiang, F.; Zhang, G.L.; Wang, J.Y.; Zhu, Y.H.; Liu, X.Y. Inhibition of Hafnia alvei H4 Biofilm Formation by the Food Additive Dihydrocoumarin. *J. Food Prot.* **2017**, *12*, 842–847. [CrossRef] [PubMed]
24. D'Almeida, R.E.; Molina, R.D.I.; Viola, C.M.; Luciardi, M.C.; Nieto Peñalver, C.; Bardón, A.; Arena, M.E. Comparison of seven structurally related coumarins on the inhibition of Quorum sensing of *Pseudomonas aeruginosa* and *Chromobacterium violaceum*. *Bioorg. Chem.* **2017**, *73*, 37–42. [CrossRef] [PubMed]
25. Girennavar, B.; Cepeda, M.L.; Soni, K.A.; Vikram, A.; Jesudhasan, P.; Jayaprakasha, G.K.; Pillai, S.D.; Patil, B.S. Grapefruit juice and its furocoumarins inhibits autoinducer signaling and biofilm formation in bacteria. *Int. J. Food Microbiol.* **2008**, *125*, 204–208. [CrossRef] [PubMed]
26. Inchagova, K.S.; Duskaev, G.K.; Deryabin, D.G. The suppression of the "quorum sensing" *Chromobacterium violaceum* when exposed to combinations of amikacin with activated carbon or small molecules of plant origin (pyrogallol and coumarin). *Microbiology* **2019**, *88*, 72–82. [CrossRef]
27. Deryabin, D.G.; Inchagova, K.S. Inhibitory effect of aminoglycosides and tetracyclines on Quorum sensing in Chromobacterium violaceum. *Microbiology* **2018**, *87*, 1–8. [CrossRef]
28. Abate, A.; Dimartino, V.; Spina, P.; Costa, P.L.; Lombardo, C.; Santini, A.; Del Piano, M.; Alimonti, P. Hymecromone in the treatment of motor disorders of the bileducts: A multicenter, double-blind, placebo-controlled clinical study. *Drugs Exp. Clin. Res.* **2001**, *27*, 223–231.
29. Božič-Mijovski, M. Advances in monitoring anticoagulant therapy. *Adv. Clin. Chem.* **2019**, *90*, 197–213. [CrossRef]

30. Dávila-Fajardo, C.L.; Díaz-Villamarín, X.; Antúnez-Rodríguez, A.; Fernández-Gómez, A.E.; García-Navas, P.; Martínez-González, L.J.; Dávila-Fajardo, J.A.; Barrera, J.C. Pharmacogenetics in the treatment of cardiovascular diseases and its current progres regarding implementation in the clinical routine. *Genes* **2019**, *10*, 261. [CrossRef]
31. Opherk, D.; Schuler, G.; Waas, W.; Dietz, R.; Kubler, W. Intravenous carbochromen: A potent and effective drug for estimation of coronary dilatory capacity. *Eur. Heart J.* **1990**, *11*, 342–347. [CrossRef]
32. Gierlak, W.; Kuch, M. How to use cardiac drugs in everyday practice? In *Anticoagulants in Cardiology—Vitamin K Antagonists*; Czelej, Lublin 1.; 2010; Chapter 6.
33. Salvo, F.; Bezin, J.; Bosco-Levy, P.; Letinier, L.; Blin, P.; Pariente, A.; Moore, N. Pharmacological treatments of cardiovascular diseases: Evidence from real-life studies. *Pharmacol. Res.* **2017**, *118*, 43–52. [CrossRef] [PubMed]
34. Patel, D.; Patel, R.; Kumari, P.; Patel, N. In vitro antimicrobial assessment of coumarin-based s-triazinyl piperazines. *Med. Chem. Res.* **2012**, *21*, 1611–1624. [CrossRef]
35. López, F.; Deglesne, J.; Arroyo, P.A.; Ranneva, R.; Deprez, E. 4hexylresorcinol-a-new-molecule-for-cosmetic-application. *J. Biomol. Res. Ther.* **2018**, *8*, 1000170.
36. Aguedo, M.; Beney, L.; Waché, Y.; Belin, J.M. Interaction of an odorant lactone with model phospholipid bilayers and its strong fluidizing action in yeast membrane. *Int. J. Food Microbiol.* **2003**, *80*, 211–215. [CrossRef]
37. Inchagova, K.S.; Deryabin, D.G.; Duskaev, G.K.; Karimov, I.F.; Ryazanov, V.A. Plant molecules influence on luminescent Escherichia coli K12 TG1 with1constitutiveexpressed luxCDABE genes. *FEBS OpenBio* **2019**, *9*, 303–304.
38. Zhang, J.; Jiang, C.S. Synthesis and evaluation of coumarin/piperazine hybrids as acetylcholinesterase inhibitors. *Med. Chem. Res.* **2018**, *27*, 1717–1727. [CrossRef]
39. Brackman, G.; Hillaert, U.; Van Calenbergh, S.; Nelis, H.J.; Coenye, T. Use of quorum sensing inhibitors to interfere with biofilm formation and development in *Burkholderia multivorans* and *Burkholderia cenocepacia*. *Res. Microbiol.* **2009**, *160*, 144–151. [CrossRef]
40. Lindsay, A.; Ahmer, B.M. Effect of sdiA on biosensors of N-acylhomoserine lactones. *J. Bacteriol.* **2005**, *187*, 5054–5058. [CrossRef]
41. Ding, X.; Yin, B.; Qian, L.; Zeng, Z.; Yang, Z.; Li, H.; Lu, Y.; Zhou, S. Screening for novel quorum-sensing inhibitors to interfere with the formation of *Pseudomonas aeruginosa* biofilm. *J. Med. Microbiol.* **2011**, *60*, 1827–1834. [CrossRef]
42. Durig, A.; Kouskoumvekaki, I.; Vejborg, R.M.; Klemm, P. Chemoinformatics-assisted development of new anti-biofilm compounds. *Appl. Microbiol. Biotechnol.* **2010**, *87*, 309–317. [CrossRef]
43. Todorov, L.; Traykova, M.; Traykov, T. *In Vitro Evaluation of the Antioxidant Effect of Yohimbine, Proceedings of the IV European Congress on Toxicology*; EUROTOX: Krakow, Poland, 2005; Poster No. 27.
44. Tyagi, Y.K.; Kumar, A.; Raj, H.G.; Vohra, P.; Gupta, G.; Kumari, R.; Kumar, P.; Gupta, R.K. Synthesis of novel amino and acetyl amino-4-methylcoumarins and evaluation of their antioxidant activity. *Eur. J. Med. Chem.* **2005**, *40*, 413–420. [CrossRef] [PubMed]
45. Yang, L.; Ding, W.; Xu, Y.Q.; Wu, D.S.; Li, S.L.; Chen, J.N.; Guo, B. New insights into the antibacterial activity of hydroxycoumarins against *Ralstonia Solanacearum*. *Molecules* **2016**, *21*, 468. [CrossRef] [PubMed]
46. Lee, J.H.; Kim, Y.G.; Cho, H.S.; Ryu, S.Y.; Cho, M.H.; Lee, J. Coumarins reduce biofilm formation and the virulence of *Escherichia coli* O157:H7. *Phytomedicine* **2014**, *21*, 1037–1042. [CrossRef] [PubMed]
47. Deryabin, D.G.; Kamayeva, A.A.; Tolmacheva, A.A.; El-Registan, G.I. The effects of alkylhydroxybenzenes on homoserine lactone-induced manifestations of quorum sensing in bacteria. *Appl. Biochem. Microbiol.* **2014**, *50*, 353–358. [CrossRef]
48. Stauff, D.L.; Bassler, B.L. Quorum sensing in *Chromobacterium violaceum*: DNA recognition and gene regulation by the CviR receptor. *J. Bacteriol.* **2011**, *193*, 3871–3878. [CrossRef]
49. Tallarida, R.J. An overview of drug combination analysis with isobolograms: Perspectives in pharmacology. *Pharmacol. Exp. Ther.* **2006**, *3*, 1–7. [CrossRef]
50. Landers, T.F.; Cohen, B.; Wittum, T.E.; Larson, E.L. A review of antibiotic use in food animals: Perspective, policy, and potential. *Public Health Rep.* **2012**, *127*, 4–22. [CrossRef]

Article

Synthesis of Ferulenol by Engineered *Escherichia coli*: Structural Elucidation by Using the In Silico Tools

Anuwatchakij Klamrak [1], Jaran Nabnueangsap [2], Ploenthip Puthongking [1] and Natsajee Nualkaew [1,*]

[1] Faculty of Pharmaceutical Sciences, Khon Kaen University, Khon Kaen 40002, Thailand; anuwat_kla@yahoo.com (A.K.); pploenthip@kku.ac.th (P.P.)

[2] Salaya Central Instrument Facility RSPG, Research Management and Development Division, Office of the President, Mahidol University, Nakhon Pathom 73170, Thailand; jaran.nab@mahidol.ac.th

* Correspondence: nnatsa@kku.ac.th; Tel.: +66-43-202178

Abstract: 4-Hydroxycoumarin (4HC) has been used as a lead compound for the chemical synthesis of various bioactive substances and drugs. Its prenylated derivatives exhibit potent antibacterial, antitubercular, anticoagulant, and anti-cancer activities. In doing this, *E. coli* BL21(DE3)pLysS strain was engineered as the in vivo prenylation system to produce the farnesyl derivatives of 4HC by coexpressing the genes encoding *Aspergillus terreus* aromatic prenyltransferase (AtaPT) and truncated 1-deoxy-D-xylose 5-phosphate synthase of *Croton stellatopilosus* (CstDXS), where 4HC was the fed precursor. Based on the high-resolution LC-ESI(±)-QTOF-MS/MS with the use of in silico tools (e.g., MetFrag, SIRIUS (version 4.8.2), CSI:FingerID, and CANOPUS), the first major prenylated product (named compound-1) was detected and ultimately elucidated as ferulenol, in which information concerning the correct molecular formula, chemical structure, substructures, and classifications were obtained. The prenylated product (named compound-2) was also detected as the minor product, where this structure proposed to be the isomeric structure of ferulenol formed via the tautomerization. Note that both products were secreted into the culture medium of the recombinant *E. coli* and could be produced without the external supply of prenyl precursors. The results suggested the potential use of this engineered pathway for synthesizing the farnesylated-4HC derivatives, especially ferulenol.

Keywords: *Escherichia coli*; biotransformation; 4-hydroxycoumarin; ferulenol; structural annotation; in silico tools

Citation: Klamrak, A.; Nabnueangsap, J.; Puthongking, P.; Nualkaew, N. Synthesis of Ferulenol by Engineered *Escherichia coli*: Structural Elucidation by Using the In Silico Tools. *Molecules* **2021**, *26*, 6264. https://doi.org/10.3390/molecules26206264

Academic Editors: Maria João Matos and Valeria Costantino

Received: 11 July 2021
Accepted: 13 October 2021
Published: 16 October 2021

1. Introduction

Prenylation is one of the post-structural modifications essential for the increasing biological activities of several natural products [1]. Transferring the prenyl moieties onto the aromatic acceptor molecules often leds to prenylated derivatives with greatly improved therapeutic potency [2]. This process has become a new frontier for developing novel drugs and lead compounds in the pharmaceutical industry, especially for antimicrobial, antioxidant, anti-inflammatory, and anti-cancer agents [3–5]. Nevertheless, plant-based production of these valuable products is limited by finite resources, low yields, slow growth rates, seasonal dependency, and rare, or completely absent in some regions [6,7]. Moreover, the chemoenzymatic and total synthesis of the prenylated products is rather difficult, as it is challenged by the structural complexity which requires multiple steps of the uncontrollable regio- and stereoselective prenylation, and expensive starting materials [8,9].

Ferulenol (2), a C-3 farnesylated 4-hydroxycoumarin, is a major constituent in *Ferula communis* (Giant fennel) which possesses many biological activities [5,10]. According to previous studies, this compound exhibits cytotoxicity against various lines of cancer cells including human breast (MCF-7), colon (Caco-2), ovarian (SK-OV-3), and leukemic (HL- 60) in a dose-dependent manner, with the mode of action resembling paclitaxel (taxol) [11,12]. Ferulenol (2) also exerts anti-cancer activity through the downregulation of Bcl2 protein along with upregulation of Bax protein in benzo[a]pyrene-induced lung cancer in a rat model [13]. This indicated ferulenol (2) as a pro-oxidant and chemotherapeutic agent and has been recognized as an interesting lead compound for anti-cancer semi-synthesis [13,14]. In searching for the novel anticoagulant warfarin derivatives, ferulenol (2) exhibits higher activity than the warfarin drug (approximately 22 times) with lower toxicity [15,16]. This compound also possesses antimycobacterial activity against fast-growing *Mycobacterium* species [17]. Although chemical synthesis of this compound could be achieved by the reaction between 4-hydroxycoumarin sodium salt (4HCNa) and all trans-farnesylchloride based on alkylation at the C-3 of coumarin [15], these processes are quite complicated, and the starting precursors are rather expensive.

Engineering microbial cells as the in vivo prenylation system provides several advantages over the two existing methods, as microbes can be grown very fast in the noncomplex medium and can be easily extended to large-scale production [9,18]. More importantly, microbes can supply various lengths of prenyl donors, e.g., dimethylallyl pyrophosphate (DMAPP, C5), geranyl pyrophosphate (GPP, C10), farnesyl pyrophosphate (FPP, C15), and geranylgeranyl pyrophosphate (GGPP, C20), through their inherent isoprenoid pathways [7,19]. *E. coli* solely synthesizes isoprenoids via the 1-deoxy-D-xylulose 5-phosphate (DXP) pathway (also known as non-mevalonate pathway), in which 1-deoxy-D-xylulose 5-phosphate synthase (DXS) is known as the first committing step enzyme that controls the metabolic flux of isoprenoids precursors [20–24]. In wild type *E. coli*, FPP serves as the key branching point in the synthesis of the vital molecules, i.e., heme O, ubiquinone, and peptidoglycan [25–28]. Consequently, the metabolic flux of this precursor in *E. coli* has been redirected towards taxadiene, carotenoids, and amorpha-4,11-diene through heterologous expression of various terpene synthases [29–34]. In the last few years, using microorganisms as the prenylation systems for the production of hybrid molecules containing prenyl moieties has focused mainly on yeasts (e.g., *Saccharomyces cerevisiae* and *Pichia pastoris*) and *Bacillus subtilis* based on the endogenous prenyl donors supplied via either the mevalonate (MVA) pathway or the DXP pathway [35–38]. However, an attempt at synthesizing ferulenol (2), the product of C-3 farnesylation of 4HC (1), through the genetic manipulation of the DXP pathway in *E. coli* has not been reported yet. Therefore, this gives rise to our interest in establishing a new synthesis pathway for producing this product by using the engineered microbe based on feeding of 4HC (1) to minimize chemical consumption.

Aromatic prenyltransferases (aPTs) are the enzymes that catalyze the regio-selective prenylation of the prenyl groups (so-called prenyl donors), incorporating the aromatic compounds (known as aromatic acceptors) that contain the electron-rich regions through a mechanism comparable to Friedel–Crafts aromatic electrophilic substitution [39–42]. Nowadays, the genes encoded for aPTs have been isolated from plants, fungi, and bacteria and have been found to exhibit broad substrates with specificity both in terms of aromatic acceptors and prenyl donors, creating a diverse range of prenylated products [1–3,43–45]. Of those enzymes characterized, *Aspergillus terreus* aromatic prenyltransferase (AtaPT) possesses the broad range substrate specificity towards various types of aromatic acceptors (e.g., coumarins, resveratrol, and naringenin) and prenyl donors, i.e., DMAPP, GPP, FPP, and GGPP, yielding prenylated products with structural diversity [46]. This enzyme also differs from formerly characterized aPTs, as it can generate mono-, di-, and/or triprenylated products via C-C- and/or C-O-bonded prenylation. Since 4-hydroxycoumarin (4HC) (1) contains the two promising regions for electrophilic alkylation, including the C3 and the oxygen atom at C4 positions [15,47], it was proposed to be utilized by AtaPT due to the unique catalytic activity of this enzyme. In wild-type *E. coli*, its native isoprenoid precursors are always insufficient for pathway engineering purposes, therefore driving metabolic fluxes via overexpression of the rate-limiting step enzymes in the DXP pathway, e.g., DXS, 1-deoxy-D-xylulose 5-phosphate reductoisomerase (DXR), and isopentenyl pyrophosphate isomerase (IDI) are required [19–24].. According to the previous findings, transcriptional profiling analysis revealed a positive correlation between CsDXS gene expression and plaunotol (acyclic diterpene alcohol) content in the young leaves of *Croton stellatopilosus*, and this enzyme has been suggested to control the metabolic flux of isoprenoid precursors in that plant [48]. Therefore, the truncated CsDXS (namely CstDXS) was chosen to co-express with the AtaPT in *E. coli* BL21(DE3)pLysS to establish a new synthetic route for the in vivo formation ferulenol (2), in which 4HC (1) was only a fed precursor.

In this study, *E. coli* was used as the bioconversion system of the newly designed pathway leading to the formation of farnesylated-4HC analogs, which includes two steps (Figure 1). The CstDXS was overexpressed for driving metabolic flux of isoprenoid precursors (e.g., GPP, FPP) in the DXP pathway [47]. Second, prenylation of those prenyls to the core structure of 4HC (1) through the catalytic function of AtaPT produced its corresponded products (2 and 3) [46]. This process requires magnesium ion (Mg^{2+}) as a cofactor to facilitate the formation of an electrophilic prenyl carbocation, which subsequently binds to 4HC (1) via the Friedel-Crafts reaction [46].

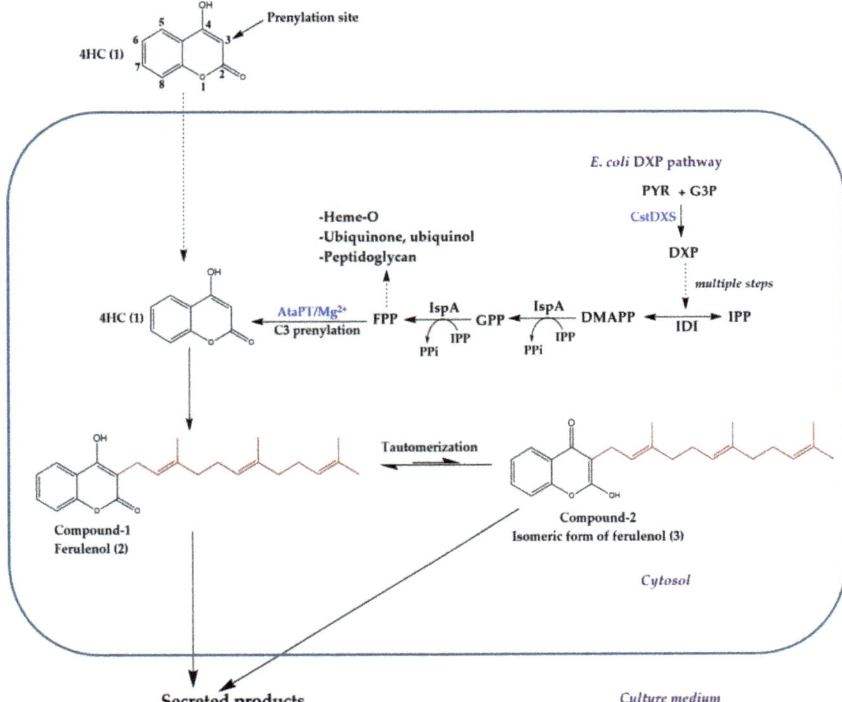

Figure 1. Proposed biosynthetic pathway of ferulenol (2) and its isomeric structure (3) reconstructed in *E. coli* BL21(DE3)pLysS carrying pCDFDuet-AtaPT-CstDXS, where 4HC (1) was the fed precursor. Note that both products can be formed without any supply of prenyl donors (i.e., FPP) and secreted into the culture medium. PYR = pyruvate; G3P = glyceraldehyde 3-phosphate; IPP = isopentenyl pyrophosphate; DMAPP = dimethylallyl pyrophosphate; GPP = geranyl pyrophosphate; FPP = farnesyl pyrophosphate; IDI = isopentenyl pyrophosphate isomerase; IspA = geranyl pyrophosphate synthase and farnesyl pyrophosphate synthase.

The structure elucidation of the target compounds by NMR techniques has been limited to the study of metabolic engineering and metabolomics, which deals with trace-level metabolites [49]. Therefore, high-resolution LC-ESI-QTOF-MS/MS has played a crucial role instead through its high sensitivity and can distinguish between the closed atomic mass numbers [49–51]. The experimental mass spectra can be interpreted into the chemical structures based on the in silico prediction tools by matching against the in silico -fragmented spectra of compounds in databases (i.e., PubChem, MetaCyc) [51–53]. In this study, a variety of the in-silico tools were chosen for the structurally assisted identification of the prenylated-4HC derivatives obtained from the bioconversion of 4HC (1) by the clones carrying pCDFDuet-AtaPT-CstDXS to verify the product formation.

Here, we report the newly artificial pathway for synthesizing ferulenol (2) established in the *E. coli* BL21(DE3)pLysS strain, where the chemical structures of this product are entirely elucidated by using in silico tools, included MetFrag [52], SIRIUS [54], CSI:FingerID web service [55], and CANOPUS [56]. The experimental mass spectra for the putative ferulenol (2) are further confirmed by the alignment to the mass spectra of the authentic ferulenol (2) established using the same LC-MS/MS condition, along with those established by Fourel and colleagues [57]. The putative mass peak corresponds to farnesylated derivative of 2-hydroxy-4-chromenone, which has been proposed to exist via the tautomerization of ferulenol (2) and is detected in this study. The results can be used to further establish the *E. coli* system as the microbial cell factory for producing the farnesylated-4HC derivatives to serve drug discovery purposes.

2. Results

2.1. Construction of Plasmid pCDFDuet-AtaPT-CstDXS

Engineering *E. coli* as the in vivo prenylation system requires at least two steps: (1) overexpression of CstDXS as the rate-limiting step enzyme in the DXP pathway to increase the available pool of prenyl-donors; and (2) transferring the prenyl-donors into the core aromatic acceptor by the catalytic function of AtaPT. Therefore, the truncated DXS (CstDXS: GenBank accession no. AB354578.1) was used for enhancing the flux of isoprenoid precursors in *E. coli*. The gene encoding for AtaPT (GenBank accession no. KP893683) was chosen for incorporating the prenyl donors into the 4HC core structure. Construction of pCDFDuet-AtaPT-CstDXS was achieved via stepwise incorporation of AtaPT, followed by CstDXS into the multiple cloning site 1 (MCS-1) and MCS-2 of pCDFDuet-1 vector, respectively. DNA sequencing confirmed that both genes were inserted into the correct regions of the pCDFDuet-1 vector with no frame-shifted insertion. Each gene was independently controlled by their T7 promotor, and gene expression was driven by adding IPTG into the culture (Figures S1 and S2). The 5′ end of AtaPT in the MCS-1 (*BamHI*/*NotI*) was joined with His-tag, and the 3′ end of CstDXS was fused with the S-tag. The map created using GenScript (https://www.genscript.com/gensmart-design/# (accessed on 26 September 2021)) is depicted in Figure 2.

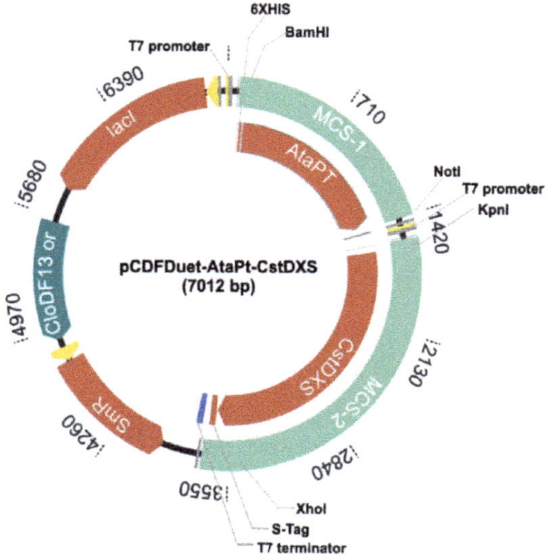

Figure 2. Map of recombinant plasmid pCDFDuet-AtaPT-CstDXS.

2.2. Identification of Prenylated 4HC Derivative Using LC-ESI(-)-QTOF-MS/MS

The high-resolution LC-ESI-QTOF-MS/MS was used for the identification of the prenylated 4HC derivative produced by clones carrying pCDFDuet-AtaPT-CstDXS. The in vivo formation for the peak corresponding to the 4HC containing farnesyl moiety was detected on the extracted ion chromatogram (EIC) at the calculated *m/z* 366.211670 $[M - H]^-$, in which the product was tentatively confirmed to be "ferulenol (2)" by comparing the MS/MS fragmentation profile against the authentic ferulenol (2) established by the previous work [57]. The results showed that there were two mass peaks (namely compound-1 and -2) which exhibited the identical *m/z* 365.21 eluted at the RT 21.3 min and 22.3 min, respectively (Figure 3A). No product formation was observed in the culture medium clones harboring pCDFDuet-AtaPT-CstDXS grown without supplementing 4HC (1) in the culture (Figure 3B). These products were therefore presumably farnesylated-4HC, where the farnesyl moiety was incorporated on the C3 position of 4HC (1) via C-C-bonded formation. Based on the direct comparison against previous mass spectra, only the compound-1 highly resembled the MS/MS profile to that of ferulenol (2) reported by Fourel et al. [57], which was characterized by the ion peak *m/z* 365 $[M - H]^-$, 228 and 174 (Figure 3C). In the MS/MS spectrum of this product, the presence of product ions with *m/z* 228.08 (base peak), 214.06, and 174.03 illustrated the partial losses of prenyl side chain attached to the C3 position of 4HC (1), which was the core structure of this prenylated product (Figure 3D). This product (compound-1) is therefore believed to be ferulenol (2), where the obtained mass data (Figure S4) were further eluicidated by many aspects of in silico tools, including MetFrag web service, SIRIUS, CSI:Finger ID web service, and CANOPUS to entirely support the structural elucidation of this prenylated product. Abdou et al. [47] revealed that 4HC (1) is able to exist in tautomeric forms including 4-hydroxy-2-chromenone (a), 2,4-chromadione (b), and 2-hydroxy-4-chromenone (c) (Figure 3E). The compound-2 (*m/z* 365.21; Rt = 22.3 min) was thus presumably the isomeric form of ferulenol (2) proceeded via the tautomerization, where the 2-hydroxy-4-chromenone (c) served as the aromatic core structure (Figure 3F). The presence of fragmented ions with *m/z* 214.06 (base peak), 228.08, and 282.12 in the resulting MS/MS spectrum also indicated the partial losses of prenyl moiety that attached to the C3 position of this prenylated product (Figure 3G). Note that 4HC (1) as the fed substrate was also detected in the culture medium of the clones carrying pCDFDuet-AtaPT-CstDXS (Figure 3F).

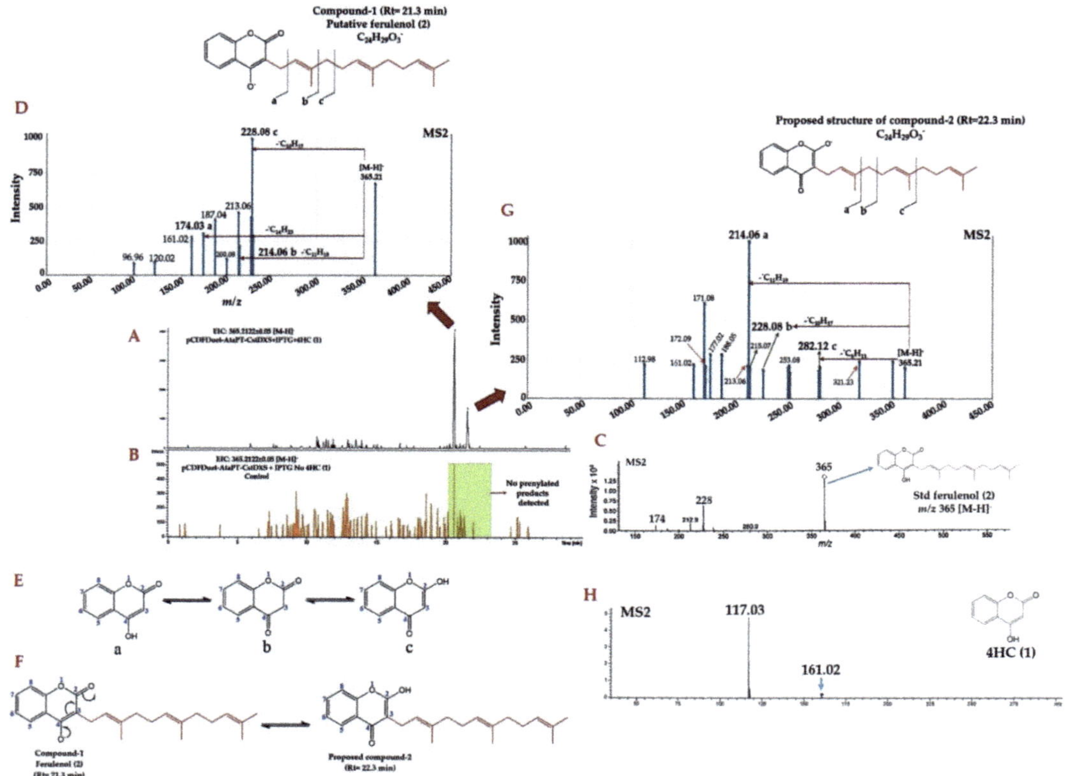

Figure 3. Identification of prenylated products using negative-ion mode LC-ESI-QTOF-MS/MS. (**A**) The EIC for molecular ion with *m/z* 365.2122 ± 0.0500 $[M - H]^-$ detected from the medium extract of clones carrying pCDFDuet-AtaPT-CstDXS grown in the presence of fed 4HC (1). (**B**) The EIC for molecular ion with *m/z* 365.2122 ± 0.0500 $[M - H]^-$ of the clones carrying pCDFDuet-AtaPT-CstDXS grown in the absence of 4HC (1) (control group). (**C**) The MS/MS spectrum of authentic ferulenol (2) (1 µg/mL) established by Fourel et al. [57] (**D**) The MS/MS spectrum of compound-1 (*m/z* 365.21; Rt = 21.3 min). (**E**) The tautomerization of 4HC (1) reported by Abdou et al. [47]. (**F**) The proposed mechanism underlying the formation of compound-2 (3), which proceeded via the tautomerization of ferulenol (2) (compound-1). (**G**) The MS/MS spectrum of compound-2 (*m/z* 365.21; Rt = 22.3 min). (**H**) The MS/MS spectrum of 4HC (1) detected from the culture medium of clones harboring pCDFDuet-AtaPT-CstDXS.

2.3. Structural Annotation of Compound-1 by Using MetFrag Web Service

The chemical structure of compound-1 was further annotated by using MetFrag web service, the in silico tool designed to elucidate chemical structure of query subjects from the experimental mass spectra [52]. Among the 2267 candidates retrieved from the PubChem database, compound-1 was best annotated as ferulenol (2), (compound CID: 54679300), with the highest score of 1.0 in which 8/11 peaks were matched with those of the in silico-generated fragmented ions of the candidated ferulenol (2) deposited in the PubChem database (Figure 4). We hence suggested that compound-1 is likely to be ferulenol (2), rather than the other prenylated 4HC derivatives exhibiting the identical mass value with *m/z* of 365.21 $[M - H]^-$.

Rank	Candidate $C_{24}H_{30}O_3$ (Mass 366.219)	Final score (8/11 peaks explained)	Coumarin core structure (Name)
1	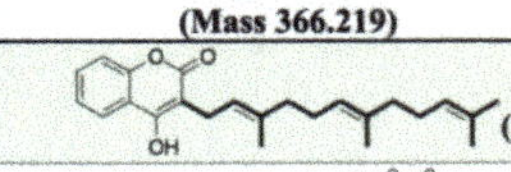 (2)	1.0	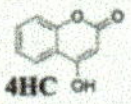 4HC
2	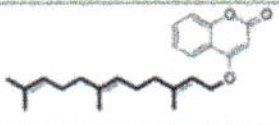	0.9995	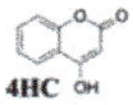 4HC
3	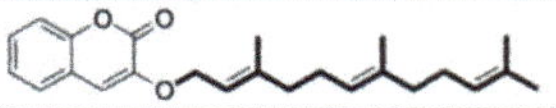	0.9995	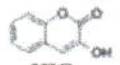 3HC
4	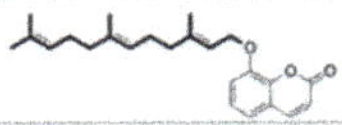	0.9924	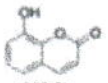 8HC
5	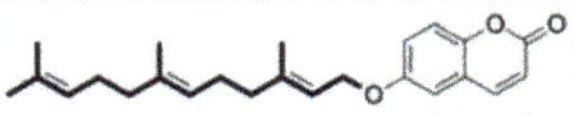	0.9924	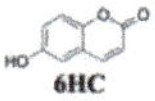 6HC
6	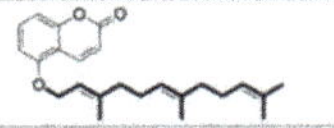	0.9924	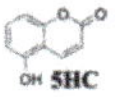 5HC
7	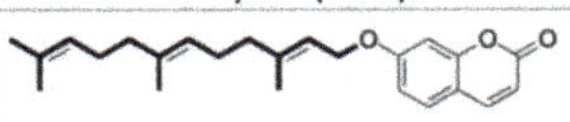	0.9924	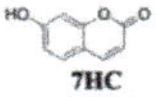 7HC

Figure 4. Structural annotation of compound-1 (*m/z* 365.21 $[M - H]^-$; Rt = 21.3 min) using MetFrag web service, where the query subject was best annotated as ferulenol (2) among 2267 structures retrieved from the PubChem database.

2.4. Computationally Assisted Identification of the Prenylated Product by Using SIRIUS (Version 4.8.2)

According to the user manual, SIRIUS requires high mass accuracy in which the ppm error is less than 20 ppm before conducting the annotation processes for the most reliable results [55]. The observed molecular ion *m/z* 365.21 $[M - H]^-$ was thus estimated by using the mass error calculation tool from the web service (https://warwick.ac.uk/fac/sci/chemistry/research/barrow/barrowgroup/calculators/mass_errors/ (accessed on 17 June 2021)), by comparing against its theoretical *m/z* (365.212218 $[M - H]^-$), and showed an error of 8.762029 ppm, permitting the elucidation of the raw mass data by using this tool.

SIRIUS utilizes the high-resolution isotopic pattern analysis to locally annotate the correct molecular formula based on the experimentally acquired MS/MS data of the query subjects. Among the potential ten elemental formulas retrieved from the PubChem database, our query subject was annotated as $C_{24}H_{30}O_3$ with the highest Sirius score of 62.243% (Figure 5A), which was identical to the native elemental formula of ferulenol (2) ($C_{24}H_{30}O_3$) deposited in the PubChem database (https://pubchem.ncbi.nlm.nih.gov/compound/Ferulenol (accessed on 17 June 2021)). SIRIUS also offers a refined search option to explore the query subjects against biological databases, e.g., Natural Products, Collection of Natural Products (COCONUTS), and NORMAN databases to specify and narrow natural molecules at user-defined cut-offs. By searching the possible structure against the aforementioned databases, our query metabolite (compound-1) was annotated as $C_{24}H_{30}O_3$ with a score that greatly improved to 100% (Figure 5A). The annotated MS/MS spectra locally computed by the SIRIUS tool revealed that 8 of 11 peaks (indicated in the green spectra) matched with their local database (Figure 5B). The result was also consistent with those fragmentation spectra predicted by MetFrag, where 8 out of 11 peaks matched with the in silico fragmented ions of the candidate ferulenol (2) in the PubChem database (Figure 4).

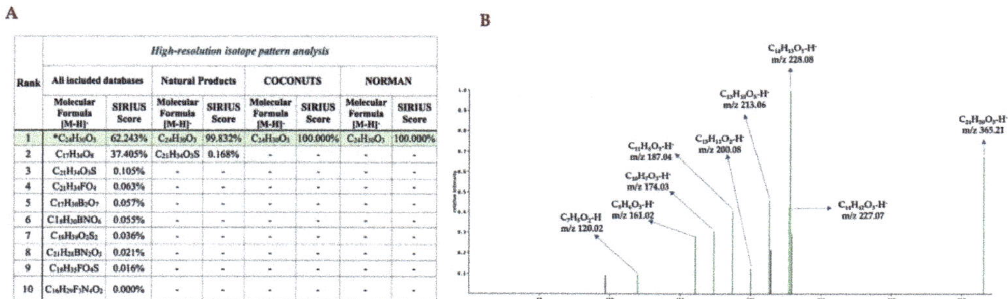

A

Rank	High-resolution isotope pattern analysis							
	All included databases		Natural Products		COCONUTS		NORMAN	
	Molecular Formula $[M-H]^-$	SIRIUS Score	Molecular Formula $[M-H]^-$	SIRIUS Score	Molecular Formula $[M-H]^-$	SIRIUS Score	Molecular Formula $[M-H]^-$	SIRIUS Score
1	*$C_{24}H_{30}O_3$	62.243%	$C_{24}H_{30}O_3$	99.832%	$C_{24}H_{30}O_3$	100.000%	$C_{24}H_{30}O_3$	100.000%
2	$C_{17}H_{34}O_8$	37.405%	$C_{21}H_{34}O_5S$	0.168%	-	-	-	-
3	$C_{21}H_{34}O_5S$	0.105%	-	-	-	-	-	-
4	$C_{21}H_{34}FO_4$	0.063%	-	-	-	-	-	-
5	$C_{17}H_{30}B_2O_7$	0.057%	-	-	-	-	-	-
6	$Cl_8H_{30}BNO_6$	0.055%	-	-	-	-	-	-
7	$C_{16}H_{38}O_2S_2$	0.036%	-	-	-	-	-	-
8	$C_{21}H_{28}BN_2O_3$	0.021%	-	-	-	-	-	-
9	$C_{18}H_{35}FO_4S$	0.016%	-	-	-	-	-	-
10	$C_{16}H_{29}F_3N_4O_2$	0.000%	-	-	-	-	-	-

Figure 5. Molecular formula annotation of the prenylated 4HC analog produced by clones carrying pCDFDuet-AtaPT-CstDXS using SIRIUS (version 4.8.2). (**A**) High-resolution isotope pattern analysis reveals the molecular formula of query subject is best annotated as $C_{24}H_{30}O_3$. (**B**) The mass spectra is annotated to match with the local databases of SIRIUS (indicated by the green spectrum with corresponding molecular formula), while the spectra with no annotations are considered to be noise peaks (indicated by the black spectrum).

2.5. Structural Annotation and Compound Classification of Compound-1 by Using CSI:FIngerID and CANOPUS

SIRIUS has recently been integrated with CSI: FingerID web service to identify the chemical structure of the query subjects [55]. In this step, the annotated mass spectrum (Figure 5B) was compared against several compounds in the chosen molecular structure databases (e.g., PubChem, MeSH, and COCONUTS). Of more than 100 possible structures retrieved from all databases and locally predicted by the tool, ferulenol (2) as the target product was ranked as the tenth candidate structure with a percentage similarity of 43.62% (Figure 6A). Since ferulenol (2) is a natural product exclusively produced by *F. communis*, we narrowed the scope of the structural elucidation by using Natural Products and COCONUTS databases and found the rank of the candidate was substantially improved from tenth to fifth (Figure 6A). By searching against the structural compounds from the NORMAN database, our prenylated 4HC analog (compound-1) was perfectly matched to that of ferulenol (2) as the top ranked candidate (Figure 6A). Besides, CSI:FingerID can provide the crucial information of so-called "molecular fingerprints" to verify various substructures that can be found in the query subjects (Figure 6B). In this instance, several molecular fingerprints belonging to ferulenol (2) were predicted to be present in this prenylated product (m/z 365.21 $[M-H]^-$). For example, substructures encoded by a SMARTS string "[#6]c1c([#8])ccc1 (Cc1c(O)ccc1)" with a score of 98% (F1 = 0.823) correspond to the benzene ring attached to the pyrone ring of 4HC (1). The basis pyrone ring of 4HC encoded by [#8]=,:[#6]-,:[#6]:[#6]-,:[#8](O=C-C:C-O) possessed 85% similarity (F1 = 0.815) and was verified to be present in same candidate structure. Biosynthetically, they were all obtained from the 4HC (1) fed in the culture medium of clones bearing pCDFDuet-AtaPT-CsTDXS. Equally important, there were several substructures belonging to the farnesyl moiety which originated from the DXP pathway, predicted to be present in the same candidate structure. Several substructures representing the basic benzene and pyrone rings along with the isoprene building block belonging to the query ferulenol (2) were predicted to be present in the trained structures of SIRIUS tool (Figure 6C). SIRIUS has also been developed to connect with CANOPUS (class assignment and ontology prediction using mass spectrometry) for logical classification of unknown metabolites based on the high-resolution MS/MS data [56]. According to the molecular properties annotated by CSI:FingerID web service, our query prenylated product m/z 365.21 $[M-H]^-$ was systematically classified as coumarins and derivatives (class), where phenylpropanoids/polyketides and organic compounds served as the superclass and kingdom, respectively (Figure 6D). CANOPUS also provides alternative classes of the query subjects. In this case, the putative ferulenol (2) was mainly classified as a coumarin derivative, but aromatic monoterpenoids, benzenoids, and lactones were recognized as the alternative classes of this product. Based on the data acquired from the direct comparison against previously established mass spectra [56] along with the molecular formula and structural elucidation by means of the in silico tools included, MetFrag, SIRIUS, CSI:FingerID, and CANOPUS strongly supported that compound-1 was ferulenol (2), which is the product of C3-farnesylation of 4HC (1).

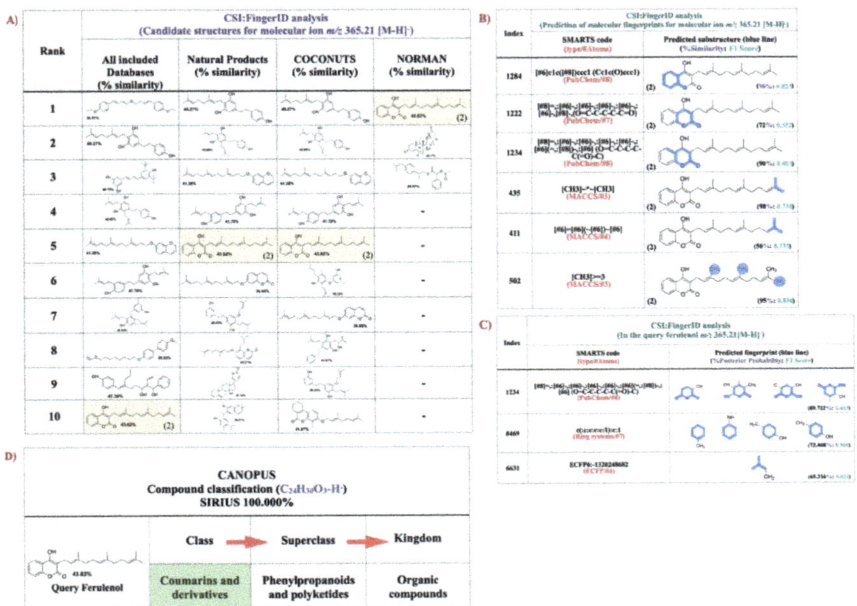

Figure 6. Structural annotation and classifications for the query ferulenol (2) (m/z 365.21 $[M-H]^-$) synthesized by the recombinant *E. coli* carrying pCDFDuet-AtaPT-CstDXS based on CSI:FingerID and CANOPUS integrated in the SIRIUS (version 4.8.2). (**A**) In all included databases, ferulenol (2) as the target metabolite was ranked as the tenth potential candidate (highlighted in yellow) for the prenylated 4HC analog m/z 365.21 $[M-H]^-$, meanwhile, the rank of this product significantly increased towards the fifth- and top-ranked candidates when structural identification was performed against the biological databases included Natural Products, COCONUTS, and NORMAN. (**B**) The examples of substructures (molecular fingerprints) predicted to exist in the query subject, including those belonging to the 4HC (1) core structure (No. 1284, 1222, and 1234), and the farnesyl moiety (No. 435, 411, and 502). (**C**) The examples of substructures belonging to ferulenol (2) found to exist in the training structures of the SIRIUS tool. (**D**) Based on the experimentally observed MS/MS data, CANOPUS showing the query compound ($C_{24}H_{30}O_3$ (SIRIUS score 100%), m/z 365.21 $[M-H]^-$) was categorized as coumarins and derivatives, describing the polycyclic aromatic compounds that contains a 1-benzopyran moiety with a ketone group at the C2 carbon atom (1-benzopyran-2-one).

2.6. Identification of Farnesylated-4HC Derivatives by Using LC-QTOF-MS/MS in Positive ESI Mode

The high-resolution LC-QTOF-MS/MS in positive ion mode was thus implemented to corroborate both prenylated products based on monitoring of the molecular ion m/z 367.227320 $[M+H]^+$ computed by the web tool (https://www.sisweb.com/referenc/tools/exactmass.htm?formula=C24H31O3 (accessed on 7 August 2021)). In the extracted ion chromatogram (EIC) of the medium extract of clones carrying pCDFDuet-AtaPT-CstDXS, there were two extracted ions (namely compound-1 and compound-2) eluted at the retention times of 21.3 and 22.3 min that exhibited identical m/z 367.22 $[M+H]^+$, where they presumably farnesylated derivatives of 4HC (Figure 7A). The resulting MS/MS spectrum of compound-1 (m/z 367.22; Rt = 21.3 min) was clearly identical to those belonging to the authentic ferulenol (2) established under the same LC-MS/MS condition (Figure 7B). In the MS/MS spectrum of this prenylated product, the presence of proposed substructures at m/z 175.04 (base peak), 217.08, 231.10, and 243.10 illustrated the partial losses of the prenyl side chain specifically attached to a C3 position of the 4HC core structure (1) (Figure 7C). The obtained evidence leds us to suggest that the compound-1 (m/z 367.22 $[M + H]^+$; Rt 21.3 min) was indeed ferulenol (2), a product derived from C3-farnesylation of 4HC (1). Hence, the compound-2 (m/z 367.22 $[M + H]^+$; Rt = 22.3 min) was presumably the tautomeric structure of ferulenol (2), where the 2-hydroxy-4-chromenone (c) served as the core structure (Figure 3E). In the MS/MS spectra along with proposed structures of this product, the existing product ions at m/z 177.08 (base peak), 215.06, 229.08, and 243.10 signify the partial elimination of prenyl moiety that incorporated the C3 region of 2-hydroxy-4-chromenone (Figure 7D). The existing of signals at m/z 189.05, which were presented in the MS/MS spectrum of both products as shown in Figure 7C,D, were presumably obtained from the prenyl moieties which were adjacent to the core structure of ferulenol (2) and its isomeric structure (3). Based on the acquired evidence, compound-2 was tentatively defined as the isomeric structure of ferulenol (2) which was formed via "tautomerization". Further elucidation (e.g., NMR) is required to verify the tentative confirmation of this prenylated product. The proposed reaction mechanisms illustrating the various chemical losses present in the MS/MS spectrum of compound-1 and compound-2 are shown in Figures S5 and S6. The postulated mechanisms underlying the formation of ions m/z 189.05 of compound-1 and compound-2 were shown in the Supplementary Materials (Figures S7 and S8).

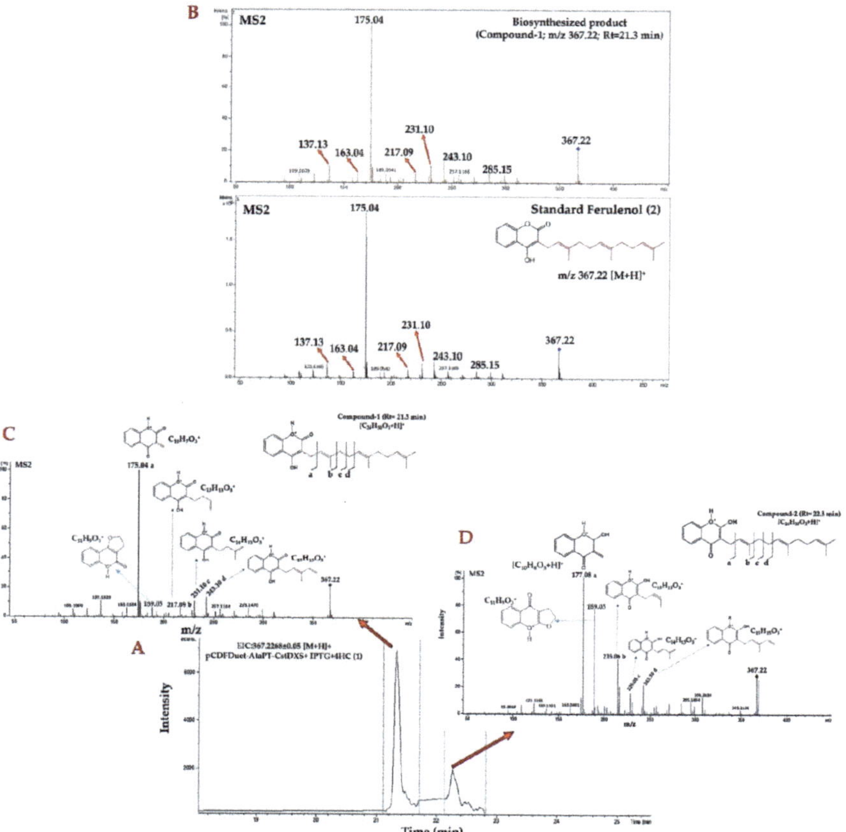

Figure 7. Identification of prenylated products produced by clones carrying pCDFDuet-AtaPT-CstDXS based on the high-resolution LC-ESI-QTOF-MS/MS in the positive-ion mode. (A) The EIC for the molecular ion m/z 367.2268 ± 0.05 $[M + H]^+$

showing two reaction products exhibited the equivalent *m/z* 367.22 (named compound-1 and 2) eluted from the HPLC colume at the retention times of 21.3 and 22.3 min, respectively. (**B**) The direct comparison between the MS/MS spectrum of compound-1 (*m/z* 367.22) and the authentic ferulenol (2) (*m/z* 367.22), which is characterized by the product ions with *m/z* 137.13, 163.04, 175.04, 217.09, 231.10, 243.10, and 285.15, respectively. (**C**) The MS/MS spectrum of compound-1 showing the partial losses of prenyl side chain from the 4HC core structure (1). (**D**) The MS/MS spectrum of compound-2 (proposed structure) signifying the partial elimination of prenyl moiety from the 2-hydroxy-4-chromenone, a core structure of the prenylated product.

3. Discussion

E. coli has the capability to supply various isoprenoid precursors through the native DXP pathway [19–24]. In the past few decades, this microorganism has been engineered to produce various bioactive terpenoids and valuable precursors such as limonene, taxadiene (taxol precursor), amorphadiene (artemisinin precursor), and carotenoids (e.g., lycopene, astaxanthin, carotenoids) by the heterologous expression of terpene synthases with the rate-limiting step enzymes in the DXP pathway [29–34]. However, few studies have focused on the use of this microorganism as the in vivo prenylation system of aromatic natural products. In wild-type *E. coli*, FPP is involved in the biosynthesis of various molecules (e.g., ubiquinone and peptidoglycan) and its level is tightly maintained, as it is essential for bacterial growth and viability [25–28]. A previous study also demonstrated that the deletion of genes encoded for FPP synthase (IspA) resulted in the observed growth retardation in the mutant strains of *E. coli* [25]. We hence speculated that FPP might become more readily available for the in vivo prenylation of 4HC (1) for ferulenol (2) than others (e.g., DMAPP, GPP, and GGPP). However, the native isoprenoid precursors of *E. coli* are always insufficient for metabolic engineering purposes, and the enhanced metabolic fluxes via overexpression of the rate-limiting step enzymes in the DXP pathway (e.g., DXS, DXR, and IDI) are required [7,19–24]. The transcriptional profile analysis in the young leaves of *C. stellatopilosus* revealed a positive association between CsDXS gene expression and plaunotol (acyclic diterpene alcohol) content, and this enzyme was hence suggested as one of the rate-limiting steps in the DXP pathway controlling the flux of isoprenoid precursors in the plant [48]. Since the active form of CsDXS could be achieved after the removal of the signaling peptide region [30,48], the truncated form of this enzyme, CstDXS, was used in this study to drive the metabolic flux of the DXP pathway in the *E. coli* BL21(DE3)pLysS strain. We demonstrated that *E. coli* BL21(DE3)pLysS harboring pCDFDuet-AtaPT-CstDXS is capable of producing the putative mass peak corresponding to ferulenol (2) from the fed 4HC (1) without relying on the external supply of prenyl-donors (e.g., DMAPP, GPP, and FPP). The synthesized product was found to be excreted into the culture medium. This means the step of breaking the cell during the downstream processes is not necessary. Furthermore, this expression system did not require the use of high-priced precursors. These reflected the potential economic system of the prenylated 4HC derivatives production by using recombinant *E. coli*. Besides, our results extend the findings of Chen et al. [46] in that AtaPT can utilize various substrates by showing that 4HC (1) could also be recognized as the aromatic acceptor, even though it has never been reported that this compound acted as the substrate of AtaPT.

Although NMR elucidation is required to verify the structures of the resulting prenylated product (compound-1), this technique is restricted to researchers in the fields of metabolic engineering and metabolomics, which have to deal with trace-level metabolites [49–51]. The structural elucidation of compound-1 was thus based on the use of in silico tools designed to annotate the structures of the query metabolites using the experimental mass spectra (MS2) (Figure S4). Having been confirmed to be highly consistent with the unique MS/MS fragmentation profile of ferulenol (2) from the previous work [56], MetFrag analysis clearly showed that the compound-1 was best annotated as "ferulenol (2)" out of the 2266 candidates presented from PubChem database. However, information regarding the molecular formula, chemical structure, substructures, and classifications of this product are still needed. SIRIUS is one of the in silico tools developed to unravel various chemical features hidden in the MS/MS data of the query subjects [52]. This tool has typically been used in the field of metabolomics and has been shown to be helpful in the field of metabolic engineering, in which it was used to identify 2,4,6-trihydroxybenzophenone produced by the engineered *E. coli* [58]. The results from SIRIUS showed that our query product (compound-1) possessed the neutral molecular formula $C_{24}H_{30}O_3$, which corresponded to that of ferulenol (2) deposited in the PubChem database (https://pubchem.ncbi.nlm.nih.gov/compound/Ferulenol (accessed on 26 September 2021)). SIRIUS also locally computed the relevant MS/MS spectra and fragmentation pathway which might be involved in the fragmentation process of compound-1 in which eight mass peaks were explained, meanwhile, the rest of the three peaks were considered as the noises. Based on the eight mass peaks explained by SIRIUS, CSI:FingerID analysis suggested that the compound-1 was annotated as ferulenol (2) as the tenth-ranked candidate after searching against all databases provided by the SIRIUS tool. Although the expected ferulenol (2) was not perfectly categorized as the first candidate for the compound-1, it might be speculated that CSI:FingerID integrated in SIRUS exhibited extremely good performance in the case of the independent MS/MS being trained and deposited in the training data, and the rate of correct identification was substantially decreased when the trained data were removed from this tool [54]. This can be confirmed by the fact that the MS/MS data for the authentic ferulenol (2) (encoded by InChI=1S/C24H30O3/c1-17(2)9-7-10-18(3)11-8-12-19(4)15-16-21-23(25)20-13-5-6-14-22(20)27-24(21)26/h5-6,9,11,13-15,25H,7-8,10,12,16H2,1-4H3/b18-11+,19-15+) has not yet been trained in the negative mode mass data of the SIRIUS tool (https://www.csi-fingerid.uni-jena.de/v1.6.0/api/fingerid/trainingstructures?predictor=2 (accessed on 23 June 2021)). Based on the LC-MS search (for the molecular ion *m/z* 365.21 $[M-H]^-$), it might also be possible to exclude the rest of candidates (ranked first to ninth retrieved from all included databases, and ranked first to fifth from the Natural Products and COCONUTS databases) as incorrect structures since they are not natural occurring in *E. coli*, and there was only 3b-allotetrahydrocortisol showing the proximal *m/z* 365.23 $[M-H]^-$ found in the *E. coli* metabolomics database (ECMDB) (https://ecmdb.ca/spectra/ms/search (accessed on 23 June 2021)). This indicated that only ferulenol (2) as the product of pathway engineering could be accepted as the correct structure. In the case of the query metabolites acquired from the biological samples, SIRIUS also provides the biological databases as the choices to narrow the scope of natural molecules. When MS/MS spectra of compound-1 were annotated against biological databases, e.g., Natural Products, COCONUTS, and NORMAN, the rank annotated as "ferulenol (2)" was greatly improved as the fifth and first candidate, respectively. Based on the substructures predicted by CSI:Finger ID and CANOPUS, it subsequently provided the vital information that the compound-1 was classified as coumarins and derivatives.

According to Abdou and colleaques [47], the 4HC (1) can exist in three different tautomeric structures: 4-hydroxy-2-chromenone (a), 2,4-chromadione (b), and 2-hydroxy-4-chromenone (c) (Figure 3E). We hence postulated that the compound-2 (eluted from the HPLC column at 22.3 min), obtained from the negative- and positive-ion mode analyses, is presumably the tautomeric form of ferulenol (2). The obtained MS/MS spectra (Figures 3G and 7D) were further elucidated to gain insight into the structural information of this prenylated product. In the negative ion mode analyses, the three chemical losses explained the daughter ions at *m/z* 214.06, 228.08, 282.12, signifying the prenyl side chain was partially removed from the 2-hydroxy-4-chromenone (c), the core structure of this prenylated product. A similar trend was observed in the positive ion mode analyses, where the partial loss of prenyl side chain attaching to the 2-hydroxy-4-chromenone could be characterized via the signals with *m/z* 177.08, 215.06, 229.08, and 243.10, respectively. The existence of product ions at *m/z* 189.05 $[M+H]^+$ in the MS/MS spectrum of both prenylated products (Figure 7C,D) were likely to be the outcome of non-enantioselective epoxidation of the double bond present in the farnesyl side chain of ferulenol (2), as recently described by Cortés et al. [59]. Based on the obtained evidence, the compound-2 (3) was interpreted as the farnesylated derivative of 2-hydroxy-4-chromenone, however further elucidation (e.g., NMR) is required to fully support a chemical point of view.

The availability of prenyl precursors influences the patterns of isoprenoid-derived natural products produced by the bacterial systems [60]. Here, ferulenol (2) and its proposed isomeric structure (3) were exclusively detected as the predominant products produced by *E. coli*-carried pCDFDuet-AtaPT-CstDXS. Although AtaPT exhibits remarkable substrate promiscuity towards a variety of prenyl-donors such as DMAPP, GPP, FPP, and GGPP [46], the presence of two farnesylated-4HC analogs (compounds-1 and -2) clearly supports our hypothesis that the FPP accumulated inside the bacterial cells is more easily accessible for synthesizing the two prenylated products rather than the others (DMAPP, GPP, and GGPP). Our result was also consistent with the previous finding, where the farnesylated-menadione was the major product of the whole-cell catalysis by *P. pastoris*, harboring the gene encoding for aromatic prenyltransferase (NovQ) [37]. Studies have shown that overexpression of multiple rate-limiting enzymes in the DXP pathway, including 1-deoxy-D-xylulose 5-phosphate reductoisomerase (DXR), isopentenyl pyrophosphate isomerase (IDI), along with FPP synthase (e.g., IspA), causes the substantially increased production of terpenoids in their engineered microbes [23–25]. Furthermore, 4HC (1) as the fed precursor is unlikely and unsustainable from an economic standpoint, as a considerable amount of this expensive substrate must be supplied to the bacterial culture to make its desired prenylated analogs. This strategy might be impracticable for industrially scaled applications, where the newly designed strains capable of de novo manufacture of the two farnesylated-4HC analogs or the usage of the lower cost precursors (e.g., sodium salicylate) should be established. Previous research has shown that *E. coli* can be engineered to produce 4HC (1) via the inherent chorismate pathway, which is accomplished through a series of reactions catalyzed by isochorismate synthase (ICS), isochorismate pyruvate-lyase (IPL), salicylate-CoA ligase (SCL), and biphenyl synthase (BIS) [61]. Based on the broad substrate specificity of benzoate-CoA ligase (BadA) and benzophenone synthase (GmBPS), our group demonstrated that *E. coli* BL21(DE3)pLysS carrying pETDuet-BadA-GmBPS was able to synthesize 4HC (1) from the fed salicylate (sodium salt) as well (data unpublished). By considering these advantages, further establishments of *E. coli* systems capable of synthesizing 4HC (1) from the inexpensive precursors (e.g., glucose, sodium salicylate) to act as the ATaPT's substrate along with enhancing the flux of isoprenoid precursors would be beneficial for large-scale production of the two farnesylated-4HC analogs reported herein.

Although we demonstrated that the *E. coli* BL21(DE3)pLysS strain could be engineered to produce ferulenol (2) and its isomeric structure (3), further optimizations are needed to improve the yields of final product, which seems to be limited by the supply of isoprenoid precursors [7–9]. Several studies have demonstrated that overexpression of the multiple enzymes regulating the flux of isoprenoid precursors led to greatly improved terpenoid-derived natural products in *E. coli* [19–24]. Future pathway engineering via overexpression of the other rate-controlling steps in the DXP pathway, such as DXR, IDI, and farnesyl pyrophosphate synthase (FPPS), will be carried out to improve the yields of the two prenylated-4HC derivatives reported herein. Since a large amount of 4HC (1) was also found to remain in the culture medium, optimizing substrate consumptions, i.e., a time course production is needed to enhance the yield of those prenyl derivatives. There are several factors affect the heterologous production of natural products in bacterial systems, such as temperature, codon usage, and plasmid copy number [62–64]. Thus, optimizing these parameters will be examined.

Our results illustrated the in vivo functional expression of AtaPT and CstDXS in the *E. coli* system, which could be seen from the formation of ferulenol (2) and its isomeric structure (3) secreted into the culture medium. These findings also shed new light on the use of the engineered *E. coli* system as the prenylation system to produce valuable secondary metabolites instead of isolation from plants and chemical synthesis. Since the AtaPT used in this study exhibits broad substrate specificity towards many types of aromatic acceptors, e.g., benzophenones, xanthones, stilbenes, and chalcones [46], the future impact of clones carrying pCDFDuet-AtaPT-CstDXS on synthesizing the prenylated products might not be restricted to 4HC (1) but could be applied to the other types of aromatic acceptors.

4. Materials and Methods

4.1. Reagents

The general reagents were analytical grade and were purchased from Sigma-Adrich (St. Louis, MO, USA), Merck (Darmstadt, Germany), and Avantor (Center Valley, PA, USA).

4.2. Construction of Plasmid pCDFDuet-AtaPT-CstDXS

pCDFDuet-1 coexpression vector (Novagen, Darmstadt, Germany) was chosen to construct the recombinant plasmid containing the two genes encoding AtaPT (GenBank accession no. KP893683) and CstDXS (GenBank accession no. AB354578.1) based on the procedure reported by Toila and Joshua-Tor [65]. The initial insertion was performed by incorporating AtaPT in the MCS-1 (*BamHI* and *NotI*) followed by insertion of CstDXS into the MCS-2 (*Kpn*I and *Xho*I) of the pCDFDuet-1 vector to yield pCDFDuet-AtaPT-CstDXS. The details were as follows: The plasmid pUC57-AtaPT obtained from the gene synthesis technology (Invitrogen, Waltham, MA, USA) was used as the DNA template to provide AtaPT. The PCR reaction consisted of Phusion High-Fidelity DNA polymerase (NEB, Ipswich, MA, USA); the forward primer was AtaPT-F: 5′-GGTGGATCCGATGCTCCCCCCATCAGACA-3′, and the reverse primer was AtaPT-R 5′-AAAGCGGCCGTCACACAGCTGCG-3′ (underlines are the recognition sites for *BamH*I and *Not*I, respectively). The PCR cycle included pre-denaturation at 98 °C for 1 min, followed by 30 cycles of 98 °C for 30 s, 60 °C for 30 s, and 72 °C for 30 s, and a final extension at 72 °C for 5 min. The obtained PCR product (~1275 bp) was purified by using a Gel Band Purification Kit (GE Healthcare, Chicago, IL, USA), digested with *BamH*I and *Not*I, and ligated into a pCDFDuet-1 vector which had been treated with the same restriction enzymes. The ligation mixture was transformed into *E. coli* DH5α and the positive clones were selected by spreading on the LB-agar-contained streptomycin (50 μg/mL). The resulting plasmids were extracted using PureYieldTM Plasmid Mini-prep System (Promega, Madison, WI, USA) and the gene insertion was confirmed by double digestion with *BamH*I and *Not*I. The resulting pCDFDuet-AtaPT was used as the DNA backbone in the next step.

Based on the previous finding, the gene-encoded 1-deoxy-D-xylose 5-phosphate synthase of *C. stellatopilosus* (CsDXS: 2163 bp) was predicted to contain the putative chloroplast transit peptide (cTP) that should be removed before the gene expression in *E. coli* systems [40]. Therefore, the truncated CsDXS (CstDXS) which was absent of the cTP coding sequence (171 bp) was cloned from the young leaves of *C. stellatopilosus*. The PCR reaction consisted of Pfu DNA Polymerase (ThermoScientific, Waltham, MA, USA), CstDXS-F: 5′-TGCGGTACCATGGCATCACTTTCAGAAA-3′, and CstDXS -R: 5′-AGCCTCGAGTGCTGACATAATTTGCAGA -3′ (underlines are the restriction sites for *Kpn*I and *Xho*I, respectively). The PCR condition was as follows: Pre-denaturation at 95 °C for 1 min, followed by 30 cycles of 95 °C for 30 s, 60 °C for 30 s, and 72 °C for 2 min, and a final extension at 72 °C for 5 min. The PCR product (1992 bp) was purified from the agarose gel by using the gel purification kit, double digested with *Kpn*I and *Xho*I, and was ligated to the pCDFDuet-AtaPT which was digested with the same restriction enzymes. The double digestion by *Kpn*I and *Xho*I was performed to confirm the insertion of CstDXS in pCDFDuet-AtaPT (Figure S3). The nucleotide sequencing (IDT, Penang, Malaysia) was performed to verify the correct bases and the in-frame arrangement of both AtaPT and CstDXS in the pCDFDuet-AtaPT-CstDXS by using two pairs of primers: ACYCDuetUP1 primer 5′-GGATCTCGACGCTCTCCT-3′, and DuetDOWN1 primer 5′-GATTATGCGGCCGTGTACAA-3′ for AtaPT in the MCS-1, and DuetUP2 primer 5′-TTGTACACGGCCGCATAATC-3′ and T7term primer 5′-GCTAGTTATTGCTCAGCGG-3′ for CstDXS in the MCS-2.

4.3. Bioconversion of 4HC (1) into the Farnesylated-4HC Derivatives

The pCDFDuet-AtaPT-CstDXS transformed into *E. coli* BL21(DE3)pLysS (Promega, Madison, WI, USA) by heat shock method. The engineered strains carrying pCDFDuet-AtaPT-CstDXS were cultured in the LB medium containing streptomycin (50 μg/mL) and chloramphenicol (34 μg/mL) at 37 °C, 200 rpm for 18 h. The 1.5 mL of culture was then inoculated into the 500 mL Erlenmeyer flask containing 150 mL of the same medium and was further cultivated at 37 °C (200 rpm) until the OD_{600} reached 1.0. The gene expression was induced by adding 1 mM IPTG (final concentration); after that, cells were grown at 18 °C, 250 rpm for 5 h. To start the bioconversion process, 3 mM 4HC (1) (TCI, Tokyo, Japan) and 3 mM $MgCl_2$ were supplied to the induced culture and the cells were further cultivated at the same condition for 18 h. After that, the culture medium was harvested by centrifugation at 4 °C, 8000 rpm for 10 min. *E. coli* BL21(DE3)pLysS carried pCDFDuet-AtaPT-CstDXS that was grown in parallel at the same condition, except without the supplement of 4HC (1) used as the control in this study. Since ferulenol (2) is light-sensitive, the bioconversion experiment throughout this study took place in the darkness to minimize the product degradation.

4.4. Extraction of Prenylated Products from the Culture Medium

The extraction process was performed in the darkness. The culture medium (150 mL) was partitioned twice with 75 mL EtOAc in the 500 mL Erlenmeyer flask by shaking at 300 rpm, 25 °C for 30 min. The EtOAc layers were then harvested after centrifugation for 5 min, 6000 rpm, at 4 °C, and concentrated until dryness using N_2 gas. The dried residues were redissolved in MeOH (HPLC grade) before identification of the prenylated-4HC derivatives by using the high-resolution LC-ESI-QTOF-MS/MS in both negative- and positive-ion mode operations.

4.5. Identification of Metabolites by LC-ESI-QTOF-MS/MS

Identification of the prenylated products were carried out by HPLC Dionex (Thermo Scientific) connected with MS Bruker Maxis (esquire 4000 Daltonics, Bremen, Germany) and the RP-18 column (Acclaim RSCL 120 C18 column, 2.1 X 100 mm, 2.2 μM, Thermo Fisher Scientific, Waltham, MA, USA). The mobile phases consisted of H_2O with 0.1% formic acid (solvent A) and acetonitrile with 0.1% formic acid (solvent B). The separation of prenylated products was achieved by using a linear gradient of solvent B as follows: 5% for 0–2 min, 5–95% for 15 min, 95% for 3 min, and back to 5% for 10 min, for a total running time of 30 min. The flow rate was set at 0.4 mL/min. The sample temperature was controlled to 10 °C. The column oven temperature was 40 °C. The injection volume was 10 μL. The product elucidation was conducted by tandem mass (MS/MS) with electrospray ionization (ESI) under the collision-induced dissociation (CID) energy 35 eV in negative ion mode analysis. Nebulizer pressure was set at 29 psi, the dry gas temperature was 180 °C, and the dry gas flow rate was 8.0 L/min. The masses were scanned over the *m/z* range of 100–1000 amu. The obtained MS/MS spectra (Figure S4) were directly compared against MS/MS spectra of ferulenol (2) [54], before being annotated by using the in silico tools to gain insight into the structure to ultimately confirm the obtained results. For positive ESI mode, the same chromatographic separation condition was applied for identifying both prenylated products (compound- 1 and 2), except the ion polarity was switched to the positive-ion mode with the CID energy of 25 eV (mass scan range 50–1500 amu). The obtained MS/MS spectra were then compared to the authentic ferulenol (2) (final concentration of 20 ppm) (Santa Cruz Biotechnology, Santa Cruz, CA, USA) run in parallel under the same positive ESI condition.

4.6. Structural Annotation of Prenylated Products by Using MetFrag

MetFrag, the freely accessible software (https://msbi.ipb-halle.de/MetFragBeta/ (accessed on 26 September 2021)), was used as a tool for structural annotation of metabolites from the high-quality MS/MS spectra of the query subjects, which plays a crucial role as the initial step of structural identification. According to the user manual provided in https://ipb-halle.github.io/MetFrag/projects/metfragweb/ (accessed on 26 September 2021), structural annotation of the target metabolites is required for two steps of data processing, including "retrieving candidates" and "processing of candidates". Based on the experimental mass spectra of the compound-1, the neutral mass of 366.219495 (relative mass deviation of 5 ppm) with the calculated molecular ion (*m/z* 365.211670 $[M - H]^-$) was defined as parameters to retrieve the candidate molecules from the PubChem database. The neutral molecular formula ($C_{24}H_{30}O_3$) and data-specific identifiers (i.e., compound ID number) could also be defined in this step to perform the candidate search. Having retrieved a large number of candidates (2267 molecular structures, in our case) from PubChem, several parameters including the relative mass deviation (10 ppm), absolute mass deviation (0.001), the certain adduct type in ($[M - H]^-$), and the MS/MS peak list (Figure S4) were then defined to match against "in silico generated fragments" of candidates from the PubChem database. After that, the score-ranked list of candidates was displayed. For each candidate in the row, the information regarding candidate image, identifier with linked database, exact mass, molecular formula, and several MS/MS peaks explained were also displayed. For a certain candidate of the list, "fragment view" also provided molecular formulas and their corresponding substructures for the matched peaks, which are highlighted in green.

4.7. Structural Annotations of Prenylated 4HC Analog Using SIRIUS Tool

SIRIUS is an in silico tool developed for the identification of molecular formula, mass spectra, and fragmentation tree annotation of the unknown metabolites based upon high-resolution raw MS/MS data [54]. This tool is also connected with CSI:FingerID web service for structural identification and prediction substructures (so-called molecular fingerprints) present in the query subject. SIRIUS can also systematically predict the classes of query metabolites via the computational tool (CANOPUS: class assignment and ontology prediction using mass spectrometry) [56]. Structural elucidation of the prenylated-4HC analogs were thus based on the prediction power of SIRIUS (version 4.8.2). According to the user manual, the raw MS/MS data in text format (Figure S4) was imported into the SIRIUS application window. The MS2 level and the collision energy (35 eV) were subsequently defined in the next dialogue. Then, two parameters, including the precursor mass ion (*m/z* 365.2154) and the specific adduct type ($[M - H]^-$), were defined in the following application window. The annotation of prenylated-4HC derivatives was accomplished by selecting "compute option". In our case, the entire tools, including SIRIUS, CSI:FingerID web service, and CANOPUS, were selected to sufficiently detail structural elucidation of the prenylated product. Certain databases (e.g., Natural Products, COCONUTS, and NORMAN) were also chosen as biological databases to improve the rate of correct identification and rank of the query metabolite.

5. Conclusions

In this study, we demonstrated that the *E coli* BL21(DE3)pLysS strain could be engineered as the prenylation system of 4HC (1) via co-expression of the genes encoding for AtaPT and CstDXS. Based on the high-resolution LC-ESI(-)-QTOF-MS/MS, feeding 4HC (1) into the engineered strain resulted in the formation of the prenylated 4HC analog (namely compound-1) showing *m/z* 365.21 $[M - H]^-$, which was detected from the culture medium. The mass fragmentation pattern of this product was highly identical to the authentic ferulenol (2) established by the previous work [66]. MetFrag analysis clearly showed that the compound was best explained as "ferulenol (2)" among 2267 candidated compounds from the PubChem database. Further annotation using SIRIUS integrated with CSI:FingerID and CANOPUS led to unraveling the chemical features hidden in compound-1, e.g., molecular formula (defined as $C_{24}H_{30}O_3$), ferulenol (2) as a potential structure, and classes (defined as coumarin derivatives), indicating the prenylate-4HC derivative was indeed ferulenol (2). Meanwhile, compound-2 (*m/z* 365.21 $[M - H]^-$ and 367.22 $[M + H]^+$) was elucidated as the farnesylated of 2-hydroxy-4-chromenone (c), which was proposed to be formed through the tautomerization of ferulenol (2). It is worthwhile to emphasize that both products could be formed without the external supply of prenyl donors (e.g., DMAPP, GPP, and FPP), which indicated the capacity of *E. coli* to supply the prenyl donors, particularly FPP, and verified the functional expression of AtaPT and CstDXS in this bacterial system. Equally important, they were secreted in the culture medium, providing a great benefit from an economic point of view since breaking cell pellets is not needed. Further experiments, e.g., overexpression of rate-limiting step enzymes in the DXP pathway along with the optimized culture conditions, are required to improve the yields of two products. Based on the substrate promiscuity of AtaPT, the in vivo prenylation by clones carrying pCDFDuet-AtaPT-CstDXS could be applied towards other types of natural products such as quercetin and mangiferin to generate diverse classes of prenylated derivatives.

Supplementary Materials: The following are available online, Figure S1: The insertion of AtaPT in MCS-1 of pCDFDuet-1 vector, Figure S2: The insertion of CstDXS in MCS-2 of pCDFDuet-1 vector, Figure S3: The verified insertion of CstDXS (~1992 bp) in pCDFDuet-AtaPT (~5020 bp) via *Kpn*I and *Xho*I digestion, Figure S4: The raw mass (MS2) data for putative ferulenol (2) (compound-1) obtained from negative-ESI mode, Figure S5: The proposed reaction mechanisms illustrating the various chemical losses present in the MS/MS spectrum of compound-1, Figure S6: The proposed reaction mechanisms illustrating the various chemical losses present in the MS/MS spectrum of compound-2, Figure S7: The postulated mechanisms underlying the formation of ion *m/z* 189.05 of compound-1, Figure S8: The postulated mechanisms underlying the formation of ion *m/z* 189.05 of compound-2.

Author Contributions: Conceptualization, N.N. and A.K.; methodology, A.K., J.N., P.P. and N.N.; validation, A.K., J.N., P.P. and N.N.; formal analysis, A.K., N.N., P.P. and J.N.; investigation, A.K. and J.N.; resources, N.N.; writing—original draft preparation, A.K. and N.N.; writing—review and editing, A.K., J.N., P.P. and N.N.; supervision, P.P. and N.N.; funding acquisition, N.N. All authors have read and agreed to the published version of the manuscript.

Funding: This work was supported by the National Research Council of Thailand-Khon Kaen University, 2018 (NRCT-KKU 2018): 610034; Faculty of Pharmaceutical Sciences, Khon Kaen University, Thailand.

Institutional Review Board Statement: Not applicable.

Informed Consent Statement: Not applicable.

Acknowledgments: A.K. wishes to thank the Graduate School, Khon Kaen University for the overseas research scholarship. The authors would like to thank Assistant Professor Tiwatt Kulja-nabhagavad, Suandusit University, Bangkok, Thailand for the valuable comments and advice on the mass spectrometry elucidation. We also thank Yutthakan Saengkun and Jatupong Sitsutheechananon for assistance.

Conflicts of Interest: The authors declare no conflict of interest.

Sample Availability: Samples of the compounds are not available from the authors.

References

1. Yazaki, K.; Sasaki, K.; Tsurumaru, Y. Prenylation of aromatic compounds, a key diversification of plant secondary metabolites. *Phytochemistry* **2009**, *70*, 1739–1745. [CrossRef]
2. Botta, B.; Vitali, A.; Menendez, P.; Misiti, D.; Monache, G. Prenylated Flavonoids: Pharmacology and Biotechnology. *Curr. Med. Chem.* **2005**, *12*, 713–739. [CrossRef] [PubMed]
3. Alhassan, A.M.; Abdullahi, M.I.; Uba, A.; Umar, A. Prenylation of aromatic secondary metabolites: A new frontier for development of novel drugs. *Trop. J. Pharm. Res.* **2014**, *13*, 307–314. [CrossRef]
4. Bartmańska, A.; Tronina, T.; Popłoński, J.; Milczarek, M.; Filip-Psurska, B.; Wietrzyk, J. Highly cancer selective antiproliferative activity of natural prenylated flavonoids. *Molecules* **2018**, *23*, 2922. [CrossRef]
5. Akaberi, M.; Iranshahy, M.; Iranshahi, M. Review of the traditional uses, phytochemistry, pharmacology and toxicology of giant fennel (Ferula communis l. subsp. communis). *Iran. J. Basic Med. Sci.* **2015**, *18*, 1050–1062. [PubMed]
6. Trantas, E.A.; Koffas, M.A.G.; Xu, P.; Ververidis, F. When plants produce not enough or at all: Metabolic engineering of flavonoids in microbial hosts. *Front. Plant Sci.* **2015**, *6*, 1–16. [CrossRef]

7. Ajikumar, P.K.; Tyo, K.; Carlsen, S.; Mucha, O.; Phon, T.H.; Stephanopoulos, G. Terpenoids: Opportunities for biosynthesis of natural product drugs using engineered microorganisms. *Mol. Pharm.* **2008**, *5*, 167–190. [CrossRef]
8. Shen, R.; Jiang, X.; Ye, W.; Song, X.; Liu, L.; Lao, X.; Wu, C. A novel and practical synthetic route for the total synthesis of lycopene. *Tetrahedron* **2011**, *67*, 5610–5614. [CrossRef]
9. Zhou, K.; Edgar, S.; Stephanopoulos, G. Engineering Microbes to Synthesize Plant Isoprenoids. *Methods Enzymol.* **2016**, *575*, 225–245.
10. Gliszczyńska, A.; Brodelius, P.E. Sesquiterpene coumarins. *Phytochem. Rev.* **2011**, *11*, 77–96. [CrossRef]
11. Bocca, C.; Gabriel, L.; Bozzo, F.; Miglietta, A. Microtubule-interacting activity and cytotoxicity of the prenylated coumarin ferulenol. *Planta Med.* **2002**, *68*, 1135–1137. [CrossRef]
12. Altmann, K.H.; Gertsch, J. Anticancer drugs from nature - Natural products as a unique source of new microtubule-stabilizing agents. *Nat. Prod. Rep.* **2007**, *24*, 327–357. [CrossRef] [PubMed]
13. Lariche, N.; Lahouel, M.; Benguedouar, L.; Zellagui, A. Ferulenol, a sesquiterpene coumarin, induce apoptosis *via* mitochondrial dysregulation in lung cancer induced by benzo[a]pyrene: Involvement of Bcl2 protein. *Anticancer. Agents Med. Chem.* **2017**, *17*, 1357–1362. [CrossRef]
14. Wiart, C. *Lead Compounds from Medicinal Plants for the Treatment of Cancer*, 1st ed.; Academic Press: Cambridge, MA, USA, 2012; ISBN 9780123983824.
15. Gebauer, M. Synthesis and structure-activity relationships of novel warfarin derivatives. *Bioorganic Med. Chem.* **2007**, *15*, 2414–2420. [CrossRef]
16. Monti, M.; Pinotti, M.; Appendino, G.; Dallocchio, F.; Bellini, T.; Antognoni, F.; Poli, F.; Bernardi, F. Characterization of anti-coagulant properties of prenylated coumarin ferulenol. *Biochim. Biophys. Acta Gen. Subj.* **2007**, *1770*, 1437–1440. [CrossRef] [PubMed]
17. Appendino, G.; Mercalli, E.; Fuzzati, N.; Arnoldi, L.; Stavri, M.; Gibbons, S.; Ballero, M.; Maxia, A. Antimycobacterial coumarins from the Sardinian giant fennel (*Ferula communis*). *J. Nat. Prod.* **2004**, *67*, 2108–2110. [CrossRef] [PubMed]
18. de Bruijn, W.J.C.; Levisson, M.; Beekwilder, J.; van Berkel, W.J.H.; Vincken, J.P. Plant Aromatic Prenyltransferases: Tools for Microbial Cell Factories. *Trends Biotechnol.* **2020**, *38*, 917–934. [CrossRef]
19. Ward, V.C.A.; Chatzivasileiou, A.O.; Stephanopoulos, G. Metabolic engineering of *Escherichia coli* for the production of isoprenoids. *FEMS Microbiol. Lett.* **2018**, *365*, 1–9. [CrossRef] [PubMed]
20. Zhao, Y.; Yang, J.; Qin, B.; Li, Y.; Sun, Y.; Su, S.; Xian, M. Biosynthesis of isoprene in *Escherichia coli via* methylerythritol phosphate (MEP) pathway. *Appl. Microbiol. Biotechnol.* **2011**, *90*, 1915–1922. [CrossRef]
21. Volke, D.C.; Rohwer, J.; Fischer, R.; Jennewein, S. Investigation of the methylerythritol 4-phosphate pathway for microbial terpenoid production through metabolic control analysis. *Microb. Cell Fact.* **2019**, *18*, 1–15. [CrossRef]
22. Niu, F.X.; Lu, Q.; Bu, Y.F.; Liu, J.Z. Metabolic engineering for the microbial production of isoprenoids: Carotenoids and isoprenoid-based biofuels. *Synth. Syst. Biotechnol.* **2017**, *2*, 167–175. [CrossRef]
23. Li, Y.; Wang, G. Strategies of isoprenoids production in engineered bacteria. *J. Appl. Microbiol.* **2016**, *121*, 932–940. [CrossRef] [PubMed]
24. Yuan, L.Z.; Rouvière, P.E.; LaRossa, R.A.; Suh, W. Chromosomal promoter replacement of the isoprenoid pathway for enhancing carotenoid production in E. coli. *Metab. Eng.* **2006**, *8*, 79–90. [CrossRef]
25. Takahashi, H.; Aihara, Y.; Ogawa, Y.; Murata, Y.; Nakajima, K.-I.; Iida, M.; Shirai, M.; Fujisaki, S. Suppression of phenotype of *Escherichia coli* mutant defective in farnesyl diphosphate synthase by overexpression of gene for octaprenyl diphosphate synthase. *Biosci. Biotechnol. Biochem.* **2018**, *82*, 1003–1010. [CrossRef]
26. Cao, Y.; Zhang, R.; Liu, W.; Zhao, G.; Niu, W.; Guo, J.; Xian, M.; Liu, H. Manipulation of the precursor supply for high-level production of longifolene by metabolically engineered *Escherichia coli*. *Sci. Rep.* **2019**, *9*, 1–10. [CrossRef]
27. Lee, P.C.; Petri, R.; Mijts, B.N.; Watts, K.T.; Schmidt-Dannert, C. Directed evolution of *Escherichia coli* farnesyl diphosphate synthase (IspA) reveals novel structural determinants of chain length specificity. *Metab. Eng.* **2005**, *7*, 18–26. [CrossRef]
28. Vallon, T.; Ghanegaonkar, S.; Vielhauer, O.; Müller, A.; Albermann, C.; Sprenger, G.; Reuss, M.; Lemuth, K. Quantitative analysis of isoprenoid diphosphate intermediates in recombinant and wild-type *Escherichia coli* strains. *Appl. Microbiol. Biotechnol.* **2008**, *81*, 175–182. [CrossRef]
29. Boghigian, B.A.; Salas, D.; Ajikumar, P.K.; Stephanopoulos, G.; Pfeifer, B.A. Analysis of heterologous taxadiene production in K- and B-derived *Escherichia coli*. *Appl. Microbiol. Biotechnol.* **2012**, *93*, 1651–1661. [CrossRef] [PubMed]
30. Huang, Q.; Roessner, C.A.; Croteau, R.; Scott, A.I. Engineering Escherichia coli for the synthesis of taxadiene, a key intermediate in the biosynthesis of taxol. *Bioorganic Med. Chem.* **2001**, *9*, 2237–2242. [CrossRef]
31. Albrecht, M.; Misawa, N.; Sandmann, G. Metabolic engineering of the terpenoid biosynthetic pathway of *Escherichia coli* for production of the carotenoids β-carotene and zeaxanthin. *Biotechnol. Lett.* **1999**, *21*, 791–795. [CrossRef]
32. Schmidt-Dannert, C. Engineering novel carotenoids in microorganisms. *Curr. Opin. Biotechnol.* **2000**, *11*, 255–261. [CrossRef]
33. Tsuruta, H.; Paddon, C.J.; Eng, D.; Lenihan, J.R.; Horning, T.; Anthony, L.C.; Regentin, R.; Keasling, J.D.; Renninger, N.S.; Newman, J.D. High-level production of amorpha-4, 11-diene, a precursor of the antimalarial agent artemisinin, in *Escherichia coli*. *PLoS ONE* **2009**, *4*, e4489. [CrossRef] [PubMed]

34. Anthony, J.R.; Anthony, L.C.; Nowroozi, F.; Kwon, G.; Newman, J.D.; Keasling, J.D. Optimization of the mevalonate-based isoprenoid biosynthetic pathway in *Escherichia coli* for production of the anti-malarial drug precursor amorpha-4,11-diene. *Metab. Eng.* **2009**, *11*, 13–19. [CrossRef] [PubMed]
35. Sasaki, K.; Tsurumaru, Y.; Yazakiy, K. Prenylation of flavonoids by biotransformation of yeast expressing plant membrane-bound prenyltransferase SfN8DT-1. *Biosci. Biotechnol. Biochem.* **2009**, *73*, 759–761. [CrossRef]
36. Luo, X.; Reiter, M.A.; d'Espaux, L.; Wong, J.; Denby, C.M.; Lechner, A.; Zhang, Y.; Grzybowski, A.T.; Harth, S.; Lin, W.; et al. Complete biosynthesis of cannabinoids and their unnatural analogues in yeast. *Nature* **2019**, *567*, 123–126. [CrossRef]
37. Li, Z.; Zhao, G.; Liu, H.; Guo, Y.; Wu, H.; Sun, X.; Wu, X.; Zheng, Z. Biotransformation of menadione to its prenylated derivative MK-3 using recombinant Pichia pastoris. *J. Ind. Microbiol. Biotechnol.* **2017**, *44*, 973–985. [CrossRef]
38. Ma, Y.; McClure, D.D.; Somerville, M.V.; Proschogo, N.W.; Dehghani, F.; Kavanagh, J.M.; Coleman, N.V. Metabolic Engineering of the MEP Pathway in *Bacillus subtilis* for increased biosynthesis of menaquinone-7. *ACS Synth. Biol.* **2019**, *8*, 1620–1630. [CrossRef] [PubMed]
39. Heide, L. Prenyl transfer to aromatic substrates: Genetics and enzymology. *Curr. Opin. Chem. Biol.* **2009**, *13*, 171–179. [CrossRef]
40. Zhou, K.; Ludwig, L.; Li, S.M. Friedel-Crafts alkylation of acylphloroglucinols catalyzed by a fungal indole prenyltransferase. *J. Nat. Prod.* **2015**, *78*, 929–933. [CrossRef]
41. Liebhold, M.; Xie, X.; Li, S.M. Expansion of enzymatic Friedel-Crafts alkylation on indoles: Acceptance of unnatural β-unsaturated allyl diphospates by dimethylallyl-tryptophan synthases. *Org. Lett.* **2012**, *14*, 4882–4885. [CrossRef]
42. Fan, A.; Xie, X.; Li, S.M. Tryptophan prenyltransferases showing higher catalytic activities for Friedel-Crafts alkylation of o- and m-tyrosines than tyrosine prenyltransferases. *Org. Biomol. Chem.* **2015**, *13*, 7551–7557. [CrossRef]
43. Miranda, C.L.; Aponso, G.L.M.; Stevens, J.F.; Deinzer, M.L.; Buhler, D.R. Prenylated chalcones and flavanones as inducers of quinone reductase in mouse Hepa 1c1c7 cells. *Cancer Lett.* **2000**, *149*, 21–29. [CrossRef]
44. Dong, X.; Fan, Y.; Yu, L.; Hu, Y. Synthesis of four natural prenylflavonoids and their estrogen-like activities. *Arch. Pharm. (Weinheim).* **2007**, *340*, 372–376. [CrossRef] [PubMed]
45. Appendino, G.; Gibbons, S.; Giana, A.; Pagani, A.; Grassi, G.; Stavri, M.; Smith, E.; Rahman, M.M. Antibacterial cannabinoids from *Cannabis sativa*: A structure-activity study. *J. Nat. Prod.* **2008**, *71*, 1427–1430. [CrossRef] [PubMed]
46. Chen, R.; Gao, B.; Liu, X.; Ruan, F.; Zhang, Y.; Lou, J.; Feng, K.; Wunsch, C.; Li, S.M.; Dai, J.; et al. Molecular insights into the enzyme promiscuity of an aromatic prenyltransferase. *Nat. Chem. Biol.* **2017**, *13*, 226–234. [CrossRef] [PubMed]
47. Abdou, M.M.; El-Saeed, R.A.; Bondock, S. Recent advances in 4-hydroxycoumarin chemistry. Part 1: Synthesis and reactions. *Arab. J. Chem.* **2019**, *12*, 88–121. [CrossRef]
48. Sitthithaworn, W.; Wungsintaweekul, J.; Sirisuntipong, T.; Charoonratana, T.; Ebizuka, Y.; De-Eknamkul, W. Cloning and expression of 1-deoxy-d-xylulose 5-phosphate synthase cDNA from *Croton stellatopilosus* and expression of 2C-methyl-d-erythritol 4-phosphate synthase and geranylgeranyl diphosphate synthase, key enzymes of plaunotol biosynthesis. *J. Plant Physiol.* **2010**, *167*, 292–300. [CrossRef]
49. Xiao, J.F.; Zhou, B.; Ressom, H.W. Metabolite identification and quantitation in LC-MS/MS-based metabolomics. *TrAC - Trends Anal. Chem.* **2012**, *32*, 1–14. [CrossRef]
50. Zhou, B.; Xiao, J.F.; Tuli, L.; Ressom, H.W. LC-MS-based metabolomics. *Mol. Biosyst.* **2012**, *8*, 470–481. [CrossRef]
51. Blaženović, I.; Kind, T.; Ji, J.; Fiehn, O. Software tools and approaches for compound identification of LC-MS/MS data in metabolomics. *Metabolites* **2018**, *8*, 31. [CrossRef] [PubMed]
52. Ruttkies, C.; Schymanski, E.L.; Wolf, S.; Hollender, J.; Neumann, S. MetFrag relaunched: Incorporating strategies beyond *in silico* fragmentation. *J. Cheminform.* **2016**, *8*, 1–16. [CrossRef] [PubMed]
53. Wolf, S.; Schmidt, S.; Müller-Hannemann, M.; Neumann, S. *In silico* fragmentation for computer assisted identification of metabolite mass spectra. *BMC Bioinform.* **2010**, *11*, 1–12. [CrossRef]
54. Dührkop, K.; Fleischauer, M.; Ludwig, M.; Aksenov, A.A.; Melnik, A.V.; Meusel, M.; Dorrestein, P.C.; Rousu, J.; Böcker, S. SIRIUS 4: A rapid tool for turning tandem mass spectra into metabolite structure information. *Nat. Methods* **2019**, *16*, 299–302. [CrossRef]
55. Dührkop, K.; Shen, H.; Meusel, M.; Rousu, J.; Böcker, S. Searching molecular structure databases with tandem mass spectra using CSI:FingerID. *Proc. Natl. Acad. Sci. USA* **2015**, *112*, 12580–12585. [CrossRef] [PubMed]
56. Dührkop, K.; Nothias, L.F.; Fleischauer, M.; Ludwig, M.; Hoffmann, M.A.; Rousu, J.; Dorrestein, P.C.; Böcker, S. Classes for the masses: Systematic classification of unknowns using fragmentation spectra. *bioRxiv* **2020**, 046672. [CrossRef]
57. Fourel, I.; Hugnet, C.; Goy-Thollot, I.; Berny, P. Validation of a new liquid chromatography-tandem mass spectrometry ion-trap technique for the simultaneous determination of thirteen anticoagulant rodenticides, drugs, or natural products. *J. Anal. Toxicol.* **2010**, *34*, 95–102. [CrossRef]
58. Klamrak, A.; Nabnueangsap, J.; Nualkaew, N. Biotransformation of benzoate to 2,4,6-trihydroxybenzophenone by engineered *Escherichia coli. Molecules* **2021**, *26*, 2779. [CrossRef]
59. Cortés, I.; Cala, L.J.; Bracca, A.B.J.; Kaufman, T.S. Furo[3,2-c]coumarins carrying carbon substituents at C-2 and/or C-3. Isolation, biological activity, synthesis and reaction mechanisms. *RSC Adv.* **2020**, *10*, 33344–33377. [CrossRef]
60. Lee, P.C.; Schmidt-Dannert, C. Metabolic engineering towards biotechnological production of carotenoids in microorganisms. *Appl. Microbiol. Biotechnol.* **2002**, *60*, 1–11.
61. Lin, Y.; Shen, X.; Yuan, Q.; Yan, Y. Microbial biosynthesis of the anticoagulant precursor 4-hydroxycoumarin. *Nat. Commun.* **2013**, *4*, 1–8. [CrossRef]

62. Gustafsson, C.; Govindarajan, S.; Minshull, J. Codon bias and heterologous protein expression. *Trends Biotechnol.* **2004**, *22*, 346–353. [CrossRef]
63. Jones, K.L.; Kim, S.W.; Keasling, J.D. Low-copy plasmids can perform as well as or better than high-copy plasmids for metabolic engineering of bacteria. *Metab. Eng.* **2000**, *2*, 328–338. [CrossRef] [PubMed]
64. Singh, A.; Upadhyay, V.; Upadhyay, A.K.; Singh, S.M.; Panda, A.K. Protein recovery from inclusion bodies of *Escherichia coli* using mild solubilization process. *Microb. Cell Fact.* **2015**, *14*, 1–10. [CrossRef]
65. Tolia, N.H.; Joshua-Tor, L. Strategies for protein coexpression in *Escherichia coli*. *Nat. Methods* **2006**, *3*, 55–64. [CrossRef] [PubMed]
66. Ridder, L.; Van Der Hooft, J.J.J.; Verhoeven, S.; De Vos, R.C.H.; Bino, R.J.; Vervoort, J. Automatic chemical structure annotation of an LC-MSn based metabolic profile from green tea. *Anal. Chem.* **2013**, *85*, 6033–6040. [CrossRef] [PubMed]

Article

New 3-Ethynylaryl Coumarin-Based Dyes for DSSC Applications: Synthesis, Spectroscopic Properties, and Theoretical Calculations

João Sarrato [1,†], Ana Lucia Pinto [1,†], Gabriela Malta [1,†], Eva G. Röck [2,3], João Pina [2], João Carlos Lima [1], A. Jorge Parola [1,*] and Paula S. Branco [1,*]

1 LAQV-REQUIMTE, Departament of Chemistry, NOVA School of Science and Technology, FCT NOVA, Universidade NOVA de Lisboa, 2829-516 Caparica, Portugal; j.sarrato@campus.fct.unl.pt (J.S.); al.pinto@campus.fct.unl.pt (A.L.P.); g.malta@campus.fct.unl.pt (G.M.); lima@fct.unl.pt (J.C.L.)
2 Department of Chemistry, Coimbra Chemistry Centre, University of Coimbra, Rua Larga, 3004-535 Coimbra, Portugal; eva.roeck@dcb.unibe.ch (E.G.R.); jpina@qui.uc.pt (J.P.)
3 Department of Chemistry, Biochemistry and Pharmaceutical Sciences, University of Bern, Freiestrasse 3, CH-3012 Bern, Switzerland
* Correspondence: ajp@fct.unl.pt (A.J.P.); paula.branco@fct.unl.pt (P.S.B.); Tel.: +351-21-294-8300 (P.S.B.)
† These authors contributed equally to this work.

Abstract: A set of 3-ethynylaryl coumarin dyes with mono, bithiophenes and the fused variant, thieno [3,2-*b*] thiophene, as well as an alkylated benzotriazole unit were prepared and tested for dye-sensitized solar cells (DSSCs). For comparison purposes, the variation of the substitution pattern at the coumarin unit was analyzed with the natural product 6,7-dihydroxycoumarin (Esculetin) as well as 5,7-dihydroxycomarin in the case of the bithiophene dye. Crucial steps for extension of the conjugated system involved Sonogashira reaction yielding highly fluorescent molecules. Spectroscopic characterization showed that the extension of conjugation via the alkynyl bridge resulted in a strong red-shift of absorption and emission spectra (in solution) of approximately 73–79 nm and 52–89 nm, respectively, relative to 6,7-dimethoxy-4-methylcoumarin (λ_{abs} = 341 nm and λ_{em} = 410 nm). Theoretical density functional theory (DFT) calculations show that the Lowest Unoccupied Molecular Orbital (LUMO) is mostly centered in the cyanoacrylic anchor unit, corroborating the high intramolecular charge transfer (ICT) character of the electronic transition. Photovoltaic performance evaluation reveals that the thieno [3,2-*b*] thiophene unit present in dye **8** leads to the best sensitizer of the set, with a conversion efficiency (η = 2.00%), best V_{OC} (367 mV) and second best J_{sc} (9.28 mA·cm^{-2}), surpassed only by dye **9b** (J_{sc} = 10.19 mA·cm^{-2}). This high photocurrent value can be attributed to increased donor ability of the 5,7-dimethoxy unit when compared to the 6,7 equivalent (**9b**).

Keywords: dye-sensitized solar cells; coumarin dyes; thieno [3,2-*b*] thiophene; charge transfer; ethynylaryl

Citation: Sarrato, J.; Pinto, A.L.; Malta, G.; Röck, E.G.; Pina, J.; Lima, J.C.; Parola, A.J.; Branco, P.S. New 3-Ethynylaryl Coumarin-Based Dyes for DSSC Applications: Synthesis, Spectroscopic Properties, and Theoretical Calculations. *Molecules* **2021**, *26*, 2934. https://doi.org/10.3390/molecules26102934

Academic Editor: Maria João Matos

Received: 31 March 2021
Accepted: 10 May 2021
Published: 14 May 2021

1. Introduction

Following O'Regan and Grätzel's seminal application [1] of an organic dye adsorbed on a mesoporous wide band gap semiconductor as a light-harvesting electrode, the field of DSSCs (Dye-Sensitized Solar Cells) has garnered much interest over the last three decades. Compared to alternative light-harvesting technologies, they possess reduced cost, ease of manufacture and low environmental impact, which combined with their wide array of possible colors and compatibility with flexible substrates, allows for a multitude of applications, such as integration into buildings [2,3] and interiors [4,5].

Although the original ruthenium(II)-polypyridyl chromophores used in DSSCs, such as N3 [6] and N719 [7], have mostly remained as gold standard dyes due to their large conversion efficiencies [8], their only moderate extinction coefficients and the use of a rare and expensive noble metal has led to the extensive search for highly efficient metal-free dyes. These metal-free dyes offer, in addition to the lower cost of production, easier and greater synthetic versatility and tunable optical and electrochemical properties through structural modification, as demonstrated by the extremely wide array of compounds explored over the years [9–11]. Despite being mostly overshadowed by their inorganic counterparts, in recent years comparable or even superior performances have been achieved by metal-free chromophores, such as bulky indoline-quinoxaline dyes (10.65%) [12], tetrathioacene dyes (10.1%) [13] and more elaborate polycyclic aromatic push-pull dyes (12.6%) [14], all employing a Co(II/III) complex as redox shuttle. Additionally, the use of co-sensitization approaches has been employed to great success with organic dyes, managing to reach efficiencies of more than 14% [15,16].

To be adequately applied in DSSCs, organic dyes must present a donor-π-acceptor (D-π-A) structure. The push-pull effect in these D-π-A dyes leads to efficient intramolecular charge transfer (ICT) from the donor to the acceptor unit through the π-bridge upon light absorption. Among the many classes of organic compounds used, coumarins are of particular interest due to their wide use as fluorescent sensors [17–19], emitting layers in Organic Light-Emitting Diodes (OLEDs) [20–22] and in laser applications [23,24], owing to their large Stokes shift, high quantum yields and good solubility. Additionally, their photophysical properties can be easily tuned through the addition of substituents, namely electron-withdrawing substituents in position 3 and electron-donating substituents in position 7 [25].

This allows for a decrease in the energy gap between the highest occupied molecular orbital (HOMO) and the lowest unoccupied molecular orbital (LUMO), making coumarins great candidates as new sensitizers for DSSCs. Hara et al. [26–28] were some of the first to successfully design and employ dyes with coumarin donor units to achieve competitive efficiencies of up to 8.2% [29] when using deoxycholic acid (DCA) as a coadsorbent. In more recent work, Jiang et al. [30] and He et al. [31] used additional indoline and triphenylamine donors (respectively) attached to the coumarin unit, while Vekariya [32] investigated the effect of various *o*-halide phenylene spacers on dye structure and device performance.

Recently we have shown that the introduction of a linear ethynyl π-bridge into coumarin-based conjugated donor–acceptor systems resulted in redshifted absorption and emission spectra relatively to the styryl counterparts [33]. Electrochemical studies revealed that this derivatization resulted in a marked decrease in the HOMO energy levels, which influenced the overall conversion efficiency with a significantly superior performance. This revealed the importance of the alignment of the substituents with the direction of the intramolecular charge transfer. This is not completely surprising since the incident photon-to-current conversion efficiency (IPCE) reveals that the electron transfer yield (Φ(v)ET) becomes larger with the introduction of a triple bond [34].

Following up on our previous experience with the synthesis of coumarin chromophores [33,35,36], several 3-ethynyl-6,7-dihydroxycoumarin-based dyes were prepared, with emphasis on the effect of varying the π-bridge on dye structure, photophysical properties and sensitizer efficiency. These groups include mono and bithiophenes, including a fused variant, thieno [3,2-*b*] thiophene, as well as a benzotriazole unit containing a long alkyl chain (Figure 1). Additionally, a 5,7-dihydroxycoumarin-based chromophore was prepared, to ascertain the effect of this alternative substitution pattern on device performance. All prepared dyes were then spectroscopically characterized, their optimized geometry was obtained from density functional theory (DFT) and their performance as sensitizers was evaluated.

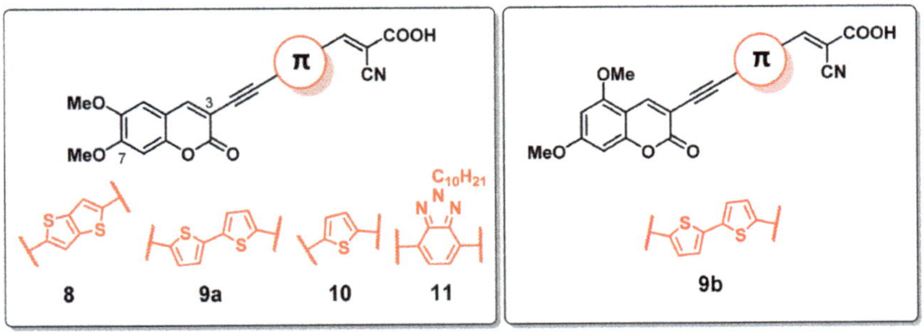

Figure 1. Structure of target dyes prepared and characterized in this work.

2. Results and Discussion

2.1. Synthesis and Characterization

The synthetic approach to obtain the coumarin-based dyes, as described in Scheme 1, initiated with the brominated derivatives **1a** and **1b**. Compound **1a** was obtained as described by Martins et al. [36], while **1b** was prepared through condensation of ethyl propiolate and 1,3,5-trihydroxybenzene [37], followed by chloroformylation, bromination [38], hydrolysis and methylation. The 3-ethynyl coumarin derivatives (**3a** and **3b**) were then obtained through a high-yield Sonogashira coupling with ethynyltrimethylsilane, followed by the removal of the trimethylsilyl group, which at first was accomplished with tetrabutylammonium fluoride (TBAF). Since this method resulted in poor yields, partly due to the possible degradation of the unprotected ethynyl derivative, a method employing K_2CO_3 in MeOH described by Wang et al. [39] was used instead, without further purification of the resulting product.

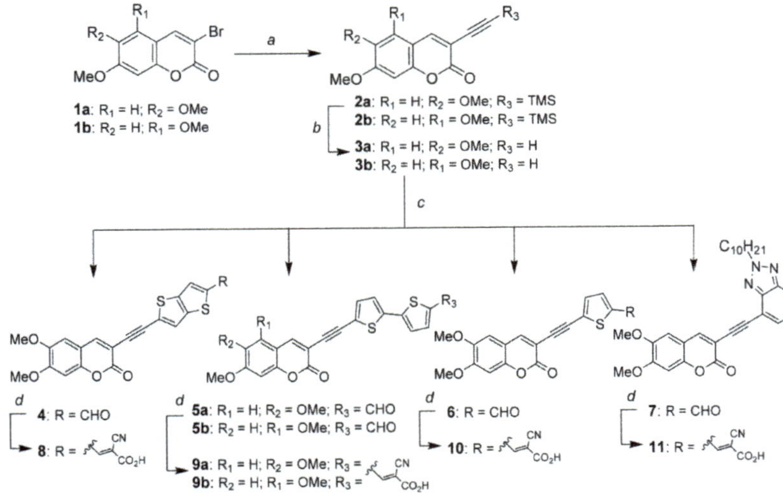

Scheme 1. Synthetic approach used for the preparation of the various chromophores: (**a**) **1a**/**1b** (1 eq.) ethynyltrimethylsilane (2 eq.), $Pd(PPh_3)_4$ (0.15 eq.), PPh_3 (0.06 eq.), CuI (0.12 eq.), (*i*-Pr) $_2$NH (2 eq.), dry dioxane, sealed tube under N_2, 40–45 °C, overnight; (**b**) **2a**/**2b** (1 eq.), K_2CO_3 (0.15 eq.), dry MeOH, r.t., 4h; (**c**) ethynylcoumarin (1 eq.), aldehyde (1 eq.), $Pd(PPh_3)_4$ (0.15 eq.), PPh_3 (0.06 eq.), CuI (0.12 eq.), (*i*-Pr) $_2$NH (2 eq.), dry dioxane, sealed tube under N_2, 40–45 °C, overnight; (**d**) aldehyde (1 eq.), cyanoacetic acid (3 eq.), piperidine (2.7 eq.), dry acetonitrile, reflux, overnight.

The heterocyclic π-bridges, in the form of brominated aldehydes, were then coupled to the ethynyl moiety through another Sonogashira coupling. In the case of compounds **4** and **7** the aldehydes were not commercially available and as such were prepared by lithiation-formylation of 2,5-dibromothieno [3,2-*b*] thiophene in the case of compound **4**, and alkylation, double bromination [40] and lithiation-formylation of benzotriazole in the case of compound **7**.

With the aldehyde groups now present in the molecules, a Knoevanagel condensation is performed as described by Martins et al. [33] in order to insert the cyanoacrylic acid group that allows the binding to the TiO_2 surface. With this, the final cromophores **8**, **9a**, **9b**, **10** and **11** were obtained with overall yields of 20.2%, 20.3%, 1.1%, 23.6% and 3.6%, respectively. The coumarin derivative with substitution at the 5 and 7-positions (as in compound **9b**) proved to be more difficult to handle then their 6,7-disubstituted analogues, due to insolubility problems. This drawback, together with the increased synthetic complexity of the 5,7-dihydroxycoumarin dye system, led us to focus on the 5,7-disubstitued derivatives.

2.2. Absorption and Fluorescence

Figure 2 presents the UV-Vis absorption and fluorescence emission spectra of the investigated samples at room temperature in acetonitrile solution. The significant Stokes shift values (in the 2201–3998 cm^{-1} range) point to a charge transfer (CT) character of the fluorescence emission band (Table 1). Indeed, when compared to 6,7-dimethoxy-4-methylcoumarin (λ_{abs} = 341 nm, λ_{em} = 410 nm in ethanol) the absorption and emission spectra of samples **8–11** are strongly redshifted [41]. This is associated with the increase in conjugation length of the chromophoric system due to overlapping of π-orbitals of the coumarin group with the π-orbitals of the cyanoacetic acid substituted benzotriazole or thienyl units. Moreover, the absorption spectra of samples **8–11** are also redshifted when comparison is made with 7-methoxy-3-acetylcoumarin (λ_{abs} = 341 nm) where intramolecular charge transfer (ICT) was found due to the introduction of the acetyl group in the 3-position of the coumarin moiety [42]. This behavior is explained by the increasing ICT character of compounds **8–11**, promoted by the strong electron-withdrawing character of the cyanoacetic acid substituted benzotriazole or thienyl units, thus leading to high ground-state dipole moments (values in the 14.560–24.230 D range, see Table 2).

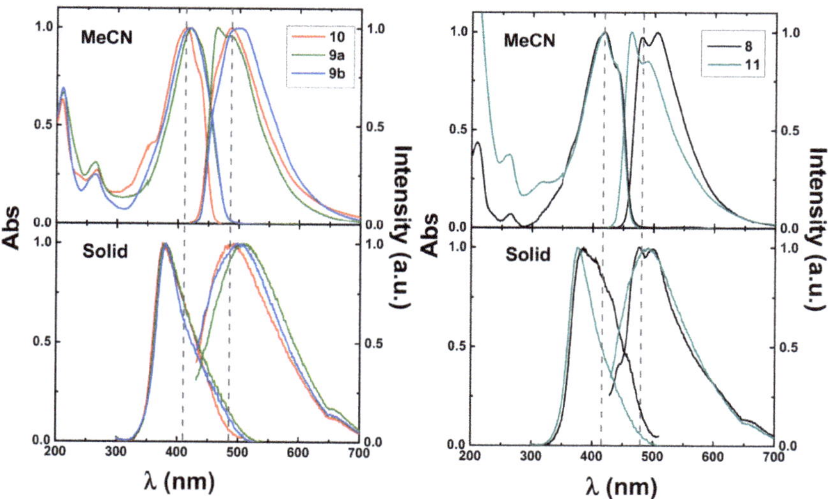

Figure 2. Normalized absorption and fluorescence emission spectra for compounds **8–11** in acetonitrile solution and in the solid state (adsorbed in TiO_2 films) at 293K.

Table 1. Spectroscopic data for compounds **8–11** in acetonitrile solution (absorption and fluorescence emission maxima, molar extinction coefficients, ε, and Stokes shift, Δ_{SS}) and absorbed in TiO_2 films (absorption and fluorescence emission maxima) at 293 K.

Dye	λ_{max}^{Abs} (mm)	λ_{max}^{Abs} Solid (nm)	ε ($cm^{-1}M^{-1}$)	λ_{max}^{Fluo} (nm)	λ_{max}^{Fluo} Solid (nm)	Δ_{SS} (nm)	Δ_{SS} (cm^{-1})
8	417	386	64,470	489	476	72	3589
9a	420	380	30,190	464	510	44	2201
9b	418	380	45,480	499	500	81	3998
10	414	375	15,300	486	488	72	3578
11	416	375	12,060	462	490	46	2451

Table 2. Experimental absorption maxima obtained in acetonitrile solution together with the relevant computed absorption properties (predicted vertical excitation energies and associated orbitals transitions major contributions together with oscillator strengths, f, and band gap, E_g) for the investigated compounds obtained by TD-DFT at the CAM-B3LYP/6-311G(d,p) level of theory after ground-state geometry optimization using the same functional and basis set.

Dye	$\lambda_{max}^{S_0 \rightarrow S_n}$ (nm)	$\lambda_{max}^{S_0 \rightarrow S_n}$ Calc. (nm)	Dipole Moment (D)	Transition and Orbitals Major Contributions	Oscillator Strength, F	E_g (eV) [a]
8	416	415	14.560	$S_0 \rightarrow S_1$, HOMO→LUMO (83%)	2.146	2.85 [2.70]
9a	421	424	19.139	$S_0 \rightarrow S_1$, HOMO→LUMO (79%)	2.156	2.77 [2.75]
9b	416	428	22.394	$S_0 \rightarrow S_1$, HOMO→LUMO (78%);	2.147	2.72 [2.73]
10	414	404	24.230	$S_0 \rightarrow S_1$, HOMO→LUMO (86%)	1.878	2.97 [2.78]
11	415	407	15.56	$S_0 \rightarrow S_1$, HOMO→LUMO (84%)	1.764	2.75 [2.76]

[a] in brackets the experimental absorption band gap obtained from the intersection between the normalized absorption and fluorescence emission spectra.

Porous TiO_2 films (about 1 µm thick) were made by spreading TiO_2 paste (ref. 30NR-D, from GreatcellSolar) onto electrically conductive Fluorine-Doped Tin Oxide (FTO) glass, using Scotch Magic tape as a spacer. The film paste was then gradually sintered up to 500 °C to reach the anatase TiO_2 phase. When the films had cooled to about 80 °C, they were placed for 1 min in concentrated solutions of the samples **8–11** in acetonitrile solution (0.1 mM) for adsorption. The absorption (here measured by the fluorescence excitation spectra) and fluorescence emission spectra are depicted in Figure 2. In general, the spectra of the investigated samples in the solid state are blue-shifted by 31–41 nm, with respect to the absorption spectra in MeCN solutions.

2.3. Theoretical Calculations

The ground state optimized geometry structures and the relevant HOMO and LUMO energy levels, together with their electron density distribution surface plots were obtained at the DFT/CAM-B3LYP/6-311G(d,p) level taking into account the bulk solvent effects of acetonitrile. Frequency analysis for each compound were also computed and did not yield any imaginary frequencies, indicating that the structure of each molecule corresponds to at least a local minimum on the potential energy surface. The geometry optimization for the investigated compound revealed that in general the thienyl and/or the benzotriazole moieties are mostly planar with the coumarin units.

The optimized ground-state molecular geometries found for the investigated compounds were used to obtain the vertical excitation energies, oscillator strengths (f) and excited state compositions in terms of excitations between the occupied and virtual orbitals using the time-dependent density functional theory (TD-DFT) approach, see Table 2. For samples **8–11** the predicted $S_0 \rightarrow S_1$ transitions are in good agreement with the observed lowest energy absorption bands in acetonitrile solution (Table 1). In general, for these transitions the major contribution arises from the HOMO→LUMO orbitals (contributions > 78%). It is worth mentioning that the calculated absorption band gap values agree with the experimental values (see Table 2), thus giving support for the predicted ground-state geometries.

The molecular orbital contours (Figure 3) show that the densities of the HOMO orbitals are, in general, spread over the entire molecules, while the LUMO shows a decrease in the electron density on the coumarin moieties and a concomitant increase in the benzotriazole or thienyl units. The electronic delocalization in the LUMO orbitals gives support for the occurrence of a charge transfer state (CT) in the singlet excited state.

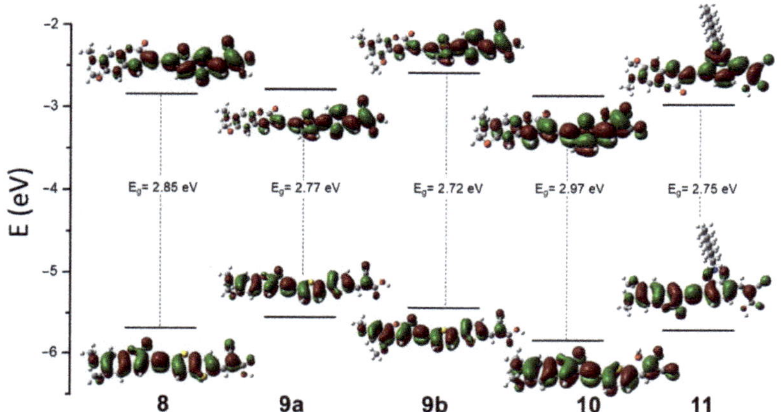

Figure 3. DFT//CAM-B3LYP/6311G(d,p) optimized ground-state geometry together with the frontier molecular orbital energy levels and the relevant electronic density contours (calculated at B3LYP/6311G(d,p) level) for the investigated compounds. Additionally, displayed are the predicted optical band gap energy values, E_g.

2.4. Electrochemical Characterization

The electrochemical properties of the dyes were determined by differential pulse voltammetry (DPV), with the main results being presented in Table 3 (See Supplementary Materials for full DPV data). From the obtained onsets of oxidation and reduction peaks, the HOMO and LUMO energies were estimated. For this purpose, the following equation was used: E [eV] = $-$(E_{onset} (V vs. SCE) + 4.44) [43].

Table 3. Electrochemical properties in dimethylformamide (DMF) obtained from DPV measurements: HOMO energy level, determined from the onset of the oxidation peak (E_{ox}); LUMO energy level, determined from the onset of the reduction peak (E_{red}). Gap energy (E_g), calculated with E_{HOMO}—E_{LUMO}.

Dye	HOMO Energy (eV)	LUMO Energy (eV)	E_g (eV)
8	−5.70	−3.57	2.13
9a	−5.64	−3.58	2.06
9b	−5.60	−3.53	2.07
10	−5.74	−3.62	2.12
11	−5.65	−3.60	2.05

All calculated band gap values are similar across the various dyes, arising from similar values in the HOMO (−5.74 to −5.60 eV) and in the LUMO energies (−3.62 to

−3.53eV) energies. These differences are within the error for electrochemical determination of the energy of frontier orbitals (~0.1 V) [43], and as such a comparison between the dyes cannot be confidently made. The determined electrochemical band gaps in DMF (~2.1 eV) are lower than the corresponding optical band gaps (~2.7 eV, Table 2), determined in acetonitrile. Solvation and coulombic effects are responsible for often observed differences between electrochemical and optical band gaps [44]. Additionally, the solvatochromic nature of substituted coumarins [45,46] leads to a shorter band gap in more polar solvent, which is the case with DMF.

When compared with other reported coumarin sensitizers (E_{HOMO}: −5.2 eV; E_{LUMO}: −2.4 eV) [30], a significant decrease in orbital energy is observed for the dihydroxycoumarin dyes. This is indicative of a lower electron-donating ability of the dihydroxy substitution pattern in comparison with the indoline moieties employed in the referenced work. On the other hand, this family of dyes presents a shorter band gap (2.0 V vs. 2.4 V), a direct consequence of the more extensive conjugation of the π-system.

2.5. Photovoltaic Performance

The prepared chromophores were tested in prototype devices and their performance was compared with reference dye N719 (in non-optimized conditions). These results are summarized in Figure 4 and Table 4.

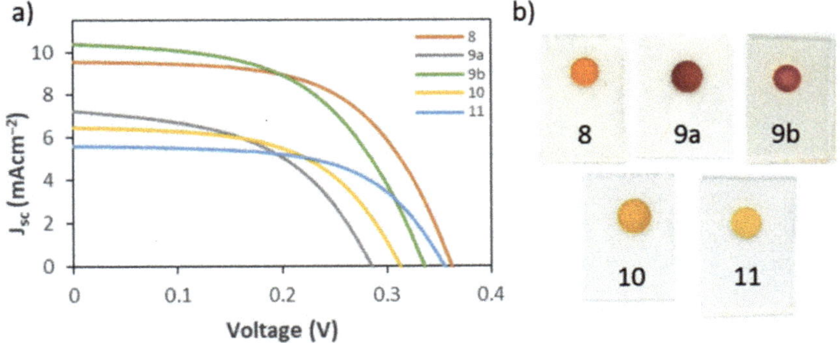

Figure 4. (**a**) I–V curves of the test cells based on the synthesized dyes under 100 mW·cm^{-2} under simulated AM 1.5 illumination. The results presented correspond to the best performing cell. (**b**) Pictures of the dyes adsorbed on the TiO_2 photoanodes.

Table 4. Performance values of the test cells based on the synthesized dyes and reference dye N719 under 100 mW·cm^{-2} AM 1.5 illumination. The results presented correspond to the average values of at least two cells per dye, each cell measured 5 times.

Dye	V_{oc} (mV)	J_{sc} (mA/cm^2)	J_{max} (mA/cm^2)	V_{max} (mV)	FF	η (%)
8	367 ± 5	9.3 ± 0.1	7.5 ± 0.2	256 ± 3	0.56 ± 0.01	2.00 ± 0.06
9a	289 ± 6	6.7 ± 0.3	4.9 ± 0.3	193 ± 4	0.49 ± 0.02	0.95 ± 0.07
9b	339 ± 3	10.2 ± 0.1	7.8 ± 0.2	227 ± 2	0.51 ± 0.01	1.78 ± 0.06
10	311 ± 5	6.4 ± 0.1	4.9 ± 0.3	214 ± 5	0.54 ± 0.02	1.07 ± 0.05
11	359 ± 2	5.4 ± 0.1	4.3 ± 0.2	258 ± 1	0.58 ± 0.02	1.13 ± 0.04
N719	440 ± 6	15.5 ± 0.4	13.1 ± 0.2	305 ± 4	0.59 ± 0.02	4.06 ± 0.05

Compound **8** was the best performing dye, with an efficiency of 2% and the highest values of V_{OC} and V_{max}, 367 and 256 mV, respectively. This marked difference from the other dyes can be attributed to the comparably higher ε value (6.4470 cm^{-1}M^{-1}), as well as the more redshifted absorption when adsorbed on the TiO_2 surface (386 nm). Dye **9a**, which contains a 2,2′-bithiophene group as π-bridge, obtained the worst efficiency value (0.95%), comparable to dye **10** (1.07%) containing only one thiophene ring. One possible explanation for this result may be the known torsion angles present between the two thiophene units [47], which are not present in the fused ring equivalent **8**, and result in hindered conjugation. This possibility seems to be further supported by the Stokes shift observed for this dye (Δ_{SS} = 2201 cm^{-1}), being the lowest of the group, which indicates a low ICT character for electronic transition.

Comparison of dyes **9a** and **9b** allows the evaluation of the effect of the position of the substituents on DSSC performance, and it is immediately apparent from the obtained efficiencies (0.95% vs. 1.78%, respectively) that position 5 (when compared to 6) is a superior choice for donor units in coumarin dyes. The superior efficiency is a consequence of the high photocurrent values (J_{sc} = 10.2 and J_{max} = 7.8 $mA \cdot cm^{-2}$), which even surpass the results for dye **8**, allowing it to have a good efficiency despite the relatively modest photovoltage (V_{oc} = 339 and V_{max} = 227 mV). The improved properties of position 5 as a donor are further supported by the obtained absorption properties, in particular the absorptivity (ε = 4.5480 $cm^{-1}M^{-1}$) and Stokes shift (Δ_{SS} = 3998 cm^{-1}), which are indicative of the dye's effective light absorption and charge transfer character. Additionally, dye **9b** possesses a higher HOMO energy than dye **9a**, which once again points to position 5 of the coumarin unit being a more suitable choice for the inclusion of electron-donating groups.

Inversely, dye **11** shows a high V_{OC}, comparable to dye **8** (359 vs. 367 mV, respectively), yet it has the worst J_{sc} (5.4 $mA \cdot cm^{-2}$) of all the synthesized dyes, which can be due to its inferior ε (12,060 $cm^{-1}M^{-1}$) and Stokes shift (Δ_{SS} = 2451 cm^{-1}). The photovoltage values can be attributed to the bulky alkyl chain present in the benzotriazole moiety, which will help suppress charge recombination between the semiconductor and the oxidized redox shuttle [48,49]. Another effect of the presence of this bulky group may be the distortion of the molecule's geometry [50], decreasing planarity and therefore leading to a less efficient electron delocalization and lower ICT character in the transition. Once again, this is reflected in the observed Stokes shift value (Δ_{SS} = 2451 cm^{-1}), which is on the lower end of the group, comparable to dye **9a**.

3. Materials and Methods

3.1. General Information and Instruments

All solvents and reagents were obtained commercially (Merck KGaA, Darmstadt, Germany) and used without further purification. The drying of the solvents was achieved with M2A molecular sieves (Merck KGaA), as described by Bradley et al. [51].

Thin-layer chromatography (TLC) was carried out on aluminum-backed Kieselgel 60 F254 silica gel plates (Merck KGaA). Plates were visualized with UV light (254 and 336 nm) and in certain cases chemical staining agents (Acidic solution of 2,4-dinitrophenylhidrazine). Preparative-layer chromatography (PLC) was performed on Keiselgel 60 F254 silica gel plates (Merck KGaA) with a thickness of 0.5 mm. Column chromatography was performed using Keiselgel 60 silica gel (Merck KGaA), 70–230 mesh and 230–400 mesh particle sizes as stationary phases, in the cases of regular and flash [52] normal-phase cromatographies, respectively.

The ^{1}H- and ^{13}C-NMR (nuclear magnetic spectroscopy) spectra were acquired with a Bruker Avance III 400 (Billerica, MA, USA), at 400 and 101 MHz, respectively. Absorption and fluorescence spectra were recorded on a Cary 5000 UV-Vis-NIR (Santa Clara, CA, USA) and Horiba–Jobin–Ivon Fluoromax4 spectrometers (Longjumeau, France), respectively. The fluorescence spectra were corrected for the wavelength response of the system.

Differential pulse voltammetry (DPV) measurements were performed on a μAutolab Type III potentiostat/galvanostat (Metrohm Autolab B. V., Utrecht, The Netherlands), controlled with GPES (General Purpose Electrochemical System) software version 4.9 (Eco-Chemie, B. V. Software, Utrecht, The Netherlands), using a cylindrical 5 mL three-electrode cell. All measurements refer to a saturated calomel electrode (SCE, saturated KCl) reference electrode (Metrohm, Utrecht, The Netherlands). A Pt wire was used as counter-electrode, a glassy carbon electrode (MF-2013, f = 1.6 mm, BAS inc., West Lafayette, IN, USA) was used as the working electrode. Prior to use, the working electrode was polished in aqueous suspensions of 1.0 and 0.3 mm alumina (Buehler, Esslingen, Germany) over 2–7/″ micro-cloth (Buehler) polishing pads, then rinsed with water and ethanol. This cleaning procedure was systematically applied before any electrochemical measurement. The electrolyte composition was 0.1 M tetrabutylammonium tetrafluoroborate in DMF, with a dye concentration of 1.5×10^{-4} M. Measurements were performed between 0 and +1.6 V for determination of oxidation potential and between 0 and −1.6 V for determination of reduction potential, with a scan rate of 10 mV/s in both cases. The samples in the electrochemical cell were de-aerated by purging with nitrogen for 10 min prior to, and during, the electrochemical measurements.

High-resolution mass spectra (HRMS) were obtained at the University of Porto, Mass Spectrometry Laboratory (LEM/CEMUP) using a mass spectrometer Linear Trap Quadrupole (LTQ) Orbitrap XL^{TM} (Thermo Fischer Scientific, Bremen, Germany) controlled by a LTQ Tune Plus 2.5.5 and Xcalibur 2.1.0 and at the University of Salamanca (Spain), Elemental Analysis, Chromatography and Mass Spectrometry Service (NUCLEUS), using a High Performance Liquid Chromatography (HPLC) Agilent 1100 coupled to a QSTAR XL Hybrid qTOF (AB Sciex, Framingham, MA, USA) mass spectrometer.

3.2. *Synthesis*

3.2.1. Synthesis of 6,7-Dimethoxy-3-((Trimethylsilyl)Ethynyl)Coumarin (**2a**) and 5,7-Dimethoxy-3-((Trimethylsilyl)Ethynyl)Coumarin (**2b**)

To a sealed tube, 0.06 eq. of PPh_3, 0.12 eq. of CuI, 0.15 eq. of $Pd(PPh_3)_4$, 1 eq. of 3-bromocoumarin (**2a**/**2b**) and 5 mL of dry dioxane were added under a N_2 atmosphere. After a few minutes, 2 eq. of ethynyltrimethylsilane and 0.35 mL 2 eq. of dry $(i\text{-Pr})_2NH$ were added and the solution was stirred at 45 °C overnight under a N_2 atmosphere. Once the reaction was confirmed to be complete by TLC (hexane/AcOEt (7:3 v/v)), the solution was cooled to room temperature, the solvent was removed under reduced pressure and the solid residue dried in vacuo before being purified by flash chromatography with hexane/AcOEt (7:3 v/v) as eluent, affording the target compounds.

6,7-dimethoxy-3-((trimethylsilyl)ethynyl)coumarin (**2a**). Starting from 347.3 mg (1.22 mmol, 1 eq.) of 3-bromo-6,7-dimethoxycoumarin (**1a**), 343.4 mg (93.2%) of 6,7-dimethoxy-3-((trimethylsilyl)ethynyl)coumarin (**2a**) were obtained. ^{1}H-NMR (400 MHz, $CDCl_3$) δ (ppm) 7.83 (s, 1H, H4), 6.82 (s, 1H, H5/H8), 6.80 (s, 1H, H5/H8), 3.95 (s, 3H, H1′/H2′), 3.91 (s, 3H, H1′/H2′), 0.26 (s, 9H, H5′); ^{13}C-NMR (101 MHz, $CDCl_3$) δ (ppm) 160.0 (C2), 153.5 (C7), 149.8 (C8a), 146.8 (C4/C6), 146.1 (C4/C6), 111.5 (C3/C4a/C5), 109.5 (C3/C4a/C5), 107.7 (C3/C4a/C5), 101.0 (C3′/C4′/C8), 99.9 (C3′/C4′/C8), 98.7 (C3′/C4′/C8), 56.6 (C1′/C2′), 56.5 (C1′/C2′), −0.1 (C5′); HRMS-ESI(+) Calculated for $C_{16}H_{19}O_4Si$ $[M + H]^+$ 303.1047; Found 303.1053.

5,7-dimethoxy-3-((trimethylsilyl)ethynyl)coumarin (**2b**). Starting from 147.1 mg (1.22 mmol, 1 eq.) of 3-bromo-5,7-dimethoxycoumarin (**1b**), 114.0 mg (72.9%) of 5,7-dimethoxy-3-((trimethylsilyl)ethynyl)coumarin (**2b**) were obtained. ^{1}H-NMR (400 MHz, $CDCl_3$) δ (ppm) 8.14 (s, 1H, H4), 6.37 (s, 1H, H6/H8), 6.25 (d, J = 2.4 Hz, 1H, H6/H8), 3.88 (s, 3H, H1′/H2′), 3.84 (s, 3H, H1′/H2′), 0.25 (s, 9H, H5′). HRMS-ESI(+) Calculated for $C_{16}H_{19}O_4Si$ $[M + H]^+$ 303.1047; Found 303.1041.

3.2.2. General Method for the Synthesis of Coupled Aldehydes (**4**–**7**)

To a sealed tube, PPh_3 (0.06 eq), CuI (0.12 eq), $Pd(PPh_3)_4$ (0.15), aldehyde (1 eq.) and 5 mL of dry dioxane were added under a N_2 atmosphere. After a few minutes ethynylcoumarin (**3**) (1 eq.) and dry $(i\text{-Pr})_2NH$ (2 eq.) were added and the solution was stirred at 45°C overnight under a N_2 atmosphere. Once the reaction was confirmed to be complete by TLC (hexane/AcOEt (7:3 v/v)), the solution was cooled to room temperature, the solvent was removed under reduced pressure and the solid residue dried in vacuo before being purified by flash chromatography with DCM/MeOH (99.8:0.02 v/v and 99.5:0.05 v/v) as eluent, affording a bright yellow solid in all cases.

5-((6,7-dimethoxy-2-oxo-2H-chromen-3-yl)ethynyl)thieno [3,2-b] tiophene-2-carbaldehyde (**4**). Starting from 104.6 mg (0.42 mmol, 1 eq.) of 5-bromothieno [3,2-*b*] thiophene-2-carbaldehyde, 89.5 mg (51.5%) of 5-((6,7-dimethoxy-2-oxo-2*H*-chromen-3-yl)ethynyl)thieno [3,2-*b*] tiophene-2-carbaldehyde (**4**) were obtained. ^{1}H-NMR (400 MHz, $CDCl_3$) δ (ppm) 9.99 (s, 1H, H7′), 7.92 (s, 1H, H4), 7.89 (s, 1H, H5′/H6′), 7.54 (s, 1H, H5′/H6′), 6.87 (s, 1H, H5/H8), 6.86 (s, 1H, H5/H8), 3.98 (s, 3H, H1′/H2′), 3.94 (s, 3H, H1′/H2′); HRMS-ESI(+) Calculated for $C_{20}H_{13}O_5S_2$ $[M + H]^+$ 397.0199; Found 397.0197.

5′-((6,7-dimethoxi-2-oxo-2H-chromen-3-yl)ethynyl)-[2,2′-bithiophene]-5-carbaldehyde (**5a**). Starting from 132.8 mg (0.49 mmol, 1 eq.) of 5-bromo-[2,2′-bithiophene]-5-carbaldehyde, 95.2 mg (46.3%) of 5′-((6,7-dimethoxi-2-oxo-2*H*-chromen-3-yl)ethynyl)-[2,2′-bithiophene]-5-carbaldehyde (**5a**) were obtained. ^{1}H-NMR (400 MHz, CD_2Cl_2) δ (ppm) 9.85 (s, 1H, H9′), 7.89 (s, 1H, H4), 7.71 (d, J = 4.8 Hz, 1H, H5′/H6′/H7′/H8′), 7.30 (s, 3H, H5′/H6′/H7′/H8′), 6.87 (s, 1H, H5/H8), 6.85 (s, 1H, H5/H8), 3.92 (s, 3H, H1′/H2′), 3.87 (s, 3H, H1′/H2′); HRMS-ESI(+) Calculated for $C_{22}H_{15}O_5S_2$ $[M + H]^+$ 423.0355; Found 423.0348.

5′-((5,7-dimethoxi-2-oxo-2H-chromen-3-yl)ethynyl)-[2,2′-bithiophene]-5-carbaldehyde (**5b**). Starting from 154.0 mg (0.56 mmol, 1 eq.) of 5-bromo-[2,2′-bithiophene]-5-carbaldehyde, 22.4 mg (14.1%) of 5′-((5,7-dimethoxi-2-oxo-2*H*-chromen-3-yl)ethynyl)-[2,2′-bithiophene]-5-carbaldehyde (**5b**) were obtained. ^{1}H-NMR (400 MHz, DMF-d_7) δ (ppm) 10.02 (s, 1H, H9′), 8.29 (s, 1H, H4), 8.08 (d, J = 4.03 Hz, 1H, H5′/H6′/H7′/H8′), 7.69–7.67 (m, 2H, H5′/H6′/H7′/H8′), 7.53 (d, J = 4.03 Hz 1H, H5′/H6′/H7′/H8′), 6.67 (s, 1H, H6/H8), 6.63 (s, 1H, H6/H8), 4.04 (s, 3H, H1′/H2′), 3.99 (s, 3H, H1′/H2′); HRMS-ESI(+) Calculated for $C_{22}H_{15}O_5S_2$ $[M + H]^+$ 423.0355; Found 423.0349.

5-((6,7-dimethoxy-2-oxo-2H-chromen-3-yl)ethynyl)-thiophene-2-carbaldehyde (**6**). Starting from 64.8 mg (0.34 mmol, 1 eq.) of 5-bromothiophene-2-carbaldehyde, 58.7 mg (50.8%) of 5-((6,7-dimethoxy-2-oxo-2*H*-chromen-3-yl)ethynyl)-thiophene-2-carbaldehyde (**6**) were obtained. ^{1}H-NMR (400 MHz, CD_2Cl_2) δ (ppm) 9.87 (s, 1H, H7′), 7.95 (s, 1H, H4), 7.71 (d, J = 3.8 Hz, 1H, H5′/H6′), 7.41 (d, J = 3.7 Hz, 1H, H5′/H6′), 6.89 (s, 1H, H5/H8), 6.87 (s, 1H, H5/H8), 3.94 (s, 3H, H1′/H2′), 3.89 (s, 3H, H1′/H2′); HRMS-ESI(+) Calculated for $C_{18}H_{13}O_5S$ $[M + H]^+$ 341.0478; Found 341.0478.

2-decyl-7-((6,7-dimethoxy-2-oxo-2H-chromen-3-yl)ethynyl)-2H-benzo[d][1,2,3]triazole-4- carbaldehyde (**7**). Starting from 113 mg (0.31 mmol, 1 eq.) of 7-dibromo-2-decyl-2*H*-benzo[*d*] [1,2,3]triazole-4-carbaldehyde, 42.5 mg (50.8%) of 2-decyl-7-((6,7-dimethoxy-2-oxo-2*H*-chromen-3-yl)ethynyl)-2*H*-benzo[*d*][1,2,3]triazole-4-carbaldehyde (**7**) were obtained. 1H-NMR (400 MHz, CD2C12) δ (ppm) 10.50 (s, 1H, H7′), 8.11 (s, 1H, H4), 7.99 (d, J = 7.3 Hz, 1H, H5′/H6′), 7.81 (d, J = 7.5 Hz, 1H, H5′/H6′), 6.96 (s, 1H, H5/H8), 6.93 (s, 1H, H5/H8), 4.90 (t, J = 7.4 Hz, 2H, H1″), 3.99 (s, 3H, H1′/H2′), 3.94 (s, 3H, H1′/H2′), 2.26–2.18 (m, 2H, H2″), 1.43–1.30 (m, 15H, H3″-H9″), 0.90 (t, J = 6.4 Hz, 3H, H10″); HRMS-ESI(+) Calculated for $C_{30}H_{34}N_3O_5$ $[M + H]^+$ 516.2493; Found 516.2509.

3.2.3. General Method for the Synthesis of Final Chromophores (8–11)

To a round-bottom flask containing aldehyde (1 eq.), cyanoacetic acid (3 eq.), 5 mL of acetonitrile (ACN) and dry piperidine (2.7 eq.) were added and the resulting solution was stirred under reflux for 24 h. Once the reaction was confirmed to be complete by TLC (DCM/MeOH (9.5:0.5 v/v)), the solvent was evaporated under reduced pressure, the solid residue was washed 3–5 times with ACN, acidified with HCl (10%) and washed 3–5 times with distilled water. After each washing step the solvent used was centrifuged (4500 rpm, 10–30 min) to recover any lost product.

2-cyano-3-(5-((6,7-dimethoxy-2-oxo-2H-chromen-3-yl)ethynyl)thieno[3,2-b]thiophen-2-yl)acrylic acid (**8**). Starting from 15 mg (0.038 mmol, 1 eq.) of 5-((6,7-dimethoxy-2-oxo-2*H*-chromen-3-yl)ethynyl)thieno[3,2-*b*]tiophene-2-carbaldehyde (**4**), 7.8 mg (44.5%) of 2-cyano-3-(5-((6,7-dimethoxy-2-oxo-2*H*-chromen-3-yl)ethynyl)thieno[3,2-*b*]thiophen-2-yl)acrylic acid (**8**) were obtained. ^{1}H-NMR (400 MHz, DMSO-d_6) δ (ppm) 8.75 (br s, 1H, OH), 8.37 (s, 1H, H4/H7′),

8.28 (s, 1H, H4/H7′), 8.09 (s, 1H, H5′/H6′), 7.89 (s, 1H, H5′/H6′), 7.26 (s, 1H, H5/H8), 7.14 (s, 1H, H5/H8), 3.90 (s, 3H, H1′/H2′), 3.82 (s, 3H, H1′/H2′). HRMS-ESI(+) Calculated for $C_{23}H_{14}NO_6S_2$ [M + H]$^+$ 464.0257; Found 464.0249.

2-cyano-3-(5′-((6,7-dimethoxy-2-oxo-2H-chromen-3-yl)ethynyl)-[2,2′-bithiophen]-5-yl)acrylic acid (**9a**). Starting from 95 mg (0.225 mmol, 1 eq.) of 5′-((6,7-dimethoxi-2-oxo-2*H*-chromen-3-yl)ethynyl)-[2,2′-bithiophene]-5-carbaldehyde (**5a**), 54.7 mg (49.7%) of 2-cyano-3-(5′-((6,7-dimethoxy-2-oxo-2*H*-chromen-3-yl)ethynyl)-[2,2′-bithiophen]-5-yl)acrylic acid (**9a**) were obtained. ^{1}H-NMR (400 MHz, DMF-d_7) δ (ppm) 8.57 (s, 1H, H4/H9′), 8.38 (s, 1H, H4/H9′), 8.08 (d, *J* = 3.5 Hz, 1H, H5′/H6′/H7′/H8′), 7.73 (d, *J* = 3.7 Hz, 1H, H5′/H6′/H7′/H8′), 7.70 (d, *J* = 4.0 Hz, 1H, H5′/H6′/H7′/H8′), 7.53 (d, *J* = 4.5 Hz, 1H, H5′/H6′/H7′/H8′), 7.37 (s, 1H, H5/H8), 7.15 (s, 1H, H5/H8), 4.02 (s, 3H, H1′/H2′), 3.92 (s, 3H, H1′/H2′). HRMS-ESI(+) Calculated for $C_{25}H_{16}NO_6S_2$ [M + H]$^+$ 490.0414; Found 490.0408.

2-cyano-3-(5′-((5,7-dimethoxy-2-oxo-2H-chromen-3-yl)ethynyl)-[2,2′-bithiophen]-5-yl)acrylic acid (**9b**). Starting from 23.2 mg (0.055 mmol, 1 eq.) of 5′-((5,7-dimethoxi-2-oxo-2*H*-chromen-3-yl)ethynyl)-[2,2′-bithiophene]-5-carbaldehyde (**5b**), 14.1 mg (52.5%) of 2-cyano-3-(5′-((5,7-dimethoxy-2-oxo-2*H*-chromen-3-yl)ethynyl)-[2,2′-bithiophen]-5-yl)acrylic acid (**9b**) were obtained. ^{1}H-NMR (400 MHz, DMF-d_7) δ (ppm) 8.53 (s, 1H, H4/H9′), 8.27 (s, 1H, H4/H9′), 8.04 (s, 1H, H5′/H6′/H7′/H8′), 7.70 (br s, 1H, H5′/H6′/H7′/H8′), 7.67 (br s, 1H, H5′/H6′/H7′/H8′), 7.52 (br s, 1H, H5′/H6′/H7′/H8′), 6.65 (s, 1H, H6/H8), 6.61 (s, 1H, H6/H8), 4.04 (s, 3H, H1′/H2′), 3.99 (s, 3H, H1′/H2′). HRMS-ESI(+) Calculated for $C_{25}H_{16}NO_6S_2$ [M + H]$^+$ 490.0414; Found 490.0406.

2-cyano-3-(5-((6,7-dimethoxy-2-oxo-2H-chromen-3-yl)ethynyl)-thiophen-2-yl)acrylic acid (**10**). Starting from 28.1 mg (0.083 mmol, 1 eq.) of 5-((6,7-dimethoxy-2-oxo-2*H*-chromen-3-yl)ethynyl)-thiophene-2-carbaldehyde (**6**), 17.7 mg (50.8%) of 2-cyano-3-(5-((6,7-dimethoxy-2-oxo-2*H*-chromen-3-yl)ethynyl)-thiophen-2-yl)acrylic acid (**10**) were obtained. ^{1}H-NMR (400 MHz, DMSO-d_6) δ (ppm) 8.50 (s, 1H, H4/H7′), 8.36 (s, 1H, H4/H7′), 7.98 (d, *J* = 3.8 Hz, 1H, H5′/H6′), 7.57 (d, *J* = 3.9 Hz, 1H, H5′/H6′), 7.22 (s, 1H, H5/H8), 7.11 (s, 1H, H5/H8), 3.89 (s, 3H, H1′/H2′), 3.81 (s, 3H, H1′/H2′). HRMS-ESI(+) Calculated for $C_{21}H_{14}NO_6S$ [M + H]$^+$ 408.0536; Found 408.0529.

2-cyano-3-(2-decyl-7-((6,7-dimethoxy-2-oxo-2H-chromen-3-yl)ethynyl)-2H-benzo[d][1,2,3]triazol-4-yl)acrylic acid (**11**). Starting from 51,7 mg (0.10 mmol, 1 eq.) of 2-decyl-7-((6,7-dimethoxy-2-oxo-2*H*-chromen-3-yl)ethynyl)-2*H*-benzo[*d*][1,2,3]triazole-4-carbaldehyde (**7**), 8.9 mg (50.8%) of 2-cyano-3-(2-decyl-7-((6,7-dimethoxy-2-oxo-2*H*-chromen-3-yl)ethynyl)-2*H*-benzo[*d*][1,2,3]triazol-4-yl)acrylic acid (11) were obtained. 1H-NMR (400 MHz, DMF-d_7) δ (ppm) 8.92 (s, 1H, H4/H7′), 8.55 (d, *J* = 8.0 Hz, 1H, H5′/H6′), 8.46 (s, 1H, H4/H7′), 7.97 (d, *J* = 7.8 Hz, 1H, H5′/H6′), 7.42 (s, 1H, H5/H8), 7.17 (s, 1H, H5/H8), 4.95 (t, *J* = 7.3 Hz, 2H, H1″), 4.04 (s, 3H, H1′/H2′), 3.93 (s, 3H, H1′/H2′), 2.22–2.15 (m, 2H, H2″), 1.44–1.14 (m, 20H, H3′-H9″), 0.85 (t, *J* = 6.4 Hz, 5H, H10″). HRMS-ESI(+) Calculated for $C_{33}H_{35}N_4O_6$ [M + H]$^+$ 583.2551; Found 583.2542.

3.3. Theoretical Calculations

The ground state molecular geometry was optimized using the density functional theory (DFT) by means of the Gaussian 09 program (Gaussian, Wallingford, CT, USA) [53], under CAM-B3LYP/6-311G(d,p) level [54,55] taking into account the bulk solvent effects of acetonitrile [56]. Optimal geometries were determined on isolated entities in acetonitrile and no conformation restrictions were imposed. For the resulting optimized geometries time-dependent DFT calculations (using the same functional and basis set as those in the previously calculations) were performed to predict the vertical electronic excitation energies. Molecular orbital contours were predicted at the B3LYP/6-311G(d,p) level of theory in vacuo and were plotted using GaussView 5.0.8 (Gaussian). The orbitals transitions percentage contributions of the predicted vertical excitation were calculated using GaussSum 2.2 (Dublin, Ireland) [57].

3.4. DSSCs Fabrication and Photovoltaic Characterization

The detailed procedure has been described elsewhere [58]. The conductive FTO-glass (TEC7, Greatcell Solar, Queanbeyan, Australia) used for the preparation of the transparent electrodes was first cleaned with detergent and then washed with water and ethanol. To prepare the anodes, the conductive glass plates (area: 15 cm × 4 cm) were immersed in a $TiCl_4$/water solution (40 mM) at 70 °C for 30 min, washed with water and ethanol and sintered at 500 °C for 30 min. This procedure is essential to improve the adherence of the subsequently deposited nanocrystalline layers to the glass plates, as well as to serve as a 'blocking-layer', helping to block charge recombination between electrons in the FTO and holes in the I^-/I_3^- redox couple. Afterwards, the TiO_2 nanocrystalline layers were deposited on these pre-treated FTO plates by screen-printing the transparent titania paste (18NR-T, Greatcell Solar) using a frame with polyester fibers with 43.80 mesh per cm^2. This procedure, involving two steps (coating and drying at 125 °C), was repeated twice. The TiO_2-coated plates were gradually heated up to 325 °C, then the temperature increased to 375 °C in 5 min, and afterwards to 500 °C. The plates were sintered at this temperature for 30 min, and finally cooled down to room temperature. A second treatment with the same $TiCl_4$/water solution (40 mM) was performed, following the procedure described previously. This second $TiCl_4$ treatment is also an optimization step that enhances the surface roughness for dye adsorption, thus positively affecting the photocurrent produced by the cell under illumination. Finally, a coating of reflective titania paste (WER2-O, Greatcell Solar) was deposited by screen-printing and sintered at 500 °C. This layer of 150–200 nm sized anatase particles functions as a 'photon-trapping' layer that further improves the photocurrent. Each anode was cut into rectangular pieces (area: 2 cm × 1.5 cm) with a spot area of 0.196 cm^2 and a thickness of 15 μm. The prepared anodes were soaked for 16 h in a 0.5 mM solution of the dye in dichloromethane:methanol:H_2O (65:20:2), at room temperature in the dark. The excess dye was removed by rinsing the photoanodes with the same solvent as that employed for the dye solution.

Each counter-electrode consisted of an FTO-glass plate (area: 2 cm × 2 cm) in which a hole (1.0 mm diameter) was drilled. The perforated substrates were washed and cleaned with water and ethanol to remove any residual glass powder and organic contaminants. The transparent Pt catalyst (PT1, Greatcell Solar) was deposited on the conductive face of the FTO-glass by doctor blade: one edge of the glass plate was covered with a strip of an adhesive tape (3 M Magic) both to control the thickness of the film and to mask an electric contact strip. The Pt paste was spread uniformly on the substrate by sliding a glass rod along the tape spacer. The adhesive tape strip was removed, and the glasses heated at 550 °C for 30 min. The photoanode and the Pt counter-electrode were assembled into a sandwich type arrangement and sealed (using a thermopress) with a hot melt gasket made of Surlyn ionomer (Meltonix 1170-25, Solaronix SA, Aubonne, Switzerland). The electrolyte was prepared by dissolving the redox couple, I^-/I_3^- (0.8 M LiI and 0.05 M I_2), in an acetonitrile/valeronitrile (85:15, % v/v) mixture. The electrolyte was introduced into the cell via backfilling under vacuum through the hole drilled in the back of the cathode. Finally, the hole was sealed with adhesive tape.

For each compound, at least two cells were assembled under the same conditions, and the efficiencies were measured 5 times for each cell resulting in a minimum of 10 measurements per compound.

Current-Voltage curves were recorded with a digital Keithley SourceMeter multimeter (PVIV-1A) (Newport, M. T. Brandão, Porto, Portugal) connected to a PC. Simulated sunlight irradiation was provided by an Oriel solar simulator (Model LCS-100 Small Area Sol1A, 300 W Xe Arc lamp equipped with AM 1.5 filter, 100 mW/cm^2) (Newport, M. T. Brandão). The thickness of the oxide film deposited on the photoanodes was measured using an Alpha-Step D600 Stylus Profiler (KLA-Tencor, Milpitas, CA, USA).

4. Conclusions

In the current work, five new 3-ethynylaryldimethoxycoumarin-based chromophores were synthesized by previously reported methods, with the aim of investigating not only the effect of varying the heterocyclic π-bridge on the device performance, but also assess the difference in properties between the 6,7- and 5,7-substitution patterns on the coumarin moiety. Through absorption and fluorescence emission spectroscopy, both in solution and adsorbed onto porous TiO_2, it was determined that dyes **8** and **9b** present the most redshifted absorption, highest absorptivity, and most pronounced Stokes shift, which help explain their high photocurrent values. Additionally, the difference in the theoretically calculated HOMO orbitals of dyes **9a** and **9b** further support the more electron-donating nature of the 5,7-substitution pattern. Despite this, this set of dyes demonstrates comparatively lower conversion efficiencies than other coumarin dyes [26–32], which can be attributed to the superior donor ability of nitrogen-based donors there used. Overall, we can see that the increased planarity of the thieno [3,2-*b*] thiophene π-bridge (dye **8**) and the superior donor ability of the 5,7-disubstituted coumarin are promising avenues for the synthesis of more efficient coumarin-based donors.

Supplementary Materials: The following are available online, the detailed synthetic procedures, the NMR and HRMS spectral data and the full differential pulse voltammograms of dyes **8–11**.

Author Contributions: Conceptualization, P.S.B. and A.J.P.; methodology, P.S.B., A.J.P., J.C.L. and J.P.; validation, P.S.B., J.C.L., A.J.P. and J.P.; formal analysis, A.L.P., J.S., G.M. and E.G.R.; investigation, J.S., G.M., A.L.P. and E.G.R.; resources, P.S.B., A.J.P., J.C.L. and J.P.; data curation, P.S.B., A.J.P., J.C.L. and J.P.; writing—original draft preparation, J.S., P.S.B., J.P.; writing—review and editing, P.S.B., A.J.P., J.C.L. and J.P.; visualization, J.S., A.L.P. and J.P.; supervision, P.S.B., J.C.L., A.J.P. and J.P.; project administration, P.S.B.; funding acquisition, P.S.B. All authors have read and agreed to the published version of the manuscript.

Funding: This work was performed under the projects PTDC/QUI-QOR/7450/2020 "Organic Redox Mediators for Energy Conversion" through FCT—Fundação para a Ciência e a Tecnologia I. P. and POCI-01-0145-FEDER-016387 "SunStorage—Harvesting and storage of solar energy", funded by European Regional Development Fund (ERDF), through COMPETE 2020—Operational Programme for Competitiveness and Internationalisation (OPCI). This work was also supported by the Associate Laboratory for Green Chemistry—LAQV which is financed by national funds from FCT/MCTES (UIDB/50006/2020 and UIDP/50006/2020). FCT/MCTES is also acknowledged for the National NMR Facility (RECI/BBB-BQB/0230/2012 and RECI/BBB-BEP/0124/2012,) and PhD grants 2020.09047.BD (J.S.), PD/BD/135087/2017 (A.L.P.) and PD/BD/145324/2019/ (G.M.).

Institutional Review Board Statement: Not applicable.

Informed Consent Statement: Not applicable.

Data Availability Statement: The data provided in this study is available in the article and Supplementary Material file submitted.

Acknowledgments: The authors acknowledge Hugo Cruz for his support with CV and DPV measurements. The authors acknowledge as well their respective, current institutions of affiliation: Universidade NOVA de Lisboa and University of Coimbra.

Conflicts of Interest: The authors declare no conflict of interest.

Sample Availability: Samples of the compounds are not available from the authors.

References

1. O'Regan, B.; Gratzel, M. A Low-Cost, High-Efficiency Solar-Cell Based on Dye-Sensitized Colloidal TiO_2 Films. *Nature* **1991**, *353*, 737–740. [CrossRef]
2. Selvaraj, P.; Ghosh, A.; Mallick, T.K.; Sundaram, S. Investigation of semi-transparent dye-sensitized solar cells for fenestration integration. *Renew. Energy* **2019**, *141*, 516–525. [CrossRef]
3. Kim, J.H.; Han, S.H. Energy Generation Performance of Window-Type Dye-Sensitized Solar Cells by Color and Transmittance. *Sustainability* **2020**, *12*, 8961. [CrossRef]

4. Devadiga, D.; Selvakumar, M.; Shetty, P.; Santosh, M.S. Dye-Sensitized Solar Cell for Indoor Applications: A Mini-Review. *J. Electron. Mater.* **2021**. [CrossRef]
5. Ghosh, A.; Selvaraj, P.; Sundaram, S.; Mallick, T.K. The colour rendering index and correlated colour temperature of dye-sensitized solar cell for adaptive glazing application. *Sol. Energy* **2018**, *163*, 537–544. [CrossRef]
6. Nazeeruddin, M.K.; Kay, A.; Rodicio, I.; Humphrybaker, R.; Muller, E.; Liska, P.; Vlachopoulos, N.; Gratzel, M. Conversion of light to eletricity by *cis*-X_2bis(2,2′-bipyridyl-4,4′-dicarboxylate)ruthenium(II) charge-transfer sensitizers (X = Cl^-, Br^-, I^-, CN^-, and SCN^-) on nanocrystalline TiO_2 electrodes. *J. Am. Chem. Soc.* **1993**, *115*, 6382–6390. [CrossRef]
7. Nazeeruddin, M.K.; Zakeeruddin, S.M.; Humphry-Baker, R.; Jirousek, M.; Liska, P.; Vlachopoulos, N.; Shklover, V.; Fischer, C.H.; Gratzel, M. Acid-base equilibria of (2,2′-bipyridyl-4,4′-dicarboxylic acid)ruthenium(II) complexes and the effect of protonation on charge-transfer sensitization of nanocrystalline titania. *Inorg. Chem.* **1999**, *38*, 6298–6305. [CrossRef]
8. Aghazada, S.; Gao, P.; Yella, A.; Marotta, G.; Moehl, T.; Teuscher, J.; Moser, J.-E.; De Angelis, F.; Grätzel, M.; Nazeeruddin, M.K. Ligand Engineering for the Efficient Dye-Sensitized Solar Cells with Ruthenium Sensitizers and Cobalt Electrolytes. *Inorg. Chem.* **2016**, *55*, 6653–6659. [CrossRef]
9. Punitharasu, V.; Kavungathodi, M.F.M.; Singh, A.K.; Nithyanandhan, J. pi-Extended cis-Configured Unsymmetrical Squaraine Dyes for Dye-Sensitized Solar Cells: Panchromatic Response. *ACS Appl. Energ. Mater.* **2019**, *2*, 8464–8472. [CrossRef]
10. Dai, P.P.; Zhu, Y.Z.; Liu, Q.L.; Yan, Y.Q.; Zheng, J.Y. Novel indeno 1,2-b indole-spirofluorene donor block for efficient sensitizers in dye-sensitized solar cells. *Dyes Pigment.* **2020**, *175*, 7. [CrossRef]
11. Li, S.Z.; He, J.W.; Jiang, H.Y.; Mei, S.; Hu, Z.G.; Kong, X.F.; Yang, M.; Wu, Y.Z.; Zhang, S.H.; Tan, H.J. Comparative Studies on the Structure-Performance Relationships of Phenothiazine-Based Organic Dyes for Dye-Sensitized Solar Cells. *ACS Omega* **2021**, *6*, 6817–6823. [CrossRef] [PubMed]
12. Yang, J.B.; Ganesan, P.; Teuscher, J.; Moehl, T.; Kim, Y.J.; Yi, C.Y.; Comte, P.; Pei, K.; Holcombe, T.W.; Nazeeruddin, M.K.; et al. Influence of the Donor Size in D-pi-A Organic Dyes for Dye-Sensitized Solar Cells. *J. Am. Chem. Soc.* **2014**, *136*, 5722–5730. [CrossRef] [PubMed]
13. Zhou, N.J.; Prabakaran, K.; Lee, B.; Chang, S.H.; Harutyunyan, B.; Guo, P.J.; Butler, M.R.; Timalsina, A.; Bedzyk, M.J.; Ratner, M.A.; et al. Metal-Free Tetrathienoacene Sensitizers for High-Performance Dye-Sensitized Solar Cells. *J. Am. Chem. Soc.* **2015**, *137*, 4414–4423. [CrossRef] [PubMed]
14. Ren, Y.M.; Sun, D.Y.; Cao, Y.M.; Tsao, H.N.; Yuan, Y.; Zakeeruddin, S.M.; Wang, P.; Gratzel, M. A Stable Blue Photosensitizer for Color Palette of Dye-Sensitized Solar Cells Reaching 12.6% Efficiency. *J. Am. Chem. Soc.* **2018**, *140*, 2405–2408. [CrossRef] [PubMed]
15. Ji, J.M.; Zhou, H.R.; Eom, Y.K.; Kim, C.H.; Kim, H.K. 14.2% Efficiency Dye-Sensitized Solar Cells by Co-sensitizing Novel Thieno 3,2-b indole-Based Organic Dyes with a Promising Porphyrin Sensitizer. *Adv. Energy Mater.* **2020**, *10*, 12. [CrossRef]
16. Kakiage, K.; Aoyama, Y.; Yano, T.; Oya, K.; Fujisawa, J.; Hanaya, M. Highly-efficient dye-sensitized solar cells with collaborative sensitization by silyl-anchor and carboxy-anchor dyes. *Chem. Commun.* **2015**, *51*, 15894–15897. [CrossRef]
17. Zehra, S.; Khan, R.A.; Alsalme, A.; Tabassum, S. Coumarin Derived "Turn on" Fluorescent Sensor for Selective Detection of Cadmium (II) Ion: Spectroscopic Studies and Validation of Sensing Mechanism by DFT Calculations. *J. Fluoresc.* **2019**, *29*, 1029–1037. [CrossRef]
18. Li, H.Q.; Sun, X.Q.; Zheng, T.; Xu, Z.X.; Song, Y.X.; Gu, X.H. Coumarin-based multifunctional chemosensor for arginine/lysine and Cu^{2+}/Al^{3+} ions and its Cu^{2+} complex as colorimetric and fluorescent sensor for biothiols. *Sens. Actuator B-Chem.* **2019**, *279*, 400–409. [CrossRef]
19. Li, X.Q.; Huo, F.J.; Yue, Y.K.; Zhang, Y.B.; Yin, C.X. A coumarin-based "off-on" sensor for fluorescence selectivily discriminating GSH from Cys/Hcy and its bioimaging in living cells. *Sens. Actuator B-Chem.* **2017**, *253*, 42–49. [CrossRef]
20. Peng, S.; Zhao, Y.H.; Fu, C.X.; Pu, X.M.; Zhou, L.; Huang, Y.; Lu, Z.Y. Acquiring High-Performance Deep-Blue OLED Emitters through an Unexpected Blueshift Color-Tuning Effect Induced by Electron-Donating-OMe Substituents. *Chem-Eur. J.* **2018**, *24*, 8056–8060. [CrossRef] [PubMed]
21. Kim, S.; Lee, K.J.; Kim, B.; Lee, J.; Kay, K.Y.; Park, J. New Ambipolar Blue Emitting Materials Based on Amino Coumarin Derivatives with High Efficiency for Organic Lightemitting Diodes. *J. Nanosci. Nanotechnol.* **2013**, *13*, 8020–8024. [CrossRef]
22. Zhang, H.; Liu, X.C.; Gong, Y.X.; Yu, T.Z.; Zhao, Y.L. Synthesis and characterization of SFX-based coumarin derivatives for OLEDs. *Dyes Pigment.* **2021**, *185*, 6. [CrossRef]
23. Esnal, I.; Duran-Sampedro, G.; Agarrabeitia, A.R.; Banuelos, J.; Garcia-Moreno, I.; Macias, M.A.; Pena-Cabrera, E.; Lopez-Arbeloa, I.; de la Moya, S.; Ortiz, M.J. Coumarin-BODIPY hybrids by heteroatom linkage: Versatile, tunable and photostable dye lasers for UV irradiation. *Phys. Chem. Chem. Phys.* **2015**, *17*, 8239–8247. [CrossRef] [PubMed]
24. Wang, C.; Feng, C.; Lu, Z.Z. Tunable co-doped dye laser of coumarin 440 and coumarin 460. *Laser Phys.* **2021**, *31*, 3. [CrossRef]
25. Liu, X.G.; Cole, J.M.; Waddell, P.G.; Lin, T.C.; Radia, J.; Zeidler, A. Molecular Origins of Optoelectronic Properties in Coumarin Dyes: Toward Designer Solar Cell and Laser Applications. *J. Phys. Chem. A* **2012**, *116*, 727–737. [CrossRef]
26. Hara, K.; Wang, Z.S.; Sato, T.; Furube, A.; Katoh, R.; Sugihara, H.; Dan-Oh, Y.; Kasada, C.; Shinpo, A.; Suga, S. Oligothiophene-containing coumarin dyes for efficient dye-sensitized solar cells. *J. Phys. Chem. B* **2005**, *109*, 15476–15482. [CrossRef] [PubMed]
27. Wang, Z.S.; Cui, Y.; Hara, K.; Dan-Oh, Y.; Kasada, C.; Shinpo, A. A high-light-harvesting-efficiency coumarin dye for stable dye-sensitized solar cells. *Adv. Mater.* **2007**, *19*, 1138–1141. [CrossRef]

28. Wang, Z.S.; Cui, Y.; Dan-Oh, Y.; Kasada, C.; Shinpo, A.; Hara, K. Molecular Design of Coumarin Dyes for Stable and Efficient Organic Dye-Sensitized Solar Cells. *J. Phys. Chem. C* **2008**, *112*, 17011–17017. [CrossRef]
29. Wang, Z.S.; Cui, Y.; Dan-Oh, Y.; Kasada, C.; Shinpo, A.; Hara, K. Thiophene-functionalized coumarin dye for efficient dye-sensitized solar cells: Electron lifetime improved by coadsorption of deoxycholic acid. *J. Phys. Chem. C* **2007**, *111*, 7224–7230. [CrossRef]
30. Jiang, S.L.; Chen, Y.Q.; Li, Y.J.; Han, L. Novel D-D-pi-A indoline-linked coumarin sensitizers for dye-sensitized solar cells. *J. Photochem. Photobiol. A-Chem.* **2019**, *384*, 7. [CrossRef]
31. He, J.; Liu, Y.; Gao, J.R.; Han, L. New D-D-pi-A triphenylamine-coumarin sensitizers for dye-sensitized solar cells. *Photochem. Photobiol. Sci.* **2017**, *16*, 1049–1056. [CrossRef]
32. Vekariya, R.L.; Vaghasiya, J.V.; Dhar, A. Coumarin based sensitizers with ortho-halides substituted phenylene spacer for dye sensitized solar cells. *Org. Electron.* **2017**, *48*, 291–297. [CrossRef]
33. Martins, S.; Avo, J.; Lima, J.; Nogueira, J.; Andrade, L.; Mendes, A.; Pereira, A.; Branco, P.S. Styryl and phenylethynyl based coumarin chromophores for dye sensitized solar cells. *J. Photochem. Photobiol. A* **2018**, *353*, 564–569. [CrossRef]
34. Teng, C.; Yang, X.C.; Yang, C.; Tian, H.N.; Li, S.F.; Wang, X.N.; Hagfeldt, A.; Sun, L.C. Influence of Triple Bonds as pi-Spacer Units in Metal-Free Organic Dyes for Dye-Sensitized Solar Cells. *J. Phys. Chem. C* **2010**, *114*, 11305–11313. [CrossRef]
35. Gordo, J.; Avo, J.; Parola, A.J.; Lima, J.C.; Pereira, A.; Branco, P.S. Convenient Synthesis of 3-Vinyl and 3-Styryl Coumarins. *Org. Lett.* **2011**, *13*, 5112–5115. [CrossRef] [PubMed]
36. Martins, S.M.A.; Branco, P.C.S.; Pereira, A. An Efficient Methodology for the Synthesis of 3-Styryl Coumarins. *J. Braz. Chem. Soc.* **2012**, *23*, 688–693. [CrossRef]
37. Leao, R.A.C.; de Moraes, P.D.; Pedro, M.; Costa, P.R.R. Synthesis of Coumarins and Neoflavones through Zinc Chloride Catalyzed Hydroarylation of Acetylenic Esters with Phenols. *Synthesis* **2011**, *2011*, 3692–3696. [CrossRef]
38. Su, J.L.; Zhang, Y.; Chen, M.R.; Li, W.M.; Qin, X.W.; Xie, Y.P.; Qin, L.X.; Huang, S.H.; Zhang, M. A Copper Halide Promoted Regioselective Halogenation of Coumarins Using N-Halosuccinimide as Halide Source. *Synlett* **2019**, *30*, 630–634. [CrossRef]
39. Wang, L.; Zhao, K.; Yao, F.; Xiong, J.; Tian, H.; Han, J. Fluorescent Nucleoside and Preparation Method Thereof. 2017. Available online: https://patents.google.com/patent/CN107325141A/en (accessed on 2 May 2019).
40. Icli, M.; Pamuk, M.; Algi, F.; Onal, A.M.; Cihaner, A. Donor-Acceptor Polymer Electrochromes with Tunable Colors and Performance. *Chem. Mat.* **2010**, *22*, 4034–4044. [CrossRef]
41. Pina, J.; de Castro, C.S.; Delgado-Pinar, E.; Sérgio Seixas de Melo, J. Characterization of 4-methylesculetin and of its mono- and di-methoxylated derivatives in water and organic solvents in its ground, singlet and triplet excited states. *J. Mol. Liq.* **2019**, *278*, 616–626. [CrossRef]
42. Donovalová, J.; Cigáň, M.; Stankovičová, H.; Gašpar, J.; Danko, M.; Gáplovský, A.; Hrdlovič, P. Spectral Properties of Substituted Coumarins in Solution and Polymer Matrices. *Molecules* **2012**, *17*, 3259–3276. [CrossRef] [PubMed]
43. Cardona, C.M.; Li, W.; Kaifer, A.E.; Stockdale, D.; Bazan, G.C. Electrochemical Considerations for Determining Absolute Frontier Orbital Energy Levels of Conjugated Polymers for Solar Cell Applications. *Adv. Mater.* **2011**, *23*, 2367–2371. [CrossRef]
44. Holze, R. Optical and Electrochemical Band Gaps in Mono-, Oligo-, and Polymeric Systems: A Critical Reassessment. *Organometallics* **2014**, *33*, 5033–5042. [CrossRef]
45. Moreira, L.M.; de Melo, M.M.; Martins, P.A.; Lyon, J.P.; Romani, A.P.; Codognoto, L.; dos Santos, S.C.; de Oliveira, H.P.M. Photophysical Properties of Coumarin Compounds in Neat and Binary Solvent Mixtures: Evaluation and Correlation Between Solvatochromism and Solvent Polarity Parameters. *J. Braz. Chem. Soc.* **2014**, *25*, 873–881. [CrossRef]
46. Bhagwat, A.A.; Sekar, N. Fluorescent 7-Substituted Coumarin Dyes: Solvatochromism and NLO Studies. *J. Fluoresc.* **2019**, *29*, 121–135. [CrossRef] [PubMed]
47. Almenningen, A.; Bastiansen, O.; Svendsas, P. Electron Diffraction Studies of 2,2′-Dithienyl Vapour. *Acta Chem. Scand.* **1958**, *12*, 1671–1674. [CrossRef]
48. Koumura, N.; Wang, Z.S.; Mori, S.; Miyashita, M.; Suzuki, E.; Hara, K. Alkyl-functionalized organic dyes for efficient molecular photovoltaics. *J. Am. Chem. Soc.* **2006**, *128*, 14256–14257. [CrossRef] [PubMed]
49. Forneli, A.; Planells, M.; Sarmentero, M.A.; Martinez-Ferrero, E.; O'Regan, B.C.; Ballester, P.; Palomares, E. The role of para-alkyl substituents on meso-phenyl porphyrin sensitised TiO_2 solar cells: Control of the e(TiO_2)/electrolyte(+) recombination reaction. *J. Mater. Chem.* **2008**, *18*, 1652–1658. [CrossRef]
50. Kim, S.; Choi, H.; Baik, C.; Song, K.; Kang, S.O.; Ko, J. Synthesis of conjugated organic dyes containing alkyl substituted thiophene for solar cell. *Tetrahedron* **2007**, *63*, 11436–11443. [CrossRef]
51. Bradley, D.; Williams, G.; Lawton, M. Drying of Organic Solvents: Quantitative Evaluation of the Efficiency of Several Desiccants. *J. Org. Chem.* **2010**, *75*, 8351–8354. [CrossRef]
52. Still, W.C.; Kahn, M.; Mitra, A. Rapid chromatographic technique for preparative separations with moderate resolution. *J. Org. Chem.* **1978**, *43*, 2923–2925. [CrossRef]
53. Frisch, M.J.; Trucks, G.W.; Schlegel, H.B.; Scuseria, G.E.; Robb, M.A.; Cheeseman, J.R.; Scalmani, G.; Barone, V.; Mennucci, B.; Petersson, G.A.; et al. *Gaussian 09, Revision A.02*; Gaussian, Inc.: Wallingford, CT, USA, 2009.
54. Becke, A.D. A New Mixing of Hartree-Fock and Local Density-Functional Theories. *J. Chem. Phys.* **1993**, *98*, 1372–1377. [CrossRef]
55. Francl, M.M.; Pietro, W.J.; Hehre, W.J.; Binkley, J.S.; Gordon, M.S.; Defrees, D.J.; Pople, J.A. Self-Consistent Molecular-Orbital Methods. 23. A Polarization-Type Basis Set for 2nd-Row Elements. *J. Chem. Phys.* **1982**, *77*, 3654–3665. [CrossRef]

56. Cammi, R.; Corni, S.; Mennucci, B.; Tomasi, J. Electronic excitation energies of molecules in solution: State specific and linear response methods for nonequilibrium continuum solvation models. *J. Chem. Phys.* **2005**, *122*, 104513. [CrossRef]
57. O'boyle, N.M.; Tenderholt, A.L.; Langner, K.M. cclib: A library for package-independent computational chemistry algorithms. *J. Comput. Chem.* **2008**, *29*, 839–845. [CrossRef]
58. Pinto, A.L.; Oliveira, J.; Araujo, P.; Calogero, G.; de Freitas, V.; Pina, F.; Parola, A.J.; Lima, J.C. Study of the multi-equilibria of red wine colorants pyranoanthocyanins and evaluation of their potential in dye-sensitized solar cells. *Sol. Energy* **2019**, *191*, 100–108. [CrossRef]

Article

Structural Characterization of Mono and Dihydroxylated Umbelliferone Derivatives

Rubén Seoane-Rivero [1,*], Estibaliz Ruiz-Bilbao [2], Rodrigo Navarro [3,4,*], José Manuel Laza [5], José María Cuevas [1], Beñat Artetxe [2], Juan M. Gutiérrez-Zorrilla [2], José Luis Vilas-Vilela [5] and Ángel Marcos-Fernandez [3,4]

1 GAIKER Technology Centre, Basque Research and Technology Alliance (BRTA), Parque Tecnológico de Bizkaia, edificio 202, E-48170 Zamudio, Spain; cuevas@gaiker.es

2 Departamento de Química Inorgánica, University of the Basque Country UPV/EHU, Apartado 644, 48080 Bilbao, Spain; estibaliz.ruiz@ehu.eus (E.R.-B.); benat.artetxe@ehu.eus (B.A.); juanma.zorrilla@ehu.eus (J.M.G.-Z.)

3 Instituto de Ciencia y Tecnología de Polímeros (CSIC), Juan de la Cierva 3, 28006 Madrid, Spain; amarcos@ictp.csic.es

4 Interdisciplinary Platform for "Sustainable Plastics towards a Circular Economy" (SUSPLAST-CSIC), 28006 Madrid, Spain

5 Departamento de Química Física, University of the Basque Country UPV/EHU, Apartado 644, 48080 Bilbao, Spain; josemanuel.laza@ehu.eus (J.M.L.); joseluis.vilas@ehu.eus (J.L.V.-V.)

* Correspondence: seoane@gaiker.es (R.S.-R.); rnavarro@ictp.csic.es (R.N.)

Academic Editor: Maria João Matos

Received: 23 June 2020; Accepted: 28 July 2020; Published: 31 July 2020

Abstract: Coumarin derivatives are a class of compounds with a pronounced wide range of applications, especially in biological activities, in the medicine, pharmacology, cosmetics, coatings and food industry. Their potential applications are highly dependent on the nature of the substituents attached to their nucleus. These substituents modulate their photochemical and photophysical properties, as well as their interactions in their crystalline form, which largely determines the final field of application. Therefore, in this work a series of mono and dihydroxylated coumarin derivatives with different chemical substituents were synthesized and characterized by UV-Visible spectroscopy, thermal analysis (differential scanning calorimetry (DSC) and TGA), ^{1}H NMR and X-Ray Diffraction to identify limitations and possibilities as a function of the molecular structure for expanding their applications in polymer science.

Keywords: coumarin; hydroxyl-modified coumarin; photophysical; thermal and structural characterization

1. Introduction

Coumarins (chromen-2-ones) are a family of benzopyrones widely distributed in nature. Since 1902, when Ciamician and Silber found that coumarin had the ability to be photoreactive, the photo-cyclodimerization and photo-cleavage of this product has received a lot of attention by several investigation groups [1–4]. Their structure are a class of lactones based on a benzene ring fused to α-pyrone ring, as can be seen in Scheme 1A [5,6]. Coumarins represent an important family of naturally occurring and/or synthetic oxygen-containing heterocycles, bearing a typical benzopyrone framework. One of the most important characteristics of coumarin derivatives is that they can undergo reversible photo-responsible reactions; depending on the type of irradiated wavelength, these moieties can yield a cyclobutane through dimerization or they can cleavage, reforming the double bond C=C. Thus, when irradiated at >300 nm, a [2 + 2] cycloaddition reaction takes place, forming a cyclobutane ring; in contrast, irradiating at 254 nm a photo-scission reaction leads to the original coumarin structures (Scheme 1B) [7–10].

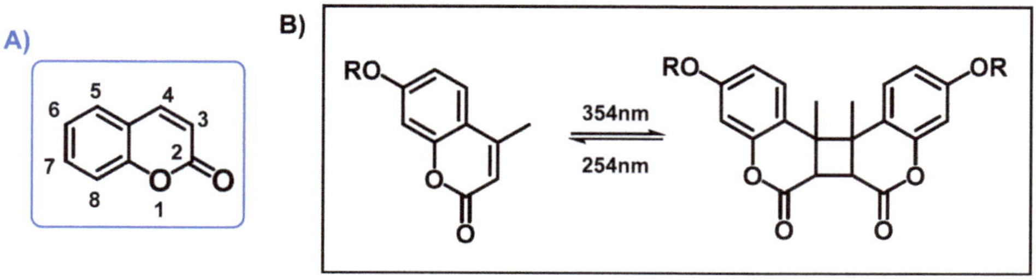

Scheme 1. (**A**) Coumarin core structure and (**B**) photo-reversibility of coumarin moieties.

Umbelliferone is a benzopyrone (also known as 7-hydroxycoumarin, hydrangine, skimmetine, and beta-umbelliferone) and belongs to the Coumarin family which is commonly found in plants [11]. The word "Umbelliferone" was originated from the plant which belongs to the Umbelliferae family. It includes significant herbs such as celery, carrot, garden angelica, sanicle, parsley, cumin, alexanders, big leaf hydrangea, fennel, asafoetida, Justicia pectoralis and giant hogweed. The phenolic coumarins, which are derived from plants, have been supposed to play a vital role in our daily life, due to their antioxidant property and are taken in the human diet in the form of vegetables and fruit [12].

Under acidic conditions and low temperatures, highly activated phenols, such as resorcinol and β-carbonyl ester, easily yield the desired coumarins; this synthetic approach is based on the biosynthesis of umbelliferone (Scheme 2).

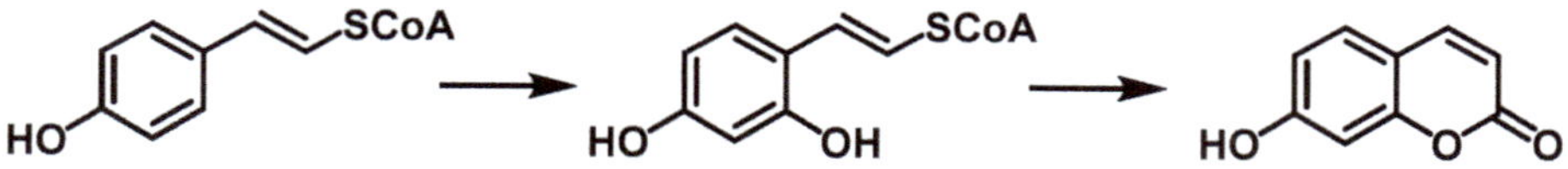

Scheme 2. Biosynthetic scheme of umbelliferone.

Moreover, coumarin and its derivatives are small molecular weight compounds that have demonstrated very interesting physical, chemical and pharmacological properties with broad applicability as biochemicals (drugs, cosmetics, dyes, antibacterials, etc.) [13]. The diverse oriented synthetic routes have led to very different derivatives with usefulness not only as biologically active agents [14–17] or optical materials [18–21], but in macromolecular chemistry they also can be seen as very diverse polymer backbones with photoreactive properties [7,22–24]. Owing to the strong demand imposed by various government organizations for the design of polymeric sustainable materials in a more circular economic model, nowadays coumarin derivatives have acquired relevant importance in this field.

To elicit photoactive polymers with self-healing properties, different coumarin derivatives have been previously incorporated in various polymeric backbones [25–28]. In the particular case of polyurethanes, some research groups have recently introduced different hydroxylated derivatives either within the hard segment (chain end or chain extender) or within the soft segment (coumarin functionalized polycaprolactone diols) [2,29–35]. More recently, we reported an outstanding three times increment increase in the tensile strength of polyurethanes with difunctional hydroxy-coumarins, which led to new irradiated polyurethanes having mechanical properties superior to any coumarin containing materials described in literature [35]. Motivated by this issue, other hydroxylated coumarins, with a high structural analogy, were also introduced into the polyurethane matrices [2,33]. Comparing the experimental data derived from those works, uneven behaviors were observed, for instance, some coumarin hydroxyl-derivatives could not be introduced into the soft segment or the photo-dimerization yields varied considerably, to name a few.

Owing to the coumarin contents in the polyurethane formulations being too low, a systematic study has been focused on the isolated coumarins. Therefore, in the present work those isolated hydroxy-coumarin compounds with different functionalities and chemical structure were synthesized and exhaustively characterized to shed light and understanding on those differences. Despite the structural similarity and considering the variability of substituents in the C7 position of the coumarins, an understanding at the atomistic level of the electronic interactions that determine the spectroscopic properties of coumarins has also been carried out. Single-crystal X-ray diffraction analyses revealed the key role of weak supramolecular forces in the self-assembly of molecular species within the crystal packing. The oxygen-rich molecular structures allow O–H···O hydrogen bonds and C–H···O type contacts to be established, which is demonstrated in their capacity to undergo edge-to-edge self-association [36]. Moreover, the presence of two fused aromatic rings in the coumarin structure contributes, with π-π stacking interactions, to the robustness of the system. These structural studies were essential to relate their different arrangements with the thermal behavior observed in DSC analyses.

2. Results and Discussion

2.1. Synthetic Approach

The chemical structures of the monohydroxy and dihydroxy-derived coumarins have been collected in Figure 1.

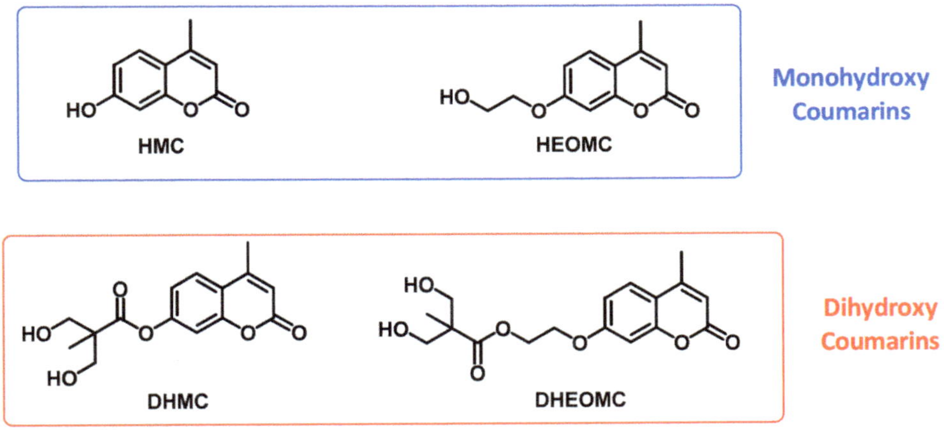

Figure 1. Chemical structures of functionalized coumarins bearing hydroxyl groups.

All studied coumarins studied are based on the umbelliferone core. Firstly, the HMC product was prepared and later different chemical reactions (etherification or esterification) were carried out on this product to achieve the rest of presented coumarins. These reactions were focused on the lateral group located at C7 carbon, which was varied to assess its impact on the performance and features of these photo-reactive systems. In essence, two types of coumarins have been proposed, monohydroxy and dihydroxy, these hydroxyl-functional groups could, in turn, be directly bound to the umbelliferone ring or separated by a spacer. The synthetic routes used for the preparation of these coumarin derivatives have been included in the Supporting Information. Likewise, a complete spectroscopic characterization of each coumarin has also been included in the Supporting Information. Additionally, it is important to note that the synthetic procedures were simple, easily scalable and yields were good in most cases. Furthermore, due to the absence of post-purification processes, this set of features are key elements for the application of these coumarins in Industry. Indeed, the quality of the obtained products following these synthetic routes were high enough in order to incorporate them directly into the polyurethane formulations, leading to polymer coatings with interesting performances [33].

On the other hand, despite the high purity of these products during their synthesis, additional recrystallization steps were required for structural analysis by X-ray diffraction. Crystals suitable were obtained by dissolving the final products in their corresponding hot solvents (ca. 90 °C) and leaving the resulting solutions to slowly evaporate in an open container. First attempts were carried out by using the solvents of the synthetic procedure, but some of the cases did not yield crystals of enough quality. Thus, mixtures of solvents with different polarities were employed for the recrystallization. The longer the aliphatic chain of the substituent, the lower the polarity of the molecule. Due to this fact, coumarins with shorter substituents (HMC and DHMC) were found to crystallize better in solvents with higher polarity than coumarins with longer substituents (HEOMC and DHEOMC). That is, single crystals of HMC were obtained in the most polar solvent mixture, EtOAc:EtOH (3:1), whereas those of DHEOMC were isolated from the most nonpolar solvent mixture, EtOAc:diethyl ether (1:1). The remaining coumarins were easily recrystallized from EtOAc.

2.2. UV Experiments

Owing to strong absorption of UV light shown by coumarins, in the first set of experiments, the UV spectra of prepared coumarins in aqueous solution were recorded for the first time.

In Figure 2, the absorption UV-spectra of pristine coumarin compounds (without irradiation) are shown. The concentration of these solutions ranged from 0.2 to 0.4 mM. In all cases, absorption of coumarins showed a π-π^* transition between 260 and 300 nm attributed to electrons of the conjugated benzene nucleus and another π-π^* transition between 310–340 nm assigned to pyrone nucleus [30]. Only in the case of DHMC, were these two transitions completely distinguishable, for the rest of coumarins the transition of the benzene ring appeared as a shoulder on the heterocycle transition band. However, the pyrone-associated transition is much more severely affected by UV radiation, while the transition of the benzene ring is practically unaltered.

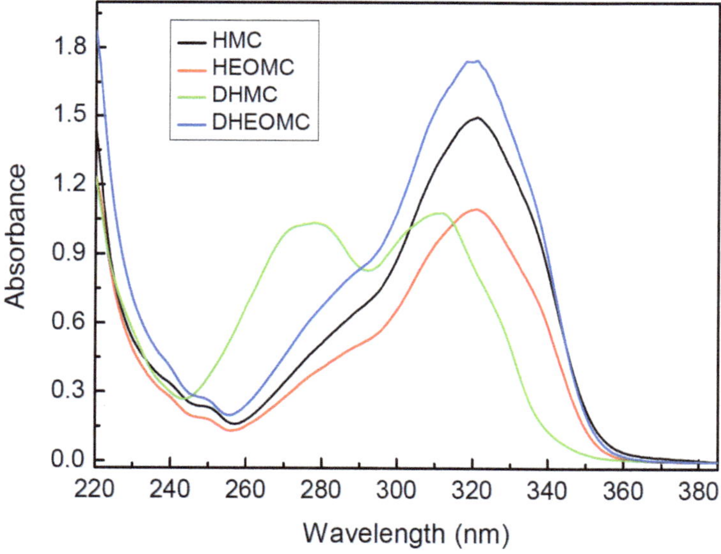

Figure 2. Absorption spectra of the coumarin compounds.

It is well known that UV-radiation markedly affects the reversible dimerization process of coumarin (Scheme 1). Thus, depending on the wavelength used as the source of excitation, this equilibrium shifts to one side or the other. When the aqueous solutions were irradiated with a set of five lamps of 354 nm, photo-dimerization reaction was performed, however, with a set of five lamps of 254 nm, photo-cleavage reaction was induced.

The absorption UV-Vis spectra of HEOMC at 354 nm (A) and 250 nm (B) are depicted in Figure 3. In the photo-dimerization reaction (Figure 3A), the maximum transition at 320 nm gradually decreased in intensity with the irradiation time. As this band was associated with the heterocycle ring, the double bond of the pyrone core progressively disappeared, leading to the formation of a cyclobutene ring by cycloaddition [2 + 2]. In contrast, during photo-cleavage (Figure 3B), the double bond was restored and the absorption peak with the maximum recovered. Additionally, as the intensity of the band at 320 nm gradually decreased, the transition of the benzene ring was easy to detect, because it remained invariant with UV radiation.

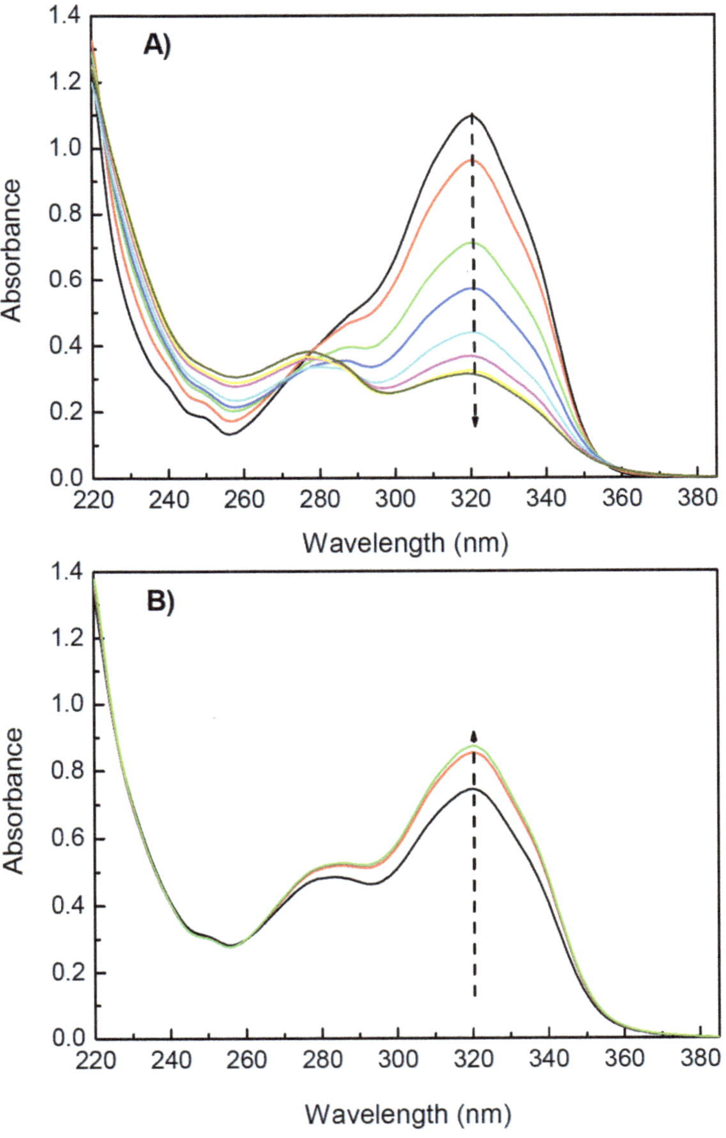

Figure 3. Photo-dimerization (**A**) and photo-cleavage (**B**) spectra of HEOMC derivative.

The photoreactivity feature of characterized solutions can be quantitatively described by the time dependence of the maximum peak height [30]. Photo-dimerization degree was estimated from Equation (1), where A_t shows the absorbance at maximum peak at time t, and A_o the original absorbance at maximum wavelength prior to 354 nm exposure. When aqueous solutions were exposed to 254 nm, in order to characterize the recovery percentage, it used Equation (2); where A_∞ denotes the absorbance after the solution exposed to 254 nm, A_∞ shows the minimum absorbance at maximum peak after exposure to 350 nm UV light, and A_o has the same meaning as that in Equation (1).

$$\%\text{ dimerization degree} = \left(1 - \frac{A_t}{A_0}\right) \times 100 \tag{1}$$

$$\%\text{photocleavage degree} = \left(\frac{A'_\infty - A_t}{A_0 - A_\infty}\right) \times 100 \tag{2}$$

Significant differences in photo-dimerization and photo-cleavage were found in the first cycle of irradiation between studied coumarins. In Figure 4, the variation of the UV-absorbance with irradiation time of the four coumarin derivatives is shown.

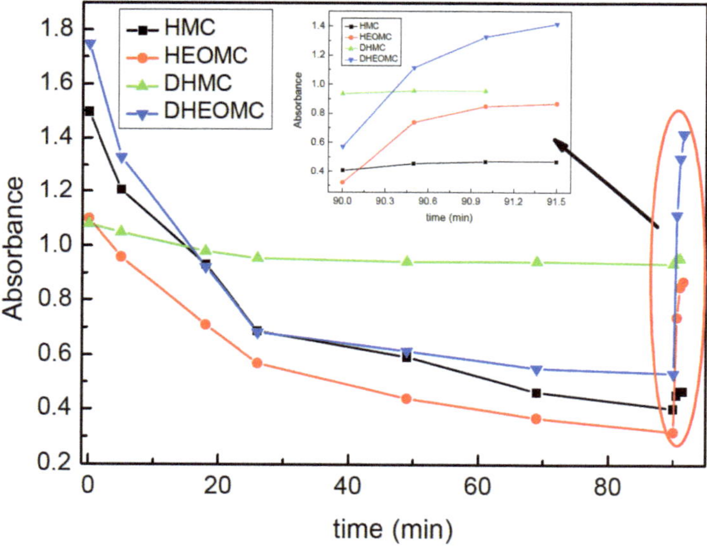

Figure 4. Photo-reversibility cycle of coumarin derivatives.

With respect to photo-dimerization, the DHMC coumarin product denoted a low dimerization degree: 13%. This could be based on the UV-spectrum of this derivative. As shown in Figure 2, the UV-spectrum of DHMC presented its absorption bands to higher wavelengths compared with its counterparts. Therefore, with an irradiation at 354 nm, the absorption of DHMC would be very weak and its dimerization would be hampered. Hence, to increase its dimerization degree, it would be necessary to irradiate at 313 nm (closer to the maximum peak of DHMC), but Seoane et al. demonstrated that irradiation at 313 nm gave a very strong irreversibility respect to photo-cleavage [2]. Optimum photoreversibility was achieved when irradiation was carried out with 354 and 254 nm sets of lamps.

Regarding the other molecules studied (HMC, HEOMC, DHEOMC), the irradiation at 354 nm led to photo-dimerization yields of about 70%. Subsequently, only the HEOMC and DHEOMC derivatives largely recovered the initial absorbance by irradiation at 254 nm. In contrast, the HMC dimer was not able to significantly cleave in aqueous solution, and the absorbance values of this solution practically did not vary with irradiation at 254 nm.

It is important to note that the HMC molecule has its functional group linked to the aromatic ring of coumarin, and this could be one reason to argue this behavior. With regard to how to improve the photoreactivity, there are some studies that denote that the addition of substituents to the coumarin dimer can improve the cleavage reaction efficiency, but Jiang et al. explained that it is not clear how the substituents modify the cleavage dynamics, or why they generally lead to enhanced efficiencies compared to the unsubstituted coumarin dimer [6].

One of the most important properties of coumarin derivatives is the photo-reversibility. This feature has been able to be studied in HEOMC and DHEOMC derivatives. In Table 1, the photo-dimerization and photo-scission yields for each cycle have been collected. Monohydroxy coumarin HEOMC progressively lost its photo-reversible capacity after the end of each cycle; in fact, after the third dimerization cycle (354 nm), only 30% of its coumarins had dimerized. In contrast, the dimers of dihydroxylated coumarin (DHEOMC) were easily cleaved after being irradiated with light at 254 nm. Although, its dimerization capacity was also depressed after each dimerization cycle, but the decrease was less pronounced than its monohydroxylated counterpart.

Table 1. Dimerization and cleavage conversions of coumarin derivatives.

Coumarin Derivatives	Dimerization (1st Cycle)	Recovery (1st Cycle)	Dimerization (2nd Cycle)	Recovery (2nd Cycle)	Dimerization (3th Cycle)
HMC	73	-	-	-	-
HEOMC	70	70	54	40	31
DHMC	13	-	-	-	-
DHEOMC	66	72	40	63	55

photo-dimerization (irradiation at 354 nm) and photo-cleavage (irradiation at 254 nm) for the HEOMC derivative is depicted in Figure 5. At the end of the photo-cleavage cycle, the absorbance value was lower than the starting point, so that the coumarin photo-reversibility was not perfect. The same effect was shown after the next cycle of photo-cleavage (end of cycle 2), where the absorbance was still even lower. Our findings suggested that photo-dimerization and photo-cleavage lost efficiency cycle by cycle. Indeed, the dimerization degree was slightly reduced whilst the cleavage dropped substantially as the number of cycles were repeated. Some authors attributed that the decrease in the photo-reversibility of coumarins could be due to the existence of an equilibrium between coumarin and its dimers, as well as the formation of non-cleavable dimers, because the lactone ring had probably opened [37].

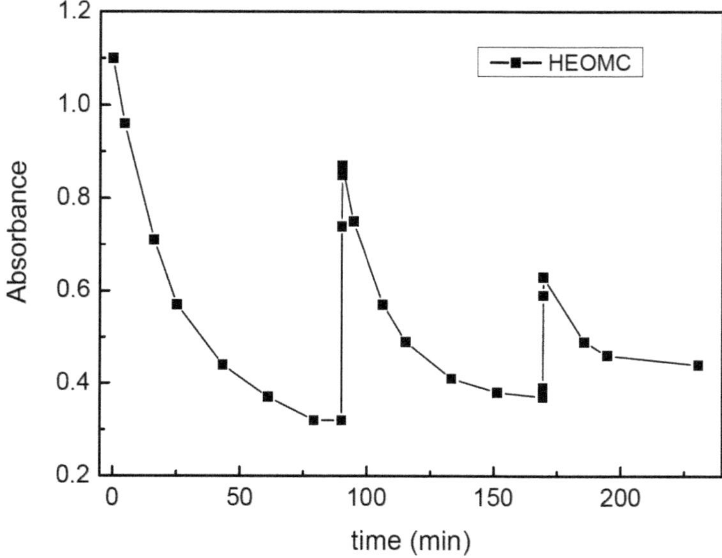

Figure 5. Photo-dimerization and photo-cleavage absorptions of HEOMC derivative cycle by cycle.

2.3. Theory: Absorption UV-Vis Spectra

As studied coumarins have exhibited disparate UV-vis behavior and been previously used as photochemical crosslinking agents to obtain high-performance coatings, the need has been raised to understand the impact of the substituent on the electron densities of coumarins.

Considering that the experimental UV-vis spectra were acquired in aqueous solution, the theoretical model should also include this effect. For this, the polarizable continuum model (PCM) was used, because this approach reproduced the experimental data with high precision at a low computational cost. Hence, we started with the optimized geometries of the ground state in the gas phase, which were then used as starting configurations for a geometric optimization of the molecules in aqueous solution using the PCM. Finally, these optimized settings served to determine the absorption properties of coumarins.

For all electrostatic potential surfaces (Figure S9), the carbonyl group (C=O) of the pyrone is the region with the highest concentration of electrons, making it the area with the lowest potential (in red). Furthermore, the coumarin family bearing two hydroxyl groups (DHMC, DHEOMC) presented a second region of low potential. This region was located in the other ester group and remained almost perpendicular to the coumarin ring. On the other hand, in all the coumarins studied, the regions with high potential correspond to hydroxyl groups, so that they will be the reactive centers during, for example, polymerization reactions.

In general, the maximum absorption for a molecule usually approximates the energy difference between the frontier orbitals (HOMO and LUMO). Eventually, the relative order of the calculated values of the HOMO–LUMO gap follows the same order with respect to the measured absorption peaks (Figure 6). Therefore, this observation could indicate that the dominant transition in UV-Vis spectra corresponds to the HOMO–LUMO transition for the C7-substituted coumarin family. However, in order to improve the description of the maximum UV-Vis absorption peaks, Time-dependent density functional theory (TD-DFT) calculations were performed for all coumarins, keeping the coumarin configuration frozen in the ground state according to the PCM. In this case, the theoretical results within the TD-DFT framework follow the same trend described above, but the maximum absorption peak is closer to the experimental one.

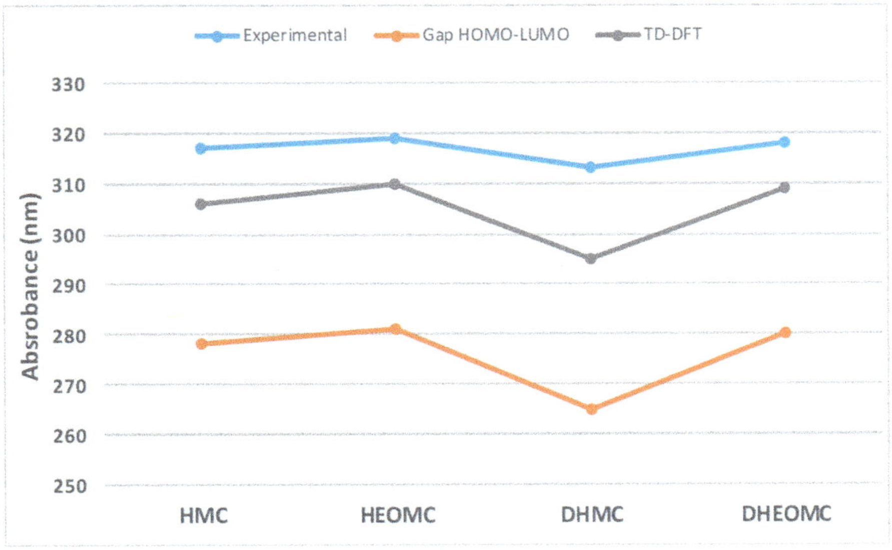

Figure 6. Comparison between experimental UV-Vis absorption peaks and calculated HOMO–LUMO gap at gas-phase and aqueous phase for C7-substituted coumarins.

Additionally, a deeper analysis of the TD-DFT transitions yields very interesting information (Table S1). For example, for all the cases studied, the first excited singlet state has the highest oscillator strength, except for coumarin DHMC, where the second excited singlet state also exhibits a significant contribution from the oscillator strength. As described above, the transition to the first singlet excited state corresponds mainly to the HOMO–LUMO transition. This transition has the same character for all C7-substituted coumarins except for DHMC, which also presents lower contributions than other transitions, such as HOMO-1 -> LUMO (2 -> 1) and HOMO-1 -> LUMO+1. (2 -> 2′).

On the other hand, the second singlet excitation state (S2) presents a very similar nature between the coumarins HMC, HEOMC and DHEOMC. For these cases, the transition is mainly governed by the HOMO-1 -> LUMO (2 -> 1′) transition and to a lesser extent by the HOMO -> LUMO +1 (1 -> 2′) transition. However, for DHMC, additionally in this transition to S2 there is also a slight contribution from the HOMO–LUMO (1 -> 1′) transition. Therefore, DHMC coumarin has a UV-Vis absorption spectrum with more significant differences compared to its counterparts, and the shape of this spectrum determines its behavior against UV radiation.

Through the morphological analysis of the frontier orbitals, which participate in the absorption processes, it should be possible to understand the optical properties of UV light absorption. Figure 7 depicts the frontier orbitals for the four C7-substituted coumarins. The two main frontier orbitals (HOMO and LUMO) of the coumarins studied are mainly extended along the heterocyclic ring defining delocalized π-orbitals. Only in the case of DHMC, is delocalization extended to the ester group located at C7; as for the rest of coumarin counterparts, the extension of the π-orbital is reduced to the C7 oxygen atom. This extension may be due to the aromatic ester nature of DHMC, while for the other ester coumarin (DHEOMC) it has a two-carbon spacer that breaks this conjugation. For the other two main orbitals (HOMO-1 and LUMO+1), they are preferentially concentrated in the benzene ring of coumarins, but only in the specific case of DHMC does the transition between these orbitals have a significant contribution.

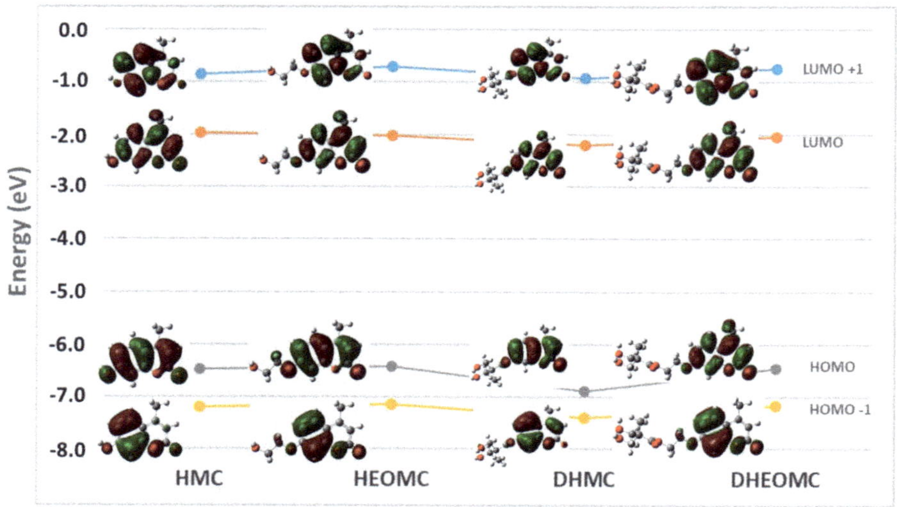

Figure 7. Distribution and schematic representation of the frontier orbitals of the coumarins studied.

2.4. Crystal Structures

Crystallographic data for compounds HMC, HEOMC, DHMC and DHEOMC are compiled in Table 2.

$$^{a}R(F) = \sum \|F_0 - F_c\| / \sum \left|F_0\right|. \; ^{b}wR(F^2) = \left\{ \sum \left[w\left(F_0^2 - F_c^2\right)^2 \right] / \sum \left[w\left(F_0^2\right)^2 \right] \right\}^{1/2}$$

Thermal vibrations of all non-hydrogen atoms were refined anisotropically. Crystals of the DHMC derivative were systematically of much poorer quality than their analogues. Measurements on several crystals were made from several different batches in quest of a better set of crystallographic data but without any success. Although completeness was as low as 91%, and the used restrain/parameter ratio was considerably high, the best structural model was obtained solving the structure in the chiral Pc space group. The two coumarin molecules that form the asymmetric unit of DHMC were found to be involved in crystallographic disorder. The C and O atoms were refined with free population factors, resulting in a chemical occupancy of ca. 0.8 for the main form and 0.2 for the minor one. In order to model the disorder, several restrictions were applied for the thermal ellipsoids of C and O atoms belonging to the minor phase (ISOR). Some C–C and C–O bond lengths were also restricted to 1.54(2) and 1.43(2), respectively (DFIX). In all cases, hydrogen atoms of the organic molecules were placed in calculated sites using standard SHELXL parameters, whereas those from hydration water molecules were located in Fourier maps and restrained to O–H bond lengths of 0.84(2)Å. For DHMC, H atoms belonging to hydroxyl groups were placed in the Fourier map and bond lengths and angles were restricted using DFIX and DANG commands.

Table 2. Crystallographic data for HMC, HEOMC, DHMC and DHEOMC.

	HMC	HEOMC	DHMC	DHEOMC
Formula	$C_{10}H_{10}O_4$	$C_{12}H_{14}O_5$	$C_{15}H_{16}O_6$	$C_{17}H_{20}O_7$
M_W (g mol^{-1})	194.18	238.23	292.31	336.33
Crystal system	Monoclinic	Monoclinic	Monoclinic	Triclinic
Space group (N°)	$P2_1/n$ (14)	$P2_1/n$ (14)	*Pc* (7)	*P*-1 (2)
λ (Å)	0.71073	0.71073	1.54184	1.54184
a (Å)	6.9524(2)	7.1794(9)	12.6290(4)	6.5260(5)
b (Å)	11.3058(4)	21.811(3)	6.9580(2)	10.7751(8)
c (Å)	11.7715(5)	7.2972(12)	16.7226(6)	12.1148(7)
α (deg.)	90	90	90	107.949(6)
β (deg.)	105.674(4)	105.052(15)	109.292(4)	93.774(5)
γ (deg.)	90	90	90	102.412(6)
V (Å^3)	890.86(6)	1103.5(3)	1386.94(8)	783.46(9)
Z	4	4	4	2
D_{calc} (g cm^{-3})	1.448	1.434	1.400	1.426
μ (mm^{-1})	0.113	0.112	0.918	0.937
Collected reflns	5797	8163	23764	4893
Unique reflns (R_{int})	1844(0.0345)	2274(0.084)	4447(0.051)	2742(0.025)
Observed reflns[$I > 2\sigma(I)$]	1442	1363	4157	2376
Parameters/Restrains	132/2	159/2	681/304	297/0
$R(F)^a$ [$I > 2\sigma(I)$]	0.045	0.059	0.072	0.039
$wR(F^2)^b$ (all data)	0.098	0.113	0.183	0.045
GoF	1.073	1.052	1.109	1.085
Flack parameter	–	–	–0.06(12)	–
Hooft parameter	–	–	–0.01(10)	–

Single-crystal X-ray diffraction experiments revealed that molecular structures of the synthetized compounds are in good agreement with those proposed from ^{1}H-NMR studies. In all cases, the coumarin backbone consists of a completely planar benzolactone ring bearing a methyl group in the C4 position which displays different substituents at the C7 position (Figure S10). Our findings on the crystal structure of monohydrate HMC and the supramolecular interactions governing its crystal structure were aligned with previously reported works [38,39]. Using this system as a base, significant changes within supramolecular interactions in the crystal structure for the rest of studied coumarins were also observed, owing to the insertion of different substituents at position C7. Indeed, intermolecular forces involved in the crystal packing of HEOMC, DHMC and DHEOMC include π-π interactions established between aromatic ring, O–H···O hydrogen bonds and C–H···O-type contacts. Their geometrical parameters are compiled in Tables S2 and S3. All the bond lengths and angles are in concordance with those found in literature [36,40–42].

In a close analysis of the HEOMC crystal structure, we observed that HEOMC crystallized in the monoclinic space group $P2_1/n$ and its asymmetric unit contained one HEOMC moiety and one hydration water molecule. As shown in Figure 8, the coumarin molecules were packed antiparallelly forming columns along the crystallographic z axis through π-π stacking. These arrangements were involved in an extensive three-dimensional network of O_W–H···O and O–H···O_W hydrogen bonds established between the hydroxyl group of this monohydroxy coumarin and the hydration water molecules. Additionally, weak C–H···O-type contacts linked adjacent columns along the [100] direction.

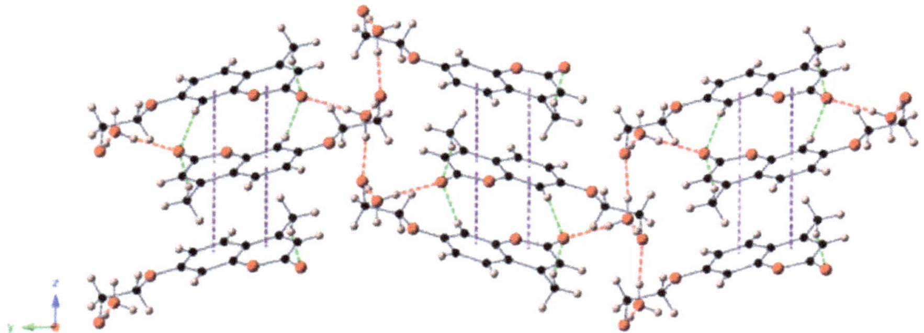

Figure 8. View of the crystal packing of HEOMC along the crystallographic x-axis. Intermolecular interactions are depicted as dashed lines: O-H···O, red; C-H···O, green; π-π purple.

On the other hand, the crystal structure of DHMC belonged to the monoclinic Pc space group, including two coumarin moieties in the asymmetric unit. Both coumarins were involved in crystallographic disorder; therefore, the forms with higher occupancy were only represented in Figure 9. DHMC molecules were packed antiparallelly, forming columns along the crystallographic y axis and interacting through π-π stacking and O–H···O hydrogen bonds. These strong interactions involved the O atoms from the ester group and both hydroxyl moieties. Owing to these four O atoms within this dihydroxy-coumarin, one-dimensional arrangements were connected to each other by creating an extensive network of C–H···O contacts.

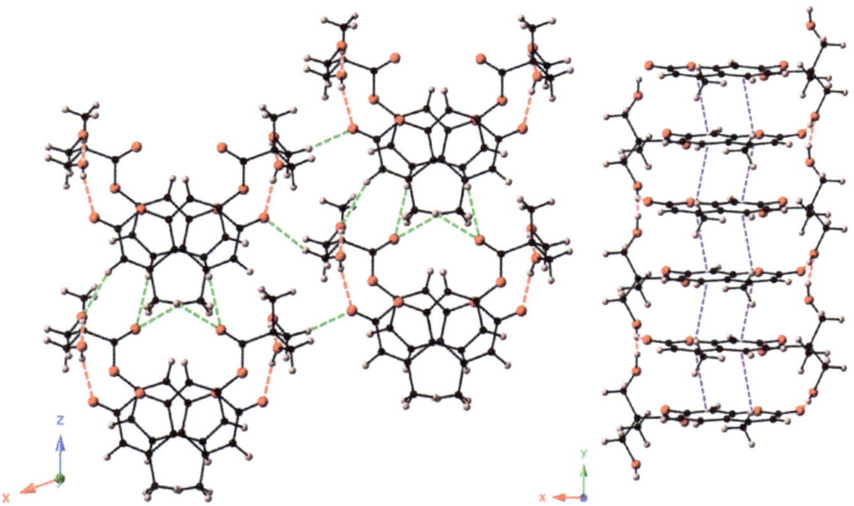

Figure 9. View of the crystal packing of DHMC along the crystallographic y axis (**left**) and detail of the supramolecular one-dimensional arrangement (**right**).

Finally, DHEOMC crystallized in the triclinic space group P-1; the lower symmetry in comparison to the other systems could come from the introduction of a flexible spacer between dihydroxyl groups and umbelliferone moiety. In the case of this dihydroxy-coumarin, the crystal structure showed a bidimensional character with ribbons that stacked along the [0–11] direction. These ribbons were constituted by double-chains of coumarins that interacted through O–H···O-type hydrogen bonds and involved O atoms from hydroxyl or ester groups of different DHEOMC moieties. Contiguous double-chains were linked to each other via π-π interactions along the crystallographic y-axis, and together with C–H···O contacts, completed the bidimensional arrangement (Figure 10).

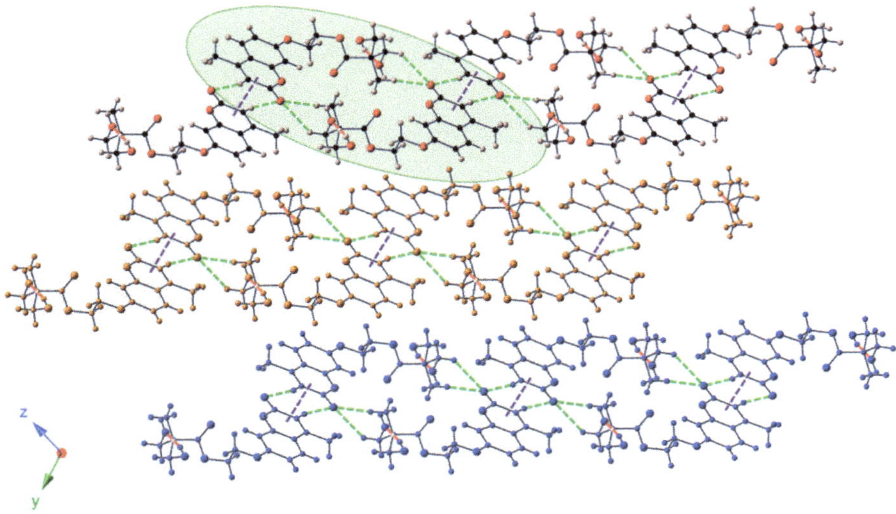

Figure 10. View of the crystal packing of DHEOMC along the crystallographic X axis indicating its bidimensional character. Hydrogen-bonded double-chains are highlighted in green.

Ultimately, despite the symmetry of the space group of the coumarins studied, it should be highlighted that one molecule of water was included within the crystalline structures of monohydroxylated coumarins, whilst their dihydroxylated counterparts showed anhydrous structures. Probably, the two hydroxyl residues and the ester functional group were able to establish interactions similar to those arranged by the water molecules, in the monohydroxylates, to stabilize the crystal structure.

2.5. Thermal Analysis

The crystal structure of coumarin derivatives was also studied by DSC. As shown in Figure 11A, initially, in the heating scan, DSC measurements for monohydroxy coumarins (HMC and HEOMC) showed a broad endothermic peak that was possibly due to the loss of the recrystallization solvents (ethyl acetate, ethanol) and hydration water that occurs at 25–80 °C and 60–110 °C, respectively. Nevertheless, the dihydroxy-coumarins (DHMC and DHEOMC) suffered this peak. This finding is in line with the crystallographic data discussed above.

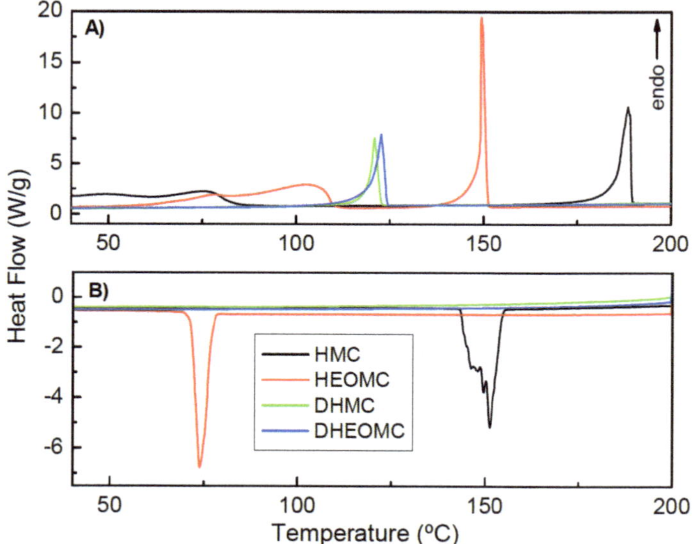

Figure 11. Differential scanning calorimetry (DSC) measurements, (**A**) heating scan, (**B**) cooling scan.

Additionally, significant differences in the melting temperatures of each compound were found in the heating scan. Coumarins bearing one hydration molecule in their asymmetric unit were found to have higher melting points than their dihydroxy partners, being 188 °C for HMC and 150 °C for HEOMC. Indeed, anhydrous crystal structures (DHMC and DHEOMC) presented a similar melting point between them, but much lower than the other hydroxylated coumarin family (121 for DHMC and 123 °C for DHEOMC). Comparing these two dihydroxy coumarins, DHEOMC presented a slightly higher melting temperature, which could be due to the flexibility of the spacer between the hydroxyl groups and umbelliferone ring favoring intermolecular interactions slightly higher.

As shown in Figure 11B, only monohydroxylated coumarins displayed an efficient crystal packing, and because of these compounds exhibited a crystallization peak during the cooling process. Moreover, this crystallization took place at lower temperatures than melting points (151 vs. 188 °C for HMC and 74 vs. 150 °C for HEOMC). This fact could be due to the kinetics of crystal nucleation and growth being very slow.

On the other hand, the enthalpies found for the melting and recrystallization processes are collected in Table S4. It can be observed that the HMC and HEOMC coumarins partially recrystallized in the cooling process as a result of the enthalpy values for this step (ΔHc) being lower than the melting process (ΔHm). Additionally, the recrystallized fraction of HMC was higher than HEOMC. This fact could be related, again, to the more efficient crystal packing shown by this monohydroxylated coumarin compared to its monohydroxy partner.

Apart from that, the study of the thermal stability of crystal structures was also completed by TGA. In Figure 12, all the TGA curves of the studied coumarins are shown. For monohydroxy coumarins (HMC and HEOMC) the thermal decomposition occurred in two main mass-loss steps, whilst for dihydroxy coumarins (DHMC and DHEOMC) the decomposition carried out in a unique mass-loss step. In HMC and HEOMC coumarins, the first decomposition step appeared around 60–100 °C, corresponding to the loss of recrystallization solvents and hydration water. This result concurs with the X-ray diffraction data and DSC measurements described above.

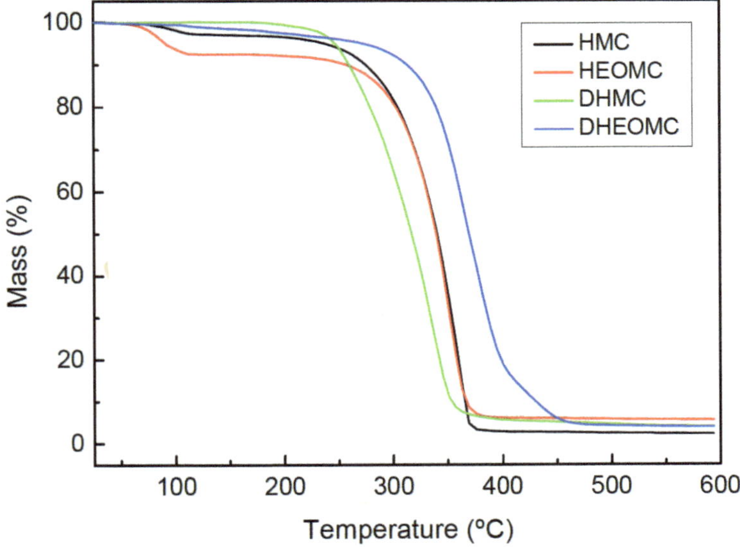

Figure 12. TGA curves for all the coumarin derivatives.

As shown in Table 3, an experimental mass loss of 2.7 and 7.5 wt % were observed for HMC and HEOMC, respectively. In the case of HEOMC, this TGA analysis agreed with the theoretically calculated value of 7.6 wt % for the loss of one water molecule within its crystal structure. Nevertheless, the theoretically calculated value for the loss of one water molecule in HMC was 9.3 wt %, which did not agree with the experimental value (2.7 wt %). This fact may be due to a weaker interaction between HMC and water that provoked a water loss at room temperature. To corroborate this issue, in a detailed analysis of the DSC curves (Figure 11), an endothermic peak was observed at room temperature for HMC, whilst for HEOMC at up 60 °C.

Table 3. TGA results.

Coumarin	T0 (°C)	Weight Loss (wt %)	T0 (°C)	Weight Loss (wt %)
HMC	63.8	−2.7	279.4	−86.0
HEOMC	65.8	−7.6	272.7	−85.9
DHMC	-	-	246.0	−95.5
DHEOMC	-	-	323.2	−95.2

After this first mass-loss, the stability of all coumarins remained unaltered until 250 °C. Hence, this high thermal stability results in suitable unalterability for bulk polymerization reactive extrusion at temperatures above 150 °C, which is an efficient and very common industrial method for manufacturing thermoplastic polyurethane with incorporated target photoreactivity [43]. At 250 °C, a second mass-loss was observed, related to the thermal degradation of the coumarin ring. Depending on the system, the temperature of this second process appeared to vary. In fact, onset degradation temperatures (T0) and the weight loss corresponding to each degradation step are collected in Table 3 for all molecules. It should be noted that the two dihydroxylated coumarins had the highest and lowest decomposition temperatures for the second stage. In this sense, the DHMC product showed the lowest decomposition temperature (246.0 °C), probably due to the aromatic nature of its ester group. Through a transesterification reaction, this aromatic ester is very susceptible to attack by any hydroxyl group. This fact has already been described previously. The thermal stability of DHEOMC increased up to 323 °C, mainly due to the aliphatic nature of its ester group.

3. Materials and Methods

3.1. Characterization

Solution 1H NMR spectra were recorded at room temperature in a Varian Unity Plus 400 instrument using deuterated chloroform ($CDCl_3$) or deuterated dimethylsulfoxide (DMSO-d^6) as solvent. Spectra were referenced to the residual solvent protons at 7.26 or 2.50 ppm, respectively.

To understand the habit of the reversible photoreaction of the coumarin derivatives, irradiations were carried out in an ultraviolet crosslinker supplied by Ultra-Violet Products equipped with two sets of 5 lamps for 354 or 254 nm irradiations (354 and 254 nm are the wavelengths on the maxima of the irradiation spectra of the lamps, respectively). The samples (from 0.2 to 0.4 mM concentration) were prepared by dissolving each coumarin derivative in distilled water. Firstly, each solution was exposed to 350 nm light in order to produce a crosslinked version via photo-dimerization. Afterwards, all the lamps were changed to give out 254 nm light and photo-cleavage of the crosslinked solution took place. In order to characterize these reversible UV kinetics, UV experiments were performed in a Perkin Elmer Lambda 35 UV-Vis spectrometer. Absorbance of the coumarin solutions were measured from 385 to 210 nm.

Thermal properties of all samples (5–10 mg) were measured (in duplicate) by differential scanning calorimetry (DSC, 822e from Mettler Toledo) in aluminum pans under constant nitrogen flow (20 $mL{\cdot}min^{-1}$). Each sample was subjected to a heating/cooling cycle from 25 to 200 °C with a heating rate of 2 $°C{\cdot}min^{-1}$. Thermogravimetric analysis measurements were also performed twice using TGA/DSC1 from Metter Toledo, under nitrogen (50 mL min^{-1}) from room temperature to 600 °C with a heating rate of 10 °C min^{-1}, where approximately 20 mg of sample was required.

3.2. Single-Crystal X-Ray Diffraction

Intensity data were collected on an Agilent Technologies SuperNova diffractometer equipped with monochromated Mo Kα radiation (λ = 0.71073 Å) and an Eos Charge-couple device CCD detector in the case of HMC and HEOMC, and monochromated Cu Kα radiation (λ = 1.54184 Å) and an Atlas CCD detector for DHMC and DHEOMC. The collection was performed at 100(2) K for HMC and HEOMC and 150(2) K for DHMC and DHEOMC. Data frames were processed (unit cell determination, mathematical absorption correction, intensity data integration and correction for Lorentz and polarization effects) using the CrysAlis Pro software package [44]. The structures were solved using the OLEX2 program [45] and refined by full-matrix least-squares using SHELXL-2014/6 [46]. Final geometrical calculations were carried out with PLATON [47] as integrated in WinGX software package [48].

3.3. Computational Methods

Quantum chemical calculations were performed with Gaussian 09 software [49]. Initially, the ground states were obtained by geometrical optimizations with the hybrid functional B3LYP and 6-311+(2d,p) as the basis set chosen for all the atoms.

As the experimental UV-Vis spectra of coumarins have been obtained in aqueous solution, the solvent effects of water were considered with the conductor-like polarizable continuum model (PCM). The calculated absorption energy was defined as the energy difference between the ground state and the excited state at the optimized ground state geometry.

4. Conclusions

Coumarin derivatives are widely distributed within the plant families and/or synthetic analogs for different applications. In this work, two types of hydroxy-derivative coumarins have been extensively analyzed by different techniques. Firstly, it is important to note that the synthetic routes presented were simple, easily scalable and with high yields, which could arouse high interest in the industry.

DHMC coumarin showed an uneven behavior compared to its counterparts, showing two strong absorption bands by UV-Vis. This performance conditioned its photophysical behavior to the dimerization reaction at 365 nm. Through a detailed atomistic study, these two absorption bands were attributed to the transitions between the frontier orbitals (HOMO, LUMO) and the two closest (HOMO-1, LUMO+1). Nevertheless, for the rest of the coumarins, the main excitation strongly corresponds to a π-π transition between the frontier orbitals.

Single-crystal X-ray diffraction analysis also demonstrated how hydroxyl groups allowed weak supramolecular forces to be established within the crystal structure, which were key elements in describing the ability to experience edge-to-edge self-association. Additionally, despite the symmetry of the space group of each coumarin, monohydroxy-derivated coumarins (HMC and HEOMC) presented one water molecule within each of their crystal structures, while their dihydroxylated counterparts (DHMC and DHEOMC) showed anhydrous structures.

Moreover, significant differences in the melting temperatures of each compound were found in the heating scan, which were consistent with an efficient crystal packing described by X-ray analysis.

Supplementary Materials: The following are available online at http://www.mdpi.com/1420-3049/25/15/3497/s1, Figure S1: A) 1H-NMR and B) 13C-NMR spectra of HMC, Figure S2: A) ATR-FTIR and B) Mass spectra of HMC, Figure S3: A) 1H-NMR and B) 13C-NMR spectra of HEOMC, Figure S4: A) ATR-FTIR and B) Mass spectra of HEOMC, Figure S5: A) 1H-NMR and B) 13C-NMR spectra of DHMC, Figure S6: A) ATR-FTIR and B) Mass spectra of DHMC, Figure S7: A) 1H-NMR and B) 13C-NMR spectra of DHEOMC, Figure S8: A) ATR-FTIR and B) Mass spectra of DHEOMC. Figure S9: Electrostatic potential surfaces for the coumarins studied. Contour values range from - 0.080 to 0.080 Hartree/e, Figure S10: ORTEP view of the asymmetric units in (A) HEOMC, (B) DHMC and (C) DHEOMC depicted at the 50% probability level, together with atom labelling. Colour code: C, black, O, red, H, white. Table S1: Experimental and theoretical vertical excitation energies and oscillator strength for the first two excited states. The labels 1, 2, 1′ and 2′ correspond to HOMO, LUMO, HOMO-1 and LUMO+1, respectively, Table S2: Geometrical parameters (Å, °) of intermolecular π-π interactions in HEOMC, DHMC and DHEOMC, Table S3: Geometrical parameters for O-H··O hydrogen bonds and C-H··O-type contacts in HEOMC, DHMC and DHEOMC, Table S4: DSC results. Scheme S1: Synthetic route for the preparation of HMC, Scheme S2: Synthetic route for the preparation of HEOMC, Scheme S3: Synthetic route for the preparation of DHMC, Scheme S4: Synthetic route for the preparation of DHEOMC.

Author Contributions: Methodology, R.S.-R. and R.N.; validation, J.M.C., B.A., J.M.G.-Z., J.L.V.-V., and Á.M.-F.; formal analysis, R.S.-R., E.R.-B., R.N. and J.M.L.; investigation, R.S.-R., E.R.-B., R.N. and J.M.L; resources, R.S.-R., J.M.C., R.N. and Á.M.-F.; writing—original draft preparation R.S.-R., E.R.-B., R.N. and J.M.L.; writing—review and editing, R.S.-R., E.R.-B., R.N. and J.M.L.; supervision, J.M.C., B.A., J.M.G.-Z., J.L.V.-V., and Á.M.-F. All authors have read and agreed to the published version of the manuscript.

Funding: This research was funded by the Basque Government within the framework ELKARTEK through the research project KK-2018/00108 and KK-2019/00077. This work was also funded by the Ministry of Economy and Competitiveness—Spain (MINECO) through the research Projects RTC-2016-4887-4 and RTI2018-096636-J-100 within the framework of the National Programme for Research Aimed at the Challenges of Society.

Acknowledgments: The authors would like to thank the Basque Government for the financial support of this work within the framework ELKARTEK through the research project KK-2018/00108 and KK-2019/00077. Additionally, this work has been supported by the research Projects RTC-2016-4887-4 and RTI2018-096636-J-100 of the Ministry of Economy and Competitiveness—Spain (MINECO) within the framework of the National Programme for Research Aimed at the Challenges of Society. Authors thank Dr Leire San Felices (Molecules and Material unit of the General X-ray Service, SGIker, UPV/EHU) for technical and human support.

Conflicts of Interest: These authors have declared no conflict of interest.

References

1. Zhang, D.; Li, D.; Li, X.; Jin, W. Post-assembly polymerization of discrete organoplatinum (II) metallacycles via dimerization of coumarin pendants. *Dye. Pigment.* **2018**, *152*, 43–48. [CrossRef]
2. Seoane Rivero, R.; Bilbao Solaguren, P.; Gondra Zubieta, K.; Peponi, L.; Marcos-Fernández, A. Synthesis, kinetics of photo-dimerization/photo-cleavage and physical properties of coumarin-containing branched polyurethanes based on polycaprolactones. *Express Polym. Lett.* **2016**, *10*, 84. [CrossRef]
3. Motoyanagi, J.; Fukushima, T.; Ishii, N.; Aida, T. Photochemical stitching of a tubularly assembled hexabenzocoronene amphiphile by dimerization of coumarin pendants. *J. Am. Chem. Soc.* **2006**, *128*, 4220–4221. [CrossRef] [PubMed]

4. Mohamed, M.G.; Hsu, K.-C.; Kuo, S.-W. Bifunctional polybenzoxazine nanocomposites containing photo-crosslinkable coumarin units and pyrene units capable of dispersing single-walled carbon nanotubes. *Polym. Chem.* **2015**, *6*, 2423–2433. [CrossRef]
5. Murray, R.D.H.; Mendez, J.; Brown, S.A. *Chemistry and Biochemistry*; John Wiely Sons: New York, NY, USA, 1982.
6. Murray, R.D.H. Naturally occurring plant coumarins. In *Progress in the Chemistry of Organic Natural Products*; Springer: Berlin/Heidelberg, Germany, 1997; pp. 1–119.
7. Trenor, S.R.; Shultz, A.R.; Love, B.J.; Long, T.E. Coumarins in polymers: from light harvesting to photo-cross-linkable tissue scaffolds. *Chem. Rev.* **2004**, *104*, 3059–3078. [CrossRef]
8. Chen, Y.; Jean, C.-S. Polyethers containing coumarin dimer components in the main chain. II. Reversible photocleavage and photopolymerization. *J. Appl. Polym. Sci.* **1997**, *64*, 1759–1768. [CrossRef]
9. Zhao, D.; Ren, B.; Liu, S.; Liu, X.; Tong, Z. A novel photoreversible poly (ferrocenylsilane) with coumarin side group: synthesis, characterization, and electrochemical activities. *Chem. Commun.* **2006**, 779–781. [CrossRef]
10. Sinkel, C.; Greiner, A.; Agarwal, S. A polymeric drug depot based on 7-(2′-methacryloyloxyethoxy)-4-methylcoumarin copolymers for photoinduced release of 5-fluorouracil designed for the treatment of secondary cataracts. *Macromol. Chem. Phys.* **2010**, *211*, 1857–1867. [CrossRef]
11. Singh, D.; Rahman, M. Umbelliferone loaded nanocarriers for healthcare applications. *Curr. Biochem. Eng.* **2020**, *6*, 25–33. [CrossRef]
12. Mazimba, O. Umbelliferone: Sources, chemistry and bioactivities review. *Bull. Fac. Pharm. Cairo Univ.* **2017**, *55*, 223–232. [CrossRef]
13. Penta, S. *Advances in Structure and Activity Relationship of Coumarin Derivatives*; Academic Press: Cambridge, MA, USA, 2015; ISBN 978-0-12-803797-3.
14. Morsy, S.A.; Farahat, A.A.; Nasr, M.N.A.; Tantawy, A.S. Synthesis, molecular modeling and anticancer activity of new coumarin containing compounds. *Saudi Pharm. J.* **2017**, *25*, 873–883. [CrossRef] [PubMed]
15. Hassan, M.Z.; Osman, H.; Ali, M.A.; Ahsan, M.J. Therapeutic potential of coumarins as antiviral agents. *Eur. J. Med. Chem.* **2016**, *123*, 236–255. [CrossRef] [PubMed]
16. Srinivasa, H.T.; Harishkumar, H.N.; Palakshamurthy, B.S. New coumarin carboxylates having trifluoromethyl, diethylamino and morpholino terminal groups: synthesis and mesomorphic characterisations. *J. Mol. Struct.* **2017**, *1131*, 97–102. [CrossRef]
17. Dandriyal, J.; Singla, R.; Kumar, M.; Jaitak, V. Recent developments of C-4 substituted coumarin derivatives as anticancer agents. *Eur. J. Med. Chem.* **2016**, *119*, 141–168. [CrossRef]
18. Sarkar, D.; Ghosh, P.; Gharami, S.; Mondal, T.K.; Murmu, N. A novel coumarin based molecular switch for the sequential detection of Al3+ and F-: Application in lung cancer live cell imaging and construction of logic gate. *Sens. Actuators B Chem.* **2017**, *242*, 338–346. [CrossRef]
19. Wang, Z.; Wu, Q.; Li, J.; Qiu, S.; Cao, D.; Xu, Y.; Liu, Z.; Yu, X.; Sun, Y. Two benzoyl coumarin amide fluorescence chemosensors for cyanide anions. *Spectrochim. Acta Part A Mol. Biomol. Spectrosc.* **2017**, *183*, 1–6. [CrossRef]
20. García, S.; Vázquez, J.L.; Rentería, M.; Aguilar-Garduño, I.G.; Delgado, F.; Trejo-Durán, M.; García-Revilla, M.A.; Alvarado-Méndez, E.; Vázquez, M.A. Synthesis and experimental-computational characterization of nonlinear optical properties of triazacyclopentafluorene-coumarin derivatives. *Opt. Mater. (Amst.)* **2016**, *62*, 231–239. [CrossRef]
21. Li, C.; Qin, J.; Wang, B.; Bai, X.; Yang, Z. Fluorescence chemosensor properties of two coumarin-based compounds for environmentally and biologically important Al3+ ion. *J. Photochem. Photobiol. A Chem.* **2017**, *332*, 141–149. [CrossRef]
22. Imit, M.; Imin, P.; Adronov, A. π-Conjugated polymers with pendant coumarins: design, synthesis, characterization, and interactions with carbon nanotubes. *Can. J. Chem.* **2016**, *94*, 759–768. [CrossRef]
23. Gindre, D.; Iliopoulos, K.; Krupka, O.; Evrard, M.; Champigny, E.; Sallé, M. Coumarin-containing polymers for high density non-linear optical data storage. *Molecules* **2016**, *21*, 147. [CrossRef]
24. He, J.; Tremblay, L.; Lacelle, S.; Zhao, Y. Preparation of polymer single chain nanoparticles using intramolecular photodimerization of coumarin. *Soft Matter* **2011**, *7*, 2380–2386. [CrossRef]
25. Kiskan, B.; Yagci, Y. Self-healing of poly (propylene oxide)-polybenzoxazine thermosets by photoinduced coumarine dimerization. *J. Polym. Sci. Part A Polym. Chem.* **2014**, *52*, 2911–2918. [CrossRef]

26. Fiore, G.L.; Rowan, S.J.; Weder, C. Optically healable polymers. *Chem. Soc. Rev.* **2013**, *42*, 7278–7288. [CrossRef] [PubMed]
27. Binder, W.H. *Self-Healing Polymers: From Principles to Applications*; John Wiley & Sons: Hoboken, NJ, USA, 2013.
28. Urban, M.W. *Stimuli-Responsive Materials: From Molecules to Nature Mimicking Materials Design*; Royal Society of Chemistry: London, UK, 2019.
29. Ling, J.; Rong, M.Z.; Zhang, M.Q. Photo-stimulated self-healing polyurethane containing dihydroxyl coumarin derivatives. *Polymer (Guildford)* **2012**, *53*, 2691–2698. [CrossRef]
30. Ling, J.; Rong, M.Z.; Zhang, M.Q. Coumarin imparts repeated photochemical remendability to polyurethane. *J. Mater. Chem.* **2011**, *21*, 18373–18380. [CrossRef]
31. Aguirresarobe, R.H.; Irusta, L.; Fernández-Berridi, M.J. UV-light responsive waterborne polyurethane based on coumarin: synthesis and kinetics of reversible chain extension. *J. Polym. Res.* **2014**, *21*, 505. [CrossRef]
32. Aguirresarobe, R.H.; Martin, L.; Aramburu, N.; Irusta, L.; Fernandez-Berridi, M.J. Coumarin based light responsive healable waterborne polyurethanes. *Prog. Org. Coat.* **2016**, *99*, 314–321. [CrossRef]
33. Seoane Rivero, R.; Bilbao Solaguren, P.; Gondra Zubieta, K.; Gonzalez-Jimenez, A.; Valentin, J.L.; Marcos-Fernandez, A. Synthesis and characterization of a photo-crosslinkable polyurethane based on a coumarin-containing polycaprolactone diol. *Eur. Polym. J.* **2016**, *76*, 245–255. [CrossRef]
34. Salgado, C.; Arrieta, M.P.; Peponi, L.; Fernández-García, M.; López, D. Influence of poly (epsilon-caprolactone) molecular weight and coumarin amount on photo-responsive polyurethane properties. *Macromol. Mater. Eng.* **2017**, *302*, 1600515. [CrossRef]
35. Seoane Rivero, R.; Navarro, R.; Bilbao Solaguren, P.; Gondra Zubieta, K.; Cuevas, J.M.; Marcos-Fernández, A. Synthesis and characterization of photo-crosslinkable linear segmented polyurethanes based on coumarin. *Eur. Polym. J.* **2017**, *92*, 263–274. [CrossRef]
36. Seth, S.K.; Sarkar, D.; Jana, A.D.; Kar, T. On the possibility of tuning molecular edges to direct supramolecular self-assembly in coumarin derivatives through cooperative weak forces: crystallographic and Hirshfeld surface analyses. *Cryst. Growth Des.* **2011**, *11*, 4837–4849. [CrossRef]
37. Cuevas, J.M.; Seoane-Rivero, R.; Navarro, R.; Marcos-Fernández, Á. Coumarins into polyurethanes for smart and functional materials. *Polymers* **2020**, *12*, 630. [CrossRef]
38. Jasinski, J.P.; Woudenberg, R.C. 7-Hydroxy-4-methylcoumarin monohydrate. *Acta Crystallogr. Sect. C Cryst. Struct. Commun.* **1994**, *50*, 1952–1953. [CrossRef]
39. Panini, P.; Venugopala, K.N.; Odhav, B.; Chopra, D. Quantitative analysis of intermolecular interactions in 7-hydroxy-4-methyl-2H-chromen-2-one and its hydrate. *Proc. Natl. Acad. Sci. India Sect. A Phys. Sci.* **2014**, *84*, 281–295. [CrossRef]
40. Gomes, L.R.; Low, J.N.; Fonseca, A.; Matos, M.J.; Borges, F. Crystal structures of three 6-substituted coumarin-3-carboxamide derivatives. *Acta Crystallogr. Sect. E Crystallogr. Commun.* **2016**, *72*, 926–932. [CrossRef] [PubMed]
41. Ouédraogo, M.; Abou, A.; Djandé, A.; Ouari, O.; Zoueu, T.J. 2-Oxo-2H-chromen-7-yl 4-tert-butylbenzoate. *Acta Crystallogr. Sect. E Crystallogr. Commun.* **2018**, *74*, 530–534. [CrossRef]
42. Abou, A.; Yoda, J.; Djandé, A.; Coussan, S.; Zoueu, T.J. Crystal structure of 2-oxo-2H-chromen-7-yl 4-fluorobenzoate. *Acta Crystallogr. Sect. E Crystallogr. Commun.* **2018**, *74*, 761–765. [CrossRef]
43. Stachak, P.; Hebda, E.; Pielichowski, K. Foaming extrusion of thermoplastic polyurethane modified by POSS nanofillers. *Compos. Theory Pract.* **2019**, *19*, 23–29.
44. Technologies, A. *CrysAlisPro Software System*; Version 1.171. 37.34; Rigaku Oxford Diffraction: Oxford, UK, 2014.
45. Dolomanov, O.V.; Bourhis, L.J.; Gildea, R.J.; Howard, J.A.K.; Puschmann, H. OLEX2: a complete structure solution, refinement and analysis program. *J. Appl. Crystallogr.* **2009**, *42*, 339–341. [CrossRef]
46. Sheldrick, G.M. A short history of SHELX. *Acta Crystallogr. Sect. A Found. Crystallogr.* **2008**, *64*, 112–122. [CrossRef]
47. Spek, A.L. Structure validation in chemical crystallography. *Acta Crystallogr. Sect. D Biol. Crystallogr.* **2009**, *65*, 148–155. [CrossRef] [PubMed]

48. Farrugia, L.J. WinGX suite for small-molecule single-crystal crystallography. *J. Appl. Crystallogr.* **1999**, *32*, 837–838. [CrossRef]
49. Frisch, J.; Trucks, G.W.; Schlegel, H.B.; Scuseria, G.E.; Robb, M.A.; Cheeseman, J.R.; Scalmani, G.; Barone, V.; Mennucci, B.; Petersson, A.; et al. *Gaussian 09 Program*; Revision C. 01; Gaussian, Inc.: Wallingford, CT, USA, 2010.

MDPI
St. Alban-Anlage 66
4052 Basel
Switzerland
Tel. +41 61 683 77 34
Fax +41 61 302 89 18
www.mdpi.com

Molecules Editorial Office
E-mail: molecules@mdpi.com
www.mdpi.com/journal/molecules

www.ingramcontent.com/pod-product-compliance
Lightning Source LLC
LaVergne TN
LVHW071615170726
843515LV00009B/2397

* 9 7 8 3 0 3 6 5 2 7 7 5 8 *